D1483413

Fungi Pathogenic for Humans and Animals

MYCOLOGY SERIES

Edited by

Paul A. Lemke

Department of Botany,
Plant Pathology, and Microbiology
Auburn University
Auburn, Alabama

Other Volumes in Preparation

Fungi Pathogenic for Humans and Animals

(IN THREE PARTS)

PART B

Pathogenicity and Detection: I

EDITED BY
Dexter H. Howard
Department of Microbiology and Immunology
School of Medicine, University of California
Los Angeles, California

with the assistance of
Lois F. Howard
Department of Medicine
School of Medicine, University of California
Los Angeles, California

MARCEL DEKKER, INC.   New York and Basel

Library of Congress Cataloging in Publication Data
(Revised for vol. 3 pt. B)
Main entry under title:

Fungi pathogenic for humans and animals.

   (Mycology series ; v. 3)
   Includes indexes.
   Contents: pt. A. Biology—pt. B. Pathogenicity &
detection: I.
   1. Fungi, Pathogenic. I. Howard, Dexter H.,
[date]. II. Howard, Lois F. III. Series.
[DNLM: 1. Fungi—Pathogenicity. QZ 65 F981]
QR245.F86   589.2'0469   82-18240
ISBN 0-8247-1875-5 (pt. A)
      0-8247-1144-0 (pt. B)

MARCEL DEKKER, INC.
270 Madison Avenue, New York, New York 10016

Current printing (last digit):
10 9 8 7 6 5 4 3 2 1

PRINTED IN THE UNITED STATES OF AMERICA

Introduction to the Series

Mycology is the study of fungi, that vast assemblage of microorganisms which includes such things as molds, yeasts, and mushrooms. All of us in one way or another are influenced by fungi. Think of it for a moment—the good life without penicillin or a fine wine. Consider further the importance of fungi in the decomposition of wastes and the potential hazards of fungi as pathogens to plants and to humans. Yes, fungi are ubiquitous and important.

Mycologists study fungi either in nature or in the laboratory and at different experimental levels ranging from descriptive to molecular and from basic to applied. Since there are so many fungi and so many ways to study them, mycologists often find it difficult to communicate their results even to other mycologists, much less to other scientists or to society in general.

This Series establishes a niche for publication of works dealing with all aspects of mycology. It is not intended to set the fungi apart, but rather to emphasize the study of fungi and of fungal processes as they relate to mankind and to science in general. Such a series of books is long overdue. It is broadly conceived as to scope, and should include textbooks and manuals as well as original and scholarly research works and monographs.

The scope of the Series will be defined by, and hopefully will help define, progress in mycology.

Paul A. Lemke

Foreword

The occurrence of mycosis as primary or secondary disease is rising rapidly. This is due, on the one hand, to increased clinical awareness and improved diagnostic skills and, on the other, to increased use of antineoplastic and immunosuppressive drugs. The diagnostic and therapeutic problems posed by mycoses strongly influence medicine today. To combat these problems, a better understanding of the ability of fungi to invade susceptible hosts, the immunological and serological changes evoked by fungi, and currently available antifungal drugs is essential. This book, the second of three parts, meets that need exactly. It contains the latest and most authoritative information on pathogenic mechanisms, immunology, serology, antifungal drugs, and fungal toxins.

The authors are prominent figures in their fields with outstanding research records. Professor Howard is a leading mycologist who has made important contributions to our knowledge of host-parasite interactions. He is also a noted teacher who has trained many young scientists in the field.

Microbiologists and clinicians alike will find this book most useful and appreciate its thorough coverage of theory and practice in the mycoses.

K. J. Kwon-Chung
Clinical Mycology Section
Laboratory of Clinical Investigation
National Institute of Allergy and Infectious Diseases
National Institutes of Health
Bethesda, Maryland

Preface

The attribute of the zoopathogenic fungi that makes them especially interesting—their ability to evoke pathological changes in a susceptible host—will be the subject of the second volume of this series. Because there are so many topics of interest under such a heading, a third part is being planned.

The overall format of Part B has chapters on theoretical treatments of a subject followed by a consideration of practical consequences. For example, the chapter on Humoral Responses of the Host is followed by one on Serodiagnosis, and the chapter on Mode of Action of Antifungal Drugs is accompanied by one on Measurement of the Activity of Antifungal Drugs.

For the most part, I have tried to be internally consistent in matters of nomenclature. Thus, the terminology set forth in Part A has been adopted for Part B. There may be some disagreement with the choices. I have adopted the recommendations set forth in Memorandum No. 23 of the Medical Research Council, London. [Medical Research Council. 1977. *Nomenclature of Fungi Pathogenic to Man and Animals,* Memorandum No. 23 (4th ed.), Her Majesty's Stationery Office, London, 26 pp.] In certain especially controversial areas I have indicated synonymous alternatives. I trust each individual author's splendid coverage will overshadow any undetected nomenclatural oversights by the editor.

Again, as in Part A of this series, I gratefully acknowledge the indispensable assistance of Mrs. Lois F. Howard. Her insistence on the accuracy of textural detail was again unfaltering. I am grateful to her, to Ms. Judy Fung, and Ms. Bette Y. Tang for the numerous typings and retypings a work of this sort requires. Some support for this venture has been supplied by research grant AI 16252 from the National Institutes of Allergy and Infectious Diseases, National Institutes of Health, which is used to fund the Collaborative California Universities-Mycology Research Unit (CCU-MRU).

I am most grateful to the splendid group of authors who collaborated with me to produce this volume.

Dexter H. Howard

Contents

Contributors

Glenn S. Bulmer, PhD, Department of Microbiology and Immunology, The University of Oklahoma Health Sciences Center, Oklahoma City, Oklahoma

H. R. Burmeister, PhD, Northern Regional Research Center, U.S. Department of Agriculture, Peoria, Illinois

A. Ciegler, PhD, Southern Regional Research Center, U.S. Department of Agriculture, New Orleans, Louisiana

Rebecca A. Cox, PhD, Department of Research Immunology, San Antonio State Chest Hospital, San Antonio, Texas

Richard D. Diamond, MD, Department of Medicine and Evans Memorial Department of Clinical Research, University Hospital and Boston University Medical Center, Boston, Massachusetts

Robert A. Fromtling, PhD,* Department of Microbiology and Immunology, The University of Oklahoma Health Sciences Center, Oklahoma City, Oklahoma

Milton Huppert, PhD, Audie L. Murphy Memorial Veterans Medical Center and University of Texas Health Science Center, San Antonio, Texas

George S. Kobayashi, PhD, Washington University School of Medicine, St. Louis, Missouri

Paul F. Lehmann, PhD, Medical College of Ohio, Toledo, Ohio

Donald W. R. Mackenzie, PhD, Mycological Reference Laboratory, London School of Hygiene and Tropical Medicine, London, England

Gerald Medoff, MD, Washington University School of Medicine, St. Louis, Missouri

Paul P. Vergeer, Richmond, California

R. F. Vesonder, Northern Regional Research Center, U.S. Department of Agriculture, Peoria, Illinois

**Present affiliation:* Merck Institute for Therapeutic Research, Rahway, New Jersey

Contents of Part A

1

Pathogenic Mechanisms of Mycotic Agents

Glenn S. Bulmer and Robert A. Fromtling* / The University of Oklahoma Health Sciences Center, Oklahoma City, Oklahoma

*Present affiliation: Merck Institute for Therapeutic Research, Rahway, New Jersey

I. INTRODUCTION

During the past 140 years, which encompass the modern history of medical mycology, considerable information has been acquired about human mycoses. In this time period all the major diseases were described and, especially to the casual observer, substantial knowledge of these diseases in the areas of etiology, epidemiology, and diagnosis was gathered. During the past three to four decades major progress has been made in understanding these diseases in such areas as immunology, pathology, predisposing factors, serology, and therapy. With this reservoir of available information, medical personnel often gain the impression that the once poorly understood mycoses can now be discussed at length by any competent medical mycologist. However, when confronted with the seemingly simple question, "How do these organisms cause disease?", the medical mycologist agrees that little is known and that much more must be learned of disease processes to help potential or existing patients.

Before writing this chapter we decided that it should not be a depressing thesis on how little is known in the area of fungal pathogenesis but, instead, it should be from the point of view that this is an area that is just beginning to be investigated. It is an area that *must* be examined in greater depth and, indeed, an area to which many investigators have already made substantial contributions.

In some regards medical mycologists are in a better position than most people currently perceive. We are fortunate to have colleagues who, in related fields such as medical microbiology and plant pathology, have investigated problems and gathered considerable information for decades on the pathogenesis of diseases caused by other microbes [259, 358-360]. Of course, not all of the accumulated concepts are directly applicable to our field, but many methods, approaches, and hypotheses can be used to assist the medical mycologist in elucidating similar problems today.

One purpose of this chapter is to present a comprehensive list of references of classical and recent contributions in the field of pathogenesis. By so doing we hope that it will encourage others to delve further into this field. New ideas, approaches, and models, and more concerted efforts are needed in this area; perhaps some of our thoughts will act as incentives to future workers in this field.

To understand pathogenicity one must first place it in the proper framework of symbiosis. *Symbiosis* is a Greek word which means living together: *sym* means together and *bios* means life. There are three generally accepted manifestations of symbiotic relationships. One is *commensalism,* in which one organism benefits but neither is dependent metabolically upon the other. A second form of symbiotic relations is *mutualism,* where both organisms are dependent metabolically upon one another. The third form of symbiosis is *parasitism.* A parasite is an organism that is dependent metabolically upon a host and contributes nothing to the host; in fact, destruction of cells usually results. A parasite may be nonpathogenic or pathogenic. The symbiotic relationships mentioned are diagrammed in Fig. 1.

Pathogenicity is the capacity of an organism to damage, i.e., to produce disease in another animal or plant. This process is considered to be the result of direct interaction between the pathogen and host. Although there are some microbes that produce materials which can alter or potentially destroy cells in another organism, it is not considered that such organisms are pathogens unless they directly interact with a host. For example, some bacteria (e.g., *Clostridium botulinum*) and some fungi (e.g., *Aspergillus* spp.) produce toxins independently of a host, and thus are not genuine pathogens. However, should it eventually be demonstrated that *Aspergillus* spp., for example, in fact produces myotoxins in vivo, then such organisms would have to be reclassified as genuine pathogens. Such distinctions are worthy of note in order to realize that a disease process cannot be fully understood if one studies only a host or only a given pathogen, i.e., as separate entities. In fact, a disease process is the culmination of an interaction between at least two organisms and the resulting interaction may be totally different from the independent

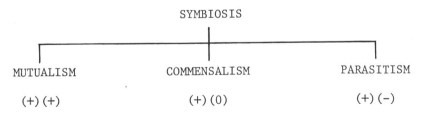

Figure 1   Potential modes of symbiosis: (+), organism benefits from the relationship; (-), organism is harmed by the relationship; (0), organism neither benefits nor is harmed by the relationship.

actions of either of the organisms. This is why, for example, it is not enough to demonstrate that a pathogenic fungus can produce materials in vitro which can harm, alter, or destroy cells in a potential host. If such a material is produced while the organism is actively metabolizing *in* another organism (i.e., in vivo), then there may be a disease process. Additionally, it is not sufficient to demonstrate that a host can respond to an organism unless such a response can be shown to be either destructive or protective.

For an organism to cause disease it must (1) enter the host, (2) multiply in host tissues, (3) resist or not stimulate host defense mechanisms, and (4) damage the host. Products of organisms which produce factors that assist in the accomplishment of one or several of these processes are called *virulence factors.* Since a pathogen must accomplish all four of these processes, and each process is complex, several determinates (virulence factors) are usually necessary to create the overall effect, i.e., to initiate a disease process. It is these properties of animal mycotic agents that will be examined in this chapter; responses to these and other factors on the part of the host are covered as separate entities in this book under sections dealing with immunology and pathology. Thus, this chapter is limited to properties of fungi that are known, or speculated to be, virulence factors in the disease process in animals. The limitation of this chapter to these factors is for practical purposes and must not be considered as a restriction of the overall goal, namely, to elucidate mechanisms and interactions that result in the final process: disease.

This chapter is divided into two areas. The first, consisting of Secs. II to XIV, covers all of the major and some of the minor human mycoses with reference to published reports on factors produced by pathogenic fungi which may play a role in the virulence of the etiologic agent. In many areas reports on virulence factors of the pathogens do not exist. By revealing the paucity of information on a given organism, we hope that investigators will accept this as a challenge to investigate such areas. In the second portion of this chapter, Sec. XV, we make brief mention of numerous factors that are not *classically* considered to be virulence factors. However, in our opinion, they do represent properties that potentially pathogenic fungi must possess in order to function in a disease process. Also, included are a few views of our colleagues in related fields, with the hope that perhaps some of these ideas will stimulate the reader to investigate pathogenesis in the human mycoses.

II. MYCETOMA, ACTINOMYCOSIS, AND NOCARDIOSIS

A. Description and General Pathogenesis

Mycetoma is a clinical syndrome characterized by a chronic, granulomatous pseudotumor, involving the cutaneous and subcutaneous tissues as well as bone. In advanced cases, the affected body area becomes deformed and multiple, draining sinuses or fistulae develop in the tumor. Pustular exudate containing pigmented granules or colonies of the infecting agent(s) is discharged from the tracts. The triad of tumefaction, draining sinuses, and grains or granules is used to define the disease mycetoma [314].

There are at least six actinomycetes and 16 species of true fungi that are capable of causing mycetoma [103]. The disease is considered to be actinomycotic or eumycotic depending on the nomenclature of the etiologic agents demonstrated. Agents of mycetoma have been isolated from soils everywhere, but North Africa, the Mediterranean region, and Mexico appear to be the main endemic areas.

The disease is believed to be acquired by traumatic implantation of the infectious agent into skin or subcutaneous tissues. The time required for the development of myce-tomas is variable and is dependent on the infecting agent and host response. Initial symp-toms may appear several months to years after actual inoculation, and the mycetoma may continue to develop slowly for as long as 25 years [252].

Cases of mycetoma have been reported worldwide [1,103,252,314]. Eumycotic mycetomas reported from the temperate areas of the world are most commonly caused by *Petriellidium boydii,* formerly named *Allescheria boydii* [251]. The imperfect or conidial phase of this fungus is known as *Monosporium (Scedosporium) apiospermum* [103]. Strains of this fungus as well as some species of the higher bacteria that cause actinomycotic mycetoma display a wide range of virulence and have been isolated from a variety of clinical syndromes [238,252,257,318,395].

Actinomycosis is a chronic, suppurative, or granulomatous disease involving the cervicofacial, thoracic, or abdominal areas. The disease is characterized by the develop-ment of firm, board-like lesions that develop draining sinus tracts. Characteristic "sulfur" granules or grains are discharged from the drainage tracts. The cervicofacial form of the disease may be described as a mycetoma of the jaw [44]. The higher bacteria *Actino-myces israelii* and *A. bovis* are the accepted etiologic agents. The disease is found world-wide due to the endogenous nature of the infecting agents, which are found as normal oral flora. Factors that convert an infection into a disease state are unknown.

Nocardiosis is an acute or chronic, suppurative disease characterized by primary pul-monary involvement with possible hematogenous dissemination to various organs. It has a predilection for the central nervous system and may manifest itself as an encephalitis or meningitis. The accepted etiologic agent is the soil actinomycete *Nocardia asteroides,* al-though *N. brasiliensis* [314] and *N. farcinica* [103] also have been reported in a few cases. Recently, Mahajan et al. [248] reported a reproducible, reliable method of producing pulmonary nocardiosis in monkeys by inoculating the animals via the lower canaliculus. Consistent morbidity, lung pathology, and mortality were reported, and the authors pro-posed using this model for further studies in the pathogenesis of this disease.

Although nocardiosis is generally recognized as a primary pulmonary infection pro-duced following inhalation of the organism, some species of *Nocardia* cause actinomycotic mycetomas. In 1973, Gonzalez-Ochoa [142] studied the virulence of three species of *Nocardia* by inoculating them into the feet of mice. He reported that the mouse footpad route yielded chronic, progressive, fistulous tumors that affected bone and produced granules. The lesions did not heal spontaneously as did the nodules produced by other inoculation routes: intraperitoneal, intramuscular, intravenous, intradermal, or subcu-taneous. This work was later substantiated by testing additional strains in a similar man-ner [143]. Stretton and Bulman [380] also have reported the experimental induction of mycetoma. Using rabbits, these workers were able to induce granule-producing tumors, typical of human mycetoma, by subcutaneous or intramuscular inoculations of a variety of actinomycetes.

Attempts to produce experimental eumycotic mycetomas have been frustrating. Murray et al. [273] claimed success in inducing murine mycetoma and granules with *Madurella mycetomatis,* but Cavanagh [57] was unable to reproduce the results. Repeated intracutaneous and subcutaneous inoculation of a monkey's palmar aspect resulted in only a transient lesion, with no granule formation and eventual spontaneous healing [57]. Recently, however, Mahgoub [249] reported the successful development of mycetoma tumors with *M. mycetomatis* in athymic nude mice. Grains appeared as early as 9 days after footpad inoculation, were well developed by 21 days and were similar to human grains in pigment and tissue reaction. The author also noted that immunologically competent mice were resistant to infection by actimomycotic or eumycotic agents, and that infection could be produced only in athymic mice.

B. Pathogenic Mechanisms

A number of pathogenic factors in this array of organisms have been delineated. The mycetoma agents do not form randomly wandering hyphal filaments in vivo; they form thick-walled grains or granules. The mycetoma grain is not a disorganized collection of hyphal filaments but has a consistent specific shape, composition, and color [83]. It is composed of an agglomeration of actinomycotic filaments or fungal hyphae of differing density. Many mycetoma grains are characteristic of specific etiologic agents [44]. The ability of mycetoma-inducing organisms to grow as grains in vivo is related to the fact that both host and parasite take part in their development; thus, the presently unknown host-parasite interaction produces conditions conducive to development of granules rather than the growth of free actinomycotic filaments or eumycotic hyphae by these pathogenic organisms [252].

In 1963, Zamora et al. [428] reported the isolation of two immunologically active polysaccharides from *Nocardia asteroides* and *N. brasiliensis*. The first somatic polysaccharides isolated from each species were chemically similar and thus were defined as group-specific; however, the second group of polysaccharides (composed of arabinose, galactose, and mannose, but in different molar ratios) reacted only with homologous serum. The polysaccharides were species-specific and aided the diagnosis of the organism as well as the stimulation of host responses.

Numerous "aggressins," such as spreading factors, toxins, and enzymes, have been implicated in the pathogenicity of bacterial diseases. While testing numerous fungi for the presence of collagenase, an enzyme capable of degrading collagen and therefore a potentially key virulence factor, Rippon and Lorincz [319] found that *Actinomadura madurae*, an agent of actinomycotic mycetoma, produced this enzyme in vitro. This was the first report of collagenolytic activity derived from a pathogenic actinomycete. In a later study Rippon and Peck [320] using mutant strains of *A. madurae*, demonstrated in vivo a direct relationship between the amount of collagenase production and the virulence of the organism. A year later Rippon and Varadi [321] reported the first isolation of the enzyme elastase from cultures of several dermatophytes, but were not able to detect any elastase activity in tested mycetoma-causing actinomycetes. Staat and Schachtele [368] reported the detection of dextranase-producing strains of *Actinomyces* in human dental plaque. This enzyme may influence the synthesis and metabolism of carries-related dextrans.

Another potential virulence factor of some mycetoma-inducing agents is a capsule. Mackinnon et al. [247] have reported the production of a capsule around very young cultured cells of seven strains of *Phialophora (Exophiala) spinifera* and *P. (E.) jeanselmei*. The capsule showed a variable thickness and could not be removed by washing the cells in water. The chemical composition and structure of this capsule as well as its role in the pathogenesis of mycetoma are yet to be discovered.

Another area investigated recently has been the immunologic activity and specificity of a water-soluble extract from *N. opaca*. Adam et al. [2] reported the isolation of a soluble immunoadjuvant extracted from the cell walls of *N. opaca* following lysozyme treatment of lipid-free cells. The main biological activity of this material was reported to stimulate an increase in the amount of circulating antibodies to ovalbumin in guinea pigs. Further studies of this extract by Bona et al. [39] revealed that it acted on bone marrow-dependent lymphocytes and was a nonspecific activator of B-lymphocytes, similar to the lipopolysaccharide of gram-negative bacteria. Further analysis revealed that stimulation of B-lymphocytes by this extract produced IgM synthesis and polyclonal proliferation that was not dependent on previous exposure to or infection by the organism.

III. DERMATOPHYTOSES

A. Description and General Pathogenesis

Dermatophytoses, superficial infections caused by dermatophytes, are characterized by itching, flaking, and sometimes inflamed patches of skin, loss of hair, or nail invasion. The etiologic agents are members of a closely related group of fungi which are classified into three genera: *Microsporum, Epidermophyton,* and *Trichophyton* [305]. As a group, dermatophytes are found worldwide, even though many species are limited geographically [3,4,293,399]. Although the diseases caused by dermatophytes are not fatal, they comprise one of the most common infections of humankind. One recent study revealed that 73.5% of almost 4000 mycoses diagnosed and analyzed in a 1-year period in India were dermatophytoses [194].

The dermatophytes can be divided into three main groups depending on their natural habitats: anthropophilic, when humans are the natural host; zoophilic, when a variety of animals act as natural hosts; and geophilic, when soil is the natural habitat [23]. Dermatophytic zoonoses (transmission of infection from animal reservoirs to humans) have been reported [106,148,199,288,355,406] and are of increasing concern to those involved with public health and dermatophyte containment and control.

Dermatophytes usually invade and parasitize only the nonliving, keratinized layers of skin, nails, and hair. This highly developed host-parasite relationship is responsible for a multitude of clinical manifestations. For example, the prominent feature of dermatophytic infections of the skin, apart from broken hairs and dystrophic nails, is the varying degrees of inflammatory and eczematous reactions which these infectious agents provoke in the host [23,148]. Many of these host-parasite interactions are dependent on specific moieties and enzymes produced by many dermatophytes, a subject discussed in greater detail later in this section.

As mentioned earlier, the dermatophytes are classified into three genera. Each of the genera is characterized by having specific "target tissues" which it infects. Species of *Microsporum* infect hair and skin, *Epidermophyton floccosum* infects skin and nails, and species of *Trichophyton* attack hair, skin, and nails [44]. The reasons for this observed tissue specificity are unknown, but may prove to be related to specific nutritional requirements or enzyme production by individual organisms. Although reliable animal models for the study of dermatophytoses are few [72,200,343], studies of dermatophyte pathogenesis using models may answer some of these pressing questions.

B. Pathogenic Mechanisms

One of the more thoroughly investigated areas of fungal pathogenesis has been the dermatophytoses. Beginning around 1962, two groups of investigators started definitive studies on the structures of antigens from dermatophytes and attempted to establish a relationship between fungus structure and immunological activity. Barker and his associates initially studied glycopeptides, discovering that immediate hypersensitivity reactions were induced by the carbohydrate moiety and delayed hypersensitivity was elicited by the protein moiety [22,27]. Recently this group also has examined the structure of an isolated galactomannan-peptide allergen [22]. Bishop, Blank, and their associates have studied both polysaccharides and proteolytic enzymes from dermatophytes [149], and most recently, Nozawa's group has examined the immunochemical characteristics of purified polysaccharide-peptide complexes [285]. This area of research is reviewed by Grappel et al. [148] and Gander [132].

Lipid moieties of microorganisms have been known to cause some occasional host responses. Numerous phospholipids have been isolated from *Trichophyton rubrum* [77], and some investigations of the sensitizing properties of dermatophytic lipids have appeared in the literature. Andersson et al. [8] have reported that lipid fractions of trichophytin, a cell wall extract of dermatophytes, elicited positive allergic delayed skin reaction in sensitized guinea pigs. The free fatty acid fraction proved the most allergenic but was significantly less reactive than the polysaccharide-peptide fraction. The middle-chain fatty acids (C_{10}-C_{12}) showed the greatest allergenic activity among the fatty acid group [166]. The authors suggested that fatty acids may act as contact sensitizers and may be responsible for the skin reactivity induced by some dermatophytoses.

Self-synthesized enzymes may serve fungi in many ways. They may enhance survival in tissues by chemically or physically altering the immediate environment, or they may act directly by digesting host proteins, thus providing a source of nutrition. Rippon and Varadi [321] reported that one strain of *Microsporum gypseum* and several species of *Trichophyton* produced the enzyme elastase which permitted these fungi to utilize elastin, a component of human tissue. An additional survey by Rippon [312] revealed several more species of dermatophytes that had elastase activity. One strain of *Trichophyton schoenleinii* produced enzymes capable of solubilizing the three scleroproteins: keratin, elastin, and collagen [319]. In general, the elastase enzyme was produced by organisms isolated from clinical cases which were characterized by marked inflammation. This concept was later substantiated and enlarged upon with some modification by Hopsu-Havu and co-workers [172,173]. They noted the production of urease and sulfatase in strains of *E. floccosum.* Interestingly, pleomorphic cultures were more active enzyme producers than were granular forms.

Recent investigations by Yu et al. [426,427] have revealed the production of three keratinases (I, II, III) by *T. mentagrophytes* var. *granulosum.* Keratinase I has been shown to digest unautoclaved white guinea pig hair by removing the medulla of the hair and producing cortical fissures. Keratinases II and III are cell bound and are immunologically dissimilar from keratinase I and each other. The pathogenic role of the two cell-bound enzymes is unknown at present.

Kunert [214,215] studied the proteolytic activity and products of keratin degradation by a strain of *M. gypseum.* He noted three phases of degradation: the initial phase, characterized by rapid growth, digestion, and increase in proteolytic activity; the middle phase, characterized by a decrease in the rate of phase one activities; the final phase, defined by a new increase in substrate decomposition but with a continuing decrease in other parameters examined [214]. The principal products of the keratinolysis were simple peptides. Further study by Kunert [214] on *M. gypseum* and by Ruffin et al. [330] on *Keratinomyces ajelloi* revealed the production of additional enzymes by dermatophytes. Sulfite is capable of splitting disulfide bonds of keratin and of producing S-sulfocysteine and possibly cysteic acid [330]. Kunert [215] proposed that keratin may be denatured and made digestible to exoproteases by such "sulfitolysis." Ruffin et al. [330] further proposed that proteolytic digestion and sulfitolysis may occur simultaneously. The issue is still under investigation.

In 1972, Nobre and Viegas [283] reported that over 75% of freshly isolated strains of *M. gypseum, M. canis, E. floccosum,* and *T. mentagrophytes* were lipase producers. Although the authors failed to relate lipase production and activity to pathogenesis, Das and Banerjee [78] reported that a strain of *T. rubrum* secreted lipase and phospholipase A when grown in Sabouraud's broth. They noted that phospholipases may play an important role in maintaining the function of the cell membrane and may aid dermatophytic invasion of host cells.

In addition to purely enzymatic studies of dermatophytes, the pattern of host invasion by these fungi has been observed microscopically. English [104,105] has described the in vitro invasion of keratin by dermatophytes. Using a light microscope, she noted that the invading mycelia were flattened, branched fronds that forced themselves beneath cuticular scales. A more recent electron microscopic study by Baxter and Mann [31] revealed that eroding dermatophytic hyphae first digested the poorly resistant endocuticular region of hair. The exocuticle layer was eroded next, leaving the resistant epicuticle. The observations indicated a concerted enzymic and physical attack on host tissues. Most recently, a report by Farley et al [110] examined the origin and ultrastructure of intrahyphae of *T. terrestre* and *T. rubrum.* Further research is required in this area, but present data at least suggest that the emergence of intrahyphal hyphae during dermatophytic infection may favor fungal survival in vivo in spite of a host-protective immune response, and thus act as an additional virulence factor in many dermatophytic infections.

IV. ASPERGILLOSIS

A. Description and General Pathogenesis

Aspergillosis may exist in many diverse disease forms. The true spectrum of syndromes has been described by Conant et al. [63] who state: "Aspergillosis . . . is characterized by the presence of inflammatory, granulomatous lesions in the skin, external ear, nasal sinuses, orbit, eye, bronchi or lung, and occasionally in the nasopharynx, vagina, uterus, heart valve, pleural cavity, mediastinum, bones, brain, and meninges." All forms of aspergillosis are found worldwide in humans. The etiologic agents are members of the ubiquitous, large genus *Aspergillus.* The vast majority of these organisms are soil and air saprophytes. The most common etiologic agents of human infections are *Aspergillus fumigatus* and *A. niger,* although *A. flavus, A. oryzae, A. glaucus,* and *A. nidulans* also have been implicated, as well as numerous other species.

Aspergillosis actually represents a group of diseases that has been divided into three categories: (1) allergic, (2) colonizing, and (3) invasive. Rippon [314] has further separated the diversity of forms into the following syndromes: pulmonary, disseminated, central nervous system, cutaneous, nasal orbital, and iatrogenic. Numerous reports in the literature support this classification as well as the premise that aspergillosis, in general, is an opportunistic infection primarily affecting debilitated, immunologically altered patients [20,50,94,258,270,310,328,402]. Allergic aspergillosis is a syndrome that has acquired more attention recently [167], but we will not discuss it in this chapter.

Animal models have proven to be very useful in studies of disease pathogenesis in aspergillosis. These studies include tumor production by aflatoxins [25,236,239], pathology of the disease [294], strain virulence and distribution in vivo [112,224,302,317], and pathogenicity in immunologically altered animals [223,410].

B. Pathogenic Mechanisms

One of the more thoroughly investigated areas of fungal pathogenicity has been the mycotoxins, and more specifically, the aflatoxins. One of the more widely studied organisms has been *A. flavus,* which is known to produce several toxic substances, including oxalic acid, kojic acid, tremorgenic acid, aspergillic acid, β-nitroproprionic acid, and aflatoxins [413]. Generally, the aflatoxins represent a group of secondary metabolites which were discovered as contaminants of certain lots of animal feed. These compounds have demonstrated a high order of acute toxicity in many animal species and humans, and cause what has been termed "mycotoxicoses," mycotically caused disease syndromes resulting from the ingestion of preformed toxic metabolites. These compounds have been shown to possess potent carcinogenic properties in several animal and fish species [25,236,239,268, 280,344]. It is not our purpose to review aflatoxins or mycotoxins in this chapter, as we are concerned with mechanisms of invasive pathogenesis rather than toxicoses. Several excellent papers and reviews on the synthesis, degradation, and general or specific characteristics of mycotoxins have been published [95,96,170,325,413,418]. See also Chap. 10.

In 1939, Henrici [168] described a toxin derived from young cultures of *A. fumigatus* and *A. flavus.* He demonstrated that the crushed mycelial "cell sap" or "endotoxin" was toxic for rabbits, guinea pigs, and chickens by subcutaneous, intraperitoneal, or intravenous routes. Oral administration of the endotoxin to animals gave no response, thus eliminating the possibility that the endotoxin was a mycotoxin. Henrici also noted that this toxin exhibited hemotoxic, neurotoxic, and histotoxic effects. Salvin [333] later demonstrated endotoxin activity in numerous fungi, including *A. fumigatus.*

Tilden and her associates [304,392,393,422] reported a series of exacting experiments involving the extraction, fractionation, purification, and determination of pathogenic characteristics of *Aspergillus* endotoxin. They clearly demonstrated that the toxin in question was cell-wall related (and thus is an endotoxin, if the term is broadly defined) and not a fungal by-product or exotoxin. The results showed that the toxic moiety was a glycopeptide and that the *A. flavus* and *A. fumigatus* toxins were electrophoretically dissimilar [422]. Further analysis showed that for dogs and rats the *A. flavus* toxin was less potent than the *A. fumigatus* toxin, although rabbits were killed by both toxins in similar doses. Additionally, both toxins were very nephrotoxic, leading to necrosis of the renal tubular epithelium with subsequent fatal uremia. Pulmonary hemorrhage and edema, chronic passive congestion with fatty infiltration of the liver, necrotic foci in the spleen and heart, and focal areas of cellular exudate in several visceral organs also were reported [392]. These early investigations contributed greatly to the study of pathogenic mechanisms and encouraged other investigators to assay for additional toxins or enzymes that may be produced by the aspergilli.

In the course of their cultivation, fungi are known to produce substantial quantities of enzymes that are secreted into the culture medium [141,220]. Nordwig and Jahn [284] reported the purification of a collagenolytic enzyme from *A. oryzae.* The isolated protease was shown to be free of nonspecific activities and it preferentially split protein at the peptide bonds and, at times, acted like an esterase. Its role in collagen degradation in vivo is still disputed. Azuma and his associates [14-16] isolated two kinds of polysaccharide, a glucan and glycopeptide (galactomannan-peptide), from *A. fumigatus* by a pyrimidine extraction procedure. The glucan did not show any immunological activity, but the glycopeptide showed precipitation, complement fixation, and passive hemagglutination reaction with specific rabbit antisera. The glycopeptide also elicited both Arthus and delayed-type skin reactions in sensitized rabbits and guinea pigs. Recently, Blythe [38] isolated several immunologically nonreactive polysaccharides and a very reactive glycoprotein from *A. clavatus.* The protease-resistant glycoprotein was shown to be an alveolitis-inducing substance when administered intranasally to mice.

Several experimental studies of the pathogenicity of aspergilli have been performed using murine models. Ford and Friedman [112] studied the relative virulence of 14 species of aspergilli for mice. A wide range of fungal virulence was demonstrated without any correlation of virulence with any defined spore or toxin characteristics. Pore and Larsh [294] studied the experimental pathology of murine infection with *Aspergillus* spp. They noted the appearance of cerebral aspergillomas, hyphal invasion of nonnecrosed tissues, and spore formation in intact tissues and vessels. The predilection for vein localization and sporulation in vivo was strain dependent.

Rippon et al. [317] noted that significant differences existed between soil and human isolates of *A. terreus* with regard to mouse virulence and in vitro growth rate. The results suggested that particular strains of potentially pathogenic fungi may be well adapted to the host's tissue environment, thus enhancing their ability to survive, multiply, and cause disease.

Tissue invasion and localization by aspergilli also have been analyzed. Purnell [302] quantitated histologically the extent of fungal invasion in brain, heart, and kidneys of intravenously inoculated mice. He noted that a direct relationship between strain virulence and the extent of tissue invasion by *A. nidulans* occurred in the brain but not in the heart or kidney. Since the results were strain dependent, he suggested that invasion of murine brain tissue may serve as a phenotypic marker of murine virulence for *A. nidulans.* Recently, Miyaji and Nishimura [263] correlated the amount of proteolytic activity of *A. fumigatus* with this organism's hyphal invasion of murine brain tissue. They noted that proteolytic activity increased in the mouse brain in proportion to the growth of hyphae within the cerebral tissue.

Corticosteroid therapy may render humans susceptible to systemic infections from opportunistic fungi [44,103,314]. Mice treated with cortisone have been shown to have lowered resistance to fungal infection in general [235,354]. Lehmann and White [223] demonstrated renal localization and enhanced growth of *A. fumigatus* hyphae in cortisone-treated mice. These investigators used a chitin assay which quantitated the total amount of mycelium that developed in susceptible organs. Lehmann and White suggested that certain host responses to mycelial development naturally exist in the kidney but are inhibited by cortisone treatment, thus permitting fungal invasion and multiplication. This hypothesis was later supported by data involving spore germination in murine lungs [223] and additional work with murine kidneys in vivo and murine kidney extract [224]. At present the reasons for renal localization of aspergilli are unknown, but recent data suggest that either the inefficient host defenses or an as yet undefined chemical germination stimulant may be major factors that determine the initial rapid spore germination in the kidney. Answering this question will draw us closer to a more complete understanding of the pathogenesis of aspergillosis.

V. PHYCOMYCOSIS

A. Description and General Pathogenesis

Phycomycosis is an acute, usually fulminating mycosis affecting the wall and lumina of blood vessels. It may be also a chronic, granulomatous subcutaneous syndrome characterized by marked deformation of the infected tissues [63]. A variety of fungi are the etiologic agents, including members of the genera *Mucor, Rhizopus, Absidia, Basidiobolus, Entomophthora,* and *Conidiobolus.* In terms of modern fungal nomenclature the class Phycomycetes has been rearranged. However, there is considerable sentiment for retaining the disease term of phycomycosis. The organisms are soil and water saprophytes that occur worldwide. Some species preferentially infect insects. Most of the etiologic agents are considered opportunists rather than primary pathogens. The spores, which become airborne, are the infectious particles. Infection can be initiated by inhalation, ingestion, or traumatic inoculation of spores [314].

Infections caused by the opportunistic phycomycetes are usually associated with a predisposing factor which has altered the host's immunological competence. Host modifications include primary or secondary immunodeficiency states, e.g., diabetes mellitus, leukemia, lymphoma, and a variety of drug-induced immunodeficiencies [103]. Generally, the acute phycomycotic infections of the upper respiratory tract, caused by species of the genera *Absidia, Mucor,* and *Rhizopus,* are described as opportunistic because they occur in patients who have been on prolonged steroid or antibiotic therapy and whose metabolism is altered and resistance lowered by debilitating disease, particularly diabetes [292,385]. It is questionable whether the subcutaneous, chronic infections caused by the Entomophthorales are truly opportunistic [254].

Clinically, a wide variety of manifestations have been described in the literature. Bovine cerebral phycomycosis associated with vasculitis, perivascular necrosis and septic thrombosis of the brain [208], and bovine mycotic abortion [181] have been reported. Cerebral phycomycosis is an occasionally reported syndrome in humans which has been reviewed by Hale [156]. Cases of ocular and orbital phycomycosis have been reported [191,345], as have incidents of chronic upper respiratory tract phycomycosis characterized by progressive nasal obstruction and severe facial disfigurement [254]. Reports of localized subcutaneous phycomycosis caused by a variety of organisms are numerous [158, 192,306,309]. However, recent reports by Koshi et al. [211] and Kamalam and Thambiah [193,195] have shown that deeper tissues, such as muscle and lymph nodes, are involved in some cases. Thus, the disease is not limited solely to the dermal and subdermal tissues.

B. Pathogenic Mechanisms

Few works have been published in the area of specific mechanisms of pathogenesis of phycomycotic agents. It is apparent that the responsible organisms have a predilection for growth in blood and lymph vessels [44], subsequently leading to vasculitis, perivascular necrosis, and septic thrombosis. As a result of this observation, the route of dissemination of many systemic phycomycotic infections is believed to be hematogenous. In addition, it has been hypothesized that if the number of spores entering the blood is large enough to overcome the ability of the reticuloendothelial system to clear them, infection may result in the most vulnerable or target tissues [208]. Kamalam and Thambiah [195] have reported in vivo sporulation of *Conidiobolus coronatus* in the lymph nodes of a naturally infected patient. Theorizing that free conidial production in the lymph nodes may permit the disease to be infectious, the investigators suggested that the pathogenic significance of in vivo formation of conidial and germ tubes should be pursued further.

Baker and Linares [19] studied experimental prednisolone-induced phycomycosis in rhesus monkeys. It was noted that the normal monkey, like the normal human, displayed a natural resistance to *Rhizopus* infection, but that the prednisolone-treated animals easily succumbed to severe systemic disease. The lesions were fulminating in the kidneys and gastric mucosa, but were localized or minimal in other viscera. The authors suggested that the mechanism of induced spread and proliferation of *Rhizopus* in prednisolone-treated monkeys may be related to the inhibition of the inflammatory reaction and the reduction of polymorphonuclear cell enzymes.

Studying phycomycotic infections in athymic nude mice [66] and in New Zealand Black mice [65], which become thymus dependent and lymphocyte deficient with age, Corbel and Eades discovered that resistance of mice to experimental infection with several phycomycotic agents was independent of effective T-lymphocyte function and probably was dependent on nonspecific host defenses such as phagocytosis. Recently, these researchers have examined the role of spore distribution in vivo and the presence of tissue-specific fungal growth stimulants or inhibitors in mice experimentally infected with *Absidia corymbifera* [67]. They observed that intravenous inoculation of spores resulted in generalized distribution of spores in vivo, with the liver and spleen acting as spore concentrating organs. Fungal growth-stimulating factors were present in extracts of all tissue, although only the brain and kidney preparations maintained stimulatory activity as concentrations approached those of whole tissue. The authors concluded that a combination of local factors, as yet undetermined, were responsible for the localization and proliferation of *A. corymbifera.*

Finally, the only studies concerned with enzyme production by the phycomycotic agents have been performed with the entomopathogens and a pathogen of crayfish. Gabriel [126] demonstrated lipase and protease activities in vitro in germ tubes of *Conidiobolus coronata.* Generally, fungi parasitic on arthropods are known to produce extracellular enzymes which permit them to penetrate their hosts [324]. For example, Soderhall and his associates [363,364] have demonstrated chitinase and protease activity in the crayfish infecting organism, *Aphanomyces astaci.* Further investigations in the area of phycomycotic extracellular enzymes, particularly with the human pathogens, may prove immensely helpful in the elucidation of phycomycotic mechanisms of pathogenesis.

VI. BLASTOMYCOSIS

A. Description and General Pathogenesis

Blastomycosis, historically known as North American blastomycosis, is a chronic granulomatous and suppurative disease with a clinical spectrum ranging from primary pulmonary to disseminated multisystem involvement. The dimorphic fungus, *Blastomyces dermatitidis,* is the only etiologic agent of the disease. The major endemic region is the Mississippi River Valley basin in the United States. The complete geographic distribution of the organism has not been elucidated since the fungus has not often been recovered from saprophytic environs and a reliable skin test antigen has not been developed [314].

Three clinical forms of blastomycosis are recognized: pulmonary, disseminated systemic, and cutaneous [44]. The dimorphic character of this organism is thought to be an important factor in the pathogenic potential of this fungus. The small conidiospores (less than 5 μm in diameter) produced by the mycelial phase of this fungus have been shown to readily undergo transformation to the in vivo yeast-like form [136]. This favors the view that the spore is the primary infectious unit of this pathogenic fungus, and that pulmonary disease results from inhalation of these entities [101,136]. Pulmonary blastomycosis may be acute and fulminating [341] or chronic [101,222,303]. Acute forms are rare and are the easiest to diagnose. Chronic blastomycosis has proven to be more troublesome [222]. Following spore inhalation a pneumonitis occurs, usually resulting in hilar lymphadeno-pathy that is not diagnosed frequently since it becomes symptomatic in only one-third to one-half of the patients [101]. It appears that this pulmonary form of blastomycosis is a benign, self-limiting infection, and its severity is dependent on host defenses [136]. The remaining pulmonary forms of this disease, which includes cavitation, fibronodular infil-tration, and mass-like lesions, are similar in roentgenographic appearance to other chronic pulmonary diseases [303]. Recently, it has been reported that reactivation, or endo-genous activation, of "cured" pulmonary blastomycosis may occur months to years after all signs of clinical disease have disappeared [222].

A murine model of blastomycosis using nasal instillation as a route of infection was reported in 1978 [160]. The authors reported the production of acute, progressive, con-solidated fungal pneumonia in mice analogous to that described in humans. Further de-velopment of this and similar animal models may pave the way to further understanding this syndrome.

Disseminated systemic blastomycosis is believed to be an extension of the pulmonary form. The actual mechanism of dissemination from a primary pulmonary focus to various extrapulmonary structures is unknown, although the accepted routes are the lymphatics and bloodstream. The most common sites of involvement are the liver and spleen, and systemic lesions may vary from frank abscesses to granulomatous nodules. Severe osseous involvement, most commonly of the long bones, may result also in some disseminated forms [137].

Cutaneous lesions are a common form of extrapulmonary blastomycosis. Their ap-pearance is often characteristic enough to reach a tentative diagnosis. A lesion develops slowly, sometimes taking months to years to evolve fully. It becomes characteristically ulcerated, verrucous, and has a well-defined, raised border [44,314]. Satellite colonies may appear at the periphery of a lesion.

B. Pathogenic Mechanisms

A few characteristics of *B. dermatitidis* may be considered as possible mechanisms of pathogenesis. Many of the tissues infected with yeast cells of *B. dermatitidis* are usually infiltrated by granulocytes [103]. Sequential steps in the process by which granulocytes locate and kill infecting microorganisms have been elucidated and include chemotaxis, phagocytosis, metabolic activation, and intracellular killing [378,379]. Sixbey et al. [357] examined the in vitro interaction between human granulocytes and the yeast phase of *B. dermatitidis*. They observed that granulocytes readily phagocytized the yeast cells, but fungicidial assays revealed that only 29% of the inoculum was killed. The authors sug-gested that the difference between phagocytosis and intracellular killing of yeast cells may be due to fungal resistance to the antimicrobial mechanisms of the granulocytes. In addi-tion, the investigators speculated that this resistance to killing may be related to the "re-lentless progression" of untreated systemic blastomycosis. In light of this, it is interesting to note that delayed hypersensitivity has been reported to be protective in murine blasto-mycosis and can be passively transferred in the same system [346,367]. It is apparent that additional research is needed in the area of host immune response to fungal infection, particularly in blastomycosis.

In 1942, Baker [17] reported that mice inoculated intraperitoneally with large doses of heat-killed yeast cells of *B. dermatitidis* developed a histologic reaction similar to one produced in mice by viable, virulent cells. This response was described as an endotoxin-like activity. He extracted a yeast cell phospholipid fraction which produced a mono-nuclear (granulomatous) response in mice. DiSalvo and Denton [90] noted that increased total lipid content of yeast cells of *B. dermatitidis* correlated directly with increased viru-lence of those cells for mice, and by inference, humans. A correlation with phospholipids, however, was not shown. While studying the "endotoxin" of *B. dermatitidis,* Taylor [388] noted that treatment of yeast cells with trypsin or hydrochloric acid greatly en-hanced the endotoxic activity. Taylor concluded that the "endotoxin" of *B. dermatitidis* was a cell wall rather than cytoplasmic fraction, and that the enhanced endotoxicity was the result of trypsin or HCl removal of cell surface proteins, with subsequent exposure of the previously masked toxic moiety.

Cox and Best [70] determined that more chitin and protein were in the cell walls of a virulent strain of *B. dermatitidis,* whereas a higher polysaccharide content was present in the cell wall of a less virulent strain. In addition, on the basis of organic phosphorus content, the authors reported that considerably more phospholipid was associated with the more virulent strain. In a later study, it was noted that cells, trypsin-treated cell walls, and alkali-extracted yeast cell walls were lethal for mice, and that on a dry weight basis, cell walls were more lethal than whole yeast cells [71]. The results suggested that the cell walls indeed possessed endotoxin activity. It was further reported that the lipid moiety of the *B. dermatitidis* yeast cells was probably responsible for the observed granu-lomatous reactions in mice.

In addition to lipid-associated endotoxin activity of *B. dermatitidis* yeast cells, a water-soluble, serologically active polysaccharide has been isolated from mycelial cells of the fungus [13]. It was determined immunologically that this polysaccharide is a common mycelial antigen in *Histoplasma, Paracoccidioides,* and *Blastomyces* species. Its signifi-cance in the pathogenicity of blastomycosis or other fungus diseases has not been deter-mined, but studies are in progress.

VII. PARACOCCIDIOIDOMYCOSIS

A. Description and General Pathogenesis

Paracoccidioidomycosis, or South American blastomycosis, is a chronic, progressive, fungus disease endemic to Central and South America. The disease is characterized by primary pulmonary involvement with subsequent lymphadenitis and pulmonary cavita-tion, and by mucocutaneous ulcerations of the buccal and nasal areas. The organism may become disseminated with systemic involvement of several organs.

The dimorphic fungus, *Paracoccidioides brasiliensis,* is the sole etiologic agent. Little is known about the epidemiology of this organism since it has been isolated only sporadic-ally from nature [5,274]; however, it is thought to be a wood or soil saprophyte. The role of bats in the epidemiology of the disease has been investigated to some extent. In 1965, Grose and Tamsitt [155] reported the isolation of *P. brasiliensis* from the intestine of a frugivorous bat; however, a very complete study by Greer and Bolanos [152] indicated that mycelial particles or yeast cells of *P. brasiliensis* could not survive in the digestive tract of the same species of bat for more than 8 h. They suggested that the bat is not a common natural reservoir for *P. brasiliensis* and thus may not play a role in the epidemiology of the disease.

Paracoccidioidomycosis is the most common systemic mycosis in Latin America, where more than 5000 cases have been reported [244]. The pathogenesis of this disease is still heavily debated. The long incubation period between initial exposure to the fungus and clinical manifestation of disease as well as problems in determining the organism's natural habitat have contributed to the indecision. Early reports suggested that the oropharynx or intestines may be the sites of primary infection [124]. However, primary cutaneous lesions are rare in paracoccidioidomycosis [125]. Thus, inhalation of spores is now accepted as the primary route of infection. Two facts seem to favor the lungs as the main means of entry. First, many cases with systemic manifestations without evidence of skin or mucosal lesions have been reported, and second, lung involvement is the most common finding, varying from 56 to 94% in different reports [42,125,139].

On the basis of these and other data, Giraldo et al. [139] proposed a model for the pathogenesis of paracoccidioidomycosis which includes an infection phase and a disease phase. According to these authors, *Paracoccidioides* infections involve an initial asymptomatic pulmonary form which subsequently transforms into a quiescent latent form, which is also asymptomatic and may be long standing. The disease state may be an acute or subacute progressive form, a chronic progressive form, or an initial symptomatic form in which patients would develop clinical manifestations shortly after initial contact with the fungus. Negroni and Negroni [275] 11 years earlier proposed a classification scheme of a simpler format which divided the disease syndromes into (1) asymptomatic or subclinical infection, (2) acute pulmonary disease, (3) chronic pulmonary disease, (4) acute disseminated disease, and (5) chronic disseminated disease.

Further study is necessary to define completely the pathogenesis of paracoccidioidomycosis. A few recent animal models of paracoccidioidomycosis have been reported which may help clarify the situation. Restrepo and Guzman [308] reported that the mycelial phase of *P. brasiliensis* can give rise to active murine infection by the respiratory route. Following nasal instillation, pulmonary infection was obtained in 38% of cortisone-treated mice, with 14.3% of the animals experiencing dissemination of the fungus to the spleen and liver. It is interesting to note that no deaths occurred, although one strain of fungus caused a more severe pulmonary infection in comparison to another tested strain. Most recently, Hay and Chandler [165] examined the dissemination of *P. brasiliensis* inoculated intravenously into mice. The lung was the earliest and most consistently infected organ, and a secondary dissemination from the lung to the he d region occurred via the blood stream. Yeast cells were found around the blood vesse   in the posterior nasal spaces. The authors also noted that after invasion of the nasal iucosa, cells of *P. brasiliensis* were subsequently discharged into the nasal cavity, from whence they might be inhaled or spread to other areas of the mucosa or skin. This possibly could lead to a self-perpetuating mucocutaneous disease.

A recent immunologic study by Mok and Greer [264] revealed that skin test positivity to paracoccidioidin was lowest in symptomatic patients with the longest duration of disease. This observation implied a decrease in specific cell-mediated immunity with prolonged active infection. However, further studies [264] by these authors also indicated that clinically recovered patients regained their immunocompetence. It is apparent that additional clinical surveys and studies are required to help elucidate this aspect of the disease.

B. Pathogenic Mechanisms

Most of the investigations on the ways in which *P. brasiliensis* may cause disease in humans and animals have been recent. Admittedly, some early work was reported using both mycelial-phase [54,93] and yeast-phase culture filtrates [218,219] as skin test antigens in normal and infected patients. Most of these studies used crudely extracted materials, some of which were contaminated with culture medium protein, and results were variable and difficult to interpret.

In 1967, Restrepo and Schneidau [309] reported the extraction and purification of skin test reactive fractions from both mycelial and yeast-phase culture filtrates of *P. brasiliensis*. The extracts were precipitated by ethyl alcohol, and chemical analysis of the reactive fractions revealed the presence of seven aliphatic amino acids: leucine, methionine, alanine, glutamic acid, serine, threonine, and glycine, and the sugars glucose, galactose, arabinose, and glucosamine. Mycelial and yeast-phase extracts were similar chemically, and they exhibited equal reactivity in sensitized guinea pigs. The authors suggested that a strong link existed between the carbohydrate and peptide portions of the extract, leading them to propose that a glycopeptide was the material primarily responsible for skin reactivity.

Using specific immunoadsorption, Yarzabal et al. [424] isolated and purified an antigen from a metabolic extract of *P. brasiliensis* that was shown to have cationic electrophoretic mobility and, more important, an alkaline phosphatase activity. The role of this enzyme in the pathogenesis of paracoccidioidomycosis is not clear at present.

San-Blas and co-workers [338-340] have proposed that the presence of α-1,3-glucan in the cell wall of the yeast phase of *P. brasiliensis* may play an important role in the pathogenicity of this fungus. The authors noted that a laboratory subcultured strain of *P. brasiliensis* developed decreased pathogenicity for mice. Chemical analysis of the less virulent organism revealed a substantial decrease in the amount of cell wall-related α-1,3-glucan. The virulence of this strain was regained upon passage through animals [339] or when grown in medium supplemented with fetal calf serum [340], as was the synthesis of α-1,3-glucan. These investigators suggested that the α-1,3-glucan could act as a protective layer for the fungus and may act by inhibiting phagocytosis of the yeast cells or by making the organism more resistant to a phagocyte's digestive enzymes. If the α-1,3-glucan is an "aggressin" or virulence factor, it may be included with well-defined factors such as the capsule or polysaccharide of *Streptococcus pneumoniae* [99], the Vi antigen of *Salmonella typhi* [62], and the poly-D-glutamic acid of *Bacillus anthracis* [205]. This area of fungal research shows great promise in the attempt to learn more about mycotic mechanisms of pathogenesis.

VIII. LOBOMYCOSIS

A. Description and General Pathogenesis

Lobomycosis, also known as Jorge Lobo's disease or keloidal blastomycosis, is a chronic, localized dermal infection apparently limited to tropical rain forests and bush country in South America, and in particular, the Amazon basin area [314]. The disease is characterized by isolated or grouped, hard, smooth to verrucous dermal nodules that generally are painless. The etiologic agent has been named *Loboa loboi,* although it has not been successfully cultured from nature or from clinical materials.

Little is known about the pathogenesis of the disease. Infection is believed to be by traumatic inoculation of the organism into skin. Patients usually suffer from very slow growing, nodular tumors that never seem to heal spontaneously and may take years to develop fully. There is no evidence of animal-to-human or human-to-human transmission; however, many patients demonstrate multiple lesions, which suggests that tumor distribution may be caused by dissemination by auto-reinfection [412]. Histopathologically, lobomycotic lesions contain fungal yeast-like cells that are either single or in chains of up to four, five, or six cells. Some budding also may be seen. The skin tumors are composed of foreign body-type granulomatous tissue in which numerous Langhan and foreign body giant cells can be found. Many of the latter contain yeast cells. One report has mentioned the presence of a pink-staining capsule in a periodic acid-Schiff stain [12]; however, this has not been substantiated.

A few reports of human lobomycosis [12,89,266,296,297,405,412] as well as a natural animal infection [51] have appeared in the literature. Azulay et al. [12] reported a human case of lobomycosis with lymphatic involvement in which fungal cells were seen in a lymphocytic exudate. Wiersema and Niemel [412] reported on 13 cases of human lobomycosis that occurred in Surinam. Because of the mild nature of the disease, these authors proposed that the skin lesions should be termed "Lobo's mycotic skin granuloma" rather than "Lobo's disease." Caldwell et al. [51] reported a case of lobomycosis in an Atlantic bottle-nosed dolphin captured off the Florida coast in 1974. The numerous lesions were histologically and microbiologically similar to those seen in human cases, and all attempts to isolate the responsible agent or transmit it to experimental animals were unsuccessful.

Experimental lobomycosis is a developing field. Borelli [40] inoculated the knee of a volunteer with a scraped tissue suspension taken from a patient with lobomycosis and reported that a large lesion developed at the infected site over a 4-year period. Numerous cells of *L. loboi* were seen in the resulting exudate 2 years after inoculation.

Animal experimental models also have been attempted. Successful infection has been reported in the footpads [412] and cheek pouches [334] of hamsters. Sampaio et al. [336] have reported the induction of *L. loboi* containing lesions in three species of inoculated tortoises as well as in an armadillo [335]. The authors were able to produce, in 11 months postinoculation, a 4 × 3 × 3 cm lesion identical grossly and microscopically to human lobomycotic lesions. This quick development suggests that the armadillo may be a suitable experimental animal for studying lobomycosis.

B. Pathogenic Mechanisms

No specific mechanisms of pathogenesis of *L. loboi* have been reported. Until the fungus can be cultured on laboratory media, elucidation of further characteristics of the organism and disease may prove difficult.

IX. COCCIDIOIDOMYCOSIS

A. Description and General Pathogenesis

Coccidioidomycosis is a benign, subclinical upper respiratory infection which may spontaneously resolve or progress into a severe systemic disease that may prove fatal. Disseminated disease may include fungal colonization of any visceral organ or the meninges, bones, joints, subcutaneous or cutaneous tissues [18,103,314]. Dermal involvement is characterized by the formation of "burrowing abscesses" [314].

Coccidioides immitis, a dimorphic soil saprophyte, is the sole etiologic agent. The organism has a geographical distribution limited to Lower Sonoran biomes [44], areas that are hot and semiarid. Known highly endemic areas include the southwestern United States and northern Mexico, as well as a few isolated areas in South America.

The morbidity from coccidioidomycosis far outweighs the mortality. It has been estimated that there are 100,000 cases of coccidioidomycosis a year in the United States, with 70 deaths annually [382]. Research on this disease has continued at an increasing pace for the past 20 years and has been reviewed excellently by Drutz and Catanzaro [97,98].

C. immitis is considered to be one of the more virulent mycotic agents. The saprophytic mycelial phase produces relatively small (4 X 6 μm), thick-walled, barrel-shaped cells called arthroconidia, which are specialized segments of hyphae. When disturbed, the arthroconidia break away from the hypha and become airborne. It is believed that inhalation of a single arthroconidium can initiate infection [44]. Following inhalation, the arthroconidia convert into the in vivo spherule or sporangium phase, and the subject becomes skin test-positive. This initial infection may resolve, leaving residual foci or progress to a disseminated state, causing extensive pulmonary or extrapulmonary involvement. Reinfection has been defined as the development of activity at some time after the occurrence of primary infection and its associated hypersensitivity. This has been clinically documented in the literature [135,416]. Reinfection can be by an endogenous source or by exogenous organisms. The former is a more common form and may occur years after a clinical cure; the latter may occur on top of a previously active, healing, or healed infection [332]. Dissemination may occur from any form of the disease.

A wide variety of clinical manifestations of coccidioidomycosis has been reported. Pulmonary coccidioidomycosis is the most common form of the disease and can be inapparent or very severe. Secondary pulmonary coccidioidomycosis may be benign and chronic, and is eventually accompanied by residual cavitation [81]. Thadepalli et al. [391] have reported a rare case of "pulmonary mycetoma" due to *C. immitis.* The authors found mycelial elements in residual pulmonary coccidioidal lesions and reported concern regarding the exhalatory release of infectious particles by such patients, although this phenomenon has not been documented. Coccidioidal meningitis [56] and spondylitis [417] have been reported as more severe forms of dissemination from a primary pulmonary focus. Even a minor syndrome, otomycosis, has been attributed to hematogenously disseminated *C. immitis* [159]. The chief concern with this manifestation is the possible neurological consequences that could result.

Experimental animal models have been used to study the adverse influence of corticosteroids [82], the infectivity and communicability of *C. immitis* [213], the development of immunity [225,226], and the range of virulence and pathogenicity of the fungus [116, 129] in coccidioidomycosis. These models have provided numerous answers, but despite the large amount of research that has been directed toward coccidioidomycosis, many questions remain unanswered.

B. Pathogenic Mechanisms

Most research concerned with coccidioidomycosis has been directed toward the epidemiology, diagnosis, and general pathogenesis of the disease rather than a search for specific mechanisms of pathogenesis. An early study by Friedman et al. [116] revealed a wide range in virulence among 27 strains of *C. immitis* inoculated into mice. A later study by Gale et al. [129] demonstrated a variation in virulence among three strains of *C. immitis* inoculated into four strains of mice. These workers determined that the dissemination rate was higher in pigmented than in albino mice, and concluded that there were too many factors involved in virulence to use animal mortality as a sole criterion of pathogenicity.

Additional studies by Gale et al. [128,129] revealed the presence of a water-soluble polysaccharide toxin. This toxin was isolated from culture filtrates and protected treated mice for a limited time against a lethal intraperitoneal challenge with *C. immitis*. However, when suspended in hog gastric mucin, the polysaccharide was toxic to mice, and the authors concluded that the mucin and culture filtrate polysaccharide shared a common antigen.

A more recent study by Anderes et al. [6] compared the lipid content of a virulent wild-type strain and an avirulent strain of *C. immitis*. They noted a major difference in the lipid composition between the virulent and avirulent strains in the arthroconidia stage. The virulent strain contained a higher total lipid content and lower C_{18} fatty acid content than the avirulent mutant. The higher lipid content in the infectious arthroconidia of the virulent strain may have been one factor in the capacity of this strain to initiate infection. Increased lipid content also has been associated with increased virulence in strains of *Histoplasma capsulatum* [281] and *Blastomyces dermatitidis* [70,90]. In vitro experiments have suggested that a high lipid content may help to protect microorganisms against attack by host defense mechanisms, thus enhancing disease-causing capabilities [7].

An extracellular protease isolated from *C. immitis* culture filtrates has been reported [327]. The synthesis of the protease was dependent on available sources of nitrogen and was most active when provided with peptone as a substrate. Its role in the pathogenesis of coccidioidomycosis has not been delineated, but if produced in vivo it has the potential of being an important virulence factor.

Much time and effort has been devoted to the immunology of coccidioidomycosis, some aspects of which appear to be related to pathogenic mechanisms of the disease. In 1961, Huppert [182] noted that delayed-type hypersensitivity in human cases of coccidioidomycosis was associated with resistance to disease spread. He hypothesized that delayed-type hypersensitivity initially functioned as a host defense mechanism, but that this process physically contained a tissue-destructive component which, in the presence of excess antigen and serum antibody, could prove fatal to the host. More recently, Galgiani and Stevens [130] reported that soluble mycelial antigens from *C. immitis* act as chemotaxinogens, which attract leukocytes indirectly by a complement-mediated pathway. Other reactive antigens derived from extracted cell wall fractions have been reported [409]. Recently, Huppert et al. [183] have emphasized the importance of standardizing and referencing many of the antigens isolated from cultures and cells of

C. immitis. These investigators presented applications of a reference system for standardization of reagents, for detecting common antigens, for monitoring successive steps during fractionation of crude preparations, and for fingerprinting strains for ecological and epidemiological studies. It is hoped that this approach will include studies on the pathogenesis of coccidioidomycosis.

X. HISTOPLASMOSIS

A. Description and General Pathogenesis

Histoplasmosis may be an acute, subacute, chronic pulmonary, or systemic mycosis. The organism has a predilection for the reticuloendothelial system and may involve any organ, including the central nervous system. The dimorphic soil saprophyte *Histoplasma capsulatum* is the sole etiologic agent. It has been isolated worldwide, mostly from soils enriched by the excreta of bats, chickens, starlings, and other birds [429]. Microconidiospores (2-3 μm in diameter) produced by the saprophytic phase in soil are believed to be the infectious particles, and infection is initiated by inhalation of these spores.

The majority of histoplasmosis cases are asymptomatic and inapparent except for a positive skin test reaction to histoplasmin. Although many variations of the disease occur, the most common clinical syndromes are (1) acute pulmonary, (2) chronic cavitary pulmonary, (3) disseminated systemic, and (4) presumed ocular.

Acute pulmonary histoplasmosis is generally a benign disease, with most patients experiencing transient, flu-like symptoms. Although this is the most minor form of the disease, it can disseminate and cause a fulminating fatal illness.

Chronic cavitary histoplasmosis is comparable to cavitary tuberculosis and was misdiagnosed as such for many years. This syndrome is characterized by fibrosis and calcification [29], and like tuberculosis, the formation of a "Ghon complex" may be induced. This syndrome may be an extension of the primary acute pulmonary form, or it may occur by reinfection either from an exogenous or latent endogenous source [29,145].

Disseminated systemic histoplasmosis may be acute or chronic. The acute form usually occurs in young children who probably have inhaled massive numbers of spores. This disease form, however, does not localize in the lungs but rapidly disseminates throughout the body and is usually fatal. The chronic form often appears to be an opportunistic invasion of immunosuppressed patients. A clinical survey by Kauffman et al. [203] showed that the most common predisposing factors were Hodgkin's disease, chronic lymphocytic leukemia, and acute lymphocytic leukemia. One unique case reported by Couch et al. [68] concerned a patient with common variable hypogammaglobulinemia who manifested histoplasma meningitis without disseminated disease. If untreated, progressive chronic histoplasmosis also may manifest itself with mucocutaneous or cutaneous lesions [28]. These lesions are an indication of the severity of the disease and are useful as sources of clinical material [44]. Further characteristics of disseminated systemic histoplasmosis are being elucidated with experimental animal models [30,92,256].

Presumed ocular histoplasmosis is a disease of the eye which often involves the macular as well as the peripapillary regions and frequently leads to blindness [113,420]. Early investigators were not able to demonstrate *H. capsulatum* in affected eyes; thus, the term "presumed ocular histoplasmosis" was applied to the disease syndrome. However, there are now numerous reports in the literature of both histopathologic and cultural proof that *H. capsulatum* is present in many cases of ocular disease [73,117,123,133,329,331,

362,421]. Additional evidence that *H. capsulatum* may cause ocular lesions has been provided by experimental animal models. Wong et al. [419] reported the production of focal histoplasmic choroiditis in rabbits by intravenous inoculation of viable yeast cells of *H. capsulatum.* Smith et al. [361,362], Woolsey et al. [421], and Fromtling et al. [123] also have produced choriodal lesions in rabbits. O'Connor [286] has listed the problems of using a rabbit model in relation to human disease and has proposed the future use of anthropoid species as a more adaptable animal model.

B. Pathogenic Mechanisms

Research reports concerning the specific mechanisms of pathogenesis in histoplasmosis are few, but several unique and promising ideas in the realm of fungal pathogenesis have been presented. In 1968, Berliner [33] reported the discovery of two morphologic types of *H. capsulatum* obtained from primary subcultures: an albino (A) and brown (B) type. These types were indistinguishable in yeast phase but were easily separated by mycelial phase morphology. Daniels et al. [76] and Tewari and Berkhout [389] compared the virulence of these two types of *H. capsulatum* in rabbits and mice. Daniels et al. were unable to reach a firm conclusion regarding differences in the pathogenicity between these strains in rabbits, but Tewari and Berkhout noted that type B cells were more pathogenic in mice than were type A. Their data suggested in vivo conversion of one morphologic type into the other. The significance of this finding is still unknown, and it is hoped further investigation in this area will be initiated.

The role of protoplasts in human and animal disease has been the subject of considerable speculation and some research [404]. In a series of papers Berliner and Reca [34-36] reported the induction of protoplasts from cells of *H. capsulatum.* These workers were able to cause the release of protoplasts from the yeast phase without the use of snail gut enzymes, which were previously reported as a requirement for protoplast induction [35]. They noted that yeast-phase protoplasts emerged from only one location on the cell wall. This area represented the weakest part of the existing glucan cell wall. They suggested that this area may be the site of new bud development. The significance of this work in relation to the pathogenesis of histoplasmosis is unclear; however, the formation of protoplasts provides a possible tool for additional studies in fungal pathogenicity.

The invasive ability of many fungi is related to the adaptation of their metabolic functions to the new host environment [313]. Howard [175,176] has reported the multiplication of yeast cells of *H. capsulatum* within mammalian histiocytes maintained in tissue culture. In addition, host temperature, nutritional state [177,178,180], and tissue oxidation reduction potential (Eh) are important factors. The ability of an invading organism to adapt to host conditions is a measure of the invader's tissue tolerance. In 1968, Rippon [313] devised a monitored environment system to study the growth, morphology, and metabolic rate of fungi by measuring oxidation-reduction potentials in vitro. In a later study Rippon and Andersson [316] assayed the metabolic rate of several fungi as a function of temperature and oxidation-reduction potential. Most interestingly, they found that cells of *H. capsulatum* grew well at a final Eh (a summation of several metabolic systems) that was very similar to that of human tissue.

The cell wall and cell products of *H. capsulatum* have been examined for toxic properties. In 1966, Nielson [281] studied the lipid content and virulence of six isolates of *H. capsulatum.* He noted that there was no apparent quantitative correlation between virulence and total extractable lipid, acetone-soluble lipid, or phospholipid. However, these data suggested a possible relationship between quantities of phosphatidylcholine and virulence.

San-Blas et al. [337] analyzed four strains of *H. capsulatum* for cell wall chemical variability and virulence. Although they confirmed the two chemotypes proposed by Domer [91], they were unable to relate cell wall composition to virulence. In particular, they concluded that the α-glucan obtained from *H. capsulatum* yeast-phase cells did not correlate with virulence, as does the α-1,3-glucan found in the cell wall of *Paracoccidioides brasiliensis* [339].

In 1978, Watson and Lee [407] reported the isolation and partial purification of two aminopeptidases from cells of *H. capsulatum.* The higher molecular weight enzyme was a proline aminopeptidase, while the second enzyme was less substrate specific. Although both enzymes were shown to be cytoplasmic, it cannot be ruled out that upon cellular lysis or degradation these enzymes may act on the host, leading to possible tissue destruction and stimulation of host defenses.

Note: African histoplasmosis is a clinically distinct form of histoplasmosis. The disease is characterized by suppurative to granulomatous lesions found primarily on the skin, although lymph nodes and bone may be affected [314]. The disease is limited to North Africa and is caused by the dimorphic fungus *Histoplasma duboisii.* The disease is probably respiratory in origin. Little is known about the epidemiology or pathogenesis of the disease, and no specific mechanisms of pathogenesis have been proposed.

XI. SPOROTRICHOSIS

A. Description and General Pathogenesis

Sporotrichosis is a chronic subcutaneous mycosis with eventual lymphatic involvement. Generalized infection involving bone, joints, lungs, central nervous system, and other internal organs may occur in some cases [44]. The dimorphic fungus *Sporothrix schenckii* is the sole etiologic agent. It is widely distributed in nature and has been isolated from soil, timbers, and a variety of plants [52,246].

Sporotrichosis may be lymphocutaneous, pulmonary, mucocutaneous, or extracutaneous (disseminated). Lymphocutaneous sporotrichosis comprises up to 75% of reported cases [314]. This form of the disease is characterized by a single initial lesion which develops on the skin following local trauma. Generally, a spore-contaminated object such as a rose thorn or a wood splinter acts as the traumatic agent. A nodular to pustular lesion forms at the site of inoculation and eventually spreads up the lymphatics, forming granulomatous to draining lesions that are painless [45,243,414]. The pulmonary form of sporotrichosis is considered to be very rare; however, an increasing number of clinical reports have appeared recently in the literature [74,207,240,311,396,430]. Mucocutaneous and extracutaneous forms of the disease remain rare.

Experimental animal models have been used to study both the pulmonary and lymphocutaneous forms of the disease. Sethi [349] failed to produce primary intestinal infection in hamsters after feeding them yeast cells of *S. schenckii.* He did note, however, that some of the experimental animals developed pulmonary disease. He suggested that the feeding of the yeast cells to the animals may have acted as an aerosol, thus permitting entrance of the yeast cells into the animals' lungs, with the subsequent development of pulmonary disease. Sethi et al. [351] induced pulmonary sporotrichosis in mice following intranasal instillation with yeast cells. Sethi [350] also produced progressive and often fatal infection in mice following intrathoracic inoculation of yeast cells. Barbee et al. [21] have suggested that adult domestic cats may be good experimental models for the lymphocutaneous form of sporotrichosis. The investigators noted that the experimentally induced cutaneous nodules were very similar to those seen in human cases.

B. Pathogenic Mechanisms

Specific pathogenic mechanisms of *S. schenckii* have not been defined. However, there are a few observations and hypotheses concerning some possible pathogenic aspects of this fungus. Investigators have reported the possible presence of a capsule around yeast cells of *S. schenckii;* however, Lurie and Still [242] reported that the spores of this fungus did not possess a capsule in vitro or in vivo. The authors showed that the cells of the fungus in tissue, when stained with hematoxylin and eosin, are surrounded by a halo resembling a capsule, but that this is an artifact due to cytoplasmic shrinkage produced by formalin fixation and histochemical processing. The halo is actually within the yeast cell and its outer surface is formed by part of the cell wall. The authors concluded that the observations of earlier investigators concerning the presence of an apparent capsule were accurate but that the interpretation was incorrect.

Occasionally, a variable amount of a homogeneous eosinophilic material surrounding some cells of *S. schenckii* has been observed in vivo. This unusual appearing element has been termed an "asteroid body." This apparently consists of a central yeast cell which is usually spherical. Its diameter may vary from 5 to 10 μm and occasionally a single bud or a cluster of three or four cells may be observed. The pink-staining material surrounding these cells has a stellate shape due to angular projections of varying length [241].

There has been much speculation about the nature of the eosinophilic material comprising the asteroid body, and there have been some suggestions that this is a mechanism of fungal pathogenesis. Lurie [241] reported that histochemical reactions indicated that the amorphous material consisted of a glycoprotein. Further studies have indicated that the asteroid body is formed by the precipitation of an antigen-antibody complex which is deposited in vivo on the surface of the fungal cell [242]. The authors confirmed this hypothesis by inducing the formation of an identical appearing asteroid body on yeast cells in vitro by suspending cells in specific antiserum. The authors also proposed that the stellate shape of the asteroid body is simply the result of the "yeast-like conidia" changing into chlamydospores, upon which a precipitate of antigen and antibody forms on the surface of the cell. It is questionable whether this antigen-antibody complex may induce the chronic inflammatory response which is seen sometimes in certain forms of sporotrichosis. In addition, it is unknown if portions of this antigen-antibody complex are shed from the asteroid body into tissues, eventually entering the blood or lymphatic vessels, where they can lodge in the alveolar or renal capillaries, thus inducing acute and fulminating immune complex disease.

Various nutritional and environmental factors, including pH and temperature, have a marked influence on the in vitro growth of dimorphic pathogenic fungi. In certain cases such studies help explain the in vivo viability of these pathogens. For example, Mariat et al. [253] have reported that certain strains of clinical isolates of *S. schenckii* require the pyrimidine moiety of thiamine. Also, *S. schenckii* isolates have been reported to be unusually temperature sensitive [217,245], and this characteristic appears to be a major factor in the epidemiology and pathogenesis of sporotrichosis. Recently, a report by Catchings and Guidry [55] showed that under experimental conditions pH and temperature both influenced the in vitro growth of *S. schenckii* yeast cells. They noted that the maximum growth of the cultures was obtained at 35°C at a pH of 7. Under these conditions the five strains tested had a minimum generation time of only 2-6 h. Earlier, Howard and Orr [178] reported that pathogenic strains of *S. schenckii* grew readily and quickly converted to the yeast phase at 37°C, while only a few strains which were isolated from nature would grow and convert to the yeast phase at the same temperature. Only the saprophytic strains capable of growing at 35-37°C showed any invasive capacity for mice. This result suggested that the pathogenic strains may have adapted to growth at higher ambient temperatures than the saprophytes and that temperature may be a determining factor in the ability of natural isolates to initiate infection in animal hosts.

This temperature-sensitivity hypothesis also has been supported by some clinical trials involving the treatment of cutaneous sporotrichosis by the application of damp heat compresses [131,245]. The local application of heat over a period of 2 months to a patient intolerant to iodides resulted in a clinical cure. The authors also combined the application of local heat and tetrahydrofurfuryl nicotinic acid (Trafuril) in the treatment of some of the cases. They reported a clinical cure of sporotrichosis in each of the nine treated patients within 3-16 weeks. The authors admitted that their heating technique was poorly controlled, but they did hypothesize that it was feasible and that the high temperatures obtained with this technique may inhibit the multiplication or metabolism of the fungus and favor the action of the host's natural defense mechanisms. In addition, the authors considered that the basic antimicrobial mechanisms of phagocytosis and antigen-antibody reactions may be improved or stimulated by the moderate increase in skin temperature.

XII. CRYPTOCOCCOSIS

A. Description and General Pathogenesis

Cryptococcosis is a chronic, subacute, or acute mycosis that can be pulmonary, systemic, or meningeal. The organism has a marked predilection for the central nervous system. The sole etiologic agent is the encapsulated yeast *Cryptococcus neoformans,* a saprophyte that has been isolated most frequently from soils enriched with pigeon excreta. The infectious particle is believed to be a small, nonencapsulated yeast cell that enters the body by inhalation [44,278,295]. A primary pulmonary infection ensues with subsequent localization or frequently fatal dissemination. The true incidence of cryptococcosis is unknown, but there is increasing evidence that the disease may be a major mycosis in the world both in incidence and prevalence [197,204].

The clinical spectrum of cryptococcosis is quite broad and has been classified into five types: (1) pulmonary, (2) central nervous system, (3) dermal or cutaneous, (4) osseous, and (5) visceral [7]. The symptoms of primary pulmonary infection are not diagnostic and may range from a systemic to severe pneumonia-like manifestation. Roentgenographically, the pulmonary manifestations fall into five major groups: (1) predilection for the lower half of the lung fields, (2) rare cavitation, (3) minimal to no fibrosis or calcification, (4) inconspicuous hilar lymphadenopathy, and (5) infrequently massive pulmonary collapse [228]. Additionally, there is some evidence that *C. neoformans* can be a commensal of the tracheobronchial tree [398].

Central nervous system (CNS) infection is characterized by severe headache, mental and visual changes, nausea and vomiting, neck pain, fever, and weakness [389]. Cryptococcosis of the CNS is the most severe form of the disease and frequently is fatal if un-treated [108]. The primary manifestation of CNS cryptococcosis is an aseptic meningitis; however, a secondary characteristic may be ocular disease. Ocular cryptococcosis has been demonstrated experimentally [372] and clinically [64,289]. It can occur as a retinitis [206] or a chorioretinitis [59] that may be unilateral or bilateral [171]. The most severe ophthalmologic complication of cryptococcal meningitis is atrophy, with resultant loss of vision [389].

Dermal or cutaneous cryptococcosis usually is accompanied by a systemic infection, and the lesions are excellent sampling sites for rapid diagnosis [342]. Although cases of primary cutaneous cryptococcosis have been reported [196,282], skin inoculation has not always been clearly established [282], leaving doubt as to the true route of infection. Experimental evidence suggests that C. neoformans may spread to deeper tissues from primary cutaneous sites and cause systemic disease [365].

Osseous involvement of the joints and bones is a severe but not uncommon manifes-tation [69,227]. The severity of this form of cryptococcosis lies in the tumor-like exten-sion to the adjacent tissues and osteolytic nature of the disease.

Visceral or disseminated cryptococcosis is many-faceted, with any organ a potential host to mucoid or granulomatous lesions which may be mistaken for tuberculosis or can-cer [18,103,314]. Heart [229], testes, prostate, and eyes are usually affected, while the kidneys, liver, spleen, and lymph nodes may escape severe invasion [314].

B. Pathogenic Mechanisms

One of the more thoroughly investigated areas of cryptococcal mechanisms of pathogenesis has been infectious particle size. For a disease organism to enter the body by the lungs, the infectious particle must be compatible with pulmonary deposition. To be clinically important, an aerosol should be composed of particles that are less than 10 μm in diam-eter. Particles less than 5 μm in diameter are more compatible with lower respiratory tract deposition [164]. In 1970, Farhi et al. [109] showed that the cell size of C. neofor-mans decreased to as small as 3 μm in diameter when incubated in soil for periods up to a year. Powell et al. [295] reported that particles of C. neoformans compatible with alveo-lar deposition could be isolated from aerosolized pigeon excreta. Recently, Neilson et al. [278] reported isolating cells of C. neoformans less than 1.1 μm in diameter from soil seeded weeks earlier with C. neoformans. The small yeast cells were capable of causing murine cryptococcosis as demonstrated by intracranial inoculations. The presence of viable cells less than 2 μm in soil aerosols indicated that under certain natural conditions cells of C. neoformans could exist in sizes that are compatible with lower respiratory tract deposition. However, now that the sexual phase of this organism has been found to pro-duce basidiospores which are 1.8-2.5 μm in diameter [216], the role of these particles in disease initiation should be examined also.

A variety of biochemical mechanisms of pathogenesis of C. neoformans has been ex-amined. Seeliger [348] reported the production of urease by C. neoformans in culture, and some correlation between the favorable growth of the yeast with some low molecular weight nitrogenous substances as substrate has been drawn [370,374]. However, urease has not been implicated clearly in the pathogenesis of the disease to date. It has been sug-gested that the urease-induced release of ammonia may destroy the host's complement function [271], and thus hinder host defenses and encourage fungal growth.

Using the auxanographic technique, Staib [369,371] concluded that C. neoformans cannot utilize human serum proteins. Recently, Müller and Sethi [271] confirmed this conclusion but also noted that cells of *C. neoformans* elicited a protease or proteases that actively degraded human fibrinogen in vitro. They suggested that if the protease activities occurred in vivo, it may explain the characteristic lack of fibrosis and hyalination around cryptococcal lesions.

Much research has been directed toward elucidating the pathogenic role of the polysaccharide capsule of *C. neoformans*. This material in vivo contributes to the characteristic gelatinous or mucinous nature of tissue lesions, and according to Bulmer et al. [46,47], it apparently is a well-defined virulence factor in cryptococcal disease. Bulmer and Sans [46,47] reported that only 24% of human leukocytes phagocytized an encapsulated strain of *C. neoformans*, whereas phagocytosis of nonencapsulated mutants ranged from 74 to 84%. They determined that cryptococcal polysaccharide acted as a specific and potent inhibitor of phagocytosis. Tacker et al. [386] noted that human peripheral leukocytes engulfed and killed cells of *C. neoformans* and the engulfing process was inhibited by small amounts of cryptococcal capsular material. Diamond and Bennett [88] noted that peripheral human macrophages did not kill engulfed cells of *C. neoformans*. Bulmer and Tacker [49] later reported that guinea pig and human pulmonary macrophages phagocytized but did not kill nonencapsulated cells of *C. neoformans*. The results of their experiments led them to speculate that pulmonary macrophages apparently do not function primarily to kill engulfed yeast cells during the first few hours following initial exposure to *C. neoformans*. This concept is not supported, however, by Mitchell and Friedman [262], who examined the intracellular fate of *C. neoformans* in rat peritoneal exudate cells. The true role of the macrophage in cryptococcosis is still unknown.

Further analysis of cryptococcal capsule by Kozel and Mastroianni [212] suggested that capsular inhibition of phagocytosis may be accounted for by polysaccharide-mediated suppression of the ability of yeast cells to attach to macrophages. If a macrophage is unable to attach itself to a foreign particle, engulfment of that particle is inhibited. These authors suggested that capsular material may sterically hinder intimate contact between the yeast cell and macrophage. According to Bulmer et al. [48]: "The degree of [yeast cell] encapsulation may not be important, but rather the presence or absence of a capsule."

Although several investigators have examined the capsule of *C. neoformans*, few have studied the organism for the presence of endotoxin. In 1964, Kobayashi and Friedman [209] reported that intravenous inoculation of 10^9 cells of *C. neoformans* into rabbits induced a febrile response which was indistinguishable from that elicited by gram-negative bacterial endotoxin. Fever induction was not dependent on the degree of encapsulation, as both slightly and heavily encapsulated cells elicited immediate endotoxic-like febrile reactions. Recently, an endotoxic-like substance has been extracted from cells of *C. neoformans* [210]. Intravenous injections of this material into rabbits induced a febrile response equivalent to that reported by Kobayashi and Friedman. Chemical analysis revealed the presence of polysaccharide and small amounts of bound lipids, glucosamine, and phosphorus. The authors suggested that based on its weak activity and chemical composition, the cell extract may resemble the cell wall component of gram-positive bacteria more than the classic endotoxin of gram-negative bacteria.

C. neoformans produces a minimal and variable immune response in humans. Host defenses against cryptococcal infections are generally considered to be inadequate [230], and numerous investigations of the role of nonspecific and specific host defenses have been reported. As stated by Gadebusch [127]: "The multi-faceted nature of the immune response makes it difficult . . . to state in precise terms the function of any one component." Some facets of host responses to infection by *C. neoformans* as they relate to mechanisms of pathogenesis have been discussed earlier in this section, and some follow in the description of experimental animal models of cryptococcosis. However, for further information on the immunology of cryptococcosis, the reader is referred to the chapters in this book concerned with the immunology of the mycoses, to recent reports [37,85-87,118-120,150], and to a review [127] on the subject.

Experimental animal models have proven to be valuable tools for studying the pathogenesis of cryptococcosis. Several models have been used to study the immunizing effect of cryptococcal polysaccharide [32,146,147], while others have been utilized to study routes of infection and dissemination. In 1963, Ritter and Larsh [322] studied the incidence of mortality and the infective pathway of *C. neoformans* in mice following intranasal instillation of yeast cells. Although the authors reported a higher mortality in mice inoculated intraperitoneally than in those inoculated intranasally, the intranasal instillation procedure was shown to be reproducible with comparable mortality results in successive trials. Using this technique, the authors were able to show primary pulmonary infection with subsequent fatal hematogenous dissemination. Recently, Karaoui et al. [201,202] examined the pathogenic and immune mechanisms in murine cryptococcosis acquired by the respiratory route. These workers emphasized that the "artificial routes" of experimental infections (i.e., intravenous, intraperitoneal, and intracerebral) may or may not simulate cryptococcosis as it occurs naturally and that their procedure may be a more valid model for the study of naturally acquired infection.

One study, in which Syrian hamsters were used, indicated that a primary lesion may occur in the nasal cavity or paranasal sinuses and that the central nervous system could become involved by direct extension [169]. Price and Bulmer [298] demonstrated the tumor-causing capabilities of at least one strain of *C. neoformans*. It was noted that young mice tended to form large abdominal tumors after intraperitoneal inoculation with a human isolate of *C. neoformans*. Older mice failed to develop tumors and succumbed to disseminated disease. Price and Bulmer proposed the term "cryptococcoma" to describe the tumor-like masses they observed. Recently, Staib and Mishra [373] noted a selective involvement of the brain in experimental murine cryptococcosis. They reported that one strain of *C. neoformans* colonized the brain without causing clinical symptoms. This phenomenon has been observed with other strains of *C. neoformans* [277,279; J. B. Neilson, personal communication] and may prove to be a new aspect of cryptococcal virulence which requires further investigation.

Although not believed to be a major form of the disease, cutaneous cryptococcosis has been examined experimentally in a variety of animal models [100,352,365]. Song [365] produced cutaneous cryptococcosis in normal and in cortisone-treated mice and guinea pigs. Cortisone-treated animals showed a severe local infection with poor production of inflammatory cells and abundant fungal multiplication. Disseminated disease developed after subcutaneous inoculation of mice, intradermal inoculation of guinea pigs, and skin scarification in both animal species. Dykstra and Friedman [100] inoculated mice subcutaneously with cells of *C. neoformans* and also noted cases of fatal systemic dissemination, as well as the appearance of cryptococcal polysaccharide in sera at certain intervals. The animal experiments described above, as well as other models dealing with unusual forms of *C. neoformans* [227], have contributed to the elucidation of crypto-coccal mechanisms of pathogenesis and should encourage further investigations.

XIII. CANDIDIASIS

A. Description and General Pathogenesis

Candidiasis or candidosis is a term used to describe a primary or secondary infection caused by yeast that belong to the genus *Candida*. Candidiasis may be manifested as a superficial skin problem, a chronic infection of nails, a disease of the mouth, throat, vagina, or a frequently fatal systemic disease that may involve the lungs, heart, gastro-intestinal tract, and other organs.

The most common etiologic agent is *Candida albicans*, although several other *Candida* spp. may be involved. *Candida* spp. are endogenous as normal flora on the mucocutaneous regions and alimentary tracts of humans and other animals [250,387,400]. The ubiquity of *C. albicans* makes it difficult to distinguish between the mucosal commensal and the etiologic agent of a disease process. *Candida albicans* is a saprophytic organism under "normal" conditions and a limited pathogen when present in large numbers, but is dis-tinctly pathogenic in a host with depressed or otherwise compromised defense mechan-isms [347]. The most important of these predisposing factors are prolonged antibiotic therapy, which disrupts the normal microbial flora; presence of underlying diseases; patient's age; surgical procedures; indwelling catheters; obesity; and drug addiction [44].

Since so many tissues and organs may be involved in candidiasis, the clinical manifes-tations are best divided into the primary organ system involved. The three main clinical conditions are mucocutaneous, cutaneous, and systemic. Allergic diseases also have been recorded but will not be discussed in this chapter. Generally, candidiasis is a great mimic of other diseases. Skin and nail infections caused by *Candida* spp. frequently resemble those caused by dermatophytes. *Candida* pulmonary disease must be differentiated from other chronic lung diseases, and *Candida* septicemia, endocarditis, and meningitis may appear similar to those caused by bacteria. Reports in the literature concerning infections by *Candida* spp. are numerous. The reader is referred to papers and reviews on the sub-ject for further insight to the vast array of clinical manifestations of this protean disease [10,250,255,323,347,376,387].

B. Pathogenic Mechanisms

Considerable attention has been directed toward defining the pathogenic mechanisms of *Candida* spp. In 1952, Salvin [333] demonstrated a toxic effect produced in mice follow-ing an intraperitoneal inoculation of killed yeast cells of *C. albicans*. He noted that the dead cells of *C. albicans*, when injected with adjuvant, had toxic properties similar to those of bacterial endotoxins. He found that the toxic moiety was not filterable. He also noted, however, that the virulence of a given isolate for mice was not necessarily related to its toxicity. In later years, endotoxin-like substances from *C. albicans* were reported by Mourad and Friedman [269], Hasenclever and Mitchell [163], Louria et al. [234], and Kobayashi and Friedman [209], although Chattaway et al. [60] failed to detect toxicity of *Candida* extracts for normal mice. A recent study by Cutler et al. [75] re-vealed that, with one exception, whole cells and cell walls of *Candida* were pyrogenic in rabbits and lethal for actinomycin D-treated mice. Cutler and co-workers emphasized that these positive tests represented only two of four accepted biological systems used to assay for bacterial endotoxin and that relatively large quantities were required to induce the reactions as opposed to microgram amounts for bacterial endotoxins.

The most definitive research on toxins of *Candida* spp. has been by Iwata and his associates [187-190]. In 1967, these investigators isolated a high molecular weight, potent, protein toxin from extracts of *C. albicans.* They designated this acidic protein "canditoxin." Enzymatic and physiochemical studies showed that canditoxin consisted of four subunits: two carboxypeptidases, a phosphomonoesterase, and an unidentified protein. Canditoxin was acutely toxic in microgram quantities for mice, and dissociation of any subunit resulted in complete loss of toxicity. Further studies by this group led to the isolation in 1974 of three high molecular weight glycoprotein toxins from another strain of *C. albicans.* Chemical analysis revealed a D-mannose moiety with varying proportions of threonine, aspartic acid, serine, histidine, and methionine. The three glycoproteins demonstrated murine toxicity comparable to that seen with canditoxin. Although the mechanism of action of the glycoproteins from *C. albicans* is not well defined, Svec [381] has proposed that these toxins may induce the release of histamine in vivo, with the subsequent pharmacologic effects being responsible for many of the clinical reactions seen in experimental animals and in human cases. Odds and Hierholzer [287] reported the isolation of an acid phosphomonoesterase which was a mannoprotein with a hexose to protein ratio of 7:1. This compound appeared to be very similar to the glycoproteins isolated by Iwata, and preliminary evidence suggested that the sugar moiety was responsible for the toxin's antigenicity.

Recent reports by Pugh and Cawson [300,301] and Price and Cawson [299] have revealed the isolation of two enzymes from cells of *C. albicans* which may play a role in the pathogenesis of the disease. Phospholipase A and lysophospholipase were isolated from developing buds and from the cell membranes and walls of nondividing cells. The authors mentioned that phospholipase A damages cell membranes, whereas lysophospholipase protects the yeast from the damaging effects of its own secreted phospholipase. It was suggested that these enzymes may assist yeast invasion of host tissues by disrupting epithelial cell membranes, thus permitting hyphal tip invasion of host cells.

Several studies concerned with the phagocytic and intracellular killing effect of various immune cell populations have been reported. *Candida albicans* has been shown to be chemotactic for polymorphonuclear (PMN) leukocytes [80]. The cell wall mannan forms a complex with serum components which induces the migration of PMN cells. Additional studies have shown that yeast cells of *C. albicans* are readily phagocytized by guinea pig [9] and human [233] PMN leukocytes and by rabbit alveolar macrophages [99]. In spite of rapid phagocytosis, the fungus was able to survive within the phagocytic cells and eventually kill the cell by germ tube and hyphae formation with subsequent rupture of the immune cell's membrane [9,120,233]. Louria and Brayton [233] proposed that the ability to circumvent host cellular defenses may permit the fungus to gain access to the renal tubular epithelium, a site where *Candida* can proliferate with minimal interference by host defenses. Studies examining the role of complement in candidiasis [267] and *Candida* related suppression of murine spleen cell blastogenesis to mitogens [189,326,375] also have been reported. Müller et al. [272] recently indicated that "asteroid bodies" (i.e., antigen-antibody precipitates on fungal cell walls which are visible by light microscopy) may occur in some cases of human candidiasis. The precipitating antibodies appeared to be those of the IgA and IgG classes.

It is disputed whether the yeast form [356,401] or the mycelial form [24,153,185,186, 221] of C. albicans is the more virulent phase. This is really a matter of definition and not for us to debate in this chapter. Barlow et al. [24] have reported factors from human serum and seminal plasma, which in the presence of glucose at 37°C, induce germination and support mycelial development of blastoconidia of C. albicans. They determined that germ tube development from yeast cells required a temperature of 37°C or above, a glucose concentration of approximately 0.1% or greater, and an albumin-like factor. Land et al. [221] later reported that morphogenesis of C. albicans was apparently correlated with a Crabree-like effect, i.e., the repression of mitochondrial activity. Ultrastructural studies of C. albicans growth in vitro [58,265] have demonstrated that the growth of Candida is intracellular and that hyphae actively penetrate cells. Discontinuous plasma membranes marked the sites of fungal penetration, and keratin-containing tonofibrils were lacking in cells invaded by the fungus. Studies by Kapica and Blank [198] have shown C. albicans grown in vitro can digest keratin. Thus it was suggested by Montes and Wilborn [265] that the invading Candida may have digested the structural tonofibrils.

Although the classic procedure for studying C. albicans in vitro was intravenous inoculation of rabbits [107], experimental animal models have been established in the mouse [58], guinea pig [397,415], dog [377], and rat [291]. In 1958, Young [425] inoculated mice intraperitoneally with cells of C. albicans and noted that within 1 h, 60% of inoculated cells developed into pseudohyphae. He also reported that the mycelial rather than the yeast phase appeared to be the infective form and that the route of invasion from the peritoneal cavity was by way of the pancreas and associated blood vessels.

Most evidence derived from experimental Candida infections in animals [140,230, 234] supports the concept that the kidney is the primary organ affected by this organism. Recent studies of systemic or disseminated candidiasis in rats [291], guinea pigs [184], and mice [397] support this basic concept and also reveal that C. albicans can affect many other organs in the body. Trnovec et al. [397] studied the distribution in mice of intravenously administered labeled cells of C. albicans and noted a rapid disappearance of the labeled cells from the blood paralleled by a transient radioactive uptake in the lungs followed by an accumulation in the liver. The spleen and kidneys did not yield very large radioactive counts, although the authors emphasized that there was no direct evidence that organ radioactivity corresponded to tissue levels of the yeast. Concerning the well-established renal pathogenesis of C. albicans, they suggested that the nephrophilic properties of the yeast may be the result of rapid multiplication of the initially deposited yeast cells in the kidney. Louria [231] has recently reviewed experimental infections with fungi and yeast, including Candida spp.

XIV. MISCELLANEOUS AND RARE MYCOSES

A. Description and General Pathogenic Mechanisms

1. Adiaspiromycosis

Adiaspiromycosis or haplomycosis is a pulmonary mycosis of humans, rodents, and some aquatic animals [103]. The etiologic agents are *Emmonsia parva* and *E. crescens*. The disease is very rare, as only 15 cases have been reported [103]. Little is known of the epidemiology or pathogenesis of these organisms. The disease is apparently secondary to a primary debilitating syndrome, and the prime characteristic is the formation of space-occupying fungal cells (as large as 600 μm in diameter in the case of *E. crescens*) which may be surrounded by tuberculoid type granulomas, although host response is usually minimal. No mechanisms of pathogenesis have been proposed.

2. Basidiomycoses

Members of the basidiomycetes are ubiquitous in nature, but very few cases of human or animal infections due to basidiomycetes have been reported. In 1954, Emmons reported the repeated isolation of *Coprinus micaceous*, a basidiomycete, from a patient's sputum [102]. Emmons presumed that basidiomycete spores had been inhaled recently or were contaminants. However, in 1971, Speller and MacIver [366] reported a confirmed case of endocarditis caused by the conidial stage of *C. cinerus*, and Restrepo et al. [307] reported the ulceration of a young girl's palate from which the basidiomycete *Schizophyllum commune* was cultured. Microscopic examination of the diseased tissue revealed thinly septate, branching hyphae that elicited a foreign body giant cell reaction.

Mechanisms of pathogenesis of basidiomycetes have not been studied since the disease potential of these organisms has been recognized only recently. A definitive study by Greer and Bolanos [151] demonstrated the pathogenic potential of *S. commune* in normal white Swiss mice. A human tissue isolate [307] was shown to invade the lungs, lymph nodes, liver, and subcutaneous tissues of intraperitoneally inoculated mice. They reported the penetration of the parenchyma by hyphae possessing clamp connections. The fungus was shown to grow well at 37°C and reportedly can tolerate temperatures up to 44°C [53].

3. Beauveriosis

Beauveriosis is a very rare mycosis of humans and animals. The etiologic agent is *Beauveria bassiana*, a well-recognized entomopathogen with worldwide distribution. Georg et al. [138] reported the isolation of *B. bassiana* from captive giant tortoises that succumbed to a pulmonary disease. Recently, Fromtling et al. [121,122] reported the isolation of *B. bassiana* from three cases of fatal pulmonary disease in captive American alligators. They noted a primary, pulmonary focus for each infection, with dissemination occurring in two of the three animals. The only reported human case of beauveriosis was by Freour et al. [114,115], who isolated the fungus from the lungs of a young female who presented with ulcerative cervical lymphadenopathy and a positive chest x-ray. Pathogenic mechanisms of *B. bassiana* have not been delineated, although Fromtling et al. [121] reported the presence of calcium oxalate crystals in the lungs of two alligators and suggested that the crystals may have physically penetrated and injured the lung parenchyma.

4. Cercosporomycosis

This disease is based on a single human case in Indonesia. The etiologic agent is the plant pathogen *Cercospora apii* [103]. No pathogenic mechanisms have been proposed.

5. Geotrichosis

Geotrichosis is an uncommon infection that is usually pulmonary in origin. *Geotrichum candidum*, the etiologic agent, is ubiquitous in nature and apparently endogenous as normal flora in humans. Some authors consider it to be dimorphic, since arthroconidia or yeast-like cells develop under certain environmental conditions. The role of *G. candidum* as a human pathogen has been disputed for more than a century, and little is known about its pathogenesis. *G. candidum* is reported to become invasive in patients with pulmonary and other diseases [111,408], and rarely disseminates [353]. Cases of oral, cutaneous, and alimentary geotrichosis have appeared in the literature [314]. The rarity of reported geotrichosis, the limited tissue invasion in most reported cases, and the evidence of rapid clearance from the blood [353] also suggest that *G. candidum* is not very virulent. No pathogenic mechanisms have been postulated to date.

6. Oidiodendronosis

The dimorphic fungus *Oidiodendron kalrai* has been the subject of several studies of fungal pathogenesis. In 1961, it was isolated from four varied clinical cases in India [390]. The organism is filamentous at room temperature and yeast-like to mycelial with occasional arthroconidia in vivo. A series of experiments concerning the pathogenesis of this organism in mice has been reported by Tewari, Swenberg, and their associates [383,384,390]. All mice showed neurological symptoms of encephalitis following intravenous or intraperitoneal inoculations of yeast cells. Arthroconidia and blastoconidia were seen in cerebral tissues, and neurological symptoms were apparently from direct invasion of neural tissue by the fungus rather than by a neurotoxin elicited by the organism [390]. Further analysis of tissues revealed that the mycelial phase represented the invasive form of the disease and that hyphae characteristically penetrated blood vessels and adjacent tissues [383]. Ultrastructural studies indicated that mechanical force exerted by the growing hyphal tips was an important factor in fungal invasion of the brain. Recent evidence has suggested that extracellular proteases are produced by *O. kalrai* and may play a role in the pathogenesis of this organism [61].

7. Otomycosis and Mycotic Keratitis

The ears and eyes may be infected by virtually any fungus, although *Fusarium* spp. and *Aspergillus* spp. are most often implicated [314]. Local injury to these organs may permit colonization and invasion of the ears and eyes by fungi, causing otomycosis or mycotic keratitis, respectively. Pathogenic mechanisms of the invading organisms are undetermined at present.

8. Penicilliosis

Penicillium spp. are ubiquitous in nature, and thus their etiologic significance when isolated from patients is questionable. Two members of this genus, *P. commune* and *P. marneffei,* have been reported as etiologic agents of human and animal disease, respectively [103]. Specific mechanisms of pathogenesis have not been elucidated.

9. Phaeohyphomycosis

Numerous dematiaceous fungi which have been implicated rarely in human infections are grouped, according to some authors, under the term "phaeohyphomycosis." These common soil saprophytes apparently can act as secondary invaders. A few species of *Curvularia, Alternaria, Chaetoconidium, Drechslera,* and *Phialophora* have been implicated in human infection [314]. The role of *Alternaria* as a possible human pathogen has been examined [41]; however, no mechanisms of pathogenesis of this fungus or the other rarely pathogenic dematiaceous fungi have been proposed.

10. Rhinosporidiosis

Rhinosporidiosis is a chronic granulomatous disease of the mucous membranes. It is characterized by polyp production with subsequent deformation of the infected tissues. *Rhinosporidium seeberi* is the etiologic agent. Its classification is uncertain, although it is considered to be a fungus despite failures to culture it. Experimental infections have failed also. Little is known of its epidemiology or pathogenesis, and no mechanisms of pathogenesis have been proposed.

11. Rhodotorulosis

Rhodotorula spp. have been isolated from sputum, urine, feces, and blood of debilitated patients [103]. Louria et al. [232] have reviewed fungemia caused by "nonpathogenic yeasts." These yeasts are not considered to be primary pathogens and no mechanisms of pathogenesis have been proposed.

12. Scopulariopsosis

Scopulariopsis brevicaulis is the only species implicated in this disease. This fungus has been associated with onychomycosis, ulcerating granuloma, and chronic granulomatous inflammation of tendon sheaths and muscle [103]. This organism is proteolytic and considered to be opportunistic. Except for its proteolytic nature, no pathogenic mechanisms have been discovered.

13. Superficial Mycoses

Superficial fungal infections such a pityriasis or tinea versicolor (*Malessesia furfur*), tinea nigra (*Exophiala werneckii*), black piedra (*Piedraia hortai*), and white piedra (*Trichosporon beigelii*) are mild syndromes that primarily constitute a cosmetic problem. No mechanisms of pathogenesis have been defined.

14. Torulopsosis

Torulopsosis is a rare, but increasingly reported, fungal infection caused by *Torulopsis glabrata,* a yeast-like fungus commonly associated with the human body as an endogenous organism. It also has been isolated from animals and soil. *T. glabrata* is an opportunistic fungus most often infecting physiologically altered patients. In 1957, Wickerham [411] noted an apparent increase in the number of *Torulopsis* infections. Several reports have appeared in the literature over the years documenting infections by *T. glabrata* that are secondary to a predisposing syndrome [154,237,260,276]. Many of these infections are reported to be fatal.

Very little is known about the mechanisms of pathogenesis of *T. glabrata.* In 1960, Hasenclever and Mitchell [161] reported that *T. glabrata* shared several antigens with seven strains of *Candida albicans* as determined by agglutination reactions. In a later study [162] these authors reported that normal mice did not develop a progressive disease when inoculated with a large infecting dose of *T. glabrata.* However, infection was induced in mice physiologically altered by cortisone, alloxan, or x-ray treatments. Howard and Otto [179] reported that *T. glabrata* was not a facultative intracellular parasite, as the fungus was unable to proliferate in mouse macrophages.

B. Concluding Remarks

There are numerous soil, water, and airborne fungi which have been implicated or proven to be etiologic agents of human or animal disease. For the most part, very few clinical cases have been recorded, and generally even less has been reported concerning their mechanisms of pathogenesis. The reader is referred to texts by Rippon [314] and Emmons et al. [103] and to some reports in the literature [290,394] for further information on the miscellaneous and rare mycoses.

XV. GENERAL DISCUSSION

The majority of papers reviewed in this chapter were directed toward elucidation of what are classically called "virulence factors." As defined previously, these are factors produced by fungi which may play a role in virulence. Certainly these factors are of prime importance when discussing infectious agents, but when taking an overview of pathogenesis, one often begins to cite other "abilities" of pathogenic fungi which may enter into how a certain organism is able to cause a disease. Although such abilities are not technically classified as virulence factors, they deserve recognition and investigation because they represent the platform from which the scenario of pathogenesis is projected. In closing this chapter it is felt that some of these properties should be presented, along with questions they evoke.

A. Chemical Composition

Considerable research has been done in this area (see specific diseases) relevant to the production of virulence factors. But, in contrast, what is known about the fate and possible disease relationship of many fungal components that appear to be inert in body fluids and tissues? Probably the most obvious example of this is the yeast cell. For example, in several of the systemic mycoses, one can see yeast cells in tissues and fluids years after a "clinical cure." Do such cells influence immunity? Can they be reactivated into multiplying infectious units? Can they predispose an individual to another disease? Why are such cells inert to catabolic enzymes? Could this phenomenon be related to the fact that the cells contain large amounts of chitin and the human body lacks chitinase?

B. Critical Mass

The literature contains little evidence (or even speculation) on the exact mechanism by which fungi cause death in animals. Perhaps in some instances we are dealing with a "critical mass" phenomenon, in which a large enough number of localized fungal cells can exert a mass effect on vital organs or systems. The best example of this is the determination of the LD_{50} for a given organism in a certain animal model. Except for *Histoplasma capsulatum* and *Coccidioides immitis*, relatively large inocula of cells are required to induce a disease. Although one may suggest that relatively large numbers of cells are required to "flood" the normal defense mechanisms of a host, this does not fully explain why, in many cases, millions of cells are required in the initial inoculum. Although it sounds nebulous, it could be worthwhile to investigate the possibilities that a large number of organisms can exert an effect on a potential host which could be totally different from a single cell multiplied by the number of cells in the inoculum. In other fields of science, critical mass phenomena are not unknown; thus this concept also may exist in the study of the mycoses.

C. Dimorphism

Dimorphism is the ability of some etiologic agents of systemic mycoses to convert from one form (usually the saprophytic phase) into another (the pathogenic or "tissue" phase). Because this conversion is associated with disease-producing potential, it is assumed that changes during and resulting from this type of morphogenesis play a key role in pathogenesis. This is probably true, except during the brief time period when the body is exposed to the as yet unconverted saprophytic infectious particles, e.g., the arthroconidia of *C. immitis* or the microconidiospores of *H. capsulatum*. The "dead-end" phenomenon of dimorphism results in many changes, notably cell wall morphology and internal metabolism. Some of these changes result in the development of what truly may be called virulence factors. However, other changes cannot be classified so simply, yet they do place the organism in a competitive status, i.e., one that did not exist when the organism was in a saprophytic phase. Changes of this type, e.g., ability to grow at body temperatures and utilize the new nutritional environment to proliferate, do not represent so-called virulence factors. However, without them, the fungus could not survive long enough to initiate a disease state. An excellent critical review on dimorphism has been published by Rippon [315]. See also Chap. 10 of Part A of this work.

D. Epidemiology

Before a disease state can be initiated, the potential host must be exposed to infectious particles. In most of the human mycoses the etiologic agents are saprophytic in nature. Despite this, it took years to accept the fact that saprophytic activity is an essential part of many diseases [11]. This sounds so simple, yet little is known about the reasons why certain pathogenic organisms grow in selective environments. For example, why is histoplasmosis diagnosed in the inhabitants of the eastern United States more than in other areas of the world? It is not enough to say that the organism proliferates in soil containing the excreta of certain birds. Or, why is coccidioidomycosis endemic in the southwestern United States? Could the etiologic agent of this, or other mycoses, exist in other areas of the world, but factors not be present for the development of the infectious phase? Perhaps we also tend to overcategorize diseases as being either endogenously or exogenously acquired. For such diseases as coccidioidomycosis, cryptococcosis, histoplasmosis, and blastomycosis, an infection initiated by the saprophytic phase of a fungus may have been acquired or initiated years before and clinical onset of the disease state evolved from the "tissue" phase of the organism. In essence, depending on one's definition, the disease state may have been endogenous in origin. Thus, in elucidating pathogenic mechanisms, the ecology of pathogens needs to be understood.

E. Fever

The development of fever is associated with numerous bacterial and viral diseases. Some pathogenic fungi induce this phenomenon in animal models [43], yet little is known about the reasons for induction and the role of fever in disease processes.

F. Genetics

We will never fully understand pathogenesis until genetic bases are understood. Surprisingly, little attention has been paid to the field of genetics of pathogenesis. The sexual phases of many of the human pathogens now are known and unlike many of our colleagues in other areas of medical microbiology, the medical mycologist has numerous models available for most of the major mycoses. Considerable information can be learned about this area from a century of investigation by plant pathologists [79,84,174].

G. Induction of Anergy

Many patients in the terminal phase of a systemic mycosis lose most of their immunologic responses—this is anergy. How is this phenomenon induced by fungi, and can it be controlled or prevented? Since anergy usually signals a poor prognosis, it would appear to represent an important segment of how a fungus causes death. Despite this, little information is available on the subject.

H. Mycotoxins

In recent years considerable knowledge has been acquired about these toxins. To date they are recognized as being poisons which are produced by fungi while outside the human body. Such chemicals exhibit enormous potential for destruction and/or alteration of animal cells [261]. Perhaps, in some instances, these chemicals could be synthesized in vivo.

I. Nutrition and Age

For decades microbiologists have known that the metabolic activity and chemical composition of a microorganism can be influenced by substrate and age [423]. Medical mycologists respect this concept as only log phase or freshly subcultured cells are used in experimenting with mycosis in the laboratory. However, most human mycoses are *not* acquired from actively growing laboratory cells; they are acquired from cells that grow in nature. The time may come when we will find it necessary to investigate the effects of some of the following conditions and/or environments on pathogenesis: the age, condition, or phase that a fungus is in during different times of the year and in different ecological niches; the diversified substrates that fungi grow upon in nature, natural predators, and competitors. For such investigations to take place we may have to sacrifice accepted laboratory equipment and procedures, such as constant-temperature incubators, synthetic media, and pure cultures.

J. Penetration of the Host

An organism cannot cause a disease unless it can enter a potential host. Despite this, it was too many years before we accepted the concept that the macroconidiospores of *H. capsulatum* were too large to enter the lungs and, in fact, the microconidiospores were the true infectious particles. In cryptococcosis it has been shown only recently that *Cryptococcus neoformans* produces airborne particles that are less than 3 μm in diameter, a size that is compatible with lung disposition [278]. Penetration also may mean entrance through intact or damaged surfaces (e.g., skin or wounds, respectively). Little emphasis has been placed on this aspect of pathogenesis for the animal mycoses. For a somewhat different and informative view on this problem, the proposals by our colleagues in plant pathology [144,403] are quite enlightening.

K. Protoplasts

Medical bacteriologists readily accept the concept that pathogenic bacteria may exist in a cell wall-less state for relatively long periods of time in a host. In this form, the cells are not susceptible to antibiotics that inhibit cell wall synthesis and the external antigenic makeup of the cell is changed. Some of these forms may cause disease or they can revert to the cell wall-containing form. Despite the publication of a text on this subject [404], medical mycology has been slow to fit this concept into mycotic pathogenesis. Do fungal protoplasts play a role in chronic disease; could they be responsible for latent disease? How effective are the antifungal drugs against them, and how does the lack of a cell wall influence the immunologic responsiveness of the host? These are just a few of the areas that require further investigation.

L. Saprophyte to Pathogen Conversion

In dealing with the systemic mycoses one normally considers that the conversion from a saprophytic to a pathogenic phase will be understood when the molecular basis of dimorphism is known. This is probably an oversimplification of the subject and it also neglects the fact that many fungi which are animal pathogens are monomorphic. This also skirts the real issue, namely, that pathogenicity is an expression of a two-component system: the host-parasite relationship [26,134,174]. Just as it will be difficult for us to break away from the pure culture concept in studying ecology and epidemiology, it will be equally difficult to study the host-parasite relationship as an integrated system.

M. Space-Occupying Lesions

In several of the mycoses large masses of tissue and organisms may be seen, e.g., crypto-coccosis, coccidioidomycosis, histoplasmosis, blastomycosis, and mycetoma. The mass of such lesions is the result of host-elicited defense mechanisms and the etiologic agent of the disease. Such masses may occupy considerable space and exert formidable pressures on surrounding cells, tissues, organs, and fluids. Histopathologically, *C. neoformans* is often suspected by pathologists by the "soap-suds" appearance of the tissue. Considerable outward pressure must have resulted from the organism's growth. Also, in this disease there is currently no explanation for the remarkably high spinal fluid pressure or the cause of death. Perhaps we have become overly concerned with finding "virulence factors," when, in fact, a huge, space-occupying lesion could by itself result in many of the disease manifestations.

N. Temperature

The statement that an organism which can grow at 37°C in vitro has one of the important parameters for pathogenicity is no longer accurate. Kwon-Chung's recent investigation [217] on several strains of *Sporothrix schenckii* represents an outstanding contribution in this area. Her studies may offer a temperature-based explanation for the differences between the development of either cutaneous or systemic sporotrichosis. Another obvious example is the dermatophytoses, which usually involve only the cooler areas of the body, such as skin, hair, and nails. Although certainly not a virulence factor, the temperature ranges of pathogenic fungi may dictate and even control the location of a disease process.

O. Summary

Until recently the diagnosis of a specific fungus disease was only of academic interest. That changed with the introduction of various antifungal drugs and an appreciation of the incidence of hospitalized cases and associated costs [157]. Medical mycology thus is placed in a "catch-up" situation. Fortunately, we have at our disposal the accumulated knowledge from several generations of bacteriologists and plant pathologists from which we can draw ideas, concepts, and techniques. We can no longer afford to be isolationists, but must direct our future studies more into the mainstream of medical microbiology and infectious diseases.

ACKNOWLEDGMENT

The authors extend their gratitude for 16 years of research support to the National Institute of Allergy and Infectious Diseases, NIH Grant AI-05022.

REFERENCES

1. Abbott, P. 1956. Mycetoma in the Sudan. Trans. R. Soc. Trop. Med. Hyg. *50*: 11-30.
2. Adam, A., R. Ciorbara, J. F. Petit, E. Lederer, L. Chedid, A Lamensans, F. Parant, M. Parant, J. P. Rosselet, and F. M. Berger. 1973. Preparation and biological properties of water-soluble adjuvant fractions from delipidated cells of *Mycobacterium smegmatis* and *Nocardia opaca*. Infect. Immun. 7: 855-861.
3. Ajello, L. 1960. Geographical distribution and prevalence of the dermatophytes. Ann. N.Y. Acad. Sci. *89*: 30-38.
4. Ajello, L. 1974. Natural history of the dermatophytes and related fungi. Mycopathol. Mycol. Appl. *53*: 93-110.
5. Albornoz, M. 1971. Isolation of *Paracoccidioides brasiliensis* from rural soil in Venezuela. Sabouraudia 9: 248-253.
6. Anderes, E. A., A. Finley, and H. A. Walch. 1973. The lipids of an auxotrophic avirulent mutant of *Coccidioides immitis*. Sabouraudia *11*: 149-157.
7. Anderes, E. A., W. E. Sandine, and P. R. Elliker. 1971. Lipids of antibiotic sensitive and resistant strains of *Pseudomonas aeruginosa*. Can. J. Microbiol. *17*: 1357-1365.
8. Andersson, B. A., L. Hellgren, and J. Vincent. 1976. Allergic delayed skin reactions from lipid fractions of trichophytin. Sabouraudia *14*: 237-241.
9. Arai, T., Y. Mikami, and K. Yokoyama. 1977. Phagocytosis of *Candida albicans* by rabbit alveolar macrophages and guinea pig neutrophils. Sabouraudia *15*: 171-177.
10. Aronson, I. K., and K. Soltani. 1976. Chronic mucocutaneous candidiasis: a review. Mycopathologia *60*: 17-25.
11. Austwick, P. K. C. 1972. The pathogenicity of fungi. In H. Smith and J. H. Pearce

(Eds.), *Microbial Pathogenicity in Man and Animals,* Cambridge University Press, Cambridge, pp. 251-268.

12. Azulay, R. D., J. A. Carneiro, M. Cunha, and L. T. Reis. 1976. Keloidal blastomycosis (Lobo's disease) with lymphatic involvement: a case report. Int. J. Dermatol. *15*: 40-42.

13. Azuma, I., F. Kanetsuna, Y. Tanaka, Y. Yamamura, and L. M. Carbonell. 1974. Chemical and immunological properties of galactomannans obtained from *Histoplasma duboisii, Histoplasma capsulatum, Paracoccidioides brasiliensis* and *Blastomyces dermatitidis.* Mycopathol. Mycol. Appl. *54*: 111-125.

14. Azuma, I., H. Kimura, F. Hirao, E. Tsubura, and Y. Yamamura. 1967. Biochemical and immunological studies on *Aspergillus.* I. Chemical and biological investigations of lipopolysaccharide, protein, and polysaccharide fractions isolated from *Aspergillus fumigatus.* Jpn. J. Med. Mycol. *8*: 210-220.

15. Azuma, I., H. Kimura, F. Hirao, E. Tsubura, Y. Yamamura, and A. Misaki. 1971. Biochemical and immunological studies on *Aspergillus.* III. Chemical and immunological properties of glycopeptide obtained from *Aspergillus fumigatus.* Jpn. J. Microbiol. *15*: 237-246.

16. Azuma, I., H. Kimura, and Y. Yamamura. 1968. Purification and characterization of an immunologically active glycoprotein from *Aspergillus fumigatus.* J. Bacteriol. *96*: 272-273.

17. Baker, R. D. 1942. Experimental blastomycosis in mice. Am. J. Pathol. *18*: 463-478.

18. Baker, R. D. 1971. *Human Infection with Fungi, Actinomycetes, and Algae,* Springer-Verlag, New York, p. 1191.

19. Baker, R. D., and G. Linares. 1974. Prednisolone-induced mucormycosis in rhesus monkeys. Sabouraudia *12*: 75-80.

20. Bambule, J., and D. Grigoriu. 1976. Sinus aspergillosis. Bull. Soc. Fr. Mycol. Med. *5*: 181-184.

21. Barbee, W. C., A. Ewert, and E. M. Davidson. 1977. Animal model of human disease: sporotrichosis. Am. J. Pathol. *86*: 281-284.

22. Barker, S. A., C. N. D. Cruickshank, and J. H. Holden. 1963. Structure of a galactomannan-peptide allergen from *Trichophyton mentagrophytes.* Biochim. Biophys. Acta *74*: 239-246.

23. Barlow, A. J. E. 1976. Recent advances in fungus diseases. Int. J. Dermatol. *15*: 418-424.

24. Barlow, A. J. E., T. Aldersley, and F. W. Chattaway. 1974. Factors present in serum and seminal plasma which promote germ tube formation and mycelial growth of *Candida albicans.* J. Gen. Microbiol. *82*: 261-272.

25. Barnes, J. M., and W. H. Butler. 1964. Carcinogenic activity of aflatoxin to rats. Nature (Lond.) *202*: 1016.

26. Barnett, H. L., and F. L. Binder. 1973. The fungal host-parasite relationship. Annu. Rev. Phytopathol. *11*: 273-292.

27. Basarab, O., M. H. How, and C. N. D. Cruickshank. 1968. Immunological relationships between glycopeptides of *Microsporum canis, Trichophyton rubrum, Trichophyton mentagrophytes* and other fungi. Sabouraudia *6*: 119-126.

28. Basler, R. S. W., and J. L. Friedman. 1974. Mucocutaneous histoplasmosis. JAMA *230*: 1434-1435.

29. Baum, G. L., and J. Schwarz. 1977. Histoplasmosis. Contrib. Microbiol. Immunol. *4*: 96-107.

30. Bauman, D. S., and E. W. Chick. 1969. An experimental model for studying extra-

pulmonary dissemination of *Histoplasma capsulatum* in hamsters. Am. Rev. Respir. Dis. *100*: 79-81.

31. Baxter, M., and P. R. Mann. 1969. Electron microscopic studies of the invasion of human hair in vitro by three keratinophilic fungi. Sabouraudia 7: 33-37.
32. Bennett, J. E., and H. F. Hasenclever. 1965. *Cryptococcus neoformans* polysaccharide: studies of serologic properties and role in infection. J. Immunol. *94*: 916-920.
33. Berliner, M. D. 1968. Primary subcultures of *Histoplasma capsulatum*. I. Macro- and micro-morphology of the mycelial phase. Sabouraudia *6*: 111-118.
34. Berliner, M. D., and M. E. Reca. 1969. Protoplasts of systemic dimorphic fungal pathogens: *Histoplasma capsulatum* and *Blastomyces dermatitidis*. Mycopathol. Mycol. Appl. *37*: 81-85.
35. Berliner, M. D., and M. E. Reca. 1970. Release of protoplasts in the yeast phase of *Histoplasma capsulatum* without added enzyme. Science *167*: 1255-1257.
36. Berliner, M. D., and M. E. Reca. 1971. Studies on protoplast induction in the yeast phase of *Histoplasma capsulatum* by magnesium sulfate and 2-deoxy-D-glucose. Mycologia *63*: 1164-1172.
37. Blumer, S. O., and L. Kaufman. 1977. Characterization of immunoglobulin classes of human antibodies to *Cryptococcus neoformans*. Mycopathologia *61*: 55-60.
38. Blyth, W. 1978. The occurrence and nature of alveolitis-inducing substances in *Aspergillus clavatus*. Clin. Exp. Immunol. *32*: 272-282.
39. Bona, C., C. Damais, and L. Chedid. 1974. Blastic transformation of mouse spleen lymphocytes by a water-soluble mitogen extracted from *Nocardia*. Proc. Natl. Acad. Sci. U.S.A. *71*: 1602-1606.
40. Borelli, D. 1962. Lobomicosis experimental. Dermatol. Venez. *3*: 72-82.
41. Botticher, W. W. 1966. *Alternaria* as a possible human pathogen. Sabouraudia *4*: 256-258.
42. Bouza, E., D. J. Winston, J. Rhodes, and W. L. Hewitt. 1977. Paracoccidioidomycosis (South American blastomycosis) in the United States. Chest *72*: 100-102.
43. Braude, A. I., J. McConnell, and H. Douglas. 1960. Fever from pathogenic fungi. J. Clin. Invest. *39*: 1266-1276.
44. Bulmer, G. S. 1976. *Lectures in Medical Mycology*, 2nd ed., University of Oklahoma Health Sciences Center, Oklahoma City.
45. Bulmer, G. S. 1978. *Medical Mycology*, Upjohn Co., Kalamazoo.
46. Bulmer, G. S., and M. D. Sans. 1967. *Cryptococcus neoformans*. II. Phagocytosis by human leukocytes. J. Bacteriol. *94*: 1480-1483.
47. Bulmer, G. S., and M. D. Sans. 1968. *Cryptococcus neoformans*. III. Inhibition of phagocytosis. J. Bacteriol. *95*: 5-8.
48. Bulmer, G. S., M. D. Sans, and C. M. Gunn. 1967. *Cryptococcus neoformans*. I. Nonencapsulated mutants. J. Bacteriol. *94*: 1475-1479.
49. Bulmer, G. S., and J. R. Tacker. 1975. Phagocytosis of *Cryptococcus neoformans* by alveolar macrophages. Infect. Immun. *11*: 73-79.
50. Burke, P. S., and C. A. Coltman. 1971. Multiple pulmonary aspergilloma in acute leukemia. Cancer. *28*: 1289-1292.
51. Caldwell, D. K., M. C. Caldwell, J. C. Woodard, L. Ajello, W. Kaplan, and H. M. McClure. 1975. Lobomycosis as a disease of the Atlantic bottle-nosed dolphin (*Tursiops truncatus*, Motangu, 1821). Am. J. Trop. Med. Hyg. *24*: 105-114.
52. Carrada-Bravo, T. 1975. New observations on the epidemiology and pathogenesis of sporotrichosis. Ann. Trop. Med. Parasitol. *69*: 267-273.
53. Cartwright, K. G., and W. P. K. Findlay. 1934. Studies in the physiology of wood-destroying fungi. II. Temperature and rate of growth. Ann. Bot. (Lond.) *48*: 481-495.

54. Carvalho, A. 1953. Sobre o emprego da paracoccidioidina na cidade do Rio de
     Janeiro. Primeiros resultados basadas no estudo de 475 individuos. Rev. Bras.
     Tuberc. *21*: 73-82.

55. Catchings, B. M., and D. J. Guidry. 1973. Effects of pH and temperature on the in
     vitro growth of *Sporothrix schenckii*. Sabouraudia *11*: 70-76.

56. Caudill, R. G. 1970. Coccidioidal meningitis. Am. J. Med. *49*: 360-365.

57. Cavanagh, L. L. 1974. Attempts to induce mycetoma in monkeys and mice using
     Madurella mycetomi. Sabouraudia *12*: 258-262.

58. Cawson, R. A., and K. C. Rajasingham. 1972. Ultrastructural features of the in-
     vasive phase of *Candida albicans*. Br. J. Dermatol. *87*: 435-443.

59. Chapman-Smith, J. S. 1977. Cryptococcal chorioretinitis: a case report. Br. J.
     Ophthalmol. *61*: 411-413.

60. Chattaway, F. W., F. C. Odds, and A. J. E. Barlow. 1971. An examination of the
     production of hydrolytic enzymes and toxins by pathogenic strains of *Candida al-
     bicans*. J. Gen. Microbiol. *67*: 255-263.

61. Cino, P. M., and W. J. Nickerson. 1975. Extracellular proteases of *Oidiodendron
     kalrai*. Abstr. Annu. Meet. Am. Soc. Microbiol., F20, p. 88.

62. Clark, W. R., J. McLaughlin, and M. E. Webster. 1958. An aminohexuronic acid
     as the principle hydrolytic component of the Vi antigen. J. Biol. Chem. *230*: 81-89.

63. Conant, N. F., D. T. Smith, R. D. Baker, and J. L. Callaway. 1971. *Manual of
     Clinical Mycology*, 3rd ed., Saunders, Philadelphia.

64. Condon, P. I., S. I. Terry, and H. Falconer. 1977. Cryptococcal eye disease. Doc.
     Ophthalmol. *44*: 49-56.

65. Corbel, M. J., and S. M. Eades. 1976. The relative susceptibility of New Zealand
     black and CBA mice to infection with opportunistic fungal pathogens. Sabouraudia
     14: 17-32.

66. Corbel, M. J., and S. M. Eades. 1977. Experimental mucormycosis in congenitally
     athymic (nude) mice. Mycopathologia *62*: 117-120.

67. Corbel, M. J., and S. M. Eades. 1978. Observations on the localization of *Absidia
     corymbifera* in vivo. Sabouraudia *16*: 125-132.

68. Couch, J. R., N. L. Abdou, and A. Sagawa. 1978. Histoplasma meningitis with
     hyperactive suppressor T cells in cerebrospinal fluid. Neurology *28*: 119-123.

69. Cowen, N. J. 1969. Cryptococcosis of bone: case report and review of the liter-
     ature. Clin. Orthop. *66*: 174-182.

70. Cox, R. A., and G. K. Best. 1972. Cell wall composition of two strains of *Blasto-
     myces dermatitidis* exhibiting differences in virulence for mice. Infect. Immun. *5*:
     449-453.

71. Cox, R. A., L. R. Mills, G. K. Best, and J. F. Denton. 1974. Histologic reactions
     to cell walls of an avirulent and a virulent strain of *Blastomyces dermatitidis*. J.
     Infect. Dis. *129*: 179-186.

72. Cox, W. A., and J. A. Moore. 1968. Experimental *Trichophyton verrucosum* in-
     fections in laboratory animals. J. Comp. Pathol. *78*: 35-41.

73. Craig, E. L., and T. Suie. 1974. *Histoplasma capsulatum* in human ocular tissue.
     Arch. Ophthalmol. *91*: 285-289.

74. Cruthirds, T. P., and D. O. Patterson. 1967. Primary pulmonary sporotrichosis.
     Am. Rev. Respir. Dis. *95*: 845-847.

75. Cutler, J. E., L. Friedman, and K. C. Milner. 1972. Biological and chemical char-
     acterization of toxic substances from *Candida albicans*. Infect. Immun. *6*: 616-
     627.

76. Daniels, L. S., M. D. Berliner, and C. C. Campbell. 1968. Varying virulence in rab-
     bits infected with different filamentous types of *Histoplasma capsulatum*. J. Bac-
     teriol. *96*: 1535-1539.

77.   Das, S. K., and A. B. Banerjee. 1974. Phospholipids of *Trichophyton rubrum*. Sabouraudia *12*: 281-286.

78.   Das, S. K., and A. B. Banerjee. 1978. Lipolytic enzymes of *Trichophyton rubrum*. Sabouraudia *15*: 313-323.

79.   Day, P. R. 1974. Genetics of host-parasite interaction. Freeman, San Francisco.

80.   Denning, T. J. V., and R. R. Davies. 1973. *Candida albicans* and the chemotaxis of polymorphonuclear neutrophils. Sabouraudia *11*: 210-221.

81.   Deppisch, L. M., and E. M. Donowho. 1972. Pulmonary coccidioidomycosis. Am. J. Clin. Pathol. *58*: 489-500.

82.   Deresinski, S. C., and D. A. Stevens. 1975. Coccidioidomycosis in compromised hosts. Medicine (Baltimore) *54*: 377-395.

83.   Destombes, P., and M. Patou. 1964. Morphologie de grains de mycétomes à *Norcardia* ou à *Cephalosporium,* reconstitués par la méthode des coupes histologiques sériées. Bull. Soc. Pathol. Exot. *57*: 393-395.

84.   De Vay, J. E., and H. E. Adler. 1976. Antigens common to hosts and parasites. Annu. Rev. Microbiol. *30*: 147-168.

85.   Diamond, R. D. 1974. Antibody-dependent killing of *Cryptococcus neoformans* by human peripheral blood mononuclear cells. Nature (Lond.) *247*: 148-150.

86.   Diamond, R. D. 1977. Effects of stimulation and suppression of cell-mediated immunity on experimental cryptococcosis. Infect. Immun. *17*: 187-194.

87.   Diamond, R. D., and A. C. Allison. 1976. Nature of the effector cells responsible for antibody-dependent cell-mediated killing of *Cryptococcus neoformans.* Infect. Immun. *14*: 716-720.

88.   Diamond, R. D., and J. E. Bennett. 1973. Growth of *Cryptococcus neoformans* within human macrophages in vitro. Infect. Immun. 7: 231-236.

89.   Dias, L. B., M. M. Sampaio, and D. Silva. 1970. Jorge Lobo's disease. Observations on its epidemiology and some unusual morphological forms of the fungus. Rev. Inst. Med. Trop. Sao Paulo *12*: 8-15.

90.   DiSalvo, A. F., and J. E. Denton. 1963. Lipid content of four strains of *Blastomyces dermatitidis* of different mouse virulence. J. Bacteriol. *85*: 927-931.

91.   Domer, J. E. 1971. Monosaccharide and chitin content of cell walls of *Histoplasma capsulatum* and *Blastomyces dermatitidis.* J. Bacteriol. *107*: 870-877.

92.   Domer, J. E. 1976. In vivo and in vitro cellular responses to cytoplasmic and cell wall antigens of *Histoplasma capsulatum* in artificially immunized or infected guinea pigs. Infect. Immun. *13*: 790-799.

93.   Douat, N. E., and V. M. Diaz. 1958. Intradermoreacoes de paracoccidioidina, coccidioidina, e histoplasmina: resultadas de testes em 30 individuos. Rev. Bras. Tuberc. *26*: 663-668.

94.   Doughten, R. M., and H. A. Pearson. 1968. Disseminated intravascular coagulation associated with *Aspergillus* endocarditis. J. Pediatr. *73*: 576-582.

95.   Doyle, M. P., and E. H. Marth. 1978. Aflatoxin is degraded by heated and unheated mycelia, filtrates of homogenized mycelia and filtrates of broth cultures of *Aspergillus parasiticus.* Mycopathologia *64*: 59-62.

96.   Doyle, M. P., and E. H. Marth. 1978. Aflatoxin is degraded by mycelia from toxigenic and nontoxigenic strains of aspergilli grown on different substrates. Mycopathologia *63*: 145-153.

97.   Drutz, D. J., and A. Catanzaro. 1978. Coccidioidomycosis. Part I. Am. Rev. Respir. Dis. *117*: 559-585.

98.   Drutz, D. J., and A. Catanzaro. 1978. Coccidioidomycosis. Part II. Am. Rev. Respir. Dis. *117*: 727-771.

99. Dubos, R. J., and J. G. Hirsch. 1965. *Bacterial and Mycotic Infections of Man,* 4th ed., Lippincott, Philadelphia.

100. Dykstra, M. A., and L. Friedman. 1978. Pathogenesis, lethality, and immunizing effect of experimental cutaneous cryptococcosis. Infect. Immun. *20*: 446-455.

101. Einstein, H. E. 1977. Coccidioidomycosis and blastomycosis. Contrib. Microbiol. Immunol. *4*: 108-112.

102. Emmons, C. W. 1954. Isolation of *Myxotrichum* and *Gymnoascus* from the lungs of animals. Mycologia *46*: 334-338.

103. Emmons, C. W., C. H. Binford, J. P. Utz, and K. J. Kwon-Chung. 1977. *Medical Mycology,* 3rd ed., Lea & Febiger, Philadelphia.

104. English, M. P. 1963. The saprophytic growth of keratinophilic fungi on keratin. Sabouraudia *2*: 115-130.

105. English, M. P. 1968. The developmental morphology of the perforating organs and eroding mycelium of dermatophytes. Sabouraudia *6*: 218–227.

106. English, M. P. 1972. The epidemiology of animal ringworm in man. Br. J. Dermatol. *86*(Suppl. 8): 78-87.

107. Evans, W. E. D., and H. I. Winner. 1954. The histogenesis of the lesions in experimental moniliasis in rabbits. J. Pathol. Bacteriol. *67*: 531-536.

108. Everett, B. A., J. A. Kusske, J. L. Rush, and H. W. Pribram. 1978. Cryptococcal infection of the central nervous system. Surg. Neurol. *9*: 157-163.

109. Farhi, F., G. S. Bulmer, and J. R. Tacker. 1970. *Cryptococcus neoformans.* IV. The not-so-encapsulated yeast. Infect. Immun. *1*: 526-531.

110. Farley, J. F., R. A. Jersild, and D. J. Niederpruem. 1975. Origin and ultrastructure of intra-hyphal hyphae in *Trichophyton terrestre* and *T. rubrum.* Arch. Microbiol. *106*: 195-200.

111. Fishback, R. S., M. L. White, and S. Feingold. 1973. Bronchopulmonary geotrichosis. Am. Rev. Respir. Dis. *108*: 1388-1392.

112. Ford, S., and L. Friedman. 1967. Experimental study of the pathogenicity of aspergilli for mice. J. Bacteriol. *94*: 928-933.

113. Francois, J., and M. Rysselaere. 1972. *Oculomycoses,* Thomas, Springfield, Ill.

114. Freour, P., M. Lahourcade, and P. Chomy. 1966. Surune mycose pulmonaire nouvelle due a *Beauveria.* J. Med. Bordeaux *143*: 823-835.

115. Freour, P., M. Lahourcade, and P. Chomy. 1966. Une mycose nouvelle: étude clinique et mycologique d'une localisation pulmonaire de "*Beauveria.*" Bull. Soc. Med. Hop. Paris *117*: 197-206.

116. Friedman, L., C. E. Smith, W. G. Roessler, and R. J. Berman. 1956. The virulence and infectivity of twenty-seven strains of *Coccidioides immitis.* Am. J. Hyg. *64*: 198-210.

117. Fromtling, R. A. 1976. Ocular histoplasmosis in immunized and in immunosuppressed dutch belted rabbits. M.S. thesis, University of Tulsa.

118. Fromtling, R. A., R. Blackstock, and G. S. Bulmer. 1981. Immunization and passive transfer of immunity in murine cryptococcosis. In E. S. Kuttin and G. L. Baum (Eds.), *Proceedings of the Seventh Congress of the International Society for Human and Animal Mycology,* Int. Congr. Ser. No. 480, Excerpta Medica, Amsterdam, pp. 122-125.

119. Fromtling, R. A., R. Blackstock, N. K. Hall, and G. S. Bulmer. 1979. Kinetics of lymphocyte transformation in mice immunized with viable avirulent forms of *Cryptococcus neoformans.* Infect. Immun. *24*: 449-453.

120. Fromtling, R. A., R. Blackstock, N. K. Hall, and G. S. Bulmer. 1979. Lymphocyte transformation of spleen cells from mice immunized with a live, avirulent, pseudohyphal form of *Cryptococcus neoformans.* Abstr. Annu. Meet. Am. Soc. Microbiol., F53, p. 371.

121. Fromtling, R. A., J. M. Jensen, B. E. Robinson, and G. S. Bulmer. 1979. Fatal mycotic pulmonary disease of captive American alligators. Vet. Pathol. *16*: 428-431.

122. Fromtling, R. A., B. E. Robinson, J. M. Jensen, and G. S. Bulmer. 1979. Fatal pulmonary beauveriosis in captive American alligators. Abstr. Annu. Meet. Am. Soc. Microbiol. F20, p. 366.

123. Fromtling, R. A., M. E. Woolsey, and M. Binstock. 1976. Ocular histoplasmosis in immunized and immunosuppressed rabbits. Proc. Okla. Acad. Sci. *56*: 26-31.

124. Furtado, T. A. 1963. Mechanism of infection in South American blastomycosis. Dermatol. Trop. *2*: 27-32.

125. Furtado, T. 1975. Infection versus disease in South American blastomycosis. Int. J. Dermatol. *14*: 117-125.

126. Gabriel, B. P. 1968. Histochemical study of the insect cuticle infected by the fungus *Entomophthora coronata*. J. Invertebr. Pathol. *11*: 82-89.

127. Gadebusch, H. H. 1979. Native and acquired resistance to infection with *Cryptococcus neoformans*. In H. H. Gadebusch (Ed.), *Phagocytes and Cellular Immunity*, CRC Press, Boca Raton, Fla., pp. 137-157.

128. Gale, O., E. A. Lockhart, and E. Kimbell. 1967. Studies of *Coccidioides immitis*. III. Further studies of toxic soluble components of *Coccidioides immitis*. In L. Ajello (Ed.), *Coccidioidomycosis*, Papers from the Second Symposium on Coccidioidomycosis, University of Arizona Press, Tucson, pp. 355-372.

129. Gale, D., E. A. Lockhart, and E. Kimbell. 1967. Studies of *Coccidioides immitis*. 1. Virulence factors of *C. immitis*. Sabouraudia *6*: 29-36.

130. Galgiani, J. M., and D. A. Stevens. 1977. Chemotaxis of human polymorphonuclear leukocytes and soluble substances of *Coccidioides immitis*. In. L. Ajello (Ed.), *Coccidioidomycosis. Current Clinical and Diagnostic Status*, Symposia Specialists, Miami, Fla., pp. 379-381.

131. Galiana, J., and I. A. Conti-Diaz. 1963. Healing effect of heat and a rubefacient on nine cases of sporotrichosis. Sabouraudia *3*: 64-71.

132. Gander, J. E. 1974. Fungal cell wall glycoproteins and peptidopolysaccharides. Annu. Rev. Microbiol. *28*: 103-119.

133. Ganley, J. P. 1975. The role of the cellular immune system in patients, with macular disciform histoplasmosis. In T. F. Schlaegel, Jr. (Ed.), *Ocular Histoplasmosis*, Little, Brown, Boston, pp. 83-91.

134. Garber, E. D. 1956. A nutrition-inhibition hypothesis of pathogenicity. Am. Nat. *90*: 183-194.

135. Gardner, P. G., and E. W. Fuller, Jr. 1969. Fatal relapse of coccidioidomycosis ten years after treatment with amphotericin B. N. Engl. J. Med. *281*: 950-952.

136. Garrison, R. G., and K. S. Boyd. 1978. Role of the conidium in dimorphism of *Blastomyces dermatitidis*. Mycopathologia *64*: 29-33.

137. Gehweiler, J. A., M. P. Capp, and E. W. Chick. 1970. Observations on the roentgen patterns in blastomycosis of bone. Am. J. Roentgenol., Radium Ther. Nucl. Med. *108*: 497-510.

138. Georg, L. K., M. Williamson, E. B. Tilden, and R. E. Getty. 1962. Mycotic pulmonary disease of captive giant tortoises due to *Beauveria bassiana* and *Paecilomyces fumoso-roseus*. Sabouraudia *2*: 80-86.

139. Giraldo, R., A. Restrepo, F. Gutierrez, M. Robledo, H. Hernandez, F. Londono, F. Sierra, and G. Calle. 1976. Pathogenesis of paracoccidioidomycosis: a model based on a study of 46 patients. Mycopathologia *58*: 63-70.

140. Goldstein, G., M. H. Grieco, and G. Fenkel, and D. B. Louria. 1965. Studies on the pathogenesis of experimental *Candida parapsilosis* and *Candida guilliermondii* infections in mice. J. Infect. Dis. *115*: 293-302.

141. Gomez, P., F. Reyes, and R. Lahoz. 1977. Effect of the level of the carbon source on the activity of some lytic enzymes released during autolysis of *Aspergillus niger*. Mycopathologia *62*: 23-30.

142. Gonzalez-Ochoa, A. 1973. Virulence of nocardiae. Can. J. Microbiol. *19*: 901-904.

143. Gonzalez-Ochoa, A., and A. Sandoval-Cuellar. 1976. Different degrees of morbidity, in the white mouse, induced by *Nocardia brasiliensis, Nocardia asteroides* and *Nocardia caviae*. Sabouraudia *14*: 255-259.

144. Goodman, R. N., Z. Kiraly, and M. Zaitlin. 1967. *The Biochemistry and Physiology of Infectious Plant Disease*, D. Van Nostrand, Princeton.

145. Goodwin, R. A., Jr., and R. M. Des Prez. 1973. Pathogenesis and clinical spectrum of histoplasmosis. South. Med. J. *66*: 13-25.

146. Goren, M. B. 1967. Experimental murine cryptococcosis: effect of hyperimmunization to capsular polysaccharide. J. Immunol. *98*: 914-922.

147. Goren, M. B., and G. M. Middlebrook. 1967. Protein conjugates of polysaccharide from *Cryptococcus neoformans*. J. Immunol. *98*: 901-913.

148. Grappel, S. F., C. T. Bishop, and F. Blank. 1974. Immunology of dermatophytes and dermatophytosis. Bacteriol. Rev. *38*: 222-250.

149. Grappel, S. F., C. A. Buscavage, F. Blank, and C. T. Bishop. 1970. Comparative serological reactivities of twenty-seven polysaccharides from nine species of dermatophytes. Sabouraudia *9*: 50-55.

150. Graybill, J. R., and R. H. Alford. 1974. Cell-mediated immunity in cryptococcosis. Cell. Immunol. *14*: 12-21.

151. Greer, D. L., and B. Bolanos. 1974. Pathogenic potential of *Schizophyllum commune* isolated from a human case. Sabouraudia *12*: 233-244.

152. Greer, D. L., and B. Bolanos. 1977. Role of bats in the ecology of *Paracoccidioides brasiliensis:* the survival of *Paracoccidioides brasiliensis* in the intestinal tract of frugivorous bat, *Artibeus lituratus*. Sabouraudia *15*: 273-282.

153. Gresham, G. A., and C. H. Whittle. 1961. Studies of the invasive, mycelial form of *Candida albicans*. Sabouraudia *1*: 30-33.

154. Grimley, P. M., L. D. Wright, and A. E. Jennings. 1965. *Torulopsis glabrata* infection in man. Am. J. Clin. Pathol. *43*: 216-223.

155. Grose, E., and J. R. Tamsitt. 1965. *Paracoccidioides brasiliensis* recovered from the intestinal tract of three bats (*Artibeus lituratus*) in Columbia, S.A. Sabouraudia *4*: 124-125.

156. Hale, L. M. 1971. Orbital-cerebral phycomycosis. Report of a case and a review of the disease in infants. Arch. Ophthalmol. *86*: 39-43.

157. Hammerman, K. J., K. E. Powell, and F. E. Tosh. 1974. The incidence of hospitalized cases of systemic mycotic infections. Sabouraudia *12*: 33-45.

158. Harmon, R. R. M., J. Hughlings, and A. J. P. Willis. 1964. Subcutaneous phycomycosis in Nigeria. Br. J. Dermatol. *76*: 408-420.

159. Harvey, R. P., D. Pappagianis, J. Cochran, and D. A. Stevens. 1978. Otomycosis due to coccidioidomycosis. Arch. Intern. Med. *138*: 1434-1435.

160. Harvey, R. P., E. S. Schmid, C. C. Carrington, and D. A. Stevens. 1978. Mouse model of pulmonary blastomycosis: utility, simplicity, and quantitative parameters. Am. Rev. Respir. Dis. *117*: 695-703.

161. Hasenclever, H. F., and W. O. Mitchell. 1960. Antigenic relationships of *Torulopsis glabrata* and seven species of the genus *Candida*. J. Bacteriol. *79*: 677-681.

162. Hasenclever, H. F., and W. O. Mitchell. 1962. Pathogenesis of *Torulopsis glabrata* in physiologically altered mice. Sabouraudia *2*: 87-95.

163. Hasenclever, H. F., and W. O. Mitchell. 1962. Production in mice of tolerance to the toxic manifestations of *Candida albicans*. J. Bacteriol. *84*: 402-409.

164. Hatch, T. F. 1961. Distribution and deposition of inhaled particles in respiratory tract. Bacteriol. Rev. *25*: 237-240.

165. Hay, R. J., and F. W. Chandler, Jr. 1978. Experimental paracoccidioidomycosis: cranial and nasal localization in mice. Br. J. Exp. Pathol. *59*: 339-344.

166. Hellgren, L., and J. Vincent. 1976. Contact sensitizing properties of some fatty acids in dermatophytes. Sabouraudia *14*: 243-249.

167. Henderson, A. H. 1968. Allergic aspergillosis. Thorax *23*: 501-512.

168. Henrici, A. T. 1939. An endotoxin from *Aspergillus fumigatus*. J. Immunol. *36*: 319-338.

169. Herrold, K. M. 1965. *Cryptococcus neoformans:* pathogenesis of the disease in Syrian hamsters (abstr.). Part 1. Fed. Proc. *24*: 492.

170. Hesseltine, C. W., O. L. Shotwell, J. J. Ellis, and R. D. Stubblefield. 1966. Aflatoxin formation by *Aspergillus flavus*. Bacteriol. Rev. *30*: 795-805.

171. Hiles, D. A., and R. L. Font. 1968. Bilateral intraocular cryptococcosis with unilateral spontaneous regression: report of a case and review of the literature. Am. J. Ophthalmol. *65*: 98-108.

172. Hopsu-Havu, V. K., C. E. Sonck, and E. Tunnela. 1972. Production of elastase by pathogenic and non-pathogenic fungi. Mykosen *15*: 105-110.

173. Hopsu-Havu, V. K., and E. Tunnela. 1976. Production of elastase, urease and sulphatase by *Epidermophyton floccosum* (Harz) Langeron et Milochevitch (1930). Mykosen *20*: 91-96.

174. Horsfall, J. G., and A. E. Dimond. 1960. Prologue—the pathogen: the concept of causality. In J. G. Horsfall and A. E. Dimond (Eds.), *Plant Pathology, an Advanced Treatise,* Academic, New York, pp. 1-19.

175. Howard, D. H. 1964. Intracellular behavior of *Histoplasma capsulatum*. J. Bacteriol. *87*: 33-38.

176. Howard, D. H. 1965. Intracellular growth of *Histoplasma capsulatum*. J. Bacteriol. *89*: 518-523.

177. Howard, D. H. 1967. Effect of temperature on the intracellular growth of *Histoplasma capsulatum*. J. Bacteriol. *93*: 438-444.

178. Howard, D. H., and G. F. Orr. 1963. Comparison of strains of *Sporotrichum schenckii* isolated from nature. J. Bacteriol. *85*: 816-821.

179. Howard, D. H., and V. Otto. 1967. The intracellular behavior of *Torulopsis glabrata*. Sabouraudia *5*: 235-239.

180. Howard, D. H., and V. Otto. 1969. Protein synthesis by phagocytized yeast cells of *Histoplasma capsulatum*. Sabouraudia *7*: 186-194.

181. Hugh-Jones, M. E., and P. K. C. Austwick. 1967. Epidemiological studies in bovine mycotic abortion. I. The effect of climate on incidence. Vet. Rec. *81*: 273-276.

182. Huppert, M. 1961. Role of hypersensitivity in pathogenesis of coccidioidomycosis. Mycopathol. Mycol. Appl. *14*: 233.

183. Huppert, M., N. S. Spratt, K. R. Vukovich, S. H. Sun, and E. H. Rice. 1978. Antigenic analysis of coccidioidin and spherulin determined by two-dimensional immunoelectrophoresis. Infect. Immun. *20*: 541-551.

184. Hurley, D. L., and A. S. Fauci. 1975. Disseminated candidiasis. I. An experimental model in the guinea pig. J. Infect. Dis. *131*: 516-521.

185. Hurley, R., and H. I. Winner. 1963. Experimental renal moniliasis in the mouse. J. Pathol. Bacteriol. *86*: 75-82.

186. Ionnini, P. B., G. D. Arai, and F. M. LaForce. 1977. Vascular clearance of blastospore and pseudomycelial phase *Candida albicans*. Sabouraudia *15*: 201-205.

187. Iwata, K. 1977. Fungal toxins and their role in the etiopathology of fungal infections. In K. Iwata (Ed.), *Recent Advances in Medical and Veterinary My-*

cology, Proceedings of the Sixth Congress of the International Society for Human and Animal Mycology, University of Tokyo Press, Tokyo, pp. 15-34.

188. Iwata, K. 1977. Toxins produced by *Candida albicans.* Contrib. Microbiol. Immunol. *4*: 77-85.

189. Iwata, K., and K. Uchida. 1978. Cellular immunity in experimental fungus infections in mice: the influence of infections and treatment with a *Candida* toxin on spleen lymphoid cells. Mykosen (Suppl.) *1*: 72-81.

190. Iwata, K., and Y. Yamamoto. 1978. Glycoprotein toxins produced by *Candida albicans.* In *The Black and White Yeasts,* Proceedings of the Fourth International Conference on the Mycoses, Pan Am. Health Organ. Sci. Publ. No. 356, Washington, D.C., pp. 246-257.

191. Jampol, L. M. 1978. Ocular and orbital phycomycosis (letter). Surv. Ophthalmol. *22*: 353.

192. Kamalam, A., and A. S. Thambiah. 1972. Subcutaneous phycomycosis in Madras. Report of a case treated medically and surgically. Antiseptic *69*: 437-441.

193. Kamalam, A., and A. S. Thambiah. 1975. Basidiobolomycosis with lymph node involvement. Sabouraudia *13*: 44-48.

194. Kamalam, A., and A. S. Thambiah. 1976. A study of 3891 cases of mycoses in the tropics. Sabouraudia *14*: 129-148.

195. Kamalam, A., and A. S. Thambiah. 1978. Lymph node invasion by *Conidiobolus coronatus* and its spore formation in vivo. Sabouraudia *16*: 175-184.

196. Kanan, W. 1970. Torulosis of the skin. Dermatologia *141*: 15-20.

197. Kanda, M., M. Moriyama, M. Ikeda, S. Kojima, M. Tokunaga, and G. Watanabe. 1974. A statistical survey of deep mycoses in Japan, with particular reference to autopsy cases of cryptococcosis. Acta Pathol. Jpn. *24*: 595-609.

198. Kapica, L., and F. Blank. 1957. Growth of *Candida albicans* on keratin as sole source of nitrogen. Dermatologia *115*: 81-105.

199. Kaplan, W. 1967. Epidemiology and public health significance of ringworm in animals. Arch. Dermatol. *81*: 714-723.

200. Kaplan, W., and L. K. Georg. 1957. A device to aid in the development of mycotic and other skin infections in laboratory animals. Mycologia *49*: 604-605.

201. Karaoui, R. M., N. K. Hall, and H. W. Larsh. 1977. Role of macrophages in immunity and pathogenesis of experimental cryptococcosis induced by the airborne route. I. Pathogenesis and acquired immunity of *Cryptococcus neoformans.* Mykosen *20*: 380-388.

202. Karaoui, R. M., N. K. Hall, and H. W. Larsh. 1977. Role of macrophages in immunity and pathogenesis of experimental cryptococcosis induced by the airborne route. II. Phagocytosis and intracellular fate of *Cryptococcus neoformans.* Mykosen *20*: 409-422.

203. Kauffman, C. A., K. S. Israel, J. W. Smith, A. C. White, J. Schwartz, and G. F. Brooks. 1978. Histoplasmosis in immunosuppressed patients. Am. J. Med. *64*: 923-932.

204. Kaufman, L., and S. Blumer. 1977. Cryptococcosis: the awakening giant. In *The Black and White Yeasts,* Proceedings of the Fourth International Conference on the Mycoses, Pan Am. Health Organ. Sci. Publ. No. 356, Washington, D.C., pp. 176-182.

205. Keppie, J., H. Smith, and P. W. Harris-Smith. 1955. The chemical basis of the virulence of *Bacillus anthracis.* III. The role of the terminal bacteraemia in death of guinea pigs from anthrax. Br. J. Exp. Pathol. *36*: 315-322.

206. Khodadoust, A. A., and J. W. Payne. 1969. Cryptococcal (Torular) retinitis. Am. J. Ophthalmol. *67*: 745-750.

207. Kinas, H. Y., and J. J. Smulewicz. 1976. Primary pulmonary sporotrichosis. Respiration *33*: 468-474.

208. Knudtson, W. U., K. Wohlgemuth, and R. J. Bury. 1973. Bovine cerebral mucormycosis: report of a case. Sabouraudia *11*: 256-258.

209. Kobayashi, G. S., and L. Friedman. 1964. Characterization of the pyrogenicity of *Candida albicans, Saccharomyces cerevisiae,* and *Cryptococcus neoformans.* J. Bacteriol. *88*: 660-666.

210. Kobayashi, T., I. Nakashima, and N. Kato. 1974. Endotoxic substance of *Cryptococcus neoformans.* Mycopathol. Mycol. Appl. *54*: 391-404.

211. Koshi, G., K. Thankamma, D. Sudarsanam, A. J. Selvapandian, and K. E. Mammen. 1972. Subcutaneous phycomycosis caused by *Basidiobolus.* A report of three cases. Sabouraudia *10*: 237-243.

212. Kozel, T. R., and R. P. Mastroianni. 1976. Inhibition of phagocytosis by cryptococcal polysaccharide: dissociation of the attachment and ingestion phases of phagocytosis. Infect. Immun. *14*: 62-67.

213. Kruse, R. H., T. D. Green, and W. D. Leeder. 1967. Infection of control monkeys with *Coccidioides immitis* by caging with inoculated monkeys. In L. Ajello (Ed.), *Coccidioidomycosis,* Papers from the Second Symposium on Coccidioidomycosis, University of Arizona Press, Tucson, pp. 387-395.

214. Kunert, J. 1972. The digestion of human hair by the dermatophyte *Microsporum gypseum* in a submerged culture. Mykosen *15*: 59-71.

215. Kunert, J. 1972. Thiosulphate esters in keratin attacked by dermatophytes in vitro. Sabouraudia *10*: 6-13.

216. Kwon-Chung, K. J. 1975. A new genus, *Filobasidiella,* the perfect state of *Cryptococcus neoformans.* Mycologia *67*: 1197-1200.

217. Kwon-Chung, K. J. 1979. Comparison of isolates of *Sporothrix schenckii* obtained from fixed cutaneous lesions with isolates from other types of lesions. J. Infect. Dis. *139*: 424-431.

218. Lacaz, C. S. 1948. Blastomicose sul-americana. Reacoes intradermicas com a paracoccidioidina, coccidioidina, e blastomicetina. Rev. Hosp. Clin. *3*: 11-18.

219. Lacaz, C. S., M. C. Rocho Passos, C. Fava Netto, and B. Macarron. 1959. Contribucao para o estudo de "Blastomicose-Infeccao." Inquerito com a paracoc cidiodina. Estudo serologico e clinico radiologico dos paracoccidiodina-positivos. Rev. Inst. Med. Trop. Sao Paulo *1*: 245-259.

220. Lahoz, R., F. Reyes, and M. I. Perez Leblic. 1976. Lytic enzymes in the autolysis of filamentous fungi. Mycopathologia *60*: 45-49.

221. Land, G. A., W. C. McDonald, R. L. Stjernholm, and L. Friedman. 1975. Factors affecting filamentation in *Candida albicans:* relationship of the uptake and distribution of proline to morphogenesis. Infect. Immun. *11*: 1014-1023.

222. Laskey, W., and G. A. Sarosi. 1978. Endogenous activation in blastomycosis. Ann. Intern. Med. *88*: 50-52.

223. Lehmann, P. F., and L. O. White. 1975. Chitin assay used to demonstrate renal localization and cortisone—enhanced growth of *Aspergillus fumigatus* mycelium in mice. Infect. Immun. *12*: 987-992.

224. Lehmann, P. F., and L. O. White. 1978. Rapid germination of *Aspergillus fumigatus* conidia in mouse kidneys and a kidney extract. Sabouraudia *16*: 203-209.

225. Levine, H. B., J. M. Cobb, and C. E. Smith. 1960. Immunity to coccidioidomycosis induced in mice by purified spherule, arthrospore, and mycelial vaccines. Trans. N.Y. Acad. Sci. *22*: 436-449.

226. Levine, H. B., Y. M. Kong, and C. E. Smith. 1965. Immunization of mice to *Coccidioides immitis.* J. Immunol. *94*: 132-142.

227. Levinson, D. J., D. C. Silcox, J. W. Rippon, and S. Thomsen. 1974. Septic arthritis due to nonencapsulated *Cryptococcus neoformans* with coexisting sarcoidosis. Arthritis Rheum. *17*: 1037-1047.

228. Lewis, J. L., and S. Rabinovich. 1972. The wide spectrum of cryptococcal infection. Am. J. Med. *53*: 315-322.

229. Littman, M. L., and L. E. Zimmerman. 1956. *Cryptococcosis,* Grune & Stratton, New York.

230. Louria, D. B. 1961. The pathogenesis of experimental infections due to *Cryptococcus* and *Candida.* Mycopathol. Mycol. Appl. *14*: 233-234.

231. Louria, D. B. 1977. Experimental infections with fungi and yeasts. Contrib. Microbiol. Immunol. *3*: 31-47.

232. Louria, D. B., A. Blevins, D. Armstrong, R. Burdick, and P. Lieberman. 1967. Fungemia caused by "nonpathogenic" yeasts. Arch. Intern. Med. *119*: 247-252.

233. Louria, D. B., and R. G. Brayton. 1964. The behavior of *Candida* cells within leukocytes. Proc. Soc. Exp. Biol. Med. *115*: 93-98.

234. Louria, D. B., R. G. Brayton, and G. Finkel. 1963. Studies on the pathogenesis of experimental *Candida albicans* infections in mice. Sabouraudia *2*: 271-283.

235. Louria, D. B., and H. G. Browne. 1960. The effect of cortisone on experimental fungal infections. Ann. N.Y. Acad. Sci. *89*: 39-46.

236. Louria, D. B., G. Finkel, J. K. Smith, and M. Buse. 1974. Aflatoxin-induced tumors in mice. Sabouraudia *12*: 371-375.

237. Louria, D. B., S. M. Greenberg, and D. W. Molander. 1960. Fungemia caused by certain non-pathogenic strains of the family Cryptococcaceae. Report of two cases due to *Rhodotorula* and *Torulopsis glabrata.* N. Engl. J. Med. *263*: 1281-1284.

238. Louria, D. B., P. H. Liebermann, H. S. Collins, and A. Blevins. 1966. Pulmonary mycetoma due to *Allescheria boydii.* Arch. Intern. Med. *117*: 748-751.

239. Louria, D. B., J. K. Smith, and G. C. Finkel. 1970. Mycotoxins other than aflatoxins: tumor-producing potential and possible relation to human disease. Ann. N.Y. Acad. Sci. *174*: 583-591.

240. Lowenstein, M., S. M. Markowitz, H. C. Nottebart, and S. Shadomy. 1978. Existence of *Sporothrix schenckii* as a pulmonary saprophyte. Chest *73*: 419-421.

241. Lurie, H. I. 1963. Histopathology of sporotrichosis. Arch. Pathol. *75*: 421-437.

242. Lurie, H. I., and W. J. S. Still. 1969. The "capsule" of *Sporotrichum schenckii* and the evolution of the asteroid body. Sabouraudia *7*: 64-70.

243. Lynch, P. J., J. J. Voorhees, and E. R. Harrell. 1970. Systemic sporotrichosis. Ann. Intern. Med. *73*: 23-30.

244. Mackinnon, J. E. 1970. On the importance of South American blastomycosis. Mycopathologia *41*: 187-193.

245. Mackinnon, J. E., and I. A. Conti-Diaz. 1962. The effect of temperature on sporotrichosis. Sabouraudia *2*: 56-59.

246. Mackinnon, J. E., I. A. Conti-Diaz, E. Gezuele, E. Civila, and S. DaLuz. 1969. Isolation of *Sporothrix schenckii* from nature and considerations on its pathogenicity and ecology. Sabouraudia *7*: 38-46.

247. Mackinnon, J. E., E. Gezuele, I. A. Conti-Diaz, and A. C. deGimenez. 1973. Production of capsule and conidia by yeast-like cells of *Phialophora spinifera* and *Phialophora jeanselmei.* Sabouraudia *11*: 33-38.

248. Mahajan, V. M., S. C. Padhy, Y. Dayal, I. M. Bhatia, and K. S. Ratnakar. 1977. Experimental pulmonary nocardiosis in monkeys. Sabouraudia *15*: 47-50.

249. Mahgoub, E. S. 1978. Experimental infection of athymic nude New Zealand mice, *nu/nu* strain with mycetoma agents. Sabouraudia *16*: 211-216.

250. Male, O. 1977. Pathogenesis of mucocutaneous mycoses caused by yeasts. Contrib. Microbiol. Immunol. *3*: 66-80.

251. Malloch, D. 1970. New concepts in the *Microascaceae* illustrated by two new species. Mycologia *62*: 727-739.

252. Mariat, F., P. Destombes, and G. Segretain. 1977. The mycetomas: clinical features, pathology, etiology and epidemiology. Contrib. Microbiol. Immunol. *4*: 1-39.

253. Mariat, F., P. Lavalle, and P. Destombes. 1962. Recherches sur la sporotrichose. Sabouraudia *2*: 60-79.

254. Martinson, F. D. 1971. Chronic phycomycosis of the upper respiratory tract. Rhinophycomycosis entomophthorae. Am. J. Trop. Med. Hyg. *120*: 449-455.

255. Masur, H., P. P. Rosen, and D. Armstrong. 1977. Pulmonary disease caused by *Candida* species. Am. J. Med. *63*: 914-924.

256. McMurray, D. N., M. E. Thomas, D. L. Greer, and N. L. Tolentino. 1978. Humoral and cell-mediated immunity to *Histoplasma capsulatum* during experimental infection in neotropical bats (*Artibeus lituratus*). Am. J. Trop. Med. Hyg. *27*: 815-821.

257. Meyer, E., and R. D. Harrold. 1961. *Allescheria boydii* isolated from a patient with chronic prostitis. Am. J. Pathol. *35*: 155-159.

258. Meyer, R. D., L. S. Young, D. Armstrong, and B. Yu. 1973. Aspergillosis complicating neoplastic disease. Am. J. Med. *54*: 6-15.

259. Mims, C. A. 1976. *The Pathogenesis of Infectious Disease,* Grune & Stratton, New York.

260. Minkowitz, S., D. Koffier, and F. G. Zak. 1963. *Torulopsis glabrata* septicemia. Am. J. Med. *34*: 252-255.

261. Mirocha, C. J., and C. M. Christensen. 1974. Fungus metabolites toxic to animals. Annu. Rev. Phytopathol. *12*: 303-330.

262. Mitchell, T. G., and L. Friedman. 1972. In vitro phagocytosis and intracellular fate of variously encapsulated strains of *Cryptococcus neoformans*. Infect. Immun. *5*: 491-498.

263. Miyaji, M., and K. Nishimura. 1977. Relationship between proteolytic activity of *Aspergillus fumigatus* and the fungus invasiveness of mouse brain. Mycopathologia *62*: 161-166.

264. Mok, D. W. Y., and D. L. Greer. 1977. Cell-mediated immune responses in patients with paracoccidioidomycosis. Clin. Exp. Immunol. *28*: 89-98.

265. Montes, L. F., and W. H. Wilborn. 1968. Ultrastructural features of host-parasite relationships in oral candidiasis. J. Bacteriol. *96*: 1349-1356.

266. Moraes, M. A. P. 1962. Blastomicose tipo. Jorge Lobo (6 casos novos encontrados no Estada do Amazonas, Brasil). Rev. Inst. Med. Trop. Sao Paulo *4*: 187-191.

267. Morelli, R., and L. T. Rosenbert. 1971. Role of complement during experimental *Candida* infection in mice. Infect. Immun. *3*: 521-523.

268. Moss, M. O. 1974. Fungal toxins acting on mammalian cells. Proc. Soc. Gen. Microbiol. *1*: 52.

269. Mourad, S., and L. Friedman. 1961. Pathogenicity of *Candida*. J. Bacteriol. *81*: 550-556.

270. Mukoyama, M., K. Gimple, and C. M. Poser. 1969. Aspergillosis of the central nervous system. Neurology *19*: 967-974.

271. Müller, H. E., and K. K. Sethi. 1972. Proteolytic activity of *Cryptococcus neoformans* against human plasma proteins. Med. Microbiol. Immunol. *158*: 129-134.

272. Müller, J., H. Takamiya, and R. Jaeger. 1977. Elektronenmikroskopische Dar-

stellung von Immunreaktionen an *Candida* Zellen: Asteroid Bodies bei *Candida albicans* im Urin von Nephritis-Patienten. Sabouraudia *15*: 87-93.

273. Murray, I., E. Spooner, and J. Walker. 1960. Experimental infection of mice with *Madurella mycetomi*. Trans. R. Soc. Trop. Med. Hyg. *54*: 335-371.

274. Negroni, P. 1966. El *Paracoccidioides brasiliensis* vive saprofiticamente en el suelo Argentino. Prensa Med. Argent. *53*: 2831-2832.

275. Negroni, P., and R. Negroni. 1965. Nuestra experiencia de la blastomicosis sud-americana en la Argentina. Mycopathol. Mycol. Appl. *26*: 264-272.

276. Negroni, R., C. Obrutzky, and R. Gonzalez. 1966. A case of septicemia by *Torulopsis glabrata*. Sabouraudia *4*: 244-249.

277. Neilson, J. B., and G. S. Bulmer. 1975. The infectious particles of *Cryptococcus neoformans*. Abstr. Annu. Meet. Am. Soc. Microbiol. F45, p. 93.

278. Neilson, J. B., R. A. Fromtling, and G. S. Bulmer. 1977. *Cryptococcus neoformans:* size range of infectious particles from aerosolized soil. Infect. Immun. *17*: 634-638.

279. Neilson, J. B., M. H. Ivey, and G. S. Bulmer. 1978. *Cryptococcus neoformans:* pseudohyphal forms surviving culture with *Acanthamoeba polyphaga*. Infect. Immun. *20*: 262-266.

280. Newberne, P. M., G. N. Wogan, W. W. Carlton, and M. M. Abdel Kader. 1964. Histopathologic lesions in ducklings caused by *Aspergillus flavus* cultures, culture extracts, and crystalline aflatoxins. Toxicol. Appl. Pharmacol. (Suppl.) *6*: 542-556.

281. Nielson, H. S., Jr. 1966. Variation in lipid content of strains of *Histoplasma capsulatum* exhibiting different virulence properties for mice. J. Bacteriol. *91*: 273-277.

282. Noble, R. C., and L. F. Fajardo. 1972. Primary cutaneous cryptococcosis: review and morphologic study. Am. J. Clin. Pathol. *57*: 13-22.

283. Nobre, G., and M. P. Viegas. 1972. Lipolytic activity of dermatophytes. Mycopathol. Mycol. Appl. *46*: 319-323.

284. Nordwig, A., and W. F. Jahn. 1968. A collagenolytic enzyme from *Aspergillus oryzae*. Purification and properties. Eur. J. Biochem. *3*: 519-529.

285. Nozawa, Y., T. Noguchi, Y. Ita, N. Suda, and S. Watanabe. 1971. Immunochemical studies on *Trichophyton mentagrophytes*. Sabouraudia *9*: 129-138.

286. O'Connor, G. R. 1975. Experimental ocular histoplasmosis. In T. F. Schlaegel, Jr. (Ed.), *Ocular Histoplasmosis*, Little, Brown, Boston, pp. 93-104.

287. Odds, F. C., and J. C. Hierholzer. 1973. Purification and properties of a glycoprotein acid phosphatase from *Candida albicans*. J. Bacteriol. *114*: 257-266.

288. O'Grady, K. J., M. P. English, and R. P. Warin. 1972. *Microsporum equinum* infection of the scalp in an adult. Br. J. Dermatol. *86*: 175-176.

289. Okun, E., and W. T. Butler. 1964. Ophthalmologic complications of cryptococcal meningitis. Arch. Ophthalmol. *71*: 52-57.

290. Oleniacz, W. S., and M. A. Pisano. 1968. Proteinase production by a species of *Cephalosporium*. Appl. Microbiol. *16*: 90-96.

291. Otero, R. B., N. L. Goodman, and J. C. Parker, Jr. 1978. Predisposing factors in systemic and central nervous system candidiasis: histopathological and cultural observations in the rat. Mycopathologia *64*: 113-120.

292. Parkhurst, G. F., and G. D. Vlahides. 1967. Fatal opportunistic fungus disease. JAMA *202*: 279-281.

293. Philpot, C. M. 1978. Geographical distribution of the dermatophytes: a review. J. Hyg. (Camb.) *80*: 301-313.

294. Pore, R., and H. W. Larsh. 1968. Experimental pathology of *Aspergillus terrus-flavipes* group species. Sabouraudia *6*: 89-93.

295. Powell, K. E., B. A. Dahl, R. J. Weeks, and F. E. Tosh. 1972. Airborne *Cryptococcus neoformans:* particles from pigeon excreta compatible with alveolar deposition. J. Infect. Dis. *125*: 412-415.

296. Pradinaud, R., E. Grosshans, and M. Basset. 1969. Un nouveau cas de maladie de Jorge Lobo en Guyane Francaise. Bull. Soc. Fr. Dermatol. Syphiligr. *76*: 837-840.

297. Pradinaud, R., F. Joly, M. Basset, A. Basset, and E. Grosshans. 1969. Les chromomycoses et la maladie de Jorge Lobo en Guyane Francaise. Bull. Soc. Pathol. Exot. *62*: 1054-1063.

298. Price, J. T., and G. S. Bulmer. 1972. Tumor induction by *Cryptococcus neoformans.* Infect. Immun. *6*: 199-205.

299. Price, M. F., and R. A. Cawson. 1977. Phospholipase activity in *Candida albicans.* Sabouraudia *15*: 179-185.

300. Pugh, D., and R. A. Cawson. 1975. The cytochemical localization of phospholipase A and lysophospholipase in *Candida albicans.* Sabouraudia *13*: 110-115.

301. Pugh, D., and R. A. Cawson. 1977. The cytochemical localization of phospholipase in *Candida albicans* infecting the chick chorio-allantoic membrane. Sabouraudia *15*: 29-35.

302. Purnell, D. M. 1975. Quantitative tissue invasion of the murine brain as a phenotypic marker of strain virulence in *Aspergillus nidulans.* Sabouraudia *13*: 209-216.

303. Rabinowitz, J. G., J. Busch, and W. R. Buttram. 1976. Pulmonary manifestations of blastomycosis. Radiology *120*: 25-32.

304. Rau, E. M., E. B. Tilden, and V. L. Koenig. 1961. Partial purification and characterization of the endotoxin from *Aspergillus fumigatus.* Mycopathol. Mycol. Appl. *14*: 347-358.

305. Rebell, G., and D. Taplin. 1974. *Dermatophytes: Their Recognition and Identification,* 2nd ed., University of Miami Press, Coral Gables.

306. Restrepo, A., D. L. Greer, M. V. Robledo, G. Constanza Diaz, R. M. Lopez, and C. R. Bravo. 1967. Subcutaneous phycomycosis: report of first case observed in Columbia, South America. Am. J. Trop. Med. *16*: 34-39.

307. Restrepo, A., D. L. Greer, M. Robledo, O. Osorio, and H. Mondragon. 1971. Ulceration of the palate caused by a basidiomycete *Schizophyllum commune.* Sabouraudia *9*: 201-204.

308. Restrepo, A., and G. E. Guzman. 1976. Paracoccidioidomicosis experimental del ration inducida por via aerogena. Sabouraudia *14*: 299-311.

309. Restrepo, A., and J. D. Schneidau, Jr. 1967. Nature of the skin-reactive principle in culture filtrates prepared from *Paracoccidioides brasiliensis.* J. Bacteriol. *93*: 1741-1748.

310. Ribner, B., G. T. Keusch, B. A. Hanna, and M. Perloff. 1976. Combination amphotericin B-rifampin therapy for pulmonary aspergillosis in a leukemic patient. Chest *70*: 681-683.

311. Ridgeway, N. A., F. C. Witcomb, E. E. Erickson, and S. W. Law. 1962. Primary pulmonary sporotrichosis. Report of two cases. Am. J. Med. *32*: 153-160.

312. Rippon, J. W. 1967. Elastase: production by ringworm fungi. Science *157*: 947.

313. Rippon, J. W. 1968. Monitored environment system to control cell growth, morphology, and metabolic rate in fungi by oxidation-reduction potentials. Appl. Microbiol. *16*: 114-121.

314. Rippon, J. W. 1974. *Medical Mycology. The Pathogenic Fungi and the Pathogenic Actinomycetes,* Saunders, Philadelphia.

315. Rippon, J. W. 1980. Dimorphism in pathogenic fungi. Crit. Rev. Microbiol. *8*: 49-97.

316. Rippon, J. W., and D. N. Anderson. 1970. Metabolic rate of fungi as a function of temperature and oxidation-reduction potential (Eh). Mycopathol. Mycol. Appl. 40: 349-352.

317. Rippon, J. W., D. N. Anderson, and M. SooHoo. 1971. Aspergillosis: comparative virulence, metabolic rate, growth rate and ubiquinone content of soil and human isolates of *Aspergillus terreus*. Sabouraudia 12: 157-161.

318. Rippon, J. W., and J. W. Carmichael. 1976. Petriellidiosis (Allescheriosis): four unusual cases and review of literature. Mycopathologia 58: 117-124.

319. Rippon, J. W., and A. L. Lorincz. 1964. Collagenase activity of *Streptomyces (Nocardia) madurae*. J. Invest. Dermatol. 43: 483-486.

320. Rippon, J. W., and G. L. Peck. 1967. Experimental infection with *Streptomyces madurae* as a function of collagenase. J. Invest. Dermatol. 49: 371-378.

321. Rippon, J. W., and D. P. Varadi. 1968. The elastase of pathogenic fungi and actinomycetes. J. Invest. Dermatol. 50: 54-58.

322. Ritter, R. C., and H. W. Larsh. 1963. The infection of white mice following an intranasal instillation of *Cryptococcus neoformans*. Am. J. Hyg. 78: 241-246.

323. Robboy, S. J., and J. Kaiser. 1975. Pathogenesis of fungal infection on heart valve prosthesis. Hum. Pathol. 6: 711-715.

324. Roberts, D. W., and W. G. Yendol. 1971. Use of fungi for microbial control of insects. In H. O. Burgess and N. W. Hussey (Eds.), *Microbial Control of Insects and Mites*, Academic, New York, pp. 125-149.

325. Rodricks, J. V., C. W. Hesseltine, and M. A. Mehlman (Eds.). 1977. *Mycotoxins in Human and Animal Health*, Pathotox Publishers, Park Forest South, Ill.

326. Rogers, T. J., and E. Balish. 1978. Suppression of lymphocyte blastogenesis by *Candida albicans*. Clin. Immunol. Immunopathol. 10: 298-305.

327. Rogozhkina, N. M., I. M. Klimova, and L. N. Zelenskaya. 1977. Izuchenie nekotorykh uslovii abrazovaniya proteazy *Coccidioides immitis*. Zh. Mikrobiol. Epidemiol. Immunobiol. 8: 87-90.

328. Roser, S. M., R. F. Canalis, and C. J. Hanna. 1976. Aspergillosis of the maxillary antrum. J. Oral Med. 31: 91-93.

329. Roth, A. M. 1977. *Histoplasma capsulatum* in the presumed ocular histoplasmosis syndrome. Am. J. Ophthalmol. 84: 293-298.

330. Ruffin, P., S. Andrieu, G. Biserte, and J. Biguet. 1976. Sulphitolysis in keratinalysis. Biochemical proof. Sabouraudia 14: 181-184.

331. Ryan, S. J. 1975. Histopathological correlates of presumed ocular histoplasmosis. In T. F. Schaegel, Jr. (Ed.), *Ocular Histoplasmosis*, Little, Brown, Boston, pp. 125-135.

332. Salkin, D. 1967. Clinical examples of reinfection in coccidioidomycosis. Am. Rev. Respir. Dis. 95: 603-611.

333. Salvin, S. B. 1952. Endotoxin in pathogenic fungi. J. Immunol. 69: 89-99.

334. Sampaio, M. M., and L. B. Dias. 1970. Experimental infection of Jorge Lobo's disease in the cheekpouch of the golden hamster *(Mesocricetus auratus)*. Rev. Inst. Med. Trop. Sao Paulo 12: 115-120.

335. Sampaio, M. M., and L. B. Dias. 1977. The armadillo *Euphractus sexcinctus* as a suitable animal for experimental studies of Jorge Lobo's disease. Rev. Inst. Med. Trop. Sao Paulo 19: 215-220.

336. Sampaio, M. M., L. B. Dias, and L. Scaff. 1971. Bizarre forms of the aetiologic agent in experimental Jorge Lobo's disease in tortoises. Rev. Inst. Med. Trop. Sao Paulo 13: 191-193.

337. San-Blas, G., D. Ordaz, and F. J. Yegres. 1978. *Histoplasma capsulatum:* chemical variability of the yeast cell wall. Sabouraudia 16: 279-284.

338. San-Blas, G., and F. San-Blas. 1977. *Paracoccidioides brasiliensis:* cell wall structure and virulence. Mycopathologia *62*: 77-86.

339. San-Blas, G., F. San-Blas, and L. E. Serrano. 1977. Host-parasite relationships in the yeast-like form of *Paracoccidioides brasiliensis* strain IVIC Pb9. Infect. Immun. *15*: 343-346.

340. San-Blas, G., and D. Vernet. 1977. Induction of the synthesis of cell wall α-1,3-glucan in the yeast-like form of *Paracoccidioides brasiliensis* strain IVIC Pb9 by fetal calf serum. Infect. Immun. *15*: 897-902.

341. Sarosi, G. A., K. J. Hammerman, F. E. Tosh, and R. S. Kronenberg. 1974. Clinical features of acute pulmonary blastomycosis. N. Engl. J. Med. *290*: 540-543.

342. Sarosi, G. A., P. M. Silberfarb, and F. E. Tosh. 1971. Cutaneous cryptococcosis. A sentinel of disseminated disease. Arch. Dermatol. *104*: 1-3.

343. Schmitt, J. A., and R. G. Miller. 1967. Variation in susceptibility to experimental dermatomycosis in genetic strains of mice. Mycopathologia *32*: 306-312.

344. Schoental, R. 1961. Liver changes and primary liver tumours in rats given toxic guinea pig diet (M.R.C. diet 18). Br. J. Cancer *15*: 812-815.

345. Schwartz, L. K., L. M. Loignon, and R. G. Webster, Jr. 1978. Posttraumatic phycomycosis of the anterior segment. Arch. Ophthalmol. *96*: 860-863.

346. Scillian, J. J., G. C. Cozad, and H. D. Spencer. 1974. Passive transfer of delayed hypersensitivity to *Blastomyces dermatitidis* between mice. Infect. Immun. *10*: 705-711.

347. Seelig, M. S. 1966. The role of antibiotics in the pathogenesis of *Candida* infections. Am. J. Med. *40*: 887-917.

348. Seeliger, H. P. R. 1959. Das Kulturell-biochemische und serologische Verhalten der *Cryptococcus*-Gruppe. Ergeb. Mikrobiol. Immunitaetsforsch. Exp. Ther. *32*: 23-72.

349. Sethi, K. K. 1967. Attempts to produce experimental intestinal cryptococcosis and sporotrichosis. Mycopathologia *31*: 245-250.

350. Sethi, K. K. 1972. Experimental sporotrichosis in the normal and modified host. Sabouraudia *10*: 66-73.

351. Sethi, K. K., V. L. Kneipp, and J. Schwarz. 1966. Pulmonary sporotrichosis in mice following intranasal infection. Am. Rev. Respir. Dis. *93*: 463-464.

352. Sethi, K. K., K. Sanfelder, and J. Schwarz. 1965. Experimental cutaneous primary infection with *Cryptococcus neoformans* (Sanfelice) Vuillemin. Mycopathol. Mycol. Appl. *27*: 357-368.

353. Sheehy, T. W., B. K. Honeycutt, and J. T. Spencer. 1976. *Geotrichum* septicemia. JAMA *235*: 1035-1037.

354. Sidransky, H., S. M. Epstein, E. Verney, and C. Horowitz. 1972. Experimental visceral aspergillosis. Am. J. Pathol. *69*: 55-70.

355. Simic, L., and S. Perisic. 1969. Microsporosis caused by *M. canis* in humans and a dog transmitted by an imported cat. Mykosen *12*: 699-703.

356. Simonetti, N., and V. Strippoli. 1973. Pathogenicity of the Y form as compared to the M form in experimentally induced *Candida albicans* infections. Mycopathol. Mycol. Appl. *51*: 19-28.

357. Sixbey, J. W., B. T. Fields, C. N. Sun, R. C. Clark, and C. M. Nolan. 1979. Interactions between human granulocytes and *Blastomyces dermatitidis*. Infect. Immun. *23*: 41-44.

358. Smith, H. 1968. Biochemical challenge of microbial pathogenicity. Bacteriol. Rev. *32*: 164-184.

359. Smith, H. 1972. Mechanisms of virus pathogenicity. Bacteriol. Rev. *36*: 281-310.

360. Smith, H. 1977. Microbial surfaces in relation to pathogenicity. Bacteriol. Rev. *41*: 475-500.

361. Smith, R. E., G. R. O'Connor, C. J. Halde, M. A. Scalarone, and W. M. Easterbrook. 1973. Clinical course in rabbits after experimental induction of ocular histoplasmosis. Am. J. Opthalmol. *76*: 284-293.

362. Smith, R. E., M. A. Scalarone, G. R. O'Connor, and C. J. Halde. 1973. Detection of *Histoplasma capsulatum* by fluorescent antibody techniques in experimental histoplasmosis Am. J. Ophthalmol. *76*: 375-380.

363. Soderhall, K., E. Svensson, and T. Unestam. 1978. Chitinase and protease activities in germinating zoospore cysts of a parasitic fungus, *Aphanomyces astaci, Oomycetes*. Mycopathologia *64*: 9-11.

364. Soderhall, K., and T. Unestam. 1975. Properties of extracellular enzymes from *Aphanomyces astaci* and their relevance in the penetration process of crayfish cuticle. Physiol. Plant. (Suppl.) *35*: 140-146.

365. Song, M. M. 1974. Experimental cryptococcosis of the skin. Sabouraudia *12*: 133-137.

366. Speller, D. C. E., and A. G. MacIver. 1971. Endocarditis caused by a *Coprinus* species: a fungus of the toadstool group. J. Med. Microbiol. *4*: 370-374.

367. Spencer, H. D., and G. C. Cozad. 1973. Role of delayed hypersensitivity in blastomycosis of mice. Infect. Immun. *7*: 329-334.

368. Staat, R. H., and C. F. Schachtele. 1974. Dextranase-producing bacteria from human dental plaque (abstr.). J. Dent. Res. *53*: 107.

369. Staib, F. 1964. Das Serum-Reststickstoff-Auxanogramm (mit Spross-und Schimmelpilzen). Zentralbl. Bakteriol. Parasitenkd. Infektionskr. Hyg., Abt. 1: Orig. *194*: 379-406.

370. Staib, F. 1964. Saprophytic life of *Cryptococcus neoformans*. Its relation to low molecular nitrogen substances in nature and the human. Ann. Soc. Belge Med. Trop. *44*: 611-618.

371. Staib, F. 1965. Serum-proteins as nitrogen source for yeast-like fungi. Sabouraudia *4*: 187-193.

372. Staib, F., S. K. Mishra, G. Grosse, and T. Abel. 1977. Ocular cryptococcosis-experimental and clinical observations. Zentralbl. Bakteriol. Parasitenkd. Infektionskr. Hyg., Abt. 1: Orig. *A237*: 378-394.

373. Staib, F., and S. K. Mishra. 1975. Selective involvement of the brain in experimental murine cryptococcosis. Zentralbl. Bakteriol. Parasitenkd. Infektionskr. Hyg., Abt. 1: Orig. *A232*: 355-364.

374. Staib, F., and J. Zissler. 1963. Über die Verwertbarkeit von Rest-Stickstoff-Substanzen menschlicher Seren für *Cryptococcus neoformans*. Zentralbl. Bakteriol. Parasitenkd. Infektionskr. Hyg., Abt. 1: Orig. *189*: 117-119.

375. Stobo, J. D., S. Paul, R. E. Van Scoy, and P. E. Hermans. 1976. Suppressor thymus-derived lymphocytes in fungal infection. J. Clin. Invest. *57*: 319-328.

376. Stone, H. H. 1974. Studies in the pathogenesis, diagnosis, and treatment of *Candida* sepsis in children. J. Pediatr. Surg. *9*: 127-133.

377. Stone, H. H., L. O. Kolb, C. A. Currie, C. E. Geheber, and J. Z. Cuzzell. 1974. *Candida* sepsis. Pathogenesis and principles of treatment. Ann. Surg. *179*: 697-711.

378. Stossel, T. P. 1974. Phagocytosis. N. Engl. J. Med. *290*: 717-723.

379. Stossel, T. P. 1974. Phagocytosis. N. Engl. J. Med. *290*: 774-780.

380. Stretton, R. J., and R. A. Bulman. 1974. Experimentally induced actinomycotic mycetoma. Sabouraudia *12*: 245-257.

381. Svec, P. 1974. On the mechanism of action of glycoprotein from *Candida albicans*. J. Hyg. Epidemiol. Microbiol. Immunol. *18*: 373-376.

382. Swatek, F. E. 1970. Ecology of *Coccidioides immitis.* Mycopathol. Mycol. Appl. *41*: 3-12.
383. Swenberg, J. A., A. Koestner, and R. P. Tewari. 1969. Experimental mycotic encephalitis. Acta Neuropathol. (Berl.) *13*: 75-90.
384. Swenberg, J. A., A. Koestner, and R. P. Tewari. 1969. The pathogenesis of experimental mycotic encephalitis. An ultrastructural study. Lab. Invest. *21*: 365-373.
385. Symmers, W. St. C. 1964. The occurrence of deep-seated fungal infections in general hospital practice in Britain today. Proc. R. Soc. Med. *57*: 405-411.
386. Tacker, J. R., F. Farhi, and G. S. Bulmer. 1972. Intracellular fate of *Cryptococcus neoformans.* Infect. Immun. *6*: 162-167.
387. Taschdjian, C. L. 1970. Opportunistic yeast infections, with a special reference to candidiasis. Ann. N.Y. Acad. Sci. *174*: 606-622.
388. Taylor, J. J. 1964. Enhanced toxicity in chemically altered *Blastomyces dermatitidis.* J. Bacteriol. *87*: 748-749.
389. Tewari, R. P., and F. J. Berkhout. 1972. Comparative pathogenicity of albino and brown types of *Histoplasma capsulatum* for mice. J. Infect. Dis. *125*: 504-508.
390. Tewari, R. P., and C. R. Macpherson. 1968. Pathogenicity and neurological effects of *Oidiodendron kalrai* for mice. J. Bacteriol. *95*: 1130-1139.
391. Thadepalli, H., F. A. Salem, A. K. Mandal, K. Rambhatla, and H. E. Einstein. 1977. Pulmonary mycetoma due to *Coccidioides immitis.* Chest *71*: 429-430.
392. Tilden, E. B., S. Freeman, and L. Lombard. 1963. Further studies of the *Aspergillus* endotoxins. Mycopathol. Mycol. Appl. *20*: 253-271.
393. Tilden, E. B., E. H. Hatton, S. Freeman, W. M. Williamson, and V. L. Koenig. 1961. Preparation and properties of the endotoxins of *Aspergillus fumigatus* and *Aspergillus flavus.* Mycopathol. Mycol. Appl. *14*: 325-346.
394. Tomsikova, A., J. Dura, and D. Novackova. 1973. Pathogenic effects of *Cladosporium herbarum* and *Penicillium decumbens.* Sabouraudia *11*: 251-255.
395. Tong, J. L., E. H. Valentine, J. R. Durrance, G. M. Wilson, and D. A. Fischer. 1958. Pulmonary infection with *Allescheria boydii:* report of a fatal case. Am. Rev. Tuberc. *78*: 604-609.
396. Trevathan, R. D., and S. Phillips. 1966. Primary pulmonary sporotrichosis. JAMA *195*: 965-967.
397. Trnovec, T., D. Sikl, M. Zemanek, V. Faberova, S. Bezek, A. Gajdosik, and V. Koprda. 1978. The distribution in mice of intravenously administered labelled *Candida albicans.* Sabouraudia *16*: 299-306.
398. Tynes, B., K. N. Mason, A. E. Jennings, and J. E. Bennett. 1968. Variant forms of pulmonary cryptococcosis. Ann. Inter. Med. *69*: 1117-1125.
399. Vanbreuseghem, R., and C. deVroey. 1970. Geographic distribution of dermatophytes. Int. J. Dermatol. *9*: 102-109.
400. VanUden, N. 1960. The occurrence of *Candida* and other yeasts in the intestinal tracts of animals. Ann. N.Y. Acad. Sci. *89*: 59-68.
401. Vaughn, V. J., and E. D. Weinberg. 1978. *Candida albicans* dimorphism and virulence: role of copper. Mycopathologia *64*: 39-42.
402. Veoder, J. S., and W. F. Schoor. 1969. Primary disseminated pulmonary aspergillosis with skin nodules. Successful treatment with inhalation nystatin therapy. JAMA *209*: 1191-1195.
403. Verhoeff, K. 1974. Latent infections by fungi. Annu. Rev. Phytopathol. *12*: 99-110.
404. Villaneuva, J. R., I. Garcia-Acha, S. Gascon, and F. Urubura (Eds.). 1973. *Yeast, Mould and Plant Protoplasts,* Academic, New York.

405. Villegas, M. R. 1965. Enfermedad de Jorge Lobo (blastomicosis queloidiana) presentacion de un nuevo caso colombiano. Mycopathol. Mycol. Appl. *25*: 373-380.

406. Walker, J. 1955. Possible infection of man by indirect transmission of *Trichophyton discoides*. Br. Med. J. *2*: 1430-1433.

407. Watson, R. R., and K. L. Lee. 1978. Isolation of aminopeptidases from *Histoplasma capsulatum*. Sabouraudia *16*: 69-78.

408. Webster, B. H. 1957. Pulmonary geotrichosis. Am. Rev. Tuberc. Pulm. Dis. *76*: 286-290.

409. Wheat, R. W., and K. S. Suchung. 1977. Antigenic fractions of *Coccidioides immitis*. In L. Ajello (Ed.), *Coccidioidomycosis. Current Clinical and Diagnostic Status,* Symposia Specialists, Miami, Fla., pp. 453-460.

410. White, L. O. 1977. Germination of *Aspergillus fumigatus* conidia in the lungs of normal and cortisone treated mice. Sabouraudia *15*: 37-41.

411. Wickerham, L. J. 1957. Apparent increase in frequency of infections involving *Torulopsis glabrata*. JAMA *165*: 47-48.

412. Wiersema, J. P., and P. L. A. Niemel. 1965. Lobo's disease in Surinam patients. Trop. Geogr. Med. *17*: 89-111.

413. Wilson, B. J. 1966. Toxins other than aflatoxins produced by *Aspergillus flavus*. Bacteriol. Rev. *30*: 478-484.

414. Wilson, D. E., J. J. Mann, J. E. Bennett, and J. P. Utz. 1967. Clinical features of subcutaneous sporotrichosis. Medicine *46*: 265-279.

415. Winner, H. I. 1960. Experimental moniliasis in the guinea pig. J. Pathol. Bacteriol. *79*: 420-423.

416. Winter, W. G., Jr., and R. K. Larson. 1975. Disseminated coccidioidomycosis in a child (letter). Am. J. Dis. Child. *129*: 1237-1238.

417. Winter, W. G., Jr., R. K. Larson, J. P. Zettas, and R. Libke. 1978. Coccidioidal spondylitis. J. Bone J. Surg. *60*: 240-244.

418. Wogen, G. N. 1966. Chemical nature and biological effects of the aflatoxins. Bacteriol. Rev. *30*: 460-470.

419. Wong, V. G., K. J. Kwon-Chung, and W. B. Hill. 1975. Koch's postulates and experimental ocular histoplasmosis. In T. F. Schlaegel, Jr. (Ed.), *Ocular Histoplasmosis,* Little, Brown, Boston, pp. 139-145.

420. Woods, A. D., and H. D. Whalen. 1959. The probable role of benign histoplasmosis in the etiology of granulomatous uveitis. Trans. Am. Ophthalmol. Soc. *57*: 318-343.

421. Woolsey, M. E., M. Binstock, and M. J. McDonald. 1974. Clinical and experimental ocular histoplasmosis. Proc. Okla. Acad. Sci. *54*: 44-47.

422. Wynston, L. K., and E. B. Tilden. 1963. Chromatographic purification of *Aspergillus* endotoxins. Mycopathol. Mycol. Appl. *20*: 272-283.

423. Yarwood, C. E. 1959. Predisposition. In J. G. Horsfall and A. E. Dimond (Eds.), *Plant Pathology, and Advanced Treatise,* Academic, New York, pp. 521-562.

424. Yarzabal, L. A., S. Andrieu, D. Bout, F. Naquira. 1976. Isolation of a specific antigen with alkaline phosphatase activity from soluble extracts of *Paracoccidioides brasiliensis*. Sabouraudia *14*: 275-280.

425. Young, G. 1958. The process of invasion and the persistence of *Candida albicans* injected intraperitoneally into mice. J. Infect. Dis. *102*: 114-120.

426. Yu, R. J., S. R. Harmon, S. F. Grappel, and F. Blank. 1971. Two cell bound keratinases of *Trichophyton mentagrophytes*. J. Invest. Dermatol. *56*: 27-32.

427. Yu, R. J., S. R. Harmon, P. E. Wachter, and F. Blank. 1969. Hair digestion by a keratinase of *Trichophyton mentagrophytes*. J. Invest. Dermatol. *53*: 166-171.

428. Zamora, A., L. F. Bojalil, and F. Bastarrachea. 1963. Immunologically active polysaccharides from *Nocardia asteroides* and *Nocardia brasiliensis*. J. Bacteriol. *85*: 549-555.

429. Zeidberg, L. D., and L. Ajello. 1954. Environmental factors influencing the occurrence of *Histoplasma capsulatum* and *Microsporum gypseum* in soil. J. Bacteriol. *68*: 156-159.

430. Zvetina, J. R., J. W. Rippon, and V. Daum. 1978. Chronic pulmonary sporotrichosis. Mycopathologia *64*: 53-57.

2

Cell-Mediated Immunity

Rebecca A. Cox / San Antonio State Chest Hospital, San Antonio, Texas

I. INTRODUCTION

The role of humoral and cellular immune mechanisms in host resistance to fungal diseases is not yet fully understood. The temporal relationship between disease susceptibility (and severity) and depressed T (thymus)-derived lymphocyte function suggests that cell-mediated immunity (CMI) contributes significantly to host defense in most fungal diseases. Humoral antibodies are readily demonstrable during the course of disease, but

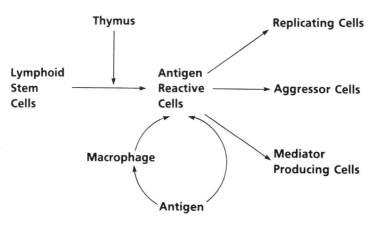

Figure 1 Model of events leading to generation of cellular immunity in humans. (From Ref. 92.)

their protective effect, if any, has not been demonstrated. This chapter reviews studies that support the role of CMI in fungal disease, and briefly describes immunological techniques used to evaluate and monitor T-cell function in experimentally and naturally infected hosts.

T-lymphocyte stimulation requires antigen presentation on the surface of a macrophage (Fig. 1). Specifically sensitized T-cells respond to antigen by proliferation and release of soluble lymphokines, which, in turn, function to (1) localize and activate macrophages (migration inhibitory factor, macrophage activating factor), (2) act directly on lymphocytes (blastogenic factor, lymphocyte chemotactic factor), or (3) mediate killing or inactivation of target cells (lymphotoxin). Expression of T-cell-mediated immunity includes delayed-type hypersensitivity (DTH), contact allergy, chronic granulomas, allograft rejection, and experimental "autoallergic" diseases [186].

It is now recognized that there exist subpopulations of T-lymphocytes which differ in function. A recent review [186] lists these into four distinct populations:

1. T-cells that produce cytotoxic or suppressor effects by elaboration of mediators such as lymphotoxin or suppressor factors
2. T helper cells that elicit an infiltrate of antigen-specific lymphocytes either cytotoxic T-cells or B-lymphocytes, the latter for antibody production
3. T-cells that attract and activate nonspecific inflammatory cells
4. T-cells that stimulate proliferation and/or increased function of normal tissue cells such as hemopoietic cells in the spleen

A diverse spectrum of T-cell defects has been demonstrated in patients with various fungal diseases. Whether or not these defects predispose patients to these diseases or are acquired during the course of disease is not yet known. Elucidation of the nature, duration, and specificity of T-cell defects will be needed to resolve this question and to provide a basis for immunotherapeutic approaches.

II. DETECTION AND ASSAY OF CMI RESPONSES

A. In Vivo Assays

1. Skin Tests

Skin testing remains one of the most valuable tools for evaluating CMI responses. Development of a DTH response to antigen establishes that the afferent, central, and efferent limbs of the immune response are intact and that the host is capable of mounting an inflammatory response. Specific anergy or failure to respond to antigen denotes a depressed T-cell function and, in most fungal diseases, skin test anergy is associated with an unfavorable prognosis for recovery. The specificity of cutaneous anergy is determined by application of skin test antigens that are antigenically unrelated to the specific etiologic agent but are derived from organisms that are ubiquitous in nature. Failure to exhibit a DTH response to any of a battery of recall antigens is consistent with a generalized anergy.

Skin tests are performed by intradermal inoculation of antigen in a physiological solution. An immediate wheal and flare response (with or without pseudopods) may develop within 10-30 min. This immediate hypersensitivity occurs with a number of fungal antigens and is presumably mediated by immunoglobulin (Ig) E antibodies [33,40,84,95,109]. An Arthus reaction may be observed within 2-4 h of skin testing. This response is mediated by precipitable antibodies which aggregate, fix complement, and excite polymorphonuclear cell infiltration. Such reactions generally subside within 10 h; however, they may persist for up to 24 h. DTH responses are read at 24 and 48 h as millimeters induration with or without erythema. Subjects who fail to respond to first-strength antigen dilutions should be retested with higher concentrations.

Recall antigens and their doses vary among laboratories but most often consist of the following: streptokinase-streptodornase (SK-SD, 4 units/1 unit); dermatophytin O (*Candida* extract, 1:10); trichophytin (50 protein nitrogen units); and tuberculin purified protein derivative (PPD, 5 units). Subjects who fail to manifest induration responses at 24 or 48 h should be skin tested with higher concentrations of the antigens listed above.

2. DNCB Sensitization

Since it is conceivable, although unlikely, that a subject may never have been exposed to any of the recall antigens commonly used, contact sensitization with a primary antigen such as dinitrochlorobenzene (DNCB) is warranted in patients who are skin test-negative to all recall antigens. Because DNCB sensitization can produce a severe local reaction in immunologically competent persons, it should be reserved for persons suspected of having a CMI deficit. The procedure described is that of Epstein and Kligman [46] as modified by Spitler [166]. Two-tenths milliliter of DNCB in acetone (10 mg/ml) is applied to the skin in an area measuring 2-3 cm in diameter. The site is examined for primary contact sensitization 10-14 days after application. If negative, challenge doses of 10 to 100 μg of DNCB in acetone are applied on day 14. The challenged sites are examined at 2, 4, and 6 days and read for erythema only (1+); erythema with induration (2+); vesicles, erythema, and induration (3+); or bullae and/or ulceration (4+). A positive reaction occurs if the subject develops a 2+ or greater response to either the sensitizing or challenge dose.

B. In Vitro Assays

Despite the usefulness of skin tests in evaluating CMI, in vivo assessment of immunological reactivity to antigen has definite disadvantages. Repeated skin tests can elicit serum antibody production and augment cutaneous hypersensitivity responses to antigen. This problem can be circumvented by use of in vitro assays of CMI responses. Moreover, in vitro assays can be used to dissect and evaluate critically the interaction of T-cells and their subpopulations with antigens, macrophages, B-lymphocytes, and serum components.

Among the most commonly used in vitro methods for assaying T-cell function are lymphocyte transformation (LT), macrophage migration inhibitory factor (MIF), and cytotoxicity assays. Since the methodology and interpretation of these assays differ among laboratories, it is imperative that healthy subjects of known skin test reactivity be included for comparisons.

1. LT Assays

Sensitized T-lymphocytes respond to specific antigen in vitro by undergoing blast transformation (LT) as measured by increased DNA, RNA, or protein synthesis, or by microscopic enumeration of blast cells [185]. LT assays can be performed on peripheral blood, spleen, lymph nodes, or any specimen that contains sufficient numbers of lymphocytes. When peripheral blood is used, mononuclear cells are separated from neutrophils and red blood cells by centrifugation of blood on a Ficoll-Hypaque medium as described by Böyum [16]. After centrifugation, the mononuclear cells (interface layers) are washed in tissue culture medium, enumerated by hemacytometer counts, and suspended to the desired concentration in tissue culture medium supplemented with serum or plasma. The cell suspension is dispensed into tissue culture tubes (macroassay) or microtiter chambers (microassay). Antigen, without preservative, is added or, for controls, tissue culture medium without antigen. Mitogens (phytohemagglutinin, PHA; concanavalin A, Con A; and pokeweed mitogen, PWM) are added to appropriate cultures to assess the overall reactivity of the lymphocyte population, irrespective of antigen specificity. The cultures are incubated at $37°C$ under a 5%-CO_2-humidified atmosphere. Optimal responses to antigen and mitogens will vary depending on the cell source, viability, and culture conditions, but generally optimal antigen responses occur at 5 days, whereas mitogen responses peak 3 days after culture. LT responses can be measured by a number of methods. The method most commonly used is addition of a radioisotopically labeled DNA precursor ($[^3H]$ thymidine) 8-18 h prior to termination of cultures. After incubation with the isotope, the cultures are harvested by successive centrifugations in saline, trichloroacetic acid, and methanol (macroassay) or by use of a multiple sample harvester which collects the labeled DNA onto glass fiber strips (microassay). The labeled DNA is transferred to vials containing a toluene-based scintillation fluid and counted for radioactivity in a liquid scintillation counter. Results of LT assays are expressed as: (1) total counts per minute (cpm); (2) Δ cpm, which is defined as the difference in the cpm of antigen- or mitogen-stimulated cultures and that of nonstimulated controls; or (3) blastogenic or stimulation indices (BI, SI), which are calculated as the cpm of stimulated cultures divided by the cpm of controls.

2. MIF Assays

Sensitized lymphocytes respond to specific antigen in vitro by production of a soluble lymphokine termed MIF, which acts to retard or inhibit migration of macrophages. MIF production in response to a specific antigen correlates strongly with in vivo skin test reactivity [36]. The direct MIF assay is limited primarily to studies of infected (or sensitized) guinea pigs, mice, or other species of experimental animals in which a sufficient number of macrophages can be obtained. Induction of macrophages is achieved by injecting 15-20 ml of sterile, lightweight mineral oil into the peritoneum. Seventy-two hours later, the peritoneal-exudate (PE) cells are harvested in Hanks' balanced salt solution and washed by centrifugation. MIF assays can be performed using either the capillary migration method [57] or the agarose-droplet assay [70]. In the former, PE cells are drawn into capillaries which are then sealed, centrifuged, and cut at the cell-fluid interface. The capillary ends containing packed PE cells are placed into tissue culture chambers to which is added tissue culture medium with fetal calf serum (FCS), guinea pig serum (GPS), and antigen. In the microagarose method, PE cells are suspended in agarose-containing medium, FCS, and GPS. The cell-agarose droplets are dispensed into microtiter wells and allowed to solidify at $4°C$. Medium with serum and antigen is then added to the agarose droplets. In both methods, macrophage cultures are incubated for 18-24 h at $37°C$ under a 5%-CO_2-humidified atmosphere. After incubation, the area or distance of migration of macrophages is measured and the results are calculated as the percent migration inhibition in wells containing antigens versus migration in controls as follows:

$$\% \text{ Inhibition} = 100 - \left[100 \times \frac{\text{distance of migration in presence of antigen}}{\text{distance of migration in absence of antigen}} \right]$$

It is of utmost importance that any toxicity of antigens is ruled out in assays of macrophages from healthy, nonimmunized animals.

MIF assays of human subjects are performed using an indirect assay [145,178] because of the difficulty in obtaining sufficient numbers of macrophages from peripheral blood. In the first step, peripheral blood lymphocytes (usually 4×10^6 in 1 ml of medium) are cultured with antigen for 48 h at $37°C$ under 5% CO_2. After incubation, the cultures are centrifuged and the supernatants are reconstituted with GPS and FCS and added to macrophages obtained from nonimmunized guinea pigs. Controls consist of assaying macrophage migration in the presence of supernatants from lymphocytes cultured in the absence of antigen. Additional controls should include assays of macrophage migration in the presence of antigen.

3. Cytotoxicity Assays

Cytotoxic T-lymphocytes can be generated by specific stimulation (i.e., incubation of sensitized T-cells with antigen) or by nonspecific stimulation (i.e., incubation of sensitized T-cells with mitogen or allogeneic cells). Once activated, cytotoxic lymphocytes release lymphokines (lymphotoxin, proliferation inhibition factor, and colony inhibition factor) which lyze or inhibit growth of target cells [72,105].

Lymphocytes (4×10^6 per ml of tissue culture medium) are cultured with the stimulating agent (antigen, mitogen, or allogeneic cells) for 4-5 days at $37°C$ under 5% CO_2. The supernatants are collected by centrifugation and assayed for lymphotoxin activity using a target cell. With the exeption of mitogen-induced cytotoxic lymphocytes, the target cell line must be antigenically related to the stimulant used to activate cytotoxic T-cells. A critical determinant in the success of cytotoxic assays is the choice of a target cell line. Because of its sensitivity, mouse fibroblast L-cells are commonly used to detect lymphotoxin activity. The assay is performed by incubating supernatants from stimulated lymphocyte cultures with target cells for 24-48 h at 37°C under CO_2. Lymphotoxic activity is measured by (1) release of radioisotope from ^{51}Cr-labeled target cells, (2) quantitating isotope uptake of viable cells remaining after incubation with lymphotoxin, or (3) visual counts of viable or dead cells.

III. EVIDENCE FOR CMI IN MYCOTIC DISEASES

A. Dermatophytosis

Dermatophytosis is a collective term used to denote infections of the skin, hair, and nails caused by species of the genera *Microsporum, Trichophyton,* and *Epidermophyton.* The dermatophytes are primary pathogens and can produce infections in otherwise healthy persons. However, widespread chronic dermatophytosis has been associated with patients having underlying disorders, such as malignant lymphomas [104] and Cushing's syndrome [122].

Primary experimental infection in humans and in animals is characterized by an intense inflammatory reaction at the site of infection, regional lymphadenopathy, spontaneous healing, and resistance to reinfection. The acquired resistance is accompanied by development of a DTH to trichophytins. The DTH response can be diminished or abrogated by repeated intradermal injections of antigen and, coincident with this, an immediate urticarial reaction develops [82]. The latter response may play an important role in the pathogenesis of dermatophyte infections, as suggested in the studies of Sulzberger and Wise [173], and later by Jillson and Huppert [82], Hanifin et al. [69], and Jones et al. [83,84]. These investigators noted that immediate wheal and flare responses were inversely correlated with the frequency and extent of DTH responses; i.e., those patients with the largest immediate wheal and flare reaction tended to have the least DTH response [82-84,173]. Conversely, patients with significant DTH responses usually did not manifest immediate hypersensitivity or manifested weak wheal and flare responses to trichophytin. This inverse relationship was further examined by Jones et al. [83,84]. Injection of an antihistamine admixed with trichophytin antigen into patients who manifested immediate responses (without DTH) reduced the size of the IgE-mediated wheal and flare response by 50%, which, in turn, resulted in development of a DTH response in three of eight patients tested. A similar intracutaneous injection of an antihistamine-antigen mixture in 17 patients who showed both an immediate and a DTH response to trichophytin diminished wheal and flare responses and augmented DTH responses from a mean induration of 5.3 mm to 9.5 mm. Atopic responses in patients with chronic dermatophytosis were not limited to dermatophytins, but rather were manifested against a number of common extrinsic allergens [83]. That atopy might predispose certain individuals to dermatophytosis was evidenced by experimental infection of three healthy, atopic subjects with *T. mentagrophytes.* Two subjects developed only DTH responses and went on to heal their infection spontaneously. The third subject developed an immediate hypersensitivity response, and coincident with this, a chronic progressive dermatophytosis. Passive transfer of serum from a subject with immediate sensitivity to three recipients who manifested DTH to trichophytin resulted in the transfer of immediate wheal and flare responses (Prausnitz-Kustner reaction), which, in turn, suppressed existing DTH. Thus, by a mechanism not yet understood, atopic responses to dermatophytins are thought to diminish or block DTH (and perhaps other CMI responses), thereby reducing immunity

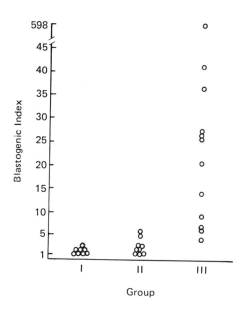

Figure 2 LT responses to *Trichophyton mentagrophytes.* Group I contained nine healthy trichophytin skin test-negative persons; group II contained nine patients with chronic *T. rubrum* infections who were DTH skin test-negative but showed immediate hypersensitivity responses to trichophytin; group III contained 12 patients with *T. mentagrophytes* or *T. rubrum* infections who were DTH positive to trichophytin with or without immediate wheal reactions. (From Ref. 69.)

against chronic dermatophytosis. The resulting CMI anergy appears to be specific for dermatophytins as skin test responses to unrelated antigens remain intact [69].

Depression of CMI responses among patients with chronic dermatophytosis occurs not only with cutaneous DTH, but also with LT responses [9,69]. Among all of nine patients (Fig. 2, group II) with chronic infections who lacked DTH but had immediate wheal and flare reactions to trichophytin, the LT responses were significantly lower than those of 12 patients (group III) who manifested DTH responses with or without immediate hypersensitivities. The defect in LT responses was specific for trichophytin antigens, was most pronounced among patients with chronic *T. rubrum* infection, and was not affected by incubation of lymphocytes in normal serum versus serum of patients with chronic disease. Thus, a pattern of defective CMI is associated with chronic dermatophytosis, although a causal relationship has not been established.

B. Candidiasis

Clinical candidiasis is a diverse disease with a spectrum ranging from a superficial infection to a severe, systemic disease. The latter most commonly involves the kidneys, lungs, and gastrointestinal tract. Chronic mucocutaneous candidiasis (CMC) is a unique form of the disease, both clinically and immunologically, and will be discussed separately.

Candida albicans and other species of *Candida* comprise part of the normal flora of the skin, mucous membranes, and gastrointestinal tract. Infections are almost invariably endogenous and, by even the most rigid criteria, candidiasis can be considered an opportunistic disease. Predisposing factors include mechanical trauma, i.e., paronychial infection and urinary tract infection resulting from catheterization; hormonal changes, i.e., *Candida* vaginitis; prolonged steroid therapy; and underlying immunologically compromising diseases. The latter most commonly include diseases associated with depressed T-cell function. Consequently, defense against candidiasis is presumed to be T-cell-mediated. This observation is supported indirectly by the finding that cortisone-treated mice [79, 108] or guinea pigs [78] were more susceptible to systemic candidiasis than were nontreated controls.

On the other hand, direct evidence has been obtained to support the role of antibody in protection against systemic candidiasis. Mourad and Friedman [118] reported that mice could be protected against *Candida* infection by passively transferred serum of immunized animals. This finding was confirmed by Pearsall et al. [134], who assessed the size of thigh lesions that developed in mice after an intramuscular infection with viable *Candida*. Whereas passively transferred immune serum significantly reduced lesion size (Fig. 3A), transfer of lymphocytes from immunized mice failed to protect against challenge (Fig. 3B), even though DTH responses were transferred with the latter.

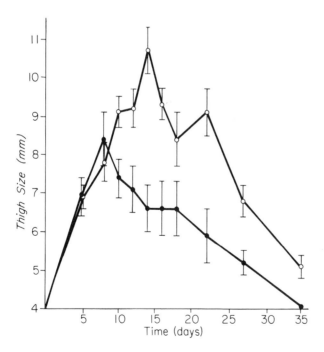

Figure 3A Results of passive transfer of immune serum to recipients given an intramuscular injection of 5×10^8 *Candida*. Each mouse received 0.5 ml of serum on day 0, a few hours before initiation of infection, and 0.1 ml each on days 1, 2, 5, 8, and 12, for a total of 1.0 ml of serum per recipient. The differences between thigh sizes in mice receiving hyperimmune serum (open circles) as compared with normal serum (filled circles) are highly significant ($P < 0.001$) by the F test. Each point represents the mean value ± S.E.M. for 8 mice. (From Ref. 134.)

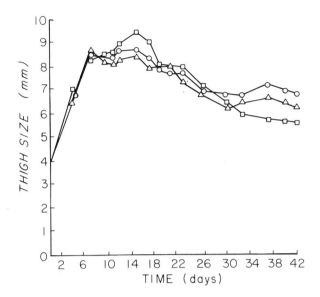

Figure 3B Lack of protection after passive transfer of "immune lymphocytes" to recipients given experimental candidiasis. No significant difference in lesion sizes was observed between any of the groups of mice, whether they received "immune lymphocytes," normal lymphocytes, or no lymphocytes (medium alone) before challenge with 5×10^8 *Candida* yeast cells. △, immune LY; □, normal LY; ○, medium. (From Ref. 134.)

Studies by Cutler [35] and Rogers et al. [146] established that congenitally athymic nude (*nu/nu*) mice were more resistant to infection with *Candida* than were their heterozygous (*nu/+*) littermates. This is in direct contrast to studies of other fungi, i.e., *Coccidioides immitis* [12], *Histoplasma capsulatum* [188], and *Cryptococcus neoformans* [26, 67], and implies that T-lymphocytes are not critical in limiting systemic *Candida* infection. Giger et al. [59] extended studies of T-cell-deficient mice by comparing the susceptibility of normal mice versus thymectomized, irradiated mice to primary and secondary cutaneous infection with *C. albicans*. Their results did not show a significant difference in the size or duration of primary lesions in thymectomized mice. A second cutaneous challenge, however, resulted in lesions that were significantly greater in size in normal mice than in thymectomized animals. The histopathology of cutaneous lesions did not differ in normal versus thymectomized mice. That reinfected, thymectomized mice were able to limit lesion size more effectively than nonthymectomized mice may be explained on the basis (1) that DTH responses, detected in normal mice but absent in T-cell-depleted mice, accelerated inflammatory responses at the site of injection; and/or (2) that antibody, either alone or complexed with antigen, may have potentiated cutaneous lesions via activation of complement. Antibody production was markedly reduced in thymectomized mice. In agreement with previous reports, Giger et al. [59] showed that T-cell-depleted mice were not more susceptible to primary intravenous infection with *Candida* than were nonthymectomized mice. However, T-cell-depleted mice which had previously been infected by a cutaneous route were significantly more susceptible to an intravenous infection. The latter finding established that primed T-lymphocytes are necessary for acquired immunity

Table 1   Immunological Abnormalities in CMC

Skin test anergy specific for *Candida albicans* [73,90,92,93,95,184].

Skin test anergy to *C. albicans* and recall antigens [18,21,92,94,95,136,144,152,182, 184].

Failure to respond to contact sensitization [21,92-95,129,136,144,152,182-184].

Negative MIF response, normal LT response [92,94,95,182,184].

Positive MIF response, negative LT response [61,183,184].

Negative MIF response, negative LT response [39,73,92-95,136,184].

Negative MIF responses to recall antigens [95].

Negative LT responses to mitogens and/or recall antigens [21,95].

Serum (or plasma)-mediated suppression of LT responses [21,131,152,184].

Elevated serum IgM [92,93,95], IgG [39,92,93,95,131]; IgA [92,93,95,109]; and IgE [95,109].

Deficiency of serum IgA and/or parotid IgA [92,93,95,109].

---

in candidiasis. The data do not, however, distinguish whether they are required as effector cells (i.e., lymphokine production) or as helper cells for antibody response, which, in turn, may protect against systemic candidiasis.

CMC is a relatively uncommon manifestation of candidiasis characterized by the development of chronic, often widespread, infections of the skin, nails, and mucous membranes. Onset is usually in infancy or during early childhood but may occur at a much later age. Approximately 70% of patients with chronic mucocutaneous candidiasis have an associated disorder (reviewed in Ref. 92). Included among these are DiGeorge's syndrome, Nezelof's syndrome, Swiss-type agammaglobulinemia and thymus dysplasia, and endocrinopathies (hypothyroidism, Addison's disease, and diabetes mellitus). Genetic predisposition to CMC has been observed in a limited number of patients with endocrine disorders.

A heterogeneous array of both cellular and humoral immune defects has been documented in patients with CMC. These are summarized in Table 1.

In general, the most profound immunological defects are demonstrable in patients who had early-onset CMC, i.e., in infancy or early childhood [95]. Late-onset CMC seemingly is associated with acquired immunological abnormalities and depressed T-cell function may be reversed after chemotherapy and/or immunotherapy [95].

The frequent dissociation between MIF production and LT responses to *Candida* (and other antigens) is consistent with the hypothesis that the population of T-cells which mediates MIF production (and perhaps skin test reactivity) differs from that which undergoes a blastogenic response to antigen [143]. Hence, a defect in one subpopulation of T-cells, i.e., mediator producing, but not the other, i.e., replicating lymphocytes, could and does occur in diseases associated with a depressed CMI. Similarly, hyperproduction of immunoglobulins, particularly IgE, may be attributed to a defect in yet another subpopulation of T-cells, i.e., those which regulate immunoglobulin synthesis.

In summary, there exists extensive clinical evidence that links T-cell deficiencies with CMC and with systemic candidiasis. However, these are two distinctly different manifestations of candidiasis, and the contributory role of T- and B-cell-mediated immunity in each remains controversial. The ability to prevent colonization by *Candida* and the ability to clear *Candida* from mucocutaneous lesions is probably dependent on CMI, as suggested by Kirkpatrick and Smith [95]. Resistance to systemic candidiasis, on the other hand, appears to be dependent on antibody, as suggested by Mourad and Friedman [118] and Pearsall et al. [134].

C. Cryptococcosis

Cellular immune studies of patients with cryptococcosis have been limited in part by lack of available antigens. This coupled with the ubiquity of the fungus in nature has hampered efforts to define control groups of previously infected versus noninfected persons. Although cryptococcosis does occur in otherwise healthy persons, at least 50% of patients with this disease have an underlying, predisposing condition, e.g., Hodgkin's disease, sarcoidosis, lupus, diabetes, or have been immunologically compromised by steroid therapy [22,43,62,106,165].

Skin test reactivity to cryptococcin varies from 50 to 58% in patients with active disease or those in short-termed (<6 months) remission and from 33 to 77% in patients in long-termed remission [8,151]. Statistically, the difference between the skin test responses of patients with active versus those with inactive disease has not been significant. In regard to the specificity of depressed skin test responses, a generalized cutaneous anergy has been documented in patients who do not have coexisting diseases [151] and to an even greater extent in patients who have underlying disorders [65,151].

Studies of LT responses in cryptococcosis have been limited for the most part to patients who were in clinical remission [8,42,65,151]. In general, the LT responses of patients in long-term remission were comparable to those of healthy, cryptococcin skin test-positive persons. Patients with active disease or in short-term remission showed depressed LT responses to cryptococcin. As a group, patients with coexisting diseases were the weakest responders. The depressed LT responses were specific for *Cryptococcus neoformans* as responses to SK-SD, PPD, and mumps antigens were comparable to those of healthy subjects [8,42,151].

Experimental studies in *Cryptococcus*-infected mice have established that cortisone [56] and antilymphocyte serum [2,67] accelerate death and, second, that *nu/nu* mice are more susceptible to cryptococcal infection than are their heterozygous littermates [26, 67]. Although *nu/nu* mice do not develop DTH responses to cryptococcal antigens, they are capable of producing anticryptococcal antibody. Cauley and Murphy [26] inoculated phenotypically normal BALB/c mice and congenitally athymic nude mice from a BALB/c background with sublethal doses of viable *C. neoformans*. Development of DTH responses, anticryptococcal antibody, and circulating cryptococcal polysaccharide were followed in the two animal groups during the course of disease. Fewer viable cryptococci were recovered from nude mice than from *nu/+* mice at day 7 and to a lesser extent at day 14 (Fig. 4). Thereafter, cryptococcal growth was significantly increased in nude mice as compared with phenotypically normal mice. This early resistance to *C. neoformans* infection is consistent with increased macrophage function in nude mice [27]. DTH responses, detected by footpad sensitivity to a soluble cryptococcal antigen were not demonstrable in nude mice (Fig. 4, bottom). DTH responses in *nu/+* mice were demonstrable 14 days after infection and gradually increased thereafter. Thus, appearance of DTH was coincident with inhibition of fungal growth in normal mice. Cryptococcal antibody production was detected in 13 of 30 nude mice and 9 of 27 *nu/+* mice during the course of infection.

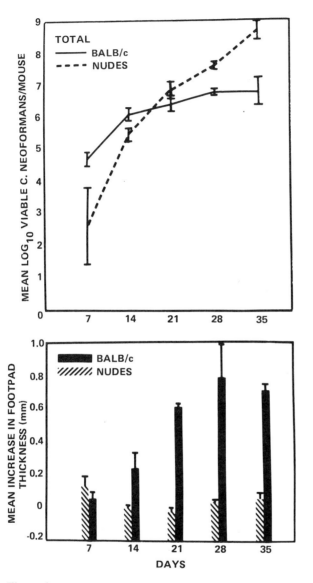

Figure 4 Top: Mean $\log_{10}$ viable *Cryptococcus neoformans* as colony forming units cultured per mouse over a 35-day experiment after infecting nude *nu/nu* or BALB/c, *nu/+* mice with 6.4×10^3 viable *C. neoformans* cells. Each point represents the mean results for four mice. Vertical bars designate standard errors of the means. Bottom: Mean increase in footpad thickness 24 h after footpad testing the same animals described above with a cryptococcal culture filtrate antigen. (From Ref. 26.)

Antibody levels did not correlate with susceptibility to fungal dissemination. Significant levels of circulating cryptococcal polysaccharide were detected in both nude and *nu*/+ mice throughout the 35-day period. In contrast to results obtained in *nu*/+ mice, circulating polysaccharide increased from day 14 on in nude mice, a finding that paralleled fungus growth in the latter animals.

The data presented by Cauley and Murphy [26] provide compelling evidence that T-cell function is critical in limiting growth of *C. neoformans.* Since cryptococcal polysaccharide is a T-independent antigen, i.e., antibody production does not require T-cell cooperation, the nude mouse is an ideal model for the study of host defense in this disease.

D. Coccidioidomycosis

Coccidioidomycosis is a fungus disease which readily lends itself to the study of the role of cellular and humoral immune responses in host defense. This statement is based on the following points. *Coccidioides immitis* is a primary pathogen. Immunologically compromising illnesses, although present in some patients [38,130,148,153,162], are not prerequisite to infection. Second, the disease is geographically limited to the southwestern United States, thereby allowing selection of control subjects with and without prior exposure to the fungus. Third, antigens available for studies of *C. immitis* infection are highly sensitive and have an acceptable range of specificity. Finally, there exists a consistent and predictable immunological pattern in patients in the various stages of this disease.

Humans acquire coccidioidomycosis by inhalation of mycelial-phase arthroconidia, which convert in tissue to the morphologically distinct spherule/endospore phase. The disease presents a diverse clinical spectrum, which includes inapparent infection, primary respiratory disease usually with uncomplicated resolution, chronic pulmonary disease either stabilized or progressive, and extrapulmonary dissemination either acute, chronic, or progressive. The degree of severity varies considerably within each category.

A fundamental feature of the disease is the direct relationship between disease severity and increased serum complement-fixing (CF) antibody titers and an inverse relationship between CF titers and CMI reactivity to *C. immitis* antigens [24,32,128,156-158]. Typically, patients with mild infection manifest strong in vivo (skin test) and in vitro (LT and MIF) responses to coccidioidin (CDN). Conversely, disease severity is accompanied by depressed CMI responses and high serum levels of CF antibody. The immunological profile of CMI responses and serum antibody titers, as they relate to disease status, evolved from the celebrated work of Smith et al. [156-158]. Depressed skin test responses were not only related to the extent of disease, but were prognostically significant in predicting the outcome of the disease. As shown in Table 2, 75% of patients who recovered from disseminated coccidioidomycosis were skin test reactive to CDN 1:100 during their disease. Only 17% of the patients who were negative to this dilution went on to cure their disease. A reciprocal relationship was observed in the patients who died of coccidioidomycosis, i.e., 83% were skin test-negative to CDN 1:100; 25% were skin test-positive.

It is pertinent to point out that spherulin, prepared as an aqueous extract of spherules of *C. immitis,* detects skin test reactivity in a significantly larger proportion of patients with active coccidioidomycosis than does CDN [169]. In a study of 23 patients with dis-

Table 2 Fate of Patients with Disseminated Coccidioidomycosis According to Level of Known Minimal Coccidioidin Sensitivity

Fate of patient	Minimal known coccidioidin sensitivity							
	1:100 dilution				1:10 dilution			
	Positive		Negative		Positive		Negative	
	Number	Percent	Number	Percent	Number	Percent	Number	Percent
Recovery	15	75	11	17	19	54	3	10
Death	5	25	52	83	16	46	26	90
Total fate known	20	100	63	100	35	100	29	100

Source: Ref. 157.

Table 3 CMI Responses in Coccidioidomycosis

Study group	Number of subjects	Percent positive in:	
		MIF assays	LT assays
Healthy subjects			
CDN skin test-positive	56	100[a]	82
CDN skin test-negative	10	0	0
Patients			
Active, pulmonary			
CDN skin test-positive	11	90	63
CDN skin test-negative	23	10	56
Active, disseminated			
CDN skin test-positive	10	33	50
CDN skin test-negative	34	13	25
Inactive			
CDN skin test-positive	13	82	69
CDN skin test-negative	3	0	0

[a]Not all subjects were assayed in each test.

Source: Modified from Ref. 25.

seminated disease, 35% reacted to both CDN and spherulin; 39% responded to spherulin only. The significance of skin test reactivity to spherulin, in the absence of skin test reactivity to CDN, as it relates to disease prognosis, is not yet known.

A cumulative report by members of the *Coccidioides* Cooperative Treatment Group (CCTG) [25], in which data were collected from clinicians and investigators at 12 institutions, showed that depressed skin test responses to CDN correlated with depressed MIF responses and, to a lesser extent, with depressed LT responses, as shown in Table 3.

The results of these studies are consistent with individual reports by a number of independent investigators [24,31,32,128,190]. With regard to specificity of the T-cell defect, Catanzaro et al. [24] reported that 2 of 13 patients with pulmonary disease and 4 of 8 patients with disseminated coccidioidomycosis showed a generalized cutaneous anergy to a battery of skin test antigens, including CDN. Our own studies [32] established a generalized cutaneous anergy in 2 of 16 patients, both of whom had disseminated disease. Anergy was also demonstrable to recall antigens in MIF assays of one of the two patients. Of particular note, this patient had significant blastogenic responses to recall antigens. More recently, Rea et al. [138] reported that 13 (52%) of 25 patients with disseminated coccidioidomycosis failed to respond to contact sensitization with DNCB. Inability to respond to DNCB sensitization correlated directly with serum CF antibody titers. In criticism of this study, it should be pointed out that 6 of the 13 patients who failed to exhibit contact sensitivity to DNCB were skin test reactive to CDN. These results conflict with the accepted principal that failure to respond to primary or secondary challenge with contact agents denotes inability of T-cells to mount a response to antigen.

The exact nature of the T-cell defect in coccidioidomycosis is not yet known. Patients seem to have sufficient numbers of peripheral blood T-lymphocytes to mediate CMI responses [24,32]. Failure of these cells to recognize and/or respond to *C. immitis* may be attributed to functionally defective T-cells, macrophages, or regulatory mechanisms, or possibly to negative feedback by antigen, antibody, or immune complexes. That excess antigen can result in T-cell anergy in coccidioidomycosis was documented by Ibrahim and Pappagianis [80]. Guinea pigs that were sensitized by footpad inoculation of killed mycelia in complete Freund's adjuvant (CFA) and subsequently injected with killed hyphae (subcutaneously) of CDN (intraperitoneally) gradually lost delayed sensitivity to CDN. Of particular interest, the resulting anergy was selective for *C. immitis* as responses to PPD were not impaired.

Whether the immune defect is the cause or the effect of the disease is not yet known. Filipinos, blacks, Mexican-Americans, and other dark-skinned races are reported to have a higher morbidity rate for disseminated coccidioidomycosis [58,156], suggesting predisposition for this disease. In support of this possibility, Scheer et al. [150] found a high frequency of the HLA-A9 phenotype in studies of patients with disseminated coccidioidomycosis. The significance of their studies is strengthened by the reported high frequency of the HLA-A9 phenotype in Filipinos and blacks [3]. Huppert [77] recently questioned the meaning of increased mortality of dark-skinned races for disseminated disease. He argued that the data did not emphasize occupational exposure and that an increased risk of disseminated disease should be accompanied by an increased morbidity rate. It is noteworthy that in a recent outbreak of windborne coccidioidomycosis, Flynn et al. [50] detected a significantly higher incidence of disseminated coccidioidomycosis in blacks than in Caucasians. This did not appear to be related to occupational exposure or socioeconomic status. Unfortunately, no attempts have been made to correlate T-cell dysfunction with racial predisposition to the disease. In this regard, it is noted that dissemination, when it occurs, appears to be an early event, an observation that may be consistent with a defect in alveolar macrophage function.

Support for the role of CMI in host resistance to coccidioidomycosis has been obtained in studies of animals experimentally infected with *C. immitis* [12,13,71,85,97,126]. An early report by Kong et al. [97] showed that spleen cell transfers from spherule-immunized mice conferred some degree of protection in mice challenged intranasally with viable arthroconidia. Beaman et al. [12,13] recently extended these studies and provided direct evidence that resistance to infection with *C. immitis* is dependent on functional T-lymphocytes. Irradiated mice that had received spleen cells from mice immunized with killed spherules were challenged with 400 viable arthroconidia. Immune spleen cells conferred resistance to infection (93% survival rate); spleen cells from nonimmune mice failed to protect against challenge (7% survival rate). Selective depletion of T-lymphocytes using antitheta serum and complement abrogated immune transfer. In additional experiments, these investigators showed that adult, thymectomized mice, which had been irradiated and reconstituted with bone marrow from normal mice, were more susceptible to infection with arthroconidia than were nonthymectomized, irradiated, bone marrow-reconstituted controls. Finally, athymic nude mice were highly susceptible to infection with 50 arthroconidia (0% survival rate at day 21), whereas their heterozygous littermates survived infection with graded doses up to 350 arthroconidia. In contrast to the above, adoptive transfer using serum from spherule-vaccinated mice has consistently failed to protect recipients against challenge with *C. immitis* [13,97].

E. Histoplasmosis

Clinical disease with *Histoplasma capsulatum* can range from a primary acute infection to a progressive disseminated disease. Despite a high infection rate in endemic areas, symptomatic illness occurs in less than 5% of the primary infections [44]. Symptomatic primary infections are usually limited to infants and children, but may occur in adults who have been exposed to a heavy inoculum of *H. capsulatum* spores. Progressive disseminated histoplasmosis is primarily associated with underlying immunological disorders. Included among these are malignancies of the lymphatic and hematopoietic systems [29,99,123] and immunosuppressive drugs [37,88]. Chronic pulmonary disease is almost invariably associated with chronic obstructive lung disease [63]. For reasons not yet understood, there is a higher incidence of the disease in males than in females with ratios of 2:1 in primary acute infection, 4:1 in progressive disseminated disease, and 10:1 in chronic cavitary histoplasmosis [181].

Cellular immune studies of patients with histoplasmosis have been limited. Skin test reactivity approaches 100% in patients with acute pulmonary histoplasmosis and decreases to less than 55% of patients with disseminated disease [54,139,149,161]. Approximately 83% of patients with chronic cavitary histoplasmosis show cutaneous anergy to histoplasmin [55]. The specificity of the cutaneous anergy has not been examined in a sufficient number of patients. In a limited study of 13 patients [30], two of seven patients who failed to respond to histoplasmin skin testing also failed to manifest a cutaneous DTH to any of five antigens in a skin test panel.

Newberry et al. [125] and, in a continuation of that study, Kirkpatrick et al. [91] reported that LT responses were depressed in patients who had recovered from active histoplasmosis. In general, depressed LT responses to histoplasmin correlated with disease severity, i.e., responses of patients with mild disease were comparable to those of healthy, histoplasmin skin test-positive persons, whereas responses of patients with severe acute or chronic histoplasmosis were significantly depressed. A more recent study of patients with active histoplasmosis confirmed these findings [30]. As shown in Table 4, LT responses of patients were depressed to *Histoplasma* antigens. Mitogen responses were reduced, but not significantly so when compared with healthy donors. The depressed responses were, in terms of antigen recognition, specific for *Histoplasma* antigens since responses to *Candida* antigen were not impaired. There was no correlation between depressed LT responses and cutaneous anergy to histoplasmin in any of the preceding reports [30,91,125]. In contrast, Alford and Goodwin [4] reported that LT responses of patients with chronic pulmonary histoplasmosis were not, as a group, depressed to histoplasmin. The only exception was noted in patients who were skin test-negative to this antigen. This disparity is difficult to explain, but may be attributed to the patient populations i.e., pulmonary versus disseminated, active versus inactive, or to differences in the techniques used to measure blastogenic responses. In regard to the latter, it is perhaps noteworthy that the lymphocyte population assayed by Alford and Goodwin [4] had been passed through a nylon column, thereby removing adherent suppressor cells. The possibility exists, therefore, that adherent suppressor cells were removed from their lymphocyte cultures (see Sec. IV.B).

Support for the role of CMI in resistance to histoplasmosis has been provided by studies in experimentally infected animals. Mice treated with antilymphocyte serum or cytotoxic agents such as cyclohexamide are far more susceptible to infection with *H. capsulatum* [2,34]. Passive transfer of spleen cells or peritoneal-exudate cells from C_3H

Table 4 LT Responses of Healthy Subjects and Patients with Histoplasmosis to Mitogens, Histoplasmin, and *Candida* Antigen

Group designation[a]	LT responses (Δ cpm)[b]			
	PHA	ConA	Histoplasmin	*Candida*
I	16,183 ± 3,937[c]	16,950 ± 3,128	2,336 ± 1,109	12,954 ± 3,274
II	23,882 ± 4,820	25,229 ± 1,963	18,339 ± 3,061	12,758 ± 1,708
III	24,263 ± 4,820	25,423 ± 4,196	1,582 ± 535	10,200 ± 1,943

[a]Group I contained 13 patients with active histoplasmosis, group II contained 14 healthy histoplasmin skin test-positive persons, and group III contained 14 healthy, histoplasmin skin test-negative persons.
[b]LT assays were performed with 15% autologous serum. Cpm of control cultures were 485 ± 131, 522 ± 132, and 503 ± 258 in groups I, II, and III, respectively.
[c]Mean ± S.E.M.
Source: Ref. 30.

mice which had been immunized with sublethal doses of viable *H. capsulatum* yeast (or with ribosomes) protected nonimmunized mice against challenge with 4×10^6 viable yeast cells [176]. Pretreatment of immune spleen cells with antitheta serum abrogated immune transfer [177].

More recently, studies have shown that congenitally athymic nude mice are significantly more susceptible to intraperitoneal infection with *H. capsulatum* than are their heterozygous littermates [188]. Thymic transplants from *nu/+* or homozygous normal BALB/c mice enhanced the survival time of nude mice as shown in Fig. 5.

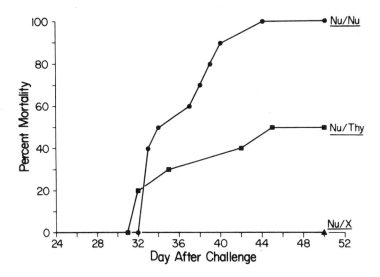

Figure 5 Cumulative percent mortality of *nu/nu, nu/thy* (thymic transplant *nu/nu* recipients), and *nu/X* mice to 10^5 *Histoplasma capsulatum* injected intraperitoneally. (From Ref. 188.)

F. Blastomycosis

Studies of immunological responses in blastomycosis have been severely limited by the
lack of suitable antigens. At present, a skin test antigen is not available commercially because of the low specificity and sensitivity associated with past preparations [10,11,15].
Antigens are available for detecting CF antibodies, but cross reactions occur with *Histoplasma* and *Coccidioides* [20]. Consequently, the incidence of exposure versus the frequency of symptomatic disease is unknown. It is generally accepted that the disease is
acquired via inhalation of the mycelial-phase microconidia, which convert to yeast cells
within host tissue. Pulmonary blastomycosis is the most common form, but there is a
high frequency of dissemination to other tissues [15]. The disease has been reported in
patients with neoplasms [15,98], Addison's disease [48,81], diabetes [15], and tuberculosis [15,137], but most patients are without coexisting diseases.

Immunological studies of patients with blastomycosis suggest that CMI responses contribute to host defense, whereas humoral antibodies do not. Patients who manifest skin
test reactivity to *Blastomyces* antigens and have low or negative serum CF antibody titers
tend to have the most favorable prognosis for recovery [160]. Skin test anergy in the
presence of high CF antibody levels is associated with a poor prognosis [160].

A number of investigators have utilized laboratory animals to study the pathogenesis
of *B. dermatitidis* and to evaluate antigen preparations (see Chap. 1). However, these
studies have not evaluated immune responses, as they relate to host resistance or susceptibility. An exception to this was a study by Spencer and Cozad [164]. These investigators showed that a subcutaneous inoculation of mice with yeast *B. dermatitidis* protected
against intraperitoneal challenge. Development of DTH, as determined by footpad sensitivity, paralleled resistance to challenge. These studies, however, did not establish whether
DTH per se was required for resistance; i.e., no attempt was made to assess the effect of
desensitization on resistance to challenge.

G. Paracoccidioidomycosis

This fungus disease is geographically limited to South and Central America. The route of
infection has not, as yet, been clearly defined but is presumed to be via inhalation. The
immunology of paracoccidioidomycosis is similar to that of coccidioidomycosis in many
aspects. Based on skin test reactivity, the infection rate ranges from 3.3 to 34.1% in
endemic areas [68,114,117,127,140]. As in coccidioidomycosis, these persons seldom
have demonstrable CF antibody titers. Chronic or progressive disease, on the other hand,
is associated with depressed T-cell function [111-113,120,141] and high CF titers [47,113,
141]. And, like coccidioidomycosis, clinical remission is often accompanied by a reacqui-
sition of CMI responses and reduced antibody titers [47,113,141].

The spectrum of immunological responses in 19 patients with paracoccidioidomycosis
was evaluated by Musatti et al. [120]. These investigators found that the percentages (but
not absolute numbers) of peripheral blood T-lymphocytes were depressed in patients.
Skin test reactivity to paracoccidioidin was demonstrable in 8 (42%) of the 19 patients.
As a group, the skin test responses of these patients were depressed to oidiomycin, *Candida*
antigen, PHA, and DNCB. Responses to trichophytin were comparable in patients and
healthy donors. As expected, those patients who responded to paracoccidioidin tended
to respond to one or more recall antigens, whereas those patients who showed anergy to

paracoccidioidin predominated in the group that failed to respond to recall antigens. In general, skin test reactivity correlated with in vitro leukocyte migration inhibition to paracoccidioidin.

In regard to LT responses, patients showed a reduced LT response to PHA, paracoccidioidin, and *Candida* antigen [120]. In vitro LT responses did not correlate with skin test reactivity. Plasma-mediated suppression of LT responses was noted in assays using autologous versus homologous serum and is discussed in Sec. IV.A.

In general, depressed CMI responses correlated with disease severity; i.e., the four patients who eventually died of progressive paracoccidioidomycosis showed depressed T-cell responses in vivo and in vitro to paracoccidioidin and to recall antigens.

A subsequent study by Mok and Greer [113] essentially confirmed the foregoing findings, in that CMI responses were depressed in patients with this disease. Their study also addressed the question of whether anergy was preexisting or acquired during the course of the disease. The results showed that the frequency of skin test reactivity to paracoccidioidin decreased with duration of active disease remission. These data suggest that anergy is acquired during the course of disease. Longitudinal studies were performed on six patients prior to and during therapy. Two of the patients who were skin test-negative to paracoccidioidin at the onset of therapy converted their responses within 6-9 months and, during this same period, decreased their serum CF antibody titers by fourfold or greater. The pattern in the immunological profile of the remaining four patients throughout the nine-month period was not consistent. Fluctuations in serum CF antibody levels did not correlate with in vitro skin test reactivity.

IV. SUPPRESSION OF CMI RESPONSES

A. Serum Factors

Cellular immune responses can be potentiated or suppressed by serum factors. Within the past decade, a number of reports have emerged sporadically, indicating that sera of patients with CMC [21,131,152,184], coccidioidomycosis [128], histoplasmosis [30,91, 125], and paracoccidioidomycosis [120] suppressed in vitro LT responses. The serum factor(s) responsible for suppression has not been characterized but is presumed to be excess serum antibody, antigen, or immune complexes, inasmuch as all three have been shown to suppress T-cell function in other systems [147]. It is pertinent to point out, however, that other serum factors which may be elevated in chronic disease processes, such as α-fetoprotein [119], α-globulin [52,60], and C-reactive protein [116], may suppress in vitro LT responses.

The first report to establish suppressive effects of serum from patients with a mycotic disease was that by Newberry et al. [125]. Sera of patients who had recovered from histoplasmosis suppressed LT responses of healthy, histoplasmin-reactive donors to histoplasmin. Conversely, sera of healthy, histoplasmin-reactive persons augmented histoplasmin responses of patients. This study was later extended by Kirkpatrick et al. [91] with essentially the same findings. Again the patient population was limited to those in clinical remission. More recently, serum-mediated suppression of LT responses has been documented in patients with active histoplasmosis [30]. In the latter study, suppression was specific for *H. capsulatum* antigens in most, but not all, patients since responses to mitogens and *Candida* antigen were unaffected by the source of serum. Suppression of LT responses correlated with CF antibody titers to histoplasmin, a finding that is consistent with and supported by the specificity of the suppression. However, the serum CF antibody levels were low (≤ 1:16) in the patient population and, second, sera of 2 of 13 patients suppressed in vitro responses to histoplasmin yet neither of these patients had demonstrable CF antibody. Thus, the direct effect of serum CF antibody on LT responses remains questionable and raises the possibility that serum CF antibody titers, in the absence of demonstrable circulating antigen, may reflect immune complex levels which, in turn, mediate suppression of LT responses.

Following Newberry's original study [125], there appeared a series of reports showing serum (or plasma) of patients with CMC-suppressed LT responses [21,131,152,184]. The inhibition was specific for *Candida* antigen in some patients [131,184] but not in others [21,152]. Inhibition of LT responses did not correlate with anti-*Candida* antibody. Of particular note, Paterson et al. [131] reported that the inhibitory factor(s) disappeared in a patient's serum during the course of therapy. Following this, the patient converted her skin test and in vitro LT response to *Candida* antigen. A similar temporal relationship between clinical remission and loss of serum-mediated suppressor activity was reported in studies of patients with paracoccidioidomycosis [120].

In a study of patients with disseminated coccidioidomycosis, Opelz and Scheer [128] reported serum-mediated suppression of LT responses to CDN in only 2 of 38 patients tested. Our own studies indicate a much higher incidence of serum suppression in patients with this disease [33a]. Eleven of 15 patients with active disease showed a twofold or greater increase in their LT response to CDN in the presence of serum from healthy subjects. In addition, sera of these same patients suppressed LT responses of healthy subjects. In contrast to patients with active disease, sera of only 1 of 13 patients in clinical remission exhibited suppression of LT responses. The suppression was specific for *C. immitis* in most, but not all, patients. Elution of sera of patients through a column containing *Staphylococcus* Protein A, which binds the Fc portion of IgG [89] and preferentially binds oligomeric or complexed IgG [110], abrogated suppressor activity as shown in Table 5. Adsorption of the healthy donor's serum neither augmented nor decreased his LT responses to PHA or *Coccidioides* antigens. Prior to adsorption of the patient's serum on Protein A, his serum suppressed LT responses of the healthy, CDN-reactive donor. Suppression was not demonstrable after adsorption of free or immune complexed IgG. These results are consistent with but do not distinguish between antibody-mediated and immune complex-mediated suppression of LT responses.

In the preceding studies, suppression was demonstrated in LT assays. No attempts were made to assess the effect of patient's sera on other in vitro T-cell responses. The exception to this was a study by Walters et al. [187] of patients with dermatophytosis. Their assay system measured inhibition of leukocyte adherence in the presence of dermatophyte antigens. Addition of sera from patients with chronic dermatophytosis specifically blocked antigen-induced inhibition of leukocyte adherence. Serum of a patient with an acute dermatophyte infection did not exhibit blocking activity.

B. Suppressor Cells

Regulation of immune responses is dependent on a series of complex interactions between T- and B-lymphocytes and macrophages. T-cell responses can be potentiated or suppressed by T-lymphocytes [51,53,135], B-cells [87,124], and macrophages [7,96,107,175]. Similarly, humoral antibody responses are regulated by T helper and T suppressor cells [86,115,174].

Table 5 LT Responses of a Healthy, CDN-Reactive Donor Prior to and After Adsorption of Serum with *Staphylococcus* Protein A[a]

Lymphocyte stimulant	Serum source[b]			
	Autologous		Patient	
	Unadsorbed	Adsorbed	Unadsorbed	Adsorbed
PHA	63,991	67,916	53,997	66,479
CDN	36,225	31,271	16,268	35,474
Spherulin	31,734	40,542	21,268	46,978

[a]2×10^6 lymphocytes from a healthy, CDN skin test-positive donor were cultured in tissue culture medium containing 15% autologous serum or 15% serum from a patient with active coccidioidomycosis. Sera were assayed for their effect on LT responses prior to (unadsorbed) and after (adsorbed) elution through a column containing *Staphylococcus* Protein A.
[b]Expressed as a change in the counts per minute.

A defect in the regulatory function of T suppressor cells has been implicated in a number of immunologically deficient diseases. Lack of suppressor T-cell modulation has been alluded to in systemic lupus erythematosus [1] and in juvenile rheumatoid arthritis [172]. Conversely, abnormally increased T suppressor cell function has been associated with agammaglobulinemia [14]. In addition, increased suppressor cell activity in non-T-cell populations has been documented in multiple myeloma [17], Hodgkin's disease [180], tuberculosis [45], schistosomiasis [179], and in chronic fungal diseases [170,171].

The latter studies consisted of 14 patients having fungal diseases of different etiology. The patients were divided into three groups on the basis of their immune responses. Group I contained seven patients who responded to one or more antigens in a skin test battery and in LT assays. The three patients who comprised group II failed to exhibit cutaneous DTH to any of a panel of skin test antigens, but had normal in vitro LT responses. Group III consisted of four patients who exhibited a generalized in vivo and in vitro anergy. Two of the four patients had disseminated histoplasmosis and two had disseminated cryptococcosis. As opposed to patients in groups I and II, three of four patients in group III had undergone two to three relapses of their disease. In pursuit of characterizing the T-cell defect in these four patients, Stobo et al. [171] showed that prior culture of peripheral blood mononuclear cells in vitro for 7 days in the absence of stimulant resulted in a lymphocyte population that would subsequently respond to mitogens and antigens. Addition of freshly isolated mononuclear cells to precultured lymphocytes abrogated LT responses in the latter. These results, coupled with the finding that patients in group III had normal numbers of peripheral blood T-cells, suggested that the depressed LT responses were not attributable to lack of functional T-cells, but rather, the reactivity of these T-cells was suppressed by a population of cells that was present in freshly isolated mononuclear cells. This regulatory cell population was shown to be non-Ig-bearing, nonphagocytic, and E-rosette forming, thus presumably T-lymphocytes. In a subsequent study, Stobo [170] showed that suppression was mediated via a soluble factor released from macrophages which, when added to freshly isolated mononuclear cells of healthy donors, suppressed transformation responses. Addition of the macrophage-soluble factor to precultured mononuclear cells of these same healthy persons failed to suppress blastogenic responses; hence it was hypothesized that the soluble factor interacted with a T-lymphocyte population, which, in turn, suppressed blastogenic responses of other T-cells. The regulatory T-cell that interacted with the macrophage-released soluble factor was considered to be short-lived and, therefore, lost during the 7-day preculture period. In addition, this subpopulation of regulatory T-cells could be separated from T-lymphocytes that responded in LT assays by density gradient centrifugation.

In related studies, Artz and Bullock [5,6] monitored T-cell function, humoral antibody responses, and the histopathology of experimental histoplasmosis in mice infected intravenously with sublethal doses of *H. capsulatum* yeast. These authors observed that during acute and severe disseminated disease [1-3 weeks postinfection) spleens were enlarged three- to fourfold in size, whereas there was a marked involution of the thymus. Infections had spontaneously cleared in most mice by week 8 and in all mice by week 18. Concomitant with eradication of infection, spleen size decreased and thymic size increased to near normal size. Parallel with severe disease, mice failed to respond to histoplasmin and a challenging dose of sheep red blood cells (SRBC) in footpad testing. The animals also failed to respond to ConA and histoplasmin in LT assays as compared to mice that had been immunized with heat-killed yeast. In addition, in vitro co-cultures of unfractionated splenic lymphocytes with allogeneic mouse spleen cells failed to generate cytotoxic T-cells. Cytotoxicity responses were also depressed in a T-lymphocyte-enriched cell population, but to a lesser extent, a result that suggests suppressor cell activity in unfractionated spleen cells. In agreement with suppressor cell activity, primary antibody response of spleen cells in vitro to SRBC (a T-dependent antigen) was markedly depressed at weeks 1 and 3 in infected mice; in vitro co-cultures of these spleen cells with lymphocytes of normal mice significantly suppressed the primary antibody responses of the latter. Resolution of active disease by 8 weeks postinfection was concomitant with acquisition of T-cell responses and loss of suppressor cell activity. The relationship between suppressor cell function, thymic involution, and splenic hyperplasia during acute disease will require further exploration but was hypothesized to represent a migration of regulatory T-lymphocytes from the thymus to an area of large antigen deposition, i.e., the spleen.

V. INTERACTION OF LYMPHOCYTES AND MACROPHAGES IN CMI

While depressed skin test reactivity, LT, and MIF responses may parallel disease progression and severity in many fungal diseases, it has not yet been clarified whether these DTH responses contribute directly to host defense. In regard to this, Youmans and associates [19,121,132,189] have shown that resistance to tuberculosis can occur in the absence of demonstrable DTH responses to PPD. Of particular importance, lymphocytes of mice immunized with viable cells of *Mycobacterium tuberculosis* H37Ra were shown to elaborate a lymphokine, termed *Mycobacteria* growth inhibitory factor, which activated macrophages to inhibit intracellular growth of tubercle bacilli.

That immune lymphocytes arm macrophages to inhibit intracellular growth of *H. capsulatum* was documented in studies by Howard and co-workers [74-76]. Freshly isolated peritoneal macrophages from mice immunized with yeast of *H. capsulatum* restricted intracellular growth of *Histoplasma* yeast. However, in vitro maintenance of peritoneal-exudate cells for 48 h or removal of nonadherent cells diminished the ability of macrophages to inhibit fungus growth. Addition of lymphocytes from immunized mice to 48-h monolayers of macrophages from immune or nonimmune mice resulted in significant inhibitory activity (Table 6). Lymphocytes from nonimmunized mice were only slightly

Table 6 Effect of Immune Lymphocytes on Growth of *Histoplasma capsulatum* Within Normal and Immune Macrophages

Type of macrophage[a]	Lymphocytes added[b]	Mean number of yeast per infected macrophage[c]	Percent reduction of intracellular growth[d]
Normal	–	10.0	0
Normal	+	3.9	62
Immune	–	10.8	0
Immune	+	3.3	70

[a]Macrophages in cell culture 48 h before parasitization.

[b]Lymphocytes were peritoneal cells from immunized animals passes through cotton-wool columns and added with inoculum of *H. capsulatum*. Number of lymphocytes added was 10^7 per milliliter.

[c]After 18 h of incubation.

[d]Compared to controls.

Source: Ref. 75.

effective in activating macrophages (Table 7). Incubation of lymphocytes with *H. capsulatum* in the absence of macrophages did not reduce fungal viability. Thus, the inhibition was not attributed to direct lymphocyte cytotoxicity.

In subsequent studies, Howard and Otto [76] explored the possibility that lymphocyte activation of macrophages was via a soluble lymphokine. The results of these studies are summarized in lines 3 and 7 in Table 8. As in earlier reports cited above, addition of immune lymphocytes to monolayers of immune or nonimmune macrophages resulted in a significant inhibition of intracellular growth of *H. capsulatum* (lines 2 and 6, 60% reduction as compared with controls). In an attempt to demonstrate lymphokine activity, supernatants from *Histoplasma*-stimulated lymphocyte cultures were added to washed peritoneal macrophages of immune and nonimmune mice. Addition of supernatants to immune macrophages was slightly effective in activating the latter (line 7, 44% reduction as compared to line 5, 33% reduction) but failed to activate nonimmune macrophages

Table 7 Effect of Immune Lymphocytes on Intracellular Growth of *Histoplasma capsulatum* Within Normal Macrophages[a]

Type of lymphocyte added[b]	Mean number of yeast per infected macrophage[c]	Percent reduction of intracellular growth[d]
None	12.7	0
Normal	10.6	17
Immune	3.3	74

[a]Macrophages harvested from normal animals and maintained in cell culture for 48 h.

[b]Peritoneal cells from immunized or normal animals passed through cotton-wool columns and added after parasitization of cultures with *H. capsulatum*. Number of lymphocytes was 10^7 per milliliter.

[c]After 21 h of incubation.

[d]Compared to controls.

Source: Ref. 75.

Table 8 Intracellular Growth of *Histoplasma capsulatum* in Macrophages Treated with Immune Lymphoid Cells or Culture Medium from Immune Lymphoid Cells and Antigen

Type of macrophage[a]	Wash in cell culture[b]	Lymphoid cells added[c]	Medium from lymphoid cells added[d]	Mean number yeast/ infected macrophage[e]	Reduction (%) of intracellular growth
Normal	−	−	−	9.0	0
Normal	−	+	−	3.6	60
Normal	−	−	+	10.1	0
Immune	−	−	−	3.6	60
Immune	+	−	−	6.0	33
Immune	+	+	−	3.6	60
Immune	+	−	+	4.0	44

[a]Normal macrophages harvested from the peritoneal cavity of unstimulated normal mice; immune macrophages harvested from mice immunized by sublethal infection.

[b]Cell cultures washed three times with HBSS 1 h after cells had settled onto coverslips.

[c]Peritoneal cells from immunized animals passed through cotton-wool columns and added after parasitization of cultures with *H. capsulatum*.

[d]Lymphoid cells recovered as in footnote c maintained in a rotator culture for 2 h at 37°C in the presence of heat-killed yeast cells (2×10^7 per milliliter) of *H. capsulatum*. After incubation, cultures were centrifuged and supernatant medium was added to parasitized macrophage cultures. The addition of antigen (heat-killed yeast cells) was not essential. The lymphoid cell populations contained viable yeast cells remaining from the sublethal dose used for immunization.

[e]After 18 h at 37°C.

Source: Ref. 76.

(line 3, 0% reduction). Stimulation of lymphocytes with PHA and ConA was also ineffective in inducing lymphokine activity. These results suggest that macrophage activation requires direct lymphocyte-macrophage interaction. However, the authors caution that their failure to demonstrate lymphokine activity in the mouse model does not preclude the possibility that lymphokine activation exists in other animal species.

A direct cytotoxic effect of supernatants from lymphocytes stimulated with PHA and allogeneic mouse spleen cells was demonstrated with *Candida albicans* and *Saccharomyces cerevisia* by Pearsall et al. [133]. Incubation of 3×10^5 *Candida* per milliliter of supernatant markedly decreased the number of viable yeast. The effect was temperature dependent with essentially no inhibition at 4°C. The cytotoxic effect of supernatants was considered to be attributed to lymphokine activity, a possibility that was supported by the toxicity of the supernatants for L-cells.

In similar studies, Kashkin et al. [85] reported that supernatants of spherule-stimulated lymphocytes from immunized guinea pigs were fungicidal for *Coccidioides immitis* arthroconidia (52% reduction in viability as compared to supernatants of nonimmune lymphocytes). Comparative studies of guinea pigs infected intramuscularly with 10^4 arthroconidia indicated that fungicidal activity of supernatants inversely correlated with disease severity, i.e., supernatants of lymphocytes from guinea pigs that developed mild disease exhibited fungicidal activity; those of lymphocytes from guinea pigs that developed severe disseminated coccidioidomycosis were only slightly fungicidal.

VI. TRANSFER OF CMI

A. Transfer with Lymphocytes

The ability of passively transferred spleen cells, peritoneal-exudate cells, or thymocytes to confer resistance to challenge with *C. immitis* [12,13,97] and *H. capsulatum* [176, 188] in experimental animals was discussed earlier in this chapter. These studies did not, however, assess the adoptive transfer of CMI responses. In the first of such attempts, Scillian et al. [155] showed that spleen cells of mice immunized with a subcutaneous inoculation of viable *B. dermatitidis* yeast transferred DTH responses (footpad sensitivity and MIF) to nonimmune mice. No effort was made to assess the protective effect afforded by the cell transfers. Rifkind et al. [142] transferred CDN sensitivity to inbred mice using immune spleen cells. Again, the protective effect of adoptive transfer on resistance to challenge was not measured. More recently, Diamond [41] showed that lymph node and splenic lymphocytes from guinea pigs immunized with killed cryptococci transferred cryptococcin skin test reactivity in two of four guinea pigs. Although the author emphasized the preliminary nature of these studies, he noted that the survival rate after challenge with viable *Cryptococcus neoformans* was increased in recipients of immune versus nonimmune lymphocytes.

The ability to transfer CMI responses with lymphocytes of immunized animals and the association of depressed T-cell responses in patients with severe fungal diseases prompted a number of investigators to evaluate the effect of lymphocyte transfers in patients with these diseases. Because of the potential risk of graft versus host reaction, only a limited number of studies were made using intact leukocytes or lysed leukocyte preparations [28,90,92,93,103]. For the most part, these studies met with little success. However, Buckley et al. [18] reported an apparent cure of a patient with CMC who was given paternal bone marrow following chemotherapy.

B. Transfer Factor

The use of transfer factor (TF), a dialyzable extract of immune leukocytes, evolved from the studies of Lawrence and co-workers [100-102]. Lysed preparations of peripheral blood leukocytes of healthy, immune donors were shown to transfer skin test reactivity to nonimmune recipients. The biologically active material was dialyzable and resistant to DNase, RNase, and proteolytic enzymes. Of primary importance, dialyzable TF was nonantigenic, thereby circumventing the risk of graft versus host reactions.

Numerous studies have been made to evaluate TF as an immunotherapeutic agent in fungal diseases. Unfortunately, most of the studies have been limited to a small number of patients and, as yet, there has not been a definitive clinical trial to evaluate the efficacy of TF in any fungal disease. In an attempt to summarize the results obtained in TF therapy, selected reports of independent investigators are presented in Table 9.

It is important to emphasize that in summarizing the results described above, no attempts were made to distinguish between patients who received a combination of TF with chemotherapy and those who received TF without adjunctive chemotherapy. Nor does the table indicate the duration of clinical and/or immunological responses. However, it has been the experience of most investigators that TF should be used in conjunction with chemotherapy, at least until clinical improvement has been obtained. It has also been the experience of most investigators that multiple doses of TF are required over a prolonged period of time and that frequently, upon discontinuation of immunotherapy, patients revert back to T-cell anergy. Thus, the real effect of TF on the ultimate course of the disease remains contentious and, to some extent, enthusiasm for TF as an immunotherapeutic agent has waned.

VII. SUMMARY

Induction and expression of cellular immunity in mycotic diseases is contingent upon a complex sequence of interactions among antigen, macrophages, and lymphocytes. Although much remains to be learned of the mechanisms involved, it has become evident that in the absence of T-cell function, host resistance to most fungal diseases is markedly reduced. Support for this is derived from (1) the high incidence of fungal diseases in patients with disorders that are associated with a depressed T-cell immunity, (2) the temporal relationship between disease progression and depressed T-cell functions, and (3) the increased susceptibility of T-cell-deficient animals in these diseases. However, it is crucial to distinguish those T-cell functions that directly contribute to host defense from those that do not. Depressed DTH responses generally parallel disease progression, but this relationship may be casual rather than causal. On the other hand, inhibition of fungal growth by T-cell-mediated activation of macrophages or by lymphocyte cytotoxicity would appear to offer an effective means of eradicating the organism.

Modulation of T-cell responses by serum factors and by suppressor cell populations is a relatively new field of investigation in mycotic diseases. The significance of suppressor activity is not clearly understood but would provide a basis for T-cell anergy. Clearly, further studies are needed to define the origin, nature, and specificity of T-cell suppression in these diseases.

As a final note, I would like to emphasize that T-cell function is but one of the many dynamic interactions between the host and the infecting organism, and it is the composite interaction that ultimately leads to disease progression or regression.

Table 9 Results of Transfer Factor Therapy in Mycotic Diseases

Disease	Number of patients	Immunological conversions[a]			Clinical improvement	Reference
		Skin test	MIF	LT		
CMC	2	1/6	1/2	ND	0/2	144
	6	6/6	5/5	1/4	0/6	95
	1	1/1	ND	1/1	1/1	129
	5	4/5	4/5	1/4	2/5	94
	1	1/1	?	1/1	1/1	183
	1	1/1	1/1	0/1	1/1	163
	1	1/1	1/1	0/1	0/1	73
	1	1/1	ND	?	1/1	154
Coccidioidomycosis	3	2/2	3/3	2/2	2/3	64
	4	2/2	1/2	4/4	4/4	23
	1	1/1	ND	0/1	1/1	168
	1	1/1	ND	1/1	1/1	167
	49	24/36	21/25	18/23	31/49	25
Histoplasmosis	1	1/1	ND	1/1	1/1	159
	1	0/1	1/1	0/1	1/1	66
Blastomycosis	1	ND	ND	ND	0/1	49
Paracoccidioidomycosis	5	1/4	ND	?	1/4	120

[a]ND, not done; ?, results not presented.

REFERENCES

1. Abdor, N. I., A. Sagawa, E. Pascual, J. Hebert, and S. Sadeghee. 1976. Suppressor
    T cell abnormality in idiopathic systemic lupus erythematosus. Clin. Immunol.
    Immunopathol. *6*: 192-199.
2. Adamson, D. M., and G. C. Cozad. 1969. Effect of antilymphocyte serum on ani-
    mals experimentally infected with *Histoplasma capsulatum* or *Cryptococcus neo-
    formans*. J. Bacteriol. *100*: 1271-1276.
3. Albert, E. D., M. R. Mickey, and P. I. Terasaki. 1973. Genetics of the HL-A system
    in four populations: American-Caucasians, Japanese-Americans, American-Negroes,
    and Mexican-Americans. In J. Dausaent and J. Colombani (Eds.), *Histocompatibility
    Testing*, Williams & Wilkins, Baltimore, pp. 233-240.
4. Alford, R. H., and R. A. Goodwin. 1972. Patterns of immune responses in chronic
    pulmonary histoplasmosis. J. Infect. Dis. *129*: 269-275.
5. Artz, R. P., and W. E. Bullock. 1979. Immunoregulatory responses in experimental
    disseminated histoplasmosis: lymphoid organ histopathology and serological studies.
    Infect. Immun. *23*: 884-892.
6. Artz, R. P., and W. E. Bullock. 1979. Immunoregulatory responses in experimental
    disseminated histoplasmosis: depression of T-cell-dependent and T-effector responses
    by activation of splenic suppressor cells. Infect. Immun. *23*: 893-902.
7. Asherson, G. L., and M. Zembala. 1974. T cell suppression of contact sensitivity.
    III. The role of macrophages and the specific triggering of nonspecific suppression.
    Eur. J. Immunol. *4*: 804-807.
8. Atkinson, A. J., and J. E. Bennett. 1968. Experience with a new skin test antigen
    prepared from *Cryptococcus neoformans*. Am. Rev. Respir. Dis. *97*: 637-643.
9. Balogh, E., C. Meszaros, and K. Halmy. 1971. Die anwendung des Lymphocyten-
    transformationstestes bei der Untersuchung der mykotischen Sensibilisation. Myko-
    sen *14*: 207-211.
10. Balows, A. 1963. Comparison of *Blastomyces dermatitidis* vaccine and blastomycin
    as skin testing agents for the diagnosis of blastomycosis. Mycopathologia *19*: 151.
11. Balows, A., K. W. Deuschle, N. R. Neede, I. R. Mersach, and K. A. Watson. 1966.
    Skin tests in blastomycosis. Arch. Environ. Health *13*: 86-90.
12. Beaman, L., D. Pappagianis, and E. Benjamini. 1977. Significance of T cells in re-
    sistance to experimental murine coccidioidomycosis. Infect. Immun. *17*: 580-585.
13. Beaman, L., D. Pappagianis, and E. Benjamini. 1979. Mechanisms of resistance to
    infection with *Coccidioides immitis* in mice. Infect. Immun. *23*: 681-685.
14. Blaese, R. M., P. L. Weiden, I. Koski, and N. Dooley. 1974. Infectious agamma-
    globulinemia: transmission of immunodeficiency with grafts of agammaglobuline-
    mic cells. J. Exp. Med. *140*: 1097-1101.
15. Blastomycosis cooperative study of the Veterans Administration. 1964. Blasto-
    mycosis: a review of 198 collected cases in Veterans Administration hospitals. Am.
    Rev. Respir. Dis. *89*: 659-672.
16. Böyum, A. 1966. Ficoll-Hypaque method for separating mononuclear cells and
    granulocytes from human blood. Scand. J. Clin. Lab. Invest. (Suppl.) *21*: 77-89.
17. Broder, S., R. Humphrey, M. Dunn, M. Blackman, B. Meade, C. Goldman, W. Strober,
    and T. A. Waldmann. 1975. Impaired synthesis of polyclonal (non-paraprotein)
    immunoglobulins by circulating lymphocytes from patients with multiple myeloma:
    role of suppressor cells. N. Engl. J. Med. *293*: 887-892.
18. Buckley, R. H., Z. J. Lucas, B. G. Hattler, Jr., C. M. Zmijewski, and D. B. Amos.
    1968. Defective cellular immunity associated with chronic mucocutaneous monili-
    asis and recurrent staphylococcal botryomycosis: immunological reconstitution by
    allogeneic bone marrow. Clin. Exp. Med. *3*: 153-169.

19. Cahall, D. L., and G. P. Youmans. 1975. Molecular weight and other characteristics of mycobacterial growth inhibitory factor produced by spleen cells obtained from mice immunized with viable attenuated mycobacterial cells. Infect. Immun. *12*: 841-850.

20. Campbell, C. C., and G. E. Binkely. 1953. Serological diagnosis with respect to histoplasmin, coccidioidomycosis, and blastomycosis, and the problem of cross reactions. J. Lab. Clin. Med. *42*: 896-906.

21. Canales, L., J. M. Louro, R. O. Middlemas III, and M. A. South. 1969. Immunological observations in chronic mucocutaneous candidiasis. Lancet *2*: 567-571.

22. Casazza, A. R., C. P. Duall, and P. P. Carbone. 1966. Infection in lymphoma; histology, treatment, and duration in relation to incidence and survival. JAMA *197*: 710-716.

23. Catanzaro, A., L. Spitler, and K. M. Moser. 1974. Immunotherapy of coccidioidomycosis. J. Clin. Invest. *54*: 690-701.

24. Catanzaro, A., L. Spitler, and K. M. Moser. 1975. Cellular immune response in coccidioidomycosis. Cell. Immunol. *15*: 360-371.

25. Catanzaro, A., and L. Spitler for the Coccidioidomycosis Cooperative Treatment Group (CCTG). 1976. Clinical and immunologic results of transfer factor therapy in coccidioidomycosis. In M. S. Ascher, A. A. Gottlieb, and C. H. Kirkpatrick (Eds.), *Transfer Factor: Basic Properties and Clinical Applications*, Academic, New York, pp. 477-491.

26. Cauley, L. K., and J. W. Murphy. 1979. Response of congenitally athymic (nude) and phenotypically normal mice to *Cryptococcus neoformans* infection. Infect. Immun. *23*: 644-651.

27. Cheers, C., and R. Waller. 1975. Activated macrophages in congenitally athymic "nude" and in lethally irradiated mice. J. Immunol. *115*: 844-847.

28. Chilgren, R. A., H. J. Meuwissen, P. G. Quie, R. A. Good, and R. Hong. 1969. The cellular immune defect in chronic mucocutaneous candidiasis. Lancet *1*: 1286-1288.

29. Cox, F., and W. T. Hughes. 1974. Disseminated histoplasmosis in childhood leukemia. Cancer *33*: 1127-1133.

30. Cox, R. A. 1979. Immunologic studies of patients with histoplasmosis. Am. Rev. Respir. Dis. *120*: 143-149.

31. Cox, R. A., J. R. Vives, A. Gross, G. Lecara, E. Miller, and E. Brummer. 1976. In vivo and in vitro cell-mediated responses in coccidioidomycosis. I. Immunologic responses of persons with primary, asymptomatic infections. Am. Rev. Respir. Dis. *114*: 937-943.

32. Cox, R. A., and J. R. Vivas. 1977. Spectrum of in vivo and in vitro cell-mediated immune responses in coccidioidomycosis. Cell. Immunol. *31*: 130-141.

33. Cox, R. A., and D. R. Arnold. 1979. Immunoglobulin E in coccidioidomycosis. J. Immunol. *123*: 194-200.

33a. Cox, R. A., R. M. Pope, and S. Yoshinoya. 1979. Immune complexes in sera of patients with coccidioidomycosis. Abstr. Annu. Meet. Am. Soc. Microbiol., 1979, F33, p. 368.

34. Cozad, G. C., and T. J. Lindsey. 1974. Effect of cyclophosphamide on *Histoplasma capsulatum* infections in mice. Infect. Immun. *9*: 261-265.

35. Cutler, J. E. 1976. Acute systemic candidiasis in normal and congenitally thymic-deficient (nude) mice. J. Reticuloendothel. Soc. *19*: 121-124.

36. David, J. R., S. Al-Askari, H. S. Lawrence, and L. Thomas. 1964. Delayed hypersensitivity in vitro. I. The specificity of inhibition of cell migration by antigens. J. Immunol. *93*: 264-273.

37. Davies, S. F., M. Khan, and G. Sarosi. 1978. Disseminated histoplasmosis in immunologically suppressed patients. Am. J. Med. *64*: 94-100.

38. Deresinski, S. C., and D. A. Stevens. 1975. Coccidioidomycosis in compromised hosts: experience at Stanford University Hospital. Medicine (Baltimore) *54*: 377-395.

39. DeSousa, M., R. Cochran, R. Machie, D. Parratt, and M. Arala-Chaves. 1976. Chronic mucocutaneous candidiasis treated with transfer factor. Br. J. Dermatol. *94*: 79-83.

40. Dessaint, J. P., D. Bout, J. Fruit, and A. Capron. 1976. Serum concentration of specific IgE antibody against *Aspergillus fumigatus* and identification of the fungal allergen. Cell. Immunol. Immunopathol. *5*: 314-319.

41. Diamond, R. D. 1977. Effects of stimulation and suppression of cell-mediated immunity on experimental cryptococcosis. Infect. Immun. *17*: 187-194.

42. Diamond, R. D., and J. E. Bennett. 1973. Disseminated cryptococcosis in man: decreased lymphocyte transformation in response to *Cryptococcus neoformans*. J. Infect. Dis. *127*: 694-697.

43. Diamond, R. D., and J. E. Bennett. 1974. Prognostic factors in cryptococcal meningitis. A study of 111 cases. Ann. Intern. Med. *80*: 176-181.

44. Edwards, L. B., F. A. Acquaviva, V. T. Livesay, F. W. Cross, and C. E. Palmer. 1969. An atlas of sensitivity to tuberculin, PPD-B, and histoplasmin in the United States: nationwide variation in sensitivity to PPD-S, PPD-B, and histoplasmin. Am. Rev. Respir. Dis. *99*: 12-17.

45. Ellner, J. J. 1978. Suppressor adherent cells in human tuberculosis. J. Immunol. *121*: 2573-2579.

46. Epstein, W. L., and A. M. Kligman. 1958. The interference phenomenon in allergic contact dermatitis. J. Invest. Dermatol. *31*: 103-108.

47. Fava Netto, C. 1971. The serology of paracoccidioidomycosis: present and future trends. In *Paracoccidioidomycosis,* Proceedings of the First Pan American Symposium, Pan Am. Health Organ. Sci. Publ. No. 254, Washington, D.C., pp. 209-213.

48. Fish, R. G., T. Taharo, and M. Lovell. 1960. Coexistent Addison's and North American blastomycosis. Am. J. Med. *28*: 152-155.

49. FitzSimons, R. B., and A. C. Ferguson. 1978. Cellular immunity and nutrition in refractory disseminated blastomycosis. CMA J. *119*: 343-346.

50. Flynn, N. M., P. D. Hoeprich, M. M. Kawachi, K. K. Lee, R. M. Lawrence, E. Goldstein, G. W. Jordan, R. S. Kundargi, and G. A. Wong. 1979. An unusual outbreak of windborne coccidioidomycosis. N. Engl. J. Med. *301*: 358-361.

51. Folch, H., and B. H. Waksman. 1974. The splenic suppressor cell. II. Suppression of mixed lymphocyte reaction by thymus-dependent adherent cells. J. Immunol. *113*: 140-144.

52. Ford, W. H., W. A. Caspary, and B. Shenton. 1973. Purification and properties of a lymphocyte inhibition factor from human serum. Clin. Exp. Immunol. *15*: 169-179.

53. Fox, R. A., and K. Rajaraman. 1979. Macrophage migration stimulation factor and the role of suppressor cells in this production. Cell. Immunol. *47*: 69-78.

54. Furcolow, M. L. 1963. Tests of immunity in histoplasmosis. N. Engl. J. Med. *268*: 357-361.

55. Furcolow, M. L., I. L. Doto, F. E. Tosh, and H. J. Lynch, Jr. 1961. Course and prognosis of untreated histoplasmosis. JAMA *177*: 292-296.

56. Gadebusch, H. H., and P. W. Gikas. 1965. The effect of cortisone upon experimental pulmonary cryptococcosis. Am. Rev. Respir. Dis. *92*: 64-74.

57. George, M., and J. H. Vaughan. 1962. In vitro cell migration as a model for delayed hypersensitivity. Proc. Soc. Exp. Biol. Med. *111*: 514-521.

58. Gifford, M. A., W. C. Bass, and R. J. Douds. 1937. Data on *Coccidioides* fungus infection in Kern County, 1901-1936. In *Kern County Health Dept. Annu. Rev. 1936-1937,* pp. 39-54.

59. Giger, D. K., J. E. Domer, S. A. Moser, and J. T. McQuitty, Jr. 1978. Experimental murine candidiasis: pathological and immune responses in T-lymphocyte-depleted mice. Infect. Immun. *21*: 729-737.

60. Glasgow, A. H., R. B. Nimberg, J. O. Menzoien, I. Saporoschetz, S. R. Cooperband, K. Schmid, and J. A. Mannick. 1974. Association of anergy with an immunosuppressive peptide fraction in the serum of patients with cancer. N. Engl. J. Med. *291*: 1263-1267.

61. Goldberg, L. S., R. Bluestone, E. V. Barnett, and J. W. Landau. 1971. Studies on lymphocyte and monocyte function in chronic mucocutaneous candidiasis. Clin. Exp. Immunol. *8*: 37-43.

62. Goldstein, E., and O. N. Rambo. 1962. Cryptococcal infection following steroid therapy. Ann. Intern. Med. *56*: 114-120.

63. Goodwin, R. A., Jr., F. T. Owens, J. D. Snell, W. W. Hubbard, R. D. Buchanan, R. T. Terry, and R. M. Des Prez. 1976. Chronic pulmonary histoplasmosis. Medicine (Baltimore) *55*: 413-452.

64. Graybill, J. R., J. Silva, Jr., R. H. Alford, and D. E. Thor. 1973. Immunologic and clinical improvement of progressive coccidioidomycosis following administration of transfer factor. Cell. Immunol. *8*: 120-135.

65. Graybill, J. R., and R. H. Alford. 1974. Cell-mediated immunity in cryptococcosis. Cell. Immunol. *14*: 12-21.

66. Graybill, J. R., C. Ellenbogen, D. Drossman, P. Kaplan, and D. E. Thor. 1976. Transfer factor therapy of disseminated histoplasmosis. In M. S. Ascher, A. A. Gottlieb, and C. H. Kirkpatrick (Eds.), *Transfer Factor,* Academic, New York, pp. 509-514.

67. Graybill, J. R., and D. J. Drutz. 1978. Host defense in cryptococcosis. II. Cryptococcosis in the nude mouse. Cell. Immunol. *40*: 263-274.

68. Greer, D. L., D. D. de Estrada, and L. A. de Trejos. 1971. Dermal reactions to paracoccidioidin among family members of patients with paracoccidioidomycosis. In *Paracoccidioidomycosis,* Proceedings of the First Pan American Symposium, Pan Am. Health Organ. Sci. Publ. No. 254, Washington, D.C., pp. 76-83.

69. Hanifin, J. M., L. F. Ray, and W. C. Lobitz, Jr. 1974. Immunological reactivity in dermatophytosis. Br. J. Dermatol. *90*: 1-8.

70. Harrington, J. T., Jr., and P. Stastny. 1973. Macrophage migration from an agarose droplet: development of a micromethod for assay of delayed hypersensitivity. J. Immunol. *110*: 752-759.

71. Hicks, H. R., and W. T. Northey. 1967. Studies on the response of thymectomized mice to infection with *Coccidioides immitis* In L. Ajello (Ed.), *Coccidioidomycosis,* Papers from the Second Symposium on Coccidioidomycosis, University of Arizona Press, Tucson, pp. 183-187.

72. Holm, G. 1969. Cytotoxic effects of lymphoid cells in vitro. Adv. Immunol. *11*: 117-193.

73. Horsmanheimo, M., K. Krohn, M. Virolainen, and K. Blomqvist. 1979. Immunologic features of chronic granulomatous mucocutaneous candidiasis before and after treatment with transfer factor. Arch. Dermatol. *115*: 180-184.

74. Howard, D. H. 1975. The role of phagocytic mechanisms in defense against *Histoplasma capsulatum.* In *Mycoses,* Proceedings of the Third International Confer-

ence on the Mycoses, Pan Am. Health Organ. Sci. Publ. No. 356, Washington, D. C., pp. 50-59.

75. Howard, D. H., V. Otto, and R. K. Gupta. 1971. Lymphocyte-mediated cellular immunity in histoplasmosis. Infect. Immun. *4*: 605-610.

76. Howard, D. H., and V. Otto. 1977. Experiments on lymphocyte-mediated cellular immunity in murine histoplasmosis. Infect. Immun. *16*: 226-231.

77. Huppert, M. 1978. Racism in coccidioidomycosis? Am. Rev. Respir. Dis. *118*: 797-798.

78. Hurley, D. L., J. E. Balow, and A. S. Fauci. 1975. Experimental disseminated candidiasis. II. Administration of glucocorticosteroids, susceptibility to infection, and immunity. J. Infect. Dis. *132*: 393-398.

79. Hurley, R. 1966. Experimental infection with *Candida albicans* in modified hosts. J. Pathol. Bacteriol. *92*: 57-67.

80. Ibrahim, A. B., and D. Pappagianis. 1973. Experimental induction of anergy to coccidioidin by antigens of *Coccidioides immitis*. Infect. Immun. *1*: 786-794.

81. Jacobsen, C. E., and M. B. Dockerty. 1943. Blastomycosis of the epididymis: report of four cases. J. Urol. *50*: 237-248.

82. Jillson, O. F., and M. Huppert. 1949. The immediate wheal and the 24-48 hour tuberculin type edematous reactions to trichophyton. J. Invest. Dermatol. *12*: 179-185.

83. Jones, H. E., J. H. Reinhardt, and M. G. Rinaldi. 1973. A clinical, mycological, and immunological survey for dermatophytosis. Arch. Dermatol. *108*: 61-65.

84. Jones, H. E., J. H. Reinhardt, and M. G. Rinaldi. 1974. Immunologic susceptibility to chronic dermatophytosis. Arch. Dermatol. *110*: 213-220.

85. Kashkin, K. P., S. M. Kikholetov, and A. V. Lipnitsky. 1977. Studies on mediators of cellular immunity in experimental coccidioidomycosis. Sabouraudia *15*: 59-68.

86. Katz, D. H., and Benacerraf. 1972. The regulatory influence of activated T cells on B cell responses to antigen. Adv. Immunol. *15*: 1-94.

87. Katz, S. I., D. Parker, and J. L. Turk. 1974. B-cell suppression of delayed hypersensitivity reactions. Nature (Lond.) *251*: 550-551.

88. Kauffman, C., K. S. Israel, J. W. Smith, A. C. White, J. Schwarz, and G. F. Brooks. 1978. Histoplasmosis in immunosuppressed patients. Am. J. Med. *64*: 923-932.

89. Kessler, S. W. 1976. Cell membrane antigen isolation with the staphylococcal protein A antibody adsorbent. J. Immunol. *117*: 1482-1490.

90. Kirkpatrick, C. H., J. W. Chandler, and R. N. Schimke. 1970. Chronic mucocutaneous moniliasis with impaired delayed hypersensitivity. Clin. Exp. Immunol. *6*: 375-385.

91. Kirkpatrick, C. H., J. W. Chandler, Jr., T. K. Smith, and W. M. Newberry, Jr. 1971. Cellular immunologic studies in histoplasmosis. In A. Balows (Ed.), *Histoplasmosis*, Thomas, Springfield, Ill., pp. 371-379.

92. Kirkpatrick, C. H., R. R. Rich, and J. E. Bennett. 1971. Chronic mucocutaneous candidiasis: model building in cellular immunity. Ann. Intern. Med. *74*: 955-978.

93. Kirkpatrick, C. H., R. R. Rich, R. G. Graw, Jr., T. K. Smith, I. Mickenberg, and G. N. Rogentine. 1971. Treatment of chronic mucocutaneous moniliasis by immunologic reconstitution. Clin. Exp. Immunol. *9*: 733-748.

94. Kirkpatrick, C. H., R. R. Rich, and T. K. Smith. 1972. Effect of transfer factor on lymphocyte function in anergic patients. J. Clin. Invest. *51*: 2948-2956.

95. Kirkpatrick, C. H., and T. K. Smith. 1974. Chronic mucocutaneous candidiasis: immunologic and antibiotic therapy. Ann. Intern. Med. *80*: 310-320.

96. Klimpel, G. R., and C. W. Henney. 1978. BCG-induced suppressor cells. I. Demon-

stration of a macrophage-like suppressor cell that inhibits cytotoxic T cell generation in vitro. J. Immunol. *120*: 563-569.

97. Kong, Y. M., D. C. Savage, and H. B. Levine. 1966. Enhancement of immune responses in mice by a booster injection of *Coccidioides* spherules. J. Immunol. *95*:
1048-1056.

98. Krick, J. A., and J. S. Remington. 1976. Opportunistic invasive fungal infection
in patients with leukemia and lymphoma. Clin. Haematol. *5*: 249-310.

99. Land, G. A. 1976. Opportunistic mycotic infections and malignancies. J. Clin.
Oncol. Hematol. *6*: 113-140.

100. Lawrence, H. S. 1955. The transfer in humans of delayed skin sensitivity to strep-
tococcal M substance and tuberculin with disrupted leukocytes. J. Clin. Invest.
34: 219-230.

101. Lawrence, H. S. 1971. Transfer factor and cellular immune deficiency disease.
N. Engl. J. Med. *283*: 411-419.

102. Lawrence, H. S., F. T. Rapaport, J. M. Converse, and W. S. Tillett. 1960. Transfer
of delayed hypersensitivity to skin homografts with leukocyte extracts in man. J.
Clin. Invest. *39*: 185-198.

103. Levy, R. L., S. W. Huang, M. L. Bach, and F. H. Bach. 1971. Thymic transplantation in a case of chronic mucocutaneous candidiasis. Lancet *2*: 898-900.

104. Lewis, G. M., M. E. Hopper, and M. D. Scott. 1953. Generalized *Trichophyton
rubrum* infections associated with systemic lymphoblastoma. Arch. Dermatol.
Syphilol. *67*: 247-262.

105. Lewis, J. E., D. S. Fair, A. M. Prieur, R. L. Lundak, and G. A. Granger. 1976.
In vitro detection of human lymphotoxin. In N. R. Rose and H. Friedman (Eds.),
Manual of Clinical Immunology, American Society for Microbiology, Washington,
D.C., pp. 110-119.

106. Lewis, J. L., and S. Rabinovich. 1972. The wide spectrum of cryptococcal infections. Am. J. Med. *53*: 315-322.

107. Lopez, L. R., A. E. Vatter, and H. Hemmingsen. 1975. The reaction of lymphocytes with stimulated and unstimulated mouse macrophages. Fed. Proc. *34*: 958.

108. Louria, D. B., N. Fallon, and H. G. Browne. 1960. The influence of cortisone on
experimental fungus infections in mice. J. Clin. Invest. *39*: 1435-1449.

109. Mathur, S., J. M. Goust, E. O. Horger III, and H. H. Fudenberg. 1977. Immunoglobulin E anti-*Candida* antibodies and candidiasis. Infect. Immun. *18*: 257-259.

110. McDougal, J. S., P. B. Redecha, R. D. Inman, and C. L. Christian. 1979. Binding
of immunoglobulin G aggregates and immune complexes in human sera to staphylococci containing protein A. J. Clin. Invest. *63*: 627-636.

111. Mendes, E., and A. Raphael. 1971. Impaired delayed hypersensitivity in patients
with South American blastomycosis. J. Allergy *47*: 17-22.

112. Mendes, N. F., C. C. Musatti, R. C. Leao, E. Mendes, and C. K. Naspitz. 1971.
Lymphocyte cultures and skin allograft survival in patients with South American
blastomycosis. J. Allergy Clin. Immunol. *48*: 40-49.

113. Mok, P. W. Y., and D. L. Greer. 1977. Cell-mediated immune responses in patients with paracoccidioidomycosis. Clin. Exp. Immunol. *28*: 89-98.

114. Mok, P. W. Y., and C. Fava Netto. Paracoccidioidin and histoplasmin sensitivity
in Coari (State of Amazonas), Brazil. Am. J. Trop. Med. Hyg. *27*: 808-814.

115. Moretta, L., S. R. Webb, C. E. Grossi, P. M. Lydyard, and M. D. Cooper. 1977.
Functional analysis of two human T-cell subpopulations: help and suppression of
B-cell responses by T cells bearing receptors for IgM or IgG. J. Exp. Med. *146*:
184-200.

116. Mortensen, R. F., A. P. Osmand, and H. Gewurz. 1975. Effects of C-reactive

protein on the lymphoid system. I. Binding to thymus-dependent lymphocytes and alteration of their functions. J. Exp. Med. *141*: 821-839.

117. Mota, C. C. A. 1967. Contribuicao ao estudo da epidemiologia da blastomicose sulamericana no Parana. An. Fac. Med. Parana *10*: 53-92.

118. Mourad, S., and L. Friedman. 1968. Passive immunization of mice against *Candida albicans.* Sabouraudia *6*: 103-105.

119. Murgita, R. A., and T. B. Tomasi, Jr. 1975. Suppression of the immune response by α-fetoprotein. II. The effect of α-fetoprotein on mixed lymphocyte reactivity and mitogen-induced lymphocyte transformation. J. Exp. Med. *141*: 269-286.

120. Musatti, C. C., M. T. Rezkallah, E. Mendes, and N. F. Mendes. 1976. In vivo and in vitro evaluation of cell-mediated immunity in patients with paracoccidioidomycosis. Cell. Immunol. *24*: 365-378.

121. Neiburger, R. G., G. P. Youmans, and A. S. Youmans. 1973. Relationship between tuberculin hypersensitivity and cellular immunity to infection in mice vaccinated with viable attenuated mycobacterial cells or with mycobacterial ribonucleic acid preparations. Infect. Immun. *8*: 42-47.

122. Nelson, L. M., and K. J. McNeice. 1959. Recurrent Cushing's syndrome with *Trichophyton rubrum* infection. Arch. Dermatol. *80*: 700-704.

123. Nelson, N. A., H. L. Goodman, and H. L. Oster. 1957. The association of histoplasmosis and lymphoma. Am. J. Med. Sci. *233*: 56-65.

124. Neta, R., and S. B. Salvin. 1976. T and B lymphocytes in the regulation of delayed hypersensitivity: the possible presence of a suppressor B cell in the regulation of delayed hypersensitivity. J. Immunol. *113*: 1716-1725.

125. Newberry, W. M., Jr., J. W. Chandler, Jr., T. D. Y. Chin, and C. H. Kirkpatrick. 1968. Immunology of the mycoses. I. Depressed lymphocyte transformation in chronic histoplasmosis. J. Immunol. *100*: 436-443.

126. Newcomer, V. D., E. T. Wright, J. E. Rarbet, L. H. Winer, and T. H. Sternberg. 1953. The effect of cortisone on experimental coccidioidomycosis. J. Invest. Dermatol. *20*: 315-327.

127. Oliveira, P. P. 1955. Contribuicao a geografia da histoplasmose no Brasil (Prana e Santa Catarina). Hospital (Rio de Janeiro). *48*: 105-112.

128. Opelz, G., and M. I. Scheer. 1975. Cutaneous sensitivity and in vitro responsiveness of lymphocytes in patients with disseminated coccidioidomycosis. J. Infect. Dis. *32*: 250-255.

129. Pabst, H. F., and R. Swanson. 1972. Successful treatment of candidiasis with transfer factor. Br. Med. J. *2*: 442-443.

130. Pappagianis, D. 1975. Opportunism in coccidioidomycosis. In E. W. Chick, A. Balows, and M. L. Furcolow (Eds.), *Proceedings on Opportunistic Fungal Infections.* Thomas, Springfield, Ill., pp. 221-229.

131. Paterson, P. Y., R. Semo, G. Blumenschein, and J. Swelstad. 1971. Mucocutaneous candidiasis, anergy and a plasma inhibitor of cellular immunity: reversal after amphotericin B therapy. Clin. Exp. Immunol. *9*: 595-602.

132. Patterson, R. J., and G. P. Youmans. 1970. Demonstration in tissue culture of lymphocyte-mediated immunity to tuberculosis. Infect. Immun. *1*: 600-603.

133. Pearsall, N. N., J. S. Sundsmo, and R. S. Weiser. 1973. Lymphokine toxicity for yeast cells. J. Immunol. *110*: 1444-1446.

134. Pearsall, N. N., B. L. Adams, and R. Bunni. 1978. Immunologic responses to *Candida albicans.* III. Effects of passive transfer of lymphoid cells or serum on murine candidiasis. J. Immunol. *120*: 1176-1180.

135. Phanuphak, P., J. W. Moorhead, and H. N. Claman. 1974. Tolerance and contact

sensitivity to DNFB in mice. III. Transfer of tolerance with "suppressor T cells." J. Immunol. *113*: 1230-1236.

136. Provost, T. T., L. K. Garrettson, R. H. Zeschke, N. R. Rose, and T. B. Tomasi, Jr. 1973. Combined immune deficiency, autoantibody formation, and mucocutaneous candidiasis. Clin. Immunol. Immunopathol. *1*: 429-445.

137. Rabinowitz, J. G., J. Busch, and W. R. Buttram. 1976. Pulmonary manifestations of blastomycosis. Radiology *120*: 25-32.

138. Rea, T. H., H. Einstein, and N. E. Levan. 1976. Dinitrochlorobenzene reactivity in disseminated coccidioidomycosis: an inverse correlation with complement-fixing antibody titers. J. Invest. Dermatol. *66*: 34-37.

139. Reddy, P., D. F. Gorelick, C. A. Brasher, and H. Larsh. 1970. Progressive disseminated histoplasmosis as seen in adults. Am. J. Med. *48*: 629-636.

140. Restrepo, A., M. M. Robledo, S. Ospina, M. Restrepo, and A. Correa. 1968. Distribution of paracoccidioidin sensitivity in Colombia. Am. J. Trop. Med. *17*: 25-37.

141. Restrepo, A., M. Robledo, R. Gutierrez, M. Sanclemente, E. Castaneda, and G. Calle. 1970. Paracoddicioidomycosis (South American blastomycosis). A study of 39 cases observed in Medellin, Colombia. Am. J. Trop. Med. Hyg. *19*: 68-76.

142. Rifkind, D., J. A. Frey, J. R. Davis, E. A. Peterson, and M. Dinowitz. 1976. Delayed hypersensitivity to fungal antigens in mice. I. Use of the intradermal skin and footpad tests as assays of active and passive sensitization. J. Infect. Dis. *133*: 50-56.

143. Rocklin, R. E. 1973. Production of migration inhibitory factor by non-dividing lymphocytes. J. Immunol. *110*: 674-678.

144. Rocklin, R. E., R. A. Chilgren, R. Hong, and J. R. David. 1970. Transfer of cellular hypersensitivity in chronic mucocutaneous candidiasis monitored in vivo and in vitro. Cell. Immunol. *1*: 290-299.

145. Rocklin, R. E., O. L. Meyers, and J. R. David. 1970. An in vitro assay for cellular hypersensitivity in man. J. Immunol. *104*: 95-102.

146. Rogers, T. J., E. Balish, and D. D. Manning. 1976. The role of thymus-dependent cell-mediated immunity in resistance to experimental disseminated candidiasis. J. Reticuloendothel. Soc. *20*: 291-298.

147. Rowley, D. A., F. W. Fitch, E. P. Stuart, H. Kohler, and H. Consenze. 1973. Specific suppression of immune responses. Science *181*: 1133-1141.

148. Rutala, P. J., and J. W. Smith. 1978. Coccidioidomycosis in potentially compromised hosts: the effect of immunosuppressive therapy in dissemination. Am. J. Med. Sci. *275*: 283-295.

149. Sarosi, G. A., D. V. Voth, B. A. Dahl, I. L. Doto, and F. E. Tosh. 1971. Disseminated histoplasmosis: results of long-term follow up. A Center for Disease Control cooperative mycoses study. Ann. Intern. Med. *75*; 511-516.

150. Scheer, M., G. Opelz, P. Terasaki, and W. Hewitt. 1973. The association of disseminated coccidioidomycosis and histocompatibility type. In *Prog. Abstr. 13th Intersci. Congr. Antimicrobial Agents Chemother.*, No. 157, Washington, D.C.

151. Schimpff, S. C., and J. E. Bennett. 1975. Abnormalities in cell-mediated immunity in patients with *Cryptococcus neoformans* infections. J. Allergy Clin. Immunol. *55*: 430-441.

152. Schlegel, R. J., G. M. Bernier, J. A. Bellanti, D. A. Maybee, G. B. Osborne, J. L. Stewart, D. S. Pearlman, J. Ouelette, and F. C. Biehusen. 1970. Severe candidiasis associated with thymic dysplasia, IgA deficiency, and plasma antilymphocyte effects. Pediatrics *45*: 926-936.

153. Schroter, G. P. J., K. Bakshandeh, B. S. Husberg, and R. Weil, III. 1977. Coccidioidomycosis and renal transplantation. Transplantation *23*: 485-489.

154. Schulkine, M. L., W. H. Adler, W. A. Altemeis, and E. M. Ayoub. 1972. Transfer factor in the treatment of a case of chronic mucocutaneous candidiasis. Cell. Immunol. *3*: 606-615.

155. Scillian, J. J., G. C. Cozad, and H. D. Spencer. 1974. Passive transfer of delayed hypersensitivity to *Blastomyces dermatitidis* between mice. Infect. Immun. *10*: 705-711.

156. Smith, C. E., R. R. Beard, H. G. Rosenberg, and E. C. Whiting. 1946. Varieties of coccidioidal infection in relation to epidemiology and control of the disease. Am. J. Public Health *36*: 1394-1402.

157. Smith, C. E., E. G. Whiting, E. E. Baker, H. G. Rosenberg, R. R. Beard, and M. T. Saito. 1948. The use of coccidioidin. Am. Rev. Tuberc. *57*: 330-360.

158. Smith, C. E., M. T. Saito, and S. A. Simons. 1956. Pattern of 39,500 serologic tests in coccidioidomycosis. JAMA *160*: 546-552.

159. Smith, C. R., D. E. Griffin, and J. R. Graybill. 1976. Chronic pulmonary histoplasmosis: improved lymphocyte response with transfer factor. Ann. Intern. Med. *85*: 708-709.

160. Smith, D. T. 1949. Immunologic types of blastomycosis: a report on 40 cases. Ann. Intern. Med. *31*: 464-469.

161. Smith, J. W., and J. P. Utz. 1972. Progressive disseminated coccidioidomycosis. A prospective study of 26 patients. Ann. Intern. Med. *76*: 557-565.

162. Smithline, N., D. A. Ogden, A. I. Cohn, and K. Johnson. 1977. Disseminated coccidioidomycosis in renal transplant patients. In L. Ajello (Ed.), *Coccidioidomycosis: Current Clinical and Diagnostic Status,* Symposia Specialists, Miami, Fla., pp. 201-206.

163. Sousa, M. D. E., R. Cochran, R. Mackie, D. Parratt, and M. Arala-Chaves. 1976. Chronic mucocutaneous candidiasis treated with transfer factor. Br. J. Dermatol. *94*: 79-83.

164. Spencer, H. D., and G. C. Cozad. 1973. Role of delayed hypersensitivity in blastomycosis of mice. Infect. Immun. *7*: 329-334.

165. Spickard, A., W. T. Butler, V. Andriole, and J. P. Utz. 1963. The improved prognosis of cryptococcal meningitis with amphotericin B therapy. Ann. Intern. Med. *58*: 66-83.

166. Spitler, L. 1976. Delayed hypersensitivity skin testing. In N. R. Rose and H. Friedman (Eds.), *Manual of Clinical Immunology,* American Society for Microbiology, Washington, D.C., pp. 53-63.

167. Steele, R. W., B. E. Sieger, T. R. McNitt, L. O. Gentry, and W. L. Moore, Jr. 1976. Therapy for disseminated coccidioidomycosis with transfer factor from a related donor. Am. J. Med. *61*: 283-286.

168. Stevens, D. A., D. Pappagianis, V. A. Marinkovich, and T. F. Waddell. 1974. Immunotherapy in recurrent coccidioidomycosis. Cell. Immunol. *12*: 37-48.

169. Stevens, D. A., H. B. Levine, S. C. Deresinski, D. R. Ten Eyck, and A. Restrepo. 1977. Epidemiological and clinical skin testing studies with spherulin. In L. Ajello (Ed.), *Coccidioidomycosis: Current Clinical and Diagnostic Status,* Symposia Sepcialists, Miami, Fla., pp. 107-114.

170. Stobo, J. D. 1977. Immunosuppression in man: suppression by macrophages can be mediated by interactions with regulatory T cells. J. Immunol. *119*: 918-924.

171. Stobo, J. D., P. Sigrun, R. E. VanScoy, and P. E. Hermans. 1976. Suppressor thymus-derived lymphocytes in fungal infection. J. Clin. Invest. *57*: 319-328.

172. Strelkauskas, A. J., R. T. Callery, J. McDowell, and Y. Borel. 1978. Direct evidence for loss of human suppressor cells during active auto-immune disease. Proc. Natl. Acad. Sci. U.S.A. *75*: 5150-5154.

173. Sulzberger, M. B., and F. Wise. 1932. Ringworm and trichophytin: newer developments, including practical and theoretical considerations. JAMA *99*: 1759-1764.

174. Tada, T., M. Taniguchi, and K. Okumura. 1977. Regulation of antibody response by antigen specific T-cell factors bearing I-region determinants. Prog. Immunol. *3*: 369-377.

175. Talmage, D. S., and H. Hemingsen. 1975. Is the macrophage the stimulating cell? J. Allergy Clin. Immunol. *55*: 442-450.

176. Tewari, R. P., D. Sharma, M. Solotorovsky, R. Lafemina, and J. Balint. 1977. Adoptive transfer of immunity from mice immunized with ribosomes or live yeast cells of *Histoplasma capsulatum*. Infect. Immun. *15*: 789-795.

177. Tewari, R. P., D. K. Sharma, and A. Mathur. 1978. Significance of thymus-derived lymphocytes in immunity elicited by immunization with ribosomes or live yeast cells of *Histoplasma capsulatum*. J. Infect. Dis. *138*: 605-613.

178. Thor, D., R. E. Jurezis, S. R. Veach, E. Miller, and S. Dray. 1968. Cell migration inhibition factor released by antigen from human peripheral lymphocytes. Nature (Lond.) *219*: 755-757.

179. Todd, C. W., R. W. Goodgame, and D. G. Colley. 1979. Immune response during human schistosomiasis mansoni. V. Suppression of schistosome antigen-specific lymphocyte blastogenesis by adherent phagocytic cells. J. Immunol. *122*: 1440-1446.

180. Twomey, J. J., A. H. Laughter, S. Farrow, and C. C. Douglass. 1975. Hodgkin's disease: an immunodepleting and immunosuppressive disorder. J. Clin. Invest. *56*: 467-475.

181. Utz, J. P. 1977. Histoplasmosis. In P. D. Hoeprich (Ed.), *Infectious Diseases: A Modern Treatise of Infectious Processes,* Harper & Row, Hagerstown, Md., pp. 383-388.

182. Valdimarsson, H., L. Holt, H. R. C. Riches, and J. R. Hobbs. 1970. Lymphocyte abnormality in chronic mucocutaneous candidiasis. Lancet *1*: 1259-1261.

183. Valdimarsson, H., C. B. S. Wood, J. R. Hobbs, and P. J. Holt. 1972. Immunological features in a case of chronic granulomatous candidiasis and its treatment with transfer factor. Clin. Exp. Immunol. *11*: 151-163.

184. Valdimarsson, H., J. M. Higgs, R. S. Wells, M. Yamamura, J. R. Hobbs, and P. J. L. Holt. 1973. Immune abnormalities associated with chronic mucocutaneous candidiasis. Cell. Immunol. *6*: 348-361.

185. Valentine, F. 1971. Lymphocyte transformation: the proliferation of human blood lymphocytes stimulated by antigen in vitro. In B. R. Bloom and P. R. Glade (Eds.), *In vitro Methods in Cell-Mediated Immunity,* Academic, New York, pp. 443-454.

186. Waksman, B. H. 1979. Cellular hypersensitivity and immunity: conceptual changes in the last decade. Cell. Immunol. *42*: 155-169.

187. Walters, B. A. J., J. E. D. Chick, and W. J. Halliday. 1974. Cell-mediated immunity and serum blocking factors in patients with chronic dermatophytic factors. Int. Arch. Allergy Appl. Immunol. *46*: 849-857.

188. Williams, D. M., J. R. Graybill, and D. J. Drutz. 1978. *Histoplasma capsulatum* infection in nude mice. Infect. Immun. *21*: 973-977.

189. Youmans, G. P., and A. S. Youmans. 1969. Allergenicity of mycobacterial ribosomal and ribonucleic acid preparations in mice and guinea pigs. J. Bacteriol. *97*: 134-139.

190. Zweiman, B., D. Pappagianis, H. Maibach, and E. Hildreth. 1969. Coccidioidin delayed hypersensitivity: skin test and in vitro lymphocyte reactivities. J. Immunol. *102*: 1284-1289.

3

Humoral Responses of the Host

Richard D. Diamond / University Hospital and Boston University Medical Center, Boston, Massachusetts

I. INTRODUCTION

The courses and outcomes of mycoses are determined largely by interactions between the fungi and various host defense mechanisms. Exposure to fungi evokes a range of specific and nonspecific humoral responses in the normal host. Much interest in host humoral responses to fungi has centered on the detection and quantitation of specific antibody responses to mycoses. Thus, measurement of antibody responses to mycoses has indirectly influenced the courses of many of these infections by providing major criteria for the presence and activity of the diseases. More recently, there has been an increased awareness that antifungal antibodies can interact with other host defense mechanisms. Such interactions might have detrimental or positive effects on the overall host immune response, so could alter the nature, extent, and severity of mycoses. Besides antibodies, complement and other nonspecific host humoral systems can interact with other host defense mechanisms in ways that also may profoundly influence the course of mycoses. Transferrin and other normally present humoral factors may damage fungi directly or at least limit the rate of fungal growth. Still other humoral substances appear to be able to alter host cellular responses to mycoses.

Specific humoral responses to fungi remain as important as markers for the presence of mycoses as in the past. Nevertheless, the entire range of host humoral responses has potential immunologic functions that have more direct implications for host defense mechanisms. It is difficult to judge the relative importance of these humoral mechanisms in the total scheme of host defense mechanisms against fungi. However, some tentative inferences should be possible after careful review of the direct data and circumstantial evidence about host humoral responses to fungi.

II. ANTIBODY RESPONSE

A. Methods of Detection

Most of the procedures that are used to detect the specific antibody responses to fungi are widely known because of their usefulness in serodiagnosis of mycoses (see Chap. 4). These techniques include immunodiffusion, tube precipitation, complement fixation, latex agglutination, macroagglutination, and immunofluorescence. Immunofluorescence may be used to detect all types of specific antibodies, or appropriate fluorescent antisera can define subclasses of antibody that are produced in the immune response. Fractionation and concentration of fractions of serum can separate serum components for use in immunologic studies. Reviews contain details of methodology as well as potential pitfalls in performance and limitations in interpretation of these tests [77].

Although not commonly used in serodiagnosis, other methods have been useful as sensitive research tools for the quantitation of antifungal antibodies. For example, passive hemagglutination has proven to be a useful assay for antibodies to Candida mannans [118], Histoplasma capsulatum [34], Aspergillus [30a], and other fungi or fungal products. The standard hemolytic plaque technique for detection of antibody-producing host cells has been adapted to assay for cells producing anticryptococcal antibodies in an experimental animal model [67]. Newer techniques are potentially sensitive, rapid, and reproducible, but may not be standardized as well as older procedures. These include detection of specific antifungal antibodies or use of these antibodies to detect circulating fungal antigens by counterimmunoelectrophoresis [4,49,51,88a], radioimmunoassay [88, 97,118a], and enzyme-linked immunoabsorbent assay (ELISA) [3,38,96a,96b].

Without reviewing technical details which are readily available in the sources cited, all of these assays share one feature. The sensitivity, specificity, and reproducibility of all methods require that antibodies bind to standardized, well-characterized antigens. Most materials used for the detection of antifungal antibodies are complex molecules with a multiplicity of antigenic determinants. Careful controls are required to ensure that different batches of such antigen preparations behave reproducibly in tests. An extreme example of antigenic complexity is the use of merthiolate-killed whole yeasts in the complement fixation test for antibodies in histoplasmosis. Similarly, crude extracts of organisms, culture filtrates, and even more purified antigens must be carefully standardized [77]. Results using more purified antigens may not necessarily correlate with results obtained with older, cruder, but stable and standardized preparations of antigens [84]. Furthermore, specificity of antibody response must be verified by appropriate controls because antibodies commonly cross-react with multiple fungi or fungal antigens [77] and sometimes even with bacteria that share common antigens [70]. For serodiagnosis of some mycoses (e.g., blastomycosis), newer and more specific reference sera are becoming available [33a]. Since whole organisms contain multiple antigens, heterogenous antibodies may be produced by the host in response to mycotic infections. Some of these antibodies may represent nonspecific evidence of exposure to the fungi, whereas others may correlate better with tissue-invasive infections. For example, crossed immunoelectrophoresis can be used to detect anti-*Candida* antibodies, but the technique gains specificity for diagnosis of invasive disease when concanavalin A is used to eliminate antimannan antibodies, leaving antibodies to cytoplasmic antigens of *Candida* [109]. Antigenic components of germinating *Candida* may differ from detectable antigens of yeast-phase organisms. In short, validity of data on antibody responses to mycoses hinges on the use of carefully standardized older procedures or meticulously controlled new ones which can be confirmed by older assays.

B. Subclasses of Antibody

1. Immunoglobulin M

As in infections caused by other types of organisms, exposure to some fungi causes an initial IgM antibody response which is later supplanted by production of specific IgG antibodies. An early IgM response might escape detection frequently because exposure to most of these fungi is chronic, whether or not an active (and often indolent) infection ever develops. Therefore, detection of IgM antibodies to infecting fungi may be a useful marker for some early, acute mycoses. However, there is little current evidence that such IgM antibodies directly help or hinder other host defense mechanisms against fungi. Nevertheless, it remains conceivable that interaction of IgM with complement components might affect host defenses by favoring generation of chemotactic factors (see Sec. III.A), by altering lymphocyte function, or by indirectly affecting other aspects of the immune system.

A specific IgM response may occur in response to different types and degrees of mycotic infections. For example, specific anti-*Candida* IgM antibodies have been detected in sera from women with a superficial (mucocutaneous) form of candidiasis, vaginitis [52,117], and with increased frequency in patients with allergic asthma [22a]. In deep, invasive mycoses which have a definable, early, acute phase, it sometimes has been possible to detect the early specific IgM antibodies, which later wane and disappear as the infections go on to clear or progress. This is most clearly seen in coccidioidomycosis with the usefulness of tube precipitation [100] or latex agglutination [40] tests in early infections. A similar pattern of antibody response appears to occur in histoplasmosis [5, 76] and may also occur in paracoccidioidomycosis [98]. In contrast, cryptococcosis is a chronic mycosis which lacks a definable clinical illness to mark the initial exposure. Therefore, it is probably not surprising that standard immunofluorescence techniques have failed to detect specific anticryptococcal IgM in large numbers of sera from patients with cryptococcosis [6,7]. That an early IgM response might occur in cryptococcosis is suggested by the production of anticryptococcal IgM antibodies by rabbits following immunization with whole cryptococci [47].

2. Immunoglobulin G

A large body of evidence documents formation of specific IgG antibodies in response to different types of exposure to a variety of fungi. However, this characteristic response may be blunted or even absent in some severely immunosuppressed hosts. It should also be noted that antibodies within the IgG subclass may have different functional activities and different specificities. Therefore, the types and sources of antigen used in determinations of antibodies affect the interpretations of results. As one example, IgG antibodies that are reactive against fungal surface antigens would be expected to be better opsonins than would antibodies to cytoplasmic antigens not on fungal surfaces. Most studies employ antigen preparations that contain multiple antigenic determinants; from this, it follows that IgG measured in some assays may include antibodies with multiple antigenic specificities which also may differ in functional activities.

The antibody response after differing degrees of exposure to fungi and of clinical illness is illustrated in aspergillosis. In allergic bronchopulmonary disease, *Aspergillus* does not invade host tissues but is chronically present in respiratory secretions. Here, presence of IgG precipitins to the colonizing organism is an almost uniform finding which may have an important role in the pathogenesis of the disease [31,48,110,114]. Similarly, colonization by *Aspergillus* stimulates a specific IgG antibody response when fungi grow within a pulmonary cavity in the form of a "fungus ball," without invasion of host tissues [30,49,121]. Invasive systemic aspergillosis is almost exclusively limited to the immunosuppressed host. In that situation, the formation of IgG antibodies to *Aspergillus* may be below detectable levels [121], or may be detectable only in low levels in some patients [49,96]. These low levels of antibodies may be due in part to neutralization by soluble *Aspergillus* antigens during infections. Such antigenemia has been demonstrated in an experimental model in rabbits [97], and may occur in human cases of invasive aspergillosis as well.

Virtually all normal human adults have circulating IgG antibodies directed against the mannan antigen which is a major constituent of the outer cell wall of *Candida* [117,119]. These serum antibody titers tend to be higher in association with locally invasive infections as in chronic mucocutaneous candidiasis [104], or in recurrent oral [52] or vaginal candidiasis [58,117], and IgG anti-*Candida* antibodies of maternal origin appear in the fetal circulation [60a]. IgG antibodies to *Candida albicans* are elevated in association with allergic respiratory disease, with differences in levels of the four IgG subclasses comparing patients with asthma versus those with rhinitis [22a]. In addition to serum, anti-*Candida* IgG is detectable in vaginal secretions of approximately 20-25% of women with or without active vaginal candidiasis or detectable colonization with *Candida* [64]. At least a proportion of these antibodies appear to be derived from the circulation [64]. In invasive, disseminated candidiasis, the level of specific IgG antibody response depends on the capacity of host immunological mechanisms. When serologic procedures are performed using carefully characterized reagents, it appears that disseminated candidiasis results in an early mannan antigenemia [119]. After a transition phase during which immune complexes may be present, antimannan antibody titers rise. This is almost entirely an IgG response, with subclasses IgG_1 and IgG_2 predominating [119]. Although antimannan antibodies may have some pathophysiologic significance, their presence in almost all normal sera minimizes their usefulness in serodiagnosis [58,119]. Antigens other than mannans have been detected in sera from mice with experimental disseminated candidiasis. Other workers suggest that antibodies to *Candida* cytoplasmic antigens provide more specific immunologic markers for disseminated candidiasis [111].

In cryptococcosis, as in candidiasis, antibodies form in response to a surface antigen of the organism. However, in cryptococcosis, the antigen is a part of the highly charged surface capsule, some of which may dissolve and circulate during infections. Circulating cryptococcal capsular polysaccharide may have a variety of effects on the host immune response, including the formation of specific antibodies. By immunofluorescence techniques, IgG anticryptococcal antibodies are detectable in sera of approximately 40% of patients with cryptococcosis [6,7,16]. In one study, no anticryptococcal antibodies were detectable by immunofluorescence in cerebrospinal fluids from several patients with cryptococcosis [6]. However, a different patient was found to have oligoclonal IgG antibodies to cryptococcal polysaccharide in cerebrospinal fluid detected by immunoelectrophoresis [85], and low, functional levels of anticryptococcal IgG appear to be present in spinal fluids from other patients with cryptococcal meningitis [19]. Local production of anticryptococcal IgG during infections may well occur. In any case, those patients with anticryptococcal IgG in serum have a better prognosis, i.e., a higher likelihood of being cured [6,16]. Anticryptococcal antibodies may have a beneficial role to play in the immune response (see Secs. II.C and III), or may represent a minor aspect of an active and successful immunological host response to infection. Although cellular host defense mechanisms are of central importance in immunity to cryptococcosis, there is some evidence in experimental animals that infusion of anticryptococcal antibodies provides some passive protection against cryptococcosis [24]. However, in a rabbit model of central nervous system cryptococcosis, preformed anticryptococcal antibodies had no apparent effect on disease progression [82a]. The fact that most cryptococcosis patients lack detectable levels of anticryptococcal antibodies may relate to a suppressive effect by the polysaccharide on antipolysaccharide antibody production by the host. Patients with high titers of circulating polysaccharide are less likely to have antipolysaccharide antibodies [6,16]. High doses of cryptococcal polysaccharide given to mice can markedly decrease or eliminate their ability to form antibodies in response to polysaccharide injected with Freund's complete adjuvant [48]. When mice are immunologically paralyzed by injection of a large dose of cryptococcal polysaccharide, antibody-forming cells are briefly demonstrable by a hemolytic plaque assay, then disappear [67]. The cryptococcal capsular polysaccharide might also shield other antigens from the host immunologic system. Cryptococcosis is, therefore, characterized by a minimal or absent specific IgG response, presumably because the surface polysaccharide of the yeast is weakly immunogenic and may even be capable of inducing immunological tolerance.

In some other mycoses, the initial encounter of host with fungus is characteristically marked by an acute, self-limited, often asymptomatic infection. Occasionally, infections fail to clear and progress to more chronic local or disseminated forms of mycoses. In coccidioidomycosis, the IgG response can be measured by immunodiffusion [40] or complement fixation tests [100]. Sequential titers of the latter provide invaluable parameters for judging severity and possible progression or ongoing dissemination. High serum complement fixation titers in serum are highly correlated with disseminated disease, and detection of complement-fixing antibodies in cerebrospinal fluid is diagnostic of coccidioidal meningitis [100]. When serum anticoccidioidal antibody titers are high without meningeal involvement, low levels of anticoccidioidal IgG may be detectable in concentrated cerebrospinal fluid by an immunodiffusion test. This presumably represents diffusion of antibodies from serum rather than local production [78]. The complement fixation test on unconcentrated cerebrospinal fluid is less sensitive in detecting IgG in this situation, so retains its serodiagnostic specificity and usefulness. In immunosuppressed patients who develop coccidioidomycosis or who have progressive disseminated infections, titers of anticoccidioidal IgG are usually detectable, although they may be low or absent in some patients [95].

As in coccidioidomycosis, the stage and degree of infection with *Paracoccidioides brasiliensis* is correlated with relative levels of IgG antibodies in serum [90]. Specific IgG antibodies to these organisms may be detected by a variety of techniques, including counterimmunoelectrophoresis, provided that careful controls are employed [71]. In acute histoplasmosis, a specific IgM response in serum is sometimes detectable by means of passive hemagglutination or complement fixation assays, but IgG antibodies predominate in most cases of histoplasmosis when these assays are used [5,34]. Assays for precipitating antibodies to *Histoplasma* antigens, including counterimmunoelectrophoresis, appear to detect the specific IgG antibody response to the organism [83]. In contrast to coccidioidomycosis and paracoccidioidomycosis, circulating IgG antibodies are more often undetectable in histoplasmosis. Even when present, changes in serum antibody titers may not correlate well with activity of the infection. While this might reflect differences in host responses to these organisms, other factors, such as antigen preparations and methods of procedure, could affect how efficiently antibodies are measured. In histoplasmosis, crude antigen preparations are commonly used in serodiagnosis: merthiolate-killed whole yeasts and mycelial phase culture filtrates. The latter are carefully standardized for the presence of the characteristic H and M antigens. It appears that H and M antigens are not on the yeast cell surface and that complement fixation results in release of antigens active in serodiagnostic tests [33]. Understanding and improvement in methods for detecting IgG antibodies to *Histoplasma* in serum depends on continuing research to characterize antigens and standardize techniques.

From a large body of circumstantial and direct evidence, it appears that host immunity to mycoses depends less on IgG antibody responses than on cell-mediated immune responses. For example, immunity to experimental histoplasmosis can be transferred between mice using whole T-lymphocytes or extracts of them; transfer of immunity is blocked by antibody against T-cells, whereas antiserum to IgG has no effect [113]. Evidence that antibodies may have positive effects on the host response to fungi are summarized later (see Secs. II.C and III). Rather than benefiting host defense mechanisms, it has been suggested that high levels of specific antifungal IgG antibodies may sometimes block or inhibit the development of cell-mediated immunity to mycoses. For example, it has been suggested that anti-*Histoplasma* antibodies might inhibit the lymphocyte proliferative response to antigens and mitogens in some patients with histoplasmosis; sera that contain such antibodies do inhibit lymphocyte transformation, but the degree of inhibition did not correlate well with the titer of antibodies [73]. Other workers have noted a better correlation and have suggested that suppression of T-cell responses, if not due to antibody alone, may be mediated by immune complexes [11a]. In some cases, there is a suggestion that antibodies may increase or decrease lymphocyte transformation, depending on the ratio of antigen to antibody [2]. In paracoccidioidomycosis, it has been suggested that antifungal antibodies may block antigenic sites and inhibit development of delayed hypersensitivity [61], but direct evidence for this concept is lacking. As in histoplasmosis, lymphocyte transformation responses to mitogens and specific antigens are often lower when autologous sera are used in cultures than when pooled normal sera are used [40]. However, the decrease in lymphocyte responsiveness does not correlate with antibody titers against *P. brasiliensis* [40,90]. Similarly, depressed parameters of cell-mediated immunity in disseminated coccidioidomycosis are most often associated with high titers of anticoccidioidal antibodies in serum, but specific IgG antibodies cannot be implicated directly as blocking agents [22]. However, it has been suggested that circulating immune complexes are present and may suppress immune responses [12]. A similar type of pattern has been described in chronic mucocutaneous candidiasis, where most patients have impaired cell-mediated immune responses to *Candida* antigens and to mitogens, while serum titers of anti-*Candida* precipitins are high [105]. As with other mycoses, this apparent reciprocal relationship in activity of antibody and cellular host defense mechanisms is not completely consistent and cannot be causally linked on the basis of currently available evidence. It has also been suggested that antigens common to *Candida*, and to ovary and thymus tissue, which could result in autoimmune-mediated defects in ovarian function and cell-mediated immunity [56a]. Antibodies might cause a different type of deleterious effect on cellular host responses in disseminated candidiasis. High titers of anti-*Candida* IgG from sera of patients with candidiasis do not alter ingestion of *Candida* by neutrophils in vitro, whereas intracellular killing of *Candida* (but not staphylococci) is markedly depressed [50]. Further evidence is required to determine the mechanism of this observation and its importance in vivo.

3. Immunoglobulin A

During systemic mycoses such as paracoccidioidomycosis [71] or candidiasis [58,119], specific antifungal antibodies of the IgA subclass are detectable in sera from some patients, although in lesser amounts than IgG antibodies of comparable specificities. Since IgA antibodies may be secreted locally onto mucous membranes in addition to circulating systemically, it is not surprising that the specific IgA antibody response has been studied most extensively in association with mucocutaneous colonization or localized tissue invasion by *Candida*.

There are significant disparities in results when some of the studies are compared with others. However, most of these differences are attributable to differences in the populations of subjects studied or, more often, to major methodologic differences in measurements of antibodies. For example, fractionation of sera may or may not be performed, subclasses of antibody may be determined by immunofluorescence using reagents of varying specificity, and different preparations of *Candida* antigens may be used. Allowing for these limitations, one may still observe several consistent features of the specific IgA response to *Candida*. Thus, approximately one-third of subjects with oral candidiasis [52] and perhaps all women with vaginal candidiasis [117] have IgA antibodies to *Candida* in their sera. However, comparable levels of such antibodies may be found in sera from subjects who are free from locally invasive candidiasis, whether or not their gastrointestinal or genital tracts appear to be colonized with *Candida* [117], as well as in sera from patients with allergic respiratory disease [22a].

In addition to serum, vaginal secretions from some women with *Candida* vaginitis may also contain anti-*Candida* IgA antibodies [58,64,116]. Detection of secretory components in some of these vaginal secretions [57] may indicate the local production of anti-*Candida* secretory IgA, some of which may reach the systemic circulation [117]. However, this may not represent a specific local antibody response to active candidiasis, since comparable local levels of IgA anti-*Candida* antibodies are detectable in vaginal secretions from women without *Candida* vaginitis [64]. In addition, specific IgG antibodies are detectable in vaginal secretions more often than those of the IgA subclass [58,64,116], and the secretory form of IgA cannot be detected in sera [117]. Therefore, it appears that at least a proportion of the antibodies found in vaginal secretions are derived from the circulation [64].

Production of anti-*Candida* IgA antibodies as well as antibodies of other subclasses may be influenced by serum levels of steroid sex hormones [60]. In normal nonpregnant women, serum levels of anti-*Candida* IgA vary with changes in hormone levels during the menstrual cycle or with administration of oral progestins. Titers of specific IgA antibodies against *Candida* rise in association with increasing levels of progesterone or low levels of estradiol, and fall when levels of estradiol are most elevated. There is no comparable change in total serum immunoglobulin concentrations, nor in titers of specific antibodies against herpesvirus or sheep red blood cells [60].

Deficiencies in serum or salivary IgA anti-*Candida* antibodies have been noted in some patients with chronic mucocutaneous candidiasis [54]. However, IgA deficiency does not appear to predispose to more severe forms of invasive candidiasis [57], nor to vaginal candidiasis [36]. IgA obtained from bronchial lavage of rabbits could agglutinate *C. albicans* [50a]. Further studies are required to determine whether or not specific IgA antibodies have any direct functional role in host defense mechanisms against mycoses.

4. Immunoglobulin E

Specific antifungal antibodies of the immunoglobulin E subclass are detectable in serum in association with exposure to or infection with several fungi, including some dermatophytes [43], *Candida* [58], *Coccidioides immitis* [11b], *Aspergillus* [13,42,43,93], and possibly *Histoplasma* as well [11c]. In allergic bronchopulmonary aspergillosis, as well as in other allergic reactions to inhalation of fungi, high titers of IgE antifungal antibodies occur in serum [93]. Levels of these antibodies may correlate with activity of the disease [42] and may contribute to the pathogenesis of the pulmonary lesions in humans and experimental animals [99]. Patients with cystic fibrosis without apparent manifestations of allergic aspergillosis are also more likely to have serum IgE antibodies against *Aspergillus* than are normal or asthmatic subjects [26]. However, this may reflect increased antigen exposure due to alterations in mucosal surfaces and pulmonary clearance mechanisms, and need not indicate that IgE contributes to ongoing damage to the lungs. The elevated antibody levels are not specific for *Aspergillus,* as approximately three-fourths of patients with allergic bronchopulmonary aspergillosis (or allergic asthma or rhinitis) have high serum titers of anti-*Candida* IgE as well [22a]. In addition to a potential role in mediating immediate hypersensitivity reactions, IgE antibodies have been observed in chronic diseases in association with depressed cell-mediated immunity. This situation may occur in some patients with chronic dermatophyte infections [43] or disseminated coccidioidomycosis [22]. However, other than the observed association, there is no direct evidence to support a causal relationship between IgE antibodies and suppression of host cell-mediated immunity to mycoses. In some patients with the "hyper-IgE" recurrent infection syndrome, IgE antibodies to *Candida* may interact with antigen and mast cells, resulting in release of inhibitors of leukocyte chemotaxis; this interference with the normal acute inflammatory response may predispose to infections [5a].

C. Interaction of Antibody with Fungi

1. Direct Damage to Fungi

Passive transfer of specific antifungal antibodies to experimental animals may provide some protection against infections with some mycoses, including cryptococcosis [24] and candidiasis [81]. However, such protective effects appear to be mediated indirectly through augmentation of other defense mechanisms. In vitro assays using a variety of different fungi have verified that factors present in serum may directly damage fungi, but damage appears to be due to factors other than specific antibodies (see Secs. IV and V). For example, even when active complement components are present, there is no direct damage to *Cryptococcus neoformans* by anticryptococcal antibodies in human sera [17], although the antibodies bind to organisms and cause a physical change in the surface polysaccharide, or "capsular reaction" [72]. When placed in human serum, *Candida* blastoconidia form pseudohyphae, germinate, and appear to proliferate actively [53]. However, high titers of IgG antibodies against *Candida* in serum may cause decreased filamentation of blastoconidia, associated with depression of fungal respiration [32]. This represents a depression in growth and metabolic activity of the tissue-invasive form of the organism. Therefore, specific antibodies might directly damage some fungi, but current evidence suggests that this is an uncommon event.

2. Opsonization

Specific IgG antibodies against fungi have the potential to combine with cellular mechanisms in different ways to augment host defenses against mycoses. Of all of the possible functional pathways for IgG, the one likely to be of greatest value to the host involves opsonization, i.e., specific coating of surfaces of fungal particles which enhances their attachment to and ingestion by host phagocytic cells. In invasive mycoses, destruction of fungi within the host appears largely attributable to the function of phagocytic cells, initially neutrophils, monocytes, and macrophages, later supplemented by "immune" or stimulated macrophages. Most of the evidence which directly supports an opsonic role in this process for antifungal IgG antibodies is derived from in vitro studies.

When sera from patients with paracoccidioidomycosis are used to opsonize yeasts in vitro, specific antibodies present in the sera appear to increase ingestion of *P. brasiliensis* but not *Rhodotorula* yeasts by human neutrophils [91]. Sera from dogs with high titer complement-fixing antibodies against *Coccidioides immitis* enhance phagocytosis of arthroconidia by neutrophils from the animals [118]. Endospores, a growth phase of the organism which appears during tissue infections in vivo, can be ingested by phagocytic cells, but it is not known whether or not spores survive within host cells. Human and animal neutrophils and monocytes require serum factors for ingestion and killing of *Histoplasma* yeasts [37, 39]. Specific antibodies may promote phagocytosis of this organism by some cell types; however, ingestion and digestion of *H. capsulatum* yeasts by "immune" macrophages from mice appear to be independent of added or circulating anti-*Histoplasma* antibodies [35].

Serum is required for phagocytosis of cryptococci by human neutrophils. However, this process is not enhanced by sera from normal subjects in whom anticryptococcal antibodies had been raised by skin testing with cryptococcin [17]. Even so, other studies indicate that normal human sera contain low levels of IgG antibodies against cryptococci, too low to be detected by standard immunofluorescence procedures used above, but enough to enhance the rate of phagocytosis by neutrophils [19]. An opsonic role of anticryptococcal antibodies is supported by data from experimental cryptococcosis in mice. Mouse neutrophils ingest small capsule forms of *Cryptococcus neoformans* only in the presence of hyperimmune serum, although specific antibodies may not be required for ingestion of the yeasts by histiocytes [25]. Data from other studies suggest that serum IgG alone facilitates attachment of mouse macrophages to unencapsulated cryptococci and triggers ingestion of attached yeasts by macrophages [47,48a]. The presence of a capsule or addition of capsular polysaccharide masks the opsonic activity of antibodies to cell walls, preventing opsonization as well as agglutination by antibodies to IgG [60b]. However, rabbit antisera that contain specific antibodies to cryptococcal capsular polysaccharide are opsonic for encapsulated cryptococci [23a]. Cryptococcal capsules are often large enough to preclude ingestion by phagocytic cells. Instead, groups of monocytes in the presence of serum opsonins may surround and damage the yeasts. Ring formation by mouse mononuclear cells requires anticryptococcal antibodies, but rabbit or guinea pig monocytes only appear to require heat-labile opsonins [2]. However, "ring" formation by monocytes shares features in common with phagocytosis, so that low levels of specific IgG might affect the kinetics of both processes.

Although decomplemented human sera that contain anti-*Candida* IgG antibodies will opsonize *Candida* yeasts for phagocytosis by human neutrophils, complement activation accounts for most of serum opsonic activity for this organism [104]. However, specific antibodies may be necessary to provide optimum kinetics of phagocytosis [92]. Unlike neutrophils, specific IgG antibodies are opsonic, whereas complement components do not appear to be necessary for phagocytosis of yeast forms of *Candida* by human monocytes [9]. Patients who have recurrent candidiasis do not have documented deficiencies in anti-*Candida* antibodies, although humoral factors other than antibodies might interfere with opsonization of *Candida* and cause defects in phagocytosis [115].

When *Candida* invades host tissues, pseudohyphal and hyphal forms develop which are too large to be completely ingested by phagocytic cells. At least in vitro, neutrophils can attach to these large organisms, damaging (and probably killing) them. Serum opsonins are not required for this to occur, but anti-*Candida* IgG antibodies enhance the process, whereas complement components do not [20]. Neutrophils and IgG antibodies interact comparably with hyphae of other potentially invasive organisms, *Aspergillus fumigatus* and *Rhizopus oryzae* [21].

3. Antibody-Dependent Cell-Mediated Damage to Fungi

Antibodies, primarily of the IgG subclass, can coat large target cells and render them vulnerable to lysis by host effector cells. Such an antibody-dependent cell-mediated cytotoxic mechanism can damage a wide variety of cells, including normal and neoplastic mammalian cell lines and parasites (e.g., schistosomula). Several types of host effector cells are capable of cooperation with antibodies to cause this type of damage, including K lymphoid cells (which differ from typical T- or B-lymphocytes), monocytes, macrophages, neutrophils, and eosinophils. Only in cryptococcosis is there even preliminary in vitro evidence that such a mechanism may be active in host defenses against mycoses. In the presence of low dilutions of IgG anticryptococcal antibodies of rabbit or human origin, human peripheral blood leukocytes kill *Cryptococcus neoformans* in vitro by a nonphagocytic mechanism [14]. Damage requires a high ratio of leukocytes to yeasts (10:1 or more). Fungi may be damaged by leukocyte preparations that are rich in neutrophils, monocytes, or nonphagocytic lymphoid cells, but not by purified preparations of T-cells [15]. Several features distinguish this process from monocyte ring formation [2] described above, including the IgG requirement and the diversity of effector cells, which can damage cryptococci. The relative importance of this process as a defense mechanism in the intact host remains to be established. Although indirect evidence suggests that IgG anticryptococcal antibodies may be beneficial to the host (see Sec. II.B.2), the high concentration of cryptococcal polysaccharide circulating during infections is likely to block antibody effects on whole organisms. It remains possible that anticryptococcal antibodies are important where polysaccharide concentrations are low. This might occur during early or late (clearing) infections, as well as in localized sites. For example, low levels of anticryptococcal antibodies are detectable in cerebrospinal fluid from some patients with cryptococcal meningitis, but the low levels of complement components are inadequate to generate opsonins or factors which are chemotactic for phagocytic cells [19].

III. COMPLEMENT

A. Classical Pathway and Early Components

In common with other antigen-antibody complexes, antifungal antibodies with whole fungi or fungal products may activate the classical complement pathway. Early complement components (C1, C4, and C2) then form a complex which can activate the third component of complement and later components (C5-C9) to generate opsonic and chemotactic factors. Such activation may be mediated by soluble substances released in vitro, e.g., by *Coccidioides immitis* mycelia or spherules [29]. In vivo, there is also the potential for classical complement pathway activation by soluble fungal products, for example, mannan-antibody complexes are probably present in sera of at least some patients with disseminated candidiasis [119], and antigen-antibody complexes are likely to form in other mycoses as well. Soluble fungal products also may, at times, be chemotactic by themselves, without interacting with the complement system [12a].

Because the processes exist for activation of the alternative complement pathway (see Sec. III.B), activation of the classical complement pathway appears likely to be of secondary importance in host defenses against mycoses. In cryptococcosis, intact yeasts or soluble capsular polysaccharide in sera may activate the classical complement pathway [19]. In the generation of opsonins for phagocytosis of cryptococci by neutrophils, early (classical pathway) complement components are required for optimum kinetics of ingestion. However, phagocytosis of cryptococci is depressed using sera depleted of properdin factor B, a component of the alternative pathway of complement activation. Therefore, early and late components of the classical complement pathway do not appear to be adequate to mediate opsonization of cryptococci [19]. When human neutrophils and blastoconidia of *Candida* are used, opsonization appears to depend primarily on activation of the classical and alternative complement pathways [104]. Phagocytosis of *Candida* by neutrophils may appear normal when sera are used which have been depleted of the C4 component of complement [44]. However, careful studies using guinea pig sera deficient in C4 establish that an intact classical complement pathway is required for optimum kinetics of phagocytosis of *Candida* by human neutrophils [92]. Thus, improvement in the early rate of opsonization of fungi appears to be the major effect of classical complement pathway activation. It is not known whether or not this phenomenon is important to host defense mechanisms in vivo.

B. Alternative Pathway and Late Components

The alternative complement pathway may be activated directly by components of fungi in the absence of antibodies. A variety of fungi and fungal products have the capacity to do this and, in turn, generate potentially opsonic and chemotactic factors from late complement components. For example, some cases of hypersensitivity pneumonitis are related to inhalation of fungi, including *Micropolyspora faeni* and several species of *Aspergillus*. Extracts of these fungi activate factor B of the alternative complement pathway and consume late complement components in the absence of specific antibodies to the fungi. Other fungi (*Mucor racemosis, Hormodendren* spp., *Penicillium*) do not activate complement [56]. This and other work suggests that alternative pathway activation might contribute to the pathophysiology of hypersensitivity pneumonitis. Soluble substances from mycelia or spherules of *Coccidioides immitis* also activate the alternative pathway [29]. Direct complement activation by these coccidioidal extracts results in the generation of chemotactic factors from sera from unexposed normal subjects, i.e., those with no detectable anticoccidioidal antibodies and negative coccidioidin skin tests [28]. If this occurs in vivo, it would favor accumulation of neutrophils and other inflammatory cells at foci of infection. Cryptococci and cryptococcal capsular polysaccharide can activate the alternative complement pathway in human serum to generate opsonic and chemotactic factors. One reason for the localization of cryptococcosis to the central nervous system may be the inability to activate the alternative pathway in cerebrospinal fluid because of inadequate local concentrations of complement components [19]. Studies of experimental cryptococcosis in guinea pigs suggest that late complement components are critical in clearing cryptococci from extraneural foci of infection, but established infections in the central nervous system progress because local complement levels are too low to aid in destruction [18]. For phagocytosis of *Cryptococcus neoformans* by murine macrophages, heat-labile opsonins may be necessary for optimum kinetics [47], although this may vary depending on whether macrophages are stimulated prior to harvesting [110]. High concentrations of cryptococcal capsular polysaccharide can inhibit phagocytosis of cryptococci by mouse macrophages or by phagocytic cells from human blood in vitro. However, these antiphagocytic effects of the polysaccharide cannot be attributed to depletion or inactivation of heat-labile opsonins [110]. Cryptococcal polysaccharide does have the capacity to deplete opsonic complement components from sera of human patients, but this is restricted only to those unusual cases of cryptococcosis which are extensive enough to cause fungemia [55].

In opsonization of *Candida*, complement activation appears to play a greater role in phagocytosis by human neutrophils [52,104] than by monocytes [9]. When the cellular events of phagocytosis and killing by neutrophils are measured separately, it appears that the C3 component of complement is required for and participates directly in intracellular candidacidal activity [120]. When soluble surface mannans from *Candida* are added to fresh normal human sera, the alternative complement pathway is activated, as evidenced by conversion of properdin factor B and activation of C3 [119]. Some patients with candidiasis have circulating mannans, with low levels of C3 and evidence of alternative pathway activation in sera [119]. In addition to generation of opsonins, local complement activation can generate chemotactic factors at infectious foci. By themselves, neither whole *Candida* organisms nor culture supernatants are chemotactic, but *Candida* activation of the alternative pathway in serum generates factors that are chemotactic for neutrophils [86]. Complement-mediated accumulation of neutrophils in lesions appears to occur in vivo in experimental cutaneous candidiasis in guinea pigs [103]. Furthermore, lesions from human patients with chronic mucocutaneous candidiasis often contain deposits of properdin or C3, but not immunoglobulins or C4 [102]. It is suggested that generation of chemotactic factors might be responsible for the intense local inflammatory response seen in this disease; the cellular response may prevent deeper spread of the infection but is insufficient to clear all fungi from local lesions. The importance of an intact alternative complement pathway is illustrated by an experimental model of disseminated candidiasis in the guinea pig. As in experimental cryptococcosis, animals depleted of late complement components do not survive *Candida* sepsis, whereas C4-deficient guinea pigs with intact alternative pathways and late components survive infections as well as normal control animals [30].

Thus, there is strong evidence that complement activation can contribute to the pathogenesis of lesions in some mycoses, as well as to generation of chemotactic and opsonic factors essential to host defenses. In most studies of fungi and other organisms, the C3b fragment of C3 appears to be responsible for most, if not all, complement-dependent opsonic activity. However, a potential role for C5 in opsonization of yeasts is suggested by decreased phagocytosis of *Saccharomyces cerevisiae* when the yeasts are opsonized by C5-deficient murine or human sera [62], or by human sera that become deficient in C5a during storage. Purified C5a restores opsonic activity to normal [63]. However, the importance of these observations must be questioned because *Saccharomyces* are nonpathogenic yeasts. In addition, other investigators observe normal opsonization of both *S. cervisiae* and *C. albicans* by human sera deficient in C5 [93]. Complement component 5 may play a role in opsonization, but perhaps only in binding particles with limited amounts of surface C3b to neutrophils [23b].

IV. IRON-BINDING COMPOUNDS

The ability to acquire iron within a host is often an important determinant of both virulence and the nature of infections caused by microorganisms [80]. Acquisition of iron by organisms can be limited by unsaturated iron-binding proteins in the host. One such compound, transferrin, is present in serum. An even more potent iron-binding compound, lactoferrin, occurs in saliva and breast milk, as well as within neutrophil granules which may be released at foci of infection.

An increased incidence of candidiasis in patients with acute leukemia is correlated with the high concentration of iron and nearly complete saturation of transferrin in these patients' sera [8]. In vitro, normal serum inhibits growth of *Candida,* whereas *Candida* grows well in hyperferremic leukemic serum. Addition of iron reverses the growth-inhibitory effects of the former, whereas transferrin restores inhibitory activity to the latter [8]. Like transferrin, lactoferrin inhibits growth of *Candida* in vitro [46]. The concentration of iron-binding substances may be altered by other agents. For example, survival of mice infected with *C. albicans* is improved by prior administration of endotoxin. Administration of iron negates this protective effect, suggesting that endotoxin effects on these infections are mediated by variations in available unsaturated iron-binding compounds [23]. Similarly, growth-inhibitory effects by transferrin have been noted for a variety of fungi, including *Torulopsis glabrata* [75], *Histoplasma capsulatum* yeasts [108], and several dermatophytes, including species of *Trichophyton, Microsporum,* and *Epidermophyton* [45].

V. OTHER HUMORAL FACTORS

A variety of other factors have the potential to make significant contributions to host defenses against mycoses. For example, human saliva contains a heat-stable, dialyzable substance which damages cryptococci. The substance differs from inhibitory substances found in serum and is not amylase or lysozyme [41]. One potential mechanism that might be operative is analogous to intracellular microbicidal activity of leukocytes, i.e., combination of hydrogen peroxide from oral microbial flora with a peroxidase and a halide or thiocyanate.

Fungal growth may be affected by several substances present in normal human serum. A heat-stable serum factor differing from immunoglobulin or complement inhibits growth of *R. oryzae* [27], although transferrin might account for some of this effect. A factor that clumps *Candida albicans* is present in sera from normal adults and children. The presence of anti-*Candida* IgG antibodies in sera appears to reduce clumping activity in sera from patients with a variety of diseases and eliminates activity in sera from patients with mucocutaneous or systemic candidiasis [54]. At least in rabbit sera, clumping activity is due in part to a macroeuglobulin of fast beta mobility [101]. Serum factors more directly kill or inhibit growth of *Cryptococcus neoformans.* This activity is separable from the inhibitory effects of transferrin and differs from immunoglobulin, complement, or properdin [41]. Iron-independent anticryptococcal activity may be due to more than one factor present in alpha$_2$ and gamma fractions of serum [89]. One active substance is heat stable, nondialyzable, and trypsin sensitive. However, all five Cohn fractions of serum stimulate growth of cryptococci; inhibitory activity might reside in a beta-globulin lipoprotein [41]. No such inhibitory substances are found in cerebrospinal fluid, which may add to other mechanisms that favor localization of cryptococcosis to the central nervous system [41]. These or other inhibitory substances may be responsible for a dramatic serum fungicidal effect capable of killing over 60% of cryptococcal inocula during 4-h incubations in vitro [111]. However, some isolates of *C. neoformans* are completely resistant to fungicidal effects of serum [17].

Nonspecific substances in serum may also influence fungal growth. Lysozyme may decrease the rate of growth of *Coccidioides immitis* spherules in vitro, probably due to binding and damage to cell membranes [11]. As with bacteria, other surface-active agents may enhance damage by lysozyme. There is increased killing of *Candida albicans* and of *C. immitis* mycelia or spherules when amphotericin and lysozyme are combined in vitro [10]. Estrogens, especially estradiol in physiologic concentrations (1 μg/ml), suppress the growth of cryptococci in vitro [65] and lower the concentrations of amphotericin B necessary to kill the fungi [66]. Estrogens also bind to neutrophils during phagocytosis and may possibly influence the microbicidal activity of the myeloperoxidase system by competing for available H_2O_2 or by preventing a fall in activity induced by high levels of H_2O_2 [46a]. At least in vitro, *S. cerevisiae* and *Candida albicans* are inhibited by lymphokines produced after phytohemmagglutinin stimulation of murine lymphocytes [82].

In addition to direct effects on the fungi, soluble factors may be produced in response to mycoses that inhibit cell-mediated immunity, as judged by in vitro assays such as lymphocyte transformation and production of migration inhibitory factor. Such inhibitors are not immunoglobulin and are found in some patients with chronic mucocutaneous candidiasis [79], paracoccidioidomycosis [69], and other mycoses. In some patients who have disseminated mycoses with depressed parameters of cell-mediated immunity, host macrophages produce a soluble substance that acts to generate or potentiate suppressor T-lymphocytes [106,107]. Such soluble mediators may prove to influence profoundly the development of cellular defense mechanisms against mycoses in the intact host.

VI. CONCLUSIONS

Several humoral mechanisms have at least theoretical capabilities for enhancement or suppression of host defenses against mycoses. In the intact host, the relative importance of these humoral responses to fungi is difficult to establish with certainty. Nevertheless, evidence for some of these processes, such as complement activation, strongly favors a major role in overall host responses to fungi. While production of antifungal antibodies remains an important marker for active mycoses, interaction of antibodies with other host defense mechanisms may prove to be of even greater significance. Other humoral substances that directly inhibit fungal growth might provide inhospitable environments for fungi which favor successful fungicidal effects by other host defense mechanisms. Ultimately, further evidence will allow accurate classification of the relative values for the intact host of all these humoral responses to fungi. In any case, if cellular host responses are critical in controlling most mycoses, humoral factors that can modulate cellular immunity are also likely to be important. It seems safe to assume that host responses to mycoses involve humoral defense mechanisms which have complex interactions with cellular mechanisms, sometimes to the ultimate benefit and sometimes to the detriment of the host.

REFERENCES

1. Aronson, M., and J. Kletter. 1973. Aspects of the defense against a large-sized parasite, the yeast, *Cryptococcus neoformans*. In A. Zuckerman and D. W. Weiss (Eds.), *Dynamic Aspects of Host-Parasite Relationships*, Vol. 1, Academic, New York, pp. 132-162.
2. Alford, R. H., and R. A. Goodwin. 1972. Patterns of immune response in chronic pulmonary histoplasmosis. J. Infect. Dis. *125*: 269-275.
3. Ambroise-Thomas, P., P. T. Desgeorges, and D. Monget. 1978. Diagnostic immunoenzymologique (ELISA) des maladies parasitaires par un microméthode modifiée. 2. Résultats pour la toxoplasmose, l'amibiase, la trichinose, l'hydatidose et l'aspergillose. Bull. W.H.O. *56*: 797-804.
4. Axelson, N. H. 1976. Analysis of human *Candida* precipitins by quantitative immunoelectrophoresis: a model for analysis of complex microbial antigen-antibody systems. Scand J. Immunol. *5*: 177-190.
5. Baum, G. L., T. M. Daniel, and E. H. Rice. 1973. Immune globulin characterization of complement fixing antibodies in histoplasmic infections. Chest *64*: 16-21.
5a. Berger, M., C. H. Kirkpatrick, and J. I. Gallin. 1980. Pathogenic role of anti-*Candida* and anti-staphylococcal IgE in patients with the hyper IgE-recurrent infection syndrome. Clin. Res. *28*: 363A.
6. Bindschadler, D. D., and J. E. Bennett. 1968. Serology of human cryptococcosis. Ann. Intern. Med. *69*: 45-52.
7. Blumer, S. O., and L. Kaufman. 1977. Characterization of immunoglobulin classes of human antibodies to *Cryptococcus neoformans*. Mycopathologia *61*: 55-60.
8. Caroline, L., F. Rosner, and P. J. Kozinn. 1969. Elevated serum iron, low unbound transferrin and candidiasis in acute leukemia. Blood *34*: 441-451.
9. Cline, M. J., and R. I. Lehrer. 1968. Phagocytosis by human monocytes. Blood *32*: 423-434.
10. Collins, M. S., and D. Pappagianis. 1974. Lysozyme-enhanced killing of *Candida albicans* and *Coccidioides immitis* by amphotericin B. Sabouraudia *12*: 329-340.
11. Collins, M. S., and D. Pappagianis. 1974. Inhibition by lysozyme of growth of the spherule phase of *Coccidioides immitis* in vitro. Infect. Immun. *10*: 616-623.

11a. Cox, R. A. 1979. Immunologic studies of patients with histoplasmosis. Am. Rev. Respir. Dis. *120*: 143-149.

11b. Cox, R. A., and D. R. Arnold. 1979. Immunoglobulin E in coccidioidomycosis. J. Immunol. *123*: 194-200.

11c. Cox, R. A., and D. R. Arnold. 1980. Immunoglobulin E in histoplasmosis. Infect. Immun. *29*: 290-293.

12. Cox, R. A., R. M. Pope, and S. Yoshinoya. 1979. Immune complexes in sera of patients with coccidioidomycosis. Abstr. Annu. Meet. Am. Soc. Microbiol., 1979, F33, p. 368.

12a. Cutler, J. E. 1977. Chemotactic factor produced by *Candida albicans*. Infect. Immun. *18*: 568-573.

13. Dessaint, J. P., D. Bout, J. Fruit, and A. Capron. 1976. Serum concentration of specific IgE antibody against *Aspergillus fumigatus* and identification of the fungal allergen. Clin. Immunol. Immunopathol. *5*: 314-319.

14. Diamond, R. D. 1974. Antibody-dependent killing of *Cryptococcus neoformans* by human peripheral blood mononuclear cells. Nature (Lond.) *247*: 148-150.

15. Diamond, R. D., and A. C. Allison. 1976. Nature of the effector cells responsible for antibody-dependent cell-mediated killing of *Cryptococcus neoformans*. Infect. Immun. *14*: 716-720.

16. Diamond, R. D., and J. E. Bennett. 1974. Prognostic factors in cryptococcal meningitis: a study of 111 cases. Ann. Intern. Med. *80*: 176-181.

17. Diamond, R. D., R. K. Root, and J. E. Bennett. 1972. Factors influencing killing of *Cryptococcus neoformans* by human leukocytes in vitro. J. Infect. Dis. *125*: 367-376.

18. Diamond, R. D., J. E. May, M. Kane, M. M. Frank, and J. E. Bennett. 1973. The role of late complement components and the alternate complement pathway in experimental cryptococcosis. Proc. Soc. Exp. Biol. Med. *144*: 312-315.

19. Diamond, R. D., J. E. May, M. A. Kane, M. M. Frank, and J. E. Bennett. 1974. The role of the classical and alternate complement pathways in host defenses against *Cryptococcus neoformans* infection. J. Immunol. *112*: 2260-2270.

20. Diamond, R. D., R. Krzesicki, and W. Jao. 1978. Damage to pseudohyphal forms of *Candida albicans* by neutrophils in the absence of serum in vitro. J. Clin. Invest. *61*: 349-359.

21. Diamond, R. D., R. Krzesicki, B. Epstein, and W. Jao. 1978. Damage to hyphal forms of fungi by human leukocytes in vitro. Am. J. Pathol. *91*: 313-324.

22. Drutz, D. J., and A. Catanzaro. 1978. Coccidioidomycosis. I. Am. Rev. Respir. Dis. *117*: 559-585.

22a. Edge, G., and J. Pepys. 1980. Antibodies in different immunoglobulin classes to *Candida albicans* in allergic respiratory disease. Clin. Allergy *10*: 45-58.

23. Elin, R. J., and S. M. Wolff. 1974. The role of iron to non-specific resistance induced by endotoxin. J. Immunol. *112*: 737-745.

23a. Follette, J. L., and T. R. Kozel. 1980. Opsonization of *Cryptococcus neoformans* by anticapsular antibodies. Abstr. Annu. Meet. Am. Soc. Microbiol., 1980, F14, p. 321.

23b. Frank, M. M. 1980. Complement receptors: chemistry and biological activity. Fed. Proc. *39*: 361.

24. Gadebusch, H. H. 1958. Passive immunization against *Cryptococcus neoformans*. Proc. Soc. Exp. Biol. Med. *98*: 611-614.

25. Gadebusch, H. H. 1972. Mechanisms of native and acquired resistance to infection with *Cryptococcus neoformans*. Crit. Rev. Microbiol. *1*: 311-320.

26. Galant, S. P., R. W. Rucker, C. E. Groncy, I. D. Wells, and H. S. Novey. 1976. Inci-

dence of serum antibodies to several *Aspergillus* species and to *Candida albicans* in cystic fibrosis. Am. Rev. Respir. Dis. *114*: 325-331.

27. Gale, G. R., and A. M. Welch. 1961. Studies of opportunistic fungi. I. Inhibition of *Rhizopus oryzae* by human serum. Am. J. Med. Sci. *241*: 604-612.

28. Galgiani, J. N., R. A. Isenberg, and D. A. Stevens. 1978. Chemotaxigenic activity of extracts from the mycelial and spherule phases of *Coccidioides immitis* for human polymorphonuclear leukocytes. Infect. Immun. *21*: 862-865.

29. Galgiani, J. N., P. Yam, L. D. Petz, P. L. Williams, and D. A. Stevens. 1980. Complement activation by *Coccidioides immitis:* in vitro and clinical studies. Infect. Immun. *28*: 944-949.

30. Gelfand, J. A., D. L. Hurley, A. S. Fauci, and M. M. Frank. 1978. Role of complement in host defense against experimental disseminated candidiasis. J. Infect. Dis. *138*: 9-16.

30a. Gold, J. W. M., B. Fisher, B. Yu, N. Chein, and D. Armstrong. 1980. Diagnosis of invasive aspergillosis by passive hemagglutination assay of antibody. J. Infect. Dis. *142*: 87-94.

31. Gordon, M. A., E. W. Lapa, and J. Kane. 1977. Modified indirect fluorescent-antibody test for aspergillosis. J. Clin. Microbiol. *6*: 161-165.

32. Grappel, S. F., and R. A. Calderone. 1976. Effect of antibodies on the respiration and morphology of *Candida albicans*. Sabouraudia *14*: 51-60.

33. Green, J. H., W. K. Harrell, S. P. Gray, J. E. Johnson, R. C. Bolin, H. Gross, and G. B. Malcolm. 1976. H and M antigens of *Histoplasma capsulatum:* preparation of antisera and location of these antigens in yeast cells. Infect. Immun. *14*: 826-831.

33a. Green, J. H., W. K. Harrell, J. E. Johnson, and R. Benson. 1979. Preparation of reference antisera for laboratory diagnosis of blastomycosis. J. Clin. Microbiol. *10*: 1-7.

34. Hermans, P. E., and H. Markowitz. 1969. Serum antibody activity in patients with histoplasmosis as measured by passive hemagglutination. J. Lab. Clin. Med. *74*: 453-463.

35. Hill, G. A., and S. Marcus. 1960. Study of cellular mechanisms in resistance to systemic *Histoplasma capsulatum* infection. J. Immunol. *85*: 6-13.

35a. Ho, Y. M., M. H. Ng, and C. T. Huang. 1979. Antibodies to germinating and yeast cells of *Candida albicans* in human and rabbit sera. J. Clin. Pathol. *32*: 399-405.

36. Hobbs, J. R., D. Brigden, F. Davidson, M. Kahan, and J. K. Oates. 1977. Immunological aspects of candidal vaginitis. Proc. R. Soc. Med. *70*(Suppl. 4): 11-13.

37. Holland, P. 1971. Circulating human phagocytes and *Histoplasma capsulatum:* ultrastructural observations. In L. Ajello, E. W. Chick, and M. L. Furcolow (Eds.), *Histoplasmosis,* Proceedings of the Second National Conference, Thomas, Springfield, Ill., pp. 580-583.

38. Holmberg, K., M. Berdischewsky, and L. S. Young. 1980. Serological immunodiagnosis of invasive aspergillosis. J. Infect. Dis. *141*: 656-664.

39. Howard, D. H. 1973. Fate of *Histoplasma capsulatum* in guinea pig polymorphonuclear leukocytes. Infect. Immun. *8*: 412-419.

40. Huppert, M., E. T. Peterson, and S. H. Sun. 1968. Evaluation of a latex particle agglutination test for coccidioidomycosis. Am. J. Clin. Pathol. *49*: 96-102.

41. Igel, H. J., and R. P. Bolande. 1966. Humoral defense mechanisms in cryptococcosis: substances in normal human serum, saliva, and cerebrospinal fluid affecting the growth of *Cryptococcus neoformans*. J. Infect. Dis. *116*: 75-83.

42. Imbeau, S. A., D. Nichols, D. Flaherty, H. Dickie, and C. Reed. 1978. Relationships between prednisone therapy, disease activity, and the total serum IgE level in allergic bronchopulmonary aspergillosis. J. Allergy Clin. Immunol. *62*: 91-95.

43. Jones, H. E., J. H. Reinhardt, and M. G. Rinaldi. 1974. Immunologic susceptibility to chronic dermatophytosis. Arch. Dermatol. *110*: 213-220.

44. Kernbaum, S. 1975. Pouvoirs phagocytaire et fongicide envers *Candida albicans* des polynucléaires neutrophiles humains en présence de sérum dépourvis de C3 et C4. Ann. Microbiol. (Inst. Pasteur) *126A*: 75-81.

45. King, R. D., H. A. Kahn, J. C. Foye, J. H. Greenberg, and H. E. Jones. 1975. Transferrin, iron, and dermatophytes. I. Serum dermatophyte inhibitory component definitely identified as unsaturated transferrin. J. Lab. Clin. Med. *86*: 204-212.

46. Kirkpatrick, C. H., I. Green, R. R. Rich, and A. L. Schade. 1971. Inhibition of growth of *Candida albicans* by iron-unsaturated lactoferrin: relation to host-defense mechanisms in chronic mucocutaneous candidiasis. J. Infect. Dis. *124*: 539-544.

46a. Klebanoff, S. J. 1979. Effect of estrogens on the myeloperoxidase-mediated anitmicrobial system. Infect. Immun. *25*: 153-156.

47. Kozel, R. T., and R. P. Mastroianni. 1976. Inhibition of phagocytosis by cryptococcal polysaccharide: dissociation of the attachment and ingestion phases of phagocytosis. Infect. Immun. *14*: 62-67.

48. Kozel, T. R., W. F. Gulley, and J. Cazin, Jr. 1977. Immune response to *Cryptococcus neoformans* soluble polysaccharide: immunological unresponsivenss. Infect. Immun. *18*: 701-707.

48a. Kozel, T. R., and T. G. McGaw. 1979. Opsonization of *Cryptococcus neoformans* by human immunoglobulin G in phagocytosis by macrophages. Infect. Immun. *25*: 255-261.

49. Kurup, V. P., and J. N. Fink. 1978. Evaluation of methods to detect antibodies against *Aspergillus fumigatus*. Am. J. Clin. Pathol. *69*: 414-417.

50. LaForce, F. M., D. M. Mills, K. Iverson, R. Cousins, and E. D. Everett. 1975. Inhibition of leukocyte candidacidal activity by serum from patients with disseminated candidiasis. J. Lab. Clin. Med. *86*: 657-666.

50a. LaForce, F. M., R. G. Sharrar, and G. Arai. 1979. Characterization of yeast agglutinins in lavage fluid from lungs of rabbits. J. Infect. Dis. *140*: 96-104.

51. Land, G. A., J. H. Foxworth, and K. E. Smith. 1978. Immunodiagnosis of histoplasmosis in a compromised host. Infect. Immun. *8*: 558-565.

52. Lehner, T. 1970. Serum fluorescent antibody and immunoglobulin estimations in candidosis. J. Med. Microbiol. *3*: 475-481.

53. Lehrer, R. I., and M. J. Cline. 1969. Interaction of *Candida albicans* with human leukocytes and serum. J. Bacteriol. *98*: 996-1004.

54. Louria, D. B., J. K. Smith, R. G. Brayton, and M. Buse. 1972. Anti-*Candida* factors in serum and their inhibitors. I. Clinical and laboratory observations. J. Infect. Dis. *125*: 102-114.

55. Macher, A. M., J. E. Bennett, J. E. Gadek, and M. M. Frank. 1978. Complement depletion in cryptococcal sepsis. J. Immunol. *120*: 1686-1690

56. Marx, J. J., and D. K. Flaherty 1976. Activation of the complement sequence by extracts of bacteria and fungi associated with hypersensitivity pneumonitis. J. Allergy Clin. Immunol. *57*: 328-334.

56a. Mathur, S., and H. H. Fudenberg. 1979. Anti-ovarian and anti-T-lymphocyte antibodies in vaginal candidiasis: antigenic cross-reactivity between *Candida,* ovary, and T cells. Clin. Res. *27*: 753A.

57. Mathur, S., G. Virella, J. Koistinen, E. O. Horger III, T. A. Mahvi, and H. H. Fudenberg. 1977. Humoral immunity in vaginal candidiasis. Infect. Immun. *15*: 287-294.

58. Mathur, S., J. Goust, E. O. Horger III, and H. H. Fudenberg. 1977. Immunoglobulin E anti-*Candida* antibodies and candidiasis. Infect. Immun. *18*: 257-259.

59. Mathur, S., J. Koistinen, C. Y. Kyong, E. O. Horger III, G. Virella, and H. H. Fudenberg. 1977. Antibodies to *Candida albicans* in IgA deficient humans. J. Infect. Dis. *136*: 436-438.

60. Mathur, S., R. S. Mathur, H. Dowda, H. O. Williamson, W. P. Faulk, and H. H. Fudenberg. 1978. Sex steroid hormones and antibodies to *Candida albicans*. Clin. Exp. Immunol. *33*: 79-87.

60a. Mathur, S., R. S. Mathur, S. C. Landgrebe, T. S. Gramling, H. O. Williamson, and H. H. Fudenberg. 1979. Antibodies to *Candida albicans* and steroid hormones during late pregnancy and in the umbilical circulation. Clin. Immunol. Immunopathol. *12*: 335-340.

60b. McGaw, T. G., and T. R. Kozel. 1979. Opsonization of *Cryptococcus neoformans* by human immunoglobulin G--masking of immunoglobulin G by cryptococcal polysaccharide. Infect. Immun. *25*: 262-267.

61. Mendes, E., and A. Raphael. 1971. Impaired delayed hypersensitivity in patients with South American blastomycosis. J. Allergy *47*: 17-22.

62. Miller, M. E., and U. R. Nilsson. 1970. A familial deficiency of the phagocytosis-enhancing activity of serum related to a dysfunction of the fifth component of complement (C5). N. Engl. J. Med. *282*: 354-358.

63. Miller, M. E., and U. R. Nilsson. 1974. A major role of the fifth component of complement (C5) in the opsonization of yeast particles. Partial dichotomy of function and immunochemical measurement. Clin. Immunol. Immunopathol. *2*: 246-255.

64. Milne, J. D., and D. W. Warnock. 1977. Antibodies to *Candida albicans* in human cervicovaginal secretions. Br. J. Vener. Dis. *53*: 375-378.

65. Mohr, J. A., H. Long, B. A. McKown, and H. G. Muchmore. 1972. In vitro susceptibility of *Cryptococcus neoformans* to steroids. Sabouraudia *10*: 171-172.

66. Mohr, J. A., B. A. Tatem, H. Long, H. G. Muchmore, and F. G. Felton. 1973. Increased susceptibility of *Cryptococcus neoformans* to amphotericin B in the presence of steroids. Sabouraudia *11*: 140-142.

67. Murphy, J. W., and G. C. Cozad. 1972. Immunological unresponsiveness induced by cryptococcal capsular polysaccharide assayed by the hemolytic plaque technique. Infect. Immun. *5*: 896-901.

68. Musatti, C. C. 1975. Cell-mediated immunity in paracoccidioidomycosis. In *Mycoses,* Proceedings of the Third International Conference on the Mycoses, Pan Am. Health Organ. Sci. Publ. No. 304, Washington, D.C., pp. 23-29.

69. Musatti, C. C., M. T. Rezkallah, E. Mendes, and N. F. Mendes. 1976. In vivo and in vitro evaluation of cell-mediated immunity in patients with paracoccidioidomycosis. Cell. Immunol. *24*: 365-378.

70. Nakamura, Y., H. Ishizaki, and R. W. Wheat. 1977. Serological cross-reactivity between group B *Streptococcus* and *Sporothrix schenckii, Ceratocystis* species, and *Graphium* species. Infect. Immun. *16*: 547-549.

71. Negroni, R. 1972. Serologic reactions in paracoccidioidomycosis. In *Paracoccidioidomycosis,* Proceedings of the First Pan American Symposium, Pan Am. Health Organ. Sci. Publ. No. 254, Washington, D.C., pp. 203-208.

72. Neill, J. M., C. G. Castillo, R. H. Smith, and C. E. Kapros. 1949. Capsular reactions and soluble antigens of *Torula histolytica* and *Sporotrichum schenckii*. J. Exp. Med. *89*: 93-106.

73. Newberry, W. M. Jr., J. W. Chandler, Jr., T. D. Y. Chin, and C. H. Kirkpatrick. 1968. Immunology of the mycoses. I. Depressed lymphocyte transformation in chronic histoplasmosis. J. Immunol. *100*: 436-443.

74. Opelz, G., and M. I. Scheer. 1975. Cutaneous sensitivity and in vitro responsive-

ness of lymphocytes in patients with disseminated coccidioidomycosis. J. Infect. Dis. *132*: 250-255.

75. Otto, V., and D. H. Howard. 1976. Further studies on the intracellular behavior of *Torulopsis glabrata*. Infect. Immun. *14*: 433-438.

76. Oxenhandler, R. W., E. H. Adelstein, and W. A. Rogers. 1977. Rheumatoid factor: a cause of false positive histoplasmin latex agglutination. J. Clin. Microbiol. *5*: 31-33.

77. Palmer, D. F., L. Kaufman, W. Kaplan, and J. J. Cavallaro. 1977. *Serodiagnosis of Mycotic Diseases*, Thomas, Springfield, Ill., p. 191.

78. Pappagianis, D., M. Saito, and K. H. VanHoosear. 1972. Antibody in cerebrospinal fluid in non-meningitic coccidioidomycosis. Sabouraudia *10*: 173-179.

79. Paterson, P. Y., R. Semo, G. Blumenschein, and J. Swelstad. 1971. Mucocutaneous candidiasis, anergy, and a plasma inhibitor of cellular immunity: reversal after amphotericin B therapy. Clin. Exp. Immunol. *9*: 595-602.

80. Payne, S. M., and R. A. Finkelstein. 1978. The critical role of iron in host-bacterial interactions. J. Clin. Invest. *61*: 1428-1440.

81. Pearsall, N. N., and D. Lagunoff. 1974. Immunological responses to *Candida albicans*. I. Mouse-thigh lesion as a model for experimental candidiasis. Infect. Immun. *9*: 999-1002.

82. Pearsall, N. N., J. S. Sundsmo, and R. S. Weiser. 1973. Lymphokine toxicity for yeast cells. J. Immunol. *110*: 1444-1446.

82a. Perfect, J. R., S. D. R. Lang, and D. T. Durack. 1980. Influence of humoral immunity on cryptococcal meningitis in rabbits. Clin. Res. *28*: 376A.

83. Picardi, J. L., C. A. Kauffman, J. Schwarz, and J. P. Phair. 1976. Detection of precipitating antibodies to *Histoplasma capsulatum* by counterimmunoelectrophoresis. Am. Rev. Respir. Dis. *114*: 171-176.

84. Pine, L., G. B. Malcolm, H. Gross, and S. B. Gray. 1978. Evaluation of purified H and M antigens of histoplasmin as reagents in the complement fixation test. Sabouraudia *16*: 257-269.

85. Porter, K. G., D. G. Sinnamon, and R. R. Gillies. 1977. *Cryptococcus neoformans*—specific oligoclonal immunoglobulins in cerebrospinal fluid in cryptococcal meningitis. Lancet *1*: 1262.

86. Ray, T., and K. D. Wuepper. 1976. Activation of the alternative (properdin) pathway of complement by *Candida albicans* and related species. J. Invest. Dermatol. *67*: 700-703.

87. Rea, T. H., R. Johnson, H. Einstein, and N. E. Levan. 1978. Dinitrochlorobenzene responsivity: differences between patients with severe pulmonary coccidioidomycosis and patients with disseminated coccidioidomycosis. J. Infect. Dis. *139*: 353-356.

88. Reiss, E., H. Hutchinson, L. Pine, D. W. Ziegler, and L. Kaufman. 1977. Solid-phase competitive-binding radioimmunoassay for detecting antibody to the M antigen of histoplasmin. J. Clin. Microbiol. *6*: 598-604.

88a. Reiss, E., and P. L. Lehmann. 1979. Galactomannan antigenemia in invasive aspergillosis. Infect. Immun. *25*: 357-365.

89. Reiss, F., G. S. Zilagyi, and E. Mayer. 1975. Immunological studies of the anti-cryptococcal factor of normal human serum. Mycopathologia *55*: 175-178.

90. Restrepo, A., M. Restrepo, F. De Restrepo, L. H. Aristizabal, L. H. Moncada, and H. Velez. 1978. Immune responses in paracoccidioidomycosis. A controlled study of 16 patients before and after treatment. Sabouraudia *16*: 151-163.

91. Restrepo, M. A., and A. Herta Velez. 1975. Efectos de la fagocitosis in vitro sobre el *Paracoccidioides brasiliensis*. Sabouraudia *13*: 10-21.

92. Root, R. K., L. Ellman, and M. M. Frank. 1972. Bactericidal and opsonic properties of C4 deficient guinea pig serum. J. Immunol. *109*: 477-486.

93. Rosenberg, M., R. Patterson, R. Mintzer, B. J. Cooper, M. Roberts, and K. E. Harris. 1977. Clinical and immunologic criteria for the diagnosis of allergic bronchopulmonary aspergillosis. Ann. Intern. Med. *86*: 405-414.

94. Rosenfeld, S. I., J. Baum, R. T. Steigbigel, and J. P. Leddy. 1976. Hereditary deficiency of the fifth component of complement in man. J. Clin. Invest. *57*: 1635-1643.

95. Rowland, V. S., R. E. Westfall, and W. A. Hinchcliffe. 1977. Fatal coccidioidomycosis: analysis of host factors. In L. Ajello (Ed.), *Coccidioidomycosis: Current Clinical and Diagnostic Status,* Symposia Specialists, Miami, Fla., pp. 91-106.

96. Schaeffer, J. C., B. Yu, and D. Armstrong. 1976. An *Aspergillus* immunodiffusion test in the early diagnosis of aspergillosis in adult leukemia patients. Am. Rev. Respir. Dis. *113*: 325-329.

96a. Scott, E. N., F. G. Felton, and H. G. Muchmore. 1980. Development of an enzyme immunoassay for cryptococcal antibody. Mycopathologia *70*: 55-59.

96b. Segal, E., R. A. Berg, P. A. Pizzo, and J. E. Bennett. 1979. Detection of *Candida* antigen in sera of patients with candidiasis by an enzyme-linked immunoabsorbant assay inhibition technique. J. Clin. Microbiol. *10*: 116-118.

97. Shaffer, P. J., G. Medoff, and G. S. Kobayashi. 1979. Demonstration of antigenemia by radioimmunoassay in rabbits experimentally infected with *Aspergillus.* J. Infect. Dis. *139*: 313-319.

98. Slavin, R. G., V. W. Fischer, E. A. Levine, C. C. Tsai, and P. Winzenburger. 1978. A primate model of allergic bronchopulmonary aspergillosis. Int. Arch. Allergy Appl. Immunol. *56*: 325-333.

99. Singer, L. M., and C. Fava Netto. 1972. Conglutinating complement-fixation reaction in paracoccidioidomycosis. In *Paracoccidioidomycosis,* Proceedings of the First Pan American Symposium, Pan Am. Health Organ. Sci. Publ. No. 254, Washinton, D.C., pp. 227-232.

100. Smith, C. E., M. T. Saito, and S. A. Simons. 1956. Pattern of 39,500 serologic tests in coccidioidomycosis. JAMA *160*: 546-552.

101. Smith, J. K., and D. B. Louria. 1972. Anti-*Candida* factors in serum and their inhibitors. II. Identification of a *Candida*-clumping factor and the influence of the immune response on the morphology of *Candida* and on anti-*Candida* serum activity in rabbits. J. Infect. Dis. *125*: 115-122.

102. Sohnle, P. G., M. M. Frank, and C. H. Kirkpatrick. 1976. Deposition of complement components in the cutaneous lesions of chronic mucocutaneous candidiasis. Clin. Immunol. Immunopathol. *5*: 340-350.

103. Sohnle, P. G., M. M. Frank, and C. H. Kirkpatrick. 1976. Mechanisms involved in elimination of organisms from experimental cutaneous *Candida albicans* infections in guinea pigs. J. Immunol. *117*: 523-530.

104. Solomkin, J. S., E. L. Mills, G. S. Giebink, R. D. Nelson, R. L. Simmons, and P. G. Quie. 1978. Phagocytosis of *Candida albicans* by human leukocytes: opsonic requirements. J. Infect. Dis. *137*: 30-37.

105. Stiehm, E. R. 1978. Chronic mucocutaneous candidiasis: clinical aspects, pp. 96-99. In J. E. Edwards, Jr. (moderator), Severe candidal infections. Clinical perspectives, immune defense mechanisms, and current concepts of therapy. Ann. Intern. Med. *89*: 91-106.

106. Stobo, J. D. 1977. Immunosuppression in man: suppression by macrophages can be mediated by interactions with regulatory T cells. J. Immunol. *119*: 918-924.

107. Stobo, J. D., S. Paul, R. E. Van Scoy, and P. E. Hermans. 1976. Suppressor thymus-derived lymphocytes in fungal infection. J. Clin. Invest. *57*: 319-328.

108. Sutcliffe, M. C., A. S. Savage, and R. H. Alford. 1980. Transferrin-dependent growth inhibition of yeast-phase *Histoplasma capsulatum* by human serum and lymph. J. Infect. Dis. *142*: 209-219.

109. Syverson, R. E., H. R. Buckley, and J. A. Gibian. 1978. Increasing the predictive value of the precipitin test for the diagnosis of deep-seated candidiasis. Am. J. Clin. Pathol. *70*: 826-831.

110. Swenson, F. J., and T. R. Kozel. 1978. Phagocytosis of *Cryptococcus neoformans* by normal and thioglycolate-activated macrophages. Infect. Immun. *21*: 714-720.

111. Tacker, J. R., F. Farhi, and G. S. Bulmer. 1972. Intracellular fate of *Cryptococcus neoformans*. Infect. Immun. *6*: 162-167.

112. Taschdjian, C. L., P. J. Kozinn, M. B. Cuesta, and E. F. Toni. 1972. Serodiagnosis of candidal infections. Am. J. Clin. Pathol. *57*: 195-205.

113. Tewari, R. P., D. K. Sharma, and A. Mathur. 1978. Significance of thymus-derived lymphocytes in immunity elicited by immunization by ribosomes or live yeast cells of *Histoplasma capsulatum*. J. Infect. Dis. *138*: 605-613.

114. Turner-Warwick, M., K. M. Citron, K. B. Carrol, B. E. Heard, D. N. Mitchell, J. Pepys, J. G. Scadding, and C. A. Sontar. 1975. Immunologic lung disease due to *Aspergillus*. Chest *68*: 346-355.

115. Verhaegen, H., D. W. DeCock, and J. DeCree. 1976. In vitro phagocytosis of *Candida albicans* by peripheral polymorphonuclear neutrophils of patients with recurrent infections. Case reports of serum-dependent abnormalities. Biomedicine *24*: 176-180.

116. Warnock, D. W., and A. L. Hilton. 1976. Value of the indirect immunofluorescence test in the diagnosis of vaginal candidiasis. Br. J. Vener. Dis. *52*: 187-189.

117. Warnock, D. W., J. D. Milne, and A. W. Fielding. 1978. Immunoglobulin classes of human serum antibodies in vaginal candidiasis. Mycopathologia *63*: 173-175.

117a. Warren, R. C., M. D. Richardson, and L. O. White. 1978. Enzyme-linked immunoabsorbant assay of antigens from *Candida albicans* circulating in infected mice and rabbits: the role of mannan. Mycopathologia *66*: 179-182.

118. Wegner, T. N., R. E. Reed, R. J. Trautman, and C. D. Beavers. 1972. Some evidence for the development of a phagocytic response by polymorphonuclear leukocytes recovered from the venous blood of dogs inoculated with *Coccidioides immitis* or vaccinated with an irradiated spherule vaccine. Am. Rev. Respir. Dis. *105*: 845-849.

118a. Weiner, M. H. 1979. Immunodiagnosis of systemic candidiasis: antigenemia detected after dissociation of serum bound mannan with a radioimmunoassay using staphylococcal protein A as adsorbant. Clin. Res. *27*: 755A.

119. Weiner, M. H., and W. H. Yount. 1976. Mannan antigenemia in the diagnosis of invasive *Candida* infections. J. Clin. Invest. *58*: 1045-1053.

120. Yamamura, M., and H. Valdimarsson. 1977. Participation of C3 in intracellular killing of *Candida albicans*. Scand. J. Immunol. *6*: 591-594.

121. Young, R. C., and J. E. Bennett. 1971. Invasive aspergillosis: absence of detectable antibody response. Am. Rev. Respir. Dis. *104*: 710-716.

4

Serodiagnosis

Donald W. R. Mackenzie / London School of Hygiene and Tropical Medicine, London, England

I. INTRODUCTION

Fungi are complex microorganisms, with enormously varied and versatile metabolisms. Many of their cytoplasmic components are immunogenic, as are the constituents and products of their life processes. Provided that physiological contact is established between a host and an infecting fungus or its metabolites, an immune response is usually elicited. The nature and extent of this response is determined by many factors, those associated with the host as well as those of the pathogen. Man's immune system is highly evolved and complex. It is capable of recognizing and responding to an enormous range of foreign substances and it does so with exquisite precision and sensitivity. In the past 20 years, new methods have been developed and new perspectives gained on the nature of the immune response. The biggest advances have been unraveling the complex interrelationships that exist between the individual elements of the immune system. Following the introduction of a foreign substance (antigen) into the tissues of a host that is capable of recognizing its alien nature, measurable responses can almost always be detected, either in the form of acquired responsiveness of white cells to the sensitizing antigens or by the appearance in the bloodstream of antibodies. These major expressions of acquired responsiveness are designated as cellular (cell mediated) and humoral, depending on whether the expression is associated primarily with lymphoid cells or with antibodies. In both instances, the occurrence and specificity of the acquired responsiveness have been exploited by the development of diagnostic tests. Serological tests are based on the detection of specific antibodies. Since their development in the last decade of the nineteenth century, they have been widely used in the diagnostic laboratory. Following the realization nearly 40 years ago that histoplasmosis was a common and widespread disease in the United States, first skin tests and then serological tests began to be used increasingly for the recognition of individuals who had been exposed to *Histoplasma capsulatum*. Serodiagnostic tests are now widely used for detecting and measuring antibodies to a wide range of fungal pathogens. Assessment of both quantitative and qualitative aspects of the humoral response can be useful, the former in suggesting the identity of the infecting agent and the latter in providing a measurement of the response. Antibodies are produced in most mycotic infections. Establishment and proliferation of the pathogen at a site which is under surveillance by the immune system normally will be followed by the formation of antibodies 10-14 days later. To some extent, the types of antibody that are formed, and their levels (i.e., titers) are expressions of the status and progress of infection. Provided that the immune system is functioning normally, serological tests can be of great importance as diagnostic aids and in evaluating the course of disease.

Laboratory tests often provide valuable clues to etiology, but every test in vitro has its innate limitations. Serological tests are no exception. Some are better than others, in the sense that results correlate better with disease. The extent to which individual serodiagnostic tests have been developed varies considerably for each mycosis. Some are reliable and of proven merit. Others are still being developed or have not yet been widely used or accepted. The margins of usefulness will be set differently by different investigators, and competence in the execution and interpretation of serodiagnostic tests is dependent, to a large extent, on the experience gained by an individual, or group of individuals, with their own reagents and their own systems. In the sections dealing with individual mycoses, attempts have been made to indicate limitations as well as disadvantages of serodiagnostic testing.

II. ANTIGENS

A. Introduction

Fungi produce many different antigens. Most are protein or polysaccharide, but complexes with carbohydrate and peptide or protein moieties are common. The biochemical changes associated with tissue persistence or invasion are little understood in the fungi, and neither the concomitants nor determinants of pathogenicity have been defined in chemical terms. It follows that since discrimination between pathogenically significant and irrelevant antigens cannot yet be achieved, the same holds true for their serological counterparts (i.e., the antibodies). A disproportionately small number of studies have been directed toward the biological nature and function of fungal antigens. In view of the gains that can be anticipated in acquiring an understanding of the mechanisms of pathogenicity, and in being able to ascribe greater significance to the results of serological tests, this failure to link studies on the chemical nature of antigens with analyses of, e.g., their enzymatic nature, is a surprising one. There are a few exceptions, including the recognition of function for serodiagnostic antigens such as chymotrypsin and catalase in *Aspergillus* [432] and protease in *Candida albicans* [249]. Fungal cells are rich in macromolecules, some cytoplasmic, others located in membranes, organelles, and cell walls, or produced extracellularly by normal physiological processes, including catabolism, secondary metabolism, and aging. Many of the substances involved or produced by these structures or activities are antigenic. The number that can be demonstrated may be considerable. Axelsen [17] described 78 water-soluble antigens in yeast cells of *Candida albicans*. It is unlikely that this represents more than a fraction of the antigens that are present in the living fungal cell. The number of antigenic components that can be recognized in an extract is limited principally by the sensitivity of the test system used to detect them. In practice, sensitivity may have disadvantages as well as advantages, since its enhancement is usually accompanied by a loss of specificity, with corresponding increases in the difficulties of interpreting the results of analytical tests.

One of the major problems in preparing serodiagnostic antigens is the widespread occurrence of cross-reactions between related, or even dissimilar fungi. This can be particularly troublesome with polysaccharide antigens. Azuma et al. have described a galactomannan common to *Histoplasma, Blastomyces,* and *Paracoccidioides,* which is also present in *Alternaria* [21]. Similarly, serum from patients with blastomycosis often reacts better in serological tests with *Histoplasma* than with *Blastomyces* [52]. A logical approach to this problem might seem to be the elimination of common antigens from a serodiagnostic preparation, so that specificity is greatly improved. However, this introduces another difficulty, since, paradoxically, the diagnostic value of the test may be diminished rather than enhanced by removing the causes of cross-reaction. Figure 1 illustrates how this

Antigens

Species A	1	2	3	4		
Species B			3	4	5	6

Figure 1   Shared and common antigens.

could happen. Species A and B have both specific (1, 2 and 5, 6, respectively) and shared (3, 4) antigens.

By appropriate absorption procedures, it may be possible to remove antigens 3 and 4, resulting in antigens that are monospecific. Sera containing antibodies to antigens 1 or 2 may be infected by species A but not by species B. Thus, the discriminatory capacity of the test has been increased. If, however, the patient's antibodies are directed exclusively to antigens 3 or 4, serological tests with antigens 1, 2, 5 or 6 will be nonreactive and a false negative would result. Different patients may produce different antibodies to the same infecting agent. For this reason it is rarely advantageous to produce highly specific antigens. Apart from the loss in the test system of reactivity to antigenic components discarded as nonspecific, purification of the reagent by physical or chemical means may lead to a product that differs radically from anything associated with the organism in vivo.

The preparation of serodiagnostic antigens will now be considered.

B. Intact Organisms

For use with agglutination or indirect immunofluorescence tests, antibodies are detected by their effect on physically unaltered vegetative cells. Yeast cells grown on surface or submerged culture are the simplest source of particulate antigens, providing a convenient means for rendering antigen-antibody reactions visible. After harvesting, cells normally are washed to remove traces of culture media, fixed by acetone, formalin, or heat, then adjusted to the selected concentration for the agglutination, immunofluorescence, or complement fixation test. To some extent, the agglutination titer bears an inverse relationship to the number of cells present in the suspension. Lower titers are obtained with a heavy concentration of cells than with a light suspension. A suspension of yeast cells grown in broth culture or harvested from semisolid media are suitable for immunofluorescence studies.

Mycelium is not physically suitable for agglutination tests, but surface-acting antibodies are demonstrated readily by indirect immunofluorescence (IF), using intact hyphae, grown on or transferred onto glass slides in such a way that they are conveniently dispersed throughout the high-power field of the microscope. It is found in practice that young hyphae are suitable in IF, but older hyphae are not. Spores are generally unsatisfactory, since they fluoresce in the absence of any specific immunological reaction (autofluorescence). A convenient procedure for examining antibodies to molds involves the preparation of several micro slide cultures on a single microscope slide. Circles about

1.0 cm in diameter are marked with a diamond scribe, then after sterilization by dry heat or autoclaving, drops of Sabouraud's agar are placed within the marked areas. After solidification, one half is removed with a sterile scalpel blade and the cut surface inoculated with the test fungus. Slides are then incubated until growth is visible on the glass surface. This is detached from the agar by cutting with a blade at its point of attachment to the agar. The latter is then removed and discarded. Surface hyphae on the glass slide are fixed lightly by heat or acetone and stored dry until required. Sporulation should be avoided, and due attention must be given to operator safety throughout.

C. Whole Cell Extracts

Whole cell extracts are used principally in gel diffusion procedures. Harvested growth is mechanically disrupted principally by treatment with ultrasonic vibration, agitation with glass beads, or by repeated high-pressure extrusion in a French press or Hughes press. Fungi, being more robust than bacteria, are less amenable to milder extraction procedures such as osmotic shock or subjection to freeze-thaw cycles. Chemical extraction of intact cellular material may yield an active antigenic preparation, but the procedure most widely adopted is physical disruption. The resultant extract contains both cell wall and cytoplasmic components. The preparation may be used without purification or calibration. More frequently, however, cellular debris is removed by centrifugation and some of the low molecular weight substances are then eliminated by dialysis. The resultant crude extract is often distributed in small measured amounts, then lyophilized so that graded strengths of the antigen can be prepared on a dry weight basis. This type of processing also assures a measure of reproducibility between one ampul or bottle and the next, for a given batch of antigen. Antigenic material which has not been subjected to lyophilization, or which has been reconstituted from the freeze-dried state, has a shelf life that may be very limited. During storage at $-40°C$ or $-20°C$, some antigens become degraded, and the qualitative and quantitative changes that result make the preparations increasingly unreliable as serodiagnostic reagents. Curiously, long-term storage of serodiagnostic antigen to which merthiolate has been added as a preservative has been found in one instance [350] to promote formation of a novel antigen, presumably formed by some degradative process, which has markedly improved specificity.

In preparing whole cell extracts for use as serodiagnostic antigens, experience has shown that the simplest and most reliable method for comparing reactivities of different batches of antigen is to measure reactivity in the relevant serodiagnostic procedure against a reference antiserum. This is generally more satisfactory than attempting to equate serological reactivity with direct measurements of protein or carbohydrate content of the extract. In practice, new antigenic preparations are titrated against standard antiserum (see below) to determine the concentration (mg/ml) which gives comparable reactivity to the previous batch or to a reference antigen if this is available.

Many constituents of crude extracts are immunologically inert in serological tests, and if these can be eliminated without loss of reactivity or atability, there are advantages in achieving partial purification. This can be done by a variety of preparative procedures, including precipitation with ammonium sulfate, extraction with detergent, and the use of such physical procedures as ion exchange chromatography or gel filtration.

Reactivity of crude or partially purified antigens can be objectively assessed using gel electrophoretic techniques. One of the most useful of these is two-dimensional immunoelectrophoresis (2D-IE). By means of this method, it is possible to detect and distinguish individual antigens, and to determine their concentrations in relation to a standard antiserum. The versatility of this procedure allows for several different types of analyses and comparisons to be made. As a tool for providing precise measurements of complex antigenic extracts, 2D-IE has much to offer, and it should be in regular use in all laboratories engaged in production and utilization of their own serodiagnostic reagents. Standardization is a critical and integral objective of all centers engaged in immunodiagnostic work, and is considered in Chap. 5.

D. Culture Filtrates

This is one of the simplest methods available for production of antigenic extracts. Culture medium in which the fungus has been grown contains many antigenic products. These include soluble cell wall substances (usually polysaccharide), extracellular proteins (enzymes), and various products associated with secondary metabolism or autolysis. Culture filtrates often contain components which are not demonstrable in young mycelial extracts. Such filtrates constitute a rich and varied source of antigens, and their reactivity, as measured by the number of precipitin lines in a gel diffusion system, is often very marked, making them a popular source of serodiagnostic antigens. A commonly used method to process culture filtrate antigens is to precipitate immunogenic materials with acetone. Such preparations are widely used for aspergilli and two commercially available antigens for serological testing (Pasteur Institute, Paris; Bencard Allergy Unit, Brentford, England) use both mycelial and culture filtrate antigens in their double-diffusion kits for aspergillosis.

It should be noted that with one exception, international reference antigens are not available for serodiagnostic tests of the mycoses, and it is correspondingly difficult to measure specifications and reactivities of antigens in use at different centers throughout the world. As a result, each center engaged in serodiagnostic work gains experience with its own systems and reagents, but except for occasional courtesy exchanges of reagents between different laboratories, there is little coordinated effort directed toward evaluation of these reagents and their application to serodiagnostic testing. Inevitably, this slows the development of improved diagnostic antigens and reliable tests. Some antigens are known to be more specific than others, since they react principally and sometimes exclusively with sera from patients infected with homologous rather than heterologous species of fungi. Such antigens have been recognized for the agents causing the major systemic mycoses (*Histoplasma, Blastomyces, Coccidioides,* and *Paracoccidioides*), and may eventually be recognized for other mycotic agents, such as aspergilli and members of the Zygomycetes. The H and M antigens of *Histoplasma* used in immunodiffusion [162] have proved to be of great value as serodiagnostic antigens, and are the most reliable ones in current use. They are produced in culture filtrate of *H. capsulatum* and are of great diagnostic value, particularly when immunodiffusion is used in association with other serological procedures, such as complement fixation. The detection of antibodies to H and M components of *H. capsulatum* has reached a level of reliability and proven efficacy that has led to their evaluation by a World Health Organization (WHO) collaborative study, with a view to their adoption as an international biological standard—the first occasion on which any fungal product has been proposed and evaluated as an international reference immunodiagnostic reagent. Apart from constituting a tribute to the developmental work of the developmental team at the Center for Disease Control (CDC) in Atlanta, Georgia

[325], this precedent could encourage further cooperative studies directed toward improvement in the quality and diagnostic value of other reagents in current use.

III. ANTIBODIES

A. Introduction

Antibodies are protein molecules produced by antibody-forming cells (B-cells) in response to foreign proteins and other immunologically active macromolecules. Antibodies exhibit a high degree of specificity to the antigen eliciting their formation, and this feature constitutes the basis of all diagnostic serological tests. Five major classes of immunoglobulins are recognized (IgG, IgM, IgA, IgD, and IgE), of which IgG and IgM are of greatest relevance to serodiagnostic tests. These two molecules differ appreciably in their physical and biological properties. Thus, IgG antibodies have a molecular weight of 150,000 and a half-life in serum of 23 days. They constitute the most abundant of the immunoglobulins, achieving significant concentrations in both the bloodstream and extravascular spaces. Antibody responses to most infective agents which are blood-borne, or whose metabolites are released into the bloodstream, are predominantly IgG.

IgM molecules are much larger, with a molecular weight of 890,000. Unlike IgG antibody, which has two sites that specifically combine with antigen, IgM antibody has 10 such sites. It is consequently more efficient than IgG in causing particulate antigen to agglutinate. As a rule, IgM is of greater importance in the first few days of infection. When a fungal pathogen is introduced into the tissues for the first time, both IgG and IgM immunoglobulins are produced almost simultaneously. The effects associated with the IgM molecules are more obvious at this stage, probably because of greater efficiency in combining with antigens. In patients who have already been exposed to antigen, subsequent (secondary) antibody responses tend to be associated with IgG rather than IgM immunoglobulins. The role of IgA antibodies in fungal infection or serodiagnosis is unclear. As with IgM antibody, IgA antibodies tend to be produced early in infection and to disappear more rapidly than IgG.

For many infectious diseases antibodies are protective. As a rule, however, acquired resistance to mycoses owes little to the humoral branch of the immune response. In the study of mycoses the presence and quantity of antibodies are used as indirect measures of the existence of a pathogen within a patient and the duration and extent of infection. What is being measured in a serological test is not any innate quality of the infecting fungus but the qualitative and quantitative responses of the host. Provided that the host's immune mechanisms are functioning satisfactorily, and sufficient time has lapsed since the fungus became established for antibodies to be formed, serological tests can provide valuable diagnostic information. As with all serological tests, however, results must always be viewed in relation to the patient's history and condition. For example, a patient with *Candida* growing in the mouth and gastrointestinal tract may have high levels of antibody detected by routine serodiagnostic procedures. In this case, presence of antibodies may have little or no diagnostic significance. A different interpretation can be considered when a patient with a recent heart valve replacement develops elevated levels of *Candida* antibody. If there is no obvious explanation for this serological finding, such as a recent colonization of the oral mucosa, urinary, or gastrointestinal tract, the antibody formation may point to a deep-seated infection, and this possibility will have to be considered specifically.

B. Detection of Antibody

The basic approach for most serodiagnostic tests in fungal serology is to use antigens of known potency, concentration, sensitivity, and specificity for the detection of serum antibody. With the exception of immunofluorescence, evidence for antigen-antibody reactions such as agglutination or precipitation are dependent on secondary manifestations. Without a clear and measurable end product, the union of antigen with antibody could not be utilized in an in vitro laboratory procedure.

C. Reference Antisera

All serological tests should include a positive control. Moreover, the choice of antigen and the concentration at which it is used in a test is determined primarily by its reactivity against a reference antiserum. The quality of reference antisera is therefore of great importance. It is obvious that no individual antigenic component can be detected in a serological test system unless antibody to the antigen is present in the antiserum. Unless the number of antigens contained in a test extract is small (e.g., *Coccidioides* or *Histoplasma*), or the test is limited to reactions with surface components of the fungus, it is equally clear that reference antisera should react to all major antigenic components of the fungus, as well as to a substantial proportion of their minor ones. Antisera for selection of antigens to be used in precipitin tests or for positive control sera should be highly reactive, not only in terms of the numbers of antibodies present but also in the extent to which it can be diluted without loss of reactivity. The higher the titer, the more economically the antiserum can be used, and the longer it can be used without replacement. One of the simplest and most effective means of minimizing batch-to-batch variation of serodiagnostic reagents is to produce them in large quantities and to have them function satisfactorily at high dilutions. Some labile components may be lost during the process of lyophilization, but provided that these do not appreciably reduce the effectiveness of the reagent, the loss is more than compensated by the standardized performance, which is always obtained more readily when test materials are freeze-dried prior to use.

Specialist laboratories make reference antisera in different ways, and in different animals, the choice of which is dependent in part on the available volume and at times the reactivity of the animal's humoral responses. Rabbits are widely used since they are of a convenient size, easy to immunize, and reactive to a wide range of antibodies. Because of individual variations in response, it is customary to immunize several animals and pool the sera after the blood has been drawn. The route and method of immunization may vary according to the purpose for which the antiserum is required. Thus, an antiserum for use as an agglutination control may be produced by using intact cells. In contrast, when required for a precipitin control, it is more customary to use disintegrated cells or culture filtrate, to promote maximum antigenic stimulation. Animals may be immunized by administration of the antigen intravenously or by its deposition subcutaneously. In the latter case, it is mixed with an adjuvant, a semiviscous substance that localizes the antigen and releases it more slowly, thereby increasing the antibody response per unit volume or mass of antigen. The number of immunizations varies according to the nature of the antigen, the reactivity of the individual animal, the route and frequency of inoculation, and the titer of antibody required. The yield of antiserum from 10-12 rabbits will be between 1 and 2 liters. Larger quantities, if required, can be obtained by immunization of goats, sheep, or cattle. Burros have been used also with success. It must always be borne in mind that responses to different antigens vary according to the species of test animal. Reactivities to polysaccharide antigens in particular may show a very wide range of responses, so much so that the use of a particular test animal may be contraindicated. Animals are test-bled before immunization to ensure a lack of specific or cross-reacting antibodies to the antigens and at intervals thereafter. Satisfactory levels of antibodies are usually attained after 4-8 weeks, after which the animals are bled out by cardiac puncture and their sera pooled, distributed into ampuls or rubber-sealed bottles in convenient quantities (e.g., 0.2-, 0.5-, or 1.0-ml quantities), and lyophilized. Before introduction into general use, the new batch of antiserum is titrated by comparison with the previous batch, against a reference antigen. The titer by each of the methods for which the antiserum will be used as a control is then determined. Ideally, it should be used at a concentration (mg/ml dry weight) which gives comparable if not indistinguishable results from those obtained with the previous batch. Sould it be necessary, a profile of reactivity can be obtained for a new reference antiserum, by 2D-IE or by tandem 2D-IE, which allows comparisons to be made of different antigenic extracts.

On occasion immunoglobulin may be precipitated out of the antiserum by ammonium sulfate or caprylic acid [411]. Such purified preparations are of value in critical studies using 2D-IE or in preparing tests for antigen detection, such as the latex agglutination test for cryptococcosis.

Many different procedures are used to detect specific antibodies. What follows is a brief account of the most widely used tests and their mechanisms of action. Tests used in the serodiagnosis of individual mycoses are described more fully under each of the disease headings in Secs. V to VIII.

IV. TESTS FOR DETECTION OF ANTIBODY

A. The Precipitin Reaction

Antigens used in precipitin tests are protein, glycoprotein, or polysaccharide solutions of an extract derived from culture filtrate or disintegrated vegetative cells of the test fungus. When mixed in approximately equal concentrations under suitable conditions of pH and molarity, antigens and antibodies will bind together to form aggregates which are insoluble and therefore become visible as precipitates. In precipitin tests, the antibody may belong to IgG, IgM, or IgA classes of immunoglobulin. Because the antigen is of molecular size, comparatively large quantities of the reactants are required to produce a visible effect, and as a consequence the test is rather insensitive.

In immunodiffusion (ID) tests, antigens and antibodies are placed in wells and allowed to diffuse toward each other through a buffered agar or agarose gel. In this system a line representing a specific immune complex (precipitin line) will form for each antigen-antibody pair. The position of the line is determined primarily by the relative concentrations of antigen and antibody. If they are in equal quantities and migrate through the gel at equivalent rates, the line will be midway between the two wells.

It can be shown by titrating increasing quantities of antigen against a constant amount of antibody that optimal precipitation occurs at conditions of near equivalence of antigen and antibody. Immune complexes are soluble under conditions of antigen excess. It follows that unless equivalence is reached on the gel between the wells containing the reactants, no precipitate will be formed. In practice, it is often advisable to use test antigen at two different concentrations, so that the possibilities of obtaining equivalence in the area of reaction are increased.

Control antisera to any antigens used are mandatory. After developing for 2-3 days, the gel is examined directly against dark-field illumination, or washed and stained with a protein stain.

If time permits, better results are always obtained when the processed gel is washed free of unbound serum proteins, dried, then stained with a protein stain to reveal precipitin lines. The initial wash helps to eliminate nonspecific lines which can resemble antigen-antibody reactions. Lines formed in the gel by the interaction between C-substance, sometimes present in antigenic extracts, and C-reactive protein, found in sera of some patients in the early stages of many illnesses, can be eliminated by a brief wash in sodium citrate or ethylenediaminetetraacetic acid (EDTA).

The use of protein stains reveals faint antigen-antibody lines which may not be visible on direct examination of the gel in dark-field illumination. In one study of *Candida* precipitins in the sera of patients with proven candidiasis [273], lines were detected in an immunodiffusion system in 41 of 52 (78%) sera from patients in unstained gels, but in 50 of 52 (96%) of gels stained with Coomassie brilliant blue.

Gel (Ouchterlony) plates are widely used. Patterns are cut in the gel by templates or by use of jigs with pegs arranged in the requisite pattern. Many variations occur in the well patterns and sizes, in the nature of the gel, and the duration of incubation. In some precipitin tests, e.g., for coccidioidomycosis [400], the reactions were originally produced in tubes rather than on plates or slides. The immunodiffusion procedure is, nevertheless, more versatile, and allows for a more ready identification of different antigen-antibody complexes. Quantitative precipitin titers are readily obtained, and involve the testing of doubling dilutions of the patient's serum against a constant concentration of antigen. The highest dilution at which an unequivocal line can be seen represents the precipitin titer. Such tests are of value in following a patient's response to therapy.

B. Agglutination

The principal difference between precipitation and agglutination reactions lies with the physical nature of the antigen. In ID the antigen is soluble, consisting of macromolecules. In agglutination, the antigen is particulate, often consisting of intact vegetative structures such as a suspension of yeast cells or spores. Because of the larger size of the antigen, fewer molecules of antibody are required to bring about their visible aggregation. Agglutination is, therefore, a more sensitive test than ID. The antibodies primarily involved are IgM, in contrast to precipitation tests where IgG antibodies predominate. For this reason, antibody responses on a single serum may be very different when measured by agglutination and precipitin tests, respectively.

The most simple and convenient agglutination procedures involve titration of test sera with a suspension of uniform-sized particles. Doubling dilutions of a patient's serum are made with saline or buffered saline, in narrow glass tubes or a series of wells in a rigid, clear plastic plate. Quantities of reagents usually vary from 50 to 500 μl. After addition of the antigen suspensions and thorough mixing, tubes or plates are covered then stood overnight at room temperature or 37°C. After agitation, to resuspend the antigen particles, the agglutination reaction is read against dark-field illumination. The agglutination titer is the highest dilution of serum which brings about aggregation of the particles. Both positive and negative controls are included. Because cell suspensions and reference antisera become increasingly ineffectual on prolonged storage at 4°C, antigen suspensions should always be tested for spontaneous (auto-) agglutination (negative control), and control antiserum titrations should be undertaken simultaneously with each set of agglutination determinations.

In conditions of antibody excess, agglutination does not occur. This prozone phenomenon is caused by the presence in the test system of an excessive number of combining sites. Upon continued dilution of the test serum, the reduced number of antibody molecules eventually results in their individual combination with the particulate antigens, and agglutination follows. Agglutination titrations can be determined on slides, which are divided into narrow rectangular areas by wax crayon or glass-marking pen.

Yeast cells, at concentrations of about 10^7-10^8 cells/ml, make convenient cell suspensions for agglutination tests. The cells are killed with merthiolate or formalin before use and stored at -20°C until required. Conidia of molds with dry spores are unsuitable, either because of the hazard they represent to the laboratory worker or because their hydrophobic nature prevents them from forming a dispersed suspension suitable for the test. Spores of *Sporothrix,* in contrast, readily form suspensions and can be used to examine sera from patients with sporotrichosis. Mycelial fragments produced by partial disintegrations of hyphal masses tend to aggregate spontaneously and do not form sufficiently homogeneous suspensions to permit their use in agglutination tests. Reproducibility of agglutination tests is generally satisfactory. In common with other quantitative serological tests, experimental variability is considered to account for differences in titer to dilutions one above and one below the figure obtained; i.e., significant rises or falls in titer in successive examinations of sera from a patient involve fourfold or greater differences in demonstrable antibody levels.

C. Inert Particle Agglutination

Soluble antigens are not suitable for agglutination tests, but can be made so by absorption onto inert carrier particles such as red blood cells, bentonite, or latex particles. When this has been done, whether by physical or chemical bonding to the carrier particles, the coated particles can then be handled in exactly the same manner as whole cell suspensions. Immunoglobulin classes of antibodies revealed by such tests are again predominantly IgM.

Inert particle agglutination (IPA) has been most widely used in detecting antibodies to *Histoplasma, Cryptococcus,* and *Candida.* The system is sensitive and effective, and IPA is a valuable serodiagnostic procedure which can be used for detection of antibodies to a wide range of antigens. Provided that antigens can be efficiently absorbed onto the carrier particle, studies can be made on antibodies to either purified or polyvalent preparations.

IPA has also been developed to detect antigen rather than antibody. In patients with cryptococcosis, capsular polysaccharide is often produced in such abundance that it can be readily detected in serum, cerebrospinal fluid (CSF), and other body fluids. In the latex agglutination test, introduced in 1963 by Bloomfield et al., immunoglobulin from rabbits hyperimmunized with *Cryptococcus neoformans* is absorbed onto latex particles [39] ; these sensitized particles will agglutinate in the presence of capsular polysaccharide antigen. The amount of free antigen in the serum of CSF can be expressed as a titer, i.e., the greatest dilution at which agglutination of latex particles can still be detected.

D. Electrophoretic Tests

Electrophoresis is the movement of molecules carrying electrical charges through an electrolyte in the presence of direct current. In practice, the net charge carried by individual proteins is rendered negative by use of a high pH, and therefore they will migrate toward the anode. Mixtures of antigens can be effectively separated by differences in their rates of electrophoretic migration. The electrophoretic field is almost always established in a gel. Cellulose acetate membranes, long used for the separation and analysis of serum proteins, have several practical advantages but are not used widely for the detection of antibodies to pathogenic fungi. The gelling agent is usually a purified agar such as agarose. In immunoelectrophoresis (IE) the individual components of an antigenic extract are first separated electrophoretically. Then they are allowed to diffuse against a serum, which is placed in an elongated trough. Immune precipitates will be formed, their relative positions depending on the direction and extent of migration of antigenic components and the relative concentrations of antigen and antibody. This system allows a better separation of the antigens than immunodiffusion, and a corresponding increase in the number of precipitins which can be recognized. It is rarely used for routine detection of antibodies, and as a tool for detailed analysis of antigens, it has been supplanted largely by techniques based on two-dimensional immunoelectrophoresis (2D-IE). In this system, the double-diffusion phase of IE is replaced by a second electrophoresis, at right angles to the first, a procedure that causes the separated antigen to migrate into an agarose gel containing antiserum. Each antigen-antibody precipitate will appear as a peak, with a broad base. The height of each peak is dependent on the relative concentration of antigen and antibody, while the position of the peaks is determined by the electrophoretic migration qualities of the antigen. The second-dimensional gel has a pH of 8.2. Since this is the isoelectric point of immunoglobulin, the pH at which it has no net charge, antibodies in the agarose gel are unaffected by the second-dimensional electrophoresis and remain in situ. This technique is a powerful and versatile analytical tool. Axelsen [17] has demonstrated 78 water-soluble antigens in an extract of *Candida albicans* cells. A modified 2D-IE has been applied to routine serodiagnostic test [135], but a simpler and more widely used technique is counterimmunoelectrophoresis (CIE). In this system, antigens are placed in a well and migrate toward the anode. Antibodies, if present in the serum, will move by endosmosis toward the cathode. Precipitates will form for each antigen-antibody combination, provided that equivalence is reached in the area of gel between the wells. Unlike immunodiffusion, there is no "dilution" of reactants caused by radial diffusion through the gel. The net effect is, therefore, an increase in sensitivity. CIE, compared with ID, has the additional advantage rapidity. Results can be read after 60-90 min, compared with 1-3 days for ID. As with ID, however, results are improved when gels are washed, dried, and stained. CIE is used commonly now in the serodiagnosis of fungal infections. One disadvantage is that no precipitates will form if there is a marked excess of antigen or antibody. Therefore, false-negative results may occur. In circumstances where this is likely to occur, it is customary to use differing concentrations of antigen or serum. As with ID, the procedure can be used to determine precipitin titers, doubling dilutions of the patient's serum being tested against a constant concentration of antigen.

E. Complement Fixation

One of the earliest of serological tests to be applied for the diagnosis of mycoses, this procedure is still in widespread use, principally for histoplasmosis and coccidioidomycosis.

Serum complement participates in many antigen-antibody interactions. The amount of complement that combines with antigen-antibody aggregates can be measured. Since for a given quantity of antigen the amount of complement bound is proportional to the amount of antibody present in a serum, the amount of complement removed from the system indicates the titer.

If antibodies are absent, complement will not be "fixed" and therefore, will be detected by addition of an indicator system consisting of sheep erythrocytes sensitized with amboceptor. Such sensitized red cells will lyse only when complement is present. The test is performed in two stages. In the first, dilutions of test serum are added to a constant concentration of antigen. After a suitable period of incubation, the hemolytic indicator system is added. If antibodies to the test antigen are present in the serum, complement will be fixed and therefore the sensitized erythrocytes of the hemolytic indicator system will not lyse. The complement-fixing (CF) titer of a serum is the greatest dilution at which complement is bound by the specific antigen-antibody reaction in sufficient quantities to prevent lysis of the sensitized erythrocytes. Antibodies detected by the CF test are principally IgM.

F. Indirect Fluorescent Antibody

In this system, antigen is derived from growth of the pathogen in vitro. In almost every instance, the antigen *is* the vegetative form of the species in question fixed by acetone or by heat coagulation. Dried and fixed preparations of yeast or mycelial growth on a microscope slide are overlaid with dilutions of the patient's serum. If specific antibodies are present, they will bind to receptor sites present on the surface of the fungal cells. After washing the fungal cell mixtures to remove unbound antibodies, the presence of human antibodies is revealed by antihuman IgG (raised in rabbit or sheep), which is conjugated to a fluorescent marker such as fluorescein isothiocyanate (FITC).

The higher the level of antibody present in the patient's serum, the greater the range of dilution that can be made before fluorescence is no longer detectable.

In practice, both positive and negative controls are always included. When examined in a fluorescent microscope, the fungal cells are surrounded by bright, continuous green fluorescence. Difficulties may be caused by natural or primary fluorescence of the fungal cells themselves. Spores, in particular, often fluoresce markedly, even in the absence of antibody. For this reason, one of the negative controls consists of fixed fungal cells that have not been stained with FITC. Any visible fluorescence, then, is due to nonspecific autofluorescence.

Antibodies participating in the indirect fluorescent antibody (IFA) test are almost exclusively IgG. Because of the amplification effect caused by multivalency of the antigen on the one hand, and interpolation of an unlabeled immunoglobulin on the other, the test is comparatively sensitive. An advantage of this system is that the same fluorescent conjugate can be used for different antigens, since its specificity is directed at the human immunoglobulin that is attached to the fungal cell. Goudswaard and Virella [144] have modified the IFA procedure by using as a source of antigen macrophages that have taken up fungal material.

As with many serological tests, various modifications can be made which allow the procedure to be used in different ways. Thus, specificity of reaction of the patient's antibodies can be determined by absorbing the serum beforehand with heterologous fungal cells. Furthermore, presence of antigen rather than antibody in a patient's serum could be demonstrated by using this serum to inhibit the binding (and hence the immunofluorescence) of the FITC conjugate to antibodies fixed on the fungal wall surface. If addition of the unknown serum to a control system causes a reduction in IFA titer, presence of free antigen in the former serum should be considered.

For reasons that are not fully understood, young fungal cells are satisfactory for IFA tests, whereas old ones are not. Cultivation of the fungi for production of the antigen is done conveniently in broth for yeast cells. Molds are more difficult to prepare, often requiring multiple slide cultures on single slides to obtain young hyphal growth of suitable quality. In some instances, germlings, produced by brief cultivation of a suitable concentration of conidia in broth culture, provide a convenient source of antigen. The IFA test is a rapid (about 90 min) and effective means of detecting and quantitating antibodies. In some systems it is helpful to undertake examinations on the same serum with IFA and another test, such as agglutination or CF, which is mediated by IgM antibodies. Positive reactions in one system but not the other may provide a clue to the nature and kinetics of the host's antibody response.

G. Immunosorbent Assays

In this type of test antigen is adsorbed onto a solid phase such as polystyrene. Dilutions of the patient's serum are then brought into contact with the "fixed" antigen. The antibodies are detected by antiglobulins which are conjugated with a marker. In radioimmunoassays (RIA), the marker is radiolabeled iodine, which is detected and quantitated in a beta counter. In enzyme-linked immunosorbent assays (ELISA), the marker is an enzyme such as alkaline phosphatase or peroxidase. Its presence is revealed by addition of a substrate which is acted on by the enzyme to produce a color reaction (Fig. 2).

Both tests are extremely sensitive, and both are very versatile. By amending the basic procedure, it is possible to measure IgG, IgA, IgM, immune complex, or antigen. One practical problem with these tests is that many serodiagnostic antigens contain an indeterminate number of chemically uncharacterized components. Where specific antigens can be recognized, the accuracy and sensitivities of the procedures should allow precise analyses to be made of serum antibodies. At the present state of development RIA and ELISA tests will reproducibly rank a series of sera according to their degrees of reactivities. They are less effective and reliable in determining absolute levels of antibodies.

H. Ammonium Sulfate and RAST Tests

For the majority of serological tests, primary union between antigen and antibody is seldom visualized directly. More commonly, secondary reactions are involved, with the production of such phenomena as precipitation, agglutination, and complement fixation. Nevertheless, primary antibody binding may be revealed by immunofluorescence or by reacting radiolabeled antigen with specific antibody, separating unbound soluble antigen and precipitating the complexes by treatment with 50% ammonium sulfate. This method has been adapted to the study of antibodies to *Aspergillus fumigatus* but has not been adopted widely for other fungal serodiagnostic procedures.

Other procedures gaining acceptance and involving the use of radiolabeled reagents include the radioallergosorbent test (RAST) commonly used for determination of IgE antibodies. In this test, antigen is covalently coupled to a filter paper disk, which is then

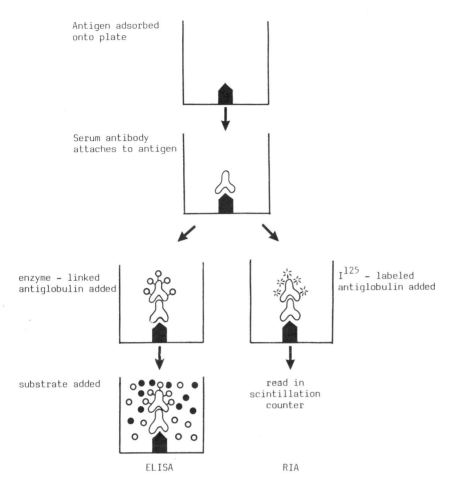

Figure 2   ELISA and RIA procedures for detection of antibody.

treated with the patient's serum. By addition of radiolabeled anti-IgE, the amount of specific IgE bound to the paper can be estimated.

V.   SYSTEMIC MYCOSES

A.   Aspergillosis

1.   Introduction

Aspergilli are widely distributed in nature. They can cause different types of disease depending on allergenic, toxigenic, or pathogenic qualities of the causal agent. Allergic bronchopulmonary aspergillosis is a common expression of respiratory hypersensitivity in atopic subjects. Sensitization to *Aspergillus* is accompanied by the appearance of several distinct clinical features and by the development of specific antibodies. Species of *Aspergillus,* notably *A. fumigatus,* may colonize devitalized areas in the lung and form one or more "fungous balls" or grow saprophytically in the diseased airways. In neither instance does the fungus invade living tissues, but in both conditions, the antigenic stimulus may be strong and protracted, and individuals with an effective immune system will therefore respond with the production of high levels of antibodies. In patients with aspergilloma (fungous ball), the *Aspergillus* forms a dense, interwoven mycelial mass usually lying within a cavity which may be several centimeters in diameter. This mass may be lined by epithelium, and may be completely sequestered, or it may connect via one or more bronchi to the outside air. The condition may be benign, being recognized as an incidental finding during routine chest x-ray or surgery. It is nevertheless, a potentially dangerous condition since in a proportion of patients an increase in size of the aspergilloma may eventually erode a blood vessel, causing either the appearance of blood in the sputum (hemoptysis) or in more severe cases, hemorrhage. In a proportion of cases, the fungus continues to spread by radial extension from the cavity, and may eventually disseminate to the kidney and other organs. *Aspergillus fumigatus* is the commonest cause of aspergilloma, but other species may be implicated, including *A. niger, A. flavus, A. terreus,* and *A. nidulans.* Invasive aspergillosis has a variety of manifestations, involving principally the lung (necrotizing pneumonitis), kidneys, liver, central nervous system, and cardiovascular system. Localized aspergillosis following implantation or deposition may be expressed as mycetoma (e.g., *A. nidulans*) infections of the eye or ear, or as paranasal aspergillosis (*A. flavus*) which affects maxillary and other paranasal sinuses.

2. Tests

  a. Gel Diffusion: The most commonly used test for detecting antibodies to aspergilli is immunodiffusion (ID). The antigen is usually prepared by concentration of culture filtrate or its extraction with acetone [63], or by mechanical disruption of the mycelium [254]. Considerable antigenic variation can be demonstrated among different isolates of *A. fumigatus* [221,430], and since no single antigenic component occurs dominantly and regularly in eliciting antibody responses in sensitized or infected subjects, many studies reported in the literature have used crude rather than purified antigens. There are advantages in using test antigens derived from more than one strain of *A. fumigatus,* and in using more than one concentration of antigen, since antigen-antibody complexes are soluble in antigen excess [319].

  Unless correctly prepared, antigenic extracts may contain C-substance, a somatic polysaccharide which will react with any serum containing C-reactive protein, and produce a line which is indistinguishable from that produced between *Aspergillus* antigen and antibody. C-substance has been demonstrated in *Epidermophyton floccosum* as well as *A. fumigatus* [244]. It can be eliminated by incorporation of a chelating substance, such as sodium citrate or EDTA in the gel, or by briefly rinsing developed gels in citrate prior to reading. C-substance appears in broth cultures after 6-7 weeks of incubation [244] but can also be demonstrated in young mycelial extracts. In either case it is likely to be a source of serodiagnostic error. Recently, studies by Kim and his associates [209,210] have been directed toward improvement of reactivity and specificity of diagnostic reagents for aspergillosis, by disintegrating mycelium mechanically and subjecting the supernatant to fractionation with ammonium sulfate, ion exchange chromatography, and gel filtration. One carbohydrate fraction comprising about 40% of the extract was found to be immunologically inert and could be discarded without reducing the antigenic reactivity of the remaining material. Similar studies elsewhere [161] have been directed at elimination of nonreactive materials, thereby improving the quality of the test antigen and in minimizing variation in content and performance between different batches.

Commercially available antigens are often prepared from both disintegrated mycelium (somatic) and culture filtrate (metabolic). These reagents, used in parallel, have different reactivities. Since serological responses of individual subjects to different antigens of *A. fumigatus* vary greatly, there is some justification for including as many antigens as possible in a serodiagnostic reagent, so that a wide range of antibody production can be detected. However, culture filtrate antigens contain many products of autolysis and secondary metabolism. Although reactivity may be superior to mycelial extracts, this might have been gained at the expense of qualitative and quantitative reproducibility of the antigen—a factor of some practical importance. A contrary argument in favor of mycelial rather than "metabolic" antigens is that in patients with invasive aspergillosis, the infecting agent is in an actively growing mycelial form, and that somatic antigens would therefore be more appropriate than diagnostic materials prepared from old or autolyzing mycelial masses.

With few exceptions, the biological nature and function of protein antigens from *Aspergillus* (or from any other pathogenic fungus) are unknown. However, the French group of investigators at Lille have studied the enzymatic nature of the principal antigenic components of *Aspergillus* extracts [433], and shown that two precipitinogens, designated C and J, have the characteristics of catalase and chymotrypsin, respectively. Furthermore, their value as serodiagnostic antigens is indicated by the demonstration of specific antibodies in the sera of 97% (C) and 100% (J) of patients with pulmonary aspergillosis. This approach has not yet been adopted outside France, but the claims for specificity and reactivity of these antigenic fractions warrant further attention from those interested in developing serodiagnostic reagents.

Antigens prepared from *A. fumigatus* may or may not react with serum from a patient infected with other species of *Aspergillus*. Biguet et al. [34], using rabbits hyperimmunized with somatic and metabolic antigens prepared from different species of *Aspergillus,* showed that specific antigens could be distinguished for each of the principal species responsible for aspergilloma (including *A. fumigatus, A. niger, A. flavus, A. terreus,* and *A. nidulans*). Common antigens were also found, however, although these were numerically less frequent. When a patient with aspergilloma is producing antibodies only to species-specific antigens, no cross-reactions will occur. If antibodies are formed against common antigens, cross-reactivity is likely to be demonstrable with heterologous antigens.

Several methods have been described for demonstrating precipitins to *A. fumigatus* by ID [63,221,244,323], differing in the nature of the gels, the pattern of wells, the volumes and concentrations of reagents, the duration of incubation of the test, and the subsequent processing and reading. Positive reactions are determined by the appearance of one or more precipitin lines. In some systems, reactivity of a serum is assessed by the number of precipitin lines which can be demonstrated. Thus, sera from patients with allergic forms of aspergillosis may have 1-5 lines, compared with up to 15 lines in patients with aspergilloma. Irrespective of the composition of the gel, the quantity and strength of reagents, and the pattern of wells used, diffusion is allowed to proceed for 2-3 days, after which the reaction is evaluated, usually after washing and staining. Coleman and Kaufman [63] reported positive ID tests in the sera of 93% of patients with aspergilloma and 88% (14 of 16) of proven cases of invasive aspergillosis. Specificity was of a high order, and precipitins were not detected in the sera of 55 patients with other mycotic or bacterial infections or neoplastic diseases, or from healthy subjects.

ID tests can be made quantitative, a constant concentration of antigen being tested against doubling dilutions of the patient's serum. This procedure can provide a means of monitoring progress following resection of an aspergilloma, or treatment of a patient with invasive aspergillosis. Following elimination of *Aspergillus,* precipitins disappear; the rate of disappearance depends on the original titer and the degree and rapidity of elimination of the nidus of infection. Halweg et al. [154] showed that antibodies present in the serum of 30 patients with aspergilloma were lost 2-13 months after resection and 1-29 months after radiographic resolution of the aspergillomata when cure was effected spontaneously or by intrapleural or intracavitary injection of nystatin.

b. Counterimmunoelectrophoresis: This procedure has been developed as a serological test for aspergillosis [80,253,439,453]. Its advantages over ID are the greater sensitivity and rapidity obtained, and the smaller volumes of reagents that are required. There are, however, some inherent disadvantages in this procedure. Reactions in the gel may be equivocal, and false negatives may occur unless antigen is used at more than one concentration [220,253]. Furthermore, antigens lacking a net negative charge will not migrate toward the anode and may not react with specific antibodies, even when they are present in the test serum. Patients with aspergilloma may have very high levels of antibody. Unless a concentrated antigen is used, highly reactive sera may be missed. For aspergillosis, CIE is an effective and reliable procedure. It is capable of quantitative as well as qualitative estimation, serum antibody levels being readily titrated by testing doubling dilutions of the patient's serum against a single concentration of antigen. Titers of 1: 128 against *A. fumigatus* are not uncommon in sera from patients with long-standing aspergilloma or heavy saprophytic colonization of diseased airways.

CIE has been adapted to cellulose acetate, as well as agarose or agarose-agar gel [277]. The use of cellulose acetate allows for a more rapid processing of the test. Since there is no gel, there is no requirement for a protracted wash to eliminate unbound serum proteins present in gels. The attractions of this method warrant further attention.

Two-dimensional immunoelectrophoresis has been used [242] to detect and monitor antibodies to *A. fumigatus* in patients with aspergilloma. This technique allows serological changes to be readily visualized, although its prodigality as measured by quantities of reagents required makes it of rather limited value as a routine serodiagnostic procedure.

c. Inert Particle Agglutination: Hipp et al. described a latex slide agglutination procedure for detection of antibodies to *Aspergillus,* using a purified galactomannan antigen obtained from *A. fischeri* [165]. Although a limited study, the results appeared encouraging, by comparison with ID. Passive hemaglutination, using red cells rather than latex particles to render the antigen particulate, has also been studied [80,187,395,431]. In general, low titers of antibody were found in healthy subjects and raised levels in patients with proven infection. In view of the rapidity and sensitivity of this type of serodiagnostic test, it might be anticipated that it will be developed further.

d. Complement Fixation: This procedure, although first used half a century ago, is now seldom used for the routine serodiagnostic examination of sera from patients with suspected aspergillosis [131,314,450]. In one study Walter and Jones [450] reported a 92.4% agreement between CF and immunoelectrophoresis procedures when both procedures were used to detect antibodies in 14 patients with active aspergillosis. Negroni and his colleagues in Argentina have shown that an extract of *A. fumigatus* inactivates serum complement in vivo and in vitro [79,291]. Complement depletion was reduced in hypogammaglobulinemic patients. "Complement deviation" has been used as a diagnostic aid in the diagnosis of *Aspergillus* infections [461]. At present, the advantage of CF tests compared with other procedures is not evident. One commercially available kit is presently on the market, but its efficacy is unknown.

    e. Indirect Immunofluorescence: First used by Drouhet and associates [92,94] in 1970, indirect immunofluorescent antibody studies have received little attention from other investigators. Drouhet and his co-workers reported high immunofluorescent titers (1:40-1:320) in all 130 sera from patients with aspergillosis in comparison with titers of 1:10-1:20 in healthy controls. Similar control serum titers were reported by Warnock [454], who also concluded that the test merited further attention. One of the difficulties encountered in IFA tests is in preparing suitable antigen. Slide-culture methods are not always satisfactory for serum titrations, being difficult to standardize. Gordon et al. [143] have recently suggested a method that obviates this difficulty, where young germlings from a conidial suspension of *A. fumigatus* are used in place of mycelial growth on a glass slide. Although more sensitive than precipitin tests, IFA reactions are less specific [92].

    f. Enzyme-Linked Immunosorbent Assay: First reported by Hommel et al. [171], this procedure has not yet been widely developed. An adaptation, enzyme-linked immuno-electrodiffusion assay, has recently been described from France [329]. In this procedure, antigen is first electrophoresed against test serum on cellulose acetate membranes. The specific immune complexes so formed are then treated with peroxidase-linked antibodies directed against the immunoglobulin heavy chains. Finally, the complexes are revealed by addition of an appropriate enzyme substrate. The test can be completed and read within 6 h and is claimed to be almost twice as sensitive as CIE.

    g. Radioimmunoassay: Marier et al. described in 1979 a solid-phase RIA procedure which was used to detect antibodies to *A. fumigatus* and *A. flavus* in the sera of 19 patients with invasive aspergillosis [264]. Significant rises in antibody levels were found in 15 patients (79%) compared with only 26% by ID and 21% by CIE. Reactions in patients with other mycoses or bacterial infections were rare (8%).

    h. Tests for Antigen Detection In Vivo: In cases of early acute invasive or disseminated aspergillosis and in patients unable to mount an effective antibody response, it might be anticipated that serum and other body fluids would contain antigen rather than antibody. Attempts have been made to develop such tests, using CIE [466] and ELISA [366] procedures. The value of these tests and the validity of this approach remain unproven. One of the problems in their development is in recognizing which antigen(s) is (are) most likely to appear in early infections. A novel approach taken by Lehmann and Reiss [233] has been the use of serum from a rabbit infected with *A. fumigatus* as a source of antigen.

3. Interpretations

Aspergilli are not members of the normal flora of humans, and infections are almost always acquired exogenously. The detection of antibodies to *Aspergillus* is therefore an indication that the subject is harboring actively growing fungal elements, or has been exposed to a sufficiently high and sustained level of inhaled *Aspergillus* material to induce a measurable antibody response.

Bardana [26], using a crude cytoplasmic extract of *A. fumigatus* showed that nonprecipitating antibodies could be detected by primary binding radioimmunodiffusion and ammonium sulfate tests, in almost all of the human sera examined. This widespread occurrence of demonstrable serum antibody was interpreted as a reflection of the ubiquity of *Aspergillus* spores in the human environment, although it was also noted that the test antigens cross-reacted with a variety of heterologous antigens from microorganisms and house dust. In this study it was not possible to distinguish between almost universal exposure to *Aspergillus* in the environment and reactivity to antigens which cross-reacted with, but were not necessarily specific for *A. fumigatus*.

With comparatively insensitive tests such as immunodiffusion, antibodies to *A. fumigatus* are rare in the serum of individuals without colonization or infection with aspergilli. Thus, Longbottom and Pepys [244] reported that detectable precipitins were not found in the sera of 60 healthy subjects, but they were found in 9% of asthmatic patients. In the serum of children with cystic fibrosis, the proportion reacting to *A. fumigatus* antigen in ID is much higher, amounting in one study in England to 31% of 112 patients [270]. In such children it is supposed that the higher proportion of subjects with *Aspergillus* antibodies is related to growth at some time of the fungus in the respiratory tract. Significantly higher levels of *Aspergillus* precipitins were also reported by Bardana et al. [27] and by Galant et al. in the United States [127].

Gerber and Jones have suggested that antibodies to *A. fumigatus* detectable by ID or CF indicate recent or active infection [131]. In patients with proven invasive aspergillosis, however, a rather different situation prevails. Predisposition, in the form of various malignancies, is an important component in the establishment of deep-seated infections by *A. fumigatus*. In one series of patients from a large cancer hospital [274], 41% of those dying with acute leukemia had evidence of aspergillosis. Young and Bennett [476] failed to detect *Aspergillus* antibodies in 15 patients with proven invasive aspergillosis, all but one having serious underlying malignant disease. This finding can perhaps be explained in part by the effects of immunosuppressive therapy that such patients would receive for their primary disorder. In one study reported from Scandinavia [412], patients on corticosteroids had marked and apparently specific reductions in the number of precipitin lines to *A. fumigatus*. It is generally found, however, that antibodies to *A. fumigatus* are present in such patients, although they may not be produced in quantity. Antibodies were reported in one study in 14 (88%) of 16 patients [63]; they were also found in 7 of 10 patients with acute leukemia who were monitored prospectively and who later developed invasive aspergillosis [382]. As with any serodiagnostic procedure, results of tests for *Aspergillus* antibodies may be uninterpretable if considered in detachment from the patient's history and current status. When antibodies are detected by whatever means, the cause must be determined. Laboratory findings may or may not be helpful. It is not always appreciated by the clinician that serodiagnostic tests for aspergilli are still some distance removed from perfection. Used with care and discrimination, however, such tests can be valuable and reliable ancillary diagnostic aids.

B. Blastomycosis

1. Introduction

Blastomycosis was first reported in 1894 by Gilchrist [133] at Johns Hopkins Hospital. The agent, *Blastomyces dermatitidis,* was described by Gilchrist and Stokes 4 years later

[134]. For 70 years it was thought that the disease was confined to the North American continent, but within the past two decades there has been realization that it is also widely distributed throughout the continent of Africa. Blastomycosis is normally acquired by inhalation of airborne spores of *B. dermatitidis*. Although the existence of an exogenous reservoir is clearly indicated by epidemiological and anatomopathological evidence and the agent has occasionally been isolated from soil, the natural habitat of *B. dermatitidis* has not yet been unequivocally established. There are some indications that dead or decaying wood may harbor the organism [14]. Blastomycosis has a wide range of clinical expression. At one end of the scale, infection may be mild or subclinical, and apparently self-limiting [377]; at the other, the disease may be disseminated, progressive, and destructive.

The fungus is dimorphic, being predominantly yeast-like in infected tissues and filamentous in culture. Interconversions from yeast to mycelial growth can be achieved by varying the temperature of incubation. Below 30°C, growth is mycelial; at 37°C it is primarily yeast-like. Antigens for serodiagnostic tests may be made from mycelial or yeast phases of growth, or by extraction from culture filtrates.

2. Tests

Serological tests were first used almost 80 years ago. Antigens derived from *B. dermatitidis* commonly cross-react with other agents, causing respiratory mycoses, and this has proven to be troublesome in developing serodiagnostic tests of acceptable specificity.

a. Agglutination: First used by Ricketts in 1901 [367], this procedure has never been widely used. The inert particle agglutination test alsi is not used widely. A microtiter latex agglutination (LA) test was developed and evaluated for histoplasmosis and blastomycosis by Gerber et al. [132]. No statistically significant differences were obtained between agglutination and CF results, but more patients were found to have antibody by CF than by LA.

b. Complement Fixation: Introduced in 1935 by Martin [265], the CF test has been widely used as a means of detecting and measuring antibodies to *B. dermatitidis*. The antigen used is a suspension of yeast phase cells. Cross-reactions with *Histoplasma* and *Coccidioides* greatly reduce the specificity of the test, and its value is further reduced by the failure of sera from culturally verified cases of blastomycosis to react in the CF test. In one series of 107 cases [52], more than 60% of the patients had negative CF tests. In a second series of 63 cases, 75% had no demonstrable CF antibodies [95]. Campbell [52] compared reactivities of 2639 sera from patients with CF antibodies to antigens from *Histoplasma, Blastomyces,* or *Coccidioides*. Reactions to antigens from one fungus only were obtained in 1790 sera. Cross-reactions involving *Blastomyces* antigen were obtained in 77.9% of the remainder. In evaluating the results of serological tests, an added source of confusion has been the finding that in patients with proven histoplasmosis or coccidioidomycosis, sera may react at higher levels to *Blastomyces* antigens than to homologous antigens. The acute, febrile nature of these two diseases generally distinguishes them from pulmonary blastomycosis, and hence this type of cross-reaction is usually recognized by the mismatch that exists between clinical and serological findings.

Skin tests are not in general use for blastomycosis. It has been suggested, however, that the prognosis is more favorable in patients with positive skin tests to blastomycin and no demonstrable CF antibodies than in patients with no cutaneous reactivity and a high CF titer [403].

c. Immunodiffusion: The use of immunodiffusion (ID) tests was reported by Abernathy and Heiner in 1961 who found precipitins to *B. dermatitidis* by ID in the sera of 14 of 22 patients with blastomycosis [1]. Correlation with CF tests was satisfactory, although discrepancies were noted. Busey and Hinton also used ID and found that the correlation obtained between positive test results and disease was superior to that obtained with CF [50]. In this latter study, based on 80 proven cases, 71 (88.7%) were positive by ID compared with only 33 (50%) of the 65 sera tested by CF. The ID test was more fully evaluated and developed by Kaufman et al. [204]. Two precipitin bands designated A and B were considered to be specific for *B. dermatitidis*. Using reference sera containing A and B precipitins, about 80% of sera from 113 proven cases of blastomycosis had demonstrable antibodies by ID. A positive reaction by this test is considered to be indicative of recent or current infection with *B. dermatitidis*. A method for preparation of reference antiserum by immunization of rabbits with precipitin arcs separated by immunoelectrophoresis was described in 1979 by Green et al. [151].

An interesting application of the ID test is its use in identifying clinical isolates of *B. dermatitidis*. In this procedure, soluble "exoantigens" produced by the agent in vitro are reacted against specific antisera. If a positive result is obtained, presumptive identification of the agent is established. First described in 1966 by Manych and Sourek in Czechoslovakia [261], the procedure was subsequently developed by Standard and Kaufman at the CDC, Atlanta [205,408,409]. The procedure has an exceptionally high degree of specificity for all isolates, tested to date, of *Histoplasma capsulatum, Coccidioides immitis, Blastomyces dermatitidis,* and *Paracoccidioides brasiliensis* [206]. Difficulties in identifying clinical isolates of many representatives of these species are common. The "exoantigen" test is an elegant, effective, and welcome innovation; on the basis of published evidence to date, it is both rapid and reliable. With antigens of such specificity now recognized for each of the major causes of deep mycoses, it would seem reasonable to anticipate that serodiagnostic tests, probably based on ID or ELISA, will become of comparable value in serodiagnosis as the histoplasmosis ID tests (see Sec. V.F.2.b).

d. Counterimmunoelectrophoresis: At the time of writing no reports have yet appeared in the literature describing the detection of precipitins to *B. dermatitidis* by this procedure.

e. Immunofluorescence: Direct immunofluorescence has been used to detect and identify yeast phase *B. dermatitidis* in tissue sections [309]. Indirect immunofluorescent procedures for the detection of antibodies have not been routinely used.

3. Interpretations

Complement fixation tests with yeast-phase antigens of *B. dermatitidis* may be difficult to interpret. Although titers of 1:8 or greater provide presumptive evidence of infection, it is necessary to appreciate that a substantial proportion of sera from infected patients fail to react in CF tests (false negatives), and that CF antibodies may be formed in subjects with proven histoplasmosis or coccidioidomycosis, or without any evidence of a systemic mycosis. It is a basic requirement—and always a responsible practice—to review serological findings in the light of the patient's history and present clinical status. The goal is to recognize and eliminate misleading findings. With CF tests it may be necessary to undertake serial examinations of sera over a period of weeks, and to pursue vigorously both cultural and histological studies for the presence of *B. dermatitidis*, before a confident laboratory diagnosis of blastomycosis can be made.

With immunodiffusion the improved specificity of the present reagents permits a more confident interpretation to be placed on a positive result. Providing that specific A and/or B lines can be detected in a patient's serum, a diagnosis of blastomycosis can be regarded as highly probable. The exact incidence of false-positive ID tests for blastomycosis is not yet known, but it would appear to be insignificant. False negatives where precipitins to *B. dermatitidis* cannot be detected are likely to occur at early stages of infection, possibly in immunosuppressed patients, in patients with subclinical infection, or in subjects who have recovered. The execution of any laboratory procedure is always followed by an evaluation of the results obtained. Recognition of "false-negative" results as such can be an important component of diagnosis.

As with any serodiagnostic test, any marked alteration in antibody level could be an expression of altered host-parasite relationship. Thus, high and rising titers carry a poor prognosis, in contrast to low and diminishing titers, where the outlook is more favorable.

C. Candidiasis

1. Introduction

Candida infections assume many clinical forms. In virtually every instance overt candidiasis is an expression of some primary underlying predisposition. The location, duration, and severity of infection owe more to localized or generalized debilitation of the host than to inherent pathogenic qualities of the associated agents. The most common cause by far is *Candida albicans,* although other species of *Candida* are occasionally implicated, notably *C. parapsilosis, C. tropicalis, C. guilliermondii, C. (Torulopsis) glabrata,* and *C. krusei. Candida albicans* becomes established in the gastrointestinal tract of humans at an early stage and persists apparently for life. Contact with *C. albicans* is therefore a rule rather than an exception, and acquired immunological reactivity to this yeast is so common that an antigenic extract is widely used to locate abnormalities of cellular immunity.

Candida commonly colonizes mucosal and other body sites without producing obvious pathological changes. In such instances the yeast has a commensal relationship with its host. When host factors controlling the *Candida* population become compromised, the balance may be tilted in favor of the yeast, and its status then shifts from commensal to pathogen. This change is usually achieved by an overall increase in numbers of yeast cells that compensates in part for their feeble pathogenicity, or by a breach in the host defense mechanisms allowing entry of the *Candida* into tissues to which access is normally denied.

Exceptionally, among the fungi causing disease in humans, *C. albicans* is endogenous; disease almost always originates from cells that are already present on or in the body rather than from any outside reservoir of infection.

In its commonest form candidiasis is manifested as superficial, localized cream-colored vegetations on the mucous membranes of the mouth or vagina (thrush). Infections of the nail plate and tissues surrounding the nail are sometimes seen in individuals whose hands are frequently wet. Cutaneous candidiasis may also occur in intertriginous areas, particularly in opposing skin surfaces that are constantly moist (e.g., diaper rash). A distinctive, intractable, and distressing form of *Candida* infection affecting mucosa, skin, and sometimes nails is chronic mucocutaneous candidiasis, a disease of immunologically deficient subjects which has attracted much investigative attention. Deep-seated candidiasis is comparatively uncommon and is often difficult to diagnose. Laboratory findings may not always be helpful. Thus, *Candida* may sometimes be isolated from the blood of subjects without any evidence of systemic infection. Moreover, it is not uncommon for blood cultures to be negative in patients with proven deep-seated candidiasis. Isolation of *C. albicans* from urine, throat swab, sputum, or feces may have little significance, since the agent occurs so frequently in such specimens in the absence of significant disease.

Serodiagnostic tests have a role to play in the recognition and management of candidiasis, but considerable confusion and controversy exist concerning their validity and reliability. For candidiasis, more than for any other systemic mycosis, care and caution are necessary not so much in the execution of a test as in the interpretation of its result. The significance of a serological test is sometimes difficult and at times impossible to assess with any measure of confidence. The literature contains many reports on the application of serological tests in the diagnosis of *Candida* infections. Data from different reports are often contradictory, and this has led to confusion and exasperation, sometimes to the extent that *Candida* serology is regarded by some as a snare and a delusion, of no real value as a diagnostic tool. This attitude is understandable but unnecessarily pessimistic. A more realistic viewpoint is to recognize that tests for the detection and measurement of antibodies to *Candida,* while having limitations, can also be of real value. Provided that the tests are used with care, common sense, and discrimination, information can often be gained which helps to implicate *Candida* as a cause of disease. On occasions serological tests can provide information which if disregarded may delay the diagnosis and perhaps imperil the patient's life.

Candida cells produce a wide range of antigens, but the two principal chemical types in serodiagnostic tests involved are proteins derived from the cell cytoplasm and cell wall polysaccharide (mannan). Glycoprotein and other macromolecules are also present in the cell sap or secreted extracellularly, but the major focus of attention in serological tests involving soluble antigens has been on cytoplasmic constituents. Using two-dimensional immunoelectrophoresis, Axelsen [17] has demonstrated 78 soluble antigens associated with blastoconidia of *C. albicans.* There is little reason to suppose that this represents the total number of antigenic substances present in the cell, or that insoluble antigens are of little or no relevance as elicitors of an immune response. For the majority of serological tests currently used, the antigens are soluble and usually consist of a mixture of many different uncharacterized components. When particulate antigens are required, either intact, killed yeast cells or carrier particles such as polystyrene latex or red cells coated with *Candida* antigen are used.

A widely used convention is to describe *Candida* antigens as "cytoplasmic" when derived from yeast cells which have been disrupted. This is misleading, for cell wall mannan is invariably present in extracts prepared in this way. Mannan is also readily demonstrable in culture filtrate ("metabolic") antigen since it is highly soluble in solution and readily produced in broth media in which the yeast cells have been cultivated. The literature contains a large number of studies on antibodies to *Candida* in the sera of infected and healthy subjects. A commensurately wide range of antigens has been used. Their chemical compositions have usually not been defined and their test concentrations have varied widely. In the absence of internationally recognized standard reagents, it is presently impossible to make accurate comparisons of results reported from different centers, and it is not altogether surprising that *Candida* serodiagnostic tests have been viewed with scepticism in some quarters [119].

2. Tests

a. Agglutination: Antibodies in human serum which cause specific agglutination of yeast cells were first detected in the early years of this century [470]. Agglutination tests are still widely used today since they provide a simple and sensitive means of detecting and measuring *Candida* antibodies. In practice, yeast cells are cultivated, washed, and killed by formalin or heat. Dilutions of test serum are mixed with a constant volume and concentration of yeast cells using rigid, molded plastic trays with wells, glass tubes, or microscope slides and allowed to react for a predetermined time. Agglutination is read against dark-field illumination.

Agglutination tests are almost always directed toward the detection of antibodies to *C. albicans*. Since this species accounts for an overwhelming majority of deep-seated *Candida* infections, the practice of excluding other species from test protocols is both rational and defensible. Different species of *Candida,* however, can be shown to carry different antigenic determinants on their cell surfaces [284,435]. This may be an important consideration in the selection of appropriate reagents and procedures for the serological testing of patients infected with *C. parapsilosis*. Agglutination reactions have been used in a series of studies by Tsuchiya et al. in Japan to study the surface antigens of a wide range of yeast genera [436]. The focus of interest in these investigations has been on the taxonomic perspectives revealed by qualitative and quantitative comparisons of the antigenic components of the yeast cell wall. Nevertheless, the sheer abundance of antigens detected by these meticulous cross-absorption studies provides a salutary reminder of the conceptual dangers inherent in imagining the yeast cell wall to be antigenically homogeneous.

The range of agglutinin titers to *C. albicans* obtained by different investigators is very wide. Agglutination will occur only within a defined range of antigen (cells) and antibody concentrations. The sensitivities of different agglutination procedures is dependent to a large extent on the concentration of cells used in the test suspension. In the agglutination system in use at the author's laboratory, the majority of sera are negative at the initial serum dilution (1:4). Titers of 1:16 or greater are uncommon. In contrast, other systems, using fewer cells in the test suspension, have raised thresholds of possible significance. Everett et al. [111], for example, regard "high" levels of agglutinins to be ≥1: 640. There are advantages and disadvantages in using tests at either end of the sensitivity scale. Their ultimate value is dependent partly on the experience gained and partly on the success with which they are intercalated with other test systems within the laboratory. In view of the widespread occurrence in humans of antibodies that react to *C. albicans,* sera from subjects with no evidence of current or previous infection with *Candida* are frequently found to contain *Candida* agglutinins. In a literature analysis made by Odds [302] the percentages of sera from healthy subjects containing *Candida* agglutinins ranged from 0 to 96%. IgG, IgA, and IgM immunoglobulin classes have all been shown to agglutinate *C. albicans* blastoconidia [239]. In common with all other serological tests, inclusion of a control serum with a known titer is mandatory.

b. Inert Particle Agglutination (1) Hemagglutination: Erythrocytes coated with *Candida* antigens have commonly been used as a convenient means for detecting and measuring antibodies. The antigen absorbed onto the surface of suitable and suitably prepared red cells is normally polysaccharide rather than protein. The test is sensitive, and titers up to 1:80 are often present in healthy subjects. Unlike the agglutination test using intact yeast cells, the immunoglobulin class involved appears to be primarily IgM. Müller [281] has suggested that elevated levels of IgM antibodies to *C. albicans* are found exclusively or principally in sera of healthy subjects or in the early stages of systemic candidiasis. In later stages of infection IgG antibody classes predominate. For this reason, an indirect hemagglutination (IHA) test using polysaccharide-coated sheep erythrocytes has been made available commercially for the serodiagnosis of deep-seated *Candida* infections. The dominant role for IgM antibody in IHA reactions was confirmed by Königer and Adam [214]. Their studies revealed, however, that IgG antibody also had hemagglutinating properties.

Human sera sometimes contain naturally occurring globulins which react with foreign antigens, including serum from animals used to raise control antisera, or erythrocytes from unimmunized animals. For this reason a preliminary absorption of test human sera with normal erythrocytes from the species of animal used to produce sensitized red cells is a necessary first step in the IHA test.

Indirect hemagglutination is more rapid to perform than agglutination of intact yeast cells and the end point may be easier to read. A practical difficulty with this test is the apparent instability of anti-*Candida* IgM antibody when maintained in the laboratory. Müller [281] reported that 14 of 20 sera showed reduction in IHA titers of three or more doubling dilutions within 2 weeks of storage. If confirmed, the IHA test may have value only in the laboratory examination of freshly drawn sera.

(2) Latex Agglutination: First described in 1972 by Stickle et al. [415], this procedure uses polystyrene latex particles coated with a soluble antigenic extract prepared by disintegration of formalin-killed blastoconidia of *C. albicans*. The resultant suspension of sensitized particles behaves in a comparable manner to a suspension of intact yeast cells. In one collaborative study by Merz et al. [273], the suspension was shown to be more sensitive than an agglutination test using yeast cells (89% compared with 64%). In the slide latex agglutination test used at the CDC [309], sensitized latex particles may be stored without loss of reactivity for up to 3 months at 4°C.

   c. Complement Fixation: CF tests were first used to differentiate between different types of yeasts. Tests were subsequently used to detect and measure antibodies in the sera of patients with candidiasis, and are still occasionally used today. Pospíšil [339] tested the blood of 300 blood donors and in 1954 reported that CF antibodies were present in only five samples (1.7%). The corresponding figure obtained by Gargani [129] in his study of 850 sera from healthy subjects was 16%, and by Andersen [8] 3.5% in 819 sera from normal subjects. Peck et al. [317] reported a similar (13.5%) prevalence of CF antibodies among 793 patients with skin infections or dermatoses. In 48 patients with superficial candidiasis, however, the figure was 75%. The range of CF positive sera in different surveys is thus comparatively wide ranging [302]. In 17 patients with systemic candidiasis reported by Dolan and Stried in 1973 [91], CF antibodies were not detected at all. Örley and Flórián detected CF antibodies in the serum of 9 (16.7%) of 54 girls with *Candida* vulvovaginitis [305]. Frisk [123] has suggested that rises of CF antibody titers equal or greater than four-fold provide strong supporting evidence for visceral candidiasis.

Early studies attested to the widespread occurrence of CF antibodies among the normal population and to the lack of a correlation between their presence and infection [129, 340]. In comparison with other serological procedures, CF tests are complicated to perform, and they provide no better correlation with disease than do other serodiagnostic methods. With few exceptions [294] their use has been discontinued for the serodiagnosis of candidiasis.

   d. Immunodiffusion: The most widely used serodiagnostic methods tests for precipitins have attracted much attention since their introduction more than a quarter of a century ago. Tube tests were used by Chew and Theus [59] to detect *Candida* precipitins in human sera, but most investigators use a double-diffusion procedure in buffered agar or agarose gel. The test is simple to perform and results are usually read without difficulty. A great deal of variation exists in the pattern and size of individual wells, the nature and concentration of the antigens, the chemical and physical properties of the gel, the duration of incubation, and the method of reading results. By manipulation of test conditions, the sensitivity of a precipitin test can be altered appreciably. The proportion of sera containing *Candida* precipitins is, therefore, largely a function of the test's sensitivity, and this is often the cause for many of the contradictory findings and much of the controversy appearing in the literature. Precipitin lines, i.e., complexes of antigen and specific antibody, may appear after as little as 2 hours of diffusion, although development time is generally extended to 1, 2, or 3 days. Precipitins to antigens rich in polysaccharide tend to be more diffuse than lines produced by reactions between antibodies and protein antigens. It is usually helpful to distinguish between the types of precipitin line since antiprotein precipitins are less commonly seen than antimannan lines. Sera may contain one to several precipitins as shown by double-diffusion (immunodiffusion) tests. Since visible precipitates are formed only when antigen and antibody are at equivalent concentrations, it may be necessary and is usually desirable to use test antigens at more than one concentration. If equivalence occurs outside the area of anticipated reaction (i.e., that part of the gel between the wells), precipitins may be absent or, more commonly, may be seen distal to reagent wells. In some systems, greater precision in identifying the type of antibody is achieved by using cytoplasmic extract, cell wall mannan having been removed by means of concanavalin A [243].

Quantitative precipitin tests may be of value in following the course of infection. Titers can be readily obtained by testing dilutions of the patient's serum against constant concentrations of antigen. Serum from subjects with precipitins to *C. albicans* may react to antigens derived from other species of *Candida,* although reactions to homologous antigens are generally much stronger. Thus, precipitin titers in a patient with endocarditis caused by *C. parapsilosis* (D. W. R. Mackenzie, unpublished observation) were established at 1:128 against *C. parapsilosis* antigen, but only 1:8 when *C. albicans* antigen was used. Precipitin lines are readily seen when the plate or slide overlaid with gel is examined against dark-field illumination. Artifacts resembling lines of precipitate are sometimes formed between antigen and antibody wells and may at times be confused with true antigen-antibody complexes. For this reason, many laboratories remove all unconjugated serum and antigen proteins from the gels by washing, leaving only the immune precipitates. Washed gels are then dried and precipitins revealed by the use of a protein stain.

As with all precipitin tests, sensitivity can be markedly influenced by altering well sizes and disposition, concentrations of reagents, buffers, gel thicknesses, and the conditions of incubation. Sensitivity per se is not necessarily desirable since the proportion of sera containing *Candida* antibodies increases together with the sensitivity of the test. The optimal level of sensitivity is that which elicits the minimal number of uninterpretable negative and positive results. In practice, this is determined largely by trial and error; different laboratories use different levels of sensitivity, and this is one of the principal reasons for the wide divergence or results obtained when precipitin tests are used to detect antibodies to *Candida.*

e. Counterimmunoelectrophoresis: First used for the detection of *Candida* antibodies by Gordon and his colleagues in 1971 [140], CIE is a rapid, convenient, and versatile method which has gained widespread use in serodiagnostic laboratories. Its use for the detection of *Candida* antibodies has been described in several studies [81,82,93,163, 217,253,273,303,353]. It is generally accepted that the practical advantages of CIE compated with ID are increased rapidity and sensitivity. The commonest type of antigen used in CIE is an extract of whole cells, containing material derived from cell walls and cytoplasm. By use of a discontinuous system, where buffer in the CIE gel is at a lower ionic strength than the tank buffer, the sensitivity of the procedure can be increased [263]. A consequence of increased sensitivity, however, is an increase in the proportion of sera reacting positively in the test system, and this may, in turn, add to the difficulties in interpreting the results. CIE has been used by Kerkering et al. [208] to detect antigenemia in the sera of patients with candidiasis (see Chap. 6).

  f. Two-Dimensional Immunoelectrophoresis: This method has also been applied to the detection of *Candida* antibodies. In this procedure, an antigenic extract is separated into its individual components by electrophoresis through an agarose gel. Distance of migration by each component is determined by its electrostatic charge. Antigens separated in this manner are then electrophoresed in a second dimension at right angles to the first run into agarose containing antiserum. Each antigen will complex with its homologous antibody to form a broad-based precipitin "peak." Since these peaks are separated in the second dimension run according to the relative concentrations of antigen and antibody, the net result is a clear separation of the individual antigen-antibody reactions. Since the height of each peak is determined by the relative concentrations of antigen and antibody, the procedure can be used to provide qualitative measurement of antibody responses by noting numbers of precipitin peaks and their heights when the second dimension run is made into patient's serum. Svendsen and Axelsen [418] examined precipitins in human serum contained in an intermediate gel interpolated in the second dimension run and devised a system where precipitin scores were estimated by adding together the number of precipitins detected and their titers, as determined by comparison with a standard rabbit antiserum. It was found that precipitin scores above 17 were found only in patients with candidiasis. Axelsen et al. used 2D-IE to demonstrate precipitins in the sera of 15 patients with chronic mucocutaneous candidiasis [18]. From 2 to 39 peaks were detected. All patients had antibodies to a mannan-protein complex. By the use of 2D-IE with an intermediate gel, Axelsen and Kirkpatrick [19] demonstrated the coexistence of free antigen and antibody in the serum of a patient with chronic mucocutaneous candidiasis and defective cellular immunity. The potential of 2D-IE in production of standardized serodiagnostic reagents was investigated by Axelsen [17]. An extract was prepared from cells of *C. albicans* and shown by 2D-IE to contain many water-soluble antigenic components. When tested against a standard rabbit antiserum, quantitative determinations were made by 2D-IE on the quantities and proportions of 30 major antigens in different batches of polyvalent extract. The method had a satisfactory degree of reproducibility. This approach provides a means of assessing batch-to-batch variation in diagnostic or reference antigenic preparations, and the method could be widely used in the production of "standard" serodiagnostic antigens from a range of fungi. The greater sensitivity and resolving capacity of 2D-IE has been applied to the routine serological examination of patients undergoing hyperalimentation for antibodies to *C. albicans* by Glew et al. in 1978 [135]. In this study, precipitins to protein rather than mannan antigens were found in the sera of all five patients who developed systemic candidiasis. In a later development of the technique, Syverson et al. [423] incorporated concanavalin A into the intermediate gel. This had the effect of eliminating mannan and ensuring that any precipitins detected in the test were directed to cytoplasmic protein antigens only.

g. Immunofluorescence: Indirect fluorescent antibody procedures have been commonly and widely used for the detection of antibodies to *C. albicans*. Blastoconidia are simply obtained and are a convenient size for immunofluorescence. The test is both simple to perform and sensitive. Lehner [237] reported in 1966 that IFA titers up to 1:8 were present in 80 control subjects, between 1:1 and 1:16 in 88 asymptomatic carriers of *C. albicans*, and >1:16 in 75 (64%) of 76 patients with oral candidiasis. In an earlier study [236], antibodies were detected in serum and saliva of patients with oral infection. It was suggested that clinical infection is indicated when serum titers were ≥1:32 and saliva titers ≥1:1. Coudert et al. reported in 1967 [72] that antibody levels up to 1:80 were present in 17 patients with superficial candidiasis. Esterly [105] examined 207 sera by an IFA procedure and suggested that normal subjects and carriers had titers ≤1:64, and patients with candidiasis had titers ≥1:64. Titers and immunoglobulin classes of antibodies to *C. albicans* were studied in sera from 65 subjects by Lehner, using an IFA test [238]. Significant levels of IgG were found in 78% of sera from patients with candidiasis. Corresponding figures for IgA and IgM were 30% and 51%, respectively. Concentrations of the serum immunoglobulin classes in patient and control groups did not differ significantly. The principal immunoglobulin class of antibody reacting in the IFA test is IgG, although IgA and IgM subclasses have also been shown to react [238]. Taschdjian et al. studied serum antibody levels to *C. albicans* in 36 high-risk patients using immunodiffusion, agglutination, and IFA [426]. IFA titers in 9 subjects who proved free of *Candida* at autopsy ranged from 0 to 1:320. Of the 24 patients in whom candidiasis was confirmed at autopsy, 16 (67%) had IFA titers ≥1:320. Within the same group, agglutinins and precipitins were detected in 70% and 89% of the patients, respectively. Negroni and Flores [292] have also studied IFA titers in subjects with and without candidiasis. Antibodies were detected in sera of 51 of 100 individuals without candidiasis at titers ranging from 1:10 to 1:60. Levels of antibody in 35 patients with superficial candidiasis ranged from 1:10 to 1:320. Sera from 10 patients with *Candida* septicemia or endocarditis contained high levels of antibody, giving IFA titers of 1:2560-1:20,480. Studies with IFA have been used in the assessment of *Candida* antibodies in women with vaginal candidiasis [371,455]. Warnock et al. [456] showed that levels of IgG and IgM antibodies to *C. albicans* in cervicovaginal secretions were similar in women with and without candidiasis colonization or vaginal candidiasis. No secretory IgA antibodies to *C. albicans* were found in the serum of women with vaginal candidiasis. These investigators concluded that IgG *Candida* antibody present in the secretions originated from the serum and was not produced locally in the vagina [478].

The two different morphological phases of *C. albicans* are known to differ antigenically [422], and differences in intensity of fluorescence between pseudohyphae and blastoconidia have been observed by Haneke [156]. The brighter immunofluorescent staining suggested that more antibodies were bound to filaments than to budding cells.

Poulain and Vernes in France have recently drawn attention to a potential cause of irreproducibility in the *Candida* IFA test [341,342]. Clones of *C. albicans* derived from a single colony by micromanipulation behaved in a different manner in the IFA test. Some clones produced variable IFA titers when tested on different occasions, whereas others maintained their immunofluorescent characteristics and titers. This variation was reflected in up to 32-fold differences in IFA titer. One clone, designated 32, appeared to be stable and provided uniformly higher titers than other clones studied. This phenomenon, which is associated with a "parietal" or p antigen in the yeast cell wall, is of relevance in the choice of a suitable strain for immunofluorescence studies. In the past it has been generally assumed that different strains of *C. albicans* were essentially interchangeable. These studies illustrate the dangers of making such assumptions and the absolute need to examine all components and parameters of a serodiagnostic system before any confidence can be felt that these have been "standardized."

Studies at the University of Colorado by Everett et al. [110] were directed to an examination of yeasts seen in the urinary tract, with the intention of establishing if the presence of immunoglobulins on the yeast surfaces could be correlated with evidence for tissue invasion. Of the 18 patients examined, 15 had *Candida* blastoconidia or filaments in the urine which fluoresced when stained by IFA for human immunoglobulin. No association was found with upper urinary tract invasion, and the procedure did not afford a useful indication that treatment should be initiated. Similar studies by Harding and Merz [157] showed that IgG was present in all five species of yeasts recovered directly from urine or other body sites. These authors also concluded that the presence of surface antibody had no value in identifying the site of urinary tract infection with yeasts.

h. Enzyme-Linked and Radioimmunoassays: The use of an enzyme-linked immunosorbent assay (ELISA) to detect *Candida* antibodies was reported by Hommel et al. [171] in 1976. The method was found to correlate well with IFA and immunoelectrophoretic procedures. In the same year, Carlsson and Bernander [56] showed in a system using experimental animal serum that the ELISA procedure was sensitive and specific. This was followed up by a study of human sera, using a purified mannan as antigen [57]. A good correlation was observed between results obtained by ELISA and passive hemagglutination. The simplicity of ELISA procedures, together with their sensitivity and versatility, make it likely that they will be used increasingly in the serodiagnosis of fungal diseases, and it is known to be under development in several laboratories with a special interest in serodiagnosis. Holmberg [170] has briefly described an ELISA procedure for the detection of antibody to cell wall mannan, and suggested that this procedure may be of some diagnostic value since specific IgA antibody to mannan appears to rise early in the course of invasive candidiasis in leukemic patients. Bull and Mackenzie [49] used a radioimmunoassay (RIA) and in examining immunoglobulin classes of antibody to different *Candida* antigens, also concluded that antimannan antibody may be an earlier and more reliable indicator of invasive candidiasis than antibodies to cytoplasmic proteins.

The striking sensitivity of ELISA has been applied to the search for antigen rather than antibody in the serum of infected subjects. Warren et al. [459,460] detected *Candida* antigen in the sera of mice and rabbits experimentally infected with *C. albicans*. Positive reactions were also noted in the sera of three patients with candidiasis. Although it might have been anticipated that the antigen detected by the ELISA procedure was mannan, the authors provide evidence to suggest that a line detectable in CIE when experimentally infected mouse serum is tested against convalescent rabbit serum could not be absorbed from the latter serum by mannan, and that its nature remains unknown. Stevens et al. [413], using a solid-phase radioimmunoassay, found *Candida* antigen in the sera of 12 of 19 patients with disseminated candidiasis. These workers also concluded that the antigens detected were unlikely to be mannan. Lehmann and Reiss reported similar findings, using an ELISA procedure [234].

Circulating mannan was, nevertheless, detected by Weiner and Yount in the sera of four patients with invasive candidiasis and in two with candidiasis of the gastrointestinal tract by a system involving hemagglutination inhibition [463]. Mannanemia was not detected in 48 patients with noninvasive candidiasis or in 99% of 234 patients in other control groups. Of interest was the finding that antibodies to mannan were present in those patients with invasive candidiasis where circulating mannan could not be detected. In a later study using RIA [462], Weiner confirmed the presence of mannan antigenemia during the course of systemic candidiasis and demonstrated the utility of fungal antigen immunoassays for early, specific immunodiagnosis of invasive *Candida* and *Aspergillus* disease.

Segal et al. used an ELISA-inhibition procedure and showed that quantitative differences in inhibition were obtained when sera were tested from patients proven at autopsy to have systemic candidiasis compared with sera from patients with clinical evidence only of *Candida* infection or from patients proven at autopsy to be free from invasive candidiasis [394]. At present, ELISA procedures using rabbit immunoglobulin absorbed onto the solid phase for detection of mannan require the careful use of controls to eliminate false-positive reactions. Sera sometimes react against immunoglobulins from normal, nonimmunized rabbits, rheumatoid factor [480], and also IgM immunoglobulins [252,480], sensitive to dithiothreitol degradation.

An enzyme-associated immunodiagnostic test using magnetic polyacrylamide agarose was described recently by Othman et al. [306]. The authors tested 39 sera from patients with various forms of candidiasis and from healthy subjects. Results compared well with those obtained by CIE and passive HA and the technique is described as being simple and very sensitive.

i. Gas-Liquid Chromatography: Miller et al. [275] described in 1974 the use of gas-liquid chromatography to detect circulating products of *C. albicans* in the serum of six patients with *Candida* septicemia. The peaks seen in these sera were not observed in normal sera but could be demonstrated in serum previously inoculated with *C. albicans* and culture filtrates and cells of *C. albicans*. Peaks were not seen in patients with bacteremia. The chromatographs of patients with mucosal candidiasis differed little from those obtained with serum from normal subjects. In two of four patients with deep-seated invasive candidiasis, chromatograph profiles were indistinguishable from those obtained from the subjects with *Candida* septicemia. In both cases the abnormal peaks disappeared following treatment. Preliminary studies with co-chromatography suggested that the peaks were derivations of mannan or mannose. This procedure has not yet been widely adopted. If shown to be reliable, the addition of a new serological procedure to the repertoire of laboratory procedures already available would be of great value.

j. Radioallergosorbent Tests: Cells and products of *C. albicans* may evoke hypersensitivity reactions in atopic subjects. Demonstration of IgE subclass of antibody to *Candida* by RAST tests has been reported by Baldo et al. [23] in Australia and Edge and Pepys [96] in the United Kingdom.

k. Germ Tube Dispersion: Described first by Katsura and Uesaka in Japan [196], this test is based on the observation that germ tubes produced by blastoconidia may not be clumped by serum of mice experimentally infected with *C. albicans*. A correlation was seen between the degree of germ tube dispersion for a serum and the extent of pathological changes in the animal. Oblack et al. [301] examined the germ tube dispersion capabilities of serum from burned patients with different severities of infection with *C. albicans*. Dispersion of germ tubes was significantly more pronounced in these sera compared with normal controls. It was found that a positive reaction with the test correlated with the existence of candidiasis in several patients with clinical evidence of infection but who lacked agglutinins or precipitins. "Clumping" of *C. albicans* occurs in most human sera. "Clumping factor," the substance responsible, has been characterized as a euprotein of fast β-globulin mobility [404]. In some sera, clumping is suppressed by the existence of an inhibitor [60]. This feature has been identified as IgG antibody specific for *C. albicans*. It is not yet known if any relationship exists between Katsura and Uesaka's germ tube dispersion test and the IgG antibody described by Chilgren et al. [60].

l. Macrophage-Engulfed Antigen: Goudswaard and Virella described in 1978 a new serodiagnostic procedure where macrophages containing *Candida* antigen are used as a substrate for an indirect fluorescent antibody test [144]. Macrophages are harvested 24 hours after intraperitoneal injection of the antigen. Dilutions of test sera are added and their titers determined. Results suggested that the new test was more sensitive than CIE and that high levels of antibody ($\geqslant 1:64$) tended to be associated with patients having systemic candidiasis.

3. Interpretations

For no other mycosis has so much attention been focused on serodiagnostic procedures and their interpretation. The literature contains many reports where attempts have been made to detect antibodies in the sera of subjects with and without candidiasis, and to correlate their presence, nature, and quantity with the status of the host-parasite relationship. For no other mycosis has so much investigative attention led to so little overall agreement on the merits and limitations of serological tests. There are several reasons why a consensus view has not emerged from the wealth of information and experience which has been amassed particularly over the past decade and a half. Some of these problems are related to the endogenous origin of candidiasis. Acquired immune responses represented by type IV hypersensitivity to *Candida* skin tests [397] or lymphocyte transformation [121] in the presence of *Candida* antigen has been demonstrated in 94% and 100% of normal adults. Moreover, provided that a sufficiently sensitive technique is used, antibodies reacting to *Candida* antigen are almost invariably demonstrable in human sera. This "background" level of immune responsiveness can vary greatly from one person to the next. It is also likely to alter as an individual experiences changes in local or systemic predisposition which cause *Candida* population levels to rise or fall. Any change in any of the measurable immunological responses to *Candida* reflects some alteration in the status quo. Recognition of the alteration by change in antibody levels is usually achieved much more readily than realization of its significance. Because of its endogenous nature, serological tests for *C. albicans* are of less diagnostic value than those used in diseases caused exogenously by acquired pathogens, since the presence and levels of antibodies may correlate with neither infection nor tissue damage. More than any other single factor, the occurrence of specific antibodies in the serum of healthy subjects or those with *Candida* colonization or superficial infections has constituted a deterrent to the acceptance of serological findings as a reliable indicator of systemic candidiasis. The frequent occurrence of *C. albicans* in clinical specimens in the absence of disease may reduce the value of cultural studies. Thus, the isolation of *Candida* from blood, urine, or sputum is by itself of no reliance as a sign of systemic infection. Many predisposing causes of candidiasis are known, and a wide range exists in the relationship between *Candida* and its host. Criteria used for establishing a clinical diagnosis may not be reliable, and the demonstration of tissue invasion by examination of biopsy material, although providing reliable evidence for invasive candidiasis, may not always be practicable. The difficulty in making a firm clinical diagnosis has provided not only a stimulus for development of ancillary aids to diagnosis, but a major barrier to their accurate evaluation.

Candida albicans has many antigenic components: some associated with the cell wall, some with the cytoplasm, and some with extracellular metabolites or products. Although many antigens have been obtained from blastoconidia, no correlation has yet been made between any component or product of actively growing cells of *C. albicans* and tissue invasion or disease. There can be little doubt that if the various stages involved in *Candida* invasiveness could be recognized and the individual components of the pathogenic process defined biochemically, the way could be opened up to the selection and use of serodiagnostic antigens that correlated with invasiveness. Staib [406] reported that strains of *C. albicans* which produced extracellular protease were more virulent toward mice than strains lacking the enzyme. This line of investigation has been developed by Macdonald and Odds [249] and may lead to tests of improved sensitivity and specificity.

Depending on the conditions for growth, *C. albicans* may produce yeast or mycelial forms. The existence of antigens specific for each phase has been convincingly demonstrated [422], but the ultimate relevance of this finding to serodiagnostic tests for candidiasis remains to be determined. Almost all antigenic extracts of *C. albicans* have been made by disintegration of blastoconidia grown in vitro. The basic assumption that these forms are biochemically, physiologically, and antigenically comparable to those produced in vivo has never been seriously challenged, although its confirmation or rebuttal would materially alter the validity of existing serodiagnostic approaches and methods. Variation in the capacity of *C. albicans* blastoconidia to elicit antibodies in experimental animals is readily demonstrable and different strains—or even different clones [342]—vary in their anticipated responses for any diagnostic test. It is a self-evident truth that any source of variation in the qualities or performance of an antigen is a cause of unpredictability in the test system and consequently a lack of reliance in any results obtained. Yet another source of variation in serodiagnostic tests associated with the agent is its specific and antigenic identity. *Candida albicans* is the major cause of candidiasis but not the only species capable of causing disease. Cross-reactions do occur between *C. albicans* and other species of *Candida* implicated at times as causes of disease [284,435], but as a rule homologous antigen elicits reactions which are more distinct and which are produced at a higher titer than those against antigens derived from other species. Hasenclever and Mitchell described two serotypes (A and B) of *C. albicans* in 1961 [158]. With very few exceptions, antigens prepared for serodiagnostic studies, when the serotype has been specified, have been derived from A serotype strains. The finding of antigenic heterogenicity among different clones of the same strain reported from France [341,342] provides a salutary reminder of the inherent complexity and variability of the yeast cell, and the dangers of assuming that the behavior of strains of *Candida* grown under uniform conditions and the nature of the serodiagnostic antigens derived are entirely predictable.

Progress toward a better understanding of the role of serodiagnostic tests in candidiasis has been hindered by variations in the test procedures used. There are at present no "standard" reagents or procedures to provide reference points for the evaluation and comparison of individual serodiagnostic systems. The need for reference agents and coordination of tests was recognized by Seeliger [391] and by Mackenzie and Austwick in 1965 [251], and more recently by the inclusion of a Round Table Conference on Standardization of Serodiagnostic Reagents and Procedures in the program of the 6th International Congress of the International Society for Human and Animal Mycology (ISHAM) in Israel in 1979 [255]. Recognition of the need for reference serodiagnostic reagents and procedures for *Candida* is now being followed up by the evaluation of candidate preparations by international organizations with a special interest in standardization of biological reagents, notably by the International Union of Immunological Societies and by ISHAM. Until and unless reference reagents are designated, the value in comparing and correlating reports from different centers remains uncertain.

Other causes for misinterpretation or uncertainty in the results of serodiagnostic tests include failure to take into account the kinetics of immunoglobulin responses, and the limited capacity of serological tests to distinguish between *Candida* colonization and invasiveness. Antibodies reacting to *Candida* antigens are common in subjects without overt candidiasis. Many studies in the literature refer to such findings as false positives. This view, if taken too literally, may reduce the value of serodiagnostic tests to an unrealistically low level. All orthodox serodiagnostic procedures used as ancillary laboratory aids to diagnosis of candidiasis are directed toward the recognition and measurement of antibodies. Their detection does not necessarily imply that *Candida* is causing disease; nor does their absence preclude infection. They are, at best, indirect tests, yielding results which then have to be linked with the primary cause. The link is not always an obvious one. In many instances the results of a single test are uninterpretable, although in a proportion of these, additional clinical or laboratory studies may provide the clue to the original findings. By reference to cumulative experience with any one test, whether used to detect precipitins, agglutinins, or indirect fluorescent antibodies, some idea can be gained of the frequency with which any single level of antibodies can be found, for patients suffering from any single type of primary disorder. A notional range of anticipated results exists for each procedure and for each clinical expression of candidiasis. Whenever a result is obtained that is outside the "normal" range, the clinician at least can be alerted to the possibility of candidiasis. If this is excluded on clinical grounds, the test result is regarded as a false positive. It should be appreciated that transient or subclinical infections may not be recognized as such, and that it is failure to recognize the link between raised antibody levels and such infections which accounts for a proportion of false positives. Similarly, *Candida* colonization of oral or vaginal mucosal surfaces or the bloodstream may elicit antibodies, and these are sometimes regarded as false positives. It is misleading and at times unhelpful to regard false positives as a source of error in viewing results of serological tests. The same is true for false-negative results. The real error comes in placing too much reliance on the results of tests on a single serum and in supposing that the presence or absence of antibodies to *C. albicans* can provide information of absolute diagnostic value.

Agglutinins to *C. albicans* have been found in the serum of healthy subjects and in patients with a wide range of *Candida* infections. Odds [302] cites reports where agglutinins were found in 0-96% of healthy subjects and in 0.2-87% of patients without proven candidiasis. In general, however, infections with *Candida* lead to an increase in the number of positive agglutination tests and in the agglutination titers obtained. Preisler et al. [343] demonstrated a fourfold or greater rise in agglutinin titer in the sera of 7 of 14 leukemic patients followed prospectively who developed visceral candidiasis. Used by these investigators, the agglutination test was more reliable than precipitin tests. Holmberg [170] cited similar results in 14 of 23 leukemic patients in whom autopsy later revealed disseminated candidiasis. Precipitins were detected in only 4 of the patients. Agglutinins are directed principally to mannan components of the cell wall. This material is produced in abundance by *Candida* in vitro and in vivo, and is capable of eliciting specific antibody whether or not tissue invasion has occurred. Agglutinins with mannan specificity are thus commonly demonstrable in patients with superficial or noninvasive forms of candidiasis, such as oral or vaginal candidiasis [45,410] or chronic mucocutaneous candidiasis [18]. Raised agglutinin titers are also found in subjects with colonization of the urinary tract or impaired bladder function [109] or with transient *Candida* fungemia [428].

The situation is broadly comparable in patients where *Candida* antibodies have been studied by immunofluorescence, although the predominating subclass of immunoglobulin which is principally involved in this test is IgG rather than the mixture of IgG, IgA, and IgM that is involved in the agglutination reaction [214,239].

The precipitin test for candidiasis was used by Elinov and Zaikina in 1959 to examine sera from 10 patients with candidiasis of the viscera or skin [99]. Only the 5 patients with deep-seated candidiasis had serum precipitins. In the 1960s, Taschdjian, Kozinn, and their colleagues at Maimonides Hospital in New York City began a series of studies on *Candida* serology in general and the precipitin test in particular, which first drew attention to the merits of this test and then to the circumstances where caution is appropriate in ascribing significance to a positive result. In 1964, Taschdjian et al. [424,425] reported that *Candida* precipitins were not detected in normal human sera, but could be recognized in 5 of 8 patients with systemic candidiasis. Stallybrass, using a similar antigen prepared by sonication of blastoconidia, found no precipitins in the sera of 244 sera from healthy subjects [407]. In later studies, Taschdjian et al. reported precipitins in 79.7-89% of patients with systemic candidiasis [426,427]. In the series of 19 reports reviewed by Odds [302], the proportion of patients with serum precipitins to *Candida* having proven candidiasis ranged from 17.4 to 100%, with 13 of the 19 recorded prevalences exceeding 70%. These precipitin studies created widespread interest, and other reports appeared in the literature describing the findings obtained by other investigators at other centers. Chew and Theus [59], using a more sensitive technique with Preer tubes, and up to 15-fold concentration of sera before testing, reported that precipitins to *C. albicans* cell wall mannan were found in 15 of 31 healthy subjects in 69% of 62 patients with candidiasis. Pepys et al., [320] using gel diffusion and threefold concentration and retesting of negative sera, showed that over 72% of patients with asthma and eosinophilia had precipitins to *C. albicans* mannan. The comparable figure for healthy subjects was 22.7% and 4.5% when tested at a threefold concentration and unconcentrated, respectively. Murray et al. [285] in 1969 examined 409 sera from 68 patients undergoing open-heart surgery and observed that precipitins were developed in 14 of 62 patients 2 weeks after surgery, in the absence of any detectable *Candida* infection. Later reports have confirmed that *Candida* precipitins appear in the sera of patients subjected to open-heart [108] or abdominal surgery [457], in burned patients [167], or in patients with subacute bacterial endocarditis [22,81]. The studies by Evans and Forster in 1976 [108], suggest that these transient antibody rises, which include agglutinins as well as precipitins, are related to temporary increases in the numbers of yeasts colonizing the gut. Reduction in the overall yeast population following recovery of the patient or the use of antifungal prophylaxis is accompanied by a reduction in the proportion of sera containing detectable antibodies. Murray et al. [285] concluded that a positive precipitin test did not correlate with *Candida* endocarditis, but a negative test apparently ruled it out. Parsons and Nassau [315] also concluded that positive sero-

logical findings were probably related to transient fungemia. They suggested that serological monitoring should be continued for at least 3 years after heart surgery. Gaines and
Remington [126] reported in 1973 that 24 of 26 patients with candidiasis had precipitins
to *C. albicans.* Two false negatives were encountered among the 107 hospital patients who
lacked *Candida* precipitins. One of these had terminal leukemia and the other had undergone multiple abdominal surgery. Candidemia occurred in both patients, shortly before
death. One possibility for the false-negative results was the existence of circulating antigen rather than antibody, but this was not investigated.

With the introduction of counterimmunoelectrophoresis (CIE) in 1971 [140], the
search for precipitins to *C. albicans* continued, but with a test whose results differ in some
respects to those obtained with classical Ochterlony immunodiffusion. In 1976, Kozinn
et al. [216] analyzed results obtained with the *Candida* precipitin test by their own group
and others, and expressed its efficiency in statistical terms. Optimal results suggested a
sensitivity (proportion of diseased individuals with positive results) of 100%, a *specificity*
(proportion of nondiseased individuals with negative results) of 95.5%, and *predictive
value* (proportion of positive reactors having the disease tested for) of 90%. At the other
end of the scale, minimal values of 27%, 56.5%, and 18.9% were obtained. Some of the
variations in individual test performance can be explained by differences in methodology
and lack of suitable standardized reference reagents; others may relate to the choice of
diseased state studied and variation in the disease prevalence, which in the studies cited
ranged from 8.8 to 60%. Merz et al. [273] extended the statistical analysis to other serodiagnostic tests in a collaborative study involving six different centers. In this investigation, tests were made with 53 sera from patients with proven, presumptive, or probable
candidiasis and 47 sera from healthy subjects or patients with other diseases. Although
marked differences were sometimes noted between results obtained in the individual
laboratories, double diffusion using a prescribed tube method gave positive results in 85%
of patients' sera and 5% of control sera. Latex agglutination [415] gave positive results
($\geqslant$1:8) in 89% and 17.4%, respectively, while the corresponding figures for tube agglutination were 64% and 12.3%. By extending the study to include a larger number of unselect-
ed control subjects, Kozinn et al. [217] showed in 1978 that both ID and CIE had sensitivities, specificities, and *efficiencies* (proportion of individuals correctly identified as having or not having the disease being tested for) of about 90%. Despite the advances in understanding made by such analytical approaches, serodiagnostic tests for candidiasis have
detractors as well as supporters. Filice et al. examined sera from patients at a cancer
hospital in New York City by double diffusion and slide agglutination for antibodies to
C. albicans [119]. The study included 40 patients with various forms of candidiasis superimposed on their primary underlying condition, and 100 control subjects. Neither presence
or quantity of *Candida* antibodies were found to correlate with invasive candidiasis. ID
tests were considered to be negative unless the precipitin reaction was as strong or stronger
than the control reaction between test antigen and the control rabbit antiserum. Moreover,
weak lines of precipitate were regarded as negative unless identity with rabbit antiserum
could be demonstrated, and only one concentration of antigen was used. With such
marked divergences in procedural detail, it cannot be considered surprising that the authors
reached such a pessimistic conclusion about the ID test. It is, nevertheless, to be noted that
97% of the control subjects had agglutinin titers of <1:16. Antibody titers generally correlated with levels of immunoglobulin. Patients with solid tumors tended to have normal

or elevated levels and produced antibodies more readily than those with lymphomas, who generally had low levels of immunoglobulin.

Myerowitz et al. [287] assessed the value of *Candida* precipitins detected by CIE in a prospective study of 39 patients with leukemia. Two patients who developed disseminated candidiasis had high titers ($\geqslant 1:8$). Because of the small number of true positives in the series, the test gave a low predictive value, and the authors considered that it did not appear to be well established as a reliable diagnostic procedure for leukemia patients.

Lehmann and Reiss [235] found by an ELISA procedure that levels of IgG antibodies directed against *Candida* mannan in leukemic patients were comparable to those in the normal population. There is some evidence to suggest that leukemic patients with pre-existing antibody to *C. albicans* respond to the onset of candidiasis by production of raised specific antibody levels [36].

An unusual finding by Iwashita et al. was the occurrence of *Candida*-specific oligo-clonal IgG antibodies in the CSF of a 13-year-old boy with chronic *Candida* meningoen-cephalitis [189].

With the recognition of cell wall mannan as the major antigen against which antibodies are directed, and accompanying increases in sensitivity of precipitin tests, the view gradually emerged that antibodies to cytoplasmic proteins, being less common, are of greater value and efficiency as serodiagnostic agents. This view was originally advanced by Taschdjian et al. [424] and constituted the basis for the choice of disrupted blastoconidia ("somatic") as the antigen for use in ID. However, cells disrupted by ultrasonication or other disruptive procedures also contain appreciable quantities of cell wall mannan, so antigens prepared in this way are more accurately regarded as whole cell extracts than as cytoplasmic extracts. Mannan, however, is highly reactive in ID and CIE and can be shown to inhibit both immunofluorescence and agglutination reactions. When removed from an extract of *Candida* blastoconidia by affinity chromatography with concanavalin A linked to sepharose, the remaining mixture of individual proteins can be assessed independently of the mannan. Using this approach, Ellsworth et al. [100] compared reactivity of sera from 20 patients with candidiasis to mannan and protein. It was found that 13 (65%) reacted to mannan and 3 (15%) to protein antigens. The correlation found between response to protein antibodies and disseminated disease was 15%. However, it was found that serum from patients without infection or with superficial candidiasis also contained precipitins to protein (5 of 19) and mannan (16 of 19) antigens. In this study the specificity of antibodies to protein was of a low order, since only 3 of 8 patients in the series with invasive candidiasis contained precipitins detectable by the ID method used (not specified).

Antibodies that react to *Candida* mannan are present in virtually all human sera. For this reason, their presence is often regarded as unhelpful. It is, nevertheless, readily demonstrable that in the course of a *Candida* infection, the first antibodies that become apparent are directed to mannan rather than protein. In one patient with *Candida* endocarditis, rising antibodies to mannan were observed 8 days before antibodies to cytoplasmic proteins could be detected (D. W. R. Mackenzie and W. Waine, unpublished results). Had the rise in antimannan titer been ignored, a critical clue to the establishment of infection would have been missed.

Serodiagnostic findings are seen in better perspective for candidiasis when made over a period of time. When viewed in relation to cultural and clinical findings, their diagnostic relevance can almost always be appreciated. There is much progress to be made in this area, but the situation is already better than the most vehement of critics might allow, if not yet developed to the degree of reliance that we can eventually expect to attain.

D. Coccidioidomycosis

1. Introduction

The first case of coccidioidomycosis was described in 1892 by Posadas [337,338] and Wernicke [465] in Argentina. They considered the agent seen in infected tissues to be a protozoan, an interpretation endorsed in 1896 by Rixford and Gilchrist when reporting the second case, which occurred in the San Joaquin Valley in California [368].

The name *Coccidioides immitis* was introduced by Rixford and Gilchrist, but the fungal nature of the pathogen was recognized by Ophüls and Moffitt 4 years later [304]. The demonstration by Hirsch and Benson in 1927 that infected subjects acquire skin re- activity to antigen derived from *Coccidioides* culture filtrate [166], the discovery by Stewart and Meyer in 1932 that the natural habitat for *C. immitis* is soil [414], the reali- zation by Dickson and Gifford in 1938 that benign pulmonary coccidioidomycosis is the rule and severe, disseminated forms the exception [87], and the extensive epidemiological and immunological studies by Charles E. Smith provided the basic framework within which our present understanding of the disease and its causal agent has been systematic- ally assembled.

Coccidioides immitis is a filamentous fungus that grows readily and rapidly in vitro as a gray, cottony mold. Individual cells in the aerial hyphae differentiate into barrel-shaped arthroconidia, about $3 \times 5 \ \mu m$ interspersed with empty cells. Arthroconidia are released by fragmentation of the walls between individual spores. The primary mechanism for establishment of infection involves inhalation of arthroconidia formed by mycelium grow- ing in the upper soil layers of endemic areas.

The mold is dimorphic, the characteristic tissue phase developing from inhaled arthro- conidia being the spherule, a spherical sporangium up to $50 \ \mu m$ in diameter. When mature, the cytoplasm undergoes cleavage to form uninucleate endospores, 2-5 μm in diameter. These are released into adjoining tissues by rupture of the spherule. Liberated endospores repeat the cycle. Both spherules and endospores can be cultivated in vitro using appro- priate cultural conditions [68].

Coccidioidomycosis is characteristically a disease of arid desert areas in the south- western United States, but endemic areas also exist in parts of Central and South America [311]. The coccidioidin skin test [402] has proved to be of value both as a diagnostic procedure and as a means of obtaining information on prevalence and endemicity. Most infections (about 60%) are asymptomatic. Primary infections may, however, result in mild, moderate, or severe respiratory disease. Progressive or disseminated infections are rare, but chronic pulmonary coccidioidomycosis develops in up to 8% of subjects with symptomatic infection [104]. Like tuberculosis, which it resembles in so many ways, the disease assumes many different forms.

Much of the information that has been acquired on the natural history of the disease and its causal agents is summarized in three major symposia, published in 1958 [118], 1967 [4], and 1977 [5].

2. Tests

a. Complement Fixation: The CF test for coccidioidomycosis originated with the pioneer studies of C. E. Smith and his associates [400]. The antigen used in the procedure, coccidioidin, was prepared from sterile culture filtrate of an asparagine synthetic broth inoculated with *C. immitis* and maintained for several months at room temperature [402]. This method of production was chosen because it was already an established procedure for the manufacture of tuberculin. The validity of the rationale has been amply confirmed by the proven reliability and efficacy of the CF test for *Coccidioides* and by the fact that coccidioidin, prepared in the same way, is still in widespread use today for CF tests and for the demonstration of skin sensitization. Other antigens have been prepared for use in immunological tests. Pappagianis et al. prepared an antigen by lysis of young (3-day) mycelia of *C. immitis* in the presence of toluene [313]. Lysates contained antigens active in the CF test which were devoid of anticomplementary activity and which failed to react significantly with sera from patients with histoplasmosis. Fractional precipitation with ethanol yielded separate fractions which were reactive in CF or precipitin tests, respectively. In the past decade, increasing attention has been taken in spherulin, an extract prepared from spherules of *C. immitis* grown in vitro [240]. Devised originally for use as a skin test antigen, spherulin has also been used in CF tests. As early as 1952, Vogel and Conant prepared an antigen from spherules cultivated in egg yolk which detected CF antibodies in experimentally infected rabbits [441]. Landay studied spherule-derived antigens in the sera of patients with coccidioidomycosis [225]. Although reactive, cross-reactions were obtained with sera from patients with histoplasmosis and blastomycosis. Huppert et al. compared coccidioidin and spherulin in CF tests on 614 sera submitted for *C. immitis* serology and 159 sera from patients with mycoses other than coccidioidomycosis [182]. Although a similar proportion of positive reactions was obtained in the former group with both types of antigen, spherulin proved to be appreciably less specific, reacting against 45% of the sera from non-*Coccidioides*-infected patients, in comparison to 20% for coccidioidin.

The CF test developed by C. E. Smith and his associates provided the means to determine patterns of antibody response in different forms of coccidioidomycosis. One report [401] is based on a study of 39,500 serological tests. With the introduction of the standardized Center for Disease Control Laboratory Branch Complement Fixation (LBCF) test for CF in 1965, a comparison of the alternative procedures became necessary. Huppert et al. undertook such a comparative evaluation in 1970 [181]. It was found that results were highly reproducible within an acceptable one-tube variation, in a twofold dilution series. The microtitration adaptation of the LBCF test [308] showed greater variation, particularly when compared with Smith's original procedure. It was concluded that the LBCF should be an acceptable procedure for *Coccidioides* serology, subject to comparative testing between different laboratories.

CF tests have been used to compare antigens derived from *C. immitis*. Landay et al. found that a preparation from sonicated spherules was more reactive than sonicated arthroconidia [229]. In addition to detection of serum antibodies, sonicated spherules are also used in the study of CSF from patients with coccidioidomycosis of the central nervous system. In the study reported by Smith et al. in 1956, complement was fixed by the spinal fluid of most patients with coccidioidal meningeal infection [401].

b.   Immunodiffusion: Tests for precipitins to *C. immitis* have been developed in parallel with CF tests and have proved to be both reliable and effective. Initially, precipitins to coccidioidin were detected by a tube precipitin test, but this was later replaced with an immunodiffusion procedure, introduced by Huppert and Bailey in 1963 [177], and evaluated more fully in subsequent reports [178,179]. Results with one antigen, prepared by lysing young mycelial growth [313], correlated well with CF tests, while a second antigen prepared from culture filtrate gave results which had a close correlation with those obtained using the original tube precipitation test. One precipitin band, designated "F," appears to be *Coccidioides*-specific. The associated antigen produces ID results which correlate with CF tests but not with the tube precipitin test. By suitable titrations of reference antisera and coccidioidin lots, a standard ID system can be devised where intensities of reactions in the agar gel correlate well with CF titers [175].

The value of ID has been confirmed by Warren [458] and Hampson and Larwood [155], and the technique has been adopted by the Pan American Health Organization in one of the procedural manuals for serodiagnostic testings for systemic mycoses [12].

ID procedures have been used to study antigens produced by *C. immitis*. Rowe et al. [374,375] showed by ID that a variety of antigens could be extracted from cultures maintained under different conditions, some of which also had activity in CF tests. Antigenic differences were found between different growth phases of the fungus. Wallraff et al. studied five coccidioidins by ID and reported the presence of two to five lines [447]. Potency of rabbit antiserum prepared by injection of coccidioidin was related to its nondialyzable carbohydrate content. Landay et al. showed by ID that antigens obtained from spherules differed from those derived from arthroconidia [227]. Antibodies to spherule antigens were detected in the sera of patients with primary pulmonary disseminated and meningeal forms of coccidioidomycosis [226].

Coccidioides immitis has been shown to produce exoantigens. The specificity of these exoantigens is remarkable. This is also true for exoantigens of other agents causing systemic mycoses [205,206,408,409]. Huppert et al. [185] found no cross-reacting exoantigens in genera of molds which reproduce by formation of arthroconidia and have, therefore, a morphological similarity to *C. immitis*. Kaufman and Standard [206] tested 128 isolates of *C. immitis* and found that all strains produced specific exoantigen. The degree of precision afforded by this presumptive test for identification of *Coccidioides* and other causes of systemic mycoses is impressive. The antigens involved for *Coccidioides* appear to be identical to those responsible for the CF test (F band in ID) and tube precipitin test, respectively. A third antigen designated HL (heat labile) was also produced as an exoantigen, and this too appeared to be specific for *C. immitis*. The role of the HL antigen in serodiagnosis has not yet been fully evaluated.

Antigenic variation has been noted for *C. immitis,* and in one set of data presented by Huppert [174] five different patterns of antigenicity were recognized among nine isolates of *C. immitis,* depending on presence or absence of four different precipitinogens. In the ID test described by Huppert et al. [180], antigen was prepared by pooling culture filtrates of no fewer than 24 strains of *C. immitis* and preparing toluene lysates of the harvested mycelial growth. With any antigenic extract intended for use in immunodiagnostic testing, there are obvious advantages in using as wide a range of antigens as practicable when differentiation of diagnostically significant components is not possible. The use of reagents containing many antigens is common in fungal serology, but it is at best a makeshift and empirical approach, to be superseded whenever (and if ever) practicable by products containing only "significant" antigens. Pooling antigenic extracts to obtain the widest spread possible of antigenic components may achieve the opposite result, since the net effect might be to dilute minor antigens (i.e., antigens present in small amounts or present in only some extracts) with extracts containing common antigens. With more sophisticated analytical procedures such as 2D-IE now available, it is now feasible to prepare mixtures containing the full range of antigens in quantities suitable for the test system chosen.

Several comparisons have been made of results obtained with ID and CF procedures [180,202,381]. Titers of precipitating antibodies are sometimes found to be lower than CF antibody levels. Ray [347] and Ray and Converse [348] therefore proposed an agar-gel precipitation inhibition (AGPI) test for coccidioidomycosis, suggesting that it is more sensitive than tube precipitin, ID, and CF test in detecting antibodies to *C. immitis*. Data supplied by Huppert et al. [180], however, indicate that sensitivities of quantitative ID and CF tests are comparable, and that ID constitutes a useful screening test for detection of *Coccidioides* antibodies. Wachs and Wallraff in 1969 described a modified AGPI test with improved sensitivity and performance [444]. Nevertheless, the procedure does not appear to have been widely adopted. Huppert has pointed out that by its nature, the AGPI test can detect only one of the several antigen-antibody systems known to be present [173].

c. Electrophoretic Procedures: Immunoelectrophoresis has not been routinely used to detect *Coccidioides* antibodies. Gordon et al., however, first used counterimmunoelectrophoresis (CIE) for detection of *Coccidioides* antibodies in 1971 [140] and this was further evaluated by Aguilar-Torres et al. [2,3], who obtained positive results in 93 of 96 (97%) specimens from 34 patients with active disease. Results correlated well with CF and ID. The procedure was also found to be simple and effective by Graham and Ryan [145], who in their series of 37 patients with coccidioidomycosis found antibodies by CIE in the sera of all patients having a CF titer of $>1:4$.

Two-dimensional immunoelectrophoresis (2D-IE) has been used by Huppert et al. [176,184] to characterize soluble antigens extracted from *Coccidioides, Histoplasma,* and *Blastomyces*. Coccidioidin was found to contain at least 26 antigens, of which 10 were also found in spherulin. Two antigens were found only in spherulin. Several extracts of *H. capsulatum* and *B. dermatitidis* were analyzed by 2D-IE using a coccidioidin-anticoccidioidin reference system. It was found that 12 common antigens were present in extracts of *B. dermatitidis* and 10 in *H. capsulatum*. Qualitative and quantitative differences were also observed in the antigens contained in different extracts.

d. Latex Particle Agglutination: Huppert et al. described the latex particle agglutination test (LPA) in 1968, using particles sensitized with coccidioidin [183]. When used in combination with ID and CF tests, antibodies were detected in 93% of sera from 328 diagnosed cases of coccidioidomycosis. A slide latex agglutination test was also used by Wallraff and Wachs [445]. The procedure is sensitive and results are rapidly obtained. Positive results were, however, obtained by Pappagianis et al. with a commercially available kit not only in spinal fluids of 35 patients with coccidioidomycosis, but also in 14 patients without coccidioidal infection [312]. Strong agglutination was obtained with simple dilution of sera from all 49 subjects, and the authors concluded that the test cannot be recommended as a screening procedure. The explanation for these false-positive results with LPA appears to be instability of sensitized particles in the spinal fluid, with its low concentrations of proteins.

e. Immunofluorescence: Kaplan et al. reported in 1966 that antibodies to *C. immitis* could be detected by fluorescent antibody inhibition [194]. This procedure was used on 91 sera from confirmed cases and 15 sera from patients with other diseases, and results were compared with those obtained by CF. The overall coincidence of results was 88% with 80 sera positive for each test and 14 negative. A similar test was also used by Wallraff and Wachs in 1971, who also concluded that the procedure has value in laboratories not equipped to perform tests [446]. Their preliminary studies suggested that the method could be adapted to allow the respective contributions of IgM and IgA immunoglobulins to be evaluated.

Using an indirect fluorescent antibody (IFA) procedure, Landay et al. studied immunofluorescence of spherules and arthroconidia of *C. immitis*. Immune sera were prepared in two groups of rabbits by repeated injections of formalinized spherules or arthroconidia, respectively [228]. Sera were used to stain spherules or arthroconidia by an IFA procedure. Distinct fluorescence was observed involving all or part of the cell wall structure, and the authors suggest that the procedure may be of potential value in distinguishing between antibodies directed to "protective" antigens and those to "nonprotective" antigens.

3. Interpretations

The general pattern of immunological responses to infection with *C. immitis* was established three decades ago. Almost every infected individual will develop cutaneous hypersensitivity to coccidioidin or spherulin. Antibodies develop in most patients with symptomatic disease, but may be absent in early or asymptomatic infections. Smith et al. [400] never observed antibodies in patients with primary coccidioidomycosis in the absence of cutaneous reactivity. The combination of CF and precipitin test gives positive reactions to coccidioidin in over 90% of primary symptomatic infections. Precipitins appear early, during the early acute phase of primary infections, in many cases during the first week of clinical illness. CF antibodies also appear early in the disease but, in contrast to precipitins, may persist for long periods. Smith et al. recognized the correlation that exists between CF titers and the extent and severity of disease [401].

These tests have a high degree of specificity and are of great value as laboratory aids to diagnosis. Any reaction in the ID test, particularly one giving a line of identity with the F band in a reference system, is strong presumptive evidence of infection.

CF titers equal to or greater than 1:16 usually indicate disseminated disease, particularly when obtained with consecutive sera. Nevertheless, titers of 1:2 and 1:4 may be significant and indicative of active infections [181,401]. Such low titers are usually related to early residual or meningeal infections. On occasions, however, low CF titers to coccidioidin have been recorded in patients without coccidioidomycosis, and caution must always be exercised in interpreting such results. Whenever possible, additional sera should be obtained and tested, and attempts made to confirm the presence of the pathogen by microscopy and culture.

Skin testing with coccidioidin does not elicit or boost antibodies to *C. immitis*. A similar lack of antibody induction has been reported for spherulin by Deresinski et al. [84]. The immunodiagnostic reagents used for detection of *Coccidioides immitis* antibodies are therefore of a higher level of specificity than those used in the past for *H. capsulatum* or *B. dermatitidis*. Nevertheless, some cross-reactions do occur, particularly in sera with low antibody titers. Kaufman and Clark reported that of 87 sera studied by them which had CF antibodies to *Coccidioides,* 65 (75%) had titers of 1:8 or less [202]. Only 40 of these (62%) were from patients with proven coccidioidomycosis, and 23 of the 65 (35%) cross-reacted with antigens prepared from *Histoplasma* or *Blastomyces*. In this study the ID was found to be highly specific, only two instances being recorded where precipitins to

Coccidioides were present in sera from patients without coccidioidomycosis. Kaufman and Clark concluded that sera with CF titers between 1:2 and 1:8, which were also ID positive, invariably denoted active or recent infections.

The comparative diagnostic merits of ID, CF, and LPA tests were evaluated by Huppert in 1970 [174]. He found that LPA tests were positive in 70.6% of sera from patients with coccidioidomycosis, compared with only 13.5% for the tube precipitin test. The proportion of false-positive results was higher with LPA (3%) than with ID (0.6%), but this was more than compensated by the very high level of sensitivity of the former test. In this series, results with ID and CF were almost identical, with positive results in more than 80% of the cases. It was suggested that a combination of LPA and ID provides a better combination of tests than CF and ID, the LPA-ID combination giving positive results in more than 90% of cases.

E. Cryptococcosis

1. Introduction

Cryptococcus neoformans, the encapsulated yeast that causes cryptococcosis, commonly occurs in association with old accumulations of pigeon droppings [102]. It has been isolated from 90% of cages used to house racing pigeons for periods up to a month after they had strayed during a race [142]. Emmons [103] reported that up to 50×10^6 viable cells of *C. neoformans* have been isolated from 1 g of dried fecal material. Primary infection is considered to be acquired exogenously. In view of the frequent occurrence of *C. neoformans* in the neighborhood of human habitation, it is considered likely that primary infections are common and almost always mild and self-limiting. Exceptionally, infection leads to overt disease, with the production of localized lesions in lung or skin, or dissemination and involvement of bone, viscera, or central nervous system. Cryptococcal meningitis of meningoencephalitis is the most familiar form of cryptococcosis, and despite the coexistence of reliable laboratory diagnostic procedures and effective chemotherapeutic drugs and regimens, the mortality from this form of the disease remains disturbingly high. Several predisposing causes have been recognized, most notably malignant lymphomas (particularly Hodgkin's disease), sarcoidosis, collagen disease, and the use of corticosteroids and immunosuppressants. Infections may also affect subjects lacking a recognized cause of susceptibility.

With most mycotic infections, proliferation of the pathogen in the host tissues elicits an antibody response, which is then detected and measured by a variety of serological procedures. *Cryptococcus,* however, is unique among pathogenic fungi in having a prominent polysaccharide capsule which is usually produced in abundance in vivo, and which results in a relatively protracted state of antigen excess. Antigenemia has now been recognized as a component element of the immunopathogenic sequence of several systemic mycotic infections, including candidiasis (see Sec. V.C.3) and aspergillosis (see Sec. V.A. 3). These are usually the result of heavy and rapid proliferation of the agent. In these diseases antigen excess appears to be an infrequent and transient state. With *C. neoformans,* however, soluble capsular polysaccharide is almost always produced in abundance, so much so that laboratory evaluative tests include those based on detection and quantitation of antigen as well as antibody.

2. Tests

   a. *Antigen Detection*: Neill and Kapros reported in 1950 that antigens could be detected in the tissues of mice experimentally infected by intraperitoneal injection of *C. neoformans* or *Sporothrix schenckii* [295]. Antigen was demonstrated by precipitation with specific antiserum and shown to be present in materials obtained from animals that were seriously ill or which had died from the infection. Recognizing a potential application of their findings, the authors suggested that examination of spinal fluid from cases of suspected cryptococcal meningitis might be worthy of attention. In a later report, Neill et al. demonstrated that soluble cryptococcal antigen was present and could be recognized in spinal fluid, urine, and blood from a patient with systemic cryptococcosis [296]. Antigen shown to be polysaccharide in nature was detected by precipitation with rabbit antiserum and by complement fixation. All tests in current use are directed toward the detection of capsular material. This material is usually produced in abundance, although infections with weakly or nonencapsulated strains are by no means uncommon. The capsule is serologically reactive and four serotypes have been recognized on the basis of precipitin, capsular, and agglutination reactions [106,107,469]. Chemical analysis has shown the presence of a mannose backbone with xylose and glucuronic acid side chains. Some cross-reactivity has been demonstrated between the yeast cells of *C. neoformans* and those of *Candida albicans* and *Saccharomyces cerevisiae,* and between *Cryptococcus neoformans* polysaccharide and trichophytin, pneumococcus, and gum tragacanth. In practice, however, tests for detection of cryptococcal antigen are singularly free of cross-reactions and exhibit a high degree of specificity. Most of the tests used to detect antigen are quantitative, allowing polysaccharide to be measured as well as detected. As with tests for antibodies, quantities of antigen are most conveniently expressed as titers. In patients with cryptococcal meningitis, polysaccharide is usually present in varying amounts in the CSF. Antigen may pass readily into the blood, where it can be detected at appreciably lower titers. In patients with nonmeningeal cryptococcosis, however, serum antigen titers may be present, but capsular material does not appear to enter the spinal fluid. When cryptococcal polysaccharide was injected intracisternally into rabbits, it disappeared rapidly from the spinal fluid, but persisted for weeks in the bloodstream [32]. It also persisted in the blood of rabbits and mice injected intravenously.

   (1) Complement Fixation: First used by Neill et al. to detect polysaccharide [296], the complement fixation test for antigen (CF-Ag) was later adapted to the routine detection of capsular polysaccharide by Anderson and Beech in Australia [9]. Bennett et al. [31] used CF to detect polysaccharide in sera and spinal fluids of patients with proven cryptococcosis. Antigen was detected in 27 of 41 patients with cryptococcal meningitis (66%) and in 2 of 8 with extraneural infection. The CF-Ag test was shown by Bennett and Hasenclever to be capable of detecting capsular polysaccharide at levels of 0.25-1 μg/ml [32]. A CF-Ag procedure was also used by Walter and Jones [451], who applied several serological procedures to the detection of antigen and antibody in sera or spinal fluids of 40 patients with cryptococcosis. It was reported that the CF-Ag test was more sensitive than whole cell or inert (latex) particle agglutination tests. Bindschadler and Bennett [35] found cryptococcal antigen by this procedure in serum or CSF of 47 patients in a series of 69 patients with cryptococcal meningoencephalitis and another 6 with nonmeningeal infections. Titers ranged from 1:4 to 1:512 in serum, and from 1:1 to 1:2048 in CSF. Only 6 patients (8%), including 4 with meningoencephalitis, were negative when tested for antigen.

(2) Latex Particle Agglutination: In a brief note published in 1962, Bloomfield [38] suggested that cryptococcal antigen could be detected by its ability to agglutinate latex particles coated with immunoglobulin from antiserum raised in rabbits immunized with *C. neoformans*. The report published the following year by Bloomfield et al. [39] provided a more detailed description of the procedure, which has become one of the most reliable and effective of all serodiagnostic procedures used in medical mycology. Immunoglobulin from immunized rabbits is coated onto a standard suspension of 0.81 μm polystyrene latex particles at an optimal concentration determined by block titration. Dilutions of serum, CSF, or urine are mixed with a constant volume of the sensitized latex particles on a glass slide for a predetermined time (e.g., 3 or 5 min) and examined for agglutination. A preliminary screening test is usually undertaken first to determine if antigen is present and to indicate its approximate titer. This is followed by a more precise determination of the titer. The procedure has several advantages over the more cumbersome CF-Ag test, including greater sensitivity and rapidity and the ability to test anticomplementary sera. Bennett and Bailey [30] consider the test to be satisfactory when latex particles are agglutinated in the presence of 30-60 μg of polysaccharide.

One major disadvantage of the LA test compared with the CF-Ag procedure was the occurrence of false-positive reactions in sera from subjects without cryptococcosis. In the series reported by Bloomfield et al. [39], two weak false-positive reactions were noted in sera with rheumatoid factor (RF). Dolan [90] reported that nonspecific reactions were obtained by LA in 15% of positive CSF and 85% of positive sera examined. In a rebuttal published subsequently, the importance of careful attention to procedural detail was emphasized by Kaufman and Blumer [201], who suggested that failure of the LA test could be explained solely on this basis.

In common with any serodiagnostic procedure, the inclusion of appropriate controls is mandatory. With the realization that RF is an important source of false positives, it became necessary to include a control for RF in the LA test protocol, and this is now a standard practice. Bennett and Bailey [30] devised a control, using latex particles coated with immunoglobulin from nonimmunized rabbits. Such particles agglutinated in the presence of RF, giving titers comparable to those obtained with sensitized particles. Gordon and Lapa [142] used mercaptoethanol or dithiothreitol to eliminate RF, and by doing so, provided the means for distinguishing between specific latex agglutination mediated by cryptococcal antigen and false positives due to RF. Use of dithiothreitol (DTT) abolished the agglutination of sensitized latex particles by 25 of 52 sera found to contain RF. Specific reactivity of three sera from patients with cryptococcosis having LA titers of 1:2 or 1:4 were abolished by 0.02 M DTT, but no effect was noted in sera with titers of 1:8 or greater.

In the examination of serum for cryptococcal antigen, controls must be included for positive and negative reactivity of the sensitized latex suspension, for sensitivity of the test, and for reactivity of individual sera to latex coated with normal rabbit immunoglobulin. Rheumatoid factor has been recognized as a source of false-positive reactions in the LA test for sera, but not for spinal fluid. Agglutination of sensitized latex particles by CSF, irrespective of titer, is strong presumptive evidence for active or recent cryptococcosis.

A novel adaptation of the LA test is in the identification of *C. neoformans* by testing culture filtrates for reactivity to sensitized latex particles [280]. Of the 104 yeasts tested by this method the supernatants of only *C. neoformans* caused agglutination in the LA test. Diagnostic kits for the cryptococcal LA test are commercially available. Kaufman et al. evaluated 24 kits and reported in 1974 that 14 were unsatisfactory due to the failure of the sensitized latex suspension to agglutinate the polysaccharide control necessary to validate the test [203]. No such difficulties were encountered in a later report by Prevost and Newell [345], who concluded that the LA test used was of value in diagnosis and prognosis of cryptococcosis. Eisses et al. have used a commercially available kit to detect *Cryptococcus* antigen in tissue extracts [98].

(3) Hemagglutination Inhibition: Kozel and Cazin coupled *C. neoformans* polysaccharide to sheep erythrocytes with $CrCl_2$ [215]. The sensitized red cells were used to detect antibody, or, by using patients' CSF or serum to inhibit hemagglutination, the test could also be used to detect antigen. The method was able to detect less than 1 μg of soluble polysaccharide.

(4) Precipitation: First used by Neill et al. in 1951 [296], precipitation of rabbit antiserum by polysaccharide antigen in tissue fluids of patients with cryptococcosis was also used by Seeliger and Christ [393] to detect antigen in the spinal fluid of a patient with cryptococcal meningitis. The insensitivity of the method led to its abandonment, in favor of latex agglutination.

(5) Gas-Liquid Chromatography: Detection of *Cryptococcus* in CSF by gas-liquid chromatography was first attempted in 1976 by Schlossberg et al. [384]. By means of electron-capture gas-liquid chromatography, analyses were made of CSF from 8 patients with cryptococcosis, 10 with viral meningitis, and 4 controls without meningitis. All 8 samples of CSF from patients with cryptococcal meningitis had similar profiles, which were distinct from the other two groups. A similar pattern was also obtained in CSF inoculated with *C. neoformans*. The authors suggest that the technique is rapid, reproducible, and simple to perform. In a more detailed study, patterns obtained by GLC were compared in CSF from 46 patients with tuberculous, cryptococcal, and viral meningitis [75]. As before, patterns obtained in the CSF of 15 patients with central nervous system (CNS) cryptococcosis resembled each other and were recognizably different from those seen in the other two groups. Moreover, compounds were detected which disappeared on therapy. If the findings are confirmed and extended, the procedure could prove to be of considerable value, perhaps complementing more traditional methods for detection of antigen.

(6) Counterimmunoelectrophoresis: In 1977, Maccani described a CIE procedure to detect and measure cryptococcal antigen in tissue fluids [248]. Using rabbit antiserum, 1.25 μg/ml of capsular polysaccharide could be detected. Precipitin lines were obtained with urine, serum, and CSF of six patients with cryptococcosis, but not with materials from noninfected subjects. By prior 100-fold concentration of the fluids, a marked increase in sensitivity was obtained, and quantities as low as 25 ng/ml of antigen could be detected in CSF or urine, and 80 ng/ml of serum. No false-positive results were obtained in sera containing rheumatoid factor. Cross-reactions were observed using sera from three patients with *Klebsiella* infections, but these could be eliminated by prior absorption with bacterial cells. To date, there has been no comparison published between results obtained by CIE and latex agglutination. Kume et al. [218] used CIE to examine serum from one patient with cryptococcal meningitis, but it is not made clear if antigen or antibody was being detected.

b. Antibody Detection: A variety of techniques have been used to measure antibody levels and responses in patients with cryptococcosis. Understandably, greater attention has been focused in the past three decades on antigen detection, in view of its greater relevance as a diagnostic laboratory finding. Antibodies to *C. neoformans* are not commonly found in human sera, but they may be produced in low-grade, localized, or successfully treated infections. In monitoring the outcome of infection, serum should be simultaneously examined for cryptococcal antibody as well as antigen. High initial levels of antigen ($\geqslant 100$) in CSF carry a poor prognosis [86]. Disappearance of antigen from the blood and its replacement by antibody is a highly favorable prognostic sign. Antibodies to *C. neoformans* are not routinely sought, but Porter et al. described specific oligoclonal immunoglobulins in the CSF of a 43-year-old man with cryptococcal meningitis [335]. The immunoglobulin could be removed by absorption with acetone-killed cryptococcal cells, but not by *Candida* cells.

(1) Indirect Fluorescent Antibody: One of the first methods to be described was immunofluorescence, and an indirect fluorescent antibody (IFA) test has been used by several investigators. Vogel and Padula described an IFA test in 1958 and detected antibodies in the serum of patients with cryptococcosis [442]. In a later study, 6 of 7 infected patients were found to have antibodies [443], although 8% of normal sera were also positive. In 1966, antibodies were found by Vogel in the serum of 36 of 45 patients (80%) with cryptococcosis [440]. A significant proportion of negatively reacting sera were in patients with CNS involvement only. The principal immunoglobulin involved was recognized as IgG. Fluorescent antibody procedures in combination with CF tests were also used by Walter and Atchinson [449] to examine antibodies in the serum of pigeon breeders. Antibodies to *Cryptococcus neoformans* were found in 28 (22%) of 134 pigeon breeders, compared with 3% of 35 nonpigeon breeders, suggesting that subclinical infections may have been more prevalent in the former group. Fink et al. demonstrated antibodies by IFA in a similar proportion of apparently healthy pigeon breeders, in 5 of 6 subjects with pigeon breeders' disease, and in 5 patients with cryptococcosis [120]. Since antibodies could be removed by absorption with pigeon droppings as well as cells of *C. neoformans*, these authors concluded that presence of cryptococcal antibody in patients with pigeon breeders' disease can be explained on the basis of increased exposure to the yeast or variation in immunological reaction to pigeon feces contaminated with *C. neoformans*. Bindschadler and Bennett [35] reported that 29 of 75 patients with cryptococcosis (38.7%) had IFA titers between 1:1 and 1:32. Positive IFA tests were also recorded for 4 of 253 control subjects (1.6%). Di Salvo et al. recently described an automated procedure for the IFA test, which compared well with the manual system [88].

(2) Hemagglutination: Hemagglutination was first introduced by Pollock and Ward in 1962 [333]. When human group O cells coated with polysaccharide were used, antibodies were detected in the sera of 2 patients with cryptococcosis at levels between 1:80 and 1:320. The procedure has not been widely employed, although its sensitivity and versatility were demonstrated by Kozel and Cazin, who used both hemagglutination and hemagglutination inhibition to detect cryptococcal antibody and antigen, respectively [215], in infected subjects.

(3) Bentonite Flocculation: Bentonite flocculation, introduced by Kimball et al. in 1967 [211], detected antibodies in 15 (42%) of 35 patients with CNS cryptococcosis and 6 (50%) of 12 with nonmeningeal infection.

(4) Charcoal Particle Agglutination: Gordon and Lapa [141] reported that although sensitized bentonite particles were stable, the sensitivity of the test in their hands was inadequate, and they developed an alternative procedure, where the inert carrier was specially prepared charcoal particles. The procedure was rapid (10 min) and sensitive compared to a tube agglutination test and had the additional advantage of providing a permanent record.

(5) Whole Cell Agglutination: This has been widely used as a convenient and reproducible procedure for measurement of cryptococcal antibody. It is routinely used as an adjunct to tests for polysaccharide antigen. Although not wholly specific, tests based on agglutination of killed intact yeast cells are of considerable value in helping to establish status and probable outcome of cryptococcal infections. Kaufman and Blumer [200] showed that a tube agglutination test for cryptococcal antibodies had a high degree of specificity (95%) when used to evaluate the serological status of 66 patients with cryptococcosis. Agglutinin titers to suspensions of intact cells are usually low, irrespective of the serotype of the infecting strain. Palmer et al. state that titers $\geqslant 1:2$ are suggestive of infection and that titers $> 1:16$ are uncommon [309]. These titers match those found in the author's own laboratory, where only 12 in one series of 51 patients with proven infections caused by *C. neoformans* had demonstrable antibody. Sensitivity of the tube agglutination test for *C. neoformans* used at the Center for Disease Control is estimated to be 35% [309]. In comparison with charcoal particle agglutination titers reported by Gordon and Lapa [141], agglutinin titers obtained by a tube agglutination procedure were lower by a factor of 40. An unusual finding reported by Noble and Fajardo was the demonstration of agglutinins to *C. neoformans* in the spinal fluid at a titer of 1:4, and in serum (titer 1:1) in a patient with cryptococcosis and underlying Hodgkin's disease [298]. Antibodies to microbial agents causing infections of the CNS are rarely observed in the CSF, and this observation is at present the only one of its kind in the laboratory investigation of patients with cryptococcosis.

(6) Other Tests for Cryptococcal Antibodies: Methods that have appeared in the literature include complement fixation [440] and passive cutaneous anaphylaxis [344]. Neither method has been widely used, presumably because of their inherent technical complexities and their ready substitution by simpler and more rapid procedures. At the time of writing no report has yet appeared on the use of a radioimmunoassay, although Desgeorges et al. have recently described the use of an enzyme-linked immunoassay (ELISA) for the detection of antibodies to *C. neoformans* [85]. Antibodies were detected in the sera of two patients with cryptococcal infection; in one patient treated with amphotericin B, the level of antibodies fell. In view of its sensitivity an ELISA procedure may prove to be of some value in detecting low levels of antibodies in the serum of subjects with subclinical infection, and in allowing acquired immunological responsiveness to *C. neoformans* to be measured in terms of antibody response rather than skin or lymphocyte reactivity.

3. Interpretations

The demonstration of high levels of cryptococcal capsular polysaccharide in serum, CSF, or urine is a laboratory finding of great diagnostic significance. With the recognition of rheumatoid factor as a source of error and its elimination from test sera by appropriate control protocols, the latex test for cryptococcal antigen has become one of the most helpful and reliable of all serodiagnostic procedures. When used in conjunction with such tests for antibody as IFA and agglutination of yeast cells of *C. neoformans,* the information obtained can be of diagnostic and prognostic value. Bennett et al. [31] have calculated that for a patient with cryptococcal meningitis to maintain the levels of antigen that are detected in spinal fluid, quantities of polysaccharide ranging from 0.1 mg to several milligrams of capsular material must be produced each day.

The latex agglutination test for antigen is usually positive in patients with CNS infection. Goodman et al. reported in 1971 that 36 of 39 (92%) patients with culturally proven meningeal cryptococcosis had demonstrable antigen [138]. In a series of 69 patients studied in the United Kingdom by Hay et al. [159] a similar proportion (86%) of patients with all forms of cryptococcosis had detectable antigen in their spinal fluid. Although the majority of patients with CNS involvement have antigen present in CSF or serum, the LA test is frequently negative when the infection is extraneural. In the series studied by Hay et al., 8 proven cases had negative LA tests. Of these, 4 had localized cutaneous disease and 3 had pulmonary cryptococcomas which had been removed (R. J. Hay et al., unpublished data). Kaufman and Blumer [200] and Palmer et al. [309] recommend that the laboratory serodiagnostic evaluation of patients thought to have cryptococcosis should include concurrent use of indirect fluorescent antibody (IFA), agglutination, and latex agglutination tests. IFA and agglutination tests are used in conjunction since it has been found that antibodies may be found by one procedure but not the other. In the study by Kaufman and Blumer [200], marked differences were seen in the serological results obtained. Of the 66 culturally proved cases, 29% were positive only in the test for antigen and 47% only for antibody; 19% of the subjects were positive in tests for both antigen and antibody. Only for 3 patients (5%) were results of tests for antigen and antibody negative. Specificities of the three tests are high. For the LA test it is apparently 100%; the tube agglutination is also specific (95%), but the IFA rather less so (79%), with cross-reactions occurring in some patients with blastomycosis and histoplasmosis.

Antigen titers in patients with proven infection, being an expression of the amount of capsular antigen being synthesized, can provide information on the patient's status and progress. High ($\geqslant 200$) initial LA titers correlate with a poor prognosis. In the report by Diamond and Bennett [86] analyzing diagnostic and prognostic findings in 111 patients with cryptococcosis, it was clearly shown that treatment failure was associated with high initial levels of antigen, high levels after a course of therapy, or failure of levels to fall during treatment.

Rising titers of antigen correlate with progressive disease and declining levels with improvement in the patient's condition. Maintenance of antigen levels during chemotherapy may indicate that treatment is inappropriate or inadequate. Disappearance of antigen from serum, and its replacement by antibodies to *C. neoformans* suggests that the disease has been brought under control. In some patients with cryptococcosis of the CNS, antigen titers may not disappear, but may be sustained at low levels for long periods. In such patients there is no evidence that viable cryptococci are present, and it may be possible that antigen is being continually and gradually released from localized accumulations of dead cells.

Cryptococcus neoformans is not antigenically homogeneous. Four serotypes (A, B, C, D) have been recognized. Most infections of humans are with the A serotype, and serological tests almost always involve antigens or antisera derived from this strain. In practice, serodiagnostic tests are usually undertaken with a serotype A strain, since cross-reactions occur between the different strains.

F. Histoplasmosis

1. Introduction

Histoplasma capsulatum, the cause of histoplasmosis, was first grown in culture by de Monbreun at Vanderbilt University in Nashville, Tennessee, in 1934 [83]. Although the disease was originally described by Darling in 1906 [76], the infective agent was mistakenly thought to be a protozoan. Its fungal nature was recognized by Da Rocha Lima in 1912 [77], but more than three decades were to pass before a clear understanding emerged of the natural history of the disease and its cause. Present-day knowledge has been systematically assembled on the basis of four key discoveries. The first was the recognition by de Monbreun of the fungal and dimorphic nature of the pathogen. The second was the realization by Christie and his associates in 1944 that subjects infected with *H. capsulatum* develop a sensitized state which can be revealed by skin testing with histoplasmin [62]. The third was the isolation, by Emmons in 1949, of *H. capsulatum* from soil [101], and the fourth, the realization that antibody responses can be elicited by naturally acquired infection [125].

Before the introduction of the histoplasmin skin test, histoplasmosis was considered to be a rare and uniformly fatal disease. Skin reactivity to histoplasmin was to prove of great value in confirming the widespread occurrence of mild, self-limiting infections, and in establishing with precision both areas of endemicity and levels of prevalence.

Histoplasmosis is commonest in the Ohio-Mississippi valley regions of the United States, with up to 90% of the population in some areas reacting to histoplasmin. It occurs in many other parts of the world, including Africa, Central and South America, and Asia. Humans become infected by inhalation of airborne spores, which are released from mycelial growth in the upper soil layers. Most infections are subclinical, but in rare instances, severe disseminated or chronic forms may develop. Focal epidemic outbreaks occur when groups of individuals are exposed to environmental sources heavily contaminated with *Histoplasma.* In disseminated forms, the disease may affect many organs, including spleen, liver, kidney, central nervous system, adrenals, gastrointestinal tract, and skin.

African histoplasmosis, a disease predominantly of skin and bone, is caused by a variety of *H. capsulatum* (var. *duboisii*) which produces large (up to 20 μm diameter) budding yeast cells in infected tissues. In culture, however, *H. capsulatum* and the African variety are identical. Epizootic lymphangitis, a disease of horses and mules producing subcutaneous and ulcerated lesions of the skin, is caused by *H. farciminosum,* a species closely related to *H. capsulatum.*

In culture, *Histoplasma* is mycelial, usually producing spores with characteristic blunt, peg- or finger-like projections. In infected tissues, the organism is yeast-like, being located primarily within histiocytes. The yeast phase can be grown in vitro by provision of appropriate cultural conditions [324]. Antigens for immunological testing are prepared principally from culture filtrates of *H. capsulatum* grown in the mycelial form for several months (histoplasmin) or by extraction of yeast or filamentous growth by chemical or physical means. Intact yeast cells are also used for complement fixation and immunofluorescent studies.

Much attention has been focused on histoplasmosis and *Histoplasma* since the realization that agent and disease are widespread and common. Much of the currently available information is summarized in the published proceedings of two national conferences on histoplasmosis [6,419].

2. Tests

   a. Complement Fixation: Introduced by Furcolow et al. in 1948 [125] and by
Campbell and Saslaw in 1949 [54], the complement fixation (CF) test has proven to be
of great value in the serological diagnosis of histoplasmosis. With the introduction in
1965 of the Laboratory Branch Complement Fixation (LBCF) test, developed at the CDC
[11], a measure of reproducibility and comparability was also introduced with the method,
and this standard procedure is now widely used for serological studies of patients with suspected or proven histoplasmosis [13]. From the earliest use of CF tests for histoplasmosis
[54,125], it was recognized that no single antigen was reactive in every case. As a result,
two different types of antigen are routinely used: one is a suspension of intact yeast cells
killed with merthiolate [386] and the other, histoplasmin, is the soluble antigen derived
from mycelial phase culture filtrate. Sera from patients with histoplasmosis may react to
only one type of antigen, but the majority will react to both. Kaufman [197] reported
that in one series of 220 sera, 182 (83%) had demonstrable CF antibodies to the histoplasmin antigen used compared with 206 (97%) to yeast-phase antigen. The figure when
both antigens were used was 96%. Intact cells for use in CF tests killed by merthiolate
were found by Pine et al. [326] to give CF titers with sera from patients with blastomycosis or coccidioidomycosis which were two- to eightfold lower than titers obtained with
formalin-treated cells. When tested against homologous sera, merthiolate-killed cells gave
titers comparable to those obtained with formalin-killed cells, although with some sera, a
twofold reduction in titer was recorded.

   b. Immunodiffusion: Precipitating antibodies to experimentally induced histoplasmosis were detected in rabbits by Pates in 1948 [316], but the presence of the antibodies in human infections was first reported by Salvin and Furcolow in 1954 [376].
Precipitins to a mycelial extract of *H. capsulatum* appeared some 2 weeks after the appearance of symptoms, reached a peak titer at 20 days, then declined until no longer detectable after about 10 weeks. Since CF antibodies in patients with acute histoplasmosis are
generally not detected, or are found at such low titers that the CF test in such cases has
little diagnostic value, increasing attention began to be taken in the ID procedure. It is
to Douglas Heiner, a pediatrician working at the University of Arkansas, that credit goes
for the first realization that diagnostically useful antigens were present in mycelial culture
filtrate [162]. Using Seitz-filtered Sabouraud's broth filtrate from 3- to 6-month cultures
of *H. capsulatum*, Heiner described six antigens, designated m, h, x, y, c, and n, which
formed precipitin lines when tested by immunodiffusion against sera from patients with
histoplasmosis and other mycoses. The m band was so designated because the associated
antibody appeared after skin testing with mycelial skin test antigen. The h band was unaffected by skin testing and was consistently seen in patients with histoplasmosis. It was
found that almost all patients with h antibodies also had m antibodies, although the
corollary was not true. In most preparations containing these antigens, the h line is found
nearer the serum well, and the m adjacent to the antigen well. Two additional bands,
designated x and y, were seen only in the sera of patients with culturally proven histoplasmosis, but the bands were not present consistently. Antigens c (cross-reacting) and n
(nonspecific) were of less diagnostic potential and have attracted no further attention.
The h and m antigens, however (by convention, now referred to as H and M antigens),
have been intensively studied and now constitute the principal diagnostic antigens for
histoplasmosis. In a series of meticulous studies from the CDC, conditions were described

for optimal production of H and M antigens in shake culture [97] and for their purification, composition, and serological characterization [44,152,327]. Methods were also published for the production of reference antisera in rabbits [150] by immunization with specific H and M precipitin bands developing in ID plates. From the same group came the report of a new soluble antigen of *H. capsulatum* [350], obtained from the supernatant of a suspension of merthiolate-treated yeast cells of *H. capsulatum* maintained for 2 months at 4°C. This antigen, designated YPS (yeast phase soluble), was originally used in CF tests, but in a later study [349] it was shown that both H and M components were present, together with a third group of unknown yeast cell antigens. From these studies has emerged a considerable body of information on the occurrence and nature of H and M antigens and methods for their production. When subjected to DEAE-cellulose fractionation, six fractions were obtained from histoplasmin previously subjected to Sephadex chromatography, which were reactive in ID, capillary precipitin, and CF tests [44].

At least five separate H antigens and three M antigens have been distinguished, although serological specificity appears to be associated with a smaller number of antigens (one M and two H). It has been suggested that H and M antigens probably represent autolytic breakdown of mature hyphae rather than products excreted during growth [44]. High carbohydrate-protein ratios in these antigens suggest that they might originate by dissolution of cell wall material. Green et al. [150] have reported that H and M antigens are not located on the surface of yeast cells, but are released during the complement fixation procedure. Detailed analyses indicate that antigens with H and M specificities have a wide range of molecular weights, but Bradley et al. [44] suggest that they have a tendency to form complexes of high molecular weight in histoplasmin concentrates. Data available to date indicate that individual low molecular weight fractions with improved stability and reactivity might eventually be developed.

Progress made toward production and characterization of H and M antigens has been considerable. The ID test for histoplasmosis has been widely adopted and is one of the more successful serodiagnostic tests available for laboratory evaluation of systemic fungal diseases. Several reagents are available in kit form, and recently, a collaborative study has been organized by WHO with a view to accepting as WHO reference materials and procedures the reagents and methods developed at the CDC. The ID procedure used in this trial involves use of a clear plastic template containing 17 patterns of seven wells in the familiar hexagon arrangement. The template is overlaid on a layer of agar and remains in situ during development of the precipitin reactions [309]. For laboratories handling comparatively small numbers of sera, a more orthodox ID system using wells cut in agar or agarose gel is generally more convenient.

Pine et al. have recently described studies made with H and M antigens in CF tests [328]. In sera containing only anti-M antibody, titers obtained with the purified M antigen were 2-16 times those recorded with histoplasmin or yeast-phase antigens. It was found that precipitin lines were obtained with H or M antigens when CF titers were ≥1:8, and that the CF test for either H or M antibody was 4-32 times as reactive as ID.

ID tests have a much lower sensitivity than CF procedures. An increase in sensitivity was reported in 1967 by Ray, using a precipitin inhibition technique [346] where doubling dilutions of a patient's serum are preincubated with a constant concentration of histoplasmin and then tested in agar against antiserum. The results showed that antibodies were detected more reliably and consistently than by orthodox ID, and that titers were comparable to those obtained by CF.

c. Electrophoretic Procedures: Immunoelectrophoresis (IE) was used by Walter in 1969 to improve resolution and detection of precipitins to *H. capsulatum,* using sera from patients with chronic histoplasmosis [448]. Sweet et al. [420] used two-dimensional immunoelectrophoresis (2D-IE) to study different batches of histoplasmin, but reported better resolution using electrodiffusion ("rocket" immunoelectrophoresis), where preparations containing multiple antigens are electrophoresed into gels containing antibodies. Such a procedure allowed quantitative and qualitative comparisons to be made of different batches of antigen.

Counterimmunoelectrophoresis (CIE) was first used to detect antibodies to *Histoplasma* by Gordon et al. in 1971 [140]. The technique was found to be more sensitive than conventional ID. Kleger and Kaufman [212] introduced a modification whereby test sera alternated in the row of serum wells with reference antisera containing H and M antibodies. This arrangement allowed not only detection of *Histoplasma* antibodies in patients' sera, but also ready differentiation into H or M specificities. Antibodies can be detected on direct examination of the gel immediately after completion of electrophoresis. The greater rapidity of CIE compared to ID may in some instances be advantageous, but for greater effectiveness, gels should be washed, dried, and stained. When this is done, the differences in time between the two test procedures are not pronounced.

d. Agglutination (1) Yeast Cells: Described by Cozad in 1958 [73] specific agglutinins were detected in the sera of rabbits inoculated with *H. capsulatum.* Subsequently, the method was applied to a preliminary study of human sera [74], but it then became largely discontinued in favor of inert particle agglutination.

(2) Inert Particles: Introduced by Saslaw and Campbell in 1949 [378,379] histoplasmin was adsorbed onto collodion particles, which agglutinated in the presence of sera from patients with histoplasmosis. Results with collodion particle agglutination tests paralleled those obtained with yeast-phase antigens in CF tests. Preparation of the sensitized collodion particles proved to be both elaborate and time consuming, and the sensitized particles were found to lack stability.

The procedure was eventually supplanted by the latex agglutination (LA) test, introduced in 1958 by Carlisle and Saslaw [55,380]. This procedure was used by Bennett [29], who found that antibodies detectable by LA developed in patients with histoplasmosis 2-3 weeks after infection, and disappeared after 5-8 months, even in patients with persistent disease. Titers equal or greater than 1:32 were considered to be strongly suggestive of active or recent disease.

Gerber et al. described a microtiter LA test which compared well with CF in detecting *Histoplasma* antibodies in the sera of acute and chronic histoplasmosis [132].

Erythrocytes were used by Norden in rabbits [299] and by Hermans and Markowitz for detection of antibodies in 10 patients with histoplasmosis [164], but passive hemagglutination has never gained wide acceptance as a routine screening procedure.

e. Other Tests: Studies were reported in 1963 by Shipe et al. using a fluorescent antibody inhibition test, but results were not encouraging and the procedure has not been developed further [398]. Reiss et al. described in 1977 a solid-phase competitive-binding radioimmunoassay for detecting antibody to the M antigen of histoplasmin [352]. Sera from 22 of 29 patients with histoplasmosis reacted in this system, compared with 21 by CF and 16 by ID. At the time of writing, no reports have yet appeared on the application of an enzyme-linked immunoassay for histoplasmosis, although the anticipated sensitivity and versatility of such a procedure could offer several advantages. Larsh has reported promising results with an immunocytoadherence test for animals experimentally infected with *Histoplasma* and *Sporothrix,* based on detection and quantitation of surface antibodies produced on circulating lymphocytes [231], but further evaluation will be required before the value of this sensitive procedure can be established.

3. Interpretations

Serological tests can play an important role in the diagnosis of histoplasmosis, but in common with all laboratory procedures, they have definite limitations. That ID and CF tests can provide so much valuable diagnostic assistance is a tribute to the acumen and persistence of those who were engaged in their development.

The interpretation of test results can be made difficult at times because antigens may cross-react with sera from individuals suffering from other systemic mycotic diseases. In the CF test, titers of 1:32 or greater are strongly suggestive of histoplasmosis [52,198]. Reactions may also occur with sera from individuals with present or previous infections with *Histoplasma, Coccidioides, Blastomyces,* and *Cryptococcus,* or even in subjects with carcinoma or uveitis [198]. Titers in such instances tend to be below 1:16, but similar levels can often be demonstrated in the sera of patients with proven histoplasmosis. Bennett has reported that CF antibodies are present in 90% of chronic active cases, but in only 50% of acute pulmonary infections [29]. Campbell reported that about 80% of 620 patients hospitalized with chest diseases had CF titers of 1:8 or 1:16 [52]. With those limitations in mind, CF antibodies are, nevertheless, of value when considered in relation to clinical findings and history, particularly if several successive sera can be examined from the same patient. Rising levels of CF antibodies are usually indicative of progressive disease, and falling titers of regression [48]. The usefulness of the CF test in chronic pulmonary histoplasmosis is limited, as with any test, by false-positive and false-negative results. Lowell and Shuford have analyzed data obtained from an extensive study of U.S. Navy recruits from endemic and nonendemic areas, and devised a decision theory method to define more precisely how limits of usefulness can be established [246]. Terry et al. reviewed records of 79 patients at the Mayo Clinic, Rochester, Minnesota, with antibodies to *H. capsulatum* demonstrable by ID and CF [429]. It was concluded that serological results were falsely positive in 28 patients (35%). CF titers of 1:32 or greater were present in 12 cases (15%) without cultural or histological evidence of active infection. These findings are at variance with those reported by Bauman and Smith [28]. In their analysis of ID and CF tests, sensitivities (i.e., number of true cases correctly classified by the test) were found to be 68% for the CF test with histoplasmin antigen and 94% with yeast-phase antigen. Specificities were 99.5% and 94.5%, respectively.

In two patients with chronic *Histoplasma* meningitis described by Plouffe and Fass [331], precipitins and CF antibodies to *H. capsulatum* were present in both serum and cerebrospinal fluid (CSF). Cultural and histological findings were negative. In three patients with active extraneural histoplasmosis, antibodies to *Histoplasma* were found in serum but not in CSF. These findings suggest that presence of such antibodies in the CSF may be of diagnostic value.

Dismukes et al. reported five cases of disseminated histoplasmosis in patients receiving corticosteroid therapy [89]. CF tests were done on sera from two patients and both were negative. Evidence accumulating from this and similar reports points to an opportunistic component in the spectrum of *Histoplasma* infections. Anergy is known to be a feature of severe disseminated histoplasmosis, and this includes both cellular and humoral elements of the immune system. Patients with terminal histoplasmosis may therefore have negative CF and other serological tests.

Levels of CF antibodies may be significantly increased in histoplasmin-sensitized subjects after a single skin test [53,207]. Similar rises may occur in levels of precipitating and agglutinating antibodies. Kaufman et al. [207] reported the appearance of CF antibodies in 17 of 114 sensitized subjects (15%). The range of CF antibody responses following a single skin test appearing in the literature is from 37 to 58%, with titers ranging from 1:8 to 1:256. Kaufman et al. concluded that skin testing does not interfere significantly with antibody levels provided that the blood is drawn within 2-3 days. Increased levels elicited by skin tests are detected primarily by histoplasmin rather than by yeast-phase antigen. Serological responses are not produced in unsensitized subjects. In view of the tendency for skin tests to boost antibody levels, serum for serological tests should always be drawn before any tests for cutaneous reactivity to histoplasmin.

H and M lines are of considerable diagnostic value. Anti-M antibodies appear earlier in infection and persist longer. They develop in sera of patients with acute and chronic forms of histoplasmosis, and may appear after skin testing of sensitized individuals with histoplasmin.

In contrast, antibodies to H antigen appear later and are found in sera from patients with active and progressive disease. Anti-H antibodies may persist for 1-2 years after apparent clinical recovery [198], but they disappear more rapidly from the serum than anti-M antibodies. They are rarely affected by skin testing.

Detection of anti-M antibodies, therefore, is compatible with active or past infection, or a recent skin test. If no skin test has been administered, presence of an M band in ID may indicate an early infection. Presence of H bands usually correlates with active infection and their disappearance with clinical improvement.

Results with latex agglutination (LA) tests are usually comparable to those obtained with CF tests, but LA antibodies may appear earlier in the course of infection. Unlike CF antibodies, those detected by LA are transient, disappearing quite rapidly from a patient's serum. Land et al. examined the LA test in relation to immunosuppressive states or treatments of patients with histoplasmosis [224]. The test was positive in 9% of "normal" patients and in 33% of patients with histoplasmosis, but in 61% of noninfected immunosuppressed patients.

Some of the reagents used in serological tests for histoplasmosis (yeast phase and histoplasmin) are antigenically very heterogeneous. Although able to provide diagnostically useful information, reagents are subject to considerable variation in performance and their interpretation may be difficult. In practice, CF tests for histoplasmosis may be undertaken with yeast-phase *Histoplasma* and *Blastomyces,* as well as histoplasmin and coccidioidin [198]. This serves to differentiate specific from cross-reacting results.

G. Paracoccidioidomycosis

1. Introduction

Paracoccidioidomycosis was first recognized in Brazil, where it was described by Lutz in 1908 [247]. Clinical features of the disease were more fully described in a series of papers by Splendore, who also provided descriptions of the causal agent in vitro [405]. Recognized from the outset were the differentiation of the disease from coccidioidomycosis and the dimorphic nature of the pathogen.

In the years following its original description, the geographical range of the disease was extended, and it is now known to be endemic in most countries in South America and extending into parts of Central America and Mexico. Although commonly referred to as South American blastomycosis, its occurrence in Mexico and Central America suggests that a more accurate disease name would have been Latin American blastomycosis. Increasing international acceptance has been gained for the name paracoccidioidomycosis, and this name is now in widespread use. The disease as first described was primarily mucocutaneous, and infections of nasal, oral, or anorectal mucosa are the most common and conspicuous manifestations of the condition. Paracoccidioidomycosis, however, has a wide range of clinical expressions, including cutaneous and mucocutaneous, lymphangitic, and visceral forms.

The existence of benign, subclinical infection is suggested by the demonstration of cutaneous reactivity to paracoccidioidin in subjects without overt disease, particularly among apparently healthy family members of patients with paracoccidioidomycosis.

The causal agent was initially named *Zymonema brasiliensis* by Splendore in 1912 [405]. Almeida created the new genus *Paracoccidioides* in 1930 [7], and although reclassified in *Blastomyces* by Conant and Howell [64], the designation almost universally accepted is *P. brasiliensis*. The natural habitat of *P. brasiliensis* remains uncertain. Most infections are thought to originate by inhalation of airborne spores, implying that the agent lives as a saprophyte somewhere in nature. Convincing demonstrations that *P. brasiliensis* occurs in soil are very infrequent, and its ecology remains largely unknown. The fungus is dimorphic, being yeast-like in tissues and mycelial in cultures maintained at room temperature. On the basis of available evidence, infections may be established principally by either the respiratory route or by direct penetration of mucosal surfaces. Infection may remain quiescent for long periods, disease developing in some individuals years or decades after their return from endemic areas. On the basis of skin tests with paracoccidioidin, males and females appear to be infected in equal numbers [276]. Males are, nevertheless, more likely to develop the disease, ratios of male to female patients in some reports ranging up to 38:1 [363] or even 74:1 [288].

Much recent information about the disease and its agent is summarized in the proceedings of a symposium on paracoccidioidomycosis sponsored in 1971 by the Pan American Health Organization (PAHO) in Colombia [310].

2. Tests

a. Complement Fixation: The first use of serological tests for paracoccidioidomycosis was by Moses in 1916 [479], when he reported the presence of CF antibodies in the sera of 8 of 10 patients [222]. The test has been in routine use since that time, and has been of great value in establishing patterns of serological response in infected subjects. Earlier studies by Lacaz [222] and Fava Netto [112] confirmed that CF antibodies were commonly demonstrable in the sera of patients with paracoccidioidomycosis. Predictably, different results have been reported for the CF test with different antigens and procedures. Lacaz recorded in 1949 the presence of CF antibodies in the sera of 29 of 31 patients (97%) with active paracoccidioidomycosis. The antigen was prepared from a culture filtrate derived from 20 strains of *P. brasiliensis* grown in the mycelial form on Sabouraud's broth for 3 months [222]. Fava Netto, in contrast, used a polysaccharide antigen prepared by autoclaving yeast-phase *P. brasiliensis,* obtaining positive results in 83% or 86% of 100 patients, depending on the CF procedure used. More recently, however, Fava Netto reported that positive CF tests were positive in 97% of 1073 infected patients [116]. Negroni and Negroni in 1968 reported the comparative performances of three different CF antigens: polysaccharide derived from autoclaved yeast cells, culture filtrate of 1-month-old mycelial growth, and mechanically disintegrated yeast cells. In this study, sera from 18 patients with paracoccidioidomycosis reacted to give CF titers ranging from 1:20 to 1:2560. Culture filtrate antigen gave higher titers in 11 of the 18 sera tested [293].

   b. Immunodiffusion: Precipitins to *P. brasiliensis* were detected by Fava Netto in 1961 using the same polysaccharide antigen developed for the CF test [114]. In this study of 220 patients, antibodies were obtained by the combined use of ID and CF in 98.4% of the sera examined. Precipitins were also detected in patient's sera by Lacaz et al. in 1962 [223], but difficulties were experienced with the test relating to poor diffusion of the lysate antigens used. In 1966, Restrepo published the first of several reports on an ID test for paracoccidioidomycosis which confirmed the diagnostic value of this procedure [354]. Using pooled culture filtrate antigens from yeast or mycelial phases, precipitins were detected in sera from 16 of 21 infected patients (76%). The test was negative in sera from 3 clinically cured patients, from 50 healthy subjects, and from 100 patients with other mycotic infections or with pulmonary disorders of unknown etiology. In subsequent reports from the same author [355,358] yeast-phase antigen was found to be superior to mycelial antigen for ID. It was shown with this antigen that 58 of 61 patients (95%) had precipitins at the time of diagnosis, compared with 78.6% with CF antibodies [358]. Up to three precipitin bands were detected in this system. Individual precipitinogens involved in this ID procedure were designated 1, 2, and 3 by Restrepo and Moncada in 1974 [360]. Reactivity of sera from 54 patients with paracoccidioidomycosis was determined against each antigen. The precipitin band against antigen 1 was found to be the most specific, and persisted longest. Band 3 was least specific and least persistent. It produced a line of identity with the M line of histoplasmin, a finding that could explain the cross-reactivity noted when sera from some patients with paracoccidioidomycosis are tested against M antigen from *H. capsulatum.*

   Restrepo has also used ID to compare antigens derived from yeast cells of different strains of *P. brasiliensis* [356]. Results suggested the existence of antigenic heterogeneity. The efficacy of culture filtrate antigens in ID tests for *P. brasiliensis* has been confirmed by Negroni [289] and Yarzabal et al. [472]. In the latter study, a micromethod was described using species-specific antigen; no false-positive results were obtained in sera from patients affected by other mycosis or from healthy controls. Quantities of reagents required in this test were significantly less than those required in other reported ID procedures. A standard ID procedure for paracoccidioidomycosis has been described in a procedural manual published by the Pan American Health Organization in 1972 [12].

   c. Immunofluorescence: Preliminary studies using an indirect fluorescent antibody test to detect antibodies to *P. brasiliensis* were reported by Kaplan in 1972 [192]. With whole yeast cells as antigen, fluorescence was demonstrated with all 24 sera from patients with active paracoccidioidomycosis and in 11 of 21 patients who had been treated and who were clinically well at the time of testing. Some cross-reactivity was noted with sera from patients with histoplasmosis. Restrepo and Moncada [359] examined sera from patients with paracoccidioidomycosis for antibodies to *P. brasiliensis,* using quantitative

ID and IFA procedures, and compared the results with those obtained by a combination of quantitative ID and CF. The IFA test was found to be more sensitive than the other two procedures, but it was not regarded as suitable for routine serodiagnostic testing because of an unacceptably low level of specificity. Cross-reactions with IFA titers $\leqslant 1:32$ were obtained with sera from patients with histoplasmosis and coccidioidomycosis. Franco et al. [122], however, suggested that IFA, using cell walls of *P. brasiliensis* remaining in the sediment after preparation of Fava Netto's polysaccharide antigen for CF [112], provided a satisfactory means for detecting antibodies to *Paracoccidioides*. A significant correlation was found between results obtained with this procedure and CF.

   d. Electrophoretic Procedures: Immunoelectrophoresis (IE) was one of three procedures used by Negroni in 1968 to detect antibodies to *P. brasiliensis* [289]. Counterimmunoelectrophoresis (CIE) has also been used for this purpose in studies reported in 1973 by Barbosa et al. [25] and Galussio et al. [128]. In the latter report, precipitins were detected in all 73 sera tested from patients with paracoccidioidomycosis.

   CIE has also been used to analyze antigens of *P. brasiliensis*. Restrepo and Drouhet described five precipitin arcs, designated A-E, in the antigenic extract developed for use in ID tests [357]. One of these (band A) appeared to be identical to antigen 1, one of the three precipitinogens previously recognized by ID studies [360]. Yarzabal has also analyzed antigens of *P. brasiliensis* by IE and shown the presence of 25 arcs, one of which (arc E) elicited antibodies in all 20 patients' sera tested [471]. Antigen E was characterized by migration toward the cathode. No cross-reactions occurred with *Histoplasma* or *Blastomyces* antigens.

   Subsequent studies revealed the presence of two fractions, designated E/1 and E/2 [473,474]. The first of these, which is cationic and had alkaline phosphatase activity, was found in only a small proportion of infected patients. In contrast, antigen E/2, which has a neutral or weakly positive charge, reacted only with sera from patients with paracoccidioidomycosis. The method for production of this antigen involves fixation of monospecific rabbit anti E/2 antiserum on sepharose followed by its elution at acid pH. It was suggested that since production of E/2 antigen is both complicated and expensive, it might be suitable for use with highly sensitive serological procedures such as ELISA. Restrepo and Moncada [361] consider that Yarzabal's antigen E and the antigen A described by Restrepo and Drouhet [357] are identical. Since antigen E migrates toward the cathode in an electophoretic system, it would fail to form precipitates in orthodox CIE tests. For this reason, Conti-Diaz et al. developed a special test which combined the principles of both CIE and ID. In this procedure, described in 1973 and 1978 [66,67], an initial CIE of *Paracoccidioides* antigen and test serum is followed by ID, additional test serum being added to a well cut on the cathode side of the CIE antigen well and double diffusion being allowed to proceed overnight. Any antibody reacting to positively charged antigen is revealed by the ID component of this two-phased system. The procedure described by its originators as immunelectroosmophoresis-immunodiffusion (IEOF-ID) has been used to detect precipitins to *P. brasiliensis*. Both cathode-migrating and anode-migrating precipitin lines were detected in the sera of 16 patients with proven paracoccidioidomycosis [66].

   e. Latex Agglutination: Restrepo and Moncada described a latex slide agglutination test in 1978 [361]. Four types of paracoccidioidin-sensitized latex particles were used to detect antibodies to *P. brasiliensis* in the sera of 100 sera from infected patients and

67 sera from other groups of patients and control subjects. In general, the procedure lacked specificity, 40% of sera from patients with histoplasmosis reacting in the slide LA test using the most specific of the sensitized latex particle preparations.

   f. *Other Procedures*:   Radial immunodiffusion has been studied by Porto et al. [336] and compared with ID, using antigen derived from *P. brasiliensis* yeast cells. Test sera were placed in wells cut in gel containing antigen. Appearance of precipitation rings around the well after incubation indicates the presence of specific antibodies. In this report, published in 1977, precipitins were detected in 17 sera by radial immunodiffusion. None of these sera was positive by the ID procedure used. Further evaluation of this procedure is awaited.

   Pons et al. described in 1976 an indirect immunoperoxidase technique for detection of *P. brasiliensis* antibodies [334]. Using yeast cells as antigen and sheep anti-IPI-Ig and anti-IgM, immunoperoxidase activity was demonstrated using sera from patients with paracoccidioidomycosis. Results with IPI-Ig correlated with ID. With IPI-IgM it was reported that clinically active patients could be distinguished from healthy subjects and those with inactive infection.

   Brief references are made in the literature to agar gel inhibition [290] and the conglutinating complement fixation reaction [399], but neither procedure has yet been fully evaluated.

3. Interpretations

As reported by Fava Netto in 1955, the combined use of tube precipitation and CF tests established the presence of antibodies in more than 98% of patients with paracoccidioidomycosis [112]. Shortly thereafter, Fava Netto et al. studied changes in serum proteins, reporting that raised levels of immunoglobulins were present in 76.6% of infected subjects [117]. A close correlation was noted between disease activity and the levels of acute-phase proteins (C-substance, mucoproteins), precipitins, and patterns of serum protein electrophoresis. These findings were confirmed nearly two decades later by Restrepo et al. [362] in a study of the immune responses in 16 patients with paracoccidioidomycosis. Untreated patients had significantly reduced levels of albumin and raised levels of IgG immunoglobulin.

   The pioneer studies by Fava Netto established the existence of two distinct serological patterns which corresponded to the two major clinical expressions of paracoccidioidomycosis [113]. Sera with low levels of antibodies corresponded to mild infections with localized lesions. In contrast, high titers of antibodies were characteristic of severe, disseminated disease. Precipitins were present in all patients studied when the disease had been present for less than a year. Precipitating antibodies were the first to appear, and were also the first to disappear as the disease responded to therapy. They were demonstrable in 60% of patients affected more than a year previously. Fava Netto [115] also reported that precipitins reappear with relapse of the disease.

   The studies by Restrepo and by Yarzabal and their respective associates have revealed the existence of a species-specific precipitinogen [357,471]. The antigen described by each group, and revealed in both instances by IE, is considered by Yarzabal to be identical [475]. Although a full evaluation of this antigen remains to be made, its specificity is clearly of a very high order, and it can be anticipated that it will prove to be of value in diagnosis of the disease.

   CF tests, using cell wall polysaccharide antigen which is distinct from Yarzabal's E/2 antigen, have been of considerable value in establishing correlations between antibody responses and patterns of disease. Fava Netto reported in 1972 that almost 97% of 1073 patients with paracoccidioidomycosis had demonstrable CF antibodies [116]. The need to interpret serological findings only in the light of the clinical picture was emphasized. In an earlier study CF antibodies were found in 67% of patients whose symptoms had appeared within the previous year, but in 95% with infection of longer duration. CF antibodies appeared later than precipitins and persisted longer, sometimes for years after clinical cure. In most instances, a correlation exists between CF titers and severity of disease. Titers greater than 1:8 with yeast-phase filtrate antigen of *P. brasiliensis* are considered strong presumptive evidence for infection [309].

Correa and Giraldo studied the immunoglobulin classes present in sera from 68 patients with paracoccidioidomycosis [70]. IgG levels were well above the normal range in patients with symptoms of less than 12 months' standing. Both IgA and IgM levels were within normal limits, in contrast to the situation reported for acute and chronic histoplasmosis and chronic coccidioidomycosis, where raised levels of IgA are present. The authors suggest that although IgM may have been present in the early stages of infection, IgG had become predominant by the time of diagnosis. In the study reported by Mota and Franco in 1979, IgM antibodies were found in 68% of the infected patients examined. Presence of antibodies was not related to serum IgM levels, to clinical activity or relapse of the disease, or to the presence of precipitating antibodies [279].

The role played by serological tests in the recognition of this protean mycosis is a significant one. The specificity of the ID test appears to be 100%, and when used in combination with CF, valuable information is provided on the existence of disease and its state of development.

VI. MYCOSES OF IMPLANTATION

A. Chromomycosis

1. Introduction

Chromomycosis (chromoblastomycosis) is a chronic, localized, mycotic disease affecting skin and subcutaneous tissues. In common with mycetoma, the disease originates by traumatic implantation of the etiological agent through the skin. Once established, the lesions develop slowly, usually over a period of years, eventually becoming nodular, verrucoid, crusted, or ulcerated. Commonest in tropical and subtropical countries, the disease was first recorded in 1915 by Lane in the United States [230]. In superficial epithelial crusts, the causal fungus can normally be seen by direct microscopy of KOH mounts; in tissue sections fungal cells in the dermis are distinctive, being round or angular, thick-walled, up to 15 μm in diameter, chestnut brown, and often divided by one or more septa.

The disease has a multiple etiology, most infections being caused by species of *Phialophora (P. pedrosoi, P. verrucosa, P. compacta), Wangiella (Exophiala) dermatitidis,* and *Cladosporium carrionii.* The taxonomy and nomenclature of this group of pathogenic fungi have been the subject of much debate and controversy, not yet fully resolved. It may be noted only that considerable support exists for the generic name *Fonsecaea* for the morphologically variable species *(F. pedrosoi* and *F. compactum).*

Serological procedures have not been widely developed or used in the diagnosis of chromomycosis. The disease has a characteristic appearance and history, and generally little difficulty is experienced in recognizing the agent in infected tissues or in isolating it in culture. Identification of the causal fungus may prove difficult for laboratories unfamiliar with the cultural features of the fungi causing chromomycosis. Partly for this reason, and also as a means of comparing the degrees of similarity among the species within the group, attention has focused on their patterns of antigens. Martin, Conant, and Baker reported in 1936 [266] and 1937 [65] from Duke University the results of their comparative studies of antigens from fungi causing chromomycosis and of antigens from other dematiaceous fungi. Sera from rabbits immunized with *F. pedrosoi* and *F. compactum* produced high CF titers to both homologous and heterologous antigens. In contrast, sera from animals immunized with *P. verrucosa* reacted only to their homologous antigen. A similar conclusion was reached by Biguet et al. in 1965, using ID and IE analysis of water-soluble antigens derived from dematiaceous pathogens [34] and by Buckley and Murray in 1966 using ID [47]. Both species-specific and group-specific antigens were described by Conant and Martin [65] and confirmed in later reports. Biguet et al. reported that more common antigens were shared between *P. verrucosa* and *C. carrionii* than between either species and *F. pedrosoi* [34]. The same conclusion was reached by Cooper and Schneidau in 1970 [69] using similar serological methods (ID and IE). Gordon and Al-Doory in 1965 reported that cross-reactivity between *P. verrucosa* and *F. pedrosoi* was not demonstrated by immunofluorescence [139].

Extensive studies have been made by Seeliger [390,392] and Trejos and Seeliger [434] on interrelatedness among a broad range of dematiaceous and other fungi using precipitin, agglutination, and absorption techniques with intact fungal cells or partially purified polysaccharide antigens. Cross-precipitation was observed between antigens from *P. verrucosa, F. compactum, F. pedrosoi,* and *C. carrionii,* and antisera against *Aureobasidium pullulans, C. werneckii,* and *Sporothrix schenckii* [392]. Nielsen and Conant examined the antigenic relationships of yeast-like dematiaceous fungi using agglutination of yeast cells, agglutinin absorption, and ID procedures [297]. These investigators noted a close antigenic similarity between *W. dermatitidis* and *Exophiala (Phialophora) jeanselmei,* the latter agent being a cause of mycetoma rather than chromomycosis. In this same study, *E. jeanselmei* and *P. gougerotii,* the latter being associated with localized subcutaneous, intramuscular, or bone cysts, were found to be antigenically similar but not identical. Stone in 1930 was unable to distinguish between these two species using a CF test [416], a finding corroborated by the precipitin studies reported in 1957 by Seeliger [389]. These findings are of particular interest in view of the later reports by McGinnis and Padhye in 1977 [267] and Padhye in 1978 [307] that *P. gougerotii* is indistinguishable from *E. jeanselmei,* with which it should be regarded as a synonym.

2. Tests and Interpretations

Montpelier and Catanei in 1927 were unable to detect antibodies in the serum of patients with chromomycosis, using CF and agglutination tests [278]. Meriin in 1930 [271] and 1932 [272] obtained positive CF results, but the reactions appeared to be nonspecific. Baliña et al. [24], however, reported in 1932 from Argentina that serum from an infected patient and experimental animals contained CF antibodies to *F. pedrosoi.* Similar findings were reported in 1936 by Martin et al. [266], who demonstrated that the CF titers in their patient from North Carolina correlated with the severity of the disease and fell during treatment.

Costello et al. in 1959 reported that precipitins were not detected in the serum of a

patient with chromomycosis when they used a saline extract of *F. pedrosoi* as antigen
[71]. Contrary findings were reported by Buckley and Murray in 1966 [47]. In this
study, precipitins were found by ID in the sera of 12 of 13 patients with chromomycosis
caused by *F. pedrosoi*. All 12 reactive sera formed precipitin lines against antigen prepared from *F. pedrosoi*, but 8 sera also reacted with *F. compactum* and 2 with *P. verrucosa*. No reactions were seen with sera from 22 normal sera or with antigens prepared
from other species of fungi, including 1 isolate of *C. carrionii* and 7 of *E. jeanselmei*.

Precipitins to heterologous antigens, i.e., to *F. compactum* or *P. verrucosa*, did not
show reactions of identity to the reactions with *F. pedrosoi*. The single serum from a
patient failing to react with *F. pedrosoi* antigen was from a patient who had been treated
with amphotericin B for 4 months and who was clinically and histologically much improved. Precipitins to *F. pedrosoi* were reported in experimentally infected mice 16 days
after intravenous injection by Kurita in 1979 [219]. No other references to precipitins
in natural or experimental infections appear in the literature. ID tests, shown to be apparently both sensitive and specific in 1966 by Buckley and Murray [47] are still not
widely used in the laboratory evaluation of chromomycosis. The relative ease with which
fungal elements can be recognized by orthodox microscopical procedures has already been
noted. It may, nevertheless, eventually prove helpful to use serological tests in the early
stages of disease when the diagnosis remains in doubt. It is self-evident that for any progressive mycoses, management of the condition is likely to be more effective when based
on a diagnosis that is made promptly and accurately.

B. Mycetoma

1. Introduction

According to Chalmers and Archibald [58], mycetoma was first recognized as a disease
entity in 1842 by Gill in southern India. The same author reported that the term "mycetoma," by which the disease is now generally identified, was introduced in 1860 by
Carter. Pinoy in 1913 was the first to distinguish mycetomas caused by actinomycetes
from those with a fungal etiology [330].

Mycetoma is a chronic, progressive, destructive, subcutaneous mycosis affecting skin
and subcutaneous tissues, fascia, and bone. Lesions are characterized by tumor-like swellings and by the development of granulomas and abscesses which suppurate and drain to
the surface by sinus tracts. The agent is present in affected tissues in the form of small
granules, 0.2-5 mm in diameter, the characteristics of which vary according to the infecting agent. The disease, originally known as Madura foot, after the State in India where
it was first described, is now known to have a worldwide distribution, although infections
in the tropics far outnumber those acquired in temperate countries.

Infection originates with introduction of the agent by means of a penetrating wound.
Development of the disease is slow, but inexorable, eventually producing gross deformity
and severe loss of function of affected sites. Feet are commonly affected, but other parts
of the body liable to come into contact with contaminated thorns may be involved, including hands, face, buttocks, and back.

Many species of fungi and actinomycetes have been implicated, although the pathogenesis, clinical appearance, and histopathology are basically the same. Infections with a
fungal etiology (eumycetoma) are most commonly caused by *Madurella mycetomatis*, but
many other molds have been implicated, including *Petriellidium boydii, M. grisea, Exo-*

phiala jeanselmei, Pyrenochaeta romeroi, Leptosphaeria senegalensis, Neotestudina ros-atii, Aspergillus nidulans, and species of *Acremonium* and *Fusarium.* Infections caused by actinomycetes (actinomycetoma) are caused principally by *Streptomyces somaliensis, Actinomadura madurae, A. pelletieri, Nocardia brasiliensis, N. asteroides,* and other aerobic species. Similar infections (botryomycosis) are caused by other bacteria, such as *Proteus, Escherichia coli,* streptococci, and *Staphylococcus aureus.* Some 60% of all cases are caused by actinomycetes and 40% by fungi [262]. Different species predominate as causes of mycetoma in different parts of the world. Thus, in Mexico 90% of cases were reported by Latapi to be caused by *N. brasiliensis* [232], in contrast to Sudan, where most infections are due to *M. mycetomatis, Streptomyces somaliensis,* or *A. pelletieri* [260]. Diagnosis is based primarily on clinical appearance and history, supplemented by macroscopic or microscopic examination of exudates from sinus tracts, or tissue sections for granules of different and at times characteristic form, texture, and color. Laboratory diagnosis is based principally on isolation and identification of the causal agent, but this may be both time consuming and difficult, particularly when spore formation is minimal or lacking. Moreover, the isolation of a species of *Fusarium* or *Aspergillus nidulans* may not by itself be sufficient proof of etiology, since these may be no more than harmless contaminants from the environment. In the early stages of infection the typical clinical picture, with tumefactions and draining sinuses, may not be present. For these reasons, serological tests can have an important role in helping to establish the diagnosis and in suggesting that an agent isolated from an infected site is causing the disease.

Confirmation of etiology is of more than theoretical importance, since the prognosis for eumycetoma and actinomycetoma differs greatly. Infections caused by fungi are highly refractory to chemotherapy. In contrast, actinomycetomas normally respond well to chemotherapy (e.g., septrin, streptomycin), even when infection is well established and of long duration [259].

2. Tests and Interpretations

One of the most important considerations in developing serological tests for mycetoma is the wide range of agents involved. With the examples of histoplasmosis or paracoccidioidomycosis as precedents, careful analysis of antigens from the causal fungi has made it possible to select individual components with a high degree of specificity and with a proven value in the laboratory recognition of infected subjects. It is no coincidence that the mycoses for which serological tests have the greatest diagnostic application have a highly restricted etiology.

It has simply not been practicable to develop antigens for use in serological tests from the many individual species of fungi and actinomycetes capable of causing mycetoma. Other considerations affecting the extent to which serological tests can be developed are the marked differences in etiology in different parts of the world and the comparatively small numbers of cases readily available to those laboratories with a special interest in developing serological tests for this condition. Such laboratories, with few exceptions, are not located in countries where mycetoma has a high prevalence. Even with these inherent limitations, however, serology has a general role in helping to establish a diagnosis of mycetoma, and a more specific one in indicating the class of microorganism to which the etiological agent belongs.

Early reports on the detection of antibodies in the sera of patients with mycetoma are largely anecdotal, referring to individual patients, and without providing details of the reagents and laboratory procedures used. One of the earliest references cited by Biguet and Seeliger in 1972 [33] was to a case of mycetoma seen in 1931 and caused by *Monosporium (Scedosporium) apiospermum* (conidial *Petriellidium boydii*). The patient's serum was found to fix complement.

The first systematic use of serological tests for mycetoma was by Seeliger in Germany. A patient, first seen in 1952, had an infection with *M. apiospermum* of 26 years standing. Combined surgical and chemotherapeutic management were successful [351]. Antibodies were detected by CF, tube precipitin, and ID and spore agglutination [388]. Some cross-reactions were observed with antigens prepared from species of *Acremonium, Trichophyton,* and *Penicillium,* but these had a low titer and could be removed by absorption. Serological reactivity was not present in the sera from 20 healthy subjects. In this study, antibodies were found in some but not all of 10 African patients with eumycetoma. Seeliger concluded that their presence may be of diagnostic value.

Rey in 1961 examined the sera of 4 patients in West Africa with mycetoma caused by *Madurella mycetomatis* or *Leptosphaeria senegalensis,* but was unable to detect antibodies by either ID or CF [364]. Avram and Nicolau in Romania studied antibody responses by CF in the sera of 20 affected patients (14 with eumycetoma and 6 with actinomycetoma). Although positive results were limited to somatic antigens from homologous species and in this sense were specific, CF titers were very low and were present in sera from only 8 of the 20 patients examined [16].

CF tests have received little additional attention as diagnostic aids. Gumaa and Mahgoub in 1973 evaluated the procedure in the serodiagnosis of actinomycetoma [153] using a somatic antigen prepared from *S. somaliensis.* When applied to the study of 95 sera from 25 patients infected with *S. somaliensis,* positive CF tests were found in 83 of 95 sera (87.4%), compared with 92.6% by ID. Eight sera were anticomplementary. Titers diminished as patients responded to therapy and rose when their conditions worsened or failed to improve. Some degree of cross-reactivity was noted with other agents of actinomycetoma (*Actinomadura madurae, A. pelletieri,* and *Nocardia brasiliensis*), but this cross-reactivity was generally at a lower titer than that to *S. somaliensis.* No reactions were obtained with 50 control sera from healthy subjects.

From the results obtained it is clear that a good correlation exists between CF results and infection with *S. somaliensis.* This serological study, by providing clinical application, complemented the report published 20 years earlier by Gonzalez Ochoa and Vasquez Hoyos on the antigens of actinomycetes detected by CF in experimentally immunized rabbits [136].

Murray, in 1961, investigated the value of skin tests in the diagnosis of mycetoma using sensitized guinea pigs [282]. His results, supplemented by later studies with naturally infected subjects [283], showed that sensitized animals reacted specifically to their homologous antigens, reactions being sufficiently distinct to differentiate between, e.g., sensitivity to *M. mycetomatis* and *M. grisea.* In humans, however, although patients with actinomycetoma reacted only to actinomycete antigens, subjects with eumycetoma either failed to respond to any of the antigens, or reacted only to homologous antigen. Failure to respond in some instances, coupled with the observation that a type 3 cutaneous response was elicited, suggested that humoral antibodies detectable by precipitin tests should be present in infected subjects, and that such tests might be of value as ancillary diagnostic procedures.

First explored in 1955 by Seeliger [387], ID tests were used in Mexico by Bojalil and

Zamora in 1963 to detect precipitating antibodies to *N. brasiliensis* in the sera of patients
with actinomycetoma caused by this species [43]. Positive results were obtained with
14 sera from patients infected with this agent but not from 136 patients having tuberculosis, hepatic or renal disease, or leprosy, or from 51 healthy subjects. Sera from three
patients whose infection with *N. brasiliensis* was arrested by chemotherapy were negative.
The first systematic study of ID in the serodiagnosis of mycetoma was made by Mahgoub
in 1964 [257]. When antigens prepared from *S. somaliensis, A. madurae, A. pelletieri,*
S. canescens, and *M. mycetomatis* were used, precipitins were found in 22 of 23 different
sera from cases of mycetoma occurring in Sudan. All but one of the sera from patients
with actinomycetoma reacted to one or more of the actinomycete antigens, but in no instance were cross-reactions observed with the antigen from *M. mycetomatis.* All sera from
patients infected with *Madurella* reacted only to this antigen, and not to actinomycete
antigens. Similar results were obtained with sera from 12 experimentally infected mice.
No reactions were noted with 36 sera from healthy control subjects. These observations
were later extended by Murray and Mahgoub in 1968 [286] and by Mahgoub in 1975
[258]. In the former study, positive ID tests were recorded for all 69 sera tested from
Sudanese patients with mycetoma of varying etiology. No reactions were obtained with
59 sera from control subjects. The specificity of the procedure was impressive, particularly for sera from patients infected with *M. mycetomatis.* Cross-reactivity was obtained between antigens from different species causing actinomycetoma, although reactions to
homologous antigens were almost always appreciably stronger. No correlation was observed between intensity of the precipitin reaction and duration of disease, although sera
from patients with extensive lesions generally contained more precipitins than those with
small, localized lesions.

In the later study by Mahgoub [258] additional data are presented, confirming the
value of serological tests and describing findings with immunoelectroosmophoresis (IEOP)
and CF procedures, in addition to ID. It was noted that antigens of different quality
were obtained from different strains of test organisms. The high level of specificity of
ID was confirmed, and a species-specific antigen was described for *N. brasiliensis.* Compared with ID, IEOP had the advantages of greater rapidity and economy. In the Mycological Reference Laboratory at the London School of Hygiene and Tropical Medicine,
IEOP has been replaced with CIE, which is the routine serological procedure of choice for
detection and characterization of antibodies in sera from patients with mycetoma [254].
As a screening procedure, sera are examined against a battery of fungal and actinomycete
antigens. If a fungus has been isolated from grains recovered from draining sinus tracts or
biopsy material, crude antigen is prepared from the isolate by mechanical disintegration
of harvested mycelial growth or lyophilized culture filtrate and tested against the patient's
serum to assist in assigning an etiological role. This procedure can be of great value for
fungi that sporulate late, poorly, or not at all. Actinomycetes are not usually subjected
to a similar procedure, since their identification is normally achieved more rapidly, and
cross-reactions with the commoner causes of actinomycetoma are common rather than
exceptional.

Few other serological tests have been used in the laboratory evaluation of mycetoma.
In 1978, an ELISA test was described by McLaren et al. [268]. When cytoplasmic antigens from five species of actinomycetes or fungi associated with mycetoma were used,
positive reactions were obtained with the sera of all nine patients tested. Reactivity
against homologous antigens was noted by both CIE and ELISA, but in addition some

cross-reactivity to heterologous antigens was observed in the more sensitive ELISA test.

Serological tests, because of the large number and variety of etiological agents, have inherent limitations as diagnostic aids. The difficulties are compounded by the demonstration of antigenic variation among individual species of fungi [46,258] and actinomycetes [258]. Early diagnosis of the condition is essential to initiate appropriate management with minimal delay, and to minimize the extent of surgical intervention should this prove necessary. Biopsy followed by culture and histopathology are the conventional methods of diagnosis, but this method is both invasive and time consuming. The characteristic granules are not always readily visible and may be overlooked. Mahgoub has reported from Sudan that a diagnosis based on culture was established in 49% of biopsy specimens, and 49% of grains discharged through sinuses. By means of histopathology, species identification was obtained in 78% of sections examined [258]. In the remaining specimens, grains were either not present or could be recognized only as actinomycotic or eumycotic. In this series, ID results were positive in 82% of the cases.

Mycetoma, being primarily a disease of the tropics that affects individuals sporadically, has not received the same attention as other mycoses. In view of the severity of established cases and the intractability of eumycetomas, the need for additional studies on methods for prompt and accurate diagnosis is critical. The full potential of serological testing is far from realization, but it can be anticipated that this is one area of mycopathology where progress can and must be made in the years ahead.

C. Sporotrichosis

1. Introduction

Sporotrichosis is a chronic and usually benign subcutaneous mycosis of humans and animals caused by *Sporothrix schenckii*. The agent is present in vegetation and soil, and initiates infection following traumatic implantation through the skin. Nodular lesions are produced in cutaneous and subcutaneous tissues, which may break down to form indolent ulcers. Lymphangitic sporotrichosis is the commonest form of cutaneous sporotrichosis, the primary lesion being followed days or weeks later by development of secondary lesions along the lymphatic vessels draining the affected area. Exceptionally, dissemination may occur with involvement of musculoskeletal tissues and viscera.

Described initially in 1898 in the United States by Schenck [383], cases were subsequently reported from many temperate and tropical countries throughout the world. Many of the earlier reports originated from France, and by the time de Beurman and Gougerot's classical monograph was published in 1912, some 200 cases had been described [78]. Curiously, the disease has since become rare in that country.

In most cases of sporotrichosis there is a history of contact with soil or vegetation. Schneidau reported in 1964 that prevalence of skin hypersensitivity to sporotrichin among 349 prisoners tested in Louisiana was 11.2%, but among 34 plant nursery workers the figure was 58.3% [385]. Sporadic cases commonly affect farmers, rose growers, or those who collect or handle straw or sphagnum moss.

Sporothrix schenckii is dimorphic, being yeast-like in infected tissues and mycelial in vitro. Individual yeast cells are rounded, oval or cigar-shaped, up to 10 μm in diameter or length. In cultures maintained at room temperature, the vegetative growth consists of delicate branching hyphae, sometimes only 1-2 μm in diameter, bearing small unicellular

conidia. Conversion of mycelial to yeast phase is readily obtained by cultivation on enriched media and incubation at 37°C.

Direct microscopy has an important role in establishing the diagnosis of most mycotic
infections. In sporotrichosis, however, individual fungal cells are often sparse and readily
overlooked. In the study reported by Fukushiro et al. [124] cells of S. schenckii were
recognized in 53 of 55 cases of sporotrichosis, but in only 2 instances were they present
in abundance. Because of their paucity in infected tissues, they are often overlooked in
tissue sections stained by orthodox or special stains for fungi. The application of a direct
fluorescent antibody technique for detection of S. schenckii in tissue sections by Kaplan
and Gonzalez Ochoa [193] has been helpful in culturally negative cases, or where the
fungus cannot be recognized by conventional staining procedures.

Isolation and identification of the agent is of diagnostic value, but this may not be
practicable in cases of deep-seated or disseminated disease. Because of the limitations inherent in diagnostic laboratory procedures based on microscopy and culture, the role of
serological testing in recognition of sporotrichosis can be an important one.

2. Tests and Interpretations

The first use of serological tests for sporotrichosis was by Widal and Abrami in 1908
[468]. They used a spore suspension for both CF and agglutination tests. Although
initial reports were favorable, being positive in all cases of proven infection, later investigators found the tests to be unreliable, giving results that were positive in sera from both
healthy and infected subjects. Inefficacy of the contemporary serological tests were
clearly demonstrated by Du Toit in 1942, during investigations of the celebrated outbreak
of sporotrichosis that affected some 3000 gold miners in South Africa [10].

In 1951, Norden used agglutination, CF, and precipitin tests to detect antibody to
S. schenckii in sera from 11 patients with different forms of sporotrichosis [300]. The
antigens used were intact, autoclaved, or disintegrated yeast cells. Although the number
of patients in the series was small, the different test procedures showed different correlations with disease; thus agglutination was unsatisfactory since titers were low in 7 of the
11 cases, CF tests were positive in only 2 sera, but precipitins were demonstrated in 8 of
the 11 patients. Autoclaved antigens gave no cross-reactions with other fungi, but these
occurred to a variable degree with intact, ground, or methylated antigens. Gonzalez Ochoa
and Soto Figueroa used a tube precipitin test with an alcohol-precipitated polysaccharide
antigen and reported in 1947 that positive results were obtained in 4 of 9 patients with
sporotrichosis [137]. Kaplan and Gonzalez Ochoa used a direct fluorescent antibody
technique to reveal cells of S. schenckii in smears from lesions from patients with suspected sporotrichosis [193].

The value of CF tests in the diagnosis of sporotrichosis has varied considerably in different reports appearing in the literature. None of the 8 patients described by Califano in
1965 had CF antibodies [51]. In contrast, Magalhães Pereira et al. found 69% of sera
from patients with localized infections to be positive [256]. A similar figure (68%) was
reported by Blumer et al. in 1973 [41]. Roberts and Larsh reported in 1971 that a yeast-
phase antigen used in CF and ID tests proved to be specific for S. schenckii and of considerable value in the serological diagnosis of sporotrichosis [369]. No false-positive reactions were observed when the antigen was tested with sera from patients with other
mycotic infections. In 1969, Jones et al. described results obtained with a yeast-phase
culture filtrate antigen when sera from 18 patients with active sporotrichosis (9 cutaneous

and 9 extracutaneous) were examined by CF and ID tests [191]. Only half of the sera reacted in ID, and positive reactions were also obtained in 20 of 78 patients with other mycotic infections, and in 6 of 47 healthy controls. In contrast, CF tests were positive in all 9 patients with extracutaneous sporotrichosis, with titers ranging from 1:8 to 1:256. Three of the 9 patients with cutaneous infections were also positive. In common with all serodiagnostic procedures, results measured by sensitivity and specificity are dependent on the method of antigen production and on procedural details of the test itself. The antigen used by Jones et al. [191] was acetone-precipitated mycelial culture filtrate. McMillen and Laverty used autoclaved mycelial culture filtrate antigen [269] and reported in 1969 that CF antibodies were present in 8 of 14 infected patients (57%).

Karlin and Nielsen used intact yeast cells and demonstrated agglutinins in the sera of 13 patients with sporotrichosis [195]. In their series, 39% of the sera reacted in the CF test and 80% by ID. The diagnostic value of agglutination tests was confirmed by results reported in 1973 by Welsh and Dolan [464]. In this study agglutinins with titers ≥ 1:160 were found in all the sera of all 5 patients with sporotrichosis; agglutinins were present in low titer in only 2 of 30 sera from healthy subjects.

Blumer et al. reported in 1973 the results of their studies with different antigens and serological procedures to determine the most specific and sensitive tests for detecting antibodies to *S. schenckii* [41]. Intact yeast cells were used for slide latex agglutination, tube agglutination, and immunofluorescence. Acetone-precipitated mycelial culture filtrate antigen was used for CF and a similarly prepared yeast culture filtrate antigen for ID. Tests were done with 80 sera from cases of sporotrichosis, 77 sera from patients with other systemic mycotic and bacterial disease, and 7 sera from healthy subjects. Agglutination tests were found to be the most sensitive, being positive for all cases of disseminated and pulmonary infection and for more than 90% of patients with cutaneous sporotrichosis. Similar results were obtained by the immunofluorescence (IF) procedure. CF and ID were less sensitive, detecting antibodies in only 68% and 55% of sera, respectively. False-positive results were generally uncommon, although present in 7 of 64 sera from patients with diseases other than sporotrichosis when tested by IF. It was concluded that tube and slide latex agglutination (SLA) had the greatest diagnostic potential, but the SLA procedure was preferred because of its rapidity and high levels of sensitivity and specificity.

Immunocytoadherence, where *S. schenckii* rosettes were observed during the acute stage of infection, was described as a means of detecting antibodies to *S. schenckii* by Roberts and Larsh [370]. The procedure was effective in the experimental infections of guinea pigs, but no other reports have appeared in the literature on its development or its use in detection of antibodies in infected humans.

Romero de Contreras reported in 1976 the results of her studies with ID, CIE, CF, and IF [372]. One antigen, prepared by mechanical disintegration of harvested mycelial growth and conidia, formed precipitins in the sera of 21 of 23 patients with different forms of sporotrichosis (91%) when tested by CIE. ID tests, in contrast, were positive in only 39% of cases. Marked differences have been noted in published results with ID, ranging from 39% [372] to 85% [269]. These differences almost certainly relate to procedural differences in preparation and use of the serodiagnostic reagents employed in the tests. Specificity of antigens from *S. schenckii* has been reported by Lloyd and Travassos in 1975 to be directed to cell wall determinants which differed in material from cultures grown at 25°C and 37°C [241]. These were identified as rhamnomannans and reacted specifically against sera from infected humans and rabbits. The elegance of this work and its precision are noteworthy and serve not only to indicate how immunological specificity can be clearly defined in chemical terms, but also as a reminder of the amount of work which needs to be done in other mycoses before such clear links can be established between serologic and chemical characteristics of fungal antigens.

Results described with the serological tests used to date for the detection of antibodies to *S. schenckii* are by no means discouraging. Jee et al. consider serology to be unreliable as a screening procedure [190]. As for most mycoses, existing reagents and procedures for sporotrichosis remain to be perfected. Nevertheless, the view that serology already has a useful diagnostic role for this disease is not likely to be seriously or vigorously challenged.

VII. CUTANEOUS MYCOSES

Superficial mycoses are caused by fungi growing on or within the keratinized parts of the body. Some, such as black and white piedra, are confined to hair shafts. Since no growth occurs in skin, affected individuals are unlikely to develop any form of immunological response, and serological tests have no role in diagnosis. In other superficial mycoses, the agent grows in the horny layer of the epidermis, resulting in the development of macular, noninflammatory lesions, characterized by pigmentation (tinea nigra). Pityriasis versicolor, caused by *Malassezia furfur,* is a common and distinctive superficial mycosis which gives rise to hypo- or hyperpigmented areas on the skin surface. The causal fungus is always seen in abundance within the affected areas, and direct microscopy is the most valualbe laboratory procedure available for confirmation of the diagnosis. As with tinea nigra, tests for detection of specific immune responses have not been examined or developed, having no discernible relevance as diagnostic or investigative aids.

Ringworm infections are caused by species of dermatophytes that grow in the stratum corneum of the skin and which may also affect nails and hairs. The lesions may be mild or inflammatory. In severe cases affecting hairs of the scalp or beard, the inflammatory responses may result in the destruction of hair follicles and the development of areas of permanent baldness and scarring. The clinical appearance of ringworm lesions varies according to the dermatophyte species involved, the site infected, and the individual response of the affected subject. Infections acquired from soil or affected animals tend to be more inflammatory than those from human sources. Infection usually results in a state of relative acquired resistance. To a large extent the altered resistance to reinfection is dependent on the species of dermatophyte involved and the site and severity of the primary infection.

Most immunological studies of dermatophyte infections have been directed to the hypersensitivity that develops in infected subjects and the immunochemical nature of the dermatophyte antigens involved. The subject was reviewed in 1974 by Grappel et al. [146]. Antibodies may appear in the course of ringworm infections, but their presence is of no diagnostic value since the lesions are always superficial, and the condition is readily recognized by direct microscopy and culture.

Antibodies to dermatophytes were first detected in naturally infected animals by Verotti in 1916. Serum from a dog infected with *Microsporum canis* contained CF antibodies which reacted to antigens from *M. canis* and other species of dermatophytes [438]. The ability of dermatophytes to elicit antibodies in experimentally infected or immunized animals has been repeatedly demonstrated. In humans, antibodies were first reported in 1915 by Kolmer and Strickler [213]. In their study, CF antibodies were found in 78% of

sera from children infected with *M. audouinii.* Later studies by various authors suggested
that antibodies are more commonly present in subjects with more severe infections. Ayres
and Anderson reported in 1934 that sera from patients with cutaneous hypersensitivity
reactions (id eruptions) inhibited growth of dermatophytes in vitro. Sera from patients
without id eruptions were not inhibitory [20]. Although it was suggested that the effect
was mediated by fungicidal antibodies, no direct link has been unequivocally established
between any defined category of antibodies and resistance to infection. Normal human
serum has been shown by several investigators to be inhibitory to dermatophytes and other
fungi [245,318,373]. The factor responsible for inhibition is present in the sera of the newborn as well as adults, and has been partially characterized as an unstable, heat-labile, dialyzable component of serum and tissue fluids. Although dermatophytes are normally in-
capable of invading living tissue, Blank et al. reported in 1959 that cultivated human skin
explants were readily penetrated by hyphae, which invaded all layers of the skin when
serum was omitted from the culture system. When serum was added to the nutrient solution, however, hyphae were confined to the keratinized layers [37].

Dermatophytes have long been recognized as keratinophilic, but keratinases were first
demonstrated by Yu et al. as recently as 1968 [477]. Grappel et al. reported in 1971 that
antibodies raised in rabbits immunized with dermatophyte keratinase caused retardation
of growth of *Trichophyton mentagrophytes* in culture, and alteration of its morphology
[149]. It is not known if such antibodies play any role in affecting the course of dermatophyte infections. Austwick suggested in 1972 that antibodies with dermatophyte specificity may diffuse into hair bulbs, become incorporated in the developing hair shaft, and
bring about the degenerative changes which can be seen in hyphae in healing ringworm
lesions [15]. Dermatophyte hyphae in infected skin lesions examined by immunofluorescence were found by Hay and Saeed to be coated with neither IgG nor IgA immunoglobulins [160]. No morphological abnormalities were observed in the hyphae, and it remains
unknown if antibodies significantly influence hyphal characteristics in vivo or persistence
in infected sites.

Precipitins to dermatophytes have been demonstrated in natural [321] and experimental [188] infections of humans. Antibodies to dermatophytes have also been detected
by passive hemagglutination [365], immunofluorescence [452], charcoal agglutination
[147], and counterimmunoelectrophoresis [42]. Grappel et al. [148] noted that in patients with scalp ringworm, antibodies were not necessarily restricted to those with severe
infections. Patients having chronic infections with *T. rubrum* often had precipitins or CF
antibodies, and antibodies were usually present in the sera of patients with id eruptions.
It was noted that in some cases antibodies were detectable for a short period only.

Nearly 40 species of dermatophytes have been recognized. Their natural histories
may be very different, and the group includes species which are adapted to life in the soil
or to specific animal or human hosts. Morphological, biochemical, and physiological characteristics may also be very different, but there are, nevertheless, close similarities within
the group, even among species with widely contrasting characteristics. Common antigens
have been found among a wide range of dermatophyte species and extensive studies made
on their chemical natures and serological reactivities. The principal antigens described
have generally been glycopeptides, polysaccharides (galactomannans), and proteins (including keratinases). Such studies have been directed to the reactivity of different antigenic preparations and their value in the hypersensitivity to dermatophyte infections.
Group-specific and species-specific antigens have been described, but cross-reactivity is

nevertheless very marked. Philpot [322] described in 1978 an analysis of antigens obtained from young mycelial growth of 20 species of dermatophytes. By ID, 48 different antigens were detected. Species of *Microsporum* formed a group which tended to be distinct from *Epidermophyton* and *Trichophyton*. *Trichophyton mentagrophytes* and *T. rubrum* shared 7 of the 9 antigens detected.

Svejgaard and Christiansen reported in 1979 that precipitins were present in the sera of 25 (9%) of 262 patients with dermatophyte infections [417]. Using 2D-1E, the same authors detected 35 antigens from *T. rubrum* and 26 from *T. mentagrophytes* [61]. By means of tandem 2D-IE, only 2 antigens were found to be common to the two species, a finding which contrasts with that reported by Philpot [322]. The explanation for these disparate findings is presumably related in part to differences in methodology for preparation and testing of the reagents.

Immunization has been widely practiced in the USSR to reduce cattle ringworm in calves. The vaccine used (TF-130) is prepared from a single strain of *T. verrucosum* [332] and has now been given to several million animals. The protection afforded was estimated to be 99.2% and the curative efficacy 98.5% [437]. In 44 of 54 regions in which the vaccine was used in 1972, no infections with *T. verrucosum* were recorded. The protection afforded by TF-130 appears to be long-lasting, but there is no evidence to suggest that antibodies are involved. Since immunity is accompanied by cutaneous sensitization, it can be assumed that the basis for protection is the acquired hypersensitivity.

The role of serological tests in the study of dermatophyte infections is principally in providing the means for analysis of antigenic preparations. At present there is little need or justification for their evaluation as ancillary aids to diagnosis.

VIII. MISCELLANEOUS MYCOSES

Humans can be affected adversely not only by a variety of fungi but also by a variety of pathogenic mechanisms. Apart from those diseases resulting from growth of fungi on or within the human body, harmful consequences can follow ingestion or inhalation of fungi or fungal products. Such diseases are related to toxic or allergic properties of the fungi or their metabolites. In only a limited number of instances have serological tests been used to study such diseases or to assist with their diagnosis. As yet, there have been no reports of serology being used in the laboratory study of a human mycotoxicosis. In view of the rarity of proven human diseases of this kind, the high level of sensitivity reached by existing methods of chemical analysis and the lack of immunogenicity of known mycotoxins, there are no reasons at present for anticipating that serology will be applied as analytical or diagnostic tools in this area.

Diseases elicited by hypersensitivity to fungi or their products are by no means uncommon. In cases of allergic aspergillosis, where atopic individuals develop sensitization to *Aspergillus* or hypersensitivity pneumonitis (extrinsic allergic alveolitis) which is caused by repeated inhalation of the airborne spores of thermophilic actinomycetes or fungi such as *A. clavatus,* antibodies may be elicited and serological tests may be of some value as ancillary diagnostic aids. Extrinsic asthma or allergic rhinitis may be related to a wide range of airborne allergenic particles, many of them being fungal spores (*Cladosporium, Alternaria, Botrytis, Acremonium, Penicillium,* etc.) In some instances antibodies may be detected in the serum of sensitized subjects, but serological tests have little diagnostic value, since they reflect the degree of exposure rather than sensitization. In such individuals, the diagnosis is made by detection of skin sensitization to specific allergens.

Infections caused by actinomycetes (actinomycosis and nocardiosis) are often accompanied by the development of antibodies, and serological tests may be of value in confirming a diagnosis and in monitoring therapy. Tests that have been described for *Actinomyces* infections include ID [130], hemagglutination and CF [168], and CIE [169]. Serological procedures for the laboratory diagnosis of nocardiosis include ID [40,172] and complement fixation [396].

IX. INTERNATIONAL STANDARDIZATION

Quality control measures are essential for all laboratory procedures, but their need is particularly evident in the production and use of serodiagnostic reagents and procedures. Unless steps are taken to ensure uniformity of quality, serological tests cannot be expected to give reliable or reproducible results. Different tests have been developed to different degrees for different mycoses. In many instances the antigens used are complex, containing a variety of immunogenic components of undetermined chemical nature and unknown biological function. Minimization of batch-to-batch variation is not only desirable, but prerequisite when new reagents are introduced in place of existing ones. If a replacement antigen differs qualitatively or quantitatively from its predecessors, results will not be comparable to those obtained with the previous reagent, and the sensitivity and specificity of the procedure will have to be evaluated anew. Experience with a given test can be consolidated and extended only when variability is reduced to a minimum. Internal quality control is usually achieved in a laboratory by comparing reactivity of new batches of reagents with previous ones using a standard procedure. As yet, physicochemical analysis of complex antigens does not affort a reliable means for measuring their specificity, sensitivity, or efficacy, and the method most frequently employed is by titration of new preparations against reference antisera to determine the concentration for use which gives comparable results to the previous antigen.

In practice, methods or preparation, assay, and use of mycotic antigens are commonly developed by trial and error. Some tests are better than others in indicating the likelihood of disease. It has long been recognized that a major hindrance to realization of the full potential of any test or immunodiagnostic antigen is the difficulty or impossibility of comparing results from different centers. Once serodiagnostic reagents and procedures have been subjected to primary evaluation, further progress with their refinement and development becomes materially assisted by comparison with other tests and antigens. The number of laboratories engaged in immunodiagnostic studies throughout the world is constantly increasing, but with few exceptions, programs of comparative evaluation are seldom undertaken, and the relative significance of published results from different centers is difficult or impossible to evaluate. It is important to recognize the distinction between diagnostic reagents on the one hand, and reference or standard products on the other. By establishing reference or standard reagents and procedures, a basis is provided for comparing different products and procedures. Diagnostic reagents produced by specialist laboratories or by commercial organizations can then be evaluated and, if necessary replaced or readjusted until their performance matches that of the standard or reference product. The anticipated gains in terms of reproducibility and predictibility are substantial.

In the development of internationally accepted "standards," a number of primary requirements have to be met. Included among these are the "recognition of need," the prior existence of candidate reagents and procedures, and the devising, funding, and implementation of collaborative studies to assess performance. International standards can only be recognized and established as such following collaboration among specialist laboratories in different countries. The earliest developments in recognition of the need for reference immunodiagnostic materials and procedures were reviewed by Mackenzie in 1977 [250]. Huppert et al. in 1973 recommended that a WHO committee be created to consider the possibility of establishing one or more central laboratories to serve as depositories and distributing reagents for serological tests used in the diagnosis of mycoses [186]. Huppert et al. also suggested that working groups should be established and commissioned with the task of seeking agreement on standardized reagents, reference cultures, and recommended procedures.

It is encouraging to note that some initial steps have been taken in this direction and that some progress has been made in providing standard methods and reagents at both national and international levels. Some of the earlier collaborative precedents were established by the Antigen Study Group of the American Thoracic Society and the Mycoses Subcommittee on Diagnostic Procedures of the Pan American Health Organization. The appearance of procedural manuals from centers with wide experience in the field [12, 199,309] have by themselves provided descriptions of tried and reliable methods for examining sera for antibodies to fungi which can be readily adopted by less experienced laboratories.

The WHO completed in 1980 a collaborative study on an ID test for histoplasmosis, the first occasion on which any fungal test or serodiagnostic reagent has been evaluated by WHO. The reagent, histoplasmin containing H and M antigens, and the procedure, using a multiwell pattern template [309], were those developed at the Center for Disease Control over many years. CDC's claim for consideration and evaluation as a reference reagent and a reference procedure is based on years of meticulous developmental study and experience with fungal serology; it is to be hoped that this precedent will be followed by similar collaborative studies.

In practical terms, however, it must be realized that such studies are time consuming, complex, and costly, and that claims for "recognition of need" for any one test and reagent have to be viewed in the light of other biological substances such as allergens and the need that exists for *their* evaluation and adoption as standard or reference products.

Serological tests have been developed to a high degree of reliability: not only the ID test for histoplasmosis, but, for example, the latex agglutination test for cryptococcal capsular antigen [39]. Reagents and procedures can, in several instances, already be selected on the basis of proven efficacy. In the case of such mycoses as candidiasis and aspergillosis, however, serological tests have not been developed so successfully, and obvious or even suitable candidate reagents and procedures are lacking. A case can be made in such instances for establishing provisional reference materials, to provide a means of comparing characteristics and performances of the many products in use throughout the world. As and when reagents of greater diagnostic value are developed and recognized, they could replace the provisional materials.

International collaboration may be arranged in accordance with the guidelines set out by WHO [467] and with the objective of establishing whether or not a serodiagnostic antigen meets the exacting requirements of a WHO standard or reference biological material. Not all collaborative studies involving laboratories in different countries are as formally structured, and much valuable information can be gained from more informal links. In this general area of fact-finding, activity is being generated by such organizations as the International Society for Human and Animal Mycology and the International Union of Immunological Societies, both of which have committees concerned with immunodiagnostic antigens and tests for diseases caused by fungi.

The establishment of reference materials and procedures calls for much patience, determination, and persistence. The rewards are not always commensurate with the expenditure of time, effort, and money. The impetus for initiating and supporting such programs rests primarily with the mycological community. It is easy to accept the need for international standards and to support in principle efforts directed toward that end. It is an altogether different matter to select materials and methods worthy of advancement to reference status, and to have them evaluated within existing frameworks and constraints of international organizations.

REFERENCES

1. Abernathy, R. S., and D. C. Heiner. 1961. Precipitation reactions in agar in North American blastomycosis. J. Lab. Clin. Med. *57*: 604-611.
2. Aguilar-Torres, F. G., L. J. Jackson, J. E. Ferstenfeld, D. Pappagianis, and M. W. Rytel. 1976. Counterimmunoelectrophoresis in the detection of antibodies against *Coccidioides immitis.* Ann. Intern. Med. *85*: 740-744.
3. Aguilar-Torres, F. G., M. W. Rytel, D. Pappagianis, and L. Jackson. 1978. Counterimmunoelectrophoresis as a rapid screening test in coccidioidomycosis. Rev. Invest. Clin. *30*: 247-251.
4. Ajello, L. (Ed.). 1967. *Coccidioidomycosis,* Papers from the Second Symposium on Coccidioidomycosis. University of Arizona Press, Tucson.
5. Ajello, L. (Ed.). 1977. *Coccidioidomycosis: Current Clinical and Diagnostic Status,* Stratton, New York.
6. Ajello, L., E. W. Chick, and M. L. Furcolow. 1971. *Histoplasmosis,* Thomas, Springfield, Ill.
7. Almeida, F. P. 1930. Estudos compartivos de granuloma coccidioidico nos Estados Unidos e no Brasil: novo genero para o parasito brasileiro. An. Fac. Med. Sao Paulo *5*: 125-141.
8. Andersen, P. 1968. The occurrence of antibodies against *Candida albicans* in sera from normal subjects. Dan. Med. Bull. *15*: 277-282.
9. Anderson, K., and M. Beech. 1958. Serological tests for the early diagnosis of cryptococcal infection. Med. J. Aust. *45*: 691-694.
10. Proceedings of the Transvaal Mine Officer's Association, *Sporotrichosis Infection on Mines of the Witwatersrand.* Cape Times, Ltd., Cape Town, 1947.
11. Laboratory branch task force. Standardised diagnostic complement fixation method and adaptation to micro test. Public Health Monogr., p. 74, 1965.
12. Manual of standardised serodiagnostic procedures for systemic mycoses. I. Agar immunodiffusion tests. Pan American Health Organization, Washington, D.C.
13. Subcommittee on diagnostic procedures. PAHO coordinating committee for the mycoses. Standardised procedures for systemic mycoses. Pan American Health Organization, Washington, D.C.
14. Morbid Mortal. Wkly. Rep., p. 450, 1979.
15. Austwick, P. K. C. 1972. The pathogenicity of fungi. In H. Smith and J. H. Pearce (Eds.), *Microbicidal Pathogenicity in Man and Animals,* Cambridge University Press, Cambridge, pp. 251-268.

16. Avram, A., and G. Nicolau. 1967. Proc. 2nd Int. Symp. Med. Mycol., Poznań., pp. 253-256.

17. Axelsen, N. 1973. Quantitative immunoelectrophoretic methods as tools of a polyvalent approach to standardisation in the immunochemistry of *Candida albicans*. Infect. Immun. *7*: 949-960.

18. Axelsen, N. H., C. H. Kirkpatrick, and R. H. Buckley. 1974. Precipitins to *Candida albicans* in chronic mucocutaneous candidiasis studied by crossed immunoelectrophoresis with intermediate gel: correlation with clinical and immunological findings. Clin. Exp. Immunol. *17*: 385-394.

19. Axelsen, N. H., and C. H. Kirkpatrick. 1973. Simultaneous characterisation of free *Candida* antigens and *Candida* precipitins in a patient's serum by means of crossed immunoelectrophoresis with intermediate gel. J. Immunol. Methods *2*: 245-249.

20. Ayres, S., and N. P. Anderson. 1934. Inhibition of fungi in cultures by blood serum from patients with "phytid" eruptions. Arch. Dermatol. *29*: 537-547.

21. Azuma, I., F. Kanetsuna, Y. Tanaka, Y. Yamamura, and L. Carbonell. 1974. Chemical and immunological properties of galactomannans obtained from *Histoplasma duboisii, Histoplasma capsulatum, Paracoccidioides brasiliensis* and *Blastomyces dermatitidis*. Mycopathol. Mycol. Appl. *54*: 111-125.

22. Bacon, P. A., C. Davidson, and B. Smith. 1974. Antibodies to *Candida* and autoantibodies in sub-acute bacterial endocarditis. Q. J. Med. *43*: 537-547.

23. Baldo, B. A., E. H. Quinn, and K. J. Turner. 1975. In vitro diagnosis of allergy. Radioallergosorbent test (RAST) studied with allergens commonly encountered in Australia. Med. J. Aust. *2*: 859-863.

24. Baliña, P., B. Bosq, P. Negroni, and M. Quiroga. 1932. Un caso de cromoblastomicosis, autóctono de Argentina. Rev. Argent. Dermatol. *16*: 369-379.

25. Barbosa, W., E. Blau, J. R. Mendonca, and R. L. Oliveira. 1973. Crossing over immunoelectrophoresis applied to the study of immunology of South American blastomycosis. Rev. Patol. Trop. *2*: 73-76.

26. Bardana, E. J., Jr. 1978. Culture and antigen variants of *Asperigillus*. J. Allergy Clin. Immunol. *61*: 225-227.

27. Bardana, E. J., K. L. Sobti, F. D. Cianciulli, and M. J. Noonan. 1975. *Aspergillus* antibody in patients with cystic fibrosis. Am. J. Dis. Child. *129*: 1164-1167.

28. Bauman, D. S., and C. D. Smith. 1975. Comparison of immunodiffusion and complement fixation tests in the diagnosis of histoplasmosis. J. Clin. Microbiol. *2*: 77-80.

29. Bennett, D. E. 1966. The histoplasmin latex agglutination test: clinical evaluation and a review of the literature. Am. J. Med. Sci. *251*: 175-183.

30. Bennett, J. E., and J. W. Bailey. 1971. Control for rheumatoid factor in the latex test for cryptococcosis. Am. J. Clin. Pathol. *56*: 360-365.

31. Bennett, J. E., H. F. Hasenclever, and B. S. Tynes. 1964. Detection of cryptococcal polysaccharide in serum and spinal fluid: value in diagnosis and prognosis. Trans. Assoc. Am. Physicians *77*: 145-150.

32. Bennett, J. E., and H. F. Hasenclever. 1965. *Cryptococcus neoformans* polysaccharide: studies of serologic properties and role in infection. J. Immunol. *94*: 916-920.

33. Biguet, J., and H. P. R. Seeliger. 1972. Le diagnostic immunologique des mycétomes. Ann. Soc. Belge. Med. Trop. *52*: 251-260.

34. Biguet, J., P. Tran Van Ky, S. Andrieu, and J. Fruit. 1965. Analyse immunoélectrophorétique des antigènes fongiques et systématique des champignons. Répercussions pratiques sur le diagnostic des mycoses. Mycopathol. Mycol. Appl. *26*: 241-256.

35. Bindschadler, D. D., and J. E. Bennett. 1968. Serology of human cryptococcosis. Ann. Intern. Med. *69*: 45-52.

36. Bläker, F., K. Fischer, and H. H. Hellwege. 1973. Bedeutung humoraler Antikörper gegen *Candida albicans* für den nachweis von *Candida*-infektionen. Dtsch. Med. Wochenshr. *98*: 194–201.

37. Blank, H., S. Sagami, C. Boyd, and F. J. Roth, Jr. 1959. The pathogenesis of superficial fungus infections in cultured human skin. Arch. Dermatol. *79*: 524-535.

38. Bloomfield, N. 1962. *Annual Report of the Division of Laboratories and Research,* New York State Department of Health, pp. 111-112.

39. Bloomfield, N., M. A. Gordon, and D. F. Elmendorf. 1963. Detection of *Cryptococcus neoformans* antigens in body fluids by latex particle agglutination. Proc. Soc. Exp. Biol. Med. *114*: 64-67.

40. Blumer, S. O., and L. Kaufman. 1979. Microimmunodiffusion test for nocardiosis. J. Clin. Microbiol. *10*: 308-312.

41. Blumer, S. O., L. Kaufman, W. Kaplan, D. W. McLaughlin, and D. E. Kraft. 1973. Comparative evaluation of five serological methods for the diagnosis of sporotrichosis. Appl. Microbiol. *26*: 4-8.

42. Böhme, H., R. Grundhoff, I. Tausch, and H. Ziegler. 1978. Detection of precipitating antibodies against dermatophytes by means of counterimmunoelectrophoresis. Dermatol. Monatsschr. *164*: 416-422.

43. Bojalil, L. F., and A. Zamora. 1973. Precipitin and skin tests in the diagnosis of mycetoma due to *Nocardia brasiliensis.* Proc. Soc. Exp. Biol. Med. *113*: 40-43.

44. Bradley, G., L. Pine, M. W. Reeves, and C. W. Moss. 1974. Purification, composition and serological characterisation of histoplasmin—H and M antigens. Infect. Immun. *9*: 870-880.

45. Buck, A. A., and H. F. Hasenclever. 1963. Epidemiologic studies of skin reactions and serum agglutinins to *Candida albicans* in pregnant women. Am. J. Hyg. *78*: 232-240.

46. Buckley, H. R. 1968. A study of dematiaceous pathogenic fungi and related saprobic species. PhD. thesis, University of London.

47. Buckley, H. R., and I. G. Murray. 1966. Precipitating antibodies in chromomycosis. Sabouraudia *5*: 78-80.

48. Buechner, H. A., J. H. Seabury, C. C. Campbell, L. K. Georg, L. Kaufman, and W. Kaplan. 1973. The current status of serologic, immunologic and skin tests in the diagnosis of pulmonary mycoses. Report of the committee on fungus diseases and subcommittee on criteria for clinical diagnosis. Chest *63*: 259-270.

49. Bull, G., and D. W. R. Mackenzie. 1976. Unpublished results.

50. Busey, J. F., and P. F. Hinton. 1967. Precipitins in blastomycosis. Am. Rev. Respir. Dis. *95*: 112-113.

51. Califano, A. 1965. Rilievi e considerazioni su otto casi di sporotricosi cutanea. G. Ital. Dermatol. *106*: 551-572.

52. Campbell, C. C. 1960. The accuracy of serologic methods in diagnosis. Ann. N.Y. Acad. Sci. *89*: 163-177.

53. Campbell, C. C., and G. B. Hill. 1964. Further studies on the development of complement fixation antibodies and precipitins in healthy histoplasmin-sensitive persons following a single histoplasmin skin test. Am. Rev. Respir. Dis. *90*: 927-934.

54. Campbell, C. C., and S. Saslaw. 1949. Use of yeast-phase antigens in a complement fixation test for histoplasmosis. III. Preliminary results with human sera. Public Health Rep. *64*: 551-560.

55. Carlisle, H. N., and S. Saslaw. 1958. A histoplasmin-latex agglutination test. 1. Results with animal sera. J. Lab. Clin. Med. *51*: 793-801.

56. Carlsson, H. E., and S. Bernander. 1975. Affinity chromatography of antigens from *Candida albicans.* In H. Peeters (Ed.), *Protides of Biological Fluids,* Pergamon, Oxford, pp. 613-616.

57. Carlsson, H. E., and K. Holmberg. 1977. Acta Soc. Med. Suec. *86*: 328 (Abstr.)

58. Chalmers, A. J., and R. G. Archibald. 1916. A Sudanese maduromycosis. Ann. Trop. Med. Parasitol. *10*: 222-223.

59. Chew, W. H., and T. L. Theus. 1967. *Candida* precipitins, J. Immunol. *98*: 220-224.

60. Chilgren, R. A., R. Hong, and P. G. Quie. 1967. Human serum interactions with *Candida albicans.* J. Immunol. *101*: 128-132.

61. Christiansen, A. H., and E. Svejgaard. 1976. Studies of the antigenic structure of *Trichophyton rubrum, Trichophyton mentagrophytes, Microsporum canis* and *Epidermophyton floccosum* by crossed immuno-electrophoresis. Acta Pathol. Microbiol. Scand., Sect. C *84*: 337-341.

62. Christie, A., and J. C. Peterson. 1945. Pulmonary calcification in negative reactors to tuberculin. Am. J. Public Health *35*: 1131-1147.

63. Coleman, R. M., and L. Kaufman. 1972. Use of the immunodiffusion test in the serodiagnosis of aspergillosis. Appl. Microbiol. *23*: 301-309.

64. Conant, N. F., and A. Howell, Jr. 1942. Similarity of fungi causing South American blastomycosis (paracoccidioidal granuloma) and North American blastomycosis (Gilchrist's disease). J. Invest. Dermatol. *5*: 353-370.

65. Conant, N. F., and D. S. Martin. 1937. The morphologic and serologic relationships of the various fungi causing dermatitis verrucosa (chromoblastomycosis). Am. J. Trop. Med. *17*: 553-577.

66. Conti-Diaz, I. A., J. E. Mackinnon, L. Calegari, and S. Casserone. 1978. Estudio comparativo de la immunoelectroforesis (IEF) y de la immunoelectroosmoforesis-immunodifusión (IEOF-ID) aplicadas al diagnóstico de la paracoccidioidomycosis. Mycopathologia *63*: 161-165.

67. Conti-Diaz, I. A., R. E. Somma-Moreira, E. Gezuele, A. C. Jimenez, M. I. Peña, and J. E. Mackinnon. 1973. Immunoelectroosmophoresis-immunodiffusion in paracoccidioidomycosis. Sabouraudia *11*: 39-41.

68. Converse, J. L. 1955. Growth of spherules of *Coccidioides immitis* in a chemically defined liquid medium. Proc. Soc. Exp. Biol. Med. *90*: 709-71.

69. Cooper, B. H., and J. D. Schneidau. 1970. A serological comparison of *Phialophora verrucosa, Fonsecaea pedrosoi* and *Cladosporium carrionii* using immunodiffusion and immunoelectrophoresis. Sabouraudia *8*: 217-226.

70. Correa, L. A., and R. M. Giraldo. 1972. Study of immune mechanisms in paracoccidioidomycosis. Changes in immunoglobulin (IgG, IgM, and IgA). In *Paracoccidioidomycosis,* Proceedings of the First Pan American Symposium, Pan Am. Health Organ. Sci. Publ. No. 254, Washington, D.C., pp. 245-250.

71. Costello, M. J., C. P. de Feo, and M. L. Littman. 1959. Chromoblastomycosis treated with local infiltration of amphotericin B solution. Arch. Dermatol. *79*: 184-193.

72. Coudert, J., T. Kien Truong, P. Ambroise-Thomas, C. Douchet, and M. A. Pothier. 1967. Septicémie à *Candida albicans.* Etude sérologique par immunofluorescence, agglutination et immuno-électrophorèse. Bull. Soc. Pathol. Exot. *60*: 497–503.

73. Cozad, G. C. 1958. A study of the whole yeast cell agglutination test in rabbits experimentally infected with *Histoplasma capsulatum.* J. Immunol. *81*: 368-375.

74. Cozad, G. W., and W. H. Marsh. 1960. A capillary tube agglutination test for histoplasmosis. J. Immunol. *85*: 387-390.

75. Craven, R. B., J. B. Brooks, D. C. Edman, J. D. Converse, J. Greenlee, D. Scholls-

berg, T. Furlow, J. M. Gwaltney, and W. F. Miner. 1977. Rapid diagnosis of lympho-cytic meningitis by frequency pulsed electron capture gas-liquid chromatography: differentiation of tuberculous cryptococcal and viral meningitis. J. Clin. Microbiol. *6*: 27-32.

76. Darling, S. T. A. 1906. A protozoan general infection producing pseudotubercles in the lungs and focal necrosis in the liver and spleen and lymphnodes. JAMA *46*: 1283-1285.

77. Da Rocha Lima, H. 1912. Beitrag zur Kenntnis der blastomykosen lymphangitis Epizootica und Histoplasmosia. Zentralbl. Bakteriol. Parasitenkd. Infektionskr. Hyg., Abt. 1: Orig. *67*: 233-249.

78. de Beurman, L., and H. Gougerot. 1912. *Les Sporotrichoses.* Ancienne Librairie Germer. Baillière, Paris.

79. De Bracco, M. M. E., D. B. Budzko, and R. Negroni. 1976. Mechanisms of activa-tion of complement by extracts of *Aspergillus fumigatus.* Clin. Immunol. Immuno-pathol. *5*: 333-339.

80. Dee, T. H. 1975. Detection of *Aspergillus fumigatus* serum precipitins by counter-immunoelectrophoresis. J. Clin. Microbiol. *2*: 482-485.

81. Dee, T. H., and M. W. Rytel. 1975. Clinical application of counterimmunoelectro-phoresis in detection of *Candida* serum precipitins. J. Lab. Clin. Med. *85*: 161-166.

82. Dee, T. H., and M. W. Rytel. 1977. Detection of *Candida* serum precipitins by counterimmunoelectrophoresis: an adjunct in determining significant candidiasis. J. Clin. Microbiol. *5*: 453-457.

83. de Monbreun, W. A. 1934. Cultivation and cultural characteristics of Darling's *Histoplasma capsulatum.* Am. J. Trop. Med. *14*: 93-125.

84. Deresinski, S. C., H. B. Levine, P. C. Kelly, R. J. Creasman, and D. A. Stevens. 1977. Spherulin skin testing and histoplasmal and coccidioidal serology: lack of effect. Am. Rev. Respir. Dis. *116*: 1116-1118.

85. Desgeorges, P. T., P. Ambroise-Thomas, and A. Goullier. 1979. Cryptococcosis: serodiagnosis by immunoenzymology (ELISA). Nouv. Presse Med. *8*: 1055-1066.

86. Diamond, R. D., and J. E. Bennett. 1974. Prognostic factors in cryptococcal menin-gitis. A study in 111 cases. Ann. Intern. Med. *80*: 176-181.

87. Dickson, E. C., and M. A. Gifford. 1938. *Coccidioides* infection (coccidioidomy-cosis) the primary type of infection. Arch. Intern. Med. *62*: 853-871.

88. Di Salvo, A. F., J. Malonas, L. Kaufman, and S. Blumer. 1978. Evaluation of an automated procedure for the indirect fluorescent antibody test for cryptococcosis. In *The Black and White Yeasts,* Proceedings of the Fourth International Conference on the Mycoses, Pan Am. Health Organ. Sci. Publ. No. 356, Washington, D. C., pp. 295-299.

89. Dismukes, W. E., S. A. Roval, and B. S. Tynes. 1978. Disseminated histoplasmosis in corticosteroid-treated patients. Report of five cases. JAMA *240*: 1495-1498.

90. Dolan, C. T. 1972. Specificity of the latex-cryptococcal antigen test. Am. J. Clin. Pathol. *58*: 358-364.

91. Dolan, C. T., and R. P. Stried. 1973. Serologic diagnosis of yeast infections. Am. J. Clin. Pathol. *59*: 49-55.

92. Drouhet, E. 1970. Fluorescent antibodies and precipitin in aspergillosis. 10th Int. Congr. Microbiol.

93. Drouhet, E. 1973. Electrosynérèse (immuno-électrodiffusion) en agarose, méthode rapide et sensible pour diagnostic des candidoses septicèmiques et viscérales. Ré-sultats comparatifs avec d'autres techniques d'immunodiffusion et d'immunofluores-cence. Bull. Soc. Fr. Mycol. Med. *2*: 11-14.

94. Drouhet, E., L. Camey, and G. Segretain. 1972. Valeur de l'immunoprécipitation

et de l'immunofluorescence indirecte dans les aspergilloses broncho-pulmonaires. Ann. Inst. Pasteur (Paris) *123*: 379-395.

95. Duttera, M. J., and S. Osterhout. 1969. North American blastomycosis: survey of 63 cases. South. Med. J. *62*: 295-301.

96. Edge, G., and J. Pepys. 1976. RAST tests avec des antigènes protéiniques et poly-saccharidiques et des cellules lavées. Rev. Fr. Allergol. Immunol. Clin. *16*: 251-255.

97. Ehrhard, H. B., and L. Pine. 1972. Factors influencing the production of H and M antigens by *Histoplasma capsulatum:* development and evaluation of a shake culture procedure. Appl. Microbiol. *23*: 236-249.

98. Eisses, J. F., M. W. Lee, G. Fine, and E. Mezger. 1978. Use of a commercial latex-agglutination test to detect *Cryptococcus neoformans* antigen in tissue extracts. Am. J. Clin. Pathol. *69*: 206-207.

99. Elinov, H. P., and N. A. Zaikina. 1958. The precipitation test in the serological diagnosis of moniliasis. Zh. Mikrobiol. Epidemiol. Immunobiol. *30*: 40-44.

100. Ellsworth, J. H., E. Reiss, R. L. Bradley, H. Chmel, and D. Armstrong. 1977. Comparative serological and cutaneous reactivity of candidal cytoplasmic proteins and mannan separated by affinity for concanavalin A. J. Clin. Microbiol. *5*: 91-99.

101. Emmons, C. W. 1949. Histoplasmosis in animals. Public Health Rep. *64*: 892-896.

102. Emmons, C. W. 1955. Saprophytic sources of *Cryptococcus neoformans* associated with the pigeon (*Columba livia*). Am. J. Hyg. *62*: 227-232.

103. Emmons, C. W. 1962. Natural occurrence of opportunistic fungi. Lab. Invest. *11*: 1026-1032.

104. Emmons, C. W., C. J. Binford, J. P. Utz, and J. Kwon-Chung. 1977. *Medical Mycology,* 3rd ed., Lea & Febiger, Philadelphia.

105. Esterly, N. B. 1968. Serum antibody titers to *Candida albicans* utilizing an immunofluorescent technique. Am. J. Clin. Pathol. *50*: 291-296.

106. Evans, E. E. 1950. The antigenic composition of *Cryptococcus neoformans.* A serologic classification by means of the capsular and agglutination reactions. J. Immunol. *64*: 423-430.

107. Evans, E. E., and J. F. Kessel. 1951. The antigenic composition of *Cryptococcus neoformans.* II. Serologic studies with the capsular polysaccharide. J. Immunol. *67*: 109-114.

108. Evans, E. G. V., and R. A. Forster. 1976. Antibodies to *Candida* after operations on the heart. J. Med. Microbiol. *9*: 303-308.

109. Everall, P. H., C. A. Morris, and D. F. Morris. 1974. Antibodies to *Candida albicans* in hospital patients with and without spinal injury and in normal men and women. J. Clin. Pathol. *27*: 722-728.

110. Everett, E. D., T. C. Eickhoff, and J. M. Ehret. 1975. Immunofluorescence of yeasts in urine. J. Clin. Microbiol. *2*: 142-143.

111. Everett, E. D., F. M. La Force, and T. C. Eickhoff. 1975. Serologic studies in suspected visceral candidiasis. Arch. Intern. Med. *135*: 1075-1078.

112. Fava Netto, C. 1955. Estudos quantitativos sôbre a fixacão de complemento na blastomicose sul-americana, com antígeno polissacarídico. Argent. Circ. Clin. Exp. *18*: 197-254.

113. Fava Netto, C. 1961. Contribuicão para o estudo immunológico da blastomicose de lutz (blastomicose sul-Americana). Rev. Inst. Adolfo Lutz *21*: 99-194.

114. Fava Netto, C. 1961. Contribuicão para o estudo immunológico da blastomicose de lutz (blastomicose sul-Americana). Rev. Inst. Adolfo Lutz *21*: 99-194.

115. Fava Netto, C. 1965. The immunology of South American blastomycosis. Mycopathol. Mycol. Appl. *26*: 349-358.

116. Fava Netto, C. 1972. The serology of paracoccidioidomycosis: present and future trends. In *Paracoccidioidomycosis,* Proceedings of the First Pan American Symposium, Pan Am. Health Organ. Sci. Publ. No. 254, Washington, D.C., pp. 209-213.

117. Fava Netto, C., R. G. Ferri, and C. S. Lacaz. 1959. Proteinograma e algumas "provas da fase aguda do sôro" na blastomicose sul-americana. Astudo comparativo com as reacões de fixacão do complemento e de precipitacão. Med. Cir. Farm. *277*: 157-163.

118. Fiese, M. J. 1958. *Coccidioidomycosis,* Thomas, Springfield, Ill.

119. Filice, G., B. Yu, and D. Armstrong. 1977. Immunodiffusion and agglutination tests for *Candida* in patients with neoplastic disease: inconsistent correlation of results with invasive infections. J. Infect. Dis. *135*: 349-357.

120. Fink, J. N., J. J. Barboriak, and L. Kaufman. 1968. Cryptococcal antibodies in pigeon breeders disease. J. Allergy *41*: 297-301.

121. Foroozanfar, N., M. Yamamura, and J. R. Hobbs. 1974. Standardization of lymphocyte transformation to *Candida* immunogen. Clin. Exp. Immunol. *16*: 301-310.

122. Franco, M. F., C. Fava Netto, and L. G. Chamna. 1973. Reacão de imunofluorecenia indireta para o diagnostico serologico da blastomicose sul-americana: padronizacão da reacão e comparacão dos resultados con a reacão de fixacão do complemento. Rev. Inst. Med. Trop. Sao Paulo *15*: 393-398.

123. Frisk, Å. 1977. Serological aspects of candidosis. Curr. Ther. Res. *22*: 46-50.

124. Fukushiro, R., S. Kagawa, S. Nishiyama, S. Takahashi, and H. Ishikawa. 1965. Die pilzelemente im gewebe der hautsporotrichose des menschen. Hautarzt *16*: 18-25.

125. Furcolow, M. L., I. L. Bunnell, and D. J. Teneberg. 1948. A complement fixation test for histoplasmosis. II. Preliminary results with human sera. Public Health Rep. *63*: 169-172.

126. Gaines, J. D., and J. S. Remington. 1973. Diagnosis of deep infection with *Candida.* A study of *Candida* precipitins. Arch. Intern. Med. *132*: 699-702.

127. Galant, S. P., R. W. Rucker, C. E. Groncy, I. D. Wells, and H. Novey. 1976. Incidence of serum antibodies to several *Aspergillus* species and to *Candida albicans* in cystic fibrosis. Am. Rev. Respir. Dis. *114*: 325-331.

128. Galussio, J. C., J. L. Friedman, and R. Negroni. 1973. Rapid diagnosis of pulmonary mycosis by counterelectrophoresis. Mycopathol. Mycol. Appl. *51*: 143-146.

129. Gargani, G. 1958. La reazione di fissazione del complemento nelle candidosi esperienze sugli animali e indagine sierologica sulla popolazione normale. Sperimentale *108*: 110-127.

130. Georg, L. K., R. M. Coleman, and J. M. Brown. 1968. Evaluation of an agar gel precipitin test for the serodiagnosis of actinomycosis. J. Immunol. *100*: 1288-1292.

131. Gerber, J. D., and R. D. Jones. 1973. Immunologic significance of aspergillin antigens of six species of *Aspergillus* in the serodiagnosis of aspergillosis. Am. Rev. Respir. Dis. *108*: 1124-1129.

132. Gerber, J. D., R. E. Riley, and R. D. Jones. 1972. Evaluation of a microtiter latex agglutination test for histoplasmosis. Appl. Microbiol. *24*: 191-197.

133. Gilchrist, T. C. 1896. A case of blastomycetic dermatitis in man. Johns Hopkins Hosp. Rev. *1*: 269-298.

134. Gilchrist, T. C., and W. R. Stokes. 1898. A case of pseudo-lupus vulgaris caused by a *Blastomyces.* J. Exp. Med. *3*: 53-78.

135. Glew, R. H., H. R. Buckley, H. M. Rosen, R. C. Moellering, and J. E. Fischer. 1978. Serologic tests in the diagnosis of systemic candidiasis. Enhanced diagnostic accuracy with crossed immunoelectrophoresis. Am. J. Med. *64*: 586-591.

136. Gonzalez Ochoa, A., and A. Vasquez Hoyos. 1953. Relaciones serólicas de los principales actinomycetes patogens. Rev. Invest. Salubr. Enferm. Trop., Mexico City *13*: 177-187.

137. Gonzalez Ochoa, A., and E. Soto Figueroa. 1947. Polisaccaridos del *Sporotrichum schenckii* datos immunológicos intradermo-reacción en el diagnóstico de la esporotricosis. Rev. Inst. Salubr. Enferm. Trop., Mexico City *8*: 143-153.

138. Goodman, J. S., L. Kaufman, and M. G. Koenig. 1971. Diagnosis of cryptococcal meningitis. Value of immunologic detection of cryptococcal antigen. N. Engl. J. Med. *285*: 434-436.

139. Gordon, M. A., and Y. Al-Doory. 1965. Application of fluorescent antibody procedures to the study of pathogenic dermatiaceous fungi. J. Bacteriol. *89*: 551-556.

140. Gordon, M. A., R. E. Almy, C. H. Greene, and J. W. Fenton. 1971. Diagnostic mycoserology by immunoelectroosmophoresis: a general, rapid and sensitive microtechnic. Am. J. Clin. Pathol. *56*: 471-474.

141. Gordon, M. A., and E. Lapa. 1971. Charcoal particle agglutination test for detection of antibody to *Cryptococcus neoformans*. A preliminary report. Am. J. Clin. Pathol. *56*: 354-359.

142. Gordon, M. A., and E. W. Lapa. 1974. Elimination of rheumatoid factor in the latex test for cryptococcosis. Am. J. Clin. Pathol. *61*: 488-494.

143. Gordon, M. A., E. W. Lapa, and J. Kane. 1977. Modified indirect fluorescent antibody test for aspergillosis. J. Clin. Microbiol. *6*: 161-165.

144. Goudswaard, J., and G. Virella. 1978. New immunofluorescent antibody test for diagnosis of candidiasis using macrophage-engulfed antigens as substrate. J. Immunol. Methods *20*: 357-363.

145. Graham, A. R., and K. J. Ryan. 1980. Counterimmunoelectrophoresis employing coccidioidin in serologic testing for coccidioidomycosis. J. Immunol. Methods *20*: 574-577.

146. Grappel, S. F., C. T. Bishop, and F. Blank. 1974. Immunology of dermatophytes and dermatophytosis. Bacteriol. Rev. *38*: 222-250.

147. Grappel, S. F., F. Blank, and C. T. Bishop. 1971. Circulating antibodies in human favus. Dermatologica *143*: 271-278.

148. Grappel, S. F., F. Blank, and C. T. Bishop. 1972. Circulating antibodies in dermatophytosis. Dermatologica *144*: 1-11.

149. Grappel, S. F., A. Fethière, and F. Blank. 1971. Effect of antibodies on growth and structure of *Trichophyton mentagrophytes*. Sabouraudia *9*: 50-55.

150. Green, J. H., W. K. Harrell, S. B. Gray, J. E. Johnson, R. C. Bolin, H. Gross, and G. B. Malcolm. 1976. H. and M antigens of *Histoplasma capsulatum*. Preparation of anti-sera and location of these antigens in yeast phase cells. Infect. Immun. *14*: 826-831.

151. Green, J. H., W. K. Harrell, J. E. Johnson, and R. Benson. 1979. Preparation of reference antisera for laboratory diagnosis of blastomycosis. J. Clin. Microbiol. *10*: 1-7.

152. Gross, H., G. Bradley, L. Pine, S. Gray, J. H. Green, and W. Harrell. 1975. Evaluation of histoplasmin for the presence of H and M antigens: some difficulties encountered in the production and evaluation of a product suitable for the immunodiffusion test. J. Clin. Microbiol. *1*: 330-334.

153. Gumaa, S. A., and E. S. Mahgoub. 1973. Evaluation of the complement fixation test in diagnosis of actinomycetoma. J. Trop. Med. Hyg. *76*: 140-142.

154. Halweg, H., J. Ciszek, and P. Krakówka. 1968. The reversal of serological reactions in patients with pulmonary and pleural aspergillosis after treatment. Tubercle *49*: 404-409.

155. Hampson, C. R., and T. R. Larwood. 1967. Observations on coccidioidomycosis immunodiffusion tests with clinical correlations. In L. Ajello (Ed.), *Coccidioidomycosis,* Papers from the Second Symposium on Coccidioidomycosis, University of Arizona Press, Tucson, pp. 211-214.

156. Haneke, E. 1974. Different immunofluorescent titres of *Candida albicans* blastospores and pseudohyphae. Mycopathol. Mycol. Appl. *52*: 269-271.

157. Harding, S. A., and W. G. Merz. 1975. Evaluation of antibody coating of yeasts in urine as an indicator of the site of urinary tract infection. J. Clin. Microbiol. *2*: 222-225.

158. Hasenclever, H. F., and W. O. Mitchell. 1961. Observation of two antigenic groups in *Candida albicans*. J. Bacteriol. *82*: 570-573.

159. Hay, R. J., D. W. R. Mackenzie, C. K. Campbell, and C. M. Philpot. 1980. Cryptococcosis in the United Kingdom: an analysis of 69 cases. J. Infect. *2*: 13-22.

160. Hay, R. J., and E. N. Saeed. 1980. Immunofluorescence staining in chronic dermatophytosis. Clin. Exp. Dermatol. *6*: 155-158.

161. Hearn, V. M., E. V. Wilson, A. G. Proctor, and D. W. R. Mackenzie. 1980. Preparation of *Aspergillus fumigatus* antigens and their analysis by two-dimensional immunoelectrophoresis. J. Med. Microbiol. *13*: 451-458.

162. Heiner, D. C. 1958. Diagnosis of histoplasmosis using precipitin reactions in agar gel. Pediatrics *20*: 616-627.

163. Hellwege, H. H., K. Fischer, and F. Bläker. 1972. Diagnostic value of *Candida* precipitins. Lancet *2*: 386.

164. Hermans, P. E., and H. Markowitz. 1969. Serum antibody activity in patients with histoplasmosis as measured by passive hemagglutination. J. Lab. Clin. Med. *74*: 454-463.

165. Hipp, S., D. S. Berns, V. Tomkins, and H. E. Buckley. 1970. Latex slide agglutination test for *Aspergillus* antibodies. Sabouraudia *8*: 237-241.

166. Hirsch, E. F., and H. Benson. 1927. Specific skin test reactions with culture filtrates of *Coccidioides immitis*. J. Infect. Dis. *40*: 629-633.

167. Holder, I. A., P. J. Kozinn, and E. J. Law. Evaluation of *Candida* precipitin and agglutination tests for the diagnosis of systemic candidiasis in burn patients. J. Clin. Microbiol. *6*: 219-223.

168. Holm, P., and J. B. Kwapinski. 1959. Studies in the detection of *Actinomyces* antibodies in human sera by using pure antigenic fractions of *Actinomyces israelii*. Acta Pathol. Microbiol. Scand. *45*: 107-112.

169. Holmberg, K., C. E. Nord, and T. Wadström. 1975. Serological studies of *Actinomyces israelii* by crossed immunoelectrophoresis. Taxonomic and diagnostic applications. Infect. Immun. *12*: 398-403.

170. Holmberg, K. 1978. Serological diagnosis of systemic candidiasis and invasive aspergillosis. Scand. J. Infect. Dis. (Suppl.) *16*: 26-32.

171. Hommel, M., T. K. Truong, and D. E. Bidwell. 1976. Technique immunoenzymatique (ELISA) appliquée au diagnostic sérologique des candidoses et aspergilloses humaines: résultats préliminaires. Nouv. Presse Med. *5*: 2789-2791.

172. Humphreys, D. M., J. G. Crowder, and A. White. 1975. Serological reactions to *Nocardia* antigens. Am. J. Med. Sci. *269*: 323-326.

173. Huppert, M. 1968. Recent developments in coccidioidomycosis. Rev. Med. Vet. Mycol. *6*: 279-294.

174. Huppert, M. 1970. Serology of coccidioidomycosis. Mycopathol. Mycol. Appl. *41*: 107-113.
175. Huppert, M. 1970. Standardization of immunological reagents. In *The First International Symposium on the Mycoses,* Pan Am. Health Organ. Sci. Publ. No. 205, Washington, D.C., pp. 243-252.
176. Huppert, M., J. P. Adler, E. H. Rice, and S. H. Sun. 1979. Common antigens among systemic disease fungi analyzed by two-dimensional immunoelectrophoresis. Infect. Immun. *23*: 479-485.
177. Huppert, M., and J. W. Bailey. 1963. Immunodiffusion as a screening test for coccidioidomycosis serology. Sabouraudia 2: 284-291.
178. Huppert, M., and J. W. Bailey. 1965. The use of immunodiffusion tests in coccidioidomycosis. The accuracy and reproducibility of the immunodiffusion test which correlates with complement fixation. Am. J. Clin. Pathol. *44*: 364-368.
179. Huppert, M., and J. W. Bailey. 1965. The use of immunodiffusion tests in coccidioidomycosis. II. An immunodiffusion test as a substitute for the tube precipitin test. Am. J. Clin. Pathol. *44*: 369-373.
180. Huppert, M., J. W. Bailey, and P. Chitjian. 1967. Immunodiffusion as a substitute for complement fixation and tube precipitin tests in coccidioidomycosis. In L. Ajello (Ed.), *Coccidioidomycosis,* Papers from the Second Symposium on Coccidioidomycosis, University of Arizona Press, Tucson, pp. 221-225.
181. Huppert, M., P. A. Chitjian, and A. J. Gross. 1970. Comparison of methods for coccidioidomycosis complement fixation. Appl. Microbiol. *20*: 328-332.
182. Huppert, M., I. Krasnow, K. R. Vukovich, S. H. Sun, E. H. Rice, and L. J. Kutner. 1977. Comparison of coccidioidin and spherulin in complement fixation tests for coccidioidomycosis. J. Clin. Microbiol. *6*: 33-41.
183. Huppert, M., E. T. Peterson, S. H. Sun, P. A. Chitjian, and W. J. Derrevere. 1968. Evaluation of a latex particle agglutination test for coccidioidomycosis. Am. J. Clin. Pathol. *49*: 96-102.
184. Huppert, M., N. S. Spratt, K. R. Vukovich, S. H. Sun, and E. H. Rice. 1978. Antigenic analysis of coccidioidin and spherulin determined by two-dimensional immunoelectrophoresis. Infect. Immun. *20*: 541-555.
185. Huppert, M., S. H. Sun, and E. H. Rice. 1978. Specificity of exoantigens for identifying cultures of *Coccidioides immitis.* J. Clin. Microbiol. *8*: 346-348.
186. Huppert, M., S. H. Sun, and K. R. Vukovich. 1974. *Standardization of Mycological Reagents,* Proc. Int. Conf. Stand. Diagn. Mater., U.S. Dept. Health, Education, and Welfare, Center for Disease Control, Atlanta, pp. 187-194.
187. Ikemoto, H., and S. Shibata. 1973. Indirect haemagglutinin in pulmonary aspergilloma diagnosis. Sabouraudia *11*: 167-170.
188. Ito, K. 1963. Immunologic aspects of superficial fungus diseases. Trichophytin: skin and serologic reactions. In D. M. Pillsbury and C. S. Livingwood (Eds.), *International Congress of Dermatology, 12th,* Excerpta Medica Foundation, New York, pp. 563-567.
189. Iwashita, H., K. Araki, Y. Kuroiwa, and T. Matsumoto. 1978. Occurrence of *Candida* specific oligoclonal IgG antibodies in CSF with *Candida* meningoencephalitis. Ann. Neurol. *4*: 579-581.
190. Jee, S-H., J-S. Deng, and Y-C. Lu. 1978. Clinical studies on sporotrichosis in Taiwan. Chin. J. Microbiol. *11*: 62-67.
191. Jones, R. D., G. A. Sarosi, J. D. Parker, R. J. Weeks, and F. E. Tosh. 1969. The complement fixation test in extra-cutaneous sporotrichosis. Ann. Intern. Med. *71*: 913-918.

192. Kaplan, W. 1972. Application of immunofluorescence to the diagnosis of para-coccidioidomycosis. In *Paracoccidioidomycosis,* Proceedings of the First Pan American Symposium, Pan Am. Health Organ. Sci. Publ. No. 254, Washington, D.C., pp. 224-226.

193. Kaplan, W., and A. Gonzalez Ochoa. 1963. Application of the fluorescent antibody technique to the rapid diagnosis of sporotrichosis. J. Lab. Clin. Med. *62:* 835-841.

194. Kaplan, W., M. Huppert, D. E. Kraft, and J. W. Bailey. 1966. Fluorescent antibody inhibition test for *Coccidioides immitis* antibodies. Sabouraudia *5:* 1-6.

195. Karlin, J. V., and H. S. Nielsen. 1970. Serological aspects of sporotrichosis. J. Infect. Dis. *121:* 316-327.

196. Katsura, Y., and I. Uesaka. 1974. Assessment of germ tube dispersion activity of serum from experimental candidiosis: a new procedure for serodiagnosis. Infect. Immun. *9:* 788-793.

197. Kaufman, L. 1970. Serology: its value in the diagnosis of coccidioidomycosis, cryptococcosis and histoplasmosis. In *The First International Symposium on the Mycoses,* Pan Am. Health Organ. Sci. Publ. No. 205, Washington, D.C., pp. 96-100.

198. Kaufman, L. 1971. Serological tests for histoplasmosis: their use and interpretation. In L. Ajello, E. W. Chick, and M. L. Furcolow (Eds.), *Histoplasmosis,* Thomas, Springfield, Ill., pp. 321-326.

199. Kaufman, L. 1976. Serodiagnosis of fungal diseases. In E. H. Lennette, E. H. Spaulding, and J. P. Truant (Eds.), *Manual of Clinical Immunology.* American Society for Microbiology, Washington, D.C., pp. 363-381.

200. Kaufman, L., and S. Blumer. 1968. Value and interpretation of serological tests for the diagnosis of cryptococcosis. Appl. Microbiol. *16:* 1907-1912.

201. Kaufman, L., and S. Blumer. 1973. Latex-cryptococcal antigen test. Am. J. Clin. Pathol. *59:* 285-286.

202. Kaufman, L., and M. J. Clark. 1974. Value of the concomitant use of complement fixation and immunodiffusion tests in the diagnosis of coccidioidomycosis. Appl. Microbiol. *28:* 641-643.

203. Kaufman, L., G. Cowart, S. Blumer, A. Stine, and R. Wood. 1974. Evaluation of a commercial latex agglutination test kit for cryptococcal antigen. Appl. Microbiol. *27:* 620-621.

204. Kaufman, L., D. W. McLaughlin, M. J. Clark, and S. Blumer. 1974. Specific immunodiffusion test for blastomycosis. Appl. Microbiol. *26:* 244-247.

205. Kaufman, L., and P. G. Standard. 1978. Improved version of the exoantigen test for identification of *Coccidioides immitis* and *Histoplasma capsulatum* cultures. J. Clin. Microbiol. *8:* 42-45.

206. Kaufman, L., and P. Standard. 1978. Immunoidentification of cultures of fungi pathogenic to man. Curr. Microbiol. *1:* 135-140.

207. Kaufman, L., R. T. Terry, J. H. Schubert, and D. McLaughlin. 1967. Effects of a single histoplasmin skin test on the serological diagnosis of histoplasmosis. J. Bacteriol. *94:* 798-803.

208. Kerkering, T. M., A. Espinel-Ingroff, and S. Shadomy. 1979. Detection of *Candida* antigenemia by counterimmunoelectrophoresis in patients with invasive candidiasis. J. Infect. Dis. *140:* 659-664.

209. Kim, S. J., and S. D. Chaparas. 1978. Characterization of antigens from *Aspergillus fumigatus.* Am. Rev. Respir. Dis. *118:* 547-551.

210. Kim, S. J., S. D. Chaparas, T. M. Brown, and M. C. Anderson. 1978. Characterization of antigens from *Aspergillus fumigatus.* Am. Rev. Respir. Dis. *118:* 553-560.

211. Kimball, H. R., H. F. Hasenclever, and S. M. Wolff. 1967. Detection of circulating

antibody in human cryptococcosis by means of a bentonite flocculation technique. Am. Rev. Respir. Dis. *95*: 631-637.

212. Kleger, B., and L. Kaufman. 1973. Detection and identification of diagnostic *Histoplasma capsulatum* precipitates by counterelectrophoresis. Appl. Microbiol. *26*: 231-238.

213. Kolmer, J. A., and A. Strickler. 1915. Complement fixation in parasitic skin diseases. JAMA *64*: 800-804.

214. Königer, G., and D. Adam. 1973. Serologischer Nachweis systemischer *Candida* Infektionen mit Hilfe von Antikörpertiterrelationen. Klin. Wochenschr. *51*: 437-444.

215. Kozel, T. R., and J. Cazin. 1972. Immune response to *Cryptococcus neoformans* soluble polysaccharide. I. Serological assay for antigen and antibody. Infect. Immun. *5*: 35-41.

216. Kozinn, P. J., R. S. Galen, C. L. Taschdjian, P. L. Goldberg, W. Protzman, M. A. Kozinn. 1976. The precipitin test in systemic candidiasis. JAMA *235*: 628-629.

217. Kozinn, P. J., C. L. Taschdjian, P. K. Goldberg, W. P. Protzman, D. W. R. Mackenzie, J. S. Remington, S. Anderson, and M. S. Seelig. 1978. Efficiency of serologic test in the diagnosis of systemic candidiasis. Am. J. Clin. Pathol. *70*: 893-898.

218. Kume, H., M. Okudaira, and M. Abe. 1978. A rapid method for diagnosis of deep fungus infection by means of counterimmunoelectrophoresis. Jpn. J. Med. Mycol. *19*: 125-128.

219. Kurita, N. 1979. Cell-mediated immune responses in mice infected with *Fonsecaea pedrosoi*. Mycopathologia *68*: 9-15.

220. Kurup, V. P., and J. N. Fink. 1978. Evaluation of methods to detect antibodies against *Aspergillus fumigatus*. Am. J. Clin. Pathol. *69*: 414-417.

221. Kurup, V. P., J. N. Fink, G. H. Scribner, and J. H. Falk. 1977. Antigenic variability of *Aspergillus fumigatus* strains. Microbios *19*(77-78): 191-204.

222. Lacaz, C. S. 1949. Novas dados em relação à blastomicose sul-americana e sev agente etiológica. Rev. Med. Circ. Sao Paulo *9*: 303-341.

223. Lacaz, C. S., R. G. Ferri, G. Fava Netto, and E. Belfort. 1962. Aspectos immunoquimicos na blastomicose sul-americana e a blastomicose queloideana (doenca de J. Lobo). Med. Cir. Farm. *298*: 63-74.

224. Land, G. A., J. H. Foxworth, and K. E. Smith. 1978. Immunodiagnosis of histoplasmosis in a compromised host. J. Clin. Microbiol. *8*: 558-565.

225. Landay, M. E. 1973. Spherules in the serology of *Coccidioides immitis*. II. Complement fixation tests with human sera. Mycopathol. Mycol. Appl. *49*: 45-52.

226. Landay, M. E., R. M. Pash, and J. W. Miller. 1970. Spherules in the serodiagnosis of coccidioidomycosis. J. Lab. Clin. Med. *75*: 197-205.

227. Landay, M. E., R. W. Wheat, N. F. Conant, and E. P. Lowe. 1967. Serological comparison of the three morphological phases of *Coccidioides immitis* by the agar gel diffusion method. J. Bacteriol. *93*: 1-6.

228. Landay, M. E., R. W. Wheat, N. F. Conant, and E. P. Lowe. 1968. Indirect IFA test with spherules and arthrospores of *Coccidioides immitis*. J. Lab. Clin. Med. *71*: 294-300.

229. Landay, M. E., R. W. Wheat, N. F. Conant, and E. P. Lowe. 1968. A serological comparison of the three morphological phases of *Coccidioides immitis:* complement fixation test with a pooled antiserum obtained from rabbits with experimental coccidioidomycosis. Mycopathol. Mycol. Appl. *34*: 289-295.

230. Lane, C. C. 1915. A cutaneous disease caused by a new fungus (*Phialophora verrucosa*). J. Cutaneous Dis. *33*: 840-846.

231. Larsh, H. W. 1971. The public health importance of histoplasmosis. In L.

Ajello, E. W. Chick, and M. L. Furcolow (Eds.), *Histoplasmosis,* Thomas, Springfield, Ill., pp. 9-17.

232. Latapi, F. 1959. Micetoma: análisis de 100 casos estudiados en la ciudad de México, Mem. III Congr. Ibero-Latino Am. Dermatol., p. 203.

233. Lehman, P. F., and E. Reiss. 1978. Invasive aspergillosis: antiserum for circulating antigen produced after immunization with serum from infected rabbits. Infect. Immun. *20*: 570-572.

234. Lehmann, P. F., and E. Reiss. 1980. Detection of *Candida albicans* mannan by immunodiffusion, counterimmunoelectrophoresis and enzyme-linked immunoassay. Mycopathologia *70*: 83-88.

235. Lehmann, P. F., and E. Reiss. 1980. Comparison of ELISA of serum anti-*Candida albicans* mannan IgG levels of a normal population and in diseased patients. Mycopathologia *70*: 89-93.

236. Lehner, T. 1965. Immunofluorescent investigation of *Candida albicans.* Arch. Oral Biol. *10*: 975-980.

237. Lehner, T. 1966. Immunofluorescence study of *Candida albicans* in candidiasis carriers and controls. J. Pathol. Bacteriol. *91*: 97-104.

238. Lehner, T. 1970. Serum fluorescent antibody and immunoglobulin estimations in candidiosis. J. Med. Microbiol. *3*: 475-481.

239. Lehner, T., H. R. Buckley, and I. G. Murray. 1972. The relationship between fluorescent agglutinating and precipitating antibodies to *Candida albicans.* J. Clin. Pathol. *25*: 344-348.

240. Levine, H. B., J. M. Cobb, and G. M. Scalarone. 1969. Spherule coccidioidin in delayed dermal sensitivity reactions of experimental animals. Sabouraudia 7: 20-32.

241. Lloyd, K. O., and L. R. Travassos. 1975. Immunochemical studies on L-rhammo-D-mannans of *Sporothrix schenckii* and related fungi by use of rabbit and human antisera. Carbohydr. Res. *40*: 89-97.

242. Longbottom, J. L. 1978. Immunological aspects of infection and allergy due to *Aspergillus* species. Mykosen (Suppl.) *1*: 207-217.

243. Longbottom, J. L., W. D. Brighton, G. Edge, and J. Pepys. 1975. Antibodies mediating type 1 stain test reactions to polysaccharide and protein antigens of *Candida albicans.* Clin. Allergy 6: 41-49.

244. Longbottom, J. L., and J. Pepys. 1964. Pulmonary aspergillosis: diagnostic and immunological significance of antigens and C-substance in *Aspergillus fumigatus.* J. Pathol. Bacteriol. *88*: 141-151.

245. Lorincz, A. L., J. O. Priestly, and P. A. Jacobs. 1958. Evidence for a humoral mechanism which prevents growth of dermatophytes. J. Invest. Dermatol. *31*: 15-17.

246. Lowell, J. R., and E. H. Shuford. 1976. The value of the skin test and complement fixation test in the diagnosis of chronic pulmonary histoplasmosis. Am. Rev. Respir. Dis. *114*: 1069-1075.

247. Lutz, A. 1908. Uma micose psuedococcidica localisada na bocca e observada no Brasil. Contribuicão ao conhecimento das hyfoblastomycoses americanas. Bras. Med. *22*: 121-124.

248. Maccani, J. E. 1977. Detection of cryptococcal polysaccharide using counterimmunoelectrophoresis. Am. J. Clin. Pathol. *68*: 39-44.

249. Macdonald, F., and F. C. Odds. 1980. Purified *Candida albicans* proteinase in the serological diagnosis of systemic candidosis. JAMA *243*: 2409-2411.

250. Mackenzie, D. W. R. 1977. Reference material and procedures for serological testing of mycoses. In K. Iwata (Ed.), *Recent Advances in Medical and Veterinary*

Mycology, Proceedings of the Sixth Congress of the International Society for Humans and Animal Mycology, University of Tokyo Press, Tokyo, pp. 37-42.

251. Mackenzie, D. W. R., and P. K. C. Austwick. 1965. The study of mycotic allergens in Britain. Prog. Immunobiol. Stand. *2:* 164-169.

252. Mackenzie, D. W. R., and E. Georgakopoulos. 1976. Unpublished results.

253. Mackenzie, D. W. R., and C. M. Philpot. 1975. Counterimmunoelectrophoresis as a routine mycoserological procedure. Mycopathologia *57:* 1-7.

254. Mackenzie, D. W. R., C. M. Philpot, and A. G. J. Proctor. 1980. Basic serodiagnostic methods for diseases caused by fungi and actinomycetes. Public Health Lab. Serv. Monogr. Ser. *12.*

255. Mackenzie, D. W. R., and E. Segal. 1980. Round table: Standardization of mycoserologic reagents and procedures. In E. S. Kuttin and G. L. Baum (Eds.), *Human and Animal Mycology,* Excerpta Medica, Amsterdam, pp. 141-145.

256. Magalhães Pereira, A., A. P. Goncalves, C. S. Lacaz, C. Fava Netto, and R. M. de Castro. 1964. Estudos sobre a imunopatologia da esporotricose. An. Bras. Dermatol. Sifilogr. *39:* 34-41.

257. Mahgoub, E. S. 1964. The value of gel diffusions in the diagnosis of mycetoma. Trans. R. Soc. Trop. Med. Hyg. *58:* 560-663.

258. Mahgoub, E. S. 1975. Serological diagnosis of mycetoma. In *Mycoses,* Proceedings of the Third International Conference on the Mycoses, Pan Am. Health Organ. Sci. Publ. No. 304, Washington, D.C., pp. 154-160.

259. Mahgoub, E. S. 1976. Medical management of mycetoma. Bull. W.H.O. *54:* 303-310.

260. Mahgoub, E. S., and I. G. Murray. 1973. *Mycetoma,* William Heineman, London.

261. Manych, J., and J. Sourek. 1966. Diagnostic possibilities of utilising precipitation in agar for the identification of *Histoplasma capsulatum, Coccidioides immitis, Blastomyces dermatitidis* and *Paracoccidioides brasiliensis.* J. Hyg. Epidemiol. Microbiol. Immunol. *10:* 74-84.

262. Mariat, F. 1963. Sur la distribution géographique et la répartition des agents de mycétomes. Bull. Soc. Pathol. Exot. *56:* 35-45.

263. Marier, R., and V. T. Andriole. 1978. Comparison of continuous and discontinuous counterimmunoelectrophoresis with immunodiffusion in identification of *Candida* antibody using HS antigen. J. Clin. Microbiol. *8:* 12-14.

264. Marier, R., W. Smith, M. Jansen, and V. T. Andriole. 1979. A solid-phase radioimmunoassay for the measurement of antibody to *Aspergillus* in invasive aspergillosis. J. Infect. Dis. *140:* 771-779.

265. Martin, D. S. 1935. Complement fixation in blastomycosis. J. Infect. Dis. *57:* 291-295.

266. Martin, D. S., R. D. Baker, and N. F. Conant. 1936. A case of verrucous dermatitis caused by *Hormodendrum pedrosoi* (chromoblastomycosis) in North Carolina. Am. J. Trop. Med. *16:* 593-619.

267. McGinnis, M. R., and A. A. Padhye. 1977. *Exophiala jeanselmei,* a new combination for *Phialophora jeanselmei.* Mycotaxon *5:* 341-352.

268. McLaren, M. L., E. S. Mahgoub, and E. Georgakopoulos. 1978. Preliminary investigation on the use of the enzyme linked immunosorbent assay (ELISA) in the serodiagnosis of mycetoma. Sabouraudia *16:* 225-228.

269. McMillen, S., and E. R. Laverty. 1969. Sporotrichosis: serology as a clinical aid. Bacteriol. Proc. *1969:* 115 (Abstr.).

270. Mearns, M., J. Longbottom, and J. Batten. 1967. Precipitating antibodies to *Aspergillus fumigatus* in cystic fibrosis. Lancet *2:* 538-539.

271. Meriin, J. A. 1930. Zur Mykologie der Chromoblastomykose. (Der Erreger

des europäischen Falles der Erkrankung.) Arch. Dermatol. Syphilol. *162*: 300-310.

272. Meriin, J. A. 1932. Weitere Beobachtungen über den Erreger der europäischen Chromoblastomycosis. Arch. Dermatol. Syphilol. *166*: 722-729.

273. Merz, W. J., G. I. Evans, S. Shadomy, S. Anderson, L. Kaufman, P. J. Kozinn, D. W. R. Mackenzie, W. P. Protzman, and J. S. Remington. 1977. Laboratory evaluation of serological tests for systemic candidiasis: a co-operative study. J. Clin. Microbiol. *5*: 596-603.

274. Meyer, R. D., L. S. Young, D. Armstrong, and B. Yu. 1973. Aspergillosis complicating neoplastic disease. Am. J. Med. *54*: 6-15.

275. Miller, G. G., M. W. Witwer, A. I. Braude, and C. E. Davis. 1974. Rapid identification of *Candida albicans* septicaemia in man by gas-liquid chromatography. J. Clin. Invest. *54*: 1235-1240.

276. Mok, W. Y., and C. Fava Netto. 1978. Paracoccidioidin and histoplasmin sensitivity in Coari (state of Amazonas) Brasil. Am. J. Trop. Med. Hyg. *27*: 808-814.

277. Monjour, L., G. Niel, J. Dautry, A. Briend, and M. Gentilini. 1975. Application de l'électrosynérèse, sur membrane d'acétate de cellulose au diagnostic des aspergilloses. Bull. Soc. Fr. Mycol. Med. *4*: 225-229.

278. Montpelier, J., and A. Catanei. 1927. Mycose humaine due à un champignon du genre "*Hormodendrum*": *H. algeriensis* nov. sp. Ann. Dermatol. Syphiligr. (Paris) *8*: 626-635.

279. Mota, F. T., and M. F. Franco. 1979. Observoçes sôbre a pesquisa de anticorpos IgM anti-*Paracoccidioides brasiliensis,* por imunofluorescencia no soro de pacientes com paracoccidioidomicose. Rev. Inst. Med. Trop. Sao Paulo *21*: 82-89.

280. Muchmore, H. G., F. G. Felton, and E. N. Scott. 1978. Rapid presumptive identification of *Cryptococcus neoformans*. J. Clin. Microbiol. *8*: 166-170.

281. Müller, H. L. 1974. IgM- and IgG- Antibodies against *Candida* polysaccharides in the serodiagnosis of candidiasis. Bull. Soc. Fr. Mycol. Med. *3*: 51-52.

282. Murray, I. G. 1961. II. Skin sensitivity to *Madurella mycetomi* in guinea pigs. Trans. R. Soc. Trop. Med. Hyg. *5*: 209-215.

283. Murray, I. G. 1971. Progress in the serological diagnosis of mycetoma and other tropical fungus diseases. Z. Tropenmed. Parasitol. *22*: 326-331.

284. Murray, I. G., and H. R. Buckley. 1966. Serological studies of *Candida* species. In H. I. Winner and R. Hurley (Eds.), *Symposium on Candida infections,* E. and S. Livingstone, Edinburgh, pp. 44-51.

285. Murray, I. G., H. R. Buckley, and G. C. Turner. 1969. Serological evidence of *Candida* infection after open-heart surgery. J. Med. Microbiol. *2*: 463-469.

286. Murray, I. G., and E. S. Mahgoub. 1968. Further studies on the diagnosis of mycetoma by double diffusion in agar. Sabouraudia *6*: 106-110.

287. Myerowitz, R. L., H. Layman, S. Perursson, and R. B. Yee. 1979. Diagnostic value of *Candida* precipitins determined by counterimmunoelectrophoresis in patients with acute leukemia. A prospective study. Am. J. Clin. Pathol. *72*: 963-967.

288. Negroni, R. 1966. Micosis Profundas, III. Publ. Especial No. 2., La Plata, Argentina.

289. Negroni, R. 1968. Nuevos estudios sobre antígenos para reacciones serológicas en blastomicosis sud-americana. Ref. Dermatol. Ibero Lat. Am. *10*: 409-416.

290. Negroni, R. 1972. Serological reactions in paracoccidioidomycosis. In *Paracoccidioidomycosis,* Proceedings of the First Pan American Symposium, Pan Am. Health Organ. Sci. Publ. No. 254, Washington, D.C., pp. 203-208.

291. Negroni, R. 1974. El nivel sanguineo de complemento en la Aspergillosis pulmonar. Rev. Inst. Med. Trop. Sao Paulo *16*: 286-291.

292. Negroni, R., and C. I. Flores. 1971. Indirect immunofluorescence in the serological diagnosis of systemic candidiasis. Mycopathol. Mycol. Appl. *43*: 355-359.

293. Negroni, R., and P. Negroni. 1968. Antigens del *Paracoccidioides brasiliensis* para las reacciones serológicas. Mycopathol. Mycol. Appl. *34*: 285-288.

294. Negroni, P., R. Negroni, and P. Galimberti. 1978. Sicosis candidiásica. Rev. Argent. Micol. *1*: 17-21.

295. Neill, J. M., and C. E. Kapros. 1950. Serological tests on soluble antigens from mice infected with *Cryptococcous neoformans* and *Sporotrichum schenckii.* Proc. Soc. Exp. Biol. Med. *73*: 557-559.

296. Neill, J. M., J. Y. Sugg, and D. W. McCauley. 1951. Serologically reactive material in spinal fluid, blood and urine from a human case of cryptococcosis (torulosis). Proc. Soc. Exp. Biol. Med. *77*: 775-778.

297. Nielsen, H. S., and N. F. Conant. 1968. A new human pathogenic *Phialophora.* Sabouraudia *6*: 228-231.

298. Noble, R. C., and L. F. Fajardo. 1972. Primary cutaneous cryptococcosis: review and morphologic study. Am. J. Clin. Pathol. *57*: 13-22.

299. Norden, A. 1979. Agglutination of sheep's erythrocytes sensitised with histoplasmin. Proc. Soc. Exp. Biol. Med. *70*: 218-220.

300. Norden, A. 1951. Sporotrichosis. Acta Pathol. Microbiol. Scand. (Suppl.) *89*: 119.

301. Oblack, D., J. Schwarz, and I. A. Holder. 1976. Comparative evaluation of the *Candida* agglutinin test, precipitin test, and germ tube dispersion test in the diagnosis of candidiasis. J. Clin. Microbiol. *3*: 175-179.

302. Odds, F. C. 1979. *Candida and Candidosis,* University of Leicester Press, Leicester, England.

303. Odds, F. C., E. G. V. Evans, and K. T. Holland. 1975. Detection of *Candida* precipitins. A comparison of double diffusion and counter immunoelectrophoresis. J. Immunol. Methods *7*: 211-218.

304. Ophüls, W., and H. C. Moffitt. 1900. A new pathogenic mould (formerly described as a protozoan (*Coccidioides immitis, Coccidioides pyogenes*). Preliminary report. Pa. Med. J. *5*: 1471-1472.

305. Örley, J., and E. Flórián. 1974. Serologic aspects of pubertal vaginal candidosis: agglutination and complement fixing tests. Mykosen *17*: 107-109.

306. Othman, T., J. L. Guesdon, and E. Drouhet. 1978. Dosage enzymoimmunologique des anticorps humains anti-*Candida albicans* en utilisant des perles de polyacrylamide agarose magnétiques. Bull. Soc. Fr. Mycol. Med. *7*: 249-253.

307. Padhye, A. A. 1978. Comparative study of *Phialophora jeanselmei* and *P. gougerotii* by morphological biochemical and immunological methods. In *The Black and White Yeasts,* Proceedings of the Fourth International Conference on the Mycoses, Pan Am. Health Organ. Sci. Publ. No. 356, Washington, D.C., pp. 60-65.

308. Palmer, D. F., H. L. Casey, J. R. Olsen, V. H. Eller, and J. M. Fuller. 1969. A guide to the performance of the standardized diagnostic complement fixation method and adaption to micro test. Center for Disease Control, Atlanta.

309. Palmer, D. F., L. Kaufman, W. Kaplan, and J. J. Cavallaro. 1977. *Serodiagnosis of Mycotic Diseases,* Thomas, Springfield, Ill.

310. Pan American Health Organization. 1972. *Paracoccidioidomycosis,* Proceedings of the First Pan American Symposium, Pan Am. Health Organ. Sci. Publ. No. 254, Washington, D.C.

311. Pappagianis, D. 1970. Epidemiology of coccidioidomycosis. In *The First International Symposium on the Mycoses,* Pan Am. Health Organ. Sci. Publ. No. 205, Washington, D.C., pp. 195-201.

312. Pappagianis, D., I. Krasnow, and S. Beall. 1976. False-positive reactions of cerebrospinal fluid and diluted sera with the coccidioidal latex-agglutination test. Am. J. Clin. Pathol. *66:* 916-921.

313. Pappagianis, D., C. E. Smith, G. S. Kobayashi, and M. T. Saito. 1961. Studies of antigens from young mycelia of *Coccidioides immitis.* J. Infect. Dis. *108:* 35-44.

314. Parker, J. D., G. A. Sarosi, I. L. Doto, and F. E. Tosh. 1970. Pulmonary aspergillosis in sanatoriums in the south central United States. A national communicable disease center co-operative mycosis study. Am. Rev. Respir. Dis. *101:* 551-557.

315. Parsons, E. R., and E. Nassau. 1974. *Candida* serology in open-heart surgery. J. Med. Microbiol. *7:* 415-423.

316. Pates, A. L. 1948. Precipitin reactions in experimental histoplasmosis and blastomycosis. Science *108:* 383-385.

317. Peck, S. M., R. Bergamini, L. C. Kelcec, and C. R. Rein. 1955. The serodiagnosis of moniliasis: its value and limitations. J. Invest. Dermatol. *25:* 301-310.

318. Peck, S. M., H. Rosenfeld, and A. W. Glick. 1940. Fungistatic power of blood serum. Arch. Dermatol. *42:* 426-437.

319. Pepys, J. 1969. *Monographs in Allergy, 4: Hypersensitivity Diseases of the Lungs Due to Fungi and Organic Dusts.* S. Karger, Basel.

320. Pepys, J., J. A. Faux, J. L. Longbottom, D. S. McCarthy, and F. E. Hargreave. 1968. *Candida albicans* precipitins in respiratory disease in man. J. Allergy *41:* 305-318.

321. Pepys, J., R. W. Riddel, and Y. M. Clayton. 1959. Human precipitins against common pathogenic and nonpathogenic fungi. Nature (Lond.) *184:* 1328-1329.

322. Philpot, C. M. 1978. Serological differences among the dermatophytes. Sabouraudia *16:* 247-256.

323. Philpot, C. M., and D. W. R. Mackenzie. 1976. Detection of antibodies to *Aspergillus fumigatus* in agar gel with different antigens and immunodiffusion patterns. J. Biol. Stand. *4:* 73-79.

324. Pine, L. 1960. Morphological and physiological characteristics of *Histoplasma capsulatum.* In H. C. Sweany (Ed.), *Histoplasmosis,* Thomas, Springfield, Ill., pp. 40-75.

325. Pine, L. 1977. *Histoplasma* antigens: their production purification and uses. Contrib. Microbiol. Immunol. *3:* 138-168.

326. Pine, L., R. G. Falcone, and C. J. Boone. 1969. Effect of thimerosal on the whole yeast phase antigen of *Histoplasma capsulatum.* Mycopathol. Mycol. Appl. *37:* 1-14.

327. Pine, L., H. Gross, G. B. Malcolm, J. R. George, S. B. Gray, and C. W. Moss. 1977. Procedures for the production and separation of H and M antigens in histoplasmin: chemical and serological properties of the isolated products. Mycopathologia *61:* 131-141.

328. Pine, L., G. B. Malcolm, H. Gross, and S. B. Gray. 1978. Evaluation of purified H and M antigens of histoplasmin as reagents in the complement fixation test. Sabouraudia *16:* 257-269.

329. Pinon, J. M., J. P. Gorse, and G. Dropsy. 1977. Exploration de l'immunité humorale dans l'aspergillose: intérêt de l'E.L.I.D.A. Mycopathologia *60:* 115-120.

330. Pinoy, E. 1913. Actinomycoses et mycétomes. Bull. Inst. Pasteur (Paris), *11:* 977-984.

331. Plouffe, J. F., and R. J. Fass. 1980. *Histoplasma* meningitis: diagnostic value of cerebrospinal fluid serology. Ann. Intern. Med. *92*: 189-191.

332. Podobedov, A. I., V. M. Korolev, K. Egerson, and K. Alunurm. 1971. Spetsificheskaya profilaktika strigushchego lishaya krupnogo rogatogo skota. Veterinariya *48*: 48-51.

333. Pollock, A. O., and L. M. Ward. 1962. A hemagglutination test for *Cryptococcus*. Am. J. Med. *32*: 6-15.

334. Pons, I., M. Gimenez, C. Guillerón, and A. Szarfman. 1976. La técnica de la immunoperoxidasa en la detección de anticuerpos especificos en la infección humana por *Paracoccidioides brasiliensis*. Medicina *36*: 510-512.

335. Porter, K. G., D. G. Sinnamon, and R. R. Gillies. 1977. *Cryptococcus neoformans*-specific oligoclonal immunoglobulins in cerebrospinal fluid cryptococcal meningitis. Lancet *1*: 1262.

336. Porto, E., R. G. Ferri, and I. Irulegui. 1977. Pesquisa de anticorpos contra *P. brasiliensis* por imunodifusao vechal (immunoplacas). Rev. Inst. Med. Trop. Sao Paulo *19*: 417-421.

337. Posadas, A. 1892. Ensayo anatomopatológico sobre una neoplasia considerada como micosis fungoidea. An. Circ. Med. Argent. *15*: 8.

338. Posadas, A. 1892. Un nuevo caso de micosis fungoidea con psorospermias. An. Circ. Med. Argent. *15*: 585-597.

339. Pospišil, L. 1954. Zur Serologie der mykotischen Erkrankungen. 1. Komplementbindungsreaktion bri Monihasis. Zentralbl. Bakteriol., Infektionskr. Hyg., Abt. 1: Orig. *161*: 311-314.

340. Pospišil, L. 1955. Zur Serologie der mykotischen Erkrankungen. 2. Antigene Beziehungen einiger Arten vom genus *Candida* zuienander. Zentralbl. Bakteriol. Parasitenkd. Infektionskr. Hyg. Abt. 1: Orig. *163*: 407-409.

341. Poulain, D., G. Desseaux, and A. Vernes. 1978. Variation de la structure antigénique pariétale des blastospores de *Candida albicans*. Bull. Soc. Fr. Mycol. Med. *7*: 37-43.

342. Poulain, D., J. Fruit, A. Vernes, A. Capron, and J. Biguet. 1977. Existe-t-il chez *Candida albicans* une fraction spécifique de la pathogénicité? Bull. Soc. Fr. Mycol. Med. *5*: 17-20.

343. Preisler, H. D., H. F. Hasenclever, and E. S. Henderson. 1971. Anti-*Candida* antibodies in patients with acute leukemia—prospective study. Am. J. Med. *51*: 352-361.

344. Prest, D. B., and A. E. Riker. 1968. The passive cutaneous anaphylaxis (PCA) test in cryptococcosis. Mycopathol. Mycol. Appl. *35*: 1-9.

345. Prevost, E., and R. Newell. 1978. Commercial cryptococcal latex kit: clinical evaluation in a medical center hospital. J. Clin. Microbiol. *8*: 529-533.

346. Ray, J. G. 1967. Agar-gel precipitin-inhibition technique for histoplasmosis antibody determinations. Appl. Microbiol. *15*: 794-799.

347. Ray, J. C. 1967. Agar gel precipitin-inhibition test for coccidioidomycosis. 1. Preliminary evaluation of the complement fixation and agar-gel precipitin tests in the serodiagnosis of coccidioidomycosis. Appl. Microbiol. *15*: 1049-1053.

348. Ray, J. C., and J. L. Converse. 1967. Agar-gel precipitin-inhibition test for coccidioidomycosis. 11. Serological test antigen studies in agar gel. Appl. Microbiol. *15*: 1054-1061.

349. Reeves, M. W., L. Pine, and G. Bradley. 1974. Characterization and evaluation of a soluble antigen complex prepared from the yeast phase of *Histoplasma capsulatum*. Infect. Immun. *9*: 1033-1044.

350. Reeves, M. W., L. Pine, L. Kaufman, and D. McLaughlin. 1972. Isolation of a new soluble antigen from the yeast phase of *Histoplasma capsulatum*. Appl. Microbiol. *24*: 841-843.

351. Reifferscheid, M., and H. P. R. Seeliger. 1955. Monosporiose und maduromykose. Dtsch. Med. Wochenschr. *80*: 1841.

353. Reiss, E., H. Hutchinson, L. Pine, D. W. Ziegler, and L. Kaufman. 1977. Solid-phase competitive-binding radioimmunoassay for detecting antibody to the M antigen of histoplasmin. J. Clin. Microbiol. *6*: 598-604.

353. Remington, J. S., J. D. Gaines, and M. A. Gilmer. 1972. Demonstration of *Candida* precipitins in human sera by counterimmunoelectrophoresis. Lancet *1*: 413.

354. Restrepo, A. M. 1966. La prueba de inmunodifusión en al diagnóstico de la paracoccidioidomicosis. Sabouraudia *4*: 223-230.

355. Restrepo, A. M. 1967. Comportamiento inmulológico de 20 pacientes con paracoccidioidomicosis. Antioquia Med. *17*: 211-230.

356. Restrepo, A. M. 1970. Serological comparison of the two morphological phases of *Paracoccidioides brasiliensis*. Infect. Immun. *2*: 268-273.

357. Restrepo, A. M., and E. Drouhet. 1970. Etude des anticorps précipitants dans la blastomycose sud-américaine par l'analyse immunoélectrophorétique des antigènes de *Paracoccidioides brasiliensis*. Ann. Inst. Pasteur (Paris) *119*: 338-346.

358. Restrepo, A. M., and L. H. Moncada. 1970. Serologic procedures in the diagnosis of paracoccidioidomycosis. In *The First International Symposium on the Mycoses*, Pan Am. Health Organ. Sci. Publ. No. 205, Washington, D.C., pp. 101-110.

359. Restrepo, A. M., and L. H. Moncada. 1972. Indirect fluorescent-antibody and quantitative agar-gel immunodiffusion tests for the serological diagnosis of paracoccidioidomycosis. Appl. Microbiol. *24*: 132-137.

360. Restrepo, A. M., and L. H. Moncada. 1974. Characterization of the precipitin bands detected in the immunodiffusion test for paracoccidioidomycosis. Appl. Microbiol. *28*: 138-144.

361. Restrepo, A. M., and L. H. Moncada. 1978. Une prueba de latex en lámina para el diagnóstico de la paracoccidioidomicosis. Bol. Of. Sanit. Panam. *84*: 520-532.

362. Restrepo, A. M., M. Restrepo, F. Restrepo, L. H. Aristizabal, L. H. Moncada, and H. Velez. 1978. Immune responses in paracoccidioidomycosis. A controlled study of 16 patients before and after treatment. Sabouraudia *16*: 151-163.

363. Restrepo, A. M., M. Roblado, F. Gutierrez, M. Sanclemente, E. Castaneda, and G. Calle. 1970. Paracoccidioidomycosis (South American Blastomycosis): a study of 39 cases observed in Medellin, Colombia. Am. J. Trop. Med. *19*: 68-76.

364. Rey, M. 1961. *Les mycétomes dans l'ouest africain*, R. Foulon, Paris, p. 276.

365. Reyes, A. C., and L. Friedman. 1966. Concerning the specificity of dermatophyte-reacting antibody in human and experimental animal sera. J. Invest. Dermatol. *47*: 27-34.

366. Richardson, M. D., L. O. White, and R. C. Warren. 1979. Detection of circulating antigen of *Aspergillus fumigatus* in sera of mice and rabbits by enzyme-linked immunosorbent assay. Mycopathologia *67*: 83-88.

367. Ricketts, H. T. 1901. Oidiomycosis of the skin and its fungi. J. Med. Res. *6*: 373-547.

368. Rixford, E., and T. C. Gilchrist. 1896. Two cases of protozoan (coccidioidal) infection of the skin and other organs. Johns Hopkins Hosp. Rep. *1*: 209-268.

369. Roberts, G. D., and H. W. Larsh. 1971. The serologic diagnosis of extracutaneous sporotrichosis. Am. J. Clin. Pathol. *56*: 597-600.

370. Roberts, G. D., and H. W. Larsh. 1971. A study of the immunocytoadherence test in the serology of experimental sporotrichosis. J. Infect. Dis. *124*: 264-269.

371. Rohatiner, J. J., and A. Grimble. 1967. Fluorescent antibodies in genital candidiasis. J. Obstet. Gynaecol. Br. Commonw. *74*: 575-578.

372. Romero de Contreras, H. 1976. Master of Philosophy thesis, University of London.

373. Roth, F. J., Jr., C. C. Boyd, S. Sagami, and H. Blank. 1959. An evaluation of the fungistatic activity of serum. J. Invest. Dermatol. *32*: 549-556.

374. Rowe, J. R., V. D. Newcomer, and J. W. Landau. 1963. Effects of cultural conditions on the development of antigens by *Coccidioides immitis*. 1. Immunodiffusion studies. J. Invest. Dermatol. *41*: 343-350.

375. Rowe, J. R., V. D. Newcomer, and E. T. Wright. 1963. Studies of the soluble antigens of *Coccidioides immitis* by immunodiffusion. J. Invest. Dermatol. *41*: 225-233.

376. Salvin, S. B., and M. L. Furcolow. 1954. Precipitins in human histoplasmosis. J. Lab. Clin. Med. *43*: 259-274.

377. Sarosi, G. A., and R. A. King. 1977. Apparent diminution of the blastomycin skin test: follow-up of an epidemic of blastomycosis. Am. Rev. Respir. Dis. *116*: 785-788.

378. Saslaw, S., and C. C. Campbell. 1949. A comparison between histoplasmin and blastomycin by the collodion agglutination technique. Public Health Rep. *64*: 290-294.

379. Saslaw, S., and C. C. Campbell. 1979. A collodion agglutination test for histoplasmosis. Public Health Rep. *64*: 424-429.

380. Saslaw, S., and H. N. Carlisle. 1958. Histoplasmin-latex agglutination test II. Results with human sera. Proc. Soc. Exp. Biol. Med. *97*: 700-703.

381. Savage, J. B. 1976. A comparison of complement fixation studies and immunodiffusion procedures for *Coccidioides immitis*. J. Am. Med. Technol. *38*: 237-238.

382. Schaefer, J. C., B. Yu, and D. Armstrong. 1976. An *Aspergillus* immunodiffusion test in the early diagnosis of aspergillosis in adult leukaemia patients. Am. Rev. Respir. Dis. *113*: 325-329.

383. Schenck, B. R. 1898. On refactory subcutaneous abscesses caused by fungus possibly related to the sporotricha. Bull. Johns Hopkins Hosp. *9*: 286-290.

384. Schlossberg, D., J. B. Brooks, and J. A. Schulman. 1976. Possibility of diagnosing meningitis by gas chromatography: cryptococcal meningitis. J. Clin. Microbiol. *3*: 239-245.

385. Schneidau, J. D. 1964. Cutaneous hypersensitivity to sporotrichin in Louisiana. JAMA *188*: 371-373.

386. Schubert, J. H., and L. Ajello. 1957. Variation in complement fixation antigenicity of different yeast phase strains of *Histoplasma capsulatum*. J. Lab. Clin. Med. *50*: 304-307.

387. Seeliger, H. P. R. 1955. Zur Anwendungsmöglichkeit der Praezipitation im Agargel bei der O-antigen Analyse von Bakterien und Pilzen. Z. Hyg. Infektionskr. *141*: 110.

388. Seeliger, H. P. R. 1956. A serologic study of hyphomycetes causing mycetoma in man. J. Invest. Dermatol. *26*: 81-93.

389. Seeliger, H. P. R. 1957. Die Serologie Schwarzer und rote Hefen. Zentralbl. Bakteriol. Parasitenkd. Infektionskr. Hyg, Abt. 1: Orig. *167*: 396-408.

390. Seeliger, H. P. R. 1958. *Mykologische Serodiagnostik*, Barth, Leipzig.

391. Seeliger, H. P. R. 1965. Standardization and assay of skin test antigens for mycotic diseases. Prog. Immunobiol. Stand. *2*: 154-163.

392. Seeliger, H. P. R. 1968. Serology as an aid to taxonomy. In G. C. Ainsworth and A. S. Sussman (Eds.), *The Fungi*, Vol. 3, Academic, New York, pp. 597-674.

393. Seeliger, H. P. R., and P. Christ. 1958. Zur Schnelldiagnose der cryptococcusmeningitis Mittles der Liquorpräzipitation. Mykosen *1*: 88-92.

394.   Segal, E., R. A. Berg, P. A. Pizzo, and J. E. Bennett. 1979. Detection of *Candida* antigen in sera of patients with candidiasis by enzyme-linked immunosorbent assay-inhibition technique. Clin. Microbiol. *10*: 116-118.

395.   Senet, J. M., and C. Brisset. 1973. The diagnosis of aspergillosis by passive hemagglutination. Biomedicine *19*: 365-368.

396.   Shaninhouse, J. Z., A. C. Pier, and D. A. Stevens. 1978. Complement fixation antibody test for human nocardiosis. J. Clin. Microbiol. *8*: 516-519.

397.   Shannon, D. C., G. Johnson, F. S. Rosen, and K. F. Austen. 1966. Cellular reactivity to *Candida albicans* antigen. N. Engl. J. Med. *275*: 690-693.

398.   Shipe, E. L., M. Williams, and J. Vann. 1963. Comparison of a fluorescent antibody inhibition test with histoplasmin-complement fixation and histoplasmin latex agglutination tests for histoplasmosis. Public Health Lab. *21*: 169-177.

399.   Singer, L. M., and C. Fava Netto. 1972. Conglutinating complement-fixation reaction in paracoccidioidomycosis. In *Paracoccidioidomycosis*, Proceedings of the First Pan American Symposium, Pan Am. Health Organ. Sci. Publ. No. 254, Washington, D.C., pp. 227-232.

400.   Smith, C. E., M. T. Saito, R. R. Beard, R. M. Kepp, R. W. Clark, and B. V. Eddie. 1950. Serological tests in the diagnosis and prognosis of coccidioidomycosis. Am. J. Hyg. *52*: 1-21.

401.   Smith, C. E., M. T. Saito, and S. A. Simons. 1956. Pattern of 39,500 serologic tests in coccidioidomycosis. JAMA *160*: 546-552.

402.   Smith, C. E., E. G. Whiting, E. E. Baker, H. G. Rosenberger, R. R. Beard, and M. T. Saito. 1948. The use of coccidioidin. Am. Rev. Tuberc. *57*: 330-360.

403.   Smith, D. T. 1949. Immunologic types of blastomycosis: a report on 40 cases. Ann. Intern. Med. *31*: 463-469.

404.   Smith, J. K., and D. B. Louria. 1972. Anti-*Candida* factors in serum and their inhibitors. 11. Identification of a *Candida* clumping factor and the influence of the immune response on the morphology of *Candida* and on anti-*Candida* activity of serum in rabbits. J. Infect. Dis. *125*: 115-122.

405.   Splendore, A. 1912. Zymonematosi con localizazzione nella cavitá della bocca, osservated in Brasile. Bull. Soc. Pathol. Exot. *5*: 313-319.

406.   Staib, F. 1969. Proteolysis and pathogenicity of *Candida albicans* strains. Mycopathol. Mycol. Appl. *37*: 345-348.

407.   Stallybrass, F. C. 1964. *Candida* precipitins. J. Pathol. Bacteriol. *87*: 89-97.

408.   Standard, P. G., and L. Kaufman. 1976. Specific immunological test for the rapid identification of members of the genus *Histoplasma*. J. Clin. Microbiol. *3*: 191-199.

409.   Standard, P. G., and L. Kaufman. 1977. Immunological procedures for the rapid and specific identification of *Coccidioides immitis* cultures. J. Clin. Microbiol. *5*: 149-153.

410.   Stanley, V. C., R. Hurley, and C. J. Carroll. 1972. Distribution and significance of *Candida* precipitins in sera from pregnant women. J. Med. Microbiol. *5*: 313-320.

411.   Steinbuch, M., and R. Audran. 1969. The isolation of IgG from mammalian sera with the aid of caprylic acid. Arch. Biochem. Biophys. *134*: 279.

412.   Stevens, E. A. M., C. J. Russchen, C. Hilvering, and N. G. M. Orie. 1970. Steroid effect on *Aspergillus* antibodies. Scand. J. Respir. Dis. *51*: 55-60.

413.   Stevens, P., S. Huang, L. S. Young, and M. Berdischewsky. 1980. Detection of *Candida* antigenemia in human invasive candidiasis by a new solid phase radioimmunoassay. Infection (Suppl. 3) *8*: S334-S338.

414.   Stewart, R. A., and K. F. Meyer. 1932. Isolation of *Coccidioides immitis* (Stiles) in the soil. Proc. Soc. Exp. Biol. Med. *29*: 937-938.

415. Stickle, G., L. Kaufman, S. Blumer, and D. W. McLaughlin. 1972. Comparison of a newly developed latex agglutination test and an immunodiffusion test in the diagnosis of systemic candidiasis. Appl. Microbiol. *23*: 490-499.

416. Stone, K. 1930. A study of yeasts by the complement fixation test. Lancet *209*: 577.

417. Svejgaard, E., and A. H. Christiansen. 1979. Precipitating antibodies in dermatophytosis demonstrated by crossed immunoelectrophoresis. Acta Pathol. Microbiol. Scand. *87*: 23-27.

418. Svendsen, P. J., and N. H. Axelsen. 1972. A modified antigen-antibody crossed electrophoresis characterizing the specificity and titre of human precipitins against *Candida albicans.* J. Immunol. Methods *1*: 169-176.

419. Sweany, H. C. (Ed.). 1960. *Histoplasmosis,* Thomas, Springfield, Ill.

420. Sweet, G. H., R. S. Cimprich, A. C. Cook, and D. E. Sweet. 1969. Antibodies in histoplasmosis detected by use of yeast and mycelial antigens in immunodiffusion and electroimmunodiffusion. Am. Rev. Respir. Dis. *120*: 441-449.

421. Swinne, D. 1978. Contribution à l'étude de l'epidémiologie de la cryptococcose. Bull. Soc. Fr. Mycol. Med. *7*: 123-126.

422. Syverson, R. E., H. R. Buckley, and C. C. Campbell. 1975. Cytoplasmic antigens unique to the mycelial or yeast phase of *Candida albicans.* Infect. Immun. *12*: 1184-1188.

423. Syverson, R. E., H. R. Buckley, and J. R. Gibian. 1978. Increasing the predictive value positive of the precipitin test for diagnosis of deep-seated candidiasis. Am. J. Clin. Pathol. *70*: 826-831.

424. Taschdjian, C. L., G. B. Dobkin, L. Caroline, and P. J. Kozinn. 1964. Immune studies relating to candidiasis. II. Experimental and preliminary clinical studies on antibody formation in systemic candidiasis. Sabouraudia *3*: 129-139.

425. Taschdjian, C. L., P. J. Kozinn, and L. Caroline. 1964. Immune studies in candidiasis. III. Precipitating antibodies in systemic candidiasis. Sabouraudia *3*: 312-320.

426. Taschdjian, C. L., P. J. Kozinn, M. B. Cuesta, and E. F. Toni. 1972. Serodiagnosis of candidal infections. Am. J. Clin. Pathol. *57*: 195-205.

427. Taschdjian, C. L., P. J. Kozinn, A. Okas, L. Caroline, and M. A. Halle. 1967. Serodiagnosis of systemic candidiasis. J. Infect. Dis. *117*: 180-187.

428. Taschdjian, C. L., P. J. Kozinn, and E. F. Toni. 1970. Opportunistic yeast infections with special reference to candidiasis. Ann. N.Y. Acad. Sci. *174*: 606-622.

429. Terry, P. B., E. C. Robinow, III, and C. D. Roberts. 1978. False-positive complement fixation serology in histoplasmosis. A retrospective study. JAMA *239*: 2453-2456.

430. Thurston, J. R., J. L. Richard, and S. McMillen. 1973. Cultural and serological comparison of ten strains of *Aspergillus fumigatus* Fresenius. Mycopathol. Mycol. Appl. *51*: 327-335.

431. Tönder, O., and M. Rödsaether. 1974. Indirect haemagglutination for demonstration of antibodies to *Aspergillus fumigatus.* Acta Pathol. Microbiol. Scand., Sect. B *82*: 871-878.

432. Tran Van Ky, P., J. Biguet, and T. Vaucelle. 1968. Etude d'une fraction antigénique d'*Aspergillus fumigatus* support d'une activité catalasique. Conséquence sur le diagnostic immunologique de l'aspergillose. Rev. Immunol. *32*: 37-52.

433. Tran Van Ky, P., J. Uriel, and F. Rose. 1966. Caractérisation de types d'activités enzymatiques dans des extraits antigéniques d'*Aspergillus fumigatus* après électrophorèse et immunoélectrophorèse en agarose. Ann. Inst. Pasteur (Paris) *111*: 161-170.

434. Trejos, A., and H. P. R. Seeliger. 1957. A serologic study of some yeast like dema-
      tiaceous fungi, cited by Seeliger, H. P. R. Zentralbl. Bakteriol. Parasitenkd. Infek-
      tionskr. Hyg., Abt. 1: Orig. *167*: 396-408.
435. Tsuchiya, T., Y. Fukazawa, and S. Kawakita. 1961. Serological classification of
      the genus *Candida*. In I. Donomae (Ed.), *Studies on Candidiasis in Japan.*, Research
      Committee on Candidiasis, Education Ministry of Japan, Tokyo, pp. 34-46.
436. Tsuchiya, T., Y. Fukazawa, and S. Kawakita. 1965. Significance of serological
      studies on yeasts. Mycopathol. Mycol. Appl. *26*: 1-15.
437. Urmanov, Z. A. 1974. Spetsificheskaya profilaktika trikhoftii krupnogo rogatogo
      skota v Bashkirskoi ASSR. Tr. Vses. Konf. Obshch. Epizoot. *74*: 179-182.
438. Verotti, G. 1916. Un case di micosi microsporica del cudio cepelluto da *Micro-
      sporum lanosum*. G. Ital. Mal. Ven. *57*: 84-92.
439. Viviani, M. A., A. Pagano, and E. Drouhet. 1974. L'elettrosineresi applicata alla
      diagnosi dell aspergillosi e dell aspergilloma broncopolmonare. Studio compara-
      tivo con le altre techniche di immunoprecipitazione e con l'immunofluorescenza.
      Ann. Sclavo *15*: 361-376.
440. Vogel, R. A. 1966. The indirect fluorescent antibody test for the detection of
      antibody in the human cryptococcal disease. J. Infect. Dis. *116*: 573-580.
441. Vogel, R. A., and N. F. Conant. 1952. *Coccidioides immitis* spherule antigen in a
      complement fixation test for experimental coccidioidomycosis. Proc. Soc. Exp.
      Biol. Med. *79*: 544-572.
442. Vogel, R. A., and J. F. Padula. 1958. Indirect staining reaction with fluorescent
      antibody for detection of antibodies to pathogenic fungi. Proc. Soc. Exp. Biol.
      Med. *98*: 135-149.
443. Vogel, R. A., T. F. Sellers, and P. Woodward. 1961. Fluorescent antibody tech-
      niques applied to the study of human cryptococcosis. JAMA *178*: 921-932.
444. Wachs, E. E., and E. B. Wallraff. 1969. Recent developments in serologic methods
      for the diagnosis of coccidioidomycosis. II. Modification of an agar gel precipitin
      inhibition test. Am. J. Clin. Pathol. *51*: 370-374.
445. Wallraff, E. B., and E. E. Wachs. 1969. Recent developments in serologic methods
      for the diagnosis of coccidioidomycosis. I. Diagnostic screening tests and agar gel
      precipitin inhibition tests. Am. J. Clin. Pathol. *51*: 366–369.
446. Wallraff, E. B., and E. E. Wachs. 1971. Recent developments in serologic methods
      for the diagnosis of coccidioidomycosis. III. A soluble antigen fluorescent anti-
      body test. Am. J. Clin. Pathol. *55*: 418-423.
447. Wallraff, E. B., E. E. Wachs, and S. L. Waite. 1968. Antigenic analysis of coccid-
      ioidins by qualitative and quantitative immunodiffusion techniques. Am. Rev.
      Respir. Dis. *97*: 406-414.
448. Walter, J. E. 1969. The significance of antibodies in chronic histoplasmosis by
      immunoelectrophoretic and complement fixation tests. Am. Rev. Respir. Dis.
      99: 50-58.
449. Walter, J. E., and R. W. Atchinson. 1966. Epidemiological and immunological
      studies of *Cryptococcus neoformans*. J. Bacteriol. *92*: 82-87.
450. Walter, J. E., and R. D. Jones. 1968. Serologic tests in diagnosis of aspergillosis.
      Dis. Chest *53*: 729-735.
451. Walter, J. E., and R. D. Jones. 1968. Serodiagnosis of clinical cryptococcosis.
      Am. Rev. Respir. Dis. *97*: 275-282.
452. Walzer, R. A., and J. Einbinder. 1962. Immunofluorescent studies in dermato-
      phyte infection. J. Invest. Dermatol. *39*: 165-168.
453. Ward, G. W., and P. F. Kohler. 1973. Counterelectrophoresis as a rapid method

for the detection of *Aspergillus* precipitins in pulmonary disease. Chest *63:* 495-515.

454. Warnock, D. W. 1974. Indirect immunofluorescence test for the detection of *Aspergillus fumigatus* antibodies. J. Clin. Pathol. *27:* 911-912.

455. Warnock, D. W., and A. L. Hilton. 1976. Value of the indirect immunofluorescence test in the diagnosis of vaginal candidiasis. Br. J. Vener. Dis. *52:* 187-189.

456. Warnock, D. W., J. D. Milne, and A. M. Fielding. 1978. Immunoglobulin classes of human serum antibodies in vaginal candidiasis. Mycopathologia *63:* 173-175.

457. Warnock, D. W., D. C. E. Speller, P. J. Finan, K. D. Vellacott, and M. N. Phillips. 1979. Antibodies to *Candida* species after operations on the large intestine: observations on the association with oral and faecal yeast colonization. Sabouraudia *17:* 405-414.

458. Warren, B. 1965. An evaluation of the immunodiffusion screening tests for coccidioidomycosis. Health Lab. Sci. *2:* 242-245.

459. Warren, R. C., A. Bartlett, D. E. Bidwell, M. D. Richardson, A. Voller, and L. O. White. 1977. Diagnosis of invasive candidosis by enzyme immunoassay of serum antigen. Br. Med. J. *1:* 1183-1185.

460. Warren, R. C., M. D. Richardson, and L. O. White. 1978. Enzyme-linked immunosorbent assay of antigens from *Candida albicans* circulating in infected mice and rabbits: the role of mannan. Mycopathologia *66:* 179-182.

461. Weill, C. P., G. D. Pesle, and P. Meh. 1969. Diagnosis of *Aspergillus* infections by the complement deviation method. Ann. Biol. Clin. (Paris) *27:* 87-91.

462. Weiner, M. H. 1980. Immunodiagnosis of opportunistic mycoses: detection of fungal antigenemia by radioimmunoassays in systemic candidiasis and aspergillosis. In E. S. Kuttin and G. L. Baum (Eds.), *Human and Animal Mycology,* Excerpta Medica, Amsterdam, pp. 165-168.

463. Weiner, M. H., and W. J. Yount. 1976. Mannan antigenemia in the diagnosis of invasive *Candida* infections. J. Clin. Invest. *58:* 1045-1053.

464. Welsh, M. S., and C. T. Dolan. 1973. *Sporothrix* whole yeast agglutination test. Am. J. Clin. Pathol. *59:* 82-85.

465. Wernicke, R. 1892. Veber einen protozoenbefund bei mycosis fungoides. Zentralbl. Bakteriol. Parasitenkd. Infektionskr. Hyg. *12:* 859-861.

466. White, L. O., M. D. Richardson, H. C. Newham, E. Gibb, and R. C. Warren. 1977. Circulating antigen of *Aspergillus fumigatus* in cortisone treated mice challenged with *Candida:* detection by counterimmunoelectrophoresis. FEMS Microbiol. *2:* 153-156.

467. W.H.O. Expert Committee on Biological Standardization. 1978. 29th report, Tech. Rep. Ser. 626, W.H.O., Geneva.

468. Widal, F., and P. Abrami. 1908. Sérodiagnostic de la sporotrichose par la sporoagglutination. La coagglutination mycosique et con application au diagnostic de l'actinomycose. La réaction de fixation. Bull. Mem. Soc. Med. Hop. Paris *25:* 947.

469. Wilson, D. E., J. E. Bennett, and J. W. Bailey. 1967. Serological grouping of *Cryptococcus neoformans.* Proc. Soc. Exp. Biol. Med. *127:* 820-823.

470. Winner, H. I., and R. Hurley. 1964. *Candida albicans.* J. and A. Churchill, London.

471. Yarzabal, L. A. 1971. Anticuerpos precipitantes especificos de la blastomicosis sud-americana. Rev. Inst. Med. Trop. Sao Paulo *13:* 320-327.

472. Yarzabal, L. A., M. B. Albornoz, N. A. Cabral, and S. R. Santiago. 1978. Specific double diffusion microtechnique for the diagnosis of aspergillosis and paracoccidioidomycosis using monospecific antisera. Sabouraudia *16:* 55-62.

473. Yarzabal, L. A., S. Andrieu, D. Bout, and F. Naquira. 1976. Isolation of specific antigen with alkaline phosphatase activity from soluble extracts of *Paracoccidioides brasiliensis.* Sabouraudia *14*: 275-280.

474. Yarzabal, L. A., D. Bout, F. Naquira, J. Fruit, and S. Andrieu. 1977. Identification and purification of the specific antigen of *Paracoccidioides brasiliensis* responsible for immunoelectrophoretic band E. Sabouraudia *15*: 79-85.

475. Yarzabal, L. A., J. M. Torres, M. Josef, I. Vigna, S. Luz, and S. Andrieu. 1972. Antigenic mosaic of *Paracoccidioides brasiliensis.* In *Paracoccidioidomycosis,* Proceedings of the First Pan American Symposium, Pan Am. Health Organ. Sci. Publ. No. 254, Washington, D.C., pp. 239-244.

476. Young, R. C., and J. E. Bennett. 1971. Invasive aspergillosis. Absence of detectable antibody response. Am. Rev. Respir. Dis. *104*: 710-716.

477. Yu, R. J., S. R. Harmon, and F. Blank. 1968. Isolation and purification of an extracellular keratinase of *Trichophyton mentagrophytes.* J. Bacteriol. *96*: 1435-1436.

478. Milne, J. D., and D. W. Warnock. 1977. Antibodies to *Candida albicans* in human cervicovaginal secretions. Br. J. Vener. Dis. *53*: 375-378.

479. Moses, A. 1916. A fixacão do complemento na blastomicose Mem. Inst. O. Cruz. *8*: 68-70.

480. Warren, R. C., L. O. White, S. Mohan, and M. D. Richardson. 1979. The occurrence and treatment of false positive reactions in enzyme-linked immunosorbent assays (ELISA) for the presence of fungal antigens in clinical samples. J. Immunol. Methods *28*: 177-186.

5

Antigens Used for Measuring Immunological Reactivity

Milton Huppert / Audie L. Murphy Memorial Veterans Medical Center and University of Texas Health Science Center, San Antonio, Texas

I. INTRODUCTION

Immunological reactions are useful aids for diagnosing mycotic infections, and may also provide prognostic information (e.g., coccidioidomycosis) when used judiciously with regard to the clinical type of disease. Hence, such reactions can serve as objective guides to clinical management. The sensitivity and specificity of these immunological tests and the interpretations derived from the results depend on the antigen preparations and testing procedures. Antigens, prepared and standardized with great care, yield reliable and reproducible results, but less attention to controlled production has led to differing results among laboratories and to much confusion, in some cases even to loss of confidence in the utility of these tests. The problems caused by use of antigen preparations differing in quality and standardization are magnified by variations in testing procedures. The latter aspect, although of equal significance, is treated in depth in other chapters and so is considered here only briefly and when relevant to a particular discussion. This chapter focuses on the production, refinement, and characterization of antigen preparations derived from fungi that cause disease in humans and other animals. Before doing so, however, three generalizations are considered.

The first generalization is that the term *antigen,* within the context of the current state of the art in medical mycology, refers to preparations containing multiple antigens. The classical example of this is the coccidioidin lot 9 of Smith et al. [444]. This single preparation elicited delayed hypersensitivity (DH) by skin testing and demonstrated both complement fixing and precipitin antibodies. Multiple antigens were involved because mild heating destroyed only complement fixation (CF) reactivity, boiling reduced the capacity for eliciting DH by only 20%, and even autoclaving had an insignificant effect on the component reactive in the tube precipitation (TP) test. Thus, coccidioidin lot 9 contained at least three antigens, each involved in different immunological responses. Similarly, all fungous preparations in current use probably contain multiple antigens (e.g., Refs. 194 and 221).

A second generalization states that demonstrable reactivity of an antigen preparation may vary depending on the stage of clinical disease. For example, coccidioidin TP reactivity among patients with coccidioidomycosis occurs during the first few months of symptomatic disease and less than 10% of these patients are still reactive after 3 months [444]. Hence, assay of coccidioidin for antigen reactive in the TP test must be done with sera obtained during early primary coccidioidomycosis. This restriction emphasizes that evaluation of the immunological reactivity of an antigen preparation must not be divorced from the type of clinical disease but must be integrated into the entire picture of patient infection and response.

The third generalization has emerged from testing an antigen preparation from one species in patients proven to be infected with other species of fungi. Positive results are not uncommon. Therefore, many of these antigen preparations contain not only components specific for the homologous infecting fungus, but also antigens found among heterologous fungi because the cross-reactions cannot be explained entirely as dual sensitizations [444].

A corollary to these three generalizations is that most current immunological reactions among the mycoses may be either a response to a single antigen or a composite of responses to several antigens, some shared among several species. For example, Ehrhard and Pine [117] reported that the CF test for histoplasmosis reflected "the additive presence of both H and M antigens." The net result of composite responses can also be masking of one by another when one is suppressive and the other stimulatory [90]. This can be critical when examining the role of cell-mediated responses as the mechanism for immunity in mycotic infections.

These generalizations receive considerable attention in this chapter in conjunction with the current state of the art as reviewed recently by Grappel et al. [164] for the dermatophytes, by Salvin and Neta [416] for antigens of pathogenic fungi, by Pine [367] for *Histoplasma* antigens, and by Levine et al. [273] for *Coccidioides* and *Histoplasma* antigens. Since these excellent reviews are quite recent, some duplication is unavoidable. This will be minimized as much as possible with regard to earlier publications by summarizing from the reviews for both substantive content and citations from the literature.

Finally, we are now in an era in which modern theories of immunological competence and response plus sophisticated technology are being applied in medical mycology. A valid argument can be made for continuing use of multicomponent antigen preparations in clinical situations based on the empirical information acquired from past experience and on the increased chance of detecting an immunological response as an aid to diagnosis. The current emphasis, however, is on understanding the intricate mechanisms by which host cells learn to recognize, remember, and respond to a foreign insult containing the many antigens of an infecting microorganism. This will require monomolecular antigens since assays with multiple antigens result in a composite of responses, some possibly synergistic, additive, indifferent, or even antagonistic to borrow from experience with chemotherapy. Hence, the current trend is toward ever more refinement of antigens. Since the primary interface between host and fungus occurs at cell surfaces, purification of cell wall antigens has received the most attention. But histopathological evidence of residual "shells" of fungous cells testifies that not all fungal elements survive in tissues and implies that the host is exposed also to antigenic substances of fungal membranes and cytoplasm. In addition, persisting viable cells grow and multiply. They are, therefore, actively metabolizing and secreting or excreting potential antigens into host tissue. Thus, a full comprehension of the immunological interaction between host and fungus requires examination of the entire cell as a physiological entity capable of producing multiple antigens. This review, therefore, will consider both crude and refined antigen preparations from cells and culture filtrates.

II. PRODUCTION

Antigenic substances are produced from culture filtrates and from whole or disrupted cells. A discussion of all methods used is not practical because of variations in isolates cultured, in media, in growth conditions, and in preparation of antigens. In addition, effects of these variables on antigen composition of the final product, combined with natural variability among fungi, create a most complex problem in terms of batch reproducibility and comparability among results obtained by different investigators. Hence, a generalized outline of production methodology will be presented with specific examples for some fungi only as deemed appropriate.

A generalized scheme for preparing various types of antigens is outlined in three stages (Fig. 1). The first involves production of fungous cultures from which antigens are made. The second stage results in crude products that can be used either directly as antigens or serve as source material for the more refined products obtained in stage 3.

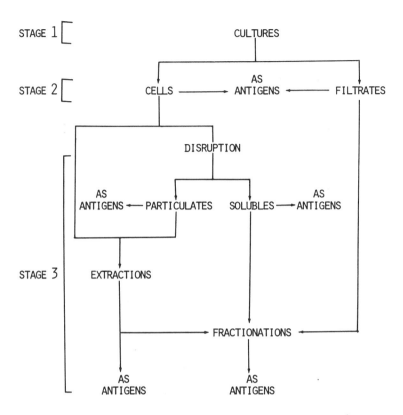

STAGE 1 [

STAGE 2 [

CULTURES

CELLS ⟶ AS ANTIGENS ⟵ FILTRATES

DISRUPTION

AS ANTIGENS ⟵ PARTICULATES   SOLUBLES ⟶ AS ANTIGENS

STAGE 3

EXTRACTIONS

FRACTIONATIONS ⟵

AS ANTIGENS   AS ANTIGENS

Figure 1   Generalized flowchart for obtaining different types of antigen preparations from fungi.

A. Stage 1—Cultures

Special reference isolates have been designated in only a few species as preferred choices for antigen production. Most investigators who produce antigens continuously use strains from their own collection based on personal experience. This may not be critical in some instances but could be in others. Immunotypes exist among some species (e.g., *Candida albicans, Cryptococcus neoformans, Histoplasma capsulatum*) and failure to select appropriate isolates could be a significant error. It is unfortunate that an international center for immunological reference cultures and reagents has not had universal acceptance.

Fungi can be grown on gelled or liquid media, with the former used primarily for obtaining yeast cells or conidia and the latter for yeast or mycelial growth forms. In the special case for the parasitic form of *Coccidioides immitis,* both types of media can be used for producing mature spherules or endospores. Nutritive substrates may be complex organic derivatives, or their dialysates, or composed of chemically defined constituents. Use of complex organic derivatives can be justified on the theoretical, but experimentally unsubstantiated, basis that fungi are more likely to produce the full spectrum of antigens to which an infected host is exposed. This judgment must be weighed against the potential for including in the final product antigens derived from the media and possibly causing nonspecific reactions. Generally, the use of complex organic substrates is restricted to the production of cells that can be separated and washed extensively to remove potential contaminating nonspecific antigens [31]. This, in turn, defeats the suppositional reason for using complex organics since any extracellular antigen (i.e., exoantigen) production is lost. The choice of chemically defined, nonantigenic nutrients eliminates the problem of nonspecific contaminating antigens and enables use of culture filtrates that may contain exoantigens. A disadvantage lies in the possibility that antigen production may not be as complete as that to which the host is exposed during infection. The choice of a dialysate from complex organic substances is a compromise that minimizes inclusion of potentially antigenic medium constituents and retains a rich nutrient source. Since the seamless cellulose dialysis tubing has a molecular weight cutoff approximating 12,000, the dialysate probably contains some peptides. The pore size of dialysis membranes can be reduced by special treatment to exclude substances with molecular weights exceeding 2000 [96]. Hollow fiber and ultrafiltration membrane systems with molecular weight cutoffs in the range 500-2000 are now available commercially. Therefore, a dialysate medium can be prepared with virtual exclusion of undesirable antigenic components.

Inoculation of the medium and incubation conditions vary according to the fungus species and growth form desired. An inoculum of conidia or yeast cells produces greater fungous mass in shorter time compared to tangled hyphae. Incubation temperature for mycelial growth is usually ambient, although the optimal temperature for most pathogenic species approximates $30°C$. Maintenance of the yeast form of dimorphic fungi usually requires incubation at $35-37°C$ and that for the spherule-endospore cycle of *C. immitis* is $37-40°C$. Liquid cultures can be incubated stationary or with continuous shaking, either reciprocating or rotary. Several advantages are gained with shake cultures. A given cell mass is attained earlier, nutrients and aeration are available equally for all cells, and successive batches are more reproducible. Mycelial fungi grow in stationary culture as a mat covering the liquid surface, and nutrient and aeration conditions vary for upper and lower layers of cells as the mat develops. Hence, liquid cultures with continuous shaking are most suitable for rapid and reproducible production of antigens, but a gelled medium would be appropriate if conidia or yeast cells are desired for use as infectious particles, for vaccines, or for analysis of antigenic composition.

B. Stage 2—Crude Antigens

Cells and culture filtrates serve as crude, multicomponent, antigenic preparations or as source material for separation into less complex fractions. Harvesting viable cells is a biohazard, especially with fungi causing respiratory infections, and extreme precautions are required with *C. immitis*. Yeast cells on gelled media are suspended directly by washing the surface growth, and conidia are obtained either in the dry state with a vacuum device or in wet suspension by adding fluid to the surface and dislodging the conidia gently (e.g., with a spinning magnetic bar). If cell viability is not a requirement, appropriate chemicals are added to suspending fluids, but heat killing is not advisable because some clinically significant antigens are heat labile. Fungi in liquid culture are harvested by filtration or centrifugation. Suspensions of cells obtained from growth on complex organic substrates should be washed repeatedly to remove all traces of media ingredients. If cells are removed from liquid cultures by centrifugation and the supernatant is to be used, it should be filtered to ensure sterility. Both live and dead cells are used as vaccines, the former usually in sublethal doses and injected via routes less likely to produce severe disease (e.g., subcutaneous). Cells are used also for tube agglutination (TA) tests, fluorescent antibody

(FA) procedures, phagocytosis studies, eliciting DH, and stimulating lymphocyte transformation (LT). Culture filtrates are used as antigenic preparations either directly or after concentration by pervaporation, concentration dialysis, ultrafiltration, or precipitation with chemicals. These preparations are used for serological tests, eliciting DH, and in vitro assays for cell-mediated immunity (CMI) responses.

C. Stage 3—Refined Antigens

Whole cells and culture filtrates are multicomponent preparations which have proved useful as vaccines and as immunological aids for diagnosis, but their value for dissecting out the separate and specific elements of the host's immunological response is very limited. Serologic and CMI assays performed with reagents containing multiple antigens must be considered composite reactions, the sum of separate responses to single antigens. Evaluation of these responses is complicated additionally by the occurrence of antigens common to several species of fungi. Some antigens are more dominant than others in terms of inducing a host response and eliciting a host reaction, and separation of these dominant antigens has been the principal objective of stage 3 operations. Specific methodology will be reserved for later sections dealing with each mycosis and only general approaches will be indicated in this section.

Fungous cells can be disrupted by several techniques, principally agitation with glass beads, high-pressure extrusion, and sonic oscillation (reviewed in Ref. 484). The most efficient rupture of cells is achieved by mechanical shaking with glass beads, resulting in the complete or almost complete absence of intact cells, depending on cell wall strength of different fungi and shaking time and speed. This efficiency, however, does have a disadvantage since subcellular particulates may be reduced to small fragments which are difficult to separate. High-pressure extrusion can achieve a similar degree of rupture, but several recycles (species dependent) through the apparatus are usually required [73]. Fragmented particulates are usually larger and more easily separated than with glass bead disruption. Sonic oscillation is less effective and can achieve comparable efficiency only with cells having weaker cell walls (e.g., some yeasts), but controlled oscillation, both power and time, is useful for dispersing clusters of cells and removing adherent membranes from cell walls after rupture by another method. All methods require operation at low temperature to protect labile antigens from the heat generated and to minimize degradation by enzymes released from intracellular compartments [97].

Suspensions of broken cells can be separated into solubles and particulates by filtration, or into several classes of solubles and particulates by differential ultrafiltration or ultracentrifugation procedures. Both particulates (e.g., ribosomes and cell walls for immunization) and solubles can serve as antigens. Cell walls deserve particular attention because they represent the primary interface between fungus and host. The study of antigens present exclusively in cell walls requires the determination that other cellular elements have been eliminated. Tests for pentose and phosphate are no longer sufficient because these have been found in some cell wall fractions from fungi. The absence of cytoplasmic components by electron microscopy and of nucleic acids by chemical tests is the most convincing evidence that contamination of cell wall preparations is insignificant.

Methods for extraction and fractionation are so numerous and varied that only examples can be presented, and the discussion of these will be reserved for later sections about separate mycoses. This approach will admittedly be selective but sufficient to illustrate the experimental approaches that have been used. In general, the most commonly successful procedures include extraction with chemicals (e.g., NaOH, phenol), precipitation (e.g., ethanol, ammonium sulfate), and various types of chromatography (e.g., gel permeation, ion exchange, affinity). Isolation of monomolecular antigen requires assays for homogeneity. Analytical physical methods (e.g., ultracentrifugation, zonal electrophoresis) do not establish antigenicity and should be supplemented with the analytical capability of two-dimensional immunoelectrophoresis.

III. SYSTEMIC MYCOSES

Several species infect susceptible hosts via the respiratory tract with nonpulmonary primary disease occasionally caused by traumatic injection. A variety of clinical syndromes occur among the infected population, ranging from mild or even inapparent respiratory symptoms, through serious pulmonary disease which may be either acute or chronic, to extrapulmonary dissemination with acute, chronic, or progressive course. Accumulated experience has established that a constant feature of nonfatal systemic mycoses is induction of a CMI response which may or may not be accompanied by demonstrable humoral antibodies. The intensity of each response varies with the several clinical syndromes: CMI with increased resistance and humoral antibody generally with reduced resistance. Recent evidence suggests that these two mechanisms are not necessarily independent but may be related in a complex manner not fully understood at present. Technical methods for step-by-step analysis of these responses are available, but their application is hampered by lack of the monomolecular antigens required to evaluate whether the host's ultimate fate is determined by successful or failed response to one or several dominant antigens. Hence, the isolation of single homogeneous antigens is the next critical step toward understanding the pathogenesis of systemic mycoses, and progress is being made.

Four mycoses will be considered in this section: blastomycosis, coccidioidomycosis, histoplasmosis, and paracoccidioidomycosis. Cryptococcosis, in recent times, occurs more frequently as an opportunistic invader associated with impairment of a host's immunological competence and is considered in the section dealing with opportunistic mycoses. This is admittedly an arbitrary but reasonable choice with which some readers may disagree for equally valid reasons. All of these fungi are dimorphic, three with mycelial and yeast forms and *Coccidioides immitis* with mycelial and spherule-endospore forms. Current belief is that the infecting fungal element derives from mycelial growth in nature, but that this differentiates rather quickly within the host into the second morphological form, with cell differentiation probably occurring before reproduction. Differences in antigenic composition among dimorphic forms might be an early factor determining ultimate outcome of the infection; therefore, investigation of antigens in both saprobic and parasitic forms is justified.

A. Blastomycosis

The etiological agent of blastomycosis is the dimorphic fungus *Blastomyces dermatitidis*. The lack of a proven endemic area with large concentrations of infected individuals and the poor sensitivity and specificity experienced with early crude preparations have hampered investigation of the antigens of *B. dermatitidis* relative to clinical applications. Experimental animal models have been studied to compensate for this deficiency.

1. Cultures

Both yeast and mycelial growth forms are used as sources for antigen preparations. A variety of isolates have been cultured by different investigators. Whether this has influenced results obtained is unknown since there has been no systematic examination of strains of *B. dermatitidis* for immunotypes, although there is some evidence indicating immunological differences among American and African isolates [460] and between "+" and "–" mating types [397]. Comparison of results reported by different groups is complicated further by the variety of culture methods used. Yeast cells have been grown on complex organic [24,62,91,229,295] or dialysate [115,397] media. Similarly, mycelia have been cultured on differing types of complex nutrients [1,62] and defined constituents [229]. Both cells and culture filtrates are used either directly as antigens or indirectly as source material for preparing antigens. Generally, harvested yeast cells are washed prior to killing by chemical, heat, or disruptive treatments [24,169,229,295,397], and culture filtrates are concentrated usually by evaporative techniques or precipitation with chemicals [1,115,229].

2. Crude Antigens

Blastomyces antigens are prepared by procedures similar to those used successfully with other fungi (reviewed in Refs. 292 and 416). Particulate suspensions (whole or fragmented cells) and soluble substances (culture filtrates or cell extracts) from both growth forms are used in a variety of immunological procedures, with variable results reported by some investigators. A valid example is found in studies by Martin [295] and by Balows et al. [24] since both used a yeast cell suspension prepared in the laboratory of N. F. Conant. Yeast cells, washed from growth on blood agar at 37°C, were made up to a 1:1,000 suspension in saline and heated at 60°C for 2 h. With this reagent Ballows' group reported 33% skin test reactors among 27 patients with proven blastomycosis, but Martin found 70% reactors among 34 similar patients. Only 30% of 20 patients yielded positive CF results among those tested by Balows et al., whereas Martin obtained positive CF titers with sera from 54% of 24 patients. Comparisons of results reported by other investigators would be meaningless because of dissimilarities in antigens employed. In general, these earlier studies indicated that all the different antigen preparations lacked both sensitivity and specificity with respect to human cases of blastomycosis, and that this was particularly evident with blastomycin, a culture filtrate from the mycelial form grown on asparagine or peptone dialysate broth.

Significant progress has been achieved since then. Abernathy and Heiner [1] studied immunodiffusion (ID) and CF reactions of sera from patients with proven blastomycosis. Antigens were filtrates, concentrated by pervaporation, from stationary cultures on Sabouraud's glucose broth for 3-6 months at ambient temperature [183]. Sensitivities of the two serological tests were equivalent, but species-specific ID reactions resolved the differential diagnosis among patients with positive CF tests for both blastomycosis and histoplasmosis. Furthermore, the specific ID reaction for blastomycosis was negative with sera from 51 healthy subjects and positive, with only 1 of 144 patients with suspected mycotic infection. Hence, blastomycin, which had been unsatisfactory in earlier studies, proved useful when applied in a serological test capable of differentiating single antigen-antibody reactions. Subsequent studies provided additional information on the use of *Blastomyces* antigens. Busey and Hinton [62] prepared culture filtrates from both growth forms and demonstrated that ID tests were more sensitive than CF, by detecting 89% of

80 proven cases with the former compared to 51% of 65 with the latter among the same patients. Kaufman et al. [229] evaluated four antigen preparations as potential diagnostic aids. Yeast cells were grown in brain heart infusion broth at 37°C with rotary mixing, and mycelial cells were grown as stationary cultures in an asparagine synthetic broth for 6 months at room temperature. Cells were killed with thimerosal, filtered, washed with thimerosal-saline, and homogenized by shaking with glass beads. The two culture filtrates were precipitated with acetone and redissolved in phosphate buffered saline, pH 7.2, at a 10-fold concentration. After preliminary determinations for optimal dilutions, the disrupted yeast cell suspension proved most satisfactory for CF and the concentrated yeast culture filtrate for ID, but the latter test resulted in the best specificity. Two precipitate bands, labeled A and B, were identified as specific for blastomycosis. These were found in 59% of 49 sera from patients with proven blastomycosis and in none of 104 sera from heterologous mycotic and actinomycotic infections. Additional tests with 113 proven blastomycosis case sera demonstrated that ID detected 79%. Thus, the ID test for blastomycosis, with the acetone-precipitated, yeast culture filtrate as antigen and appropriate reference antisera for A and B bands, demonstrates respectable sensitivity and completely eliminates nonspecific reactions which previously had discouraged use of serological reactions as a diagnostic aid for blastomycosis.

Immunological differences occur among American and African isolates and among mating type strains. When fluorescein-conjugated rabbit anti-*Blastomyces* globulins were absorbed with yeast cells of *H. capsulatum* and cells of *Geotrichum candidum,* specific staining for *B. dermatitidis* yeast cells was obtained [217]. Reciprocal absorption studies demonstrated that American isolates had at least one antigen not present in African isolates [460]. Antigenic differences among mating type cultures were demonstrated by ID tests [397]. Antigens were prepared separately from three "+" and two "–" mating types of *Ajellomyces dermatitidis.* Yeast cells were grown in a dialysate broth for 1 week at 37°C with shaking. The dialyzed and concentrated filtrate was precipitated with ethanol and the precipitate redissolved in water at a concentration of 25 mg of protein per milliliter. A cytoplasmic antigen was obtained by disruption of yeast cells in a French press, centrifuging, and adjusting the supernatant to a concentration of 25 mg of protein per milliliter. The "+" mating type antigens yielded maximum sensitivity. Only 9 of 20 stored sera were positive, but all freshly obtained sera from 20 patients with active blastomycosis were reactive. It must be noted, however, that the stored sera were tested only with a single cytoplasmic preparation, whereas the fresh specimens were tested with the battery of antigens. The mating types exhibited different α-esterase profiles by disc gel electrophoresis, and the enzymes formed immune complexes in ID tests. It is unfortunate that the antigen-antibody references of Kaufman et al. [229] were not included in these studies and no relationship between results of these two reports has been established.

A major problem limiting evaluation of antigens for routine use in ID and FA has been the difficulty of obtaining adequate volumes of specific antiserum. Green et al. [169] resolved this problem by immunizing rabbits with specific immune complexes dissected from precipitated bands in immunoelectrophoresis (IEP). The complexes were developed with antisera raised in rabbits by immunization with *B. dermatitidis* cell walls and reacted against an electrophoresed water-soluble antigen. The cell walls had been prepared from thimerosal-killed yeast cells broken by shaking with beads and washed repeatedly with saline. The water-soluble antigen was obtained also from thimerosal-killed cells which had been harvested, stored in thimerosal-saline for 1 week at room temperature, and the supernatant after centrifugation concentrated and dialyzed against distilled water. The A and B specific precipitated bands were cut out separately and pools from many IEP slides were emulsified in Freund's complete adjuvant and injected into rabbits intradermally. The antisera raised against these specific immune complexes developed not only strong precipitates with the *Blastomyces*-specific antigens in ID tests, but also at least one weak band against a third antigen identified as the c, or common, antigen reported previously by Heiner [183]. Absorption of this antiserum with *H. capsulatum* yeast cells eliminated the nonspecific c immune complex. The absorbed antiserum, however, was not completely free of cross-reacting antibody when fluorescein-conjugated immunoglobulins were tested by the more sensitive FA procedure against 19 heterologous fungus species. Cells of *G. candidum, Sporothrix schenckii,* and endospores, but not spherules, of *C. immitis* still stained weakly. Nevertheless, the absorbed antiserum serves as a satisfactory reference reagent of sufficient volume for ID tests detecting *Blastomyces*-specific reactions, and this had been the objective of the study.

3. Refined Antigens

Purification of specific antigens from *B. dermatitidis* has been attempted because of the earlier results demonstrating that crude preparations lacked sensitivity and specificity. Disrupted yeast cells or their culture broths were precipitated with ethanol in the presence of sodium acetate and acetic acid to obtain "polysaccharides" [115,241,292,352]. These procedures yielded products containing peptides; e.g., pneumococcal polysaccharides prepared similarly contained up to 7% total nitrogen and 2% amino N [210]. In fact, Dyson and Evans [115] reported positive Folin-Ciocalteau reactions with their *Blastomyces* product. The polysaccharide-enriched substances obtained from yeast cell culture filtrates generally were better than cell extracts in yield, sensitivity, and specificity. Dyson and Evans [115], for example, used fractional ethanol precipitation and the 67% ethanol fraction yielded the most satisfactory product, determined by eliciting DH in infected rabbits and guinea pigs. The preparation was positive in all 11 *Blastomyces*-infected animals and in none of the 12 *Histoplasma*-infected animals at doses of 1-10 μg. Specificity declined at higher doses and sensitivity at lower doses. After the preparation was deproteinized by chloroform treatment, all reactivity was lost, indicating an absolute requirement for the peptide moiety to elicit DH. Kashkin et al. [221] demonstrated a similar peptide requirement for eliciting DH. Lancaster and Sprouse purified antigens from blastomycin by gel chromatography followed by preparative isotachophoresis, obtaining nine fractions. Three of these elicited significant DH responses in guinea pigs infected with *B. dermatitidis,* and two of the three were completely negative with *Histoplasma*-infected animals. Only one of these two antigens elicited positive DH in all animals, with average induration matching the original crude material. Unfortunately, the low yields of purified substance precluded studies in humans.

The results reported by Cox and associates [89,91,93,99] are much more encouraging in that a specific and sensitive product was obtained reproducibly and with high yield. Yeast cells served as the source for preparing cytoplasmic and cell wall fractions after fracture of thimerosal-killed cells in a Braun homogenizer. The several cytoplasmic fractions, obtained by sequential ultrafiltration through membranes of decreasing pore diameter, demonstrated good sensitivity but poor specificity in DH tests on guinea pigs sensitized to *B. dermatitidis* or *H. capsulatum.* In contrast, fractionation of a cell wall extract yielded a preparation with good sensitivity and specificity. The cell wall pellet recovered after centrifugation of the ruptured yeast cells was washed repeatedly with distilled water, treated with trypsin and then extracted with 1 N NaOH at 25°C for 3 h. Dialysis of the filtered extract against distilled water resulted in two fractions; one was alkaline soluble, water insoluble, and the other was alkaline soluble, water soluble (B-ASWS). The insoluble fraction, an α-linked glucan [91,212], did not elicit DH, but the B-ASWS did [91]. Sequential ultrafiltration established that this reactivity was retained by a membrane with a nominal cutoff at 10,000 daltons, that addition of this ultrafiltration step improved reproducibility among successive batches, and that the ultrafiltration retentate matched the original B-ASWS in DH sensitivity and specificity. The B-ASWS exhibited good dose-response curves in DH and macrophage migration inhibition (MMI) assays within the range 10-100 μg whether the animals were sensitized by vaccination with dead cells or by infection [99,173]. In contrast to the low level (7%) of cross-reactive DH responses by B-ASWS in *Histoplasma*-sensitized guinea pigs, there were no significant nonspecific responses in LT and MMI assays [99,173]. The B-ASWS was markedly superior to standard blastomycin, which was evaluated simultaneously but yielded sensitivities no better than 66% and nonspecificity ranging from 40 to 86% [91,99]. The B-ASWS proved to be a complex solution despite the extensive purification. Disc gel electrophoresis [92] and isoelectric focusing [173] demonstrated multiple components and at least two of these elicited DH in infected animals. Huppert et al. [194] found by two-dimensional IEP that there were five common antigens among ASWS preparations from *B. dermatitidis, C. immitis,* and *H. capsulatum.* Lancaster and Sprouse [259] separated four fractions by preparative disc electrophoresis. All were reactive but only one, fraction f_4, was 100% sensitive and specific when tested in guinea pigs infected with *B. dermatitidis* or *H. capsulatum.* No additional studies of fraction f_4 have been reported. The fractions obtained by sequential ultrafiltration according to Cox and Larsh [91,92] demonstrated DH activity in the molecular weight range 10,000-50,000, but substances greater than 20,000 were less specific than the final B-ASWS, with a molecular weight of 10,000-20,000. Electrophoretic studies demonstrated that both protein and polysaccharide molecules dissociated into smaller subunits [91] and R. A. Cox (personal communication) has suggested that the native antigen dissociated into smaller molecules which retained antigenic determinants. Chemical analysis of B-ASWS demonstrated 40% amino acids and 31% polysaccharide composed of mannose, galactose, and glucose in molar ratios of 6.8:5.2:1, respectively [92,212]. Roy and Landau [404] precipitated an alkali-soluble, water-soluble extract with ammonium sulfate. This material was not fractionated by ultrafiltration as had been done in the Cox procedure. The precipitated portion represented 12.9% of the original cell wall and contained 15 amino acids. The comparable analysis of the supernatant after ammonium sulfate treatment was 16.1%, with 16 amino acids. Azuma et al. [16], working with total alkali solubles extracted from whole cells, obtained galactomannan from mycelia but only a trace from yeasts. This might indicate that yeast cell B-ASWS is located beneath the outer layer of the cell wall.

Recent work, therefore, has demonstrated considerable progress toward obtaining improved immunological reactions in blastomycosis. The acetone-precipitated A and B antigens present in the reference ID system of Kaufman et al. appear to be specific for this mycosis, with a sensitivity approximating 80% for detection of active disease. The B-ASWS preparation is significantly superior to previous antigen preparations for CMI studies in experimental animals, but has not yet been evaluated among human patients. The B-ASWS and fractions thereof may prove to be highly efficacious with respect to sensitivity and specificity, but this must await additional evaluation in mycoses of humans.

B. Coccidioidomycosis

Immunological tests for fungous infection have had the greatest use and acceptance for coccidioidomycosis among all the mycoses. This has been due almost entirely to the classical studies either directed or influenced by Charles E. Smith, a unique instance in which man, time, and place were in most opportune conjunction. The antigen preparation he produced, the immunological tests he employed, and the clinical interpretations derived from the results of his work are as valid today as when he published them three decades ago. The principal new developments are either derived from his original work or based on techniques not available to him 30 years ago: i.e., the ability to grow the parasitic form of *C. immitis* in the laboratory and to perform in vitro assays for CMI responses. Antigens now are prepared from both the saprobic and parasitic forms, and these have been fractionated to extract antigens with greater specificity and sensitivity.

1. Cultures

Although some strains of *C. immitis* have been used more frequently than others, Smith et al. found no strain differences among 27 isolates with respect to eliciting DH [444]. Huppert [192] demonstrated some strain differences by ID analyses, but these were not correlated with reactivity in CF and TP tests. Stock cultures generally are grown on Sabouraud's glucose agar or 1% glucose, 0.5% yeast extract agar at room temperature to produce inocula for initiating antigen production. When cultures are grown for 1-2 months and considerable drying has occurred, a harvest consisting almost entirely of arthroconidia can be obtained. These are an infection hazard for personnel, even as wet suspensions, and biohazard precautions must be enforced.

Several media and growth conditions have been used for antigen production. Smith et al. [442-444] modified the asparagine synthetic broth developed for tuberculin production by the Bureau of Animal Industry by reducing the glycerin content to 2.5% and substituting 0.7% ammonium chloride for one-half of the asparagine. The asparagine synthetic broth was inoculated with a suspension of mycelial cells, incubated stationary at 37°C, and sampled for activity at 6-8 weeks. If potency was inadequate, cultures were removed to room temperature and incubation continued with weekly sampling. Variability was a constant problem not only among successive batches, but also among replicate flasks. These production problems were resolved by Pappagianis et al. [349,350], who used a shake culture procedure. Flasks containing a broth of 2% glucose and 1% yeast extract were inoculated with a dense mycelial suspension and incubated at 34°C for 3-4 days with rotary shaking. The mycelial mass was collected as a thick sheet on two layers of filter paper by vacuum filtration. Other investigators have used this method successfully with some modifications [2,195].

In contrast to the ease with which the saprobic form of *C. immitis* could be grown, in vitro cultivation of the parasitic form has been difficult and inconstant (reviewed in Refs. 22 and 136). Reproducible in vitro growth of the spherule-endospore cycle was accomplished by Converse [79,80], Converse and Besemer [81], and Levine et al. [268] using a basal salts liquid medium containing ammonium acetate and glucose as the sole nutritional sources of nitrogen and carbon. Inoculated flasks were incubated at 37°C under increased CO_2 tension and with rotary shaking for several days (depending on the strain). Maximum yields of mature spherules occur when endospores are used as inoculum and these are obtained from preliminary liquid cultures seeded with arthroconidia. Mycelial elements are invariably present, but repetitive filtrations reduce this morphological contamination to less than 0.1% [268]. Thus, standardized cultural conditions have been established for growing either the mycelial or spherule-endospore forms with consistency and reproducibility.

2. Crude Antigens

Soluble substances from mycelial growth have been called coccidioidin. Smith's coccidioidin was not the first antigen preparation used for coccidioidomycosis (reviewed in Refs. 245 and 444), but it became the established reference standard. This preparation [442-444] was composed of the combined filtrates of several strains grown on the modified asparagine broth. Although 4 isolates were used initially, the number was increased to 10 with deliberate selection of cultures from diverse types of clinical cases and from different geographical areas to cover the possibility of strain-dependent immunological variation. After sampling demonstrated adequate potency, aqueous thimerosal was added to a final concentration of 1:10,000, cultures were sterilized by filtration, and filtrates pooled to constitute one lot of coccidioidin.

Pappagianis et al. [349,350] improved coccidioidin production by using a combination of shake culturing to obtain increased growth of uniform young mycelia and of toluene-induced lysis of the fungous cells for extracting antigens. The fungus was harvested from glucose yeast extract broth by filtration, washed in situ with distilled water, and suspended as a thick slurry in distilled water containing 3% toluene. After incubating at $37°C$ for 3 days, the mass of lysed cells was removed by filtration and the filtrates (i.e., lysates) preserved with thimerosal. Appropriate testing has established that the lysate contains antigens reactive in all three routine immunological tests and that the production variability is minimal. This method has been modified by using dialysates of peptone [2] or yeast extract [195] in the culture medium so that immunologically active culture filtrates can be used without contamination by extraneous high molecular weight components from nutrients.

Patients have been used to assay coccidioidin for optimal activity in DH, CF, and TP tests. Dilutions of newly prepared coccidioidin are compared with a previously standardized lot to determine potency and sensitivity, and subjects from nonendemic areas are tested to evaluate specificity. Bioassay for DH in infected guinea pigs was abandoned by the Smith group because of variable sensitivity among animals, but Sutton and Marcus [467] reported successful assays in the guinea pig model by evaluating new lots in terms of a potency ratio compared to reactions elicited by simultaneous tests with a previously standardized coccidioidin. Although the TP test as performed by Smith et al. [442] detects a *Coccidioides*-specific antigen, tests for DH and CF with coccidioidin can be positive with human and animal cases of other mycoses, particularly blastomycosis and histoplasmosis (e.g., Refs. 65, 441, and 444). Although multiple sensitizations occur, recent studies of antigen composition demonstrate that the fungi causing these mycoses have antigens in common. Nevertheless, analysis of the multiple antigen-antibody precipitates formed in ID tests with patient sera has demonstrated that appropriate reference systems enable detection of reactions specific for coccidioidomycosis [192,193,347,402,403]. One of these correlates with results of CF tests [195,196,198] and a second with results of the TP test [197,198]. In addition, the specific reactions have been adapted into a test for cultural exoantigens as a rapid method for identifying an isolate as *C. immitis* [202,230,453]. Coccidioidin is also used to sensitize latex particles for a rapid slide agglutination test reported to correlate with Smith's TP results [200], and both the filtrate and lysate preparations have been used for assaying in vitro CMI responses (e.g., Refs. 88, 90, and 222) and DH (e.g., Refs. 6, 507, and 508) in infected animals.

Some physicochemical characteristics of coccidioidin have been reported by several groups of investigators. Smith et al. [442,444] demonstrated that the DH-eliciting, CF, and TP activities in coccidioidin are stable for at least 9 years when stored at 4°C or at room temperature. Autoclaving for 10 min or steaming for 1 h has an insignificant effect on TP activity, but reduces DH potency by 20% and completely destroys substances reactive in the CF test. The heat lability of the latter is shown by a marked reduction of activity even after 56°C for 30 min [197,349]. Coccidioidin contains 60-70% polysaccharide and 14-18% protein [349], but the DH response parallels only the protein concentration [444]. Acid hydrolysis of filtrates and lysates reveals 18 amino acids and mannose, 3-O-methylmannose, glucose, and galactose as major monosaccharides in both preparations [6]. Fractionation of dialyzed preparations on a Sephadex G-25 column discloses a major and a minor carbohydrate-protein peak in the lysate but only the major peak in the filtrate. Both peaks elicit DH and exhibit only minor differences in content of three monosaccharides but an almost 10-fold increase of 3-O-methylmannose in the major peak fraction of the lysate [6]. The native structure of the carbohydrate(s) as homoglycans or heteroglycans has not been reported, although Wheat et al. [526] suggested that a β-$(1 \rightarrow 3)$-glucan might be present based on release of reducing sugar by β-$(1 \rightarrow 3)$-glucanase. Precipitates resulting from addition of ethanol up to 70% retain DH, TP, and CF activity, but CF reactivity is lost with higher concentrations of ethanol. Thus, coccidioidin contains multiple antigens based on different physicochemical characteristics, ID analyses, and clinical correlations with DH, TP, and CF reactivities.

The development of consistent and reproducible growth of the parasitic form enabled antigen production from endospores and spherules. When formalin-killed spherules, endospores, and arthroconidia are tested as vaccines in experimental animals, the spherules induce greater protection against intranasal challenge with arthroconidia (reviewed in Refs. 246 and 273), and the spherule wall appears to be the major site of the antigen [247,270]. An immunogen can be extracted from cell walls of disrupted spherules with phosphate buffered saline at pH 6.2, but full protection requires administration with alum adjuvant in contrast to intact spherules without adjuvant [346]. It was not determined whether the 2% chloroform used as an antimicrobial preservative extracted or denatured any potentially immunogenic components. Viable spherules have served as the source for a soluble antigen preparation designated spherulin [267]. Harvested spherules are washed repeatedly by centrifugation in distilled water, resuspended in distilled water, and incubated at 35°C with shaking for 40 days as the optimal time [423]. The cells, still in the distilled water, are sonicated, centrifuged, and the supernatant preserved with thimerosal. Scalarone et al. [423] reported that successive lots from a single strain are very reproducible on a dry weight basis for eliciting DH, and five of six preparations from separate isolates were comparable. All of these preparations contain protein and polysaccharide but, unlike coccidioidin, neither correlates with DH reactivity. Dialyzed culture supernatants from in vitro grown spherules contain the same four major monosaccharides found in coccidioidin [527]. Comparable studies with spherules have not been reported, but, presumably, these sugars would be present since they are found in culture filtrates and in hydrolysates of spherule walls [527]. Levine and associates [267,269,271,455,456] demonstrated that spherulin is equally specific but more sensitive than coccidioidin for eliciting DH among humans and experimental animals. The greater sensitivity with spherulin appears due to a heat-labile component since heating spherulin at 60°C for 20 min reduces its activity to that of coccidioidin, whereas heating coccidioidin has no significant effect [267]. Spherulin, like coccidioidin, does not induce sensitization and does not affect serological responses among sensitized persons [101,348] or experimental animals [423]. Spherulin has been used also for CF [199,424] and for CMI in vitro assays [70,88,90, 100,102]. In contrast to results with DH tests, spherulin is significantly more cross-reactive than coccidioidin in CF tests with sera from patients with heterologous mycoses [199], and spherulin also exhibits some nonspecific activity with cells from healthy non-sensitive persons in LT [88] and MMI [90] assays.

3. Refined Antigens

Since coccidioidin and spherulin contain multiple antigens, indicated by differential heat susceptibilities, by ID, and by both species-specific and nonspecific reactions in serological and CMI tests, efforts have been directed toward purifying immunologically active components. Cox and associates [88,90,94,512] concentrated on antigens in the fungous cell wall. Her alkali-soluble, water-soluble (C-ASWS) cell wall antigen is prepared from formalin-killed, mycelial cells disrupted in a Braun homogenizer by the same procedure as that described for B-ASWS from *B. dermatitidis*. The final C-ASWS product is immunologically reactive with antibody [194] and in CMI assays, including DH [88,90,94,95,512]. The sensitivity of C-ASWS for eliciting LT responses is comparable to or better than that of coccidioidin and spherulin, but cross-reactions occur with all three antigens among cells from subjects sensitized to *H. capsulatum* [88]. In addition, C-ASWS, like spherulin and spherules, exhibits a low level of nonspecific stimulation in LT assays with cells from non-sensitive subjects [88,90,100].

Cell walls and culture filtrates have been analyzed for immunological reactivity and chemical composition by Wheat and his associates [6,375,523-527]. Culture filtrates and toluene-induced lysates from mycelial growth, after dialysis and concentration, were fractionated by Sephadex G-25 chromatography or by sequential ultrafiltration through membranes of decreasing pore size. Reactive fractions contained carbohydrate and protein with molecular weights in the range 10,000-60,000 [6]. The molecular and antigenic complexity of the reactive substances were revealed by further fractionation with ethanol precipitation and by extraction with 44% aqueous phenol. DH was elicited with all water-soluble, ethanol-precipitated fractions, with a water-insoluble, 20% ethanol-soluble preparation, and with both aqueous and phenol phases after dialysis [524]. The sensitins were obtained also from culture filtrates and lysates and by extraction of defatted mycelia with deionized water, or perchloric acid, or 44% phenol. All fractions elicited DH in infected guinea pigs. The active material concentrated into the phenol phase of a water-phenol mix and contained both protein and the same four major monosaccharides found in coccidioidin. These investigators concluded that either multiple antigens were present in the extracts or degradation of one or more large molecules results in smaller components retaining antigenic integrity. A similar interpretation was suggested by Cox and Larsh [91] for extracts from *B. dermatitidis*. The complex antigen composition was demonstrable with antisera as well as by DH testing. Kashkin et al. [221] found four to seven antigens by IEP among β-naphthol extracts of mycelia and fractions of culture filtrates obtained by precipitation with ammonium sulfate (protein rich) or ethanol (polysaccharide rich). Huppert et al. [194,201], using two-dimensional IEP, found multiple antigens not only in coccidioidin and spherulin, but also in C-ASWS. Coccidioidin, as the recombined culture filtrate and toluene-induced lysate, developed 26 distinct antigen-antibody peaks; spherulin had 12, with 10 common to those in coccidioidin; and C-ASWS had 8, with all present in both spherulin and coccidioidin. At least two of these antigen-antibody precipitates formed shoulders on the anodal side of primary peaks, supporting the suggestion by both Cox and Wheat that some of these antigens occur in polymeric form with fragmenting subunits which retain antigenicity. Similarly, Ibrahim and Pappagianis [204] reported that a small, dialyzable, molecular component of coccidioidin was involved in DH, possibly as a hapten. It would be of some interest to determine if the coccidioidin dialysate contains substances corresponding to the polymeric antigens demonstrated by two-dimensional IEP.

It is apparent that preparations from both the saprobic and parasitic forms of *C. immitis* are complex mixtures with multiple, immunologically reactive components. The frequent demonstration of proteins (or peptides) and of carbohydrates with mannose, 3-O-methylmannose, glucose, and galactose does not resolve the molecular nature of the individual antigens, since they may exist in multiple molecular mixtures of polysaccharides, peptides, or glycopeptides. Some, possibly all, of these antigens may induce immunological responses in the infected host and these may be beneficial, detrimental, or indifferent in terms of host survival. Furthermore, present methods for measuring the host's humoral and CMI responses probably represent a composite of multiple immunological reactions. The answer to such questions must await isolation of monomolecular antigens.

C. Histoplasmosis

The natural distribution of *Histoplasma capsulatum* extends to many parts of the world. The interest in, and immunobiological studies generated by, this fungus have resulted in a wide variety of methods for preparing antigens. No attempt will be made to discuss and evaluate all of these since most of this information for histoplasmosis has been reviewed elsewhere during the past decade [64,245,367,416]. The production methods presented here will be those of investigators at the Center for Disease Control (CDC) of the U.S. Public Health Service, who have had probably the most extensive experience in the production, standardization, and use of histoplasmal antigens. This will be supplemented by more recent information as appropriate.

The principal immunological tests used in the study of histoplasmosis require antigen preparations from both the mycelial and yeast forms of *H. capsulatum*. Particulate antigens are derived primarily from yeast cultures and soluble preparations from both morphological forms with either cells or culture filtrates as source material. The term *histoplasmin* has been applied traditionally to crude, mycelial, culture filtrates, but it was originally used and has occasionally been used subsequently for soluble products from yeast cultures [272,546]. It will be convenient and more specific to differentiate between these filtrates as histoplasmin-MF for the mycelial culture filtrate and histoplasmin-YF for the yeast culture filtrate.

1. Cultures

The selection of isolates for producing antigenic materials depends on the type of immunological test to be used because variation has been reported among strains used in the production of antigens for skin tests (ST) [273,422] and serological tests [367]. For example, in his study of ID reactions with patient sera and histoplasmin-MF, Heiner [183] demonstrated that two of the several antigen-antibody precipitates bore clinical significance and he designated these antigens m and h. Subsequent studies by other investigators (e.g., Refs. 117, 118, 171, and 452) revealed strain-dependent production of either m antigen or h antigen or both. Selection is important also for yeast cells when the immunological test involves surface antigens, e.g., FA procedures. Kaufman and Blumer [226] defined five serotypes among yeast cells determined by four surface antigens and designated these serotypes: 1,2; 1,4; 1,2,3; 1,2,4; and 1,2,3,4. Ordinarily, an isolate of serotype 1,2,3,4 would be selected for maximum antigen coverage, but serotypes 1,2,3,4 and 1,2,4 are uncommon. Therefore, the best choice generally available is an isolate of serotype 1,2,3. This may not be a critical compromise but might, in fact, be the preferred choice. Fluorescein-conjugated immunglobulins raised against serotype 1,2,3 stain all serotypes strongly, but one raised against 1,4 yields much weaker reactions with all serotypes [367]. This could mean that surface orientation of antigen 4 might interfere with antibody induction by other surface antigens, but this possibility has not been examined by absorption studies with antisera against serotype 1,2,3,4.

Since antigen preparations from both mycelial and yeast forms are required for the routine immunological tests, two sets of stock cultures must be maintained. At the CDC, yeast cells are cultured on brain heart infusion agar at 37°C for use either as a source of inoculum or directly as an antigen. The seed cultures are transferred three times at 3 to 4-day intervals prior to inoculating production cultures. No strain preference is indicated for yeast cells to be used in CF tests, but a serotype 1,2,3 isolate is specified as the immunogen for raising antisera to be used in FA procedures. Since yeast cells can be grown in defined media [367], the use of complex organic nutrients might not be advisable. Bauman and Chick [31] were unable to wash out human blood group antigens adsorbed to yeast cells grown on blood agar, and three to six passages on blood-free media were required. The CDC production unit does not use a blood supplement in their media, but it is not known whether other potential antigens in complex organic media might provide a similar threat of nonspecific cross-reactivity. Other investigators have used dialysate [415,417,418] or defined [304,366,370] broths to avoid this problem. Comparison of antigenic reactivity among cultures grown on diverse media indicates that differences which occur are quantitative rather than qualitative [367].

In the past, mycelial cultures used as seed for producing histoplasmin-MF were grown on Sabouraud's glucose agar at room temperature until a dense mycelial mat formed and then several 3 to 6-mm pieces were floated on the surface of Smith's asparagine broth [444]. Cultures were incubated stationary for about 6 months. More recently, Ehrhard and Pine [117,118] inoculated defined broths with a standard dense yeast suspension which converted to mycelium during incubation as shake cultures at room temperature. The m and h antigens in these culture filtrates were produced more rapidly and consistently than in either stationary cultures or those inoculated with mycelial fragments. Differences among three media tested [304,366,444] were slight, but strain selection is of prime importance since variation occurs in quantitative and qualitative production of the Heiner antigens. Although it is not generally emphasized, *H. capsulatum* is considered an infectious hazard for personnel, and biohazard precautions should be maintained.

2. Crude Antigens

The CDC unit produces histoplasmin-MF for use in routine CF and ID tests [175]. The variable immunogenicity by serotype of the yeast cells does not extend to mycelial-derived antigens; i.e., a yeast cell serotype 1,4 may produce a histoplasmin-MF equal to that of other serotypes. Hence, the CDC unit does not recommend a specific isolate for production of histoplasmin-MF but suggests a pilot study with several isolates followed by selection of one or more isolates for a production lot. The final product from each, if satisfactory, should be combined. After killing the cultures with thimerosal (1:5,000 final concentration), the histoplasmin-MF is harvested by coarse filtration, followed by sterilizing filtration. The undiluted product is evaluated for optimal dilution in the CF test and a 10-fold concentrate (acetone precipitated) for the ID test. Histoplasmin-MF for other serological tests [67,415,421] and for ST [120,190] generally is prepared in a similar manner.

The antigen preparation specifically for use in ID tests is of special interest not only as a practical aid for diagnosis, but also for the knowledge gained about the complex of antigens present in *H. capsulatum*. Heiner [183] originally demonstrated six antigen-antibody reactions by ID with a histoplasmin-MF prepared from Sabouraud's glucose broth inoculated with mycelia and incubated stationary at room temperature for 5 months. Four of these precipitates (labeled m, h, x, and y) were found only with sera from histoplasmosis patients or healthy persons sensitized to histoplasmin-MF. The fifth band (labeled c) was cross-reactive with blastomycosis and coccidioidomycosis, and the sixth (labeled n) was nonspecific since it occurred in uninoculated medium. Subsequent IEP studies by other investigators [7,44,184] demonstrate that histoplasmin-MF contains possibly 23 antigens with at least 15 shared by other fungi and that Heiner's m antigen exhibits catalase activity. The presence of patient antibody to m or h bands has been accepted as indicating past or current infection with *H. capsulatum*.

Yeast cells have been used in serological tests (reviewed in Refs. 64, 216, 223, 225, and 245) and as vaccines (reviewed in Refs. 246, 266, and 414). At the CDC yeast cells for CF tests are obtained from a serotype 1,2,3 isolate. The fungus is harvested with saline containing thimerosal at 1:5,000 from growth on brain heart infusion agar. After verification of purity and sterility, the suspension is filtered through gauze or glass wool to remove agar particles, washed once by centrifugation, and adjusted to 10% packed cells in saline-thimerosal 1:10,000 and the optimal dilution determined. Thimerosal is preferred over formalin for killing cells [371]. Although cells killed with each of these chemicals yield comparable CF results with homologous antisera, the specificity of thimerosal-treated cells is usually better when tested against anti-*Blastomyces* and anti-*Coccidioides* patient sera. Cell walls from disrupted yeast cells have not been better reagents for CF and, in general, their use has been discontinued.

Immunity against challenge infection has been induced with viable or killed cells and with subcellular particles. Cell wall preparations and killed cells are equally effective immunogens, but live cells given in sublethal doses are the most efficient stimulators of immunity. A ribosomal preparation, admixed with Freund's incomplete adjuvant, induces immunity only slightly less than does a preparation with viable cells when each is administered subcutaneously [135,485]. The ribosomes are prepared by differential centrifugation after disrupting yeast cells by shaking with glass beads. In all reports of experimental immunization, variable but significant numbers of animals surviving the challenge infection are still infected. It is not known whether longer observation periods would reveal eventual death of all animals or complete elimination of fungi from tissues. This would be significant information for evaluating the efficacy of antigens for immunization.

Crude soluble products have been produced also from yeast cultures, i.e., histoplasmin-YF and unfractionated extracts from intact, disrupted, or autolyzed cells. Generally, the fungus is grown in nonantigenic broth [366,370,417,418] to obtain extracellular as well as cellular antigens. Reeves et al. [379,380] found that yeast cells, harvested from gelled casein hydrolysate medium [366] and suspended in 0.001 M borate buffered saline (pH 8.6) containing thimerosal, released antigens into the suspending fluid during incubation at 4°C. This product has less anticomplementary activity than intact yeast cells and contains h and m antigens plus additional antigens (labeled Y antigens) by ID tests. These new Y antigens, which are not necessarily the same as Heiner's y, react frequently (38%) with convalescent sera from human cases of histoplasmosis and appear to be specific when tested against sera from patients with blastomycosis, coccidioidomycosis, and cryptococcosis. Treatment of cells with phenol extracts additional antigens. Tompkins [489] obtained an antigen (YP) different from Heiner's but reactive by ID with sera from a few patients with histoplasmosis. Garcia and Howard [151] distinguished three antigens in a phenolic extract, one of which was identical to one of two antigens present in an ethylenediamine extract of intact cells. These antigens were not evaluated for a relationship to Heiner's antigens. Similarly, Domer's [107,109] soluble cytoplasmic substance and ethylenediamine extract from cell walls were not compared with Heiner's antigens, although they were reactive in DH and CMI assays. Levine and associates [273,422] employed their controlled lysis procedure to obtain DH-eliciting reagents from both yeast and mycelia. Cultures were grown on defined medium [304] at either 37°C or 27°C, harvested, and washed. Cells, suspended in sterile, pyrogen-free water, were allowed to autolyze during shaking 1-10 days at 37°C. Activity decreased with extended lysis times indicating enzymatic degradation of the antigen(s). Extensive testing of multiple lots in guinea pigs infected with *H. capsulatum* or *C. immitis* demonstrated that the yeast- and mycelial-derived preparations possessed equivalent sensitivity, but the former was generally more potent on a dry weight basis and completely specific, whereas the mycelial preparation elicited DH among 31% of 26 animals infected with *C. immitis.* Both autolysates were superior to two standard histoplasmin-MF reference preparations in every parameter measured. Furthermore, ST with the yeast autolysate on sensitized animals [422] did not induce any significant increase in CF titer (> twofold) compared with a commercial histoplasmin-MF which did cause increases among 6 of 9 animals. Additional tests among humans confirm the absence of CF titer displacement by ST with the yeast autolysate on sensitive subjects. The autolysis method apparently produces a ST reagent superior to current histoplasmin-MF preparations, and in this respect, is similar to the reported superiority of spherulin over coccidioidin.

3. Refined Antigens

Since crude *Histoplasma* antigens frequently react in serological and CMI assays among patients and animals with heterologous mycoses, many attempts have been directed toward isolation of antigens reacting specifically for histoplasmosis (reviewed in Refs. 245, 367, and 416). Cells and culture filtrates from both morphological forms have served as source materials for extractions and fractionations by a variety of procedures.

Earlier studies focused on isolation of polysaccharides. Products enriched in carbohydrate but still containing significant amounts of nitrogen resulted from enzymatic digestion of protein [243], from precipitation of polysaccharides with ethanol [63,116, 241,242,293,418,445], or from precipitation of protein with ammonium sulfate and collection of the carbohydrate-enriched supernatant [341,511]. Most of these preparations are reactive in serological tests and elicit DH, but attempts to remove protein completely by chemical, electrophoretic, or chromatographic procedures while retaining full immunological reactivity are generally unsuccessful [293,341,418]. Extensive reduction of nitrogen content is associated with loss of capacity for eliciting DH but not necessarily serological reactivity [116,243,511]. In contrast, protein enrichment, but again not entirely free of carbohydrate, results in products reactive in ST and, usually, serological assays [341,384,418,446,448,511]. Furthermore, most investigators report immunological or physicochemical heterogeneity in the fractions isolated. These results show that the various refined preparations contain a complex of antigenic substances, both carbohydrate and protein or glycoprotein, and that the protein or peptide moiety of undetermined size is required for eliciting DH.

More recent studies have focused on the isolation and characterization of Heiner's m, h, and related antigens, of the serotype antigens of Kaufman and Blumer, and of substances reactive in CMI assays, including DH. Antigenic complexity becomes more obvious with increasing refinement of the preparations and with the application of analytical methods with greater sensitivity; e.g., histoplasmin-YF contains at least 18 different antigens by fused rocket IEP [474,476].

Although many antigens have been found in culture filtrates, cells, and cell-derived extracts [151,221,314,474,476], the m and h antigens remain among the most significant as aids to clinical diagnosis, and attempts have been made to isolate them. Column chromatography on DEAE-cellulose followed by gradient or stepwise elution with decreasing pH (8.0-2.0) and increasing molarity (0.05-0.2 M) yields separated fractions enriched with either m or h and small amounts of other antigens [56,105,170,290,372]. The enriched m fraction generally elutes between pH 8.0-7.0 and the enriched h fraction between pH 5.0 and 4.0. The best separation is achieved with sequential procedures: preliminary concentration followed by gel permeation, DEAE-cellulose with Tris buffer and NaCl, and DEAE-cellulose with phosphate buffer [372]. The purest m fraction obtained contains 3% of the total h, and the comparable h fraction, 5% of the total m. Final purification is achieved by preparative electrophoresis in a polyacrylamide gel, cutting out the bands, and eluting with 0.1 M phosphate (pH 4.0) in 0.3 M NaCl. Although only a single band by disc electrophoresis is found with each of these two purified fractions, a rabbit anti-*B. dermatitidis* serum developed a band of nonidentity with both preparations. This is presumed to be the c antigen of Heiner. Both m and h purified fractions are reactive in CF and ID tests and elicit DH in sensitized guinea pigs with as little as 0.5 μg of m and 0.6 μg of h [372]. At present it appears that m and h are not located on the outer surface of the cell [168]. Although there is a strong association of CF results with m and h in the antigen preparations [373,529], absorption of antisera specific for m and h with yeast cells does not change the CF titer and the cells do not stain with fluorescein-conjugated antibody specific for m or h. Green et al. [168] postulated that the CF reaction must alter the surface structure of the cells causing release of the antigens into the surrounding fluid. This theory is based on evidence that when washed yeast cells are mixed and incubated with antiserum, with complement, or with antiserum and complement, and then

Table 1 Comparison of Highly Purified Antigens m, h, and Y from Yeast and Mycelial Growth of *Histoplasma capsulatum*

Antigen[a]	Molecular weights	Relative monosaccharide ratios Glucose/Mannose/Galactose			Prot/CHO[b]
Mycelial					
h	>200,000	1	14.9	7.3	1.3
m	>200,000	1	2.7	1.2	0.9
Yeast					
h	>50,000	1	1.0	Tr[c]	6.5
m	>50,000	1	0.3	Tr	14.0
h	<50,000	1	1.0	Tr	0.3
m	<50,000	1	0.3	Tr	0.4
Y	<50,000	1	0.3	Tr	1.1

[a] Data for mycelial antigens from Ref. 56 and for yeast antigens from Ref. 290. Antigens were eluted from DEAE-cellulose columns with Tris buffers of descending pH.

[b] Protein/carbohydrate ratio.

[c] Trace amounts.

the supernatants are used as antigens in the complete CF test, only the supernatant from the yeast cell-antiserum-complement mixture yields a positive test.

Physicochemical examination of these purified antigens demonstrates a complex of molecular species with determinants for m or h. These, and the Y antigen of Reeves et al. [379], appear to be peptidoheteropolysaccharides with the carbohydrate moiety composed of glucose, mannose, and galactose as indicated in Table 1 [56,290]. Molecular weights of the mycelial derivatives (histoplasmin-MF) were determined by gel permeation in Sephadex G-200. Not shown in the table are the low molecular weight m (24,000) and h (30,000) antigens which were found in addition to high molecular weight components, but yields were too low for extensive analysis [56]. The yeast derivatives were obtained from cell autolysates with preliminary separation by hollow fiber ultrafiltration into fractions of greater or less than 50,000 daltons, from which purified antigens were prepared by gel permeation and DEAE-cellulose chromatography. Earlier work by the same investigators demonstrated that Y is also a complex, with molecular weights ranging from less than 10,000 to 100,000 [379]. There are obvious differences in the relative amounts of mannose and galactose in the mycelial- compared to the yeast-derived antigens and also in the protein:carbohydrate ratios of the high molecular weight yeast-derived antigens compared to others. Although qualitative amino acid analysis reveals a generally similar composition for all preparations, quantitative differences are apparent. The mycelial m antigen contains greater amounts of proline, threonine, glutamate, and serine than the h antigen. The yeast-derived antigens are generally similar except for methionine and histidine. Methionine is absent from the high molecular weight antigens, is the amino acid with the relatively highest concentration in low molecular m and h, and is present in Y at only one-tenth the concentration in the others. Histidine is present in the high molecular weight m and h antigens, but in low concentration or absent in the low molecular weight antigens. Since stepwise degradation and terminal amino acid analyses were not done, it is not known whether these might be structural sites at which large molecules separate into antigenically reactive smaller molecules. Treatment with β-(1 → 3) glucanase or periodate has no effect on serological activity of any antigen, but protease decreases capillary precipitin activity of the h antigen by 78-96%. Boiling for 10 min destroys all serological activity. Even though there are differences in composition, the several molecular species exhibit the indicated antigenic specificities by ID and these assays do not definitively differentiate between carbohydrate and protein moieties as antigenic determinants.

The Kaufman-Blumer serotype antigens also may be peptidoheteroglycans. Reiss and associates [381,383] subjected isolated and washed yeast cell walls to serial enzymatic hydrolysis, analyzing solubilized nondialyzable products and residual cell walls at each stage by ID, IEP, CF, and FA with conjugated antiglobulins of known specificity for serotypes. Two isolates of *H. capsulatum* were chosen because of a correlation between serotypes and cell wall chemotypes [106,214,368,381,420], i.e., serotype 1, 2, 3 with chemotype I and serotype 1,4 with chemotype II. Cell walls of serotype 1,4 were extracted with α-(1 → 3)-glucanase (walls of chemotype I, serotype 1,2,3 do not contain α-glucan) and then walls of both types were extracted successively with β-(1 → 3)-glucanase, pronase, and chitinase. The original cell walls, all solubilized, nondialyzable extracts, and the residual mural cores contained glucose, mannose, galactose, and protein. Glucosamine was absent only in the α-glucanase extract from serotype 1,4, but amino sugar was still present in the mural core. All soluble fractions were serologically reactive with a maximum of three antigenic components in the β-glucanase extract of serotype 1,2,3 by IEP, and all solubilized antigens differed from m and h antigens by ID and IEP assays. Despite enzymatic removal of antigenic substances from cell walls, FA staining intensity of residual mural cores was undiminished until after pronase treatment. Since enzymatic degradation with a-(1 → 3)-glucanase and β-(1 → 3)-glucanase did not decrease FA and CF reactions, a definitive role as antigenic determinants cannot be assigned to these glucans as yet. In contrast, the presence of mannose, galactose, and protein in all immunologically active fractions is consistent with a role as antigens composed of these substances. In addition, the evidence that specific FA serotype staining of mural cores decreases in intensity after proteinase treatment implies a significant role for peptides in serotype antigens.

The fact that the appropriate chemical components are present does not prove that these are peptidogalactomannan antigens. Definitive isolation, demonstration of molecular and immunochemical homogeneity, and structural analyses would be required. Nevertheless, studies by other investigators point toward peptidoglycans as immunologically reactive components of cell walls and culture filtrates [16,107-109,214,384; reviewed in Refs. 367 and 416]. The more recent procedures for obtaining peptidoglycans are based on alkaline extraction of cells or isolated cell walls [16,151,214,369,384,420]. The alkali-soluble substances are separated by neutralization into a precipitated fraction containing peptidoglycans, or, alternatively, by direct precipitation from solution with ethanol. Although some investigators report only glucose [244] or glucose and mannose [106,108, 109], others consistently identify glucose, mannose, and galactose as the hexose components in the alkali-soluble nonprecipitable (or ethanol-precipitated) fractions along with appreciable amounts of amino acids [16,214,384,420]. Anion exchange chromatography [384] or precipitation as copper complexes with Fehling's solution [16,214] yields a product with reduced or absent glucose but with constant molar ratios for galactose:

mannose and intact immunological reactivity, indicating that the reactive component is a peptidogalactomannan [214,384] with a molecular weight approximating 100,000 by gel permeation analysis [384]. Dissociation of the copper complex by mild acid and removal of the peptide by chromatography through a DEAE-cellulose column results in a pure galactomannan which precipitates with antibody and elicits Arthus-type sensitivity but not DH in sensitized rabbits [16,384], a fact which indicates that the peptide moiety is probably a requirement for CMI responses. Purified galactomannans from *B. dermatitidis, H. capsulatum, H. duboisii,* and *Paracoccidioides brasiliensis* form single component reactions of complete identity by ID, and therefore represent a common antigenic determinant among these species [16]. Azuma et al. [16] reported that the purified galactomannan contains a main chain of $(1 \rightarrow 6)$-linked mannopyranoside with galactofuranose and mannopyranose as nonreducing terminal units. The similarity to purified galactomannan I from *Trichophyton* species prompted an immunological comparison and the galactomannan I cross-reacted with anti-*Histoplasma* serum in quantitative precipitin tests. Thus, galactomannans appear to be widespread among fungi as potentially common antigenic determinants.

The standard histoplasmin-MF also contains one or more peptidogalactomannans apparently as the DH-eliciting substance. Bartels [29] obtained two reactive fractions following disc gel electrophoresis. These contained at least 18 amino acids in addition to mannose, galactose, and an unidentified hexosamine. A histidine derivative was present by N-terminal analysis and both serine and threonine by C-terminal analysis. Sprouse and associates [446-448] also isolated two fractions by using gel permeation followed by disc gel electrophoresis. One (dIII) elicited DH in both *Histoplasma-* and *Blastomyces*-infected guinea pigs and the other (dII) in only the *Histoplasma*-infected animals. The dII appeared to be homogeneous by continuous gradient ultracentrifugation with a molecular weight approximating 12,000 and to be devoid of contaminating substances by isoelectric focusing and IEP. Spectral analysis, however, showed a peak at 219 nm with a shoulder beginning at 248 nm, indicating some lack of homogeneity, and was interpreted either as possible contamination with nucleic acid or protein, or as polydispersion of the dII molecule. Staining of analytical disc electrophoresis gels demonstrated carbohydrate and protein in both dII and dIII, while acid and heat denaturation studies indicated a covalent bond between carbohydrate and protein. Monosaccharide composition has not been determined. Since both dII and dIII were present in more than 30 separate lots of histoplasmin-MF, these could be the major antigens eliciting DH, dII being species-specific and dIII with an antigenic determinant common to *B. dermatitidis.* Comparable studies with the controlled yeast lysate of Levine et al. [272] have not been reported as yet, but it would be of considerable interest to learn whether the yeast autolysate preparation contains the dII, dIII, or additional antigens capable of eliciting DH but not yet identified.

D. Paracoccidioidomycosis

Immunological studies of paracoccidioidomycosis, like blastomycosis, have been hampered by lack of a well-defined endemic area with a high concentration of sensitized persons. The disease occurs throughout Latin America, but the distribution of cases is uneven and there have been no focal epidemics documented with respect to immunological studies of the native population. The fungus, *Paracoccidioides brasiliensis,* is dimorphic with a saprobic mycelial form and a parasitic yeast form. Both morphological forms can be grown in the laboratory with little difficulty and both have been used as source material for antigen preparations (reviewed in Refs. 328 and 389).

1. Cultures

Comparisons among antigen preparations produced by different investigators must be qualified by recognizing that there has been no uniformity in isolates or culture media used and no general agreement on standardized reference reagents. Strain-dependent variation in antigen composition has been demonstrated among yeast cells by FA procedures [215,439,440] and in culture filtrates by ID [387], but no strain preference is indicated, although Kaufman [224] reported that Restrepo's strain B-341 produces a yeast culture filtrate antigen equal to that from pooled isolates. Culture media have varied from complex organic substrates to defined chemical constituents. Since these are frequently concentrated 10-fold or more, false-positive results from reactions with media ingredients may occur. The mycelial form can be grown in defined medium, but the yeast form generally requires either enriched organic nutrients or a peptone dialysate supplemented with thiamin [326] or multiple vitamins [134,386] for optimal growth. The dialysate media with supplements support growth of either yeast cells or mycelium, depending on incubation temperatures of 35-37°C or ambient, respectively. In general, antigen production is faster when the fungus is grown with constant agitation, but 1 month may be required for this slow-growing organism [331,386].

2. Crude Antigens

The antigen used most extensively has been the "polysaccharide" of Fava Netto [130, 134] prepared by the method of Norden [339]. Washed and defatted yeast cells are suspended (15% by volume) in veronal buffer and autoclaved at 120°C for 20 min. The supernatant after centrifugation constitutes the "polysaccharide" antigen, so-called because Norden had found no protein (Biuret method) in his preparations from *Sporothrix schenckii*. Costa [85], however, working in Fava Netto's laboratory, reported that these "polysaccharide" antigens contained 3.5 mg/ml of Folin-Ciocalteau reactive material (Lowry method) and 0.8 mg/ml polysaccharide (anthrone method). This polysaccharide-protein mixture has been used extensively for CF, TP, and ST [57,85-87,130-134]. Costa and associates [86,87] chose the residual yeast cells after extraction of the polysaccharide-protein for ST in animals because the well-circumscribed and indurated reactions were easier to read. Histologically, these lesions were microabscesses surrounded by mononuclear infiltrates interpreted as consistent with a DH response. Frequent cross-reactions in CF [131] and DH [86,132] with the Fava Netto antigen among histoplasmosis patients represents a serious diagnostic problem since the endemic areas for these two mycoses overlap. Positive reactions could have been caused by dual infections or by antigens shared by both fungi.

   Other types of cell-derived antigens from both yeast [306,323,324,331,533] and mycelia [7,533,538] have been evaluated. Supernatants from disrupted cells are used in serological and CMI tests, but in general these appear no more specific or sensitive than culture filtrates. Recently, Negroni et al. [329] prepared a soluble extract from yeast cells by controlled lysis in distilled water followed by mechanical disruption, a method similar to that of Levine et al. for *C. immitis* [267]. This extract contained both polysaccharide and protein and demonstrated sensitivity and specificity equivalent to Fava Netto's antigen in comparative ST among patients with paracoccidioidomycosis and histoplasmosis. The potency of Negroni's antigen may be greater since it elicited larger dermal reactions in more patients. This could be a quantitative rather than a qualitative difference since Negroni's antigen was standardized on a dry weight basis and Fava Netto's by optimal dilution. Whole yeast cells stain in FA tests, but fluorescein-conjugated immunoglobulins require extensive absorption with heterologous fungi because of common antigens shared with many species, including *Blastomyces dermatitidis, Candida albicans, Coccidioides immitis, Histoplasma capsulatum,* and *Sporothrix schenckii* [215,439,440].

Filtrates from yeast and mycelial cultures are used as crude antigens (reviewed in Refs. 328 and 389). Generally, the cell-free filtrates are dialyzed and concentrated prior to use either by pervaporation to 5-10% of original volume or by lyophilization. These filtrates are used for evaluating CMI [315,393; reviewed in Ref. 76] and serological responses (reviewed in Refs. 328 and 390). When several different types of antigen preparations are compared, most investigators agree that the yeast cell culture filtrates are superior for serological tests [331,386-388]. Use of these preparations in ID and IEP tests reveals at least 5 antigen-antibody reactions with patient sera [388] and a minimum of 25 with rabbit hyperimmune sera [7,536]. Restrepo and associates [386,390-392] found that a yeast culture filtrate produces three precipitated bands, and rarely a fourth [392], by ID with patient sera. Band 1 is most common and persists longest among patients receiving therapy. Band 3 is identical to the *H. capsulatum* m antigen [392], explaining the frequency of cross-reactions in these two mycoses. Analysis of the same antigen preparation by IEP reveals five antigens, labeled A through E, with antigen A corresponding to the ID antigen 1 [388]. Yarzábal and colleagues [533,536,538] demonstrated by IEP that many of these antigens possess enzymatic activity and that one, labeled E, of three cathodically migrating antigens is an alkaline phosphatase which corresponds to Restrepo's antigen A. This has been confirmed by Conti-Diaz et al. [77,78]. Yarzábal et al. [536] also reported that rabbit anti-*Paracoccidioides* sera reacts with soluble extracts from *Aspergillus fumigatus, Cryptococcus neoformans,* and species of *Emmonsia.* It is obvious that the crude preparations from *P. brasiliensis* contain multiple antigens, some species-specific and many shared with other fungi.

3. Refined Antigens

Several groups have attempted refinement and characterization of immunologically reactive substances present in crude preparations derived from *P. brasiliensis.* Addition of ethanol to filtrates from either mycelial or yeast growth precipitates a fraction which, when dissolved in buffered saline, elicits DH in humans and infected guinea pigs [537]. The fractions from both growth forms are equivalent in sensitivity to the original culture filtrates, but the precipitated fraction from yeast culture filtrates exhibits better specificity. In a later evaluation among patients with paracoccidioidomycosis or sporotrichosis, the mycelial ethanol-precipitated fraction yielded better sensitivity but less specificity than the comparable product from yeast filtrates [426]. The ethanol-precipitated products contain 2-4% nitrogen, 29-37% reducing substances, 6-8% pentose, and 1-2% phosphorus. Chromatographic analysis of acid hydrolysates reveals seven amino acids and glucose, galactose, arabinose, and glucosamine. Extraction with chloroform-amyl alcohol results in almost complete loss of DH-eliciting capability in the aqueous phase, but substances recovered from the chloroform-alcohol phase after evaporation to dryness elicit DH equal to the original ethanol-precipitated fraction. This recovered antigenic material contains nitrogen and reducing substances, leading to speculation that the active substance is a peptidoglycan which must have strong bonding because the immunological reactivity withstood extraction and emulsification in chloroform-alcohol.

Purification of the alkaline phosphatase, cathodic antigen (i.e., Yarzábal's E and Restrepo's A) has been accomplished [534,535,537] and is of special interest because of its apparent species specificity and common occurrence in sera of patients with paracoccidioidomycosis. A rabbit antiserum, raised against crude mycelial culture filtrate, was absorbed with crude extracts from heterologous fungi and was then used with the crude filtrate to develop antigen-antibody precipitates of IEP. Cathodic band E was excised, washed, emulsified in Freund's complete adjuvant and injected intradermally into rabbits. The rabbit antisera differentiated band E into two precipitates, EI with alkaline phosphatase activity and EII without enzyme activity. Excision of the EI and EII precipitates and repetition of the immunization produced monospecific antisera against each. Immunoabsorbent affinity chromatography extracted EI and EII separately from the crude culture filtrate and the purified antigens were recovered by elution with glycine-HCl-NaCl buffer at pH 2.8. Although both EI and EII appeared to be species-specific, better yields with EII enabled more extensive studies. The EII antigen produced positive ID and IEP results with sera from 43 of 45 patients with paracoccidioidomycosis and did not react with specimens from 25 healthy donors or 60 patients with various heterologous mycoses, including 18 with histoplasmosis [534]. The 96% sensitivity and complete specificity mark the EII antigen as a valuable aid for the diagnosis of paracoccidioidomycosis.

Peptidogalactomannan has been demonstrated as a cell wall component of *P. brasiliensis* [16,213]. The peptidoglycan was obtained in the water-soluble supernatant following alkali extraction and neutralization. The purified galactomannan moiety (molar ratio 1:3.8) was separated from peptide by formation of copper complex (Fehling's solution), decomposition of the complex with dilute acid, dialysis, and removal of peptide by chromatography through DEAE-cellulose. Addition of ethanol precipitated the purified galactomannan. The molecule was structured on a main chain backbone of mostly $(1 \rightarrow 6)$-linked mannose residues with branches of $(1 \rightarrow 2)$-linked galactofuranose terminal residues [16]. The purified galactomannan elicited only Arthus-type dermal reactions in immunized rabbits and developed precipitated bands of identity in ID tests with similar purified galactomannans from *B. dermatitidis, H. capsulatum,* and *H. duboisii.* The demonstration that these galactomannans occur in relatively large amounts in mycelia but only in trace amounts in yeast cells may explain the greater specificity of antigen preparations derived from yeast growth, but definitive studies have not been reported.

IV. OPPORTUNISTIC MYCOSES

The achievements of modern medicine have produced a human population with unique susceptibility to microbial infection. An extended lifetime is now common among people with debilitating diseases, genetic defects, or those requiring immunosuppressive therapy. Some lifesaving therapeutic modalities are accompanied by immunologic dysfunction with suppression of CMI responses predisposing the patient to infection by fungi as well as other microorganisms. Accumulated experience has demonstrated that the former include not only fungi with known pathogenic capability, but also a broad spectrum of fungi previously considered innocuous. A presentation of antigen production and characterization from all of these is impractical, and, therefore, only *Aspergillus, Candida,* and *Cryptococcus* will be considered since these are the most common etiological agents of opportunistic mycoses.

A. Aspergillosis

Species of *Aspergillus* are ubiquitous environmental microorganisms and humans are exposed to them constantly. These fungi cause the multiple clinical entities known loosely as aspergillosis. Although demonstrable immunological reactions vary with the type of clinical disease, antigen preparations used for evaluating a patient's immunological status are the same for each species. Hence, methods for crude and refined antigens are presented in general without reference to particular species except for choice of isolates, *Aspergillus fumigatus* being the most frequent.

1. Cultures

Current evidence indicates that antigen preparations from only a few species are required to assure maximum detection of clinical cases [66,74,152,280,281,376,378]. For example, patient sera with weak or negative reactions against *A. fumigatus* antigens are reactive with comparable extracts from *A. nidulans* [280,378], *A. terreus* [280], *A. flavus* [376], and *A. niger* [74,152,332]. This is not consistent because in other cases of non-*A. fumigatus* aspergillosis, including most of the same species, patient sera did react with *A. fumigatus* antigens [74,152,280]. Since sera from patients with *A. niger* and *A. flavus* infections are the most frequent among those nonreactive with *A. fumigatus* antigens, the use of preparations from these three species are recommended as a practical minimum number for detecting antibody to *Aspergillus* species [74,152,238]. Experimental evidence comparing antigenic relationships among species of *Aspergillus* supports this recommendation (e.g., Ref. 236). An additional consideration is significant antigenic variability among isolates of *A. fumigatus* depending on preparative methods and source material [5,252,280, 360,361,376]. A reasonable approach would be two antigen preparations from *A. fumigatus*, one extracted from mycelia and the other derived from culture filtrates.

Cultural conditions have been various. The media include complex organic substrates [74,159,280,361], peptone dialysates [61,332,488], and defined constituents [152,235, 280,449,510,539]. Incubation variations are temperatures from ambient to 37°C and with stationary or mixing conditions. Three examples comparing results obtained with antigens made under differing cultural conditions should suffice. Longbottom and Pepys [280] used filtrates of *A. fumigatus* cultured on or in Sabouraud's glucose or Czapek-Dox broth, varying time, temperature, and location of growth, i.e., surface, submerged, or constant agitation. The filtrates were evaluated with patient sera in ID tests. Shake or submerged preparations were positive with the greatest number of specimens and growth with either medium at 37°C for 3-5 weeks yielded the optimum antigen content. Kurup et al. [252] compared preparations from a single isolate of *A. fumigatus* grown in two defined broths, Czapek-DOX or Association of Official Analytical Chemists (AOAC) supplemented with glucose, as surface or shake cultures at 25°C or 30°C for up to 6 weeks. These were evaluated for total yield and number of antigens detectable with hyperimmune rabbit sera by crossed IEP. Preparations from AOAC medium were superior and the optimal conditions were shake cultures at 30°C for 7 days. Kim and Chaparas [235] compared antigen production by *A. fumigatus* in three defined media incubated at 37°C with shaking. Results were better with preparations from Long's asparagine than from Czapek-Dox or Dorset-Henley broths. Optimal antigen content in mycelia occurred in young growth (4 days) which was homogenized and disrupted in water. As antigen yield from mycelia decreased with incubation time, it increased in cultrue filtrates throughout a 6-week period.

Several factors favor use of defined media and shake culture conditions. Growth is faster, more uniform, and reproducibility is better. Since *Aspergillus* species are proto-trophic and produce a broad spectrum of antigens when grown on defined media, no special advantage is gained by using complex organic substrates unless it can be shown that significantly different antigens are produced from growth in complex media.

2. Crude Antigens

Whole cells, soluble substances from disrupted cells, and culture filtrates serve as crude antigen preparations. Suspensions of conidia agglutinate spontaneously but are satisfac-tory at carefully determined optimal concentrations for CF tests by means of which some immunological differences have been demonstrated among morphological groups of *Asper-gillus* species [299,300]. Antigens for indirect fluorescent antibody (IFA) tests include hyphae grown on slide cultures [513,514] and dried suspensions of homogenized mycel-ium [111,141,330,539]. Gordon et al. [159] noted that specific FA staining concen-trates along younger hyphae and nonspecific staining along older portions. When germ-lings are used [4,159], specific and uniform fluorescence occurs along germ tubes and around the periphery of swollen conidia not yet germinating. Specificity and staining in-tensity decrease as branching begins. The IFA test with this germling antigen is highly sensitive, detecting all of 25 cases of aspergillosis, including 5 with invasive disease. The claims for specificity, however, require confirmation because specimens were selected from proven or suspected cases of aspergillosis and nonaspergillosis cases were based only on negative ID tests for aspergillosis without any information about specimens from heterologous mycoses.

Most antigens used in immunological tests for aspergillosis are soluble substances ob-tained from culture filtrates or extracted from cells. Filtrates usually require concentra-tion before determining optimal dilutions for each immunological test, and a variety of concentration methods have been used: dialysis followed by freeze drying [42,98,140, 280,376], pervaporation [301,332,510,539], flash evaporation [539], dialysis against an adsorbent bed [361,425], and precipitation with acetone [74,157,159]. The require-ment for concentrating culture filtrates argues against the use of complex organic nutrients in media. Crude extracts from washed cells are obtained following disruption by mechan-ical methods [42,153,286,361,519], by freeze pressing [61,84,400,490], or by direct alkaline extraction of cells [263,435]. Concentration methods are similar to those used for filtrates. Some investigators [42,153,316] combine mycelial extracts with concen-trated culture filtrates. The presence of hemolytic, cytotoxic, and complement-inactivat-ing substances in crude cell extracts necessitates special treatment before use in serological or CMI studies [59,400,490]. According to Longbottom and Pepys [280], the poly-saccharide C-substance appeared in culture filtrates after 6-7 weeks at 37°C but was pres-ent earlier in mycelia. The occurrence of C-substance in culture filtrates appears variable, being present at 3 weeks [539] and absent at 6 weeks [488]. Whether production of C-substance is characteristic of certain isolates or cultural conditions has not been deter-mined, and testing each preparation is advisable.

Crude antigen preparations are used in ID and IEP tests in concentrations of from 0.6 to 60 mg/ml, and they contain multiple antigens demonstrable with human or animal sera [42,98,140,153,280,360-362,376,488]. Some of these antigens occur among other fungi. Biguet et al. [42] identified 16 antigens in a cell extract from *A. fumigatus* and 15 of these were in the culture filtrate also. One, antigen 10, was present in similar prepara-tions from all of 12 additional species of *Aspergillus*, 3 species of *Penicillium*, 1 species of

Mucor, and *Epidermophyton floccosum.* Antigens numbered 5, 8, and 13 or 15 were found among other fungus species but less commonly. In addition, immunological reactions to crude antigens of aspergilli occur among patients with heterologous mycoses [152, 280,400,510], and parasitic [316] and bacterial infections [280,425]. The possibility of dual sensitization must be considered because of the ubiquity of aspergilli. Bardana et al. [25,26] reported that sera from everyone tested exhibited primary binding in the ammonium sulfate test with a nonprecipitating mucopolysaccharide-protein complex from *A. fumigatus* and *A. nidulans.* Most investigators, however, consider the presence of multiple precipitin bands by ID or EIP as strong indication of current aspergillosis because the potentially nonspecific reactors seldom demonstrate more than one antigen-antibody precipitate by these tests. The evidence that many of these antigens are enzymes [495] and that one of two chymotrypsins [45,103] and a catalase [494] are highly specific suggests the possibility that tests with appropriate enzyme substrates would be an aid for specific diagnosis.

3. Refined Antigens

Refinement of antigens with clinical significance have followed patterns similar to those for fungi causing systemic mycoses. Initial steps include extraction of acid [42]-, alkali [42,182]-, or phenol [18,42,280]-soluble substances, and preliminary separation of protein and polysaccharide-enriched fractions by precipitation with chemicals [18,25,42,237, 252,301,434] or through various types of columns [17,42,237,280,301]. Supernatants from disrupted cells, after disc electrophoresis and differential staining of the gels, contain proteins, polysaccharides or glycoproteins, lipids, and mucopolysaccharides, many of which produce precipitates when the gel is incorporated in agar and diffused against an antiserum [25]. Trichloroacetic acid precipitates most of the proteins and cathodal mucopolysaccharides but has little effect on glycoprotein, lipid, polysaccharide, and an anodal mucopolysaccharide as seen in the stained gels. Precipitation with ammonium sulfate followed by gel permeation and/or ion exchange chromatography results in protein and polysaccharide-enriched fractions both of which react serologically [236,237,434] and elicit DH or stimulate LT [237,301]. Similarly, detergent-soluble extracts from isolated cell walls as well as intracellular substances from disrupted protoplasts contain proteins and polysaccharides or glycoproteins which are immunologically reactive [182]. Hence, culture filtrates, cell walls, and cytosols as source material all contain a variety of immunologically reactive macromolecules.

Investigators differ about whether the carbohydrate or protein of glycopeptides is the reactive moiety in serological and CMI assays and which of the two is responsible for cross-reactivity. Hearn and Mackenzie [182] reported that proteolytic enzymes abolished all reactivity and carbohydrases or periodate treatment had no effect on the antigens present in either an extract from whole cells or a detergent-soluble extract from isolated cell walls. Protoplast cytosols, however, reacted differently; periodate affected one antigen and proteolysis a second one. The binding activity of the mucopolysaccharide-protein complex of Bardana et al. [25] also was reduced by periodate oxidation and borohydride reduction but not by proteolysis. Goldstein et al. [154] also reported that their antigen fractions were affected differently by heat or by reduction and alkylation. Thus, a variety of chemically complex antigens were present in these preparations with immunological determinants attributable to the carbohydrate moiety in some and to protein in others [236]. The polysaccharide-containing antigens appeared to be responsible for cross-reactions because they reacted with antisera against similar preparations from other species of *Aspergillus* and from dermatophytes, bacteria, and vegetable dusts [355,434]. Since the polysaccharide antigens are so common, the demonstration of antibody to *Aspergillus* polysaccharide does not necessarily indicate an *Aspergillus* etiology. Nevertheless, a differential response to protein or polysaccharide antigens may have some clinical significance. Patients with aspergilloma have higher levels of antibody to protein by enzyme-linked immunoabsorbent assay (ELISA) tests while those with allergic bronchopulmonary aspergillosis or asthma have higher levels of antipolysaccharide antibody [434].

A few of the many antigens in these extracts have been isolated and characterized. Azuma and associates [17-21] precipitated alkaline [18,19]- or pyridine [20,21]-soluble extracts from defatted cells with acetone obtaining two fractions, one at 0-33% acetone and the second at 33-66% acetone. Purification steps included ion exchange and gel permeation chromatography. The 33% acetone fraction was pure glucan and nonreactive in serological tests and ST. The 66% acetone fraction contained a peptidogalactomannan with 1.01% nitrogen and a galactose:mannose ratio of 1:1.5 from the pyridine soluble and 1:3 from the alkaline-soluble extracts. In contrast to the glucan, the peptidogalacto-mannan was reactive in serological tests and elicited both Arthus and DH responses in rabbits and guinea pigs immunized with acetone-dried mycelia in complete Freund's adjuvant [20]. Exposure to papain eliminated the capacity for eliciting DH without affecting the Arthus response, and the reverse occurred following periodate oxidation. These reactions to the carbohydrate moiety have been confirmed with purified galactomannan which acted as a precipitinogen and elicited only Arthus-type dermal sensitivity [20]. The estimated molecular weights of both glucan and the galactomannan exceed 50,000 [18]. Suzuki et al. [471] and Sakaguchi et al. [407,408] also isolated galactomannans from cell extracts and culture filtrates of *A. fumigatus* and *A. niger* by alcohol precipitation followed by ion exchange chromatography. These contained no nitrogen or phosphorus and gave a single symmetrical peak by ultracentrifugation. Partial acid hydrolysis showed that these contain a mannan core with many branches of galactofuranose and oligomannan. Markedly reduced TP activity following the partial acid hydrolysis indicated that these side chains were the immunologically active sites. Low levels of TP activity with antiserum to *Candida albicans* and *Saccharomyces cerevisiae*, and the negative reaction following absorption of the antisera with yeast mannan, indicated that the oligomannan branches were more likely responsible for cross-reactions than were the galactofuranose residues [471].

The *A. fumigatus* specific antigen with chymotryptic activity [45,495], designated C_2, has been isolated by affinity chromatography [55] and immunoabsorption [537]. In the former method, the α-chymotrypsin competitive inhibitor ditryptophane methyl-ester was linked to an agarose solid phase and the absorbed enzyme eluted with 0.1 M acetic acid. Monomolecular purity was confirmed with two-dimensional IEP. In the immunoabsorption method, specific antibody was raised by immunizing rabbits with antigen-antibody precipitates dissected out of IEP slides. The specific antigen C_2 reacted in ID and IEP tests with sera from patients with aspergilloma and was nonreactive with specimens from healthy donors and patients with heterologous mycoses [534]. Although not found in all patients with aspergillosis, the presence of anti-C_2 antibodies does signify an infection caused by *A. fumigatus*.

Antigenemia occurs in 3-5 days in animal models of invasive aspergillosis, and detection of circulating antigen in patients may provide an early diagnosis [263,264,382,395,

435,518,519,528]. Although the original reports of *Aspergillus* antigen in animal and human sera used counter-IEP [263,528], sensitivity has been increased greatly with radioimmunoassay (RIA) and ELISA tests [395,435,518,519]. Radiolabeling, antigen-antibody binding, and periodate oxidation studies [435,519] indicate that a carbohydrate moiety is the reactive site of at least one of the several antigens present in sera [395,519, 528]. When antigen-containing serum from a rabbit with experimental aspergillosis was injected into normal rabbits, antibody specific for a single antigen was produced [263, 264]. This antiserum detects *A. fumigatus* serum antigen (AFSA) in patient sera and the reaction is inhibited by preabsorption of the antiserum with a mycelial extract obtained with hot alkaline borohydride. The inhibiting antigen, presumably the same as AFSA, has been isolated by affinity [concanavalin A (Con A)] and ion exchange chromatography [382]. It is a galactomannan (ratio 1:1.17) with oligogalactoside side chains of three units attached to a mannan main chain and with a molecular weight in the range 25,000-75,000. This contrasts with an estimated molecular weight exceeding 125,000 for AFSA and suggests that the galactomannan occurs in serum as a complex, probably with peptide, as indicated also by Weiner and Coats-Stephen [519]. The common occurrence and immunological similarity of peptidogalactomannans from many fungi raises the specter of nonspecific results with these highly sensitive but nondiscriminating testing procedures. The presence of sterile allergic skin manifestations during mycotic infections (e.g., dermatomycoses, coccidioidomycosis) implies that antigens circulate at some time from a distant infection site. Unfortunately, the reported specificity studies [435,519] have included other *Aspergillus* species, with which cross-reactions occur, and *Candida* mannan and *Cryptococcus* polysaccharide, both of which are not peptidogalactomannans. Nevertheless, the clinical situation in invasive aspergillosis combined with additional laboratory studies would minimize the likelihood of a nonspecific positive test, and, on a practical basis, a positive test should alert the clinician to impending complications regardless of specificity.

B. Candidiasis

A vast literature about candidiasis exists and much of it includes immunological studies. It is neither practical nor expedient to cite all of these, and reference to original sources will be highly selective. Many reviews are available which include information on antigens used in immunological tests [203,355,410,411,413,416,532], on their physicochemical nature [129,411,416,531,532], and on their cellular location [129,203,358]. Many species of *Candida* and other yeasts cause opportunistic infections, but among these *Candida albicans* is the most frequent etiological agent. The methods for preparing antigens from *C. albicans* are applicable to other yeasts and, therefore, only this species will be discussed. This yeast-like fungus is unique among pathogenic fungi because it lives as a commensal organism in the gastrointestinal tract awaiting an opportunity to infect. Low levels of humoral antibody are present in most healthy persons and dermal sensitization is so common that extracts from this fungus are used frequently as one of the recall antigens to investigate the integrity of the CMI system. In this instance, an antigen preparation with high sensitivity can produce significant clinical information but specificity is of lesser concern since the objective is to demonstrate the patency of the CMI system. Consequently, preparations containing multiple antigens are desirable for this purpose and crude extracts are used extensively.

1. Cultures

The selection of cultures for antigen production is important. *Candida albicans* shares many antigens with other yeasts [502], but at least two serotypes among strains of *C. albicans,* types A and B, can be differentiated by agglutination and absorption techniques [177]. Although reciprocal reactions occur with the two serotypes, type A appears to be the stronger and more complex immunogen. Antibodies to type A strains recognize all the antigenic determinants of type B strains, but the converse is not true, as will be discussed later. Hence, if a single-strain antigen preparation is to be produced, a type A isolate should be used.

Whole cells have served as the source of antigens with few exceptions (e.g., Refs 137 and 356). Consequently, most investigators culture *C. albicans* on complex organic media with subsequent washings of harvested cells. The question of whether yeast cells, pseudo-hyphae, or hyphae are more significant in pathogenesis is not settled, but these morphological forms differ immunologically to some extent and this can influence the results of immunological studies [126,185,479,503]. Cultural conditions favoring one or the other type of growth are reviewed elsewhere [260,336,532]. Land et al. [260] devised a glucose biotin phosphate buffer medium (pH 7.2) in which addition of proline favors filamentous growth and substitution of ammonium chloride for proline induces yeast cell formation.

2. Crude Antigens

Whole cells are used for both inducing [177,179,475] and eliciting (reviewed in Refs. 155, 203, and 532) immunological responses. Immunological assays with whole cells presumably involve cell surface antigens, and some of these occur in other microorganisms. The ubiquity of *C. albicans* and other yeasts among humans and the widespread distribution of common antigens [502], even to bacteria [3,298; cited in Ref. 210], make it imperative that interpretation of positive reactions with whole cells be corroborated by clinical, pathological, and cultural studies. In contrast to the qualified value of whole cells as serological aids for diagnosis, such preparations are useful for the identification and immunological typing of isolates. Hasenclever and Mitchell [177] defined two antigenic groups among isolates of *C. albicans* by agglutination, demonstrating that serotype A is closely related to *C. tropicalis* and serotype B to *C. stellatoidea* [180]. Gordon et al. [158] differentiated each species in these two pairs, and also the antigenically related *Torulopsis glabrata,* with fluorescein-labeled anti-*C. albicans* A and anti-*T. glabrata* immunoglobulins, the former absorbed with *C. stellatoidea* and the latter with *C. albicans* A. Tsuchiya et al. (summarized in Ref. 502) developed a Kauffman-White type of system for identifying 88 species among 12 genera of yeasts based on heat-stable or heat-labile antigens differentiated by agglutination and reciprocal absorption of agglutinins. It is apparent that the yeast cell surface contains multiple antigens, some of which are distributed widely.

Solubilized antigens generally are obtained from cells rather than culture broths, although several reports indicate the presence of potentially significant antigens in culture filtrates [356,450,478,493]. Methods vary so much that comparisons of results from different investigators would be unreasonable. Cells have been disrupted by sonic energy [41,185,263,451,483,540], pressure extrusion [142,327,401], freeze-thaw cycles [127], shaking with glass beads [186,312,458,479,496,515], grinding with sand [262], heating cells [143,261,409], and by lysing protoplasts [142,478,479,505]. Fluids for extracting antigens include distilled water [127,483], physiological concentration of salts [142,143,

186,262,401,458,479,496,540], and hypertonic salt solution [291]. The relative sensitivity and specificity of some of these preparations have been considered in a few reports. Laskownicka et al. [262] compared four crude antigens, three from cells by sonic disruption, grinding with sand, or heating at 56°C for 72 h, and one from culture filtrate by acetone precipitation. These were analyzed by ID with the conclusion that differences were quantitative rather than qualitative. Syverson and Buckley [478] used two-dimensional IEP to compare cytoplasmic antigens obtained from lysed spheroplasts and from mechanically disrupted mycelial or yeast cells and made two significant observations. First, the so-called cytoplasmic antigens prepared by disruption and centrifugation (40,000 g for 1.5 h) were contaminated with antigens from the cell wall. Other investigators [505] using less sensitive analytic techniques did not detect cell wall antigens in cytoplasmic fractions. This bears on the contradictory reports about whether cytoplasmic antigens detect invasive candidiasis since antibodies to cell wall antigens are widespread in healthy populations. Second, the mycelial and yeast cell preparations contained at least six antigens unique to the respective morphological forms, suggesting that the type of cell used as source material was important for detecting significant host responses [479]. This suggestion is partially supported by a recent report [187] comparing cytoplasmic (pressure extrusion)- and cell wall-derived (citrate buffer extraction and methanol precipitation) preparations from yeast and mycelial forms. Among these four preparations, and a fifth commonly used commercial preparation, specificity is equivalent, but the yeast cytoplasmic preparation possesses the greatest sensitivity for differentiating patients with systemic candidiasis from those with superficial disease or no evidence of infection. It must be noted, however, that growth conditions, antigen preparatory methods, and the serological test (counter-IEP) differed from those used by Syverson and associates. The question of whether patients with systemic disease can be detected by responses to cytoplasmic antigens is confounded further by evidence that ethylenediamine extracts from separated cell walls as well as cytoplasmic antigens from both cells and spheroplasts produce stimulation of lymphocytes from patients and healthy donors, half of whom are negative by ST [142]. Although ST results did not correlate with LT, leucocyte migration inhibition by a similar cell wall extract showed an association with ST results [357].

Additional studies have demonstrated that many of the cytoplasmic antigens possess enzyme activity [496] and that a ribosomal fraction induces in rabbits antibodies which form precipitation bands of identity with a cytoplasmic extract and its antiserum [431].

3. Refined Antigens

Isolation of purified antigens has focused primarily on the cell wall mannans and peptidomannans. Several reasons provided the impetus for these studies. Taschdjian et al. [483], based on their own studies and those of others [46,396], postulated that antibody formation against cytoplasmic (somatic) antigens of *C. albicans* is induced by lysis of fungous cells in the infected host. Studies by other investigators have established that precipitins are found in sera from most persons without evident infection but that these are primarily reactions with cell wall mannan antigens [71,356]. Hence, precipitins to purified cell wall mannan antigens need to be distinguished from antibodies to cytoplasmic antigens. Second, the role of mannans or peptidomannans in the structure and plasticity of yeast cell walls (reviewed in Refs. 150 and 358) is of particular interest because of the question of whether hyphal or yeast forms are invasive. Third, the mannans appear to be the principal cell wall antigens involved as antigenic determinants of *C. albicans* serotypes A and

B and as the molecular components responsible for cross-reactions with other species [23, 176-180,358,461]. Finally, mannan antigenemia is an early event of invasive candidiasis in experimental candidiasis [374,516-519] and humans [15,137,430,516,518,520] and a potentially useful aid for early diagnosis. Thus, the isolation of purified mannan and its structure have been studied intensively.

The earlier extensive literature about yeast mannans has been reviewed by Phaff [358]. Most investigators prepare mannans by the method of Peat et al. [351] to avoid the degradation of the oligosaccharide that occurred with the earlier methods of extraction with hot alkali. Yeasts, suspended in neutral citrate buffer, are autoclaved, the extracted soluble materials concentrated under reduced pressure, and acetic acid added to normality. The resulting gelatinous material is removed by centrifugation, the solution neutralized with NaOH, and crude mannan precipitated with ethanol. After dissolving in water, addition of Fehling's solution separates mannan as an insoluble copper complex from glycogen. The precipitated mannan salt is suspended in water and solubilized with minimal amounts of concentrated HCl. Successive precipitations with alcohol in the presence of acetic acid yield mannan with a trace of glucose, probably as a contaminant from cell wall glucan because acid hydrolysis produces no evidence that glucose is combined with mannose in the polysaccharide. This product can be separated into neutral and acidic mannans by treatment with Cetavlon [522] or by chromatography on DEAE-Sephadex (acetate form), eluting first with water to obtain neutral mannan followed by gradient or stepwise elution up to 1 M NaCl, which yields the acidic mannan [342,406]. Both fractions contain mannose as the only sugar component and the neutral fraction is homogeneous by ultracentrifugation and moving boundary electrophoresis. The neutral mannan contains no phosphorus and insignificant amounts of nitrogen, but both phosphorus and nitrogen are present in the acidic mannan, a peptidomannan [342,406].

Several groups of investigators have analyzed the *C. albicans* mannan structure. Almost all agree that the core is composed of α-(1 → 6)-linked mannose units with side branches joined to the core by α-(1 → 2)-linkages (reviewed in Ref. 358). All mannose units occur in the α-D-mannopyranosyl form. In contrast to the uniform core, structural variations occur among the oligomannan branches and these determine immunochemical specificities [406,462,463,469,470]. The side chain mannopyranose oligomers range from mannobiose to mannohexaose in *C. albicans* type B and to mannoheptaose in type A, with some evidence for even larger oligomers in the latter [48,438,457,462,469,470]. The α-(1 → 2)-linkages predominate but a small proportion of α-(1 → 3)-linkages occur in the mannotetraose of type A and in the mannopentaose and mannohexaose of both serotypes in ratios of α-(1 → 2)-linkages to α-(1 → 3)-linkages of 2:1, 3:1, and 4:1, respectively [457,462,463,469,470]. Quantitative precipitation inhibition tests with certified oligomannans containing only α-(1 → 2)-linkages establish that inhibition power parallels branch length. Mitchell and Hasenclever [308] reported similar results with mannan from *C. stellatoidea*. Although inhibition increased rapidly beginning with mannobiose, the increments between successive larger oligomers decreased, with little difference in inhibitory power between the hexaose and heptaose, suggesting that these two larger oligomers are the dominant antigenic determinants and approach the upper limit in size for the antibody combining site [462,463]. The presence of an α-(1 → 3)-linkage represents an additional influence on antigenic determinance since the inhibitory power of the mannohexaose containing an α-(1 → 3)-linkage is greater than the mannoheptaose composed of only α-(1 → 2)-linkages [469,470]. The relative proportion of long to short chains is an additional factor influencing immunochemical specificity. Although the structure of the mannans from *C. albicans* types A and B are essentially the same, type A mannan contains a higher proportion of long oligomers, averaging 5.0 mannopyranosyl units in chain length, compared to type B mannan, with an average of 3.3 residues [462,463]. Sunayama [462] and Sunayama and Suzuki [463] postulated that the greater density of long-chain oligomers in type A mannan explained the fact that antibody to this mannan differentiated between cells of serotypes A and B whereas antibody to type B mannan did not. The greater size of the combining site on anti-type A mannan would accommodate all the antigenic determinants of both mannans and differentiate between them. In contrast, the smaller antibody combining site for type B mannan would accommodate only determinants consistent with those of type B mannan and would not combine with that portion containing the greater density of larger oligomers in type A mannan. The question of whether a combining site for a smaller antigenic determinant would fail to differentiate this from a larger cross-reacting antigenic determinant has been explored with Con A, which has a combining site for only three mannopyranosyl residues [468]. The results provide indirect evidence supporting the hypothesis because Con A precipitates virtually identical amounts of mannans from *C. albicans* serotypes A and B, *C. tropicalis, C. stellatoidea,* and *S. cerevisiae.* Therefore, these mannans are the dominant antigenic determinants differentiating the cells of *C. albicans* serotypes and responsible for many of the cross-reactions with other species [176,178,406,461,471,480,541].

Although crude extracts from *C. albicans* contain many antigens [13-15,221,478,479], the proteinaceous components and the acidic peptidomannans have not been studied as intensively as the polysaccharides. Fractions enriched in protein composition are usually obtained from cells or cell walls by alkaline extraction [110,211,221,305,385] and from culture filtrates or cytosols by precipitation with ammonium sulfate [119,221,297,305, 522]. In almost all of these reports, tests for homogeneity were not done and immunological assays revealed multiple components (e.g., Refs. 119, 221, 305, and 385). The acid peptidomannan fractions eluted with NaCl from DEAE-Sephadex (acetate form) contain phosphorus and peptides [342,406] and develop only a single precipitate identical with neutral mannan in ID tests. Since rechromatography produces a neutral fragment from the acidic one, it appears that the acidic fraction is a peptidophosphomannan with degradation splitting off a neutral mannan. Okubo et al. [342] obtained five fractions by eluting first with water and then stepwise with NaCl up to 0.2 M. The salt-eluted fractions all contained protein and increasing amounts of phosphate, with the latter being directly proportional to the antibody nitrogen precipitated. If the phosphate linkages occur in the core of the mannan moiety, as it does in most of the yeast mannans (reviewed in Ref. 358), this could explain the splitting off of neutral mannans on rechromatography.

Glycopeptides have been isolated from defatted cell walls by Reiss et al. [385]. Two soluble fractions were obtained, the first by extraction with cold 0.5 N NaOH and the second by sonicating the cell wall residue suspended in the cold dilute alkali. The first fraction contained glucose and mannose in a ratio of 2:3, plus 1.47% nitrogen and 0.05% phosphorus (i.e., a peptidoglucomannan). The second fraction contained glucose and mannose in a ratio of 6:1, plus 0.94% nitrogen but no phosphorus. This fraction was considered to be a mannoglucan probably with a peptide moiety. Permeation chromatography indicated molecular weights of $>5 \times 10^6$ for the first and $>1.3 \times 10^4$ for the second. Both fractions elicited DH and produced MMI in an immunized animal model, but each contained more than one antigen by ID and IEP assays. The extraction method differs slightly from that reported earlier by Kessler and Nickerson [233], who used 0.5 N NaOH at room temperature but also obtained two peptidoglucomannans: one with a high peptide content and precipitable with ammonium sulfate, the other with a low peptide component and nonprecipitable with saturated salt solution. Peptidoglucomannan complexes have also been obtained from cytoplasm of mechanically disrupted yeasts by precipitation with ammonium sulfate [119] or ethanol [305]. Con A-Sepharose chromatography separates a protein complex in a buffer eluate followed by mannan eluted with α-methylmannoside. Glucose is absent from both fractions when sought by gas chromatography [119]. Whether these various glycopeptides are truly peptidoglucomannans or peptidomannans mixed with glucan has not been resolved. Nickerson [336] pointed out the difficulty of separating peptide residues proximal to a polysaccharide moiety, and Phaff [358] agreed with Bishop et al. [48] that criteria for homogeneity of the "glucomannan-protein complexes" have been insufficient. Since a glucan fraction is extracted with cold alkali, these complexes could be mixtures of an alkali-soluble glucan and peptidomannans [358]. On the other hand, the facile separation of the glucose moiety in the cytoplasmic peptidoglucomannan by Con A-Sepharose chromatography could be due to assembly of the components into the intact molecule at the cell membrane or wall. The glycopeptides are present not only throughout the cell wall, but also on the cell membrane [68,294]. Farkas [129] has reviewed current knowledge about the mode of attachment of yeast mannans to peptides. In one type, the mannan is attached by an O-glycosidic linkage to serine or threonine in the peptide moiety, but in the second type the mannan is linked to an asparaginyl residue through a diacetylchitobiose bridge by an N-glycosidic bond. Weak alkali liberates the mannan from the O-glycosidic bond by β-elimination, but the N-glycosidic linkage is more resistant to alkali. This construction would explain the frequent isolation of a pure polysaccharide and a glycopeptide by the methods that have been used.

C. Cryptococcosis

Immunological responses of hosts infected with *Cryptococcus neoformans* are receiving more attention now that it is apparent that the frequency of cryptococcosis is increasing among patients with depressed CMI responses. This fungus grows as a yeast in the infected host and in the laboratory at both ambient and incubator temperatures, and is notable for producing a capsule. The quantity of capsular material produced in the host may be so great that it can be found as free antigen in body fluids. Thus, immunological tests must include assays for detecting antigenemia as well as for humoral and CMI responses. Therefore, the discussion of antigen production will include a preparation used to immunize animals for raising antisera that will detect antigen in patient specimens.

1. Cultures

Cryptococcus neoformans grows only as a yeast, usually encapsulated, under usual laboratory conditions, and both cells and culture filtrates have served as source material for preparing antigens. Selection of isolates should recognize the existence of serotypes within this species. Originally, Benham [35] demonstrated by reciprocal absorption-agglutination studies that all pathogenic isolates of *Cryptococcus* species belonged in her serogroup C. Evans [121,122] differentiated 19 of these group C isolates into three antigenic types by agglutination-absorption and by the capsular reaction described by Neill et al. [334]. Finally, Wilson et al. [530] distinguished four serotypes, A through D, among 106 isolates by whole cell agglutination and absorption procedures. Their serotypes A, B, and C correspond to isolates typed originally by Evans. The validity of these serotypes is supported by studies involving natural distribution, biochemistry, and sexual compatibility [38,39, 254-257]. Kwon-Chung described a species-linked relationship when she demonstrated that successful matings to produce a viable perfect form, genus *Filobasidiella*, were obtained with serotypes A and D for *F. neoformans* and with serotypes B and C for *F. bacillispora*.

Cultural conditions have varied so widely that it is pointless to do more than generalize. Whole cells to be used directly as antigens or as the source for antigen preparations are grown either on gelled or in liquid media varying from complex organic substrates to chemically defined constituents. Incubation temperatures range from ambient to 37°C, with cultures either stationary or shaken. Similar liquid media and cultural conditions are used for obtaining culture filtrates. According to Littman [275], optimal capsule synthesis for all of 40 isolates occurs in sodium glutamate basal salts medium with added thiamine (1 μg/ml), maltose (5 mg/ml), and sucrose (5 mg/ml). Bulmer and Sans [60] found that capsule production was improved markedly by culturing at pH 7.5 compared to pH 5.0, and that a variety of carbohydrates could be used but glucose and mannose yielded the thickest capsules.

2. Crude Antigens

Whole cells, either killed or viable, have been used as vaccines for immunization or as antigens for detecting immunological responses (reviewed in Refs. 246, 266, and 416). In general, cells with small capsules are more effective as immunogens, but cells with varying amounts of encapsulation are used in serological tests. The latter are obtained by selection [249,333], by culture on soil or special media [60,114,205], and by enzymatic [145-148] or chemical [35,506] degradation of the capsule. Antibody for detection of cryptococcal antigen in body fluids is prepared by short-term, intravenous immunization of rabbits with formalin-killed and washed suspensions of lightly encapsulated cells [51,333; for the detailed procedure, see Ref. 345]. The immunoglobulins for sensitizing latex particles are precipitated from antisera with ammonium sulfate, dialyzed, and made up to 4 g of protein per deciliter in buffered saline, pH 8.4. The cells used for immunization can serve also for TA after adjustment to optimal concentration [160,227,345,509]. Capsule size is not a critical factor in determining virulence for the same strain [113,114], or when cells are used as antigens in either direct or indirect FA procedures [138,219], although Vogel [506] reported that staining by IFA was stronger and more specific after decapsulation of cells with acid.

Whole cells are used also for determining the capsular serotypes of *C. neoformans*. Although soluble capsular substance can be found in cell-free culture supernatants, much remains bound to cells, apparently as a gel, even after prolonged storage and repeated washings [123,161]. According to Evans [123], this bound capsular substance appears to be a concentration effect because more than 10 mg/ml of a soluble preparation gels in water or saline and does not dissolve again during 100 rinses with saline. The nature of the bond holding the capsule to the cell wall is not known definitively. Kozel and colleagues [248,250,302,477] presented evidence that the capsular polysaccharide was bound to the cell wall by specific receptors, and that the presence of a capsule did not prevent binding of opsonizing immunoglobulin to the cell wall. Therefore, even though the polysaccharide occurred as a gel proximal to the cell wall, it was still permeable to large molecules. Goren and Middlebrook [161] proposed that the polysaccharide occurred as a lightly cross-linked gel and that this would be consistent with its known "physical integrity, essential insolubility in aqueous media, high permeability, and reversible collapse under the influence of simple cationic substances."

Soluble antigens, cryptococcins, are produced from culture filtrates and by extraction of cells. Murphy and associates [69,319-321] obtained culture filtrates from cells grown in a dialysate broth and killed with formalin prior to centrifugation. These filtrates are reactive in tests for humoral antibody and elicit specific DH when tested on guinea pigs infected with *C. neoformans, B. dermatitidis, H. capsulatum, Sporothrix schenckii,* or *Cryptococcus albidus.* Sequential ultrafiltration reveals that the major activity for eliciting DH occurs in the fraction with molecular weight greater than 50,000, and that this fraction contains 7.5 mg of protein per milliliter, 3.1 mg of carbohydrate per milliliter, and two bands by disc electrophoresis, one a glycoprotein and the other polysaccharide. After extraction from the electrophoresis gel, the polysaccharide is nonreactive, but the glycoprotein elicits DH equally in guinea pigs sensitized with viable *C. neoformans* types A, B, or C. This glycoprotein, therefore, is species- but not type-specific. Cryptococcin has been extracted from cells either before [12] or following disruption [12,37,166,167,279,419]. Urea extraction of viable cells followed by exhaustive dialysis against saline yields a product with sensitivity but relative nonspecificity against blastomycosis, coccidioidomycosis, and histoplasmosis [12]. In addition, ST with this preparation induces antibody formation in at least 30% of individuals who had been serologically negative earlier [47,104] and significant LT in at least 25% of healthy subjects [166]. Similarly, alkaline extraction (0.1 M glycine buffer, pH 10) of decapsulated cell walls (pressure extrusion) elicits DH in guinea pigs sensitized by infection with *C. neoformans* but not in those infected with *Candida albicans* or *H. capsulatum* [419]. The same preparation, however, elicits DH reactions among healthy people (32% of 82 persons) in an area where cases of cryptococcal meningitis have occurred [318]. Since this is also a known endemic area for histoplasmosis, with 55% reactors to histoplasmin among the same subjects and with 18 of the 82 people reacting to both antigens, the specificity of this cryptococcin remains uncertain. It is possible, although not established, that clinically inapparent infection by *Cryptococcus neoformans* may occur in this area with a frequency similar to those in endemic areas of coccidioidomycosis and histoplasmosis. Graybill and Taylor [167] obtained a soluble antigen as the supernatant from heat-killed and disrupted cells. This antigen induces and elicits DH in mice which do not respond significantly to soluble antibens from *Candida albicans, Coccidioides immitis, H. capsulatum,* and *T. glabrata.* In addition, immunization with this antigen significantly extends survival time of challenged mice but does not protect them from ultimate death.

3. Refined Antigens

The antigenic activity of the *Cryptococcus neoformans* capsule has received particular attention because it is involved in serotyping, in inhibition of phagocytosis, in immunological unresponsiveness, and in antigenemia (reviewed in Ref. 416). Several procedures are used to prepare capsular substance, an acidic heteropolysaccharide. The more common methods are modifications of the Heidelberger and Sevag method for pneumococcal polysaccharides [210]. The source material usually is culture filtrates from yeasts grown on defined components [128,144,285] or dialysate broths [36,249], although filtrates from complex organic broths [125] and supernatants from lightly sonicated cells [60] are used also. The essential steps in the Heidelberger-Sevag procedure require precipitation of polysaccharide and some protein with ethanol in the presence of sodium acetate and acetic acid, denaturation and removal of protein in an emulsion layer formed by shaking with chloroform-butanol, and reprecipitation of polysaccharide from the aqueous layer with ethanol and electrolyte. The final product should be soluble in distilled water. Removal of protein has not been included by all investigators and others have used differing volumes of ethanol which could result in precipitation of several molecular species of polysaccharides [60,128,210,249].

The molecular nature of the capsular polysaccharide has been studied extensively. Purified polysaccharide analyzed by enzymatic [148] or chemical [49,112,124,125,249, 309-311,377] degradation usually contains mannose, xylose, glucuronic acid, and galactose. In some cases, glucuronolactone is detected but only after extended hydrolysis [49, 125], and in others galactose is not found [112,309-311]. The presence or absence of galactose might be due to differences in preparatory methods or to more than one polysaccharide precipitated with ethanol. Rebers et al. [377] found an increased proportion of galactose in precipitates formed with cryptococcal polysaccharide type A and antipneumococcal type XIV serum but not with antipneumococcal type II serum, indicating that galactose is an integral part of the fungal molecule. Hay and Reiss [181] reported that the polysaccharide forms at least two precipitates in ID tests. The molar ratio reported for mannose, xylose, and glucuronic acid from an unspecified serotype [309-311] was 3:1:1 and for mannose, xylose, glucuronic acid, and galactose from a serotype B isolate [49] was 6:4:2:1. These differences could be procedural since polysaccharide was obtained in the former case from a 30% KI, 1% K_2CO_3 extract of cells followed by alcohol precipitation and deproteinization, but in the latter direct ethanol precipitation without deproteinization was used. Procedural differences may account also for differing molecular weights reported: $6.6\text{-}15 \times 10^3$ [128,309-311] and 1×10^6 [149,249]. Miyazaki [309-311] proposed a molecular structure consisting of a $(1 \rightarrow 2)$-linked mannan backbone with $(1 \rightarrow 4)$-linked mannopyranose branches and the remaining carbohydrates as terminal units. Levine et al. [274] found quantitative but not qualitative differences among polysaccharides from types A, B, and C by infrared spectroscopy and suggested that varying degrees of acetylation may be important for serological type specificity. Thus, according to current information, the capsule of *C. neoformans* contains one or more acidic heteropolysaccharides with a branched mannan structure and with other monosaccharides as terminal units and probable determinants of antigenic type specificity. The similarity of this structure to those found in polysaccharides from other yeasts and some bacteria probably accounts for the cross-reactions that have been reported.

The evidence that the polysaccharide occurs in a capsule which is not readily removed [60,123] presents a problem for studies of cell wall antigens. Decapsulated cells have been prepared by sonic oscillation [279], by pressure extrusion [82,313], and by enzymatic [145-148] or chemical [35,506] degradation. Cells without capsules can be grown by selective culturing methods. Substantial evidence (e.g., electron microscopy, specific FA staining) that the capsule has been eliminated should be provided.

The antigenicity of subcellular fractions has been examined by Hay and Reiss [181]. A type A, small capsule isolate was disrupted by shaking with glass beads and the fractions obtained by differential centrifugation. Assays for DH on mice sensitized by infection demonstrated the greatest activity in the original culture filtrate and the postmitochondrial supernatant, which contained ribosomes, protein, and carbohydrate. Other fractions in decreasing order of activity were a ribosome-enriched fraction, a turbid 100,000 *g* supernatant, and a mitochondrial-membrane fraction. All contained appreciable amounts of carbohydrate as well as protein and most elicited 4-h reactions, presumably of the Arthus type, in addition to DH. Further studies with the postmitochondrial supernatant demonstrated MMI with mouse peritoneal exudate cells and DH in immunized guinea pigs which was specific when tested on animals sensitized to *B. dermatitidis, Candida albicans,* or *H. capsulatum.*

The molecular nature of immunologically active preparations other than the capsular polysaccharide has not been studied extensively. The DH-eliciting activity in the culture filtrate studied by Murphy and Pahlavan [321] had a molecular weight greater than 50,000, was a glycoprotein, and showed no difference with respect to serotypes A, B, and C (serotype D isolates were not studied). All the subcellular fractions prepared by Hay and Reiss [181] contained both carbohydrate and protein. The 100,000 g supernatant produced two or three precipitation bands in ID tests and one of these was identical to one of the two formed by cryptococcal polysaccharide, a probable example of the advisability of using unencapsulated cells as source material for preparing subcellular fractions. At the present time, therefore, these preparations contain multiple antigens and their molecular characterization is uncertain.

V. INOCULATION MYCOSES

The three mycoses included in this group are chromomycosis, mycetoma, and sporotrichosis. The most common clinical form of all three begins with traumatic injection into or through the epidermis followed by subcutaneous pathological changes and, usually, associated dermal lesions. Thus, specimens are readily available for diagnosis by direct microscopic examination and culture, and immunological aids are secondary and frequently unnecessary. The use and value of immunological tests are compromised additionally by the multiple etiological agents causing chromomycosis and mycetoma; only sporotrichosis is caused by a single species. Even though immunological procedures have had limited use as diagnostic aids, they have been employed for several objectives. Host responses have been examined to explore the role of immunology during pathogenesis. Fungal antigens have been useful as aids for studying taxonomic relationships and may prove valuable for understanding fundamental differences between pathogens and taxonomically related nonpathogens. Finally, the organisms as biological entities with an impact on human life are valid subjects for basic studies, and immunological procedures can be useful for locating cellular sites of antigenic molecules and for studying subcellular changes during the life cycle of the fungus.

A. Chromomycosis

1. Cultures

Several species of dematiaceous fungi cause chromomycosis and most of these have been included in studies of antigenic relationships among the pathogens and related saprobic species in addition to the use of antigens for evaluating host responses to infection. Strains used for antigen production are those readily available either as stock cultures or as recent isolates from pathological specimens. These are grown usually on either complex organic [4,58,473] or partially defined media such as Czapek-Dox supplemented with a casein digest [83]. Incubation temperatures vary from ambient to 37°C [58,83,473], and growth conditions include stationary [58], shaking [473], or shaking with forced aeration [83]. Whether the variety of strains and cultural methods influence antigen production has not been determined.

2. Crude Antigens

Killed whole or disrupted cells as well as soluble antigens have been used for intradermal and serological tests [4,43,58,75,83,156,296]. In most cases the antigen preparation derives from fungous cells grown on complex organic nutrients. Conant and Martin [75], for example, scraped off the growth from Sabouraud's glucose agar, ground the cells in a mortar, and prepared a suspension in saline. This was shaken with glass beads for 1 h and used for CF tests. Al-Doory and Gordon [4] grew the fungus in Sabouraud's glucose broth. Cells were killed with formalin, washed, and suspended in distilled water as the antigen preparation for FA staining. In addition, cross-reactive antibodies were absorbed from sera with disrupted cells that had been heat killed, washed, dried, and ground in a mortar. Antigens for ID tests are prepared from broth filtrates by precipitation with acetone and dissolving in saline to one-half the original volume. Other investigators [43,58, 83] also use solubilized antigens either extracted from disrupted cells or concentrated from broth filtrates. Similar preparations from several pathogenic and taxonomically related saprobic species have been evaluated by ID and IEP. All investigators find interspecies relationships among the pathogens through common antigens and, in some cases, even with the saprobic species. In general, these studies support taxonomic relationships established by morphologic criteria. Discrepancies among these results probably are due to variations among strains used, methods for preparing antigens, standardization (or lack of it), and in procedures for analysis of antigenic content.

3. Refined Antigens

The major focus has been on carbohydrate antigens. Glucans and galactomannans are prepared from either cells or culture filtrates [472,473,480-482]. Killed cells are extracted with phosphate buffer (pH 7) at 135°C, and polysaccharides from cell extracts or culture filtrates are precipitated with ethanol. Glucan is separated from galactomannan by fractionation on a column of Sephadex G-75. Subsequent chromatography of the galactomannans on DEAE-Sephadex A-50 yields three subfractions obtained by stepwise elution with water, 0.1 M NaCl, and 0.5 M NaCl. Assays for TP activity with homologous anti-whole cell sera demonstrate that the glucans are inactive and that the galactomannan fractions eluted with 0.1 M NaCl contain the strongest activity. The latter also elicit passive cutaneous reactions in guinea pigs sensitized with the antisera. Comparative assays by ID and quantitative TP tests show that acid-labile galactofuranosyl residues are required for TP activity, but the acid-treated galactomannans still retain some activity by passive cutaneous anaphylaxis. These investigators concluded that the galactofuranosyl side chains function as antigenic determinants and are responsible for cross-reactions among these fungi because identical reactions are found by ID and similar curves are obtained with quantitative TP assays.

Chemical analysis indicates a total carbohydrate content of 90-94% with mannose: galactose:glucose ratios in the ranges 1:0.66-1.15:0.06-0.35. Hexosamine is present at less than 1% in all fungous extracts and protein content (Lowry method) varies from 1.65% to 5.50%. Mild acid hydrolysis reduces significantly the galactose content and yields higher values of specific rotation, indicating that the acid-labile galactofuranosyl residues are β-linked. The presence of some glucose and protein in the galactomannan fractions indicates that these preparations are not pure galactomannan. Digestion of the protein moiety with pronase does not change the quantitative TP curves, but mild acid hydrolysis results in loss of reactivity. Hence, Suzuki and Takeda [473] concluded that the acid-labile galactofuranose residues were solely responsible for the TP activity. Since acid hydrolysis released only 50% of the total galactose, these investigators suggested that the remaining galactose was present as galactopyranose residues which were responsible for cross-reactions with heterologous fungi. It should be noted, however, that enzyme digestion removed only 70% of the protein and the potential immunological role of the nitrogenous components was not studied.

B. Mycetoma

1. Cultures

Although this disease is similar to chromomycosis in the sense that both are clinical entities caused by many etiological agents, the problems in mycetoma are more complex because some of the microorganisms are eumycetes and others actinomycetes. When cultures are not available, immunological tests with appropriate antigens can help resolve the question of etiology [32,282]. The isolates used for antigen production have varied among investigators and no reference strains have been designated. Since both eumycetes and actinomycetes are involved, the species or the group of microorganisms used and the cultural conditions for producing antigens are presented in the following section in which crude and refined antigen preparations for each group are combined.

2. Crude and Refined Antigens

The fungus *Petriellidium boydii* has been studied most extensively among the eumycete agents of mycetoma. Lupan and Cazin [282] obtained antigenic fractions from liquid cultures. The fungus was grown stationary in glucose yeast extract dialysate, killed with formalin, and filtered. After concentration by pervaporation and dialysis against distilled water, the filtrate was brought into equilibrium with 0.15 M NaCl at an approximate 50-fold concentration. The crude concentrates from six strains were fractionated by ion exchange chromatography on DEAE-cellulose followed by gel permeation through Sephadex G-100. Four of the six strains yielded only a single major peak (antigen 1) off the ion exchange column, and two of the six had a second trailing peak (antigen 2). Antigen 1 was composed of 90-95% carbohydrate, 3-4% protein, and 2-5% nucleic acid. Antigen 2 contained approximately 70% carbohydrate, 16% protein, and 4% nucleic acid. The gel permeation chromatography separated antigen 1 into three subfractions. Although the chromatographic methods differentiated a total of four fractions (three from antigen 1 plus antigen 2), all were immunologically identical by ID evaluation with antisera from infected rabbits. Antigen 1 was a sensitive reagent for detecting antibody by indirect hemagglutination among both patients and experimental animals infected with *P. boydii.* It was also specific with respect to antigen preparations from 9 heterologous fungi and antisera from 17 patients with other mycoses [283,284]. Bell [33], using the methods of Murray and Mahgoub [322] for antigen preparation (discussed later), demonstrated at least 10 antigen-antibody reactions by ID with antisera raised by infecting rabbits. Although there were more antigens than those found in the carbohydrate-enriched antigen 1 of Lupan and Cazin, the specificity of these multiple antigens and their sensitivity for detecting naturally acquired petriellidiosis were not evaluated. Ziegler et al. [547] ex-tracted acetone-dried cells separately with ethylene glycol, glycerol, and α-naththol, re-sulting in solutions containing glycoproteins, polysaccharides, proteins, and nucleic acid. All were reactive by IFA, CF, and LT tests in which material from rabbits hyperimmunized by injecting dried mycelium intradermally was used. But sensitivity and specificity for naturally acquired infection was not investigated.

Mahgoub and associates [288,322] prepared antigens from both eumycetes and actino-mycetes by four methods. The eumycetes were grown in glucose peptone broth, the actin-omycetes on glucose peptone agar. Harvested cells were broken by shaking with glass beads, and the homogenized cells were used either directly as suspensions in saline (method 1), as a saline extract (method 2), or precipitated from the saline extract with acetone and re-dissolved in saline (method 3). The separated broth from the eumycete cultures was treated with acetone and the precipitate redissolved in saline (method 4). ID tests with patient sera and antigens from three eumycetes and five actinomycetes demonstrated that antigens prepared by methods 3 and 4 detect a few more antibodies. These methods were modified later by homogenizing the cells in phosphate buffered saline, exposing them to ultrasonication, and concentrating the extract [289]. Results with sera from patients and infected mice demonstrated species specificity with antigens from eumycetes but marked cross-reactivity among actinomycete antigen solutions. This confirmed earlier reports by Seeliger [428,429] who used alcohol-precipitated antigens. Mahgoub also reported intra-species strain variations expressed as qualitative differences in antigen preparations [289].

Cytoplasmic antigens from disrupted cells (method 3) were used to evaluate serological methods for detecting antibodies against the many etiological agents of mycetoma. Coun-terimmunoelectrophoresis (CIE) proved more sensitive and equally specific compared with ID, and ELISA equaled CIE in sensitivity but was somewhat less specific [172,303]. Some of the cross-reactivity to ELISA assays was reduced by centrifugation at 20,000 g for 30 min to remove particulate substances which may have been responsible for nonspecific adsorption.

Antigenic preparations from actinomycetes have been used by many investigators not only for mycetoma, but also for pulmonary and systemic infections by these microor-ganisms (reviewed in Refs. 240, 411, 412, and 504). Bojalil and colleagues [54,343,344, 545] used cells which were grown in asparagine-minerals broth, defatted with methanol-acetone, and ground in a mortar. Crude polysaccharide was obtained from a cell-free super-natant by precipitation with methanol. Two polysaccharide fractions were prepared by differential solubility in 0.02 M carbonate-bicarbonate buffer (pH 10) and 0.02 M citrate buffer (pH 5), yielding an alkali-soluble polysaccharide (poly I) and an acid-soluble poly-saccharide (poly II). Additional precipitation with TCA yielded material composed mainly of protein. The poly I substances from *Nocardia asteroides* and *N. brasiliensis* were very similar chemically and immunologically, containing arabinose and galactose in molar ratios of 2.1-2.6:1 for *N. asteroides* and 2.6-3.1:1 for *N. brasiliensis*. There was, however, two to four times as much protein in preparations from *N. brasiliensis*. Poly I preparations from both actinomycetes yielded identical reactions in ID tests and absorbed antibodies from both homologous and heterologous antisera raised in hyperimmunized rabbits. The poly I antigens also reacted with antibodies in sera from some patients with lepromatous leprosy and from 58 of 71 patients with tuberculosis [54], confirming earlier reports of common antigens among *Nocardia* and *Mycobacteria* (reviewed in Refs. 54 and 545). The poly II fractions contained arabinose, galactose, and also mannose in molar ratios of 1.2-

1.5:2.0-4.5:1 for *N. asteroides* and 4.3-4.9:1.7-2.0:1 for *N. brasiliensis*. In contrast to the similar arabinose and galactose content of poly I from these two actinomycetes, the molar ratios for these two sugars were reversed in the poly II substance from *N. asteroides*. Furthermore, the poly II antigens were species-specific; they neither reacted with nor absorbed antibodies from the heterologously immunized rabbits [545]. The poly II preparations also were negative in ID tests with sera from patients with mycobacterial infections while still positive in species-specific mycetoma infections [54]. The protein content of these two polysaccharide preparations from three strains each of *N. asteroides* and *N. brasiliensis* varied from 1.3 to 6.9%, so they were not homogeneous. The possible effect of protein composition on immunological reactivity was not investigated.

The potential importance of protein has been shown by other investigators: Kingsbury and Slack [239] with polypeptide antigens (N-PP) and Bojalil and Magnusson [53,287] with purified protein derivatives (N-PPD) from *N. asteroides* and *N. brasiliensis*. Cultures were grown on defined constituents liquid medium at incubator temperatures and cells were heat killed prior to separation from broth. The N-PP was prepared from defatted cells which were extracted with 0.1 N HCl, neutralized, and then precipitated with picric acid. The filtered precipitate was dissolved in distilled water by carefully adding NaOH to pH 7 and then precipitated again with acetone. The final product was dissolved in, and dialyzed against, 0.001 N HCl because solubility decreased sharply at pH 7. This product contained no carbohydrate or lipid, the molecualr weight approximated 20,000 and acid hydrolysis revealed 16 amino acids with arginine and glutamic acid predominating. As little as 0.5 μg elicited DH in guinea pigs and rabbits sensitized with *N. asteroides*. Sensitivity (100%) and species specificity was observed among guinea pigs with 2-μg doses of N-PP from both actinomycetes, but some of the *N. asteroides* animals reacted to a 10-μg dose of N-PP from *N. brasiliensis*. It is of considerable interest that specificity is lost if cells are not completely defatted or if the N-PP is not dialyzed, and that nonsensitized rabbits, in contrast to guinea pigs, react to N-PP. It may be that lipid (or polysaccharide) accounts for previously reported nonspecific reactions and that "normal" rabbits may have been sensitized to these actinomycetes, emphasizing the importance of testing representative animals prior to initiation of immunological studies.

Bojalil and Magnusson [53,287] used the culture filtrates as starting material for preparing N-PPD. These were concentrated by ultrafiltration, precipitated with TCA, and dried with ether, a procedure that usually yields a solution enriched with protein but containing small amounts of polysaccharide and nucleic acid [432,433]. ST on patients with *N. brasiliensis* mycetoma and among tuberculin positive and negative subjects demonstrated both sensitivity (100%) and species specificity for the N-PPD from *N. brasiliensis* with a 0.2-μg dose, but a few nonspecific reactors with a 2-μg dose, e.g., 15% for tuberculosis patients. The N-PPD from *N. asteroides* was considerably more cross-reactive with the 2-μg dose, e.g., 74% among tuberculosis patients and 65% among healthy subjects. This sensitization to *N. asteroides* may not be nonspecific since Hosty et al. [188] recovered this actinomycete from sputum of 134 patients among whom a diagnosis of nocardiosis could not be established. It might be that some "normals" have been sensitized by clinically inapparent exposure.

Protein-enriched preparations from these two actinomycetes have been studied also by Ortiz-Ortiz et al. [343,344]. Defatted cells were suspended in a Tris buffer (pH 7.4) and treated with deoxyribonuclease before disruption in a Ribi cell. Cells, cellular debris, and highly polymerized DNA were removed by sequential centrifugation with increasing force. The crude extract (supernatant) after centrifugation at 48,000 g was the source material for separation of ribosomal and cytoplasmic proteins. Ribosomes were obtained by repeated precipitation with ammonium sulfate, centrifugation, and suspension in Tris buffer. The ribosomal protein was isolated after precipitation of RNA with 2-chloro-ethanol and HCl. The cytoplasmic proteins in the extract were obtained in the supernatant following centrifugation at 144,000 g to remove ribosomes. These can be considered only as enriched protein solutions because they contain approximately 55% protein, 29% carbohydrate, and 8% nucleic acid [344]. Guinea pigs infected with either *N. asteroides* or *N. brasiliensis* were used for DH and MMI assays. The ribosomal and the cytoplasmic protein solutions elicited DH in both animal groups, but homologous reactions were significantly greater. The DH reactions elicited by ribosomal protein were not affected by heat (1 h, boiling water) but were abolished completely by digestion with pronase, indicating that an intact protein moiety is required. Similar cross-reactivity was found in the MMI test with ribosomal protein, but the cytoplasmic proteins were species-specific in this assay. Mycobacterial PPD, however, produced significant MMI with cells from both animal groups, indicating that sensitized animals are capable of reacting to common antigens and emphasizing the need for eliminating these common antigens from the nocardial cytoplasmic protein preparations. Sundararaj and Agarwal [464-466] demonstrated that immunization of guinea pigs with a ribonucleoprotein fraction from *N. asteroides* protected the animals against subsequent intravenous challenge. These animals were nonreactive to ST with N-PP and N-PPD even at 10-μg doses, but their peritoneal cells responded to all three antigens (N-PP, N-PPD, and ribonucleoprotein) in tests for MMI and macrophage aggregation.

A major problem complicating the use of immunological reactions as aids for diagnosing and for understanding the host response to actinomycotic mycetoma has been common antigenic relationships among microorganisms in the order *Actinomycetales*. The literature prior to 1969 has been reviewed thoroughly by Kwapinski [253], with sections devoted to the genera *Mycobacterium, Nocardia, Actinomyces, Streptomyces,* and *Dermatophilus.* The obvious complexity of the relationships among these microorganisms, the use of cytoplasmic, cell wall, and culture filtrate antigens, and the variety of procedures for making and testing antigens has led to conflicting claims for species-, genus-, and group-specific antigens. Pier and associates [231,363-365] described an easily prepared culture filtrate antigen with which they could differentiate *N. asteroides* from mycobacterial infections in cattle by DH and CF tests. Cultures were grown in a peptone dialysate glucose broth at 36°C with rotary shaking for 2 weeks. Cells were separated and thimerosal added to the filtrate. The filtrate, either undiluted or concentrated 10-fold by pervaporation and dialyzed, was satisfactory for ST, CF, and ID [364], and for serologic typing of nocardiae by ID [363]. DEAE-cellulose chromatography of the filtrate revealed five protein peaks (fractions A to E) by linear gradient elution with phosphate-NaCl buffers and a sixth fraction by subsequent stepwise elution with borate-0.5 M NaCl at pH 9.3 [364]. The sixth fraction, designated F, contained the DH, CF, and ID activity. A later study [487], however, demonstrated that fraction A contained some ID activity and also inhibited both hemagglutination and hemolysis of erythrocytes sensitized with culture filtrate. Since these tests were reactive with antisera to mycobacteria and heterologous nocardiae, it appeared that most of the common antigens were in fraction A. Further analysis of fraction F from four strains of *N. asteroides* demonstrated an average composition of 45% protein, 9% hexose, and less than 1% pentose with all strains quite similar [231]. Hexosamine, deoxypentose, heptose, and methylpentose were absent.

Paper chromatography revealed 16-18 amino acids and the sugars galactose, glucose, mannose, xylose, and ribose, with a possible trace of arabinose. The protein:nucleic acid ratio was 23:1. Although electrophoresis, amino acid end group analysis, and DEAE fractionation indicated homogeneity for fraction F, analysis by ID and polyacrylamide gel electrophoresis revealed up to three antigens and three protein peaks, respectively. Keeler and Pier [231] considered the possible cellular source of fraction F. They reasoned that the chemical composition and narrow molecular size distribution suggested an intracellular product actively transferred into the media. Since there is little, if any, arabinose and no hexosamine, fraction F probably does not derive from degraded cell wall. If cytoplasmic and cell wall components are present, there should be a much greater heterogeneity in the molecular size distribution. These studies, however, did not allow for inhibition of enzyme activity following harvest and the source of fraction F remains unanswered.

Lipid mannophosphoinositides from *Nocardia* have been reported to be antigenic in rabbits when injected with Freund's incomplete adjuvant [492]. The strongest activity in ID tests is obtained with the dimannoside. Studies of sensitivity and specificity were not done.

C. Sporotrichosis

1. Cultures

Sporotrichosis differs from chromomycosis and mycetoma in that a single fungus species, *Sporothrix schenckii,* is the etiological agent. This is a dimorphic fungus, mycelial when cultivated at ambient temperature and a yeast at 37°C or in tissue. Both morphological forms are used as source material for manufacturing antigens. Isolates are those readily available to investigators and reference strains have not been recognized. Culture media, either gelled or liquid, include complex organic substrates [174,437,454], peptone or yeast extract dialysates [207,325,337], and defined nutrients [307,491]. Mycelial growth occurs during incubation at 25-30°C and yeast cells at 35-37°C. Morphological purity is an important consideration. Mendonca et al. [307] reported that morphological cell types synthesized polysaccharides with different molecular structures: the yeasts formed rhamnomannans with monorhamnosyl side chains, conidia-forming mycelia synthesized rhamnomannans with dirhamnosyl side chains, and nonconidial mycelia formed galactomannans or mannans. Their defined constituents broth enabled growth of yeast cells at both 25°C and 37°C, so medium composition was not a factor determining the type of polysaccharide produced. Therefore, studies of antigens should include control of morphological purity in source cultures.

2. Crude Antigens

The similarity of cutaneous lesions in sporotrichosis to those of other diseases stimulated immunological studies at the beginning of this century. The earlier literature, summarized briefly here, has been reviewed by Norden [339], by Kligman and Delamater [240], and by Salvin [411,413]. Conidia and yeast cells, after exposure to heat or formalin, agglutinate in sera from individuals with sporotrichosis. The optimal concentration of conidial suspensions must be determined carefully because autoagglutination occurs when suspensions are excessively dense and no agglutination when suspensions are too dilute. Nonspecific reactions are a constant problem; i.e., conidia may agglutinate in sera from humans and animals without evidence of sporotrichosis. Yeast cells are superior to conidia and are easier to prepare, although careful standardization to optimal dilution is required, as with conidia. Antigenic reactions are unaffected by formalin, heating to $100°C$ for 1 h, or overnight treatment with up to 1 N HCl. In general, yeast cells are also better as particulate antigens for CF tests and stain more uniformly with FA, a test recommended as a potential aid for identifying young or contaminated cultures. Soluble antigens are prepared from cells or culture filtrates. Yeast or mycelial cells are broken, extracted with aqueous solvents, and reactive substances precipitated with alcohol and redissolved in lesser volumes. Broth culture filtrates are concentrated similarly. Some investigators treat these carbohydrate-enriched preparations with proteinases or chemicals to further reduce protein content. Almost all of these antigens produce precipitates when mixed with serum from humans or animals with sporotrichosis. Whole killed cells and concentrated culture filtrates elicit DH in patients and animals with sporotrichosis. Nielsen [337] tested infected guinea pigs with heat-killed yeast cells, three different cell wall preparations from disrupted yeasts, and six soluble preparations from separated cytosols. The frequency of positive DH reactions was 95% with yeast cells, 75-90% with cell wall preparations, and 60% or less with cytosol antigens.

Some degree of nonspecificity is apparent in these early studies. Schneidau et al. [427] suggests that this might not be a fault of the antigen preparations but rather an expression of more ubiquitous sensitization to S. schenckii than is apparent from clinical disease alone. The frequency of DH reactors is at least three times greater among professional horticulturists than among other occupational groups, and this increased incidence is magnified to sixfold among nursery employees with at least 10 years of experience. This possibility needs reevaluation because of the more recent finding that antigens from three of four species of Ceratocystis exhibit significant correlations with a S. schenckii antigen in DH tests on sporotrichosis patients [207]. These antigens are prepared from dialyzed culture filtrates by precipitation with ethanol, drying with acetone, and solution in saline at 100 μg/ml.

More recently, several groups of investigators evaluated a number of antigen preparations in a variety of serological tests [52,218,398,521]. For example, Blumer et al. [52] examined yeast cell suspensions and culture filtrates from both yeast and mycelial growth. Yeast cells were harvested from broth cultures after incubating 1 week at $37°C$ while mixing at 150-160 rpm. The cells in broth were heat killed, washed, and adjusted to an optimal dilution for TA and IFA tests. The yeast culture broth, filtered and preserved with thimerosal, was added to an equal volume of a standardized latex particle suspension for use in a 5-min LA test. A portion of the yeast culture filtrate was concentrated 10-fold by acetone precipitation and the redissolved precipitate used in an ID test. A similarly concentrated mycelial broth culture filtrate was used for CF testing. Sensitivity of 90% or better was achieved with TA, LA, and IFA tests, and all tests except IFA exhibited outstanding specificity ($\geqslant$97%). These investigators recommend using the yeast cell-derived antigens in TA and LA tests for clinical laboratory practice. Their conclusions agree in general with those of other investigators [218,339,398,521]. The detailed procedures for antigen preparation, determination of optimal dilutions, and performance of TA and slide LA tests have been published [345]. Although Blumer et al. [52] began dilutions of patient sera at 1:8 and found only two cases (both leishmaniasis) yielding positive results among 85 heterologous sera tested, others [11,521] report nonspecific results with serum dilutions up to 1:40. It is possible that at least some of the latter may not be false positives but residual activity from prior subclinical infections, as suggested by Schneidau et al. [427]. Roberts and Larsh [399] describe an immunocytoadherence test in which thimerosal-killed yeast cells form rosettes with lymphocytes from infected guinea pigs. The results with this test parallel those obtained with TA; i.e., maximum reactivity with equivalent sensitivity occurs at 2 weeks postinfection and declines thereafter to essentially negative as lesions heal.

3. Refined Antigens

One group of investigators has produced an elegant study of the immunologically reactive peptidorhamnomannans found in *S. Schenckii* and in morphologically similar but saprobic species of *Ceratocystis*. The glycopeptides were isolated from supernatants of autoclaved cultures by initial precipitation with sodium acetate-ethanol, followed by precipitation with Cetavlon in the presence of borate at pH 8.5 and chromatography on a DEAE-Sephadex column with NaCl gradient elution [276]. Polysaccharides were obtained from a recombined mixture of the autoclaved cells and the ethanol precipitate by extraction with 2% KOH at 100°C for 2 h, neutralization, and precipitation with ethanol [163,497]. The precipitate was dissolved in water at 100°C and the polysaccharides precipitated as copper complexes with Fehling's solution. The dried copper complexes were shaken as a suspension with Amberlite IR-120 and the final product precipitated with acidified ethanol. The peptidorhamnomannans originally isolated by Lloyd and Bitoon [276] contained 51% D-mannose, 33% L-rhamnose, 16% peptide, and 0.2% phosphate. The low content of phosphate indicates a lack of phosphodiester linkages found in other fungal polysaccharides, but a relatively high content of glutamic and aspartic acids explains the adsorption to DEAE-Sephadex. An increased proportion of serine and threonine suggests that the oligosaccharide is linked to the peptide through these amino acids. The peptidorhamnomannan is similar to a compound reported earlier by Ishizaki [206], except that the latter contains glucose and galactose, and is not homogeneous by IEP and ID.

After Toriello and Mariat [491] demonstrated that different polysaccharides are recovered from cell and culture filtrates, greater attention has been focused on products obtained from different morphological forms [34,277,307,501]. Extensive analysis of the carbohydrate moiety obtained from *S. schenckii* demonstrates a preferential synthesis of different polysaccharides by specific cell types with some variations depending on temperature of incubation, age of culture, and location of the substances, i.e., in the cell wall or excreted into the medium. The backbone structure is a linear chain of $(1 \rightarrow 6)$-linked α-D-mannopyranose units. Yeast cells form rhamnomannans with monorhamnosyl side chains as $(1 \rightarrow 3)$-linked α-L-rhamnopyranosyl units. Conidia and conidia-forming mycelium also produce rhamnomannans but with dirhamnosyl side chains containing both $(1 \rightarrow 2)$- and $(1 \rightarrow 3)$-linkages. Young nonconidial mycelium produces galactomannans and mannans which are excreted readily into the medium. The quantity of rhamnomannans increases with incubation time and eventually masks the presence of other types of polysaccharides which were not detected in the earlier studies. For example, Travassos et al. [501] mention unpublished studies indicating the presence of galactan and glucan; Mendonca et al. [307] report galactomannan and mannan from nonsporulated mycelium; and Travassos et al. [500] present evidence by nuclear magnetic resonance spectroscopy for glucuronic acid residues in polysaccharides from one strain. The rhamnomannans are located in two sublayers of the outer cell wall [34,501], apparently being shed into the supporting medium as new wall is synthesized.

There are several reports demonstrating the immunological reactivity of the rhamnomannans and the intact glycopeptide by TP and IFA reactions with sera from patients with sporotrichosis and from hyperimmunized rabbits [276-278,325,437,491]. These studies demonstrate that the rhamnomannans react as determinant groups with respect to humoral antibody [277,278] and that they exhibit reciprocal cross-reactions with other microorganisms bearing rhamnomannans of similar structure, particularly *Ceratocystis* species [162,174,207,278,325,437,491,497]. Although these oligosaccharides appear to be immunodominant in reactions with humoral antibody, their role in cell-mediated responses is less certain. Crude preparations were employed in earlier studies [207,337,427,454], and all of these probably contained peptide as well as polysaccharide. Shimonaka et al. [437] provided more relevant information with respect to the roles of polysaccharide and peptide moieties. A glycopeptide product was obtained from mycelium by phenol extraction and fractionation on a Sephadex G-100 column, which contained 37% total carbohydrate, predominantly mannose, rhamnose, and galactose, and 12% peptide with 15 amino acids identified. Aliquots of this preparation were either digested with papain, reducing the peptide content by more than 50%, or oxidized with periodate, destroying all sugars detectable by gas-liquid chromatography. These aliquots were assayed for immunological reactivity in guinea pigs sensitized by intraperitoneal injection of whole cells. Periodate treatment completely abolished reactivity in IEP and passive cutaneous anaphylaxis, but these responses were not affected by digestion with the enzyme. In contrast, papain treatment significantly reduced elicitation of MMI factor and DH which had not changed after periodate treatment. Thus, the polysaccharide moiety is immunodominant in reactions involving humoral antibody, but the peptide moiety is required for eliciting CMI responses. It cannot be said unequivocally that the polysaccharides are not reactive in CMI responses because they might be involved in the specificity, or nonspecificity, of such responses. This would require similar experiments, including animals sensitized to heterologous fungi.

VI. DERMATOMYCOSES

The dermatomycoses represent another example of a clinical entity caused by many different fungi, in this case species in the three genera *Epidermophyton, Microsporum,* and *Trichophyton.* Since these are infections of keratinized tissues, lesions are accessible for diagnosis by direct microscopic examination or culture. This ease of diagnosis, the ubiquity of infection, and common antigens shared among species in the three genera result in little value for immunological reactions as a diagnostic aid. Recently, however, the fact that these infections are so common and that they induce DH has popularized the use of extracts from these fungi as one of the recall antigens for testing the patency of CMI mechanisms in patients.

A. Cultures

A conspicuous characteristic of dermatomycotic fungi is the possession of common antigens, and therefore almost any species can be used for producing antigen. Isolates of *T. mentagrophytes* and *T. rubrum* are used most frequently, but a specified isolate of *T. verrucosum,* strain 130, is used for vaccinating farm animals in eastern Europe [9]. The fungus is usually grown in broth, either a complex organic medium, a partially defined mixture, or with completely defined constituents [232,251,353,354,436]. Cultures are incubated at ambient or 30°C temperatures either stationary for 2-3 months or on a shaker for shorter periods, depending on the growth characteristics of the species and the culture medium used.

B. Crude Antigens

Traditionally, the antigen preparation used to test for hypersensitivity or humoral antibody among persons with dermatomycoses has been called trichophytin [335]. Since infection by fungi in the three genera induces immunological responses which can be elicited or demonstrated with antigen preparations from almost any of the same group of fungi [164,353], the designation "trichophytin" is not sufficiently descriptive and should be replaced by dermatophytin.

Crude dermatophytins are prepared from fungous substance by extraction or from filtrates of cultures grown in broth [354]. Cells grown on or in broth are collected on several layers of filter paper, washed with water, and sometimes treated with nonpolar solvents to remove lipids. The cells are broken and soluble compounds obtained in saline or a portion of the initial culture filtrate. The aqueous extract (i.e., dermatophytin) is sterilized by filtration and a preservative added (e.g., phenol or thimerosal). Culture filtrates also are used as dermatophytin. If the potency is too low, a satisfactory product can be obtained by concentration procedures (e.g., precipitation with acetone or ultrafiltration). Peck [353] reported that the antigens in dermatophytin which elicit DH are found in the pellicle during the first week of stationary growth with a constant concentration thereafter. In contrast, the concentration in culture filtrates increases with time and is pH dependent. The DH-eliciting antigens appear in broth only when the pH is rising to an ultimately constant value of 8.6. It does not matter whether the culture medium is adjusted to an acid or an alkaline pH prior to inoculation. The pH decreases during early growth to below 5.0 and then increases gradually to pH 8.6, with the concentration of DH-eliciting substances roughly paralleling the increase. This fact is consistent with evidence that young, actively growing fungous cells are most productive [139,459]. Differences in species or strains used, in fungous cells or culture filtrates as source material, and in preparatory procedures have produced dermatophytins varying in complexity and potency. Hence, uniform responses should not be anticipated from the various dermatophytins available.

Some physicochemical characteristics of dermatophytins have been reported. The crude desiccated product is soluble in water as a clear brown solution. Skin test reactivity remains after boiling the aqueous solution but is destroyed in the presence of mineral acids under the same conditions. The active substances are dialyzable according to Bloch et al. [50] and Thompson [486]. The immunologically reactive components in dermatophytins are nitrogen-containing polysaccharides (reviewed in Ref. 164). Acid hydrolysis consistently yields glucose, mannose, galactose, and glucosamine, with additional carbohydrates reported irregularly and in minor amounts. Analysis of the nitrogenous component demonstrates up to 15 amino acids, with serine, threonine, proline, glycine, alanine, and aspartic acid usually present. Digestion analysis with proteolytic enzymes indicates that peptide rather than carbohydrate-amino acid linkages predominate and, therefore, the reactive material is glycopeptide. Immunologically related glycopeptides have been isolated from species in all three genera of dermatophytes.

Other kinds of immunologically reactive preparations include ribonucleotides [209], keratinases [542,543], and cells. Dead whole or disrupted cells have been used as vaccines with variable success (reviewed in Refs. 191 and 265). An attenuated culture of

T. verrucosum strain 130 administered intramuscularly is used for immunoprophylaxis, with marked reduction in the incidence of infection among farm animals (reviewed in Ref. 9).

C. Refined Antigens

Dermatophytins contain a complex mixture of many antigens. For example, Ito [208, 209] prepared 11 fractions from cells and an additional 11 from culture filtrates by a variety of methods involving chemical extraction, precipitation, and column chromatography. All 22 fractions elicited 24-h erythematous and cutaneous reactions in infected rabbits. It is not clear whether all of these were true DH reactions because responses to only five fractions were studied histologically. There was no mention of tests in nonsensitized animals to determine primary irritation, and there were no transfer experiments with lymphocytic cells or serum. The various fractions were characterized as glycopeptides, peptides, ribonucleic acid, and polysaccharides. The glycopeptides elicited the greatest dermal response. In addition, Andrieu et al. [8] demonstrated more than 10 antigens among each of 17 species by IEP, Christiansen and Svejgaard [72] found 25-35 antigens by two-dimensional EIP, the number varying with the species examined, and Philpot [359] differentiated with ID a total of 48 antigens in sodium acetate-acetone-precipitated extracts from 19 species with 13 antigens common to more than one species. Obviously, dermatophytins are multicomponent antigenic preparations.

Since the most significant immunological activity is found in the glycopeptides, these have received the most attention. Glycopeptide complexes have been extracted from cells with ethylene glycol [27,30,189], with phenol [208,340], and with TCA [10]. Several purified glycopeptides have been separated from the complex by various techniques involving combinations of chemical fractionation and column chromatography using gel permeation and ion exchange principles. Barker and associates [27,28,30; reviewed in Ref. 164] isolated five glycopeptides from ethylene glycol extracts of mycelia of several dermatophyte species. Amino acid analysis demonstrated similar ratios except for high proline among three of the glycopeptides. Glycine, alanine, and threonine occurred as N-terminal amino acid residues, indicating at least three peptide chains. Quantitative sugar analysis revealed only galactose and mannose in ratios varying from 1:3 to 1:14. These peptidogalactomannans elicited DH in sensitized guinea pigs. Arnold et al. [10] obtained two major, highly purified fractions by TCA extraction of washed mycelia, precipitation with ethanol, and chromatography through Sephadex G-200 and then through GE-52-cellulose (borate form). Both were peptidoheteropolysaccharides with mannose:galactose:glucose ratios of 7.5:0.7:1 for one and 9:0.3:1 for the second. Molecular weights exceeded 200,000 compared to the 20,000-40,000 for the peptidogalactomannans reported by Barker and associates. Although these two peptidoheteropolysaccharides eluted differently from the cellulose-borate column, they appeared as single, identical antigen-antibody precipitates by ID and two-dimensional IEP. Moser and Pollack [317] used a modified ethylene glycol procedure and obtained four glycopeptides separated first through a Sephadex G-200 column and then through a DEAE-Sephadex A-50 column with gradient NaCl elution. These also contained glucose as well as galactose and mannose but in ratios differing from those reported by Arnold et al. Only one of the four glycopeptides exceeded 200,000 in molecular weight and it had an unusually high glucose content. These discrepancies could be caused by differences in fungous strains, in methodology, or in the actual existence of more peptidoheteropolysaccharides than those isolated by any one group of investigators.

Another group of investigators (reviewed in Ref. 164) obtained three pure polysaccharides from defatted and enzymatically deproteinized mycelia by extraction with hot dilute alkali, precipitation of one as a copper complex, and separation of the other two by DEAE-cellulose chromatography. Each of nine species yielded one glucan and two galactomannans, GM-I and GM-II. Both galactomannans were branched polysaccharides with linear backbones of D-mannopyranose and branches terminating in nonreducing end-groups of D-galactofuranose and D-mannopyranose. A basic difference in molecular structure occurred in the linear portion of these two galactomannans. Both had $(1 \rightarrow 6)$-linkages but $(1 \rightarrow 2)$-linkages predominated in GM-II and were absent from GM-I. Some variations, which appear to be species-related, occurred in the relative frequency of branching through either C-6, C-3 or C-6, C-2 disubstituted D-mannopyranose units. In contrast, the glucans from all species were very similar. The linear chains contained $(1 \rightarrow 6)$- and $(1 \rightarrow 3)$-linked glucopyranose units, with all branches through C-6, C-3 positions. Other investigators, however, have isolated glucans with $(1 \rightarrow 4)$-linkages identified by infrared spectroscopy [189].

The chemical composition of the glycopeptides has been related to their immunological activity [28,30,317,340,405]. The intact molecule elicits either immediate or DH dermal reactions in experimental animals or humans with appropriate hypersensitivity. All the glycopeptides are reactive to varying degrees, and their specificity is directed to the dermatophytes as a group. Enzyme digestion of the peptide portion decreases DH without affecting the immediate response, but degradation of the polysaccharide greatly diminishes the immediate response without significant effect on DH. Furthermore, highly purified polysaccharides do not elicit DH in infected guinea pigs, but still elicit immediate responses [189,405]. Thus, the peptide portion is essential for eliciting DH but not required for the immediate dermal response.

The glycopeptides also react with humoral antibody but serological activity is determined primarily by the carbohydrate structure [340; reviewed in Ref. 164]. The galactomannans and glucan from each of five species react with antiserum raised against any single species, and are therefore characteristic of a group rather than a species relationship. Mild acid hydrolysis to remove D-galactose from GM leaves mannan I with the same serological reactivity as GM-I, but mannan II differs from the parent GM-II by ID and CF tests. Since mannans I and II are cross-reactive, it appears that the greater number of D-galactofuranose residues in GM-II induces some differential lymphocyte recognition of GM-II.

The three keratinases isolated by Yu et al. [542-544] contained the same 18 amino acids. Keratinase I was an extracellular protein with a molecular weight of 48,000. Keratinases II and III were intracellular glycoproteins with molecular weights of 440,000 and 20,300, respectively. The carbohydrate moiety contained galactose, mannose, and glucose [164]. All three enzymes possessed some antigenic determinant groups in common, elicited DH in infected guinea pigs, and reacted with circulating antibody [164]. Whether an immunological relationship exists among the keratinases and other dermatophyte antigens has not been reported.

The quantitative and qualitative variations among the many dermatophytins present an insoluble problem for clinical practice at the present time. There are obvious differences in the manner by which dermatophytins are produced, i.e., in culture medium, in time and temperature of incubation, in the way of stationary or shaking growth conditions, in harvesting fungous cells or broth as source material, and in postharvest manipulations. In addition, very little attention has been given to changes effected by the release of enzymes during growth and harvest and their potential effect on the final product in terms of antigens lost or modified [220]. Hence, bioassay in known reactors has been the only reasonable, albeit approximate, method for standardization. All the defined antigens (glycopeptides, keratinases, polysaccharides) are immunologically reactive in either CMI or serological assays or in both and, even though some may contain multiple antigens [340], one or more of the defined antigens could serve as a reproducible standard. Unfortunately, there have been no definitive studies relating reactions to these antigens with clinical disease and, therefore, it remains to be determined which would provide the most clinically significant information and would serve best as a standard reference antigen.

VII. COMMENTARY

Antigenic substances from fungi have served for a long time to elicit or demonstrate immunological reactions as an aid for diagnosis. Some (e.g., coccidioidin and histoplasmin) have survived the crucible of accumulated experience as useful medical devices, others have not, and a few of the latter have been revived for a new objective (e.g., dermatophytin and candidin). All of these are multiantigen preparations and their continued use can be justified by several valid arguments. They have had a long trial and proved their value for clinical practice. They have acceptable specificity and sensitivity within limits that are recognized. Their multiantigen content probably increases the chance of detecting immunological responses among patients. Serious problems exist. Many preparations are produced in different laboratories with variations occurring in fungous strains, growth medium, cultural conditions, cellular or extracellular source material, and antigen production methodology. The complexity of these preparations and the potential for qualitative and quantitative differences require bioassays to determine optimal concentrations for each immunological test. But even this is insufficient. Reagent standardization in one laboratory with its own product does not guarantee results matching those of another laboratory. If we are even to anticipate uniform responses with these manifold preparations, it is imperative that standardized reference reagents be available from a national, or preferably an international, agency.

These multicomponent preparations have additional limitations. The presence of similar or identical antigenic determinants among more than one species requires qualifying reservations for determining specific etiology by immunological means. The limitations are even more critical for investigators exploring cell-mediated mechanisms of immunity. A fundamental difference between serological and CMI assays is that a nonviable end product is measured in the former but a complex interaction among viable cells is involved in the latter. Cellular function may be stimulated or suppressed either directly by different antigens or indirectly by mechanisms mediated within the cell systems in response to effects elicited by antigens. Hence, if one antigen in a mixture leads to stimulation but a second to suppression, the observed result of a CMI assay might be less stimulation or less suppression than is actually present, or even no response if one cancels the other. These possibilities are compounded by qualitative and quantitative differences among antigen preparations. Resolution of this problem requires isolation of antigens which satisfy criteria for physicochemical and immunochemical homogeneity.

The technology is available for producing monomolecular antigens, but two questions need consideration. Is inclusion of all antigens from a fungus necessary? Are the isolated antigens identical to their native structure in the cell? A purist would answer both questions affirmatively. A pragmatist would qualify both answers, with "probably not" to the first question and by "consistent with observable immunological reactions" to the second, such as electron microscopy of cells treated with antibody conjugated to an electron-dense marker. Whatever their philosophy, both need a beginning. From the viewpoint of *medical* mycology, a priority selection can be chosen from the field of antigens available. A knowledgeable choice can be made for antigens reactive with patient antibody. Although a large number of antigens can be demonstrated with antisera from hyperimmunized animals, only a comparatively small number of antigen-antibody precipitates are found when patient sera are used in the analytically powerful two-dimensional IEP procedures. In my own experience, I have encountered only one patient serum producing as many as 7 antigen-antibody precipitates among 26 precipitates produced by a hyperimmune antiserum with the same antigen preparation. Axelsen [14] and Axelsen and Kirkpatrick [15] demonstrated 78 antigens in a *Candida albicans* extract with pooled hyperimmune antisera but only 13 with a patient's serum. Thus, selection of first antigens to be isolated can be based on those found reactive with patient sera. There is no assurance, however, that these would be the same antigens involved in the patient's CMI responses. For example, in coccidioidomycosis antigens reactive in TP and CF tests have heat sensitivities differing from those of at least one of the antigens eliciting DH, and we do not know whether the latter is present among those developed in the two-dimensional IEP plates. But the heat sensitivity property and preliminary fractionations by chromatography procedures can narrow the choice for priority isolations. The evidence that proteins and the peptide portion of glycopeptides are required for eliciting CMI responses is an additional factor guiding the selection.

Finally, morphological purity and changes in antigen content or composition during morphogenesis are frequently overlooked. The studies with *Sporothrix schenckii* are a fine example of preferential synthesis of molecularly different polysaccharides by specific cell types. We have noted in our electron microscopy studies of *Coccidioides immitis* that outer layers of the cell wall are shed from arthoconidia during both germination and conversion into the parasitic cycle, and that only the inner layers of the spherule wall contribute to the wall of endospores. Thus, production of some antigens depends on the morphogenetic stage harvested and establishing synchronous growth should improve reproducibility. Studies of antigens should include examination of morphological type and purity as done by Levine et al. [268] during their development of a vaccine for coccidioidomycosis.

REFERENCES

1. Abernathy, R. S., and D. C. Heiner. 1961. Precipitation reactions in agar in North American blastomycosis. J. Lab. Clin. Med. *57:* 604-611.
2. Ajello, L., K. Walls, J. C. Moore, and R. Falcone. 1959. Rapid production of complement fixation antigens for systemic mycotic diseases. I. Coccidioidin: influence of media and mechanical agitation on its development. J. Bacteriol. *77:* 753-756.
3. Aksoycan, N. 1977. Einteilung von Stämmen verschiedener *Candida*-Arten nach dem Antigen von *Salmonella cholerae-suis* 06,7. Zentralbl. Bakteriol. Parasitenkd. Infektionskr. Hyg., Abt. 1: Orig. *238:* 379-382.
4. Al-Doory, Y., and M. A. Gordon. 1963. Application of fluorescent-antibody procedures to the study of pathogenic dematiaceous fungi. I. Differentiation of

Cladosporium carrionii and *Cladosporium bantianum*. J. Bacteriol. *86*: 332-338.

5. Amos, W. M. 1970. The extraction of fungal antigens and their use in serological tests as an aid to the diagnosis of bronchial disorders. J. Med. Lab. Technol. *27*: 18-32.

6. Anderson, K. L., R. W. Wheat, and N. F. Conant. 1971. Fractionation and composition studies of skin test active components of sensitins from *Coccidioides immitis*. Appl. Microbiol. *22*: 294-299.

7. Andrieu, S., J. Biguet, L. Dujardin, and T. Vaucelle. 1969. Etude antigénique des agents des mycoses profondes par l'analyse comparée des milieux de culture. Mycopathol. Mycol. Appl. *39*: 97-108.

8. Andrieu, S., J. Biguet, and B. Laloux. 1968. Analyse immuno-électrophorétique comparée des structures antigénique de 17 espèces de dermatophytes. Mycopathol. Mycol. Appl. *34*: 161-185.

9. Specific prophylactic vaccine against bovine trichophytosis. Vreshtorgizdat., Moscow, No. 21M313: 1-27, 1975.

10. Arnold, M. T., S. F. Grappel, A. V. Lerro, and F. Blank. 1976. Peptidopolysaccharide antigens from *Trichophyton mentagrophytes* var. *granulosum*. Infect. Immun. *14*: 376-382.

11. Ashbrook, B. M. 1964. Studies on antigens of *Sporotrichum schenckii*. Master's thesis, Duke University, Durham, N.C.

12. Atkinson, A. J., Jr., and J. E. Bennett. 1968. Experience with a new skin test antigen prepared from *Cryptococcus neoformans*. Am. Rev. Respir. Dis. *97*: 637-643.

13. Axelsen, N. H. 1971. Antigen-antibody crossed electrophoresis (Laurell) applied to the study of the antigenic structure of *Candida albicans*. Infect. Immun. *4*: 525-527.

14. Axelsen, N. H. 1973. Quantitative immunoelectrophoretic methods as tools for a polyvalent approach to standardization in the immunochemistry of *Candida albicans*. Infect. Immun. *7*: 949-960.

15. Axelsen, N. H., and C. H. Kirkpatrick. 1973. Simultaneous characterization of free *Candida* antigens and *Candida* precipitins in a patient's serum by means of crossed immunoelectrophoresis with intermediate gel. J. Immunol. Methods *2*: 245-249.

16. Azuma, I., F. Kanetsuna, Y. Tanaka, Y. Yamamura, and L. M. Carbonell. 1974. Chemical and immunological properties of galactomannans obtained from *Histoplasma duboisii*, *Histoplasma capsulatum*, *Paracoccidioides brasiliensis* and *Blastomyces dermatitidis*. Mycopathol. Mycol. Appl. *54*: 111-125.

17. Azuma, I., H. Kimura, F. Hirao, E. Tsubura, and Y. Yamamura. 1967. Skin-testing and precipitation antigens from *Aspergillus fumigatus* for diagnosis of aspergillosis. Am. Rev. Respir. Dis. *95*: 305-306.

18. Azuma, I., H. Kimura, F. Hirao, E. Tsubura, and Y. Yamamura. 1967. Biochemical and immunological studies on *Aspergillus*. I. Chemical and biological investigations of lipopolysaccharide, protein and polysaccharide fractions from *Aspergillus fumigatus*. Jpn. J. Med. Mycol. *8*: 210-220.

19. Azuma, I., H. Kimura, F. Hirao, E. Tsubura, and Y. Yamamura. 1969. Biochemical and immunological studies on *Aspergillus*. II. Immunological properties of protein and polysaccharide fractions obtained from *Aspergillus fumigatus*. Mycopathol. Mycol. Appl. *37*: 289-303.

20. Azuma, I., H. Kimura, F. Hirao, E. Tsubura, Y. Yamamura, and A. Misaki. 1971. Biochemical and immunological studies on *Aspergillus*. III. Chemical and immunological properties of glycopeptide obtained from *Aspergillus fumigatus*. Jpn. J. Microbiol. *15*: 237-246.

21. Azuma, I., H. Kimura, and Y. Yamamura. 1968. Purification and characterization of an immunologically active glycoprotein from *Aspergillus fumigatus*. J. Bacteriol. *96*: 272-273.

22. Baker, O., and A. I. Braude. 1956. A study of stimuli leading to the production of spherules in coccidioidomycosis. J. Lab. Clin. Med. *47*: 169-181.

23. Ballou, C. E. 1970. A study of the immunochemistry of three yeast mannans. J. Biol. Chem. *245*: 1197-1203.

24. Balows, A., K. W. Deuschle, N. R. Nedde, I. P. Mersack, and K. A. Watson. 1966. Skin tests in blastomycosis. Arch. Environ. Health *13*: 86-90.

25. Bardana, E. J., Jr., J. K. McClatchy, R. S. Farr, and P. Minden. 1972. The primary interaction of antibody to components of aspergilli. I. Immunologic and chemical characteristics of a nonprecipitating antigen. J. Allergy Clin. Immunol. *50*: 208-221.

26. Bardana, E. J., Jr., J. K. McClatchy, R. S. Farr, and P. Minden. 1972. The primary interaction of antibody to components of aspergilli. II. Antibodies in sera from normal persons and from patients with aspergillosis. J. Allergy Clin. Immunol. *50*: 222-234.

27. Barker, S. A., O. Basarab, and C. N. D. Cruickshank. 1967. Galactomannan peptides of *Trichophyton mentagrophytes*. Carbohydr. Res. *3*: 325-332.

28. Barker, S. A., C. N. D. Cruickshank, J. H. Morris, and S. R. Wood. 1962. The isolation of *Trichophyton* glycopeptide and its structure in relation to the immediate and delayed reactions. Immunology *5*: 627-632.

29. Bartels, P. A. 1971. Partial chemical characterization of histoplasmin H-42. In L. Ajello, E. W. Chick, and M. L. Furcolow (Eds.), *Histoplasmosis*, Thomas, Springfield, Ill., pp. 56–63.

30. Basarab, O., M. J. How, and C. N. D. Cruickshank. 1968. Immunological relationships between glycopeptides of *Microsporum canis, Trichophyton rubrum, Trichophyton mentagrophytes* and other fungi. Sabouraudia *6*: 119-126.

31. Bauman, D. S., and E. W. Chick. 1967. Carry-over of blood group antigens in *Histoplasma capsulatum* grown on blood agar. Am. Rev. Respir. Dis. *96*: 823-826.

32. Baxter, M., I. G. Murray, and J. J. Taylor. 1966. A case of mycetoma with serological diagnosis of *Allescheria boydii*. Sabouraudia *5*: 138-140.

33. Bell, R. G. 1978. Comparative virulence and immunodiffusion analysis of *Petriellidium boydii* (Shear) Malloch strains isolated from feedlot manure and a human mycetoma. Can. J. Microbiol. *24*: 856-863.

34. Benchimol, M., W. de Souza, and L. R. Travassos. 1979. Distribution of anionic groups at the cell surface of different *Sporothrix schenckii* cell types. Infect. Immun. *24*: 912-919.

35. Benham, R. W. 1935. Cryptococci—their identification by morphology and by serology. J. Infect. Dis. *57*: 255-274.

36. Bennett, J. E., and H. F. Hasenclever. 1965. *Cryptococcus neoformans* polysaccharide: studies of serologic properties and role in infection. J. Immunol. *94*: 916-920.

37. Bennett, J. E., and H. F. Hasenclever. 1965. Evaluation of a skin test for cryptococcosis. Am. Rev. Respir. Dis. *91*: 616.

38. Bennett, J. E., K. J. Kwon-Chung, and D. H. Howard. 1977. Epidemiologic differences among serotypes of *Cryptococcus neoformans*. Am. J. Epidemiol. *105*: 582-586.

39. Bennett, J. E., K. J. Kwon-Chung, and T. S. Theodore. 1978. Biochemical differences between serotypes of *Cryptococcus neoformans*. Sabouraudia *16*: 167-174.

40. Biguet, J., R. Havez, P. Tran Van Ky, and R. Degaey. 1961. Etude électrophor-

étique, chromatographique et immunologique des antigènes de *Candida albicans*. Caractérisation de deux antigènes spécifiques. Ann. Inst. Pasteur (Paris) *100*: 13-24.

41. Biquet, J., R. Havez, and P. Tran Van Ky. 1959. Les possibilités d'application aux champignons pathogènes de la méthode d'Ouchterlony et de l'immunoélectrophorèse. Résultats encourageants d'une première tentative concernant l'étude des antigènes de *Candida albicans*. Mykosen 2: 115-120.

42. Biguet, J., P. Tran Van Ky, S. Andrieu, and J. Fruit. 1964. Analyse immunoélectrophorétique d'extraits cellulaires et de milieux de culture d'*Aspergillus fumigatus* par des immunsérums expérimentaux et des sérums de malades atteints d'aspergillome bronchopulmonaire. Ann. Inst. Pasteur (Paris) *107*: 72-97.

43. Biguet, J., P. Tran Van Ky, S. Andrieu, and J. Fruit. 1965. Analyse immunoélectrophorétique des antigènes fongigues et systématique des champignons. Répercussions pratiques sur le diagnostic des mycoses. Mycopathol. Mycol. Appl. *26*: 241-256.

44. Biguet, J., P. Tran Van Ky, S. Andrieu, and T. Vancelle. 1967. Premières caractérisations d'activitiés enzymatiques sur les immuno-électrophorégrammes des extraits antigéniques de *Histoplasma capsulatum*. Conséquences diagnostiques pratiques. Ann. Soc. Belge Med. Trop. *47*: 425-434.

45. Biquet, J., P. Tran Van Ky, J. Fruit, and S. Andrieu. 1967. Identification d'une activité chymotrypsique au niveau de fractions remarquables de l'extrait antigénique d'*Aspergillus fumigatus*. Répercussions sur le diagnostic, immunologique de l'aspergilloses. Rev. Immunol. Ther. Antimicrob. *31*: 317-328.

46. Biguet, J., P. Tran Van Ky, R. Havez, and S. Andrieu. 1962. Etude électrophorétique, chromatographique et immunologique des protéines sériques du lapin immunisé contre *Candida albicans*. Ann. Inst. Pasteur (Paris) *102*: 328-338.

47. Bindschadler, D. D., and J. E. Bennett. 1968. Serology of human cryptococcosis. Ann. Intern. Med. *69*: 45-52.

48. Bishop, C. T., F. Blank, and P. E. Gardner. 1960. The cell wall polysaccharides of *Candida albicans*, glucan, mannan, and chitin. Can. J. Chem. *38*: 869-881.

49. Blandamer, A., and I. Danishefsky. 1966. Investigation on the structure of the capsular polysaccharide from *Cryptococcus neoformans* type B. Biochim. Biophys. Acta *117*: 305-313.

50. Bloch, B., A. Labouchere, and F. Schaaf. 1925. Versuche einer chemischen Charakterisierung und Reindarstellung des Trichophytins. Arch. Dermatol. Syphilol. *148*: 413-424.

51. Bloomfield, N., M. A. Gordon, and D. F. Elmendorf, Jr. 1963. Detection of *Cryptococcus neoformans* antigen in body fluid by latex particle agglutination. Proc. Soc. Exp. Biol. Med. *114*: 64-67.

52. Blumer, S. O., L. Kaufman, D. W. McLaughlin, and D. E. Kraft. 1973. Comparative evaluation of five serological methods for the diagnosis of sporotrichosis. Appl. Microbiol. *26*: 4-8.

53. Bojalil, L. F., and M. Magnusson. 1963. Specificity of skin reactions of humans to *Nocardia* sensitins. Am. Rev. Respir. Dis. *88*: 409-411.

54. Bojalil, L. F., and A. Zamora. 1963. Precipitin and skin tests in the diagnosis of mycetoma due to *Nocardia brasiliensis*. Proc. Soc. Exp. Biol. Med. *113*: 40-43.

55. Bout, D., J. Fruit, and A. Capron. 1973. Isolement par chromatographie d'affinité de la fraction antigénique spécifique, à activité chymotrypsique, de *Aspergillus fumigatus*. Bull. Soc. Fr. Mycol. Med. *2*: 125-129.

56. Bradley, G., L. Pine, M. W. Reeves, and C. W. Moss. 1974. Purification, composition, and serological characterization of histoplasmin—H and M antigens. Infect. Immun. *9*: 870-880.

57. Brito, T. de, A. Raphael, C. Fava Netto, and S. de A. P. Sampaio. 1961. Histopathology of the skin test using a polysaccharide antigen of the *Paracoccidioides brasiliensis*. J. Invest. Dermatol. *37*: 29-37.

58. Buckley, H. R., and I. G. Murray. 1966. Precipitating antibodies in chromomycosis. Sabouraudia *5*: 78-80.

59. Budzko, D. B., and R. Negroni. 1975. Hemolytic, cytotoxic and complement inactivating properties of extracts of different species of *Aspergillus*. Mycopathologia *57*: 23-26.

60. Bulmer, G. S., and M. D. Sans. 1968. *Cryptococcus neoformans*. III. Inhibition of phagocytosis. J. Bacteriol. *95*: 5-8.

61. Burrell, R., and C. K. Thomas. 1977. Improved methods of producing precipitating *Aspergillus* antigens. Ann. Allergy *38*: 202-205.

62. Busey, J. F., and P. F. Hinton. 1967. Precipitins in blastomycosis. Am. Rev. Respir. Dis. *95*: 112-113.

63. Campbell, C. C. 1953. Antigenic fractions of *Histoplasma capsulatum*. Am. J. Public Health *43*: 712-717.

64. Campbell, C. C. 1971. History of the development of serologic tests for histoplasmosis. In L. Ajello, E. W. Chick, and M. L. Furcolow (Eds.), *Histoplasmosis,* Thomas, Springfield, Ill., pp. 341-357.

65. Campbell, C. C., and G. E. Binkley. 1953. Serologic diagnosis with respect to histoplasmosis, coccidioidomycosis and blastomycosis and the problem of cross reactions. J. Lab. Clin. Med. *42*: 896-906.

66. Campbell, M. J., and Y. M. Clayton. 1964. Bronchopulmonary aspergillosis. A correlation of the clinical and laboratory findings in 272 patients investigated for bronchopulmonary aspergillosis. Am. Rev. Respir. Dis. *89*: 186-196.

67. Carlisle, H. N., and S. Saslaw. 1958. A histoplasmin-latex agglutination test. J. Lab. Clin. Med. *51*: 793-801.

68. Cassone, A., E. Mattia, and L. Boldrini. 1978. Agglutination of blastospores of *Candida albicans* by concanavalin A and its relationship with the distribution of mannan polymers and the ultrastructure of the cell wall. J. Gen. Microbiol. *105*: 263-273.

69. Cauley, L. K., and J. W. Murphy. 1979. Response of congenitally athymic (nude) and phenotypically normal mice to *Cryptococcus neoformans* infection. Infect. Immun. *23*: 644-651.

70. Chaparas, S. D., H. B. Levine, G. M. Scalarone, and D. Pappagianis. 1975. Cellular responses, protective immunity, and virulence in experimental coccidioidomycosis. In *Mycoses,* Proceedings of the Third International Conference on the Mycoses, Pan Am. Health Organ. Sci. Publ. No. 304, Washington, D.C., pp. 80-85.

71. Chew, W. H., and T. L. Theus. 1967. *Candida* precipitins. J. Immunol. *98*: 220-224.

72. Christiansen, A. H., and E. Svejgaard. 1976. Studies of the antigenic structure of *Trichophyton rubrum, Trichophyton mentagrophytes, Microsporum canis,* and *Epidermophyton floccosum* by crossed immuno-electrophoresis. Acta Pathol. Microbiol. Scand., Sect. C *84*: 337-341.

73. Cole, G. T., T. Sekiya, R. Kasai, T. Yokoyama, and Y. Nozawa. 1979. Surface ultrastructure and chemical composition of the cell walls of conidial fungi. Exp. Mycol. *3*: 132-156.

74. Coleman, R. M., and L. Kaufman. 1972. Use of the immunodiffusion test in the serodiagnosis of aspergillosis. Appl. Microbiol. *23*: 301-308.

75. Conant, N. F., and D. S. Martin. 1937. The morphologic and serological relationships of the various fungi causing dermatitis verrucosa (chromoblastomycosis). Am. J. Trop. Med. *17*: 553-578.

76. Conti-Diaz, I. A., 1972. Skin tests with paracoccidioidin and their importance. In *Paracoccidioidomycosis,* Proceedings of the First Pan American Symposium, Pan Am. Health Organ. Sci. Publ. No. 254, Washington, D.C., pp. 197-202.

77. Conti-Diaz, I. A., J. E. Mackinnon, L. Calegari, and S. Casserone. 1978. Estudio comparative de la immunoelectroforesis (IEF) y de la immunoelectroosmoforesis-immunodifusión (IEOF-ID) aplicadas al diagnóstico de la paracoccidioidomicosis. Mycopathologia *63*: 161-165.

78. Conti-Diaz, I. A., R. E. Somma-Moreira, E. Gezuele, A. C. de Giménez, M. I. Pena, and J. E. Mackinnon. 1973. Immunoelectroosmophoresis-immunodiffusion in paracoccidioidomycosis. Sabouraudia *11*: 39-41.

79. Converse, J. L. 1956. Effect of physico-chemical environment on spherulation of *Coccidioides immitis* in a chemically defined medium. J. Bacteriol. *72*: 784-792.

80. Converse, J. L. 1957. Effect of surface active agents on endosporulation of *Coccidioides immitis* in a chemically defined medium. J. Bacteriol. *74*: 106-107.

81. Converse, J. L., and A. R. Besemer. 1959. Nutrition of the parasitic phase of *Coccidioides immitis* in a chemically defined liquid medium. J. Bacteriol. *78*: 231-239.

82. Cook, W. L., F. G. Felton, H. G. Muchmore, and E. R. Rhoades. 1970. Cell wall differences of patient and soil isolates of *Cryptococcus neoformans.* Sabouraudia *7*: 257-260.

83. Cooper, B. H., and J. D. Schneidau. 1970. A serological comparison of *Phialophora verrucosa, Fonsecaea pedrosoi* and *Cladosporium carrionii* using immunodiffusion and immunoelectrophoresis. Sabouraudia *8*: 217-226.

84. Corbel, M. J., G. A. Pepin, and P. G. Millar. 1973. The serological response to *Aspergillus fumigatus* in experimental mycotic abortion in sheep. J. Med. Microbiol. *6*: 539-548.

85. Costa, E. O. da. 1975. Paracoccidioidomicose em animais domésticos. Infecção experimental em bovinos, "Micose infecção" em bovinos, ovinos e equídeos. Doctoral thesis, Instituto de Ciência Biomedica, University of São Paulo, pp. 1-112.

86. Costa, E. O. da, and C. Fava Netto. 1978. Contribution to the epidemiology of paracoccidioidomycosis and histoplasmosis in the state of São Paulo, Brazil. Paracoccidioidin and histoplasmin intradermic tests in domestic animals. Sabouraudia *16*: 93-101.

87. Costa, E. O. da, C. Fava Netto, A. Rodrigues, and T. Brito. 1978. Bovine experimental paracoccidioidomycosis intradermic test standardization. Sabouraudia *16*: 103-113.

88. Cox, R. A. 1979. Cross-reactivity between antigens of *Coccidioides immitis, Histoplasma capsulatum,* and *Blastomyces dermatitidis* in lymphocyte transformation assays. Infect. Immun. *25*: 932-938.

89. Cox, R. A., and G. K. Best. 1972. Cell wall composition of two strains of *Blastomyces dermatitidis* exhibiting differences in virulence for mice. Infect. Immun. *5*: 449-453.

90. Cox, R. A., E. Brummer, and G. Lecara. 1977. In vitro lymphocyte responses of coccidioidin skin test-positive and -negative persons to coccidioidin, spherulin, and a *Coccidioides* cell wall antigen. Infect. Immun. *15*: 751-755.

91. Cox, R. A., and H. W. Larsh. 1974. Isolation of skin test-active preparations from yeast-phase cells of *Blastomyces dermatitidis.* Infect. Immun. *10*: 42-47.

92. Cox, R. A., and H. W. Larsh. 1974. Yeast- and mycelial-phase antigens of *Blastomyces dermatitidis:* Comparison using disc gel electrophoresis. Infect. Immun. *10*: 48-53.

93. Cox, R. A., L. R. Mills, G. K. Best, and J. F. Denton. 1974. Histologic reactions to

cell walls of an avirulent and a virulent strain of *Blastomyces dermatitidis*. J. Infect. Dis. *129*: 179-186.

94. Cox, R. A., and J. R. Vivas. 1977. Spectrum of in vivo and in vitro cell-mediated immune responses in coccidioidomycosis. Cell. Immunol. *31*: 130-141.

95. Cox, R. A., J. R. Vivas, A. Gross, G. Lecara, E. Miller, and E. Brummer. 1975. In vivo and in vitro cell-mediated responses in coccidioidomycosis. I. Immunologic responses of persons with primary asymptomatic infections. Am. Rev. Respir. Dis. *114*: 937-943.

96. Craig, L. C. 1968. Dialysis and ultrafiltration. In C. A. Williams and M. W. Chase (Eds.), *Methods in Immunology and Immunochemistry*, Academic, New York, pp. 119-133.

97. Davis, T. E., Jr., J. E. Domer, and Y.-T. Li. 1977. Cell wall studies of *Histoplasma capsulatum* and *Blastomyces dermatitidis* using autologous and heterologous enzymes. Infect. Immun. *15*: 978-987.

98. Dee, T. H. 1975. Detection of *Aspergillus fumigatus* serum precipitins by counterimmunoelectrophoresis. J. Clin. Microbiol. *2*: 482-485.

99. Deighton, F., R. A. Cox, N. K. Hall, and H. W. Larsh. 1977. In vivo and in vitro cell-mediated immune responses to a cell wall antigen of *Blastomyces dermatitidis*. Infect. Immun. *15*: 429-435.

100. Deresinski, S. C., R. J. Applegate, H. B. Levine, and D. A. Stevens. 1977. Cellular immunity to *Coccidioides immitis:* in vitro lymphocyte response to spherules, arthrospores, and endospores. Cell. Immunol. *32*: 110-119.

101. Deresinski, S. C., H. B. Levine, P. C. Kelly, R. J. Creasman, and D. A. Stevens. 1977. Spherulin skin testing and histoplasmal and coccidioidal serology: lack of effect. Am. Rev. Respir. Dis. *116*: 1116-1118.

102. Deresinski, S. C., H. B. Levine, and D. A. Stevens. 1974. Soluble antigens of mycelia and spherules in the in vitro detection of immunity to *Coccidioides immitis*. Infect. Immun. *10*: 700-704.

103. Dessaint, J. P., D. Bout, J. Fruit, and A. Capron. 1976. Serum concentration of specific IgE antibody against *Aspergillus fumigatus* and identification of the fungal allergen. Clin. Immunol. Immunopathol. *5*: 314-319.

104. Diamond, R. D., R. K. Root, and J. E. Bennett. 1972. Factors influencing killing of *Cryptococcus neoformans* by human leukocytes in vitro. J. Infect. Dis. *125*: 367-376.

105. Dickerson, W. H., Jr., and J. F. Busey. 1968. Chromatographic separation of h and m antigens from histoplasmin. Proc. Soc. Exp. Biol. Med. *128*: 654-658.

106. Domer, J. 1971. Monosaccharide and chitin content of cell walls of *Histoplasma capsulatum* and *Blastomyces dermatitidis*. J. Bacteriol. *107*: 870-877.

107. Domer, J. E. 1976. In vivo and in vitro cellular responses to cytoplasmic and cell wall antigens of *Histoplasma capsulatum* in artificially immunized or infected guinea pigs. Infect. Immun. *13*: 790-799.

108. Domer, J. E., J. G. Hamilton, and J. C. Harkin. 1967. Comparative study of the cell walls of the yeastlike and mycelial phases of *Histoplasma capsulatum*. J. Bacteriol. *94*: 466-474.

109. Domer, J. E., and H. Ichinose. 1977. Cellular immune responses in guinea pigs immunized with cell walls of *Histoplasma capsulatum* prepared by several different procedures. Infect. Immun. *16*: 293-301.

110. Domer, J. E., and S. A. Moser. 1978. Experimental murine candidiasis: cell-mediated immunity after cutaneous challenge. Infect. Immun. *20*: 88-98.

111. Drouhet, E., L. Camey, and G. Segretain. 1972. Valeur de l'immunoprécipitation et de l'immunofluorescence indirecte dans les aspergilloses bronchopulmonaires. Ann. Inst. Pasteur (Paris) *123*: 379-395.

112. Drouhet, E., G. Segretain, and J. P. Aubert. 1950. Polyoxide capsulaire d'un champignon pathogène *Torulopsis neoformans*. Relation avec la virulence. Ann. Inst. Pasteur (Paris) *79*: 891-900.
113. Dykstra, M. A., and L. Friedman. 1978. Pathogenesis, lethality, and immunizing effect of experimental cutaneous cryptococcosis. Infect. Immun. *20*: 446-455.
114. Dykstra, M. A., L. Friedman, and J. W. Murphy. 1977. Capsule size of *Cryptococcus neoformans:* control and relationship to virulence. Infect. Immun. *16*: 129-135.
115. Dyson, J. E., Jr., and E. E. Evans. 1955. Delayed hypersensitivity in experimental fungus infections. The skin reactivity of antigens from the yeast phase of *Blastomyces dermatitidis*. J. Invest. Dermatol. *24*: 447-454.
116. Dyson, J. E., and E. E. Evans. 1955. Delayed hypersensitivity in experimental fungus infections. The skin reactivity of antigens from the yeast phase of *Histoplasma capsulatum*. J. Lab. Clin. Med. *45*: 449-454.
117. Ehrhard, H. B., and L. Pine. 1972. Factors influencing the production of H and M antigens by *Histoplasma capsulatum:* development and evaluation of a shake culture procedure. Appl. Microbiol. *23*: 236-249.
118. Ehrhard, H. B., and L. Pine. 1972. Factors influencing the production of H and M antigens by *Histoplasma capsulatum*. Effect of physical factors and composition of medium. Appl. Microbiol. *23*: 250-261.
119. Ellsworth, J. H., E. Reiss, R. L. Bradley, H. Chmel, and D. Armstrong. 1977. Comparative serological and cutaneous reactivity of candidal cytoplasmic proteins and mannan separated by affinity for concanavalin A. J. Clin. Microbiol. *5*: 91-99.
120. Emmons, C. W., B. J. Olson, and W. W. Eldridge. 1945. Studies of the role of fungi in pulmonary disease. I. Cross reactions of histoplasmin. Public Health Rep. *60*: 1383-1394.
121. Evans, E. E. 1949. An immunologic comparison of twelve strains of *Cryptococcus neoformans (Torula histolytica)*. Proc. Soc. Exp. Biol. Med. *71*: 644-646.
122. Evans, E. E. 1950. The antigenic composition of *Cryptococcus neoformans*. I. A serologic classification by means of the capsular and agglutination reactions. J. Immunol. *64*: 423-430.
123. Evans, E. E. 1960. Capsular reactions of *Cryptococcus neoformans*. Ann. N.Y. Acad. Sci. *89*: 184-192.
124. Evans, E. E., and J. W. Mehl. 1951. A quantitative analysis of capsular polysaccharides from *Cryptococcus neoformans* by filter paper chromatography. Science *114*: 10-11.
125. Evans, E. E., and R. J. Theriault. 1953. The antigenic composition of *Cryptococcus neoformans*. IV. The use of paper chromatography for following purification of the capsular polysaccharide. J. Bacteriol. *65*: 571-577.
126. Evans, E. G. V., M. D. Richardson, F. C. Odds, and K. T. Holland. 1973. Relevance of antigenicity of *Candida albicans* growth phases to diagnosis of systemic candidiasis. Br. Med. J. *4*: 86-87.
127. Everett, E. D., F. M. LaForce, and T. C. Eickhoff. 1975. Serologic studies in suspected visceral candidiasis. Arch. Intern. Med. *135*: 1075-1078.
128. Farhi, F., G. S. Bulmer, and J. R. Taker. 1970. *Cryptococcus neoformans*. IV. The not-so-encapsulated yeast. Infect. Immun. *1*: 526-531.
129. Farkas, V. 1979. Biosynthesis of cell walls of fungi. Microbiol. Rev. *43*: 117-144.
130. Fava Netto, C. 1955. Estudos quantitativos sõbre a fixacão do complemento na blastomicose sul-americana, com antigeno polisaccaridico. Argent. Circ. Clin. Exp. *18*: 197-254.
131. Fava Netto, C. 1972. The serology of paracoccidioidomycosis: present and future trends. In *Paracoccidioidomycosis*. Proceedings of the First Pan Amer-

ican Symposium, Pan Am. Health Organ. Sci. Publ. No. 254, Washington, D.C., pp. 209-213.

132. Fava Netto, C., M. A. G. Guerra, and E. O. da Costa. 1976. Contribuição ao estudo imunológico de paracoccidioidomicose. Reações intradérmicas em pacientes com dois antígenos homólogos e dois heterólogos. Rev. Inst. Med. Trop. Sao Paulo *18*: 186-190.

133. Fava Netto, C., M. A. G. Guerra, E. O. da Costa, and P. H. Yasuda. 1976. Contribuição à serologia de paracoccidioidomicose. Comparação entre reações de fixação do complemento pelas técnicas de Wadsworth, Maltaner and Maltaner e micrométodo e reações, de precipitação em meio líquido e gel de ágar. Rev. Inst. Med. Trop. Sao Paulo *18*: 81-86.

134. Fava Netto, C., V. S. Vegas, I. M. Sciannaméa, and D. B. Guarnieri. 1969. Antígeno polissacarídico do *Paracoccidioides brasiliensis* estudo do tempo de cultivo do *P. brasiliensis,* necessário ao preparo do antígeno. Rev. Inst. Med. Trop. Sao Paulo *11*: 177-181.

135. Feit, C., and R. P. Tewari. 1974. Immunogenicity of ribosomal preparations from yeast cells of *Histoplasma capsulatum.* Infect. Immun. *10*: 1091-1097.

136. Fiese, M. J. 1958. *Coccidioidomycosis,* Thomas, Springfield, Ill.

137. Fischer, A., J.-J. Ballet, and C. Griscelli. 1978. Specific inhibition of in vitro *Candida*-induced lymphocyte proliferation by polysaccharidic antigens present in the serum of patients with chronic mucocutaneous candidiasis. J. Clin. Invest. *62*: 1005-1013.

138. Fischer, J. B., and N. A. Labzoffsky. 1955. Preparation of complement fixing antigen from *Cryptococcus neoformans.* Can. J. Microbiol. *1*: 451-454.

139. Flaherty, D. K., and R. Burrell. 1970. Further environmental factors affecting the antigenicity of *Trichophyton rubrum.* Mycopathol. Mycol. Appl. *42*: 165-175.

140. Forman, S. R., J. N. Fink, V. L. Moore, J. Wang, and R. Patterson. 1978. Humoral and cellular immune responses in *Aspergillus fumigatus* pulmonary disease. Allergy Clin. Immunol. *62*: 131-136.

141. Franck, J., and S. Dunan. 1977. Etude comparative de plusieurs types d'antigènes pour la réaction d'immunofluorescence indirecte dans l'aspergillose. Med. Malad. Infect. *7*: 73-76.

142. Frisk, A., L.-V. von Stedingk, and J. Wasserman. 1974. Lymphocyte stimulation in *Candida albicans* infections. Sabouraudia *12*: 87-94.

143. Fukazawa, Y., T. Shinoda, K. Kagaya, and A. Nishikawa. 1977. Immune response to live and killed cells of *Candida albicans* and related species. In K. Iwata (Ed.), *Recent Advances in Medical and Veterinary Mycology,* Proceedings of the Sixth Congress of the International Society for Human and Animal Mycology, University of Tokyo Press, Tokyo, pp. 67-73.

144. Gadebusch, H. H. 1958. Active immunization against *Cryptococcus neoformans.* J. Infect. Dis. *102*: 219-226.

145. Gadebusch, H. H. 1960. Decomposition of the capsular polysaccharide of *Cryptococcus neoformans* 3723 by a soil microorganism. Naturwissenschaften *47*: 329-330.

146. Gadebusch, H. H. 1960. Specific degradation of *Cryptococcus neoformans* 3723 capsular polysaccharide by a microbial enzyme. II. Biological activity of the enzyme. J. Infect. Dis. *107*: 402-405.

147. Gadebusch, H. H. 1960. Specific degradation of *Cryptococcus neoformans* 3723 capsular polysaccharide by a microbial enzyme. III. Antibody stimulation by partially decapsulated cells. J. Infect. Dis. *107*: 406-409.

148. Gadebusch, H. H., and J. D. Johnson. 1961. Specific degradation of *Cryptococcus neoformans* 3723 capsular polysaccharide by a microbial enzyme. I.

Isolation, partial purification, and properties of the enzyme. Can. J. Microbiol. *7:* 53-60.

149. Gadebusch, H. H., P. A. Ward, and E. P. Frenkel. 1964. Natural host resistance to infection with *Cryptococcus neoformans*. III. The effect of cryptococcal polysaccharide upon the physiology of the reticuloendothelial system of laboratory animals. J. Infect. Dis. *114:* 95-106.

150. Gander, J. E. 1974. Fungal cell wall glycoproteins and peptido-polysaccharides. Annu. Rev. Microbiol. *28:* 103-119.

151. Garcia, J. P., and D. H. Howard. 1971. Characterization of antigens from the yeast phase of *Histoplasma capsulatum*. Infect. Immun. *4:* 116-125.

152. Gerber, J. D., and R. D. Jones. 1973. Immunologic significance of aspergillin antigens of six species of *Aspergillus* in the serodiagnosis of aspergillosis. Am. Rev. Respir. Dis. *108:* 1124-1129.

153. Gernez-Rieux, Ch., J. Biguet, C. Voisin, A. Capron. P. Tran Van Ky, and P. Foissac-Gegoux. 1964. Intérêt diagnostique de la recherche des précipitines sériques au cours des aspergilloses bronchopulmonaires. Med. Hyg. *22:* 513-514.

154. Goldstein, G. B., H. Park, and M. Yokoyama. 1974. Studies of the precipitating and antibody response in pulmonary aspergilloma. Int. Arch. Allergy Appl. Immunol. *44:* 1-10.

155. Gordon, M. A. 1975. Current status of serology for diagnosis and prognostic evaluation of opportunistic fungus infections. In *Mycoses,* Proceedings of the Third International Conference on the Mycoses, Pan Am. Health Organ. Sci. Publ. No. 304, Washington, D.C., pp. 144-151.

156. Gordon, M. A., and Y. Al-Doory. 1965. Application of fluorescent-antibody procedures to the study of pathogenic dematiaceous fungi. II. Serological relationships of the genus *Fonsecaea*. J. Bacteriol. *89:* 551-556.

157. Gordon, M. A., R. E. Almy, C. H. Greene, and J. W. Fenton. 1971. Diagnostic mycoserology by immunoelectrophoresis. Am. J. Clin. Pathol. *56:* 471-474.

158. Gordon, M. A., J. C. Elliott, and T. W. Hawkins. 1967. Identification of *Candida albicans*, other *Candida* species and *Torulopsis glabrata* by means of immunofluorescence. Sabouraudia *5:* 323-328.

159. Gordon, M. A., E. W. Lapa, and J. Kane. 1977. Modified indirect fluorescent-antibody test for aspergillosis. J. Clin. Microbiol. *6:* 161-165.

160. Gordon, M. A., and D. K. Vedder. 1966. Serologic tests in diagnosis and prognosis of cryptococcosis. JAMA *197:* 131-137.

161. Goren, M., and G. M. Middlebrook. 1967. Protein conjugates of polysaccharide from *Cryptococcus neoformans*. J. Immunol. *98:* 901-913.

162. Gorin, P. A. J., R. H. Haskins, L. R. Travassos, and L. Mendonca-Previato. 1977. Further studies on the rhamnomannans and acidic rhamnomannans of *Sporothrix schenckii* and *Ceratocystis stenoceras*. Carbohydr. Res. *55:* 21-33.

163. Gorin, P. A. J., and J. F. T. Spencer. 1970. Structures of the L-rhamno-D-mannan from *Ceratocystis ulmi* and the D-gluco-D-mannan from *Ceratocystis brunea*. Carbohydr. Res. *13:* 339-349.

164. Grappel, S. F., C. T. Bishop, and F. Blank. 1974. Immunology of dermatophytosis. Bacteriol. Rev. *38:* 222-250.

165. Grappel, S. F., F. Blank, and C. T. Bishop. 1967. Immunological studies on dermatophytes. I. Serological reactivities of neutral polysaccharides with rabbit antiserum to *Microsporum quinckeanum*. J. Bacteriol. *93:* 1001-1008.

166. Graybill, J. R., and R. H. Alford. 1974. Cell mediated immunity in cryptococosis. Cell. Immunol. *14:* 12-21.

167. Graybill, J. R., and R. L. Taylor. 1978. Host defense in cryptococcosis. I. An

in vivo model for evaluating immune response. Int. Arch. Allergy Appl. Immunol. *57*: 101-113.

168. Green, J. H., W. K. Harrell, S. B. Gray, J. E. Johnson, R. C. Bolin, H. Gross, and G. B. Malcolm. 1976. H and M antigens of *Histoplasma capsulatum:* preparation of antisera and location of these antigens in yeast-phase cells. Infect. Immun. *14*: 826-831.

169. Green, J. H., W. K. Harrell, J. E. Johnson, and R. Benson. 1979. Preparation of reference antisera for laboratory diagnosis of blastomycosis. J. Clin. Microbiol. *10*: 1-7.

170. Greene, C. H., L. S. DeLalla, and V. N. Tompkins. 1960. Separation of specific antigens of *Histoplasma capsulatum* by ion-exchange chromatography. Proc. Soc. Exp. Biol. Med. *105*: 140-141.

171. Gross, H., G. Bradley, L. Pine, S. Gray, J. H. Green, and W. K. Harrell. 1975. Evaluation of histoplasmin for the presence of H and M antigens: some difficulties encountered in the production and evaluation of a product suitable for the immunodiffusion test. J. Clin. Microbiol. *1*: 330-334.

172. Gumaa, S. A., and E. S. Mahgoub. 1975. Counterimmunoelectrophoresis in the diagnosis of mycetoma and its sensitivity as compared to immunodiffusion. Sabouraudia *13*: 309-315.

173. Hall, N. K., F. Deighton, and H. W. Larsh. 1978. Use of an alkali-soluble water-soluble extract of *Blastomyces dermatitidis* yeast-phase cell walls and isoelectrically focused components in peripheral lymphocyte transformations. Infect. Immun. *19*: 411-415.

174. Harada, T., T. Nishikawa, and H. Hatano. 1976. Antigenic similarity between *Ceratocystis* species and *Sporothrix schenckii* as observed by immunofluorescence. Sabouraudia *14*: 211-215.

175. Harrell, W. K., H. Ashworth, L. E. Britt, J. R. George, S. B. Gray, J. H. Green, H. Gross, and J. E. Johnson. 1973. *Procedural Manual for Production of Bacterial, Fungal, Parasitic Reagents,* Center for Disease Control, Atlanta.

176. Hasenclever, H. F., and F. J. McAtee. 1977. Antigenic relationships of *Candida albicans, Saccharomyces telluris,* and *Saccharomyces serevisiae.* Contrib. Microbiol. Immunol. *3*: 126-137.

177. Hasenclever, H. F., and W. O. Mitchell. 1961. Antigenic studies of *Candida.* I. Observation of two antigenic groups in *Candida albicans.* J. Bacteriol. *82*: 570-573.

178. Hasenclever, H. F., and W. O. Mitchell. 1964. Immunochemical studies on polysaccharides of yeasts. J. Immunol. *93*: 763-771.

179. Hasenclever, H. F., and W. O. Mitchell. 1964. A study of yeast surface antigens by agglutination inhibition. Sabouraudia *3*: 288-300.

180. Hasenclever, H. F., W. O. Mitchell, and J. Loewe. 1961. Antigenic studies of *Candida.* II. Antigenic relation of *Candida albicans* group A and group B to *Candida stellatoidea* and *Candida tropicalis.* J. Bacteriol. *82*: 574-577.

181. Hay, R. J., and E. Reiss. 1978. Delayed-type hypersensitivity responses in infected mice elicited by cytoplasmic fractions of *Cryptococcus neoformans.* Infect. Immun. *22*: 72-79.

182. Hearn, V. M., and D. W. R. Mackenzie. 1979. The preparation and chemical composition of fractions from *Aspergillus fumigatus* wall and protoplasts possessing antigenic activity. J. Gen. Microbiol. *112*: 35-44.

183. Heiner, D. C. 1958. Diagnosis of histoplasmosis using precipitin reactions in agar gel. Pediatrics *22*: 616-627.

184. Hinton, P. F., and C. W. Campbell. 1967. Demonstration of six antigenic con-

stituents of histoplasmin by immunoelectrophoresis. Am. J. Med. Technol. *33*: 46-54.

185.   Ho, Y. M., M. H. Ng, and C. T. Huang. 1979. Antibodies to germinating and yeast cells of *Candida albicans* in human and rabbit sera. J. Clin. Pathol. *32*: 399-405.

186.   Hommel, M., T. K. Truong, and D. E. Bidwell. 1976. Technique immunoenzymatique (E.L.I.S.A.) appliquée au diagnostic sérologique des candidoses et aspergilloses humaines. Nouv. Presse Med. *5*: 2789-2791.

187.   Hopfer, R. L., and D. Gröschel. 1979. Detection by counterimmunoelectrophoresis of anti-*Candida* precipitins in sera from cancer patients. Am. J. Clin. Pathol. *72*: 215-218.

188.   Hosty, T. S., C. McDurmont, L. Ajello, L. K. Georg, G. L. Brumfield, and A. A. Calix. 1961. Prevalence of *Nocardia asteroides* in sputa examined by a tuberculosis diagnostic laboratory. J. Lab. Clin. Med. *58*: 107-114.

189.   How, M. J., M. T. Withnall, and C. N. D. Cruickshank. 1972. Allergenic glucans from dermatophytes. Part I. Isolation, purification, and biological properties. Carbohydr. Res. *25*: 341-353.

190.   Howell, A., Jr. 1947. Studies on fungus antigens. I. Quantitative studies on cross reactions between histoplasmin and blastomycin in guinea pigs. Public Health Rep. *62*: 631-651.

191.   Huppert, M. 1962. Immunization against superficial fungus infection. In G. Dalldorf (Ed.), *Fungi and Fungous Diseases,* Thomas, Springfield, Ill., pp. 239-253.

192.   Huppert, M. 1970. Serology of coccidioidomycosis. Mycopathol. Mycol. Appl. *41*: 107-113.

193.   Huppert, M. 1970. Standardization of immunological reagents. In *The First International Symposium on the Mycoses,* Pan Am. Health Organ. Sci. Publ. No. 205, Washington, D.C., pp. 243-252.

194.   Huppert, M., J. P. Adler, E. H. Rice, and S. H. Sun. 1979. Common antigens among systemic disease fungi analyzed by two-dimensional immunoelectrophoresis. Infect. Immun. *23*: 479-485.

195.   Huppert, M., and J. W. Bailey. 1963. Immunodiffusion as a screening test for coccidioidomycosis serology. Sabouraudia *2*: 284-291.

196.   Huppert, M., and J. W. Bailey. 1965. The use of immunodiffusion tests in coccidioidomycosis. I. The accuracy and reproducibility of the immunodiffusion test which correlates with complement fixation. Am. J. Clin. Pathol. *44*: 364-368.

197.   Huppert, M., and J. W. Bailey. 1965. The use of immunodiffusion tests in coccidioidomycosis. II. An immunodiffusion test as a substitute for the tube precipitin test. Am. J. Clin. Pathol. *44*: 369-373.

198.   Huppert, M., J. W. Bailey, and P. A. Chitjian. 1967. Immunodiffusion as a substitute for complement fixation and tube precipitin tests in coccidioidomycosis. In L. Ajello (Ed.), *Coccidioidomycosis,* Papers from the Second Symposium on Coccidioidomycosis, University of Arizona Press, Tucson, pp. 221-225.

199.   Huppert, M., I. Krasnow, K. R. Vukovich, S. H. Sun, E. H. Rice, and L. J. Kutner. 1977. Comparison of coccidioidin and spherulin in complement fixation tests for coccidioidomycosis. J. Clin. Microbiol. *6*: 33-41.

200.   Huppert, M., E. T. Peterson, S. H. Sun, P. A. Chitjian, and W. J. Derrevere. 1967. Evaluation of a latex particle agglutination test for coccidioidomycosis. Am. J. Clin. Pathol. *49*: 96-102.

201.   Huppert, M., N. S. Spratt, K. R. Vukovich, S. H. Sun, and E. H. Rice. 1978. Antigenic analysis of coccidioidin and spherulin determined by two-dimensional immunoelectrophoresis. Infect. Immun. *20*: 541-551.

202.   Huppert, M., S. H. Sun, and E. H. Rice. 1978. Specificity of exoantigens for identifying cultures of *Coccidioides immitis.* J. Clin. Microbiol. *8*: 346-348.

203. Hurley, R. 1967. The pathogenic *Candida* species. A review. Rev. Med. Vet. Mycol. *6*: 159-176.

204. Ibrahim, A. B., and D. Pappagianis. 1973. Experimental induction of anergy to coccidioidin by antigens of *Coccidioides immitis*. Infect. Immun. *7*: 786-794.

205. Ishaq, C. M., G. S. Bulmer, and F. G. Felton. 1968. An evaluation of various environmental factors affecting the propagation of *Cryptococcus neoformans*. Mycopathol. Mycol. Appl. *35*: 81-90.

206. Ishizaki, H. 1970. Some antigenic substances from culture filtrate of *Sporotrichum schenckii*. Jpn. J. Dermatol. *80*: 16-23.

207. Ishizaki, H., Y. Nakamura, H. Kariya, T. Iwatsu, and R. Wheat. 1976. Delayed hypersensitivity cross-reactions between *Sporothrix schenckii* and *Ceratocystis* species in sporotrichotic patients. J. Clin. Microbiol. *3*: 545-547.

208. Ito, Y. 1965. On the immunologically active substances of the dermatophytes. I. J. Invest. Dermatol. *45*: 275-284.

209. Ito, Y. 1965. On the immunologically active substances of the dermatophytes. II. J. Invest. Dermatol. *45*: 285-294.

210. Kabat, E. A., and M. M. Mayer. 1961. *Experimental Immunochemistry*, 2nd ed., Thomas, Springfield, Ill.

211. Kabe, J., Y. Aoki, T. Ishizaki, T. Miyamoto, H. Nakazawa, and M. Tomaru. 1971. Relationship of dermal and pulmonary sensitivity to extracts of *Candida albicans*. Am. Rev. Respir. Dis. *104*: 348-357.

212. Kanetsuna, F., and L. M. Carbonell. 1971. Cell wall composition of the yeastlike and mycelial forms of *Blastomyces dermatitidis*. J. Bacteriol. *106*: 946-948.

213. Kanetsuna, F., L. M. Carbonell, I. Azuma, and Y. Yamamura. 1972. Biochemical studies on the thermal dimorphism of *Paracoccidioides brasiliensis*. J. Bacteriol. *110*: 208-218.

214. Kanetsuna, F., L. M. Carbonell, F. Gil, and I. Azuma. 1974. Chemical and ultrastructural studies on the cell walls of the yeastlike and mycelial forms of *Histoplasma capsulatum*. Mycopathol. Mycol. Appl. *54*: 1-13.

215. Kaplan, W. 1972. Application of immunofluorescence to the diagnosis of paracoccidioidomycosis. In *Paracoccidioidomycosis*. Proceedings of the First Pan American Symposium, Pan Am. Health Organ. Sci. Publ. No. 254, Washington, D.C., pp. 224-226.

216. Kaplan, W. M. 1975. Practical application of fluorescent antibody procedures in medical mycology. In *Mycoses*. Proceedings of the Third International Conference on the Mycoses, Pan Am. Health Organ. Sci. Publ. No. 304, Washington, D.C., pp. 178-185.

217. Kaplan, W., and L. Kaufman. 1963. Specific fluorescent antiglobulins for the detection and identification of *Blastomyces dermatitidis* yeast-phase cells. Mycopathol. Mycol. Appl. *19*: 173-180.

218. Karlin, J. V., and H. S. Nielsen, Jr. 1970. Serologic aspects of sporotrichosis. J. Infect. Dis. *121*: 316-327.

219. Kase, A., and J. D. Marshall. 1960. A study of *Cryptococcus neoformans* by means of the fluorescent antibody technic. Am. J. Clin. Pathol. *34*: 52-56.

220. Kashkin, A. P., and Y. N. Voevodin. 1976. Proteoliticheskie fermenty *Trichophyton mentagrophytes* (Robin) Blanchard shtamm 69. Mikol. Fitopatol. *10*: 179-185. (Abstract in Rev. Med. Vet. Mycol. *13*: No. 228, 1978.)

221. Kashkin, K. P., A. I. Drozdov, and V. A. Ponomarenko. 1978. Comparative immunochemical studies of different antigens of deep mycosis pathogens. Ann. Immunol. (Paris) *129C*: 287-300.

222. Kashkin, K. P., S. M. Likholetov, and A. V. Lipnitsky. 1977. Studies on media-

tors of cellular immunity in experimental coccidioidomycosis. Sabouraudia *15*: 59-68.

223. Kaufman, L. 1970. Serology: its value in the diagnosis of coccidioidomycosis, cryptococcosis, and histoplasmosis. In *The First International Symposium on the Mycoses,* Pan Am. Health Organ. Sci. Publ. No. 205, Washington, D.C., pp. 96-100.

224. Kaufman, L. 1972. Evaluation of serological tests for paracoccidioidomycosis. Preliminary report. In *Paracoccidioidomycosis,* Proceedings of the First Pan American Symposium, Pan Am. Health Organ. Sci. Publ. No. 254, Washington, D.C., pp. 221-223.

225. Kaufman, L. 1977. Immunology: its value in diagnosing systemic fungal infections. Contrib. Microbiol. Immunol. *3*: 95-105.

226. Kaufman, L., and S. Blumer. 1966. Occurrence of serotypes among *Histoplasma capsulatum* strains. J. Bacteriol. *91*: 1434-1439.

227. Kaufman, L., and S. Blumer. 1968. Value and interpretation of serological tests for the diagnosis for cryptococcosis. Appl. Microbiol. *16*: 1907-1912.

228. Kaufman, L., and M. J. Clark. 1974. Value of the concomitant use of complement fixation and immunodiffusion tests in the diagnosis of coccidioidomycosis. Appl. Microbiol. *28*: 641-643.

229. Kaufman, L., D. W. McLaughlin, M. J. Clark, and S. Blumer. 1973. Specific immunodiffusion test for blastomycosis. Appl. Microbiol. *26*: 244-247.

230. Kaufman, L., and P. G. Standard. 1978. Improved version of the exoantigen test for identification of *Coccidioides immitis* and *Histoplasma capsulatum* cultures. J. Clin. Microbiol. *8*: 42-45.

231. Keeler, R. F., and A. C. Pier. 1965. Extracellular antigens of *Nocardia asteroides.* II. Fractionation and chemical composition. Am. Rev. Respir. Dis. *91*: 400-408.

232. Keeney, E. L., and N. Ericksen. 1949. The chemical isolation and biologic assay of extracellular antigenic fractions from pathogenic fungi. J. Allergy *20*: 172-184.

233. Kessler, G., and W. J. Nickerson. 1959. Glucomannan protein complexes from the cell walls of yeasts. J. Biol. Chem. *234*: 2281-2285.

234. Khuller, G. K. 1976. The mannophosphoinositides of *Nocardia asteroides.* Experientia *32*: 1371.

235. Kim. S. J., and S. D. Chaparas. 1978. Characterization of antigens from *Aspergillus fumigatus.* I. Preparation of antigens from organisms grown in completely synthetic medium. Am. Rev. Respir. Div. *118*: 547-551.

236. Kim, S. J., and S. D. Chaparas. 1979. Characterization of antigens from *Aspergillus fumigatus.* III. Comparison of antigenic relationships of clinically important aspergilli. Am. Rev. Respir. Dis. *120*: 1297-1303.

237. Kim. S. J., S. D. Chaparas, T. M. Brown, and M. C. Anderson. 1978. Characterization of antigens from *Aspergillus fumigatus.* II. Fractionation and electrophoretic, immunologic, and biologic activity. Am. Rev. Respir. Dis. *118*: 553-560.

238. Kim. S. J., S. D. Chaparas, and H. R. Buckley. 1979. Characterization of antigens from *Aspergillus fumigatus.* IV. Evaluation of commercial and experimental preparations and fractions in the detection of antibody in aspergillosis. Am. Rev. Respir. Dis. *120*: 1305-1311.

239. Kingsbury, E. W., and J. M. Slack. 1967. A polypeptide skin test antigen from *Nocardia asteroides.* I. Production, chemical, and biological characterization. Am. Rev. Respir. Dis. *95*: 827-832.

240. Kligman, A. M., and E. D. Delamater. 1950. The immunology of the human mycoses. Annu. Rev. Microbiol. *4*: 283-312.

241. Knight, R. A., S. Coray, and S. Marcus. 1959. *Histoplasma capsulatum* and *Blastomyces dermatitidis* polysaccharide skin tests on humans. Am. Rev. Respir. Dis. *80*: 264-266.

242. Knight, R. A., and S. Marcus. 1958. Polysaccharide skin test antigens derived from *Histoplasma capsulatum* and *Blastomyces dermatitidis*. Am. Rev. Tuberc. Pulm. Dis. *77*: 983-989.

243. Kobayashi, G. S. 1971. Isolation and characterization of polysaccharide of *Histoplasma capsulatum*. In L. Ajello, E. W. Chick, and M. L. Furcolow (Eds.), *Histoplasmosis*, Thomas, Springfield, Ill., pp. 38-44.

244. Kobayashi, G. S., and P. L. Guiliacci. 1967. Cell wall studies of *Histoplasma capsulatum*. Sabouraudia *5*: 180-188.

245. Kobayashi, G. S., and D. Pappagianis. 1970. Preparation and standardization of antigens of *Histoplasma capsulatum* and *Coccidioides immitis*. Mycopathol. Mycol. Appl. *41*: 139-153.

246. Kong, Y. M., and H. B. Levine. 1967. Experimentally induced immunity in the mycoses. Bacteriol. Rev. *31*: 35-53.

247. Kong, Y. M., H. B. Levine, and C. E. Smith. 1963. Immunogenic properties of nondisrupted and disrupted spherules of *Coccidioides immitis* in mice. Sabouraudia *2*: 131-142.

248. Kozel, T. R. 1977. Non-encapsulated variant of *Cryptococcus neoformans*. II. Surface receptors for cryptococcal polysaccharide and their role in inhibition of phagocytosis by polysaccharide. Infect. Immun. *16*: 99-106.

249. Kozel, T. R., and J. Cazin, Jr. 1971. Nonencapsulated variant of *Cryptococcus neoformans*. I. Virulence studies and characterization of soluble polysaccharide. Infect. Immun. *3*: 287-294.

250. Kozel, T. R., and T. G. McGaw. 1979. Opsonization of *Cryptococcus neoformans* by human immunoglobulin G: role of immunoglobulin G in phagocytosis by macrophages. Infect. Immun. *25*: 255-261.

251. Kuehn, H. H. 1961. Nutritional requirements of *Arthroderma tuberculatum*. Mycopathol. Mycol. Appl. *14*: 123-128.

252. Kurup, V. P., J. N. Fink, G. H. Scribner, and M. J. Falk. 1977. Antigenic variability of *Aspergillus fumigatus* strains. Microbios *19*: 191-204.

253. Kwapinski, J. B. G. 1969. Analytical serology of Actinomycetales. In J. B. G. Kwapinski (Ed.), *Analytical Serology of Microorganisms*, Interscience, New York, pp. 1-122.

254. Kwon-Chung, K. J., 1975. Description of a new genus, *Filobasidiella*, the perfect state of *Cryptococcus neoformans*. Mycologia *67*: 1197-1200.

255. Kwon-Chung, K. J. 1976. A new species of *Filobasidiella*, the sexual state of *Cryptococcus neoformans* B and C serotypes. Mycologia *68*: 942-946.

256. Kwon-Chung, K. J., and J. E. Bennett. 1978. Distribution of α and a mating types of *Cryptococcus neoformans* among natural and clinical isolates. Am. J. Epidemiol. *108*: 337-340.

257. Kwon-Chung, K. J., J. E. Bennett, and T. S. Theodore. 1978. *Cryptococcus bacillisporus* sp. nov.: serotype B-C of *Cryptococcus neoformans*. Int. J. Syst. Bacteriol. *28*: 616-620.

258. Lancaster, M. V., and R. F. Sprouse. 1976. Preparative isotachophoretic separation of skin test antigens from blastomycin purified derivative. Infect. Immun. *13*: 758-762.

259. Lancaster, M. V., and R. F. Sprouse. 1976. Isolation of a purified skin test antigen from *Blastomyces dermatitidis* yeast-phase cell wall. Infect. Immun. *14*: 623-625.

260. Land, G. A., W. C. McDonald, R. L. Stjernholm, and L. Friedman. 1975. Factors affecting filamentation in *Candida albicans:* relationship of the uptake and distribution of proline to morphogenesis. Infect. Immun. *11*: 1014-1023.

261. Laskownicka, Z., A. Porebska, and K. Zemburowa. 1967. The serological grouping

of *Candida albicans* strains isolated in the Krakow province in the years 1963-1965 by means of agar gel diffusion test. Mycopathol. Mycol. Appl. *33*: 10-16.

262. Laskownicka, Z., A. Porebska, and K. Zemburowa. 1969. *Candida albicans* antigens studied in complement fixation test and agar gel diffusion test. Mycopathol. Mycol. Appl. *37*: 357-368.

263. Lehmann, P. F., and E. Reiss. 1978. Invasive aspergillosis: antiserum for circulating antigen produced after immunization with serum from infected rabbits. Infect. Immun. *20*: 570-572.

264. Lehmann, P. F., and E. Reiss. 1979. *Aspergillus fumigatus* antigenaemia: detection of antigen in mice and in human patients. Bull. Soc. Mycol. Med. *8*: 57-64.

265. Lepper, A. W. D. 1969. Immunological aspects of dermatomycoses in animals and man. Rev. Med. Vet. Mycol. *6*: 435-446.

266. Levine, H. B. 1962. Immunogenicity of experimental vaccine in systemic mycoses. In G. Dalldorf (Ed.), *Fungi and Fungous Diseases,* Thomas, Springfield, Ill., pp. 254-276.

267. Levine, H. B., J. M. Cobb, and G. M. Scalarone. 1969. Spherule coccidioidin in delayed dermal sensitivity reactions of experimental animals. Sabouraudia 7: 20-32.

268. Levine, H. B., J. M. Cobb, and C. E. Smith. 1960. Immunity to coccidioidomycosis induced in mice by purified spherule, arthrospore, and mycelial vaccines. Trans. N.Y. Acad. Sci. *22*: 436-449.

269. Levine, H. B., A. Gonzalez Ochoa, and D. R. Ten Eyck. 1973. Dermal sensitivity to *Coccidioides immitis.* A comparison of responses elicited in man by spherulin and coccidioidin. Am. Rev. Respir. Dis. *107*: 379-386.

270. Levine, H. B., Y. M. Kong, and C. E. Smith. 1965. Immunization of mice to *Coccidioides immitis:* dose, regimen and spherulation stage of killed spherule vaccines. J. Immunol. *94*: 132-142.

271. Levine, H. B., M. A. Restrepo, D. R. Ten Eyck, and D. A. Stevens. 1975. Spherulin and coccidioidin: cross-reactions in dermal sensitivity to histoplasmin and paracoccidioidin. Am. J. Epidemiol. *101*: 512-516.

272. Levine, H. B., G. M. Scalarone, G. D. Campbell, J. R. Graybill, P. C. Kelly, and S. D. Chaparas. 1979. Histoplasmin-CYL, a yeast phase reagent in skin test studies with humans. Am. Rev. Respir. Dis. *119*: 629-636.

273. Levine, H. B., G. M. Scalerone, and S. D. Chaparas. 1977. Preparation of fungal antigens and vaccines: studies on *Coccidioides immitis* and *Histoplasma capsulatum.* Contrib. Microbiol. Immunol. *3*: 106-125.

274. Levine, S., E. E. Evans, and P. W. Kabler. 1959. Studies of *Cryptococcus* polysaccharides by infrared spectrophotometry. J. Infect. Dis. *104*: 269-273.

275. Littman, M. L. 1958. Capsule synthesis by *Cryptococcus neoformans.* Trans. N.Y. Acad. Sci. *20*: 623-648.

276. Lloyd, K. O., and M. A. Bitoon. 1971. Isolation and purification of a peptidorhamnomannan from the yeast form of *Sporothrix schenckii.* Structural and immunochemical studies. J. Immunol. *107*: 663-671.

277. Lloyd, K. O., L. Mendonca-Previato, and L. R. Travassos. 1978. Distribution of antigenic polysaccharides in different cell types of *Sporothrix schenckii* as studied by immunofluorescent staining with rabbit antisera. Exp. Mycol. *2*: 130-137.

278. Lloyd, K. O., and L. P. Travassos. 1975. Immunochemical studies on L-rhamno-D-mannans of *Sporothrix schenckii* and related fungi by use of rabbit and human antisera. Carbohydr. Res. *40*: 89-97.

279. Lomanitz, R., and J. M. Hale. 1963. Production of delayed hypersensitivity to *Cryptococcus neoformans* in experimental animals. J. Bacteriol. *86*: 505-509.

280. Longbottom, J. L., and J. Pepys. 1964. Pulmonary aspergillosis: diagnostic and

immunological significance of antigens and C-substance in *Aspergillus fumigatus.* J. Pathol. Bacteriol. *88*: 141-151.
281. Longbottom, J. L., J. Pepys, and F. T. Cline. 1964. Diagnostic precipitin test in aspergillus pulmonary mycetoma. Lancet *1*: 588-589.
282. Lupan, D. M., and J. Cazin, Jr. 1976. Serological diagnosis of petriellidiosis (allescheriosis). I. Isolation and characterization of soluble antigens from *Allescheria boydii* and *Monosporium apiospermum.* Mycopathologia *58*: 31-38.
283. Lupan, D. M., and J. Cazin, Jr. 1977. Serological diagnosis of petriellidiosis (allescheriosis). II. Indirect (passive) hemagglutination assay for antibody to polysaccharide antigens of *Petriellidium (Allescheria) boydii* and *Monosporium apiospermum.* Mycopathologia *62*: 87-95.
284. Lupan, D. M., and J. Cazin, Jr. 1979. Humoral response to experimental petriellidiosis. Infect. Immun. *24*: 843-850.
285. Maccani, J. E. 1977. Detection of cryptococcal polysaccharide using counterimmunoelectrophoresis. Am. J. Clin. Pathol. *68*: 39-44.
286. Mackenzie, D. W. R., and C. M. Philpot. 1975. Counterimmunoelectrophoresis as a routine mycoserological procedure. Mycopathologia *57*: 1-7.
287. Magnusson, M. 1961. Specificity of mycobacterial sensitins. I. Studies in guinea pigs with purified "tuberculin" prepared from mammalian and avian tubercle baccilli, *Mycobacterium balnei,* and other acid-fast bacilli. Am. Rev. Respir. Dis. *83*: 57-68.
288. Mahgoub, E. S. 1964. The value of gel diffusion in the diagnosis of mycetoma. Trans. R. Soc. Trop. Med. Hyg. *58*: 560-563.
289. Mahgoub, E. S. 1975. Serologic diagnosis in mycetoma. In *Mycoses,* Proceedings of the Third International Conference on the Mycoses, Pan Am. Health Organ. Sci. Publ. No. 304, Washington, D.C., pp. 154-161.
290. Malcolm, G. B., L. Pine, R. Cherniak, and C. W. Moss. 1979. Biochemical and serological characteristics of soluble yeast phase antigens of *Histoplasma capsulatum.* Mycopathologia *67*: 3-16.
291. Marconi, P., F. Bistoni, M. Pitzurra, L. Mosci, and A. Vecchiarelli. 1977. Immunogenicità in vivo di estratti 3 M KCl da *Candida albicans.* Boll. Soc. Ital. Biol. Sper. *53*: 395-399.
292. Marcus, S., G. A. Hill, and R. A. Knight. 1960. Antigens of *Blastomyces.* Ann. N.Y. Acad. Sci. *89*: 193-201.
293. Markowitz, H. 1964. Polysaccharide antigens from *Histoplasma capsulatum.* Proc. Soc. Exp. Biol. Med. *115*: 697-700.
294. Marriott, M. S. 1977. Mannan-protein location and biosynthesis in plasma membranes from the yeast form of *Candida albicans.* J. Gen. Microbiol. *103*: 51-59.
295. Martin, D. S. 1950. Practical applications of immunologic principles in the diagnosis and treatment of fungus infections. Ann. N.Y. Acad. Sci. *50*: 1376-1379.
296. Martin, D. S., R. D. Baker, and N. F. Conant. 1936. A case of verrucous dermatitis caused by *Hormodendrum pedrosoi* (chromoblastomycosis) in North Carolina. Am. J. Trop. Med. *16*: 593-620.
297. Masler, L., D. Sikl, S. Bauer, and J. Sandula. 1966. Extracellular polysaccharide-protein complexes produced by selected strains of *Candida albicans* (Robin) Berkhout. Folia Microbiol. (Praha) *11*: 373-378.
298. Matsufuji, S., M. Fukuda, and S. Nakamura. 1963. A common antigen between *Candida albicans* and *Mycobacterium tuberculosis.* Kurume Med. J. *10*: 225-231.
299. Matsumoto, T. 1928. Preliminary note on some serological studies of aspergilli. Phytopathology *18*: 691-696.
300. Matsumoto, T. 1929. The investigation of aspergilli by serological methods. Trans. Br. Mycol. Soc. *14*: 69-88.

301. May, L. K., R. A. Knight, and H. W. Harris. 1966. *Allescheria boydii* and *Aspergillus fumigatus* skin test antigens. J. Bacteriol. *91*: 2155-2157.

302. McGaw, T. G., and T. R. Kozel. 1979. Opsonization of *Cryptococcus neoformans* by human immunoglobulin G: masking of immunoglobulin G by cryptococcal polysaccharide. Infect. Immun. *25*: 262-267.

303. McLaren, M. L., E. S. Mahgoub, and E. Georgakopoulos. 1978. Preliminary investigations on the use of the enzyme linked immunosorbent assay (ELISA) in the serodiagnosis of mycetoma. Sabouraudia *16*: 225-228.

304. McVeigh, I., and D. Morton. 1965. Nutritional studies of *Histoplasma capsulatum*. Mycopathol. Mycol. Appl. *25*: 294-309.

305. Meister, H., B. Heymer, H. Schäfer, and O. Haferkamp. 1977. Role of *Candida albicans* in granulomatous tissue reactions. I. In vitro degradation of *C. albicans* and immunospecificity of split products. J. Infect. Dis. *135*: 224-234.

306. Mendes, E. 1975. Delayed hypersensitivity reactions in patients with paracoccidioidomycosis. In *Mycoses*, Proceedings of the Third International Conference on the Mycoses, Pan Am. Health Organ. Sci. Publ. No. 304, Washington, D.C., pp. 17-22.

307. Mendonça, L., P. A. J. Gorin, K. O. Lloyd, and L. R. Travassos. 1976. Polymorphism of *Sporothrix schenckii* surface polysaccharides as a function of morphological differentiation. Biochemistry *15*: 2423-2431.

308. Mitchell, W. O., and H. F. Hasenclever. 1970. Precipitin inhibition of *Candida stellatoidea* antiserum with oligosaccharides obtained from the parent mannan. Infect. Immun. *1*: 61-63.

309. Miyazaki, T. 1961. Studies on fungal polysaccharides. I. On the isolation and chemical properties of capsular polysaccharide from *Cryptococcus neoformans*. Chem. Pharm. Bull. *9*: 715-718.

310. Miyazaki, T. 1961. Studies on fungal polysaccharides. II. On the componental sugars and partial hydrolysis of the capsular polysaccharide of *Cryptococcus neoformans*. Chem. Pharm. Bull. *9*: 826-829.

311. Miyazaki, T. 1961. Studies on fungal polysaccharides. III. Chemical structure of the capsular polysaccharide from *Cryptococcus neoformans*. Chem. Pharm. Bull. *9*: 829-833.

312. Mogahed, A., L. Monjour, P. Druilhe, and M. Gentilini. 1979. Intérêt de la concanavaline A dans le diagnostic immunologique des candidosis. Bull. Soc. Fr. Mycol. Med. *8*: 51-56.

313. Mohr, J. A., H. G. Muchmore, F. G. Felton, E. R. Rhoades, and B. A. McKown. 1970. The effect of cryptococcin on leukocytic migration. J. Infect. Dis. *122*: 454-458.

314. Mok, W. Y., H. R. Buckley, and C. C. Campbell. 1977. Characterization of antigens from type A and B yeast cells of *Histoplasma capsulatum*. Infect. Immun. *16*: 461-466.

315. Mok, W. Y., and D. L. Greer. 1977. Cell-mediated immune responses in patients with paracoccidioidomycosis. Clin. Exp. Immunol. *28*: 89-98.

316. Monjour, L., G. Niel, J. Dautry, A. Briend, and M. Gentilini. 1975. Application de l'électrosynérèse, sur membrane d'acétate de cellulose, au diagnostic des aspergilloses. Bull. Soc. Fr. Mycol. Med. *4*: 225-229.

317. Moser, S. A., and J. D. Pollack. 1978. Isolation of glycopeptides with skin test activity from dermatophytes. Infect. Immun. *19*: 1031-1046.

318. Muchmore, H. G., F. G. Felton, S. B. Salvin, and E. R. Rhoades. 1968. Delayed hypersensitivity to cryptococcin in man. Sabouraudia *6*: 285-288.

319. Murphy, J. W., and G. C. Cozad. 1972. Immunological unresponsiveness induced

by cryptococcal capsular polysaccharide assayed by the hemolytic plaque technique. Infect. Immun. *5*: 896-901.

320. Murphy, J. W., J. A. Gregory, and H. Larsh. 1974. Skin testing of guinea pigs and footpad testing of mice with a new antigen for detecting delayed hypersensitivity to *Cryptococcus neoformans*. Infect. Immun. *9*: 404-409.

321. Murphy, J. W., and N. Pahlavan. 1979. Cryptococcal culture filtrate antigen for detection of delayed-type hypersensitivity in cryptococcosis. Infect. Immun. *25*: 284-292.

322. Murray, I. G., and E. S. Mahgoub. 1968. Further studies on the diagnosis of mycetoma by double diffusion in agar. Sabouraudia *6*: 106-110.

323. Musatti, C. C. 1975. Cell-mediated immunity in patients with paracoccidioidomycosis. In *Mycoses*, Proceedings of the Third International Conference on the Mycoses, Pan Am. Health Organ. Sci. Publ. No. 304, Washington, D.C., pp. 23-27.

324. Musatti, C. C., M. T. Rezkallah, E. Mendes, and N. F. Mendes. 1976. In vivo and in vitro evaluation of cell-mediated immunity in patients with paracoccidioidomycosis. Cell. Immunol. *24*: 365-374.

325. Nakamura, Y., H. Ishizaki, and R. W. Wheat. 1977. Serological cross-reactivity between group B streptococcus and *Sporothrix schenckii, Ceratocystis* species, and *Graphium* species. Infect. Immun. *16*: 547-549.

326. Negroni, R. 1968. Nuevos estudios sobre antígenos para reacciones serológicas en blastomicosis sudamericana. Rev. Dermatol. Ibero Lat. Am. *10*: 409-416.

327. Negroni, R. 1969. Inmulogía de las candidiasis. Mycopathol. Mycol. Appl. *38*: 189-197.

328. Negroni, R. 1972. Serologic reactions in paracoccidioidomycosis. In *Paracoccidioidomycosis*, Proceedings of the First Pan American Symposium, Pan Am. Health Organ. Sci. Publ. No. 254, Washington, D.C., pp. 203-208.

329. Negroni, R., M. R. I. de E. Costa, O. Bianchi, and R. Galimberti. 1976. Preparación y estudio de un antígeno celular de *Paracoccidioides brasiliensis*, útil para pruebas cutáneas. Sabouraudia *14*: 265-273.

330. Negroni, R., C. I. de Flores, and A. M. Robles. 1977. Estudio sobre el valor diagnóstico de la inmunofluorescenia indirecta en la aspergillosis pulmonar. Sabouraudia *15*: 195-200.

331. Negroni, R., and P. Negroni. 1968. Antígenos del *Paracoccidioides brasiliensis* para las reacciones serológicas. Mycopathol. Mycol. Appl. *34*: 285-288.

332. Negroni, R., A. M. Robles, and J. C. Galussio. 1972. Estudio comparativo de las reacciones serológicas cuantitativas con un antígeno metabólico de *Aspergillus fumigatus*. Mycopathol. Mycol. Appl. *48*: 275-287.

333. Neill, J. M., I. Abrahams, and C. E. Kapros. 1950. A comparison of the immunogenicity of weakly encapsulated and of strongly encapsulated strains of *Cryptococcus neoformans (Torula histolytica)*. J. Bacteriol. *59*: 263-275.

334. Neill, J. M., C. G. Castillo, R. H. Smith, and C. E. Kapros. 1949. Capsular reactions and soluble antigens of *Torula histolytica* and of *Sporotrichum schenckii*. J. Exp. Med. *89*: 93-105.

335. Neisser, A. 1902. Plato's Versuche über die Herstellung und Verwendung von "Trichophytin." Arch. Dermatol. Syphilol. *60*: 63-76.

336. Nickerson, W. J. 1963. Molecular bases of form in yeasts. Bacteriol. Rev. *27*: 305-324.

337. Neilsen, H. S., Jr. 1968. Biologic properties of skin test antigens of yeast form *Sporotrichum schenckii*. J. Infect. Dis. *118*: 173-180.

338. Nishikawa, T. T., T. Harada, S. Harada, and H. Hatano. 1975. Serologic differences in strains of *Sporothrix schenckii*. Sabouraudia *13*: 285-290.

339. Norden, A. 1951. Sporotrichosis. Clinical and laboratory features and a serologic study in experimental animals and humans. Acta Pathol. Microbiol. Scand. (Suppl.) *89*: 1-119.

340. Nozawa, Y., T. Noguchi, Y. Ito, N. Sudo, and S. Watanabe. 1971. Immunochemical studies on *Trichophyton mentagrophytes*. Sabouraudia *9*: 129-138.

341. O'Connell, E. J., P. E. Hermans, and H. Markowitz. 1967. Skin-reactive antigens of histoplasmin. Proc. Soc. Exp. Biol. Med. *124*: 1015-1020.

342. Okubo, Y., Y. Honma, and S. Suzuki. 1979. Relationship between phosphate content and serological activities of the mannans of *Candida albicans* strains NIH A-207, NIH B-792 and J-1012. J. Bacteriol. *137*: 677-680.

343. Ortiz-Ortiz, L., M. F. Contreras, and L. F. Bojalil. 1972. The assay of delayed hypersensitivity to ribosomal proteins from *Nocardia*. Sabouraudia *10*: 147-151.

344. Ortiz-Ortiz, L., M. F. Contreras, and L. F. Bojalil. 1972. Cytoplasmic antigens from *Nocardia* eliciting a specific delayed hypersensitivity. Infect. Immun. *5*: 879-882.

345. Palmer, D. F., L. Kaufman, W. Kaplan, and J. J. Cavallero. 1977. *Serodiagnosis of Mycotic Diseases*. Thomas, Springfield, Ill.

346. Pappagianis, D., R. Hector, H. B. Levine, and M. S. Collins. 1979. Immunization of mice against coccidioidomycosis with a subcellular vaccine. Infect. Immun. *25*: 440-445.

347. Pappagianis, D., E. W. Putman, and G. S. Kobayashi. 1961. Polysaccharide of *Coccidioides immitis*. J. Bacteriol. *82*: 714-723.

348. Pappagianis, D., C. E. Smith, and C. C. Campbell. 1967. Serologic status after positive coccidioidin skin reactions. Am. Rev. Respir. Dis. *96*: 520-523.

349. Pappagianis, D., C. E. Smith, G. S. Kobayashi, and M. T. Saito. 1961. Studies of antigens from young mycelia of *Coccidioides immitis*. J. Infect. Dis. *108*: 35-44.

350. Pappagianis, D., C. E. Smith, M. T. Saito, and G. S. Kobayashi. 1957. Preparation and property of a complement-fixing antigen from mycelia of *Coccidioides immitis*. U.S. Public Health Serv. Rep. Publ. No. *575*: 57-63.

351. Peat, S., W. J. Whelan, and T. E. Edwards. 1961. Polysaccharides of baker's yeast. Part IV. Mannan. J. Chem. Soc. *1961*: 29-34.

352. Peck, R. L., D. S. Martin, and C. R. Hauser. 1940. Polysaccharides of *Blastomyces dermatitidis*. J. Immunol. *38*: 449-455.

353. Peck, S. M. 1950. Fungus antigens and their importance as sensitizers in the general population. Ann. N.Y. Acad. Sci. *50*: 1362-1375.

354. Peck, S. M., and A. Glick. 1941. Trichophytin-methods of preparation with special reference to the specific skin-reactive factor. Arch. Dermatol. Syphilol. *43*: 839-849.

355. Pepys, J. 1969. *Hypersensitivity Diseases of the Lungs Due to Fungi and Organic Dusts*, S. Karger, Basel.

356. Pepys, J., J. A. Fauz, J. L. Longbottom, D. S. McCarthy, and F. E. Hargreave. 1968. *Candida albicans* precipitins in respiratory disease in man. J. Allergy *41*: 305-318.

357. Petrini, B. 1977. Association between skin-test reactivity and in vitro migration inhibition of leucocytes using a *Candida* antigen. Curr. Ther. Res. *22*: 17-19.

358. Phaff, H. J. 1971. Structure and biosynthesis of the yeast cell envelope. In A. H. Rose and J. S. Harrison (Eds.), *The Yeasts*, Vol. 2, Academic, New York, pp. 135-210.

359. Philpot, C. M. 1978. Serological differences among the dermatophytes. Sabouraudia *16*: 247-256.

360. Philpot, C. M., and D. W. R. Mackenzie. 1976. A comparison of *Aspergillus fumigatus* antigens by counterimmunoelectrophoresis. Mycopathologia *58*: 19-20.

361. Philpot, C. M., and D. W. R. Mackenzie. 1976. Detection of antibodies to *Aspergillus fumigatus* in agar gel with different antigens and immunodiffusion patterns. J. Biol. Stand. *4*: 73-79.

362. Piacentini, L., A. Lechi, C. Mengoli, E. Stepan, and U. Bonomi. 1976. Analisi immunoelectroforetica di un preparato antigene Aspergillare. Ann. Sclavo *18*: 286-290.

363. Pier, A. C., and R. E. Fichtner. 1971. Serologic typing of *Nocardia asteroides* by immunodiffusion. Am. Rev. Respir. Dis. *103*: 698-707.

364. Pier, A. C., and R. F. Keeler. 1962. Extracellular antigens of *Nocardia asteroides*. I. Production and immunologic characterization. Am. Rev. Respir. Dis. *91*: 391-399.

365. Pier, A. C., J. R. Thurston, and A. B. Larsen. 1968. A diagnostic antigen for nocardiosis: comparative tests in cattle with nocardiosis and mycobacteriosis. Am. J. Vet. Res. *29*: 397-403.

366. Pine, L. 1970. Growth of *Histoplasma capsulatum*. VI. Maintenance of the mycelial phase. Appl. Microbiol. *19*: 413-420.

367. Pine, L. 1977. *Histoplasma* antigens: their production, purification and uses. Contrib. Microbiol. Immunol. *3*: 138-168.

368. Pine, L., and C. J. Boone. 1968. Cell wall composition and serological reactivity of *Histoplasma capsulatum* serotypes and related species. J. Bacteriol. *96*: 789-798.

369. Pine, L., S. J. Boone, and D. McLaughlin. 1966. Antigenic properties of the cell wall and other fractions of the yeast form of *Histoplasma capsulatum*. J. Bacteriol. *91*: 2158-2168.

370. Pine, L., and E. Drouhet. 1963. Sur l'obtention et la conservation de la phase levure *d'Histoplasma capsulatum* et *d'H. duboisii*, en milieu chimiquement défini. Ann. Inst. Pasteur (Paris) *105*: 798-804.

371. Pine, L., R. G. Falcone, and C. J. Boone. 1969. Effect of thimerosal on the whole yeast phase antigen of *Histoplasma capsulatum*. Mycopathol. Mycol. Appl. *37*: 1-14.

372. Pine, L., H. Gross, G. B. Malcolm, J. R. George, S. B. Gray, and C. W. Moss. 1977. Procedures for the production and separation of H and M antigens in histoplasmin: chemical and serological properties of the isolated products. Mycopathologia *61*: 131-141.

373. Pine, L., G. B. Malcolm, H. Gross, and S. B. Gray. 1978. Evaluation of purified H and M antigens of histoplasmin as reagents in the complement fixation test. Sabouraudia *16*: 257-269.

374. Poor, A. H., and J. E. Cutler. 1979. Partially purified antibodies used in a solid-phase radioimmunoassay for detecting candidal antigenemia. J. Clin. Microbiol. *9*: 362-368.

375. Porter, J. F., E. R. Scheer, and R. W. Wheat. 1971. Characterization of 3-O-methylmannose from *Coccidioides immitis*. Infect. Immun. *4*: 660-661.

376. Rangaswamy, V., R. V. Kumari, and K. Seshagiri. 1977. Serological diagnosis of pulmonary aspergillosis. Indian J. Pathol. Bacteriol. *20*: 101-104.

377. Rebers, P. A., S. A. Barker, M. Heidelberger, Z. Dische, and E. E. Evans. 1958. Precipitation of the specific polysaccharide of *Cryptococcus neoformans* A by types II and XIV of antipneumococcal sera. J. Am. Chem. Soc. *80*: 1135-1137.

378. Redmond, A., I. J. Carré, J. D. Biggart, and D. W. R. Mackenzie. 1965. Aspergillosis *(Aspergillus nidulans)* involving bone. J. Pathol. Bacteriol. *89*: 391-395.

379. Reeves, M. W., L. Pine, and G. Bradley. 1974. Characterization and evaluation of a soluble antigen complex prepared from the yeast phase of *Histoplasma capsulatum*. Infect. Immun. *9*: 1033-1044.

380. Reeves, M. W., L. Pine, L. Kaufman, and D. McLaughlin. 1972. Isolation of a new soluble antigen from the yeast phase of *Histoplasma capsulatum*. Appl. Microbiol. *24*: 841-843.

381. Reiss, E. 1977. Serial enzymatic hydrolysis of cell walls of two serotypes of yeast-form *Histoplasma capsulatum* with $\alpha(1\rightarrow3)$ glucanase, $\beta(1\rightarrow3)$ glucanase, pronase, and chitinase. Infect. Immun. *16*: 181-188.

382. Reiss, E., and P. F. Lehmann. 1979. Galactomannan antigenemia in invasive aspergillosis. Infect. Immun. *25*: 357-365.

383. Reiss, E., S. E. Miller, W. Kaplan, and L. Kaufman. 1977. Antigenic, chemical, and structural properties of cell walls of *Histoplasma capsulatum* yeast-form chemotypes 1 and 2 after serial enzymatic hydrolysis. Infect. Immun. *16*: 690-700.

384. Reiss, E., W. O. Mitchell, S. H. Stone, and H. F. Hasenclever. 1974. Cellular immune activity of a galactomannan-protein complex from mycelia of *Histoplasma capsulatum*. Infect. Immun. *10*: 802-809.

385. Reiss, E., S. H. Stone, and H. F. Hasenclever. 1974. Serological and cellular immune activity of peptidomannan fractions of *Candida albicans* cell walls. Infect. Immun. *9*: 881-890.

386. Restrepo-Moreno, A. 1966. La prueba de inmunodifusión en el diagnóstico de la paracoccidioidomicosis. Sabouraudia *4*: 223-230.

387. Restrepo-Moreno, A. 1970. Serological comparison of the two morphological phases of *Paracoccidioides brasiliensis*. Infect. Immun. *2*: 268-273.

388. Restrepo, A., and E. Drouhet. 1970. Etude des anticorps précipitants dans la blastomycose sud-américaine par l'analyse immunoélectrophorétique des antigènes de *P. brasiliensis*. Ann. Inst. Pasteur (Paris) *119*: 338-346.

389. Restrepo, A., D. L. Greer, and M. Vasconcellos. 1973. Paracoccidioidomycosis: a review. Rev. Med. Vet. Mycol. *8*: 97-123.

390. Restrepo-Moreno, A., and L. H. Moncada F. 1970. Serologic procedures in the diagnosis of paracoccidioidomycosis. In *The First International Symposium on the Mycoses,* Pan Am. Health Organ. Sci. Publ. No. 205, Washington, D.C., pp. 101-110.

391. Restrepo-Moreno, A., and L. H. Moncada F. 1972. Indirect fluorescent-antibody and quantitative agar-gel immunodiffusion tests for the serological diagnosis of paracoccidioidomycosis. Appl. Microbiol. *24*: 132-137.

392. Restrepo, A., and L. H. Moncada. 1974. Characterization of the precipitin bands detected in the immunodiffusion test for paracoccidioidomycosis. Appl. Microbiol. *28*: 138-144.

393. Restrepo, A., M. Restrepo, F. de Restrepo, L. H. Aristizábal, L. H. Moncada, and H. Vélez. 1978. Immune responses in paracoccidioidomycosis. A controlled study of 16 patients before and after treatment. Sabouraudia *16*: 151-163.

394. Restrepo-Moreno, A., and J. D. Schneidau, Jr., 1967. Nature of the skin reactive principle in culture filtrates prepared from *Paracoccidioides brasiliensis*. J. Bacteriol. *93*: 1741-1748.

395. Richardson, M. D., L. O. White, and R. C. Warren. 1979. Detection of circulating antigen of *Aspergillus fumigatus* in sera of mice and rabbits by enzyme-linked immunosorbent assay. Mycopathologia *67*: 83-88.

396. Rimbaud, P., H. Harant, C. Bessière, J. A. Rioux, and J. M. Bastide. 1960. Contribution a l'étude séro-immunologique de la candidose expérimentale du lapin. Pathol. Biol. *8*: 329-335.

397. Rippon, J. W., D. N. Anderson, S. Jacobsohn, M. Soo Hoo, and E. D. Garber. 1977. Blastomycosis: specificity of antigens reflecting the mating types of *Ajellomyces dermatitidis*. Mycopathologia *60*: 65-72.

398. Roberts, G. D., and H. W. Larsh. 1971. The serologic diagnosis of extracutaneous sporotrichosis. Am. J. Clin. Pathol. *56*: 597-600.

399. Roberts, G. D., and H. W. Larsh. 1971. A study of the immuno-cytoadherence test in the serology of experimental sporotrichosis. J. Infect. Dis. *124*: 264-269.

400. Rodsether, M., and O. Tonder. 1975. Antibodies to *Aspergillus fumigatus.* Characterization of a haemagglutinogen. Acta Pathol. Microbiol. Scand., Sect. C *83*: 423-428.

401. Rogers, T. J., and E. Balish. 1978. Effect of systemic candidiasis on blastogenesis of lymphocytes from germfree and conventional rats. Infect. Immun. *20*: 142-150.

402. Rowe, J. R., V. C. Newcomer, and J. W. Landau. 1963. Effect of cultural conditions on the development of antigens by *Coccidioides immitis.* I. Immunodiffusion studies. J. Invest. Dermatol. *41*: 343-350.

403. Rowe, J. R., V. D. Newcomer, and E. T. Wright. 1963. Studies of the soluble antigens of *Coccidioides immitis* by immunodiffusion. J. Invest. Dermatol. *41*: 225-233.

404. Roy, I., and J. W. Landau. 1972. Protein constituents of cell walls of dimorphic phases of *Blastomyces dermatitidis.* Can. J. Microbiol. *18*: 473-478.

405. Saferstein, H. L., A. A. Strachan, F. Blank, and C. T. Bishop. 1968. Trichophytin activity and polysaccharides. Dermatologica *136*: 151-154.

406. Sakaguchi, O., S. Suzuki, M. Suzuki, and H. Sunayama. 1967. Biochemical and immunochemical studies of fungi. VIII. Immunochemical studies of mannans of *Candida albicans* and *Saccharomyces cerevisiae.* Jpn. J. Microbiol. *11*: 119-128.

407. Sakaguchi, O., M. Suzuki, and K. Yokota. 1968. Effect of partial acid hydrolysis on precipitin activity of *Aspergillus fumigatus* galactomannan. Jpn. J. Microbiol. *12*: 123-124.

408. Sakaguchi, O., K. Yokota, and M. Suzuki. 1969. Immunochemical and biochemical studies of fungi. XIII. On the galactomannans isolated from mycelia and culture filtrates of several filamentous fungi. Jpn. J. Microbiol. *13*: 1-7.

409. Saltarelli, C. G. 1968. Immunoelectrophoresis to detect differences between strains of *Candida albicans.* Mycopathol. Mycol. Appl. *34*: 225-233.

410. Salvin, S. B. 1959. Current concepts of diagnostic serology and skin hypersensitivity in the mycoses. Am. J. Med. *27*: 97-114.

411. Salvin, S. B. 1963. Immunologic aspects of the mycoses. Prog. Allergy 7: 210-331.

412. Salvin, S. B. 1968. Allergic reactions to pathogenic fungi. In P. A. Miescherand and H. J. Muller-Eberhard (Eds.), *Textbook of Immunopathology*, Grune & Stratton, New York, pp. 323-336.

413. Salvin, S. B. 1969. Analytical serology of microfungi. In J. B. G. Kwapinski (Ed.), *Analytical Serology of Microorganisms*, Interscience, New York, pp. 515-547.

414. Salvin, S. B. 1972. Immunity in the systemic mycoses. In F. Borek (Ed.), *Immunogenicity,* Vol. 25 of A. Neuberger and E. L. Tatum (Eds.), *Frontiers of Biology Series,* North-Holland, Amsterdam, pp. 225-229.

415. Salvin, S. B., and G. A. Hottle. 1948. Serologic studies on antigens from *Histoplasma capsulatum* Darling. J. Immunol. *60*: 57-66.

416. Salvin, S. B., and R. Neta. 1977. Antigens of pathogenic fungi. In M. Sela (Ed.), *The Antigens,* Vol. IV, Academic, New York, pp. 285-332.

417. Salvin, S. B., and E. Ribi. 1955. Antigens from the yeast phase of *Histoplasma capsulatum.* II. Immunologic properties of protoplasm vs. cell walls. Proc. Soc. Exp. Biol. Med. *90*: 287-294.

418. Salvin, S. B., and R. F. Smith. 1959. Antigens from the yeast phase of *Histo-*

plasma capsulatum. III. Isolation, properties, and activity of a protein-carbohydrate complex. J. Infect. Dis. *105*: 45-53.

419. Salvin, S. B., and R. F. Smith. 1961. An antigen for detection of hypersensitivity to *Cryptococcus neoformans.* Proc. Soc. Exp. Biol. Med. *108*: 498-501.

420. San-Blas, G., D. Ordaz, and F. J. Yegres. 1978. *Histoplasma capsulatum:* chemical variability of the yeast cell wall. Sabouraudia *16*: 279-284.

421. Saslaw, S., and C. C. Campbell. 1948. A method for demonstrating antibodies in rabbit sera against histoplasmin by the collodion agglutination technic. Proc. Soc. Exp. Biol. Med. *68*: 559-562.

422. Scalarone, G. M., H. B. Levine, and S. D. Chaparas. 1978. Delayed hypersensitivity responses of experimental animals to histoplasmin from the yeast and mycelial phases of *Histoplasma capsulatum.* Infect. Immun. *21*: 705-713.

423. Scalarone, G. M., H. B. Levine, S. D. Chaparas, and J. M. Cobb. 1973. Properties and assay of spherulins from *Coccidioides immitis* in delayed sensitivity responses of animals. Sabouraudia *11*: 222-234.

424. Scalarone, G. M., H. B. Levine, D. Pappagianis, and S. D. Chaparas. 1974. Spherulin as a complement-fixing antigen in human coccidioidomycosis. Am. Rev. Respir. Dis. *110*: 324-328.

425. Schaefer, J. C., B. Yu, and D. Armstrong. 1976. An *Aspergillus* immunodiffusion test in the early diagnosis of aspergillosis in adult leukemia patients. Am. Rev. Respir. Dis. *113*: 325-329.

426. Schneidau, J. D., Jr. 1972. A cooperative study of cross-reactivity among fungal skin test antigens in tropical Latin America. In *Paracoccidioidomycosis,* Proceedings of the First Pan American Symposium, Pan Am. Health Organ. Sci. Publ. No. 254, Washington, D.C., pp. 233-238.

427. Schneidau, J. D., L. L. Lamar, and M. A. Hairston, Jr. 1964. Cutaneous hypersensitivity to sporotrichin in Louisiana. JAMA *188*: 371-373.

428. Seeliger, H. P. R. 1956. A serologic study of hyphomycetes causing mycetoma in man. J. Invest. Dermatol. *26*: 81-94.

429. Seeliger, H. P. R. 1963. Immunobiologisch-Serologische Nachweisverfahren bei Pilserkrankungen. In *Handbuch der Haut und Geschlechtskrankheiten,* Springer-Verlag, Berlin, pp. 606-734.

430. Segal, E., R. A. Berg, P. A. Pizzo, and J. E. Bennett. 1979. Detection of *Candida* antigen in sera of patients with candidiasis by an enzyme-linked immunosorbent assay-inhibition technique. J. Clin. Microbiol. *10*: 116-118.

431. Segal, E., R. Levy, and E. Eylan. 1978. Antibody formation in experimental immunizations with *Candida albicans* ribosomal fractions. Mycopathologia *64*: 121-123.

432. Seibert, F. B. 1934. The isolation and properties of the purified protein derivative of tuberculin. Am. Rev. Tuberc. (Suppl.) *30*: 713-720.

433. Seibert, F. B. 1940. Removal of impurities, nucleic acid and polysaccharide from tuberculin protein. J. Biol. Chem. *133*: 593-604.

434. Sepulveda, R., J. L. Longbottom, and J. Pepys. 1979. Enzyme linked immunosorbent assay (ELISA) for IgG and IgE antibodies to protein and polysaccharide antigens of *Aspergillus fumigatus.* Clin. Allergy *9*: 359-371.

435. Shaffer, P. J., G. Medoff, and G. S. Kobayashi. 1979. Demonstration of antigenemia by radioimmunoassay in rabbits experimentally infected with *Aspergillus.* J. Infect. Dis. *139*: 313-319.

436. Shechter, Y., J. W. Landau, N. Dabrowa, and V. D. Newcomer. 1966. Comparative disc electrophoretic studies of proteins from dermatophytes. Sabouraudia *5*: 144-149.

437. Shimonaka, H., T. Noguchi, K. Kawai, I. Hasegawa, Y. Nozawa, and Y. Ito. 1975. Immunochemical studies on the human pathogen *Sporothrix schenckii:* effects of chemical and enzymatic modification of the antigenic compounds upon immediate and delayed reactions. Infect. Immun. *11:* 1187-1194.

438. Sikl, D., L. Masler, and S. Bauer. 1964. Mannan from the extracellular surface of *Candida albicans* Berkhout. Experientia *20:* 456.

439. Silva, M. E., and W. Kaplan. 1965. Specific fluorescein-labeled antiglobulins for the yeast form of *Paracoccidioides brasiliensis.* Am. J. Trop. Med. Hyg. *14:* 290-294.

440. Silva, M. E., W. Kaplan, and J. L. Miranda. 1968. Antigenic relationships between *Paracoccidioides loboi* and other pathogenic fungi determined by immunofluorescence. Mycopathol. Mycol. Appl. *36:* 97-106.

441. Smith, C. E., M. T. Saito, and R. R. Beard. 1949. Histoplasmin sensitivity and coccidioidal infection. I. Occurrence of cross reactions. Am. J. Public Health *39:* 722-736.

442. Smith, C. E., M. T. Saito, R. R. Beard, R. N. Kepp, R. W. Clark, and B. U. Eddie. 1950. Serological tests in the diagnosis and prognosis of coccidioidomycosis. Am. J. Hyg. *52:* 1-21.

443. Smith, C. E., M. T. Saito, and S. A. Simons. 1956. Pattern of 39,500 serologic tests in coccidioidomycosis. JAMA *160:* 546-552.

444. Smith, C. E., E. G. Whiting, E. E. Baker, H. G. Rosenberger, R. R. Beard, and M. T. Saito. 1948. The use of coccidioidin. Am. Rev. Tuberc. *57:* 330-360.

445. Sorensen, L. J., and E. E. Evans. 1954. Antigenic fractions specific for *Histoplasma capsulatum* in the complement fixation reaction. Proc. Soc. Exp. Biol. Med. *87:* 339-341.

446. Sprouse, R. F. 1969. Purification of histoplasmin purified derivative. I. Disc electrophoresis separation of skin test-reactive components. Am. Rev. Respir. Dis. *100:* 685-690.

447. Sprouse, R. F. 1977. Determination of molecular weight, isoelectric point, and glycoprotein moiety for the principal skin test-reactive component of histoplasmin. Infect. Immun. *15:* 263-271.

448. Sprouse, R. F., N. L. Goodman, and H. W. Larsh. 1969. Fractionation, isolation, and chemical characterization of skin test active components of histoplasmin. Sabouraudia *7:* 1-11.

449. Staib, F., U. Folkens, B. Tompak, T. Abel, and D. Thiel. 1978. A comparative study of antigens of *Aspergillus fumigatus* isolates from patients and soil of ornamental plants in the immunodiffusion test. Zentralbl. Bakteriol. Parasitenkd. Infektionskr. Hyg., Abt. 1: Orig. *242:* 93-99.

450. Staib, F., S. K. Mishra, and T. Abel. 1977. Serodiagnostic value of extracellular antigens of an actively proteolysing culture of *Candida albicans.* Zentralbl. Bakteriol. Parasitenkd. Infektionskr. Hyg., Abt. 1: Orig. *238:* 284-287.

451. Stallybrass, F. C. 1964. *Candida* precipitins. J. Pathol. Bacteriol. *87:* 85-97.

452. Standard, P. G., and L. Kaufman. 1976. A specific immunological test for the rapid identification of members of the genus *Histoplasma.* J. Clin. Microbiol. *3:* 191-199.

453. Standard, P. G., and L. Kaufman. 1977. Immunological procedure for the rapid and specific identification of *Coccidioides immitis* cultures. J. Clin. Microbiol. *5:* 149-153.

454. Steele, R. W., P. B. Cannady, Jr., W. L. Moore, Jr., and L. O. Gentry. 1976. Skin test and blastogenic responses to *Sporotrichum schenckii.* J. Clin. Invest. *57:* 156-160.

455. Stevens, D. A., H. B. Levine, S. Deresinski, and L. J. Blaine. 1975. Spherulin in clinical coccidioidomycosis. Chest *68*: 697-702.

456. Stevens, D. A., H. B. Levine, and D. R. Ten Eyck. 1974. Dermal sensitivity to different doses of spherulin and coccidioidin. Chest *65*: 530-533.

457. Stewart, T. S., and C. E. Ballou. 1968. A comparison of yeast mannans and phosphomannans by acetolysis. Biochemistry *7*: 1855-1863.

458. Stickle, D., L. Kaufman, S. O. Blumer, and D. W. McLaughlin. 1972. Comparison of a newly developed latex agglutination test and an immunodiffusion test in the diagnosis of systemic candidiasis. Appl. Microbiol. *23*: 490-499.

459. Stuka, A. J., and R. Burrell. 1967. Factors affecting the antigenicity of *Trichophyton rubrum*. J. Bacteriol. *94*: 914-918.

460. Sudman, M. S., and W. Kaplan. 1974. Antigenic relationship between American and African isolates of *Blastomyces dermatitidis* as determined by immunofluorescence. Appl. Microbiol. *27*: 496-499.

461. Summers, D. F., A. P. Grollman, and H. F. Hasenclever. 1964. Polysaccharide antigens of *Candida* cell wall. J. Immunol. *92*: 491-499.

462. Sunayama, H. 1970. Studies on the antigenic activities of yeasts. IV. Analysis of the antigenic determinant groups of the mannan of *Candida albicans* serotype A. Jpn. J. Microbiol. *14*: 27-39.

463. Sunayama, H., and S. Suzuki. 1970. Studies on the antigenic activities of yeasts. VI. Analysis of the antigenic determinants of the mannan of *Candida albicans* serotype B-792. Jpn. J. Microbiol. *14*: 371-379.

464. Sundararaj, T., and S. C. Agarwal. 1977. Cell-mediated immunity in experimental *Nocardia asteroides* infection. Infect. Immun. *15*: 370-375.

465. Sundararaj, T., and S. C. Agarwal. 1977. Cell-mediated immunity to *Nocardia asteroides* induced by its ribonucleic acid protein fraction. Infect. Immun. *18*: 253-256.

466. Sundararaj, T., and S. C. Agarwal. 1978. Relationship of macrophages to cell-mediated immunity in experimental *Nocardia asteroides* infection. Infect. Immun. *20*: 685-691.

467. Sutton, A., and S. Marcus. 1971. Standardization of coccidioidin by bioassay in guinea pigs. Proc. Soc. Exp. Biol. Med. *136*: 257-259.

468. Suzuki, S., H. Hatsukaiwa, H. Sunayama, and N. Honda. 1971. Studies on the antigenic activities of yeasts. VII. The reaction between concanavalin A and the mannans isolated from five species of yeasts including two serotypes of *Candida* and *Saccharomyces cerevisiae*. Jpn. J. Microbiol. *15*: 437-442.

469. Suzuki, S., and H. Sunayama. 1968. Studies on antigenic activities of yeasts. II. Isolation and inhibition assay of the oligosaccharides from acetolysate of mannan of *Candida albicans*. Jpn. J. Microbiol. *12*: 413-422.

470. Suzuki, S., and H. Sunayama. 1969. Studies on antigenic activities of yeasts. III. Isolation and inhibition assay of the oligosaccharides from acid-hydrolysate of mannan of *Candida albicans*. Jpn. J. Microbiol. *13*: 95-101.

471. Suzuki, S., M. Suzuki, K. Yokota, H. Sunayama, and O. Sakaguchi. 1967. On the immunochemical and biochemical studies of fungi. XI. Cross reactions of the polysaccharides of *Aspergillus fumigatus, Candida albicans, Saccharomyces cerevisiae*, and *Trichophyton rubrum* against *Candida albicans* and *Saccharomyces cerevisiae* antisera. Jpn. J. Microbiol. *11*: 269-273.

472. Suzuki, S., and N. Takeda. 1975. Serologic cross-reactivity of the D-galacto-D-mannans isolated from several pathogenic fungi against anti-*Hormodendrum pedrosoi* serum. Carbohydr. Res. *40*: 193-197.

473. Suzuki, S., and N. Takeda. 1977. Immunochemical studies on galactomannans

isolated from mycelia and culture broths of three *Hormodendrum* strains. Infect. Immun. *17*: 483-490.

474. Sweet, G. H., R. S. Cimprich, A. C. Cook, and D. E. Sweet. 1979. Antibodies in histoplasmosis detected by use of yeast and mycelial antigens in immunodiffusion and electroimmunodiffusion. Am. Rev. Respir. Dis. *120*: 441-449.

475. Sweet, C. E., and L. Kaufman. 1970. Application of agglutinins for the rapid and accurate identification of medically important *Candida* species. Appl. Microbiol. *19*: 830-836.

476. Sweet, G. H., D. E. Wilson, and J. D. Gerber. 1973. Application of electroimmunodiffusion and crossed electroimmunodiffusion to the comparative serology of a microorganism *(Histoplasma capsulatum)*. J. Immunol. *111*: 554-564.

477. Swenson, F. J., and T. R. Kozel. 1978. Phagocytosis of *Cryptococcus neoformans* by normal and thioglycollate-activated macrophages. Infect. Immun. *21*: 714-720.

478. Syverson, R. E., and H. R. Buckley. 1977. Cell wall antigens in soluble cytoplasmic extracts of *Candida albicans* as demonstrated by crossed immunoaffinoelectrophoresis with concanavalin A. J. Immunol. Methods *18*: 149-156.

479. Syverson, R. E., H. R. Buckley, and C. C. Campbell. 1975. Cytoplasmic antigens unique to the mycelial or yeast phase of *Candida albicans*. Infect. Immun. *12*: 1184-1188.

480. Takeda, N., Y. Okubo, T. Ichikawa, and S. Suzuki. 1979. Cross-reactions between mycelial galactomannans of three *Hormodendrum* strains and the mannans of two *Candida albicans* strains of different serotypes, A and B. Infect. Immun. *23*: 146-149.

481. Takeda, N., and S. Suzuki. 1974. (Immunochemical studies on chromoblastomycosis. I. Isolation and purification of antigenic polysaccharides from mycelia and culture filtrates of *Hormodendrum pedrosoi, H. compactum* and *H. dermatitidis.*), from English summary, Jpn. J. Microbiol. *29*: 369-378.

482. Takeda, N., and S. Suzuki. 1974. (Immunochemical studies on chromoblastomycosis. II. Serological relationships among galactomannans from three species of the genus *Hormodendrum, H. pedrosoi, H. compactum,* and *H. dermatitidis.*), from English summary, Jpn. J. Microbiol. *29*: 757-763.

483. Taschdjian, C. L., G. B. Dobkin, L. Caroline, and P. J. Kozinn. 1964. Immune studies relating to candidiasis. II. Experimental and preliminary clinical studies on antibody formation in systemic candidiasis. Sabouraudia *3*: 129-139.

484. Taylor, I. E. P., and D. S. Cameron. 1973. Preparation and quantitative analysis of fungal cell walls: strategy and tactics. Annu. Rev. Microbiol. *27*: 243-259.

485. Tewari, R. P., D. Sharma, M. Solotorovsky, R. Lafemina, and J. Balint. 1977. Adoptive transfer of immunity from mice immunized with ribosomes or live yeast cells of *Histoplasma capsulatum*. Infect. Immun. *15*: 789-795.

486. Thompson, K. W. 1942. The skin reacting antigen of *Trichophyton purpureum*. J. Invest. Dermatol. *5*: 475-480.

487. Thurstone, J. R., M. Phillips, and A. C. Pier. 1968. Extracellular antigens of *Nocardia asteroides*. III. Immunologic relationships demonstrated by erythrocytesensitizing antigens. Am. Rev. Respir. Dis. *97*: 240-247.

488. Thurston, J. R., J. L. Richard, and S. McMillen. 1973. Cultural and serological comparison of ten strains of *Aspergillus fumigatus* Fresenius. Mycopathol. Mycol. Appl. *51*: 327-335.

489. Tompkins, V. N. 1965. Soluble antigenic constituents of yeast-phase *Histoplasma capsulatum*. Am. Rev. Respir. Dis. *92* (Suppl.): 126-133.

490. Tonder, O., and M. Rodsaether. 1974. Indirect haemagglutination for demonstration of antibodies to *Aspergillus fumigatus*. Acta Pathol. Microbiol. Scand., Sect. B *82*: 871-878.

491. Toriello, C., and F. Mariat. 1974. Etude comparée des polyosides des champignons *Ceratocystis stenoceras* et *Sporothrix schenckii*. Composition chimique et analyse immunologigue. Ann. Microbiol. (Paris) A. *125*: 287-307.

492. Trana, A. K., and G. K. Khuller. 1978. Antigenicity of mannophosphoinositides of *Nocardia*. Indian J. Med. Res. *67*: 734-736.

493. Tran Van Ky, P., J. Biguet, and S. Andrieu. 1963. Etude par l'électrophorèse et la double diffusion en gélose des substances antigèniques excrétées dans le milieu de culture de *Candida albicans*. Sabouraudia *2*: 164-169.

494. Tran Van Ky, P., J. Biguet, and T. Vaucelle. 1968. Etude d'une fraction antigènique d'*Aspergillus fumigatus* support d'une activité catalasique. Conséquence sur le diagnostic immunologique de l'aspergillose. Rev. Immunol. Ther. Antimicrob. *32*: 37-52.

495. Tran Van Ky, P., J. Uriel, and F. Rose. 1966. Caractérisation de types d'activités enzymatiques dans des extraits antigèniques d'*Aspergillus fumigatus* après électrophorèse et immunoélectrophorèse en agarose. Ann. Inst. Pasteur (Paris) *111*: 161-170.

496. Tran Van Ky, P., T. Vaucelle, S. Andrieu, C. Torck, and F. Floc'h. 1969. Caractérisation de complexes enzymes-antienzymes dans les extraits de levures du genre *Candida* après immunoélectrophorégramme en agarose. Mycopathol. Mycol. Appl. *38*: 345-357.

497. Travassos, L. R., P. A. J. Gorin, and K. O. Lloyd. 1973. Comparison of the rhamnomannans from the human pathogen *Sporothrix schenckii* with those from the *Ceratocystis* species. Infect. Immun. *8*: 685-693.

498. Travassos, L. R., P. A. J. Gorin, and K. O. Lloyd. 1974. Discrimination between *Sporothrix schenckii* and *Ceratocystis stenoceras* rhamnomannans by proton and carbon-13 magnetic resonance spectroscopy. Infect. Immun. *9*: 674-680.

499. Travassos, L. R., and L. Mendonca-Previato. 1978. Synthesis of monorhamnosyl L-rhamno-D-mannans by conidia of *Sporothrix schenckii*. Infect. Immun. *19*:1-4.

500. Travassos, L. R., L. Mendonca-Previato, and P. A. J. Gorin. 1978. Heterogeneity of the rhamnomannans from one strain of the human pathogen *Sporothrix schenckii* determined by ^{13}C nuclear magnetic resonance spectroscopy. Infect. Immun. *19*: 1107-1109.

501. Travassos, L. R., W. de Sousa, L. Mendoca-Previato, and K. O. Lloyd. 1977. Location and biochemical nature of surface components reacting with concanavalin A in different cell types of *Sporothrix schenckii*. Exp. Mycol. *1*: 293-305.

502. Tsuchiya, T., Y. Fukazawa, M. Taguchi, T. Nakase, and T. Shinoda. 1974. Serologic aspects on yeast classification. Mycopathol. Mycol. Appl. *53*: 77-92.

503. Umenai, T., and H. Chiba. 1977. Evaluation of the antigen specific to the mycelial phase of *Candida albicans* in the serodiagnosis of candidiasis. Tohoku J. Exp. Med. *123*: 395-396.

504. Vanbreuseghem, R. 1967. Early diagnosis, treatment and epidemiology of mycetoma. Rev. Med. Vet. Mycol. *6*: 49-60.

505. Venezia, R. A., and R. G. Robertson. 1974. Efficacy of the *Candida* precipitin test. Verification with a protoplast antigen preparation. Am. J. Clin. Pathol. *61*: 849-855.

506. Vogel, R. A. 1966. The indirect fluorescent antibody test for the detection of antibody in human cryptococcal disease. J. Infect. Dis. *116*: 573-580.

507. Wallraff, E. B., R. M. Van Liew, and S. Waite. 1967. Skin reactivity and serological response to coccidioidin skin tests. J. Invest. Dermatol. *48*: 553-559.

508. Wallraff, E. B., and S. A. Wilson. 1965. Skin test activity of dialyzed coccidioidins. Am. Rev. Respir. Dis. *92*: 121-123.

509. Walter, J. E., and R. D. Jones. 1968. Serodiagnosis of clinical cryptococcosis. Am. Rev. Respir. Dis. *97*: 275-282.

510. Walter, J. E., and R. D. Jones. 1968. Serologic tests in diagnosis of aspergillosis. Dis. Chest *53*: 729-735.

511. Walter, J. E., and G. B. Price, Jr. 1968. Chemical, serologic, and dermal hypersensitivity activities of two fractions of histoplasmin. Am. Rev. Respir. Dis. *98*: 474-479.

512. Ward, E. R., R. A. Cox, J. A. Schmitt, Jr., M. Huppert, and S. H. Sun. 1975. Delayed-type hypersensitivity responses to a cell wall fraction of the mycelial phase of *Coccidioides immitis*. Infect. Immun. *12*: 1093-1097.

513. Warnock, D. W. 1974. Indirect immunofluorescence test for the detection of *Aspergillus fumigatus* antibodies. J. Clin. Pathol. *27*: 911-912.

514. Warnock, D. W., and G. F. Eldred. 1975. Immunoglobulin classes of antibodies to *Aspergillus fumigatus* in patients with pulmonary aspergillosis. Sabouraudia *13*: 204-208.

515. Warnock, D. W., D. C. E. Speller, J. A. Morris, and P. H. Mackie. 1976. Serological diagnosis of infection of the urinary tract by yeasts. J. Clin. Pathol. *29*: 836-840.

516. Warren, R. C., A. Bartlett, D. E. Bidwell, M. D. Richardson, A. Voller, and L. O. White. 1977. Diagnosis of invasive candidosis by enzyme immunoassay of serum antigen. Br. Med. J. *1*: 1183-1185.

517. Warren, R. C., M. D. Richardson, and L. O. White. 1978. Enzyme-linked immunosorbent assay of antigens from *Candida albicans* circulating in infected mice and rabbits: the role of mannan. Mycopathologia *66*: 179-182.

518. Warren, W. C., L. O. White, S. Mohan, and M. D. Richardson. 1979. The occurrence and treatment of false positive reactions in enzyme-linked immunosorbent assays (ELISA) for the presence of fungal antigens in clinical samples. J. Immunol. Methods *28*: 177-186.

519. Weiner, M. H., and M. Coats-Stephen. 1979. Immunodiagnosis of systemic aspergillosis. I. Antigenemia detected by radioimmunoassay in experimental infection. J. Lab. Clin. Med. *93*: 111-119.

520. Weiner, M. H., and W. J. Yount. 1976. Mannan antigenemia in the diagnosis of invasive *Candida* infections. J. Clin. Invest. *58*: 1045-1053.

521. Welsh, R. D., and C. T. Dolan. 1973. *Sporothrix* whole yeast agglutination test: low titer reactions of sera of subjects not known to have sporotrichosis. Am. J. Clin. Pathol. *59*: 82-85.

522. Westphal, H.-J., and U. Kaben. 1977. Isolierung von Antigenfraktionen aus *Candida albicans*. Mykosen *20*: 297-300.

523. Wheat, R. W., and E. Scheer. 1977. Cell walls of *Coccidioides immitis:* neutral sugars of aqueous alkaline extract polymers. Infect. Immun. *15*: 340-341.

524. Wheat, R. W., and K. S. Su Chung. 1977. Antigenic fractions of *Coccidioides immitis*. In L. Ajello (Ed.), *Coccidioidomycosis. Current Clinical and Diagnostic Status*, Symposia Specialists, Miami, Fla., pp. 453-460.

525. Wheat, R. W., K. S. Su Chung, E. P. Ornellas, and E. R. Scheer. 1978. Extraction of skin test activity from *Coccidioides immitis* mycelia by water, perchloric acid, and aqueous phenol extraction. Infect. Immun. *19*: 152-159.

526. Wheat, R. W., T. Terai, A. Kiyomoto, N. F. Conant, E. P. Lowe, and J. Converse. 1967. Studies on the composition and structure of *Coccidioides immitis* cell walls. In L. Ajello (Ed.), *Coccidioidomycosis*. Papers from the Second Symposium on Coccidioidomycosis, University of Arizona Press, Tucson, pp. 237-242.

527. Wheat, R. W., C. Tritschler, N. F. Conant, and E. P. Lowe. 1977. Comparison of *Coccidioides immitis* arthrospore, mycelium, and spherule cell walls, and influence of growth medium on mycelial cell wall composition. Infect. Immun. *17*: 91-97.

528. White, L. O., M. D. Richarson, H. C. Newham, E. Gibb, and R. C. Warren. 1977. Circulating antigen of *Aspergillus fumigatus* in cortisone-treated mice challenged with conidia: detection by counterimmunoelectrophoresis. FEMS Microbiol. *2*: 153-156.

529. Wiggins, G. L., and J. H. Schubert. 1965. Relationship of histoplasmin agar-gel bands and complement fixation titers in histoplasmosis. J. Bacteriol. *89*: 589-596.

530. Wilson, D. E., J. E. Bennett, and J. W. Bailey. 1968. Serologic grouping of *Cryptococcus neoformans*. Proc. Soc. Exp. Biol. Med. *127*: 820-823.

531. Winner, H. I. 1977. Recent advances in systemic candidosis. Contrib. Microbiol. Immunol. *4*: 64-76.

532. Winner, H. I., and R. Hurley. 1964. *Candida albicans*, Little, Brown, Boston.

533. Yarzábal, L. A. 1971. Anticuerpos precipitantes especificos de la blastomicosis sudamericana, revelados por inmunoelectroforesis. Rev. Inst. Med. Trop. São Paulo *13*: 320-327.

534. Yarzábal, L. A., M. B. de Albornez, N. A. de Cabral, and A. R. Santiago. 1978. Specific double diffusion microtechnique for the diagnosis of aspergillosis and paracoccidioidomycosis using monospecific antisera. Sabouraudia *16*: 55-62.

535. Yarzábal, L. A., S. Andrieu, D. Bout, and F. Naquira. 1976. Isolation of a specific antigen with alkaline phosphatase activity from soluble extracts of *Paracoccidioides brasiliensis*. Sabouraudia *14*: 275-280.

536. Yarzábal, L. A., J. Biguet, T. Vaucelle, S. Andrieu, J. M. Torres, and S. da Luz. 1973. Análisis inmunoquimico de extractos solubles de *Paracoccidioides brasiliensis*. Sabouraudia *11*: 80-88.

537. Yarzábal, L. A., D. Bout, F. Naquira, J. Fruit, and S. Andrieu. 1977. Identification and purification of the specific antigen of *Paracoccidioides brasiliensis* responsible for immunoelectrophoretic band E. Sabouraudia *15*: 79-85.

538. Yarzábal, L. A., J. M. Torres, M. Josef, I. Vigna, S. da Luz, and S. Andrieu. 1972. Antigenic mosaic of *Paracoccidioides brasiliensis*. In *Paracoccidioidomycosis*, Proceedings of the First Pan American Symposium, Pan Am. Health Organ. Sci. Publ. No. 254, Washington, D.C., pp. 239-244.

539. Young, R. C., and J. E. Bennett. 1971. Invasive aspergillosis: absence of detectable antibody response. Am. Rev. Respir. Dis. *104*: 710-716.

540. Yu, B. H., and D. Armstrong. 1977. Serological tests for invasive aspergillosis and *Candida* infections in patients with neoplastic disease. In K. Iwata (Ed.), *Recent Advances in Medical and Veterinary Mycology*, Proceedings of the Sixth Congress of the International Society for Human and Animal Mycology, University of Tokyo Press, Tokyo, pp. 47-58.

541. Yu, R. J., C. T. Bishop, F. P. Cooper, H. F. Hansenclever, and F. Blank. 1967. Structural studies of mannans from *Candida albicans* (serotypes A and B), *Candida parapsilosis*, *Candida stellatoidea*, and *Candida tropicalis*. Can. J. Chem. *45*: 2205-2211.

542. Yu, R. J., S. R. Harmon, and F. Blank. 1968. Isolation and purification of an extracellular keratinase of *Trichophyton mentagrophytes*. J. Bacteriol. *96*: 1435-1436.

543. Yu, R. J., S. R. Harmon, S. F. Grappel, and F. Blank. 1971. Two cell-bound keratinases of *Trichophyton mentagrophytes*. J. Invest. Dermatol. *56*: 27-32.

544. Yu, R. J., S. R. Harmon, P. E. Wachter, and F. Blank. 1969. Amino acid composition and specificity of a keratinase of *Trichophyton mentagrophytes*. Arch. Biochem. Biophys. *135*: 363-370.

545. Zamora, A., L. F. Bojalil, and F. Bastarrachea. 1963. Immunologically active polysaccharides from *Nocardia asteroides* and *Nocardia brasiliensis*. J. Bacteriol. *85*: 549-555.

546. Zarafonetis, C. J. D., and R. B. Lindberg. 1941. Histoplasmosis of Darling: observations on antigenic properties of causative agent. Univ. Hosp. Bull. (Ann Arbor) 7: 47-48.

547. Ziegler, V. H., I. Tausch, H. Böhme, P. Schulze, and H. Barthelmes. 1978. Zur Immunochemie von *Petriellidium boydii* und zur Immunantwort mit *Petriellidium boydii*-Vollantigen sensibilierter Kaninschen. Dermatol. Monatsschr. *164*: 423-431.

6

The Detection of Fungal Metabolites Including Antigens in Body Fluids

Paul F. Lehmann / Medical College of Ohio, Toledo, Ohio

I. INTRODUCTION

The diagnosis of disease from studies of the body fluids has had a long history in medicine. Although such techniques were at first used solely for pecuniary reasons [9], after the microbial origin of disease had been understood in the late nineteenth century they began to be applied to the diagnosis of infectious diseases. The first tests were largely ones for the detection of antibody (see Chap. 4), but it was soon discovered that in certain diseases, notably cryptococcosis among the mycoses, an antibody response was not necessarily detectable in infected patients. More recently, with the increasing use of immunosuppressive and cytotoxic drugs in the treatment of cancer and the maintenance of organ transplants, a number of fungal diseases have appeared with increased frequency. These diseases, which are also found in other types of immunosuppressed patients, include disseminated forms of aspergillosis, candidiasis, zygomycosis, histoplasmosis, and cryptococcosis [31,37] and have been difficult to diagnose by specific antibody or skin tests. Significant titers of antibody may be absent from these immunosuppressed patients or may be detected too late for successful therapy to be initiated. For these reasons, other methods of diagnosis have been used and these have included the detection of fungal metabolites in body fluids. Most of these studies have been of a preliminary nature, but testing for the presence of *Cryptococcus neoformans* capsular antigen is now widely used to aid the diagnosis of cryptococcosis.

This chapter includes a discussion of both soluble and selected insoluble fungal metabolites that are detected in patients, and is largely restricted to metabolites that are not produced by the host. Several fungi are not reported owing to the lack of available data. The methods used have included immunological techniques for the detection of antigenic materials and chemical techniques, especially gas-liquid chromatography, for the detection of other metabolites. Although these studies have been primarily of a diagnostic nature, it is clear that studies on the pathogenic mechanism of different fungi must include an understanding of the role that these metabolites play in inducing the symptoms of disease.

II. INSOLUBLE METABOLITES

A. Cell Wall and Chitin

Because the fungal cell wall differs both chemically and antigenically from host materials, it can be detected in tissue sections using histological reagents, such as the methenamine-silver and periodic acid Schiff stains, and labeled antibodies. Kaplan [30] has discussed the use of the fluorescent antibody technique in diagnostic mycology and the reader is referred to this paper for a review. Histological procedures have also involved the use of fluorescein-labeled lectins and aprotinin [69] which selectively bind to different terminal sugar groups in cell walls, but this method does not have as great a specificity as can be found with the antibody reagents.

These histological techniques can be used for the quantitation of fungal growth in experimental infections, but the methods, which involve the enumeration of yeasts or the measurement of hyphae in separate sections, are extremely tedious. For this reason, a chemical assay for fungal chitin [60] was used to study the growth of *Aspergillus fumigatus* in mice [42]. Chitin is a $\beta(1 \rightarrow 4)$-linked homopolymer of N-acetylglucosamine and is not produced in vertebrates. It appears as a structural component of the cell wall of all fungi that have been investigated, with the exception of the Oomycetes [4], and thus can be used as a marker to estimate fungal biomass. The assay is useful in studying the mycelial growth of filamentous fungi where the quantitation of viable organisms by plating out a tissue homogenate bears no relation to the amount of mycelium present. This lack of correlation can be due to the presence of viable ungerminated spores or to other factors. The chitin assay has been used to demonstrate the development of cortisone-resistant immunity to *A. fumigatus* in infected mice [43], and in studies of *Candida albicans* and *Absidia corymbifera* infections [81,82]. In the latter study, mycelial growth in the brain and kidney of mice was detectable at a time when no viable count could be recorded. Presumably, the homogenization procedures had killed the organism.

B. Oxalic Acid

Although oxalic acid is soluble, the calcium salt forms crystals that may be detected in tissue sections using several histological and chemical methods [27]. Deposits of calcium oxalate have been detected surrounding aspergillomas in patients infected by *Aspergillus niger* and *A. fumigatus* [38,52,68]. Crystals were not observed in numerous other infections by fungi and bacteria, and it would appear that the oxalate resulted from production of oxalic acid in the aspergilloma. *Aspergillus niger* is also known to produce oxalic acid in vitro and when causing crown rot of peanuts [21]. This contrasts with several other mycotoxins which, although present in plant materials, have with a single exception not been reported to be produced in animal infections (see Sec. III.C.1).

The report of Nime and Hutchins [52] showed the production of a metabolite in vivo which directly caused tissue damage. Here the oxalosis was considered to have caused local tissue necrosis as well as having contributed to the renal dysfunction seen in some patients. Renal oxalosis and the presence of crystals in animals infected by *Aspergillus* spp. have been reported occasionally (see Ref. 27 for review). It would be of great interest to know whether this is a common consequence of aspergillosis or of other fungal diseases which were not included in the study [52]. In this respect it is known that oxalic acid is of widespread occurrence in the fungi [13]. Although oxalic acid does not appear to be utilized by humans, the presence of oxalate crystals does not prove that these have been formed from oxalic acid produced by a fungus because the acid is also an end product in human metabolism.

III. SOLUBLE METABOLITES

A. Gas-Liquid Chromatographic and Chemical Methods

The presence of soluble metabolites in the serum, urine, and cerebrospinal fluid (CSF) can be detected by chemical techniques including the use of gas-liquid chromatography (GLC). Proper collection and storage of samples is essential because many metabolites are present only in trace amounts. These detection methods should be useful in the diagnosis of fungal diseases since several small molecules, for example, the sugar alcohols, ergosterol, and aromatic substances, are not produced by animals or may not be metabolized by them. Such molecules could act as markers for fungal infections and have the added advantage that, being nonantigenic in most cases, they would not be removed in an antibody response.

There have been two approaches to the diagnosis of fungal disease by detecting small metabolites. Either known chemicals have been detected or GLC profiles of the body fluids have been made with the object of finding characteristic peaks in infected patients. The chemical nature of such peaks could, in many instances, be determined by using mass spectroscopy, but this is rarely done; the GLC profile itself acts as a "fingerprint" for each infection.

1. Ethanol and Arabitol

Two small molecules have been studied by chemical techniques. These are ethanol and D-arabitol (= D-arabinitol). Both are detected by GLC, although, in addition, ethanol has been detected by chemical and enzymatic assays. These are not as sensitive or specific as the GLC method which is now well established and widely used for forensic purposes [15]. In this assay, the vapor phase of a heated body fluid is directly sampled as there is no need to form a volatile derivative. This contrasts with D-arabitol, which is not volatile at the temperatures used in GLC.

Ethanol has been detected at high levels in the blood of animals fed a glucose-modified milk diet [84]. These animals were heavily colonized by *Torulopsis glabrata,* which fermented the glucose present in the stomach and induced clinical signs of drunkenness in some individuals. The low blood levels of ethanol found in normal humans [15] may also result from yeast fermentations taking place in the gut, although it is difficult to ensure that it does not result from fermentations taking place outside the body. Reports of an elevated ethanol level in the CSF of patients with cryptococcal meningitis [17,71] were not confirmed [55,85] and the possibility was raised that the original findings were due to contamination with a skin disinfectant used to prepare patients for the spinal tap.

Recently serum levels of arabitol in patients with *Candida* infections have been studied by means of GLC [34]. Serum samples from patients were treated with acetic anhydride as a derivatizing agent and a peak was found on both polar and nonpolar columns which behaved as the arabitol standard. Although the serum peak was not confirmed as arabitol by mass spectroscopy, it seems almost certain to be that substance. For routine tests, trimethylsilyl derivatives of serum were used and significantly higher levels of arabitol were found in cases of invasive disease than in patients where *Candida* spp. merely colonized the intestines and mucosal surfaces. Should the result be confirmed, this assay would be useful in helping the diagnosis of invasive candidiasis. In addition, significant arabitol levels may also be found in other fungal infections, as arabitol is widely distributed among the fungi [8,13]. The L-arabitol isomer is found in the blood of persons with congenital pentosuria, but the complications that this may cause in studies on D-arabitol levels due to *Candida* infections are unknown.

Arabitol appears to be a more useful marker of infection than ethanol. Although both are rapidly cleared by the kidneys, arabitol is not broken down by humans and appears not to enter the blood at significant levels when released from yeasts colonizing the intestines [34]. Clearly, other fungal products might be sought with GLC. Such compounds, if they do not form complexes with host materials and are not metabolized, could be used to indicate fungal disease. Substances of this sort have been found in several nonclinical situations. For example, studies with the green plant-fungal symbioses of mycorrhizae and lichens, as well as with phytopathogens, have demonstrated that the sugar alcohols are major fungal metabolites which are not utilized by the plant host [67].

2. GLC Profiles

The second way that GLC has been used in diagnosis is exemplified by the use of a "fingerprint" of peaks produced after running the processed body fluid on a column. A pattern of abnormal peaks may be characteristic of certain infections. Miller et al. [48] reported the presence of such peaks for cases of *Candida* septicemia, after methanolic-HCl extraction of serum and conversion to trimethylsilyl (TMS) derivatives. This result was later confirmed and the same peaks were found in septicemia caused by *C. albicans, C. tropicalis,* and *C. parakrusei* [25]. These investigators utilized a flame ionization thermal detector, which detects almost any volatile compound that passes through it. By using volatilizing agents of greater selectivity than the TMS reagents, together with an electron capture (EC) detector which detects only those compounds and derivatives that have an electron capturing potential, Brooks [10] has shown that EC-GLC profiles can be routinely and usefully applied to the diagnosis of various microbial diseases. Figure 1 shows that cryptococcal meningitis could be distinguished from other causes of meningitis by the presence of one of two characteristic profiles obtained with patients' CSF processed with heptafluorobutyric anhydride [16,63]. These two patterns, which were seen with *Cryptococcus neoformans,* may be due to metabolic differences related either to the stage of disease or fungal genotype.

EC-GLC requires careful collection of specimens to avoid contamination by substances that interfere with the assay. Preservatives such as merthiolate, and plastic storage vessels, must not be used; freshly collected or frozen samples, stored in clean glass containers, are best. The EC-GLC profiles of such samples are likely to become increasingly important in disease diagnosis. The computerization of profile patterns allows rapid comparison of unknown samples with those from documented cases [10], and there are obvious advantages to be gained from the use of a single test to distinguish between several pathogens causing, for example, a lymphocytic meningitis [16]. However, prior to their routine use for the diagnosis of invasive infections, EC-GLC profiles of serum, urine, and other fluids must be made for several control patients who have well-characterized infections. As yet, these are unpublished. In addition, it will be necessary to demonstrate that the technique is essential in diagnosis of diseases of microbial and nonmicrobial origin. Certainly, the impressive demonstration of the possibility of diagnosing tuberculous meningitis by EC-GLC profiles [16] may result in the more common use of this powerful analytical tool.

B. Immunological Methods

Since the demonstration of antigenemia, produced in experimental infections of animals by *Streptococcus pneumoniae* [18], there have been several reports on the occurrence of antigens in the blood, urine, and CSF of patients and animals infected with other bacteria, fungi, viruses, and parasites. In the case of the fungi, much of the work has been reported only recently, and has required sensitive immunoassay procedures to detect the antigens, which may be found only at low concentration. The test for capsular polysaccharide antigens of *C. neoformans* is the only procedure that has become widely used clinically. Other tests for antigen are still experimental and will need to be compared and evaluated before being used in routine clinical tests. However, the finding of circulating antigens in fungal diseases other than cryptococcosis has demonstrated that immunodiagnosis of these is a possibility. In this section the major methods used and the reports that show the presence of soluble antigen in the different diseases are discussed.

1. Methods

The various methods that have been used to detect antigen by immunological means may be placed into two major categories: (1) direct assay of antigen by its reaction with unlabeled or labeled antibody, and (2) indirect assay of antigen by its ability to inhibit the attachment of labeled antigen to antibody. Additionally, methods involving the denaturation of immune complexes have been used.

   a. Direct Assays: Antigen detection by precipitin formation has been studied in free solution or in gels in which Ouchterlony immunodiffusion is employed. In addition, antigen has been detected by counterimmunoelectrophoresis (CIE) and by antigen-antibody

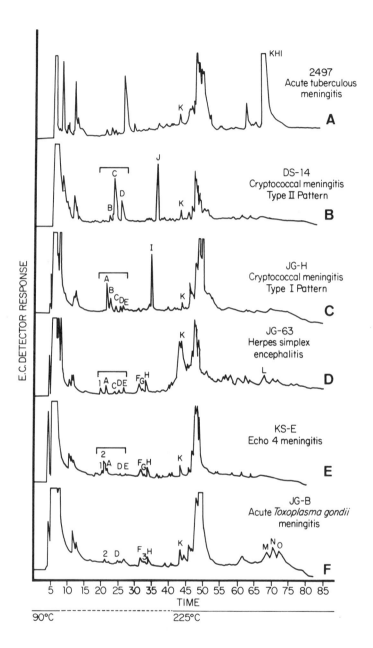

Figure 1 Electron capture, gas-liquid chromatographic profiles of amines in CSF associated with cryptococcal, viral, protozoan, and tuberculous meningitis. Note that two patterns were seen for *Cryptococcus neoformans* infections and that viral infections could be distinguished by the absence of peak B, I, or J and the presence of peaks F, G, and H. (From Ref. 16.)

crossed electrophoresis (AACE). The theoretical basis of this methods is well established and is not discussed here (the interested reader is referred to other sources [53]). In studies that use electrophoresis, it is important to record the amount of electroendomosis (EEO) that occurs, although this figure has rarely been given. The electroendosmotic flow found in agarose gels is related to the amount of sulfate and pyruvate residues bound on the matrix. An increase in these residues results in an increased flow of cations and their associated water molecules toward the cathode when an electric current is applied. The amount of sulfate and pyruvate present depends on the batch of agarose, and this makes the precise timing of a CIE test dependent on the EEO in the agarose as well as the buffer and current. Results of tests may be radically altered by using agaroses of widely different EEO, as has been found for adenovirus antigens [26] and for the detection of *Candida albicans* mannan in vitro and *A. fumigatus* galactomannan in vivo (P. F. Lehmann and E. Reiss, unpublished observation). A method for measuring EEO by comparison of the migration of a neutral molecule (dextran) with an anion (albumin) has been described [26] which uses a modification of Wieme's procedure. However, in many cases it is now possible to get this figure from the distributors of agarose products.

Antigen may also be detected by its ability to agglutinate antibody-coated latex and antibody-coated red cells [45] and by its ability to cross-link specific antibody molecules in the sandwich enzyme-linked immunosorbent assay (ELISA) test [72].

The direct assays depend on the production of a specific antiserum which does not cross-react with antigens present in normal body fluids. False-positive results due to such cross-reactivity can often be removed by adsorption of the antiserum with tissue powders or with the insoluble product of glutaraldehyde-treated normal serum [1]. Alternatively, specific antibody may be eluted from antigen bound on columns, although here the eluate must be tested for possible cross-reactivity with normal human antigens, as it is known that single immunoglobulin molecules may bind more than one antigen.

Rheumatoid factor (RF) can give false-positive reactions in direct assays. It is a human immunoglobulin, usually of the IgM class, which reacts with IgG of other species and can cross-link antibody molecules in the direct assays. Although the presence of RF does not appear to interfere with the detection of cryptococcal polysaccharide by CIE [46], RF can agglutinate antibody-coated latex or red blood cells. In these tests a control of normal rabbit IgG-coated particles may be incorporated [5], or the serum to be tested can be treated with dithiothreitol [23], which reduces the disulfide bonds in IgM and thus prevents the action of RF. RF may also give false-positive results in the sandwich ELISA test, and here dithiothreitol does not appear to be as effective as in the latex test [76]. Since the reactivity of RF is directed primarily at the Fc portion of the animal IgG molecule, the use of Fab or $F(ab')_2$ reagents should result in removal of false positives caused by RF. These reagents do not appear to have been used in mycological research.

An important drawback of direct methods of detecting antigen is that the antigen to be detected must have at least two antigenic determinants available for binding antibody. Thus, small fragments of antigen or antigen-containing soluble immune complexes may not be detected in direct assays.

b. Inhibition Assays: Hemagglutination inhibition (HAI), radioimmunoassay (RIA), and enzyme-linked immunosorbent assay inhibition (EAI) techniques are among those which have been used to detect fungal antigens in indirect assays [28,72,80]. These methods depend on the inhibition of binding of a standard amount of a "labeled" antigen with a fixed amount of specific antibody. The antigen is labeled by attachment to a surface, such as plastic or a red cell, or may be linked to an assayable material such as a radioactive atom (e.g., ^{125}I). The inhibition is seen when "unlabeled" antigen is introduced into the reaction. Several other types of label are available for use in inhibition assays, including fluorescein and enzymes.

As in the direct assays, rheumatoid factor could interfere by complexing with the antibody and causing a blockage of the antigen binding sites by steric hindrance. This would give rise to a false-positive reaction; however, the importance of RF in assays for fungal antigens by inhibition methods has not been shown.

Inhibition assays have certain advantages over direct methods. They are able to detect the presence of antigen fragments and soluble immune complexes provided that, in the latter case, one antigenic determinant is available for binding. Another advantage of the inhibition techniques is that only the reaction with specific antifungal antibody is assayed. This is because a defined fungal antigen is labeled and, provided that no cross-reactivity occurs between this antigen and normal human antigens, the assay should show no interference by background binding of the latter. Thus, exhaustive adsorption with normal human tissue antigens and insolubilized serum is not necessary in inhibition techniques.

c. *Detection of Antigen Bound to Serum Constituents and to Immune Complexes*: In several patients, antigen may be present but bound to serum constituents, including antibody. Here, the detection of antigen may be facilitated by disrupting the complex prior to the assay for antigen. The denaturing methods include high salt concentration, high molarity urea, heat, and incubation of the sera at high or low pH. The conditions must be chosen so as not to destroy the antigen. Subsequent to disruption of the complex, the antigen may be detected by either direct or inhibition assays.

That the antigen may occur in immune complexes has been demonstrated by Yoshinoya et al. [86]. Certain patients with coccidioidomycosis showed elevated levels of immune complexes, and the presence of complexed fungal antigen was demonstrated by radioimmunoassay.

2. Cryptococcus neoformans

Antigen was first detected in the blood, urine, and CSF of infected animals and a patient with cryptococcosis in the early 1950s [50,51]. The antigen, which may be only slowly removed from the circulation and CSF after a patient's recovery [7,22], was shown to be the capsular polysaccharide. In most patients, antibody to this material is not detected during active infection [24].

In the initial reports, the presence of antigen was demonstrated by precipitin formation with a rabbit anti-*C. neoformans* antiserum; however, a far more sensitive test using latex agglutination (LA) was developed and found to be specific in diagnosis [7]. With a control for rheumatoid factor [5], the LA test has been used routinely for diagnosis of cryptococcal meningitis and other forms of cryptococcosis [22,54]. Occasional reports of false-negative results are made [22], but it is not clear whether such results are due to an absence of antigen or whether the reagents fail to detect each polysaccharide serotype with equal sensitivity.

Other methods of detecting cryptococcal antigen include complement fixation [6,51] and counterimmunoelectrophoresis (CIE) [46]. The latter does not appear to give a false positive result in the presence of rheumatoid factor, which gives it an advantage over LA, but the two methods were not compared in a clinical laboratory trial. Two precipitin lines were observed with some patients' sera after testing in CIE; however, other workers have not reported these in immunodiffusion tests [32]. Double lines are occasionally seen when purified polysaccharide antigens are tested in CIE (P. F. Lehmann, unpublished observation), so the presence of two lines may not indicate two separate antigens in the serum.

The capsular antigen, an acidic polysaccharide, has antiphagocytic properties [12,36], and the capsule itself appears to be important in the establishment of infection because unencapsulated varieties have low virulence [35]. Cross-reactivity with other microbial antigens has been reported; for instance, the serotype A polysaccharide is antigenically cross-reactive with types II and XIV pneumococci [57], but problems of this sort rarely affect a clinical diagnosis. Detection of cryptococcal antigen is, at the present time, the only routinely applied test for diagnosis of a fungal disease from the presence of a soluble metabolite.

3. *Sporothrix schenckii*

The detection of soluble antigen in the peritoneal washings of mice infected by *Sporothrix schenckii* was reported. The material was detected by means of precipitin formation with an anti-*Sporothrix* antiserum [50]. No further work appears to have been done with the antigen, which was reported to be heat stable. It is not known what role, if any, it plays in the pathogenesis of sporotrichosis, nor whether more than one antigen is involved.

4. *Candida albicans*

The reports of antigens in the serum of patients and of infected animals are summarized in Table 1. Mannan has been detected in sera from patients with invasive disease using HAI [80], RIA [79], EAI [64]. Other investigators, using sandwich ELISA, have failed to detect mannan in animal models and human infections [41,75]. It is possible to explain these conflicting findings by assuming that mannan, if present, occurs in small fragments or in complexes that are unable to cross-link antibody molecules, a property necessary for sandwich ELISA although not for the other techniques. Evidence for the presence of mannan in a complexed, or hidden, state has been presented [64,79]. Preincubation of the sera in alkaline or in acidic conditions at elevated temperatures allowed mannan to be detected in several samples which otherwise showed no evidence of antigen. The nature of the material binding the mannan was not determined but may include antibody or certain complement components.

Although mannan may be detected early in infection using HAI [80], the presence of antibody and complement, which occur routinely, may bring about the rapid removal of the majority of the antigen from the plasma. Possibly, fluctuating antibody levels will be found to characterize invasive disease because all the antibody may become bound by antigen in situations where antibody production is slow or absent. The detection of mannan is further complicated by the finding that, for certain antisera raised against mannan, immunoassay sensitivity is strongly dependent on the mannan's serotype and the assay temperature [41]. This complication has not been investigated routinely in testing for circulating mannan.

Another report has demonstrated the presence of several antigens, at the same time as anti-*Candida* antibody, in the circulation of a patient [2]. These antigens may be found to be more characteristic than mannan in the diagnosis of invasive candidiasis. The structure of these and of those antigens detected by HA, CIE, RIA, or ELISA is unknown

Table 1 Detection of Fungal Antigens by Immunological Methods in Animals and Humans

Methods	Fungus	Described antigens	Reference[a]
Direct			
Precipitin formation (including CIE and AACE)	*Aspergillus fumigatus*	Galactomannan	*39, 40, 58, 59, 78, 83*
	Candida albicans		*2, 33, 74*
	Cryptococcus neoformans	Capsular polysaccharide	*32, 46, 50, 51*
	Sporothrix schenckii		*50*
Complement fixation	*Cryptococcus neoformans*	Capsular polysaccharide	*6, 51*
	Aspergillus fumigatus		*62*
	Candida albicans		*61*
Agglutination (LA and HA)	*Cryptococcus neoformans*	Capsular polysaccharide	*7, 22, 23, 24, 54*
Sandwich ELISA	*Aspergillus fumigatus*		*59, 76*
	Candida albicans		*74, 75, 76*
Inhibition assays			
HAI	*Candida albicans*	Mannan	*80*
RIA	*Aspergillus fumigatus*	Carbohydrate[b]	*65, 66, 77, 78*
	Candida albicans		*56, 79*
	Coccidioides immitis		*86*
EAI	*Candida albicans*	Mannan	*64*

[a]Italic references are those in which antigen in human body fluids is recorded.
[b]Appears to have properties similar to galactomannan [58].

(Table 1). Before any of the tests are accepted it will be necessary to show that they have a diagnostic value.

5. *Aspergillus fumigatus*

The reports of circulating antigen in invasive aspergillosis are shown in Table 1. As with the tests for antigenemia in candidiasis, the procedures have not been compared as to their sensitivity and specificity. A material with antigenic similarity to a galactomannan extracted from mycelium was detected in the serum of infected animals and humans by means of a CIE test [39,40,58]. It was not found in animals infected by several other fungi. The antigen was not detected in sera from patients with aspergilloma and allergic bronchopulmonary aspergillosis, but was found in six cases where invasive disease was either proven by histology or suspected. However, the antigen has not been detected in several proven cases of invasive aspergillosis in humans and animals infected by some strains of *A. fumigatus,* and although it is possible that the antigen is not produced by these, it is quite likely that the CIE test is not sensitive enough to detect the galactomannan at the low concentration found in many animals. The galactomannan appears to be of low antigenicity. Antibody was not detected in patients with aspergilloma or allergic bronchopulmonary aspergillosis and, indeed, it has proved difficult to raise a high titered antiserum in rabbits that are injected with antigen isolated from the serum [39].

The structure of the galactomannan seems similar to that of the purified antigen used in the RIA test to detect antigenemia in infected rabbits [78] and humans [77], and RIA, if more sensitive than the CIE test, may prove more useful in the diagnosis of human disease. The authors reported that the antigen might occur as some form of complex in the serum, as it could be detected more easily if the serum was subjected to denaturing conditions prior to assay for antigen. The nature of this complex, which could possibly be an antigen-antibody complex, is unclear.

Other reports of antigenemia have been made in experimental animals and in humans, but the precise nature of the antigens detected is not known (Table 1).

6. *Coccidioides immitis*

Soluble antigen, occurring as immune complexes, has been detected in certain individuals with chronic coccidioidomycosis [86]. The chemical nature of the antigen was not described. The biological activities of immune complexes are discussed later (Sec. III.C.2).

C. Detection by Biological Activity

1. Toxins

Although a toxin has been reported to be produced in vivo by certain strains of *Candida albicans,* there is no report of its detection in body fluids by any biological or chemical assay [29]. Other toxins have been described for both *A. fumigatus* and *C. albicans* [29]. They include low molecular weight molecules which might be detected by appropriate chemical or GLC tests; however, the high molecular weight toxins, which are highly antigenic, are likely to be rapidly cleared in an antibody response should they be produced in vivo.

Toxins such as mushroom toxins and mycotoxins, which may be absorbed from the gut, have been detected in vivo, but this is beyond the scope of this chapter; however aflatoxin, produced by *A. flavus,* has been found at autopsy in the lung of a child with severe combined immunodeficiency who had died with a pneumonia (E. M. Bernard, personal communication). The cells surrounding the lesion appeared to have been transformed, but it is not clear whether this was a result of the carcinogenic action of the aflatoxin.

2. Immune Complexes

Although fungal materials may be bound in a nonspecific manner to proteins in the plasma, on many occasions they may exist as soluble immune complexes. Such complexes are formed in conditions of antigen excess or when the antigen exists as a molecule or fragment that is too small to be bound by more than one antibody molecule. Immune complexes have been implicated as having potent immunological and pathological activities in other systems [70]. These include their capacity to block specific T-lymphocyte responses to antigen, their inhibition of antibody-dependent cell-mediated cytotoxicity, their interaction with platelets, polymorphonuclear cells, and complement, and the induction of tissue damage.

Soluble immune complex formation in fungal disease has been suggested for chronic dermatophyte infections where a species-specific blocking factor was detected in patients' sera [73]. This factor interfered with leukocyte adherence in vitro, but no evidence was presented that the blocking factor was an immune complex. Indeed, antibody or free antigen could have been the blocking factor. Weiner and Yount [80] reported that in some cases of systemic candidiasis, antimannan antibody was released from serum treated with α-mannosidase. The action of this enzyme on *C. albicans* mannan was not reported, but it was suggested that immune complexes were present in these individuals. No binding of C1q, a common property of soluble immune complexes, was observed. However, complexes containing IgG4 or IgA may have been present which would not bind C1q. Also, it is possible that the antigen-antibody binding was too weak to produce the structural modifications in the Fc of the immunoglobulin which are needed for C1q binding.

Chesney et al. [12a] reported the deposition of *Candida* antigens, both in the glomeruli and on the renal tubular basement membrane of a patient who had chronic mucocutaneous candidiasis associated with glomerulonephritis. Although there are several reports of glomerulonephritis associated with mycotic endocarditis (reviewed in Ref. 12a), the case described by Chesney et al. appears to be the first in which fungal antigen was detected in the glomerular deposits, which also were found to contain IgG and C3. Although soluble immune complexes were not sought, the presence of either antigenemia or complexes was likely, for the patient's renal function was reported to have improved markedly following treatment for his candidiasis.

One might expect soluble immune complexes in other diseases, especially when massive antigen load is present, and indeed they have been detected in several patients with persistent coccidioidomycosis [86], a case of acute allergic bronchopulmonary aspergillosis [20], and in a case of disseminated histoplasmosis with glomerulonephritis [11]. Only in coccidioidomycosis were fungal antigens demonstrated in the complexes.

Overall, it appears probable that soluble immune complexes will be found in several more cases of fungal infection. Undoubtedly, they can have a role in pathogenesis and may be important in preventing the resolution of certain chronic diseases.

3. *Candida albicans* Blocking Factor

Blocking factors are not always antigen-antibody complexes. A factor has been described in the serum of some patients having chronic mucocutaneous candidiasis. It is heat stable at 80°C, bound by concanavalin A, and soluble in ammonium sulfate solutions at concentrations that precipitate immune complexes and antibody molecules [19]. The factor was recognized by its ability to block the proliferation of normal donor lymphocytes in the presence of a *C. albicans* antigen preparation. The authors considered that the factor was possibly mannan or other carbohydrate such as glucan. Purified mannan of both serotypes was shown to block the lymphocyte responses in vitro.

IV. FUTURE DIRECTIONS

It is clear that the diagnosis of fungal diseases from the detection of metabolites in body fluids is undergoing rapid changes as new techniques become available. Chemical and immunochemical assays have been developed and it is hard to predict which approach is likely to be more useful in laboratory diagnosis in the future. For some diseases in the immunosuppressed patient, neither approach may be necessary, as, for example in histoplasmosis, biopsy can provide suitable material for an early diagnosis of disease [31]. In other situations it may be found that both chemical and immunochemical methods will be needed to provide a clear laboratory diagnosis.

Certain advantages exist in detecting metabolites by chemical methods, especially those using GLC. Even though metabolites of these sorts have only been detected in candidiasis and cryptococcosis, the production of a GLC profile could be used to distinguish among several organisms, any of which could give rise to similar symptoms in the patient. If a specific compound is characteristic of a given organism, it should be possible, theoretically at least, to assay for it directly by high-resolution mass spectroscopy [44]. Such a procedure would eliminate the GLC step and allow for more rapid analysis of samples. In this situation the benefits of a profile, which can act as a fingerprint for each organism, would be lost. In addition to GLC, it is very likely that analysis of body fluids by high-performance liquid chromatography (HPLC) will be applied to the detection of fungal metabolites because in this procedure there is no need to make the volatile derivatives necessary for GLC.

The expensive equipment and specialized technical staff required for GLC methods make it likely that the less expensive immunodiagnostic methods will continue to be used in the hospital laboratory. The success of such methods depends to some extent on the standardization of the antibody reagents and on the physical and chemical characterization of the antigens that can be detected in body fluids. Semistandardized reagents have already been shown to be useful in the LA test for cryptococcal polysaccharide [54]. Some recent developments may aid future diagnostic work. First, immunological methodology is undergoing rapid changes with the increasing use of monoclonal antibody reagents [47]; however, these have yet to be applied to fungal antigen detection. Second, stable red blood cell reagents have been developed for the detection of hepatitis B antigen (hepatest) [3]. These antibody-coated red blood cells are extremely sensitive to antigen, are easy to use, and give results rapidly. The use of stable red blood cells in inhibition tests such as HAI or inhibition of mixed reverse passive antiglobulin hemagglutination (M_R PAH) [14] could make the tests almost as sensitive as radio- or enzyme immunoassays. In contrast, direct detection of antigens in invasive fungal diseases seems less promising, mainly because of the problems of detecting small antigen fragments and soluble immune complexes (Sec. III.B.1.a). In spite of this, at present the only test used routinely for detecting fungal antigen is the LA test for cryptococcal polysaccharide, and this is a direct assay.

Table 2 Fungal Metabolites in Human Body Fluids[a]

Metabolite	Occurrence and fate in humans[b]			
	Produced	Destroyed	Blood to Urine	Reference
Ethanol	–	+	+	15
D-Arabitol	–	–	+	34
Candida mannan	–	?	?	80
Cryptococcal polysaccharide	–	?[c]	?[d]	51
Aspergillus galactomannan[e]	–	?	?[d]	58
Aspergillus carbohydrate[e]	–	?	?[d]	78
Oxalic acid[f]	+	–	+	52
Aflatoxin	–	?	?	g

[a]Table shows those metabolites discussed in this chapter. (See also Sec. VI.)
[b]Refers to metabolite's production in and breakdown by human metabolism and its transport from blood to urine. A question mark (?) indicates that data are not available.
[c]Probably not metabolized. It persists for long periods in rabbits [6].
[d]Isolation in urine of animals may be due to kidney infections.
[e]May be same material.
[f]Not truly soluble since it forms calcium oxalate crystals in vivo. Some may be present as soluble acid.
[g]Personal communication from E. M. Bernard.

Finally, further efforts to detect soluble immune complexes, small molecules, and antigenic fragments which may have important biological activities ought to be made.

V. CONCLUSIONS

Successful diagnosis of fungal infection from analysis of metabolites in body fluids depends in part on the properties of those metabolites. Most of those described (Table 2) have proven to be carbohydrates and, with the exception of *C. albicans* mannan, do not appear to be routinely cleared from the blood by a strong antibody reaction. This lack of antigenicity may be an important property of circulating metabolites, for the detection of *Candida* mannan antigenemia is transient because of the antibody response [80]. In addition, other fungal metabolites which are strongly bound to host molecules or which, like oxalic acid, react to form an insoluble material will not be routinely detected.

Further important properties of fungal metabolites must include their resistance to host enzymes. Trehalose, a fungal sugar [13], is unlikely to be detected owing to its breakdown to glucose by trehalases present in the blood and tissue. The resistance of cryptococcal polysaccharide, oxalic acid, and arabitol to degradation by animals is known, but it is not clear how long *Aspergillus* galactomannan or *Candida* mannan resist decay. Purified mannan, if injected into rabbits, is rapidly cleared from the bloodstream [75].

The rapid excretion of small molecules into the urine could make it difficult to detect high serum levels of these. Mannitol, which has not been reported as a metabolite produced by fungi during the course of infection, might be expected to have similar properties to arabitol, and this is rapidly removed in the urine. The detection of fungal products in urine would only be useful in diagnosis for those fungi that are not common colonizers of the bladder. For this reason it is unlikely that urine will prove to be useful in the diagnosis of invasive candidiasis.

The biological activity of these metabolites is not well known and in many cases only a few strains of a single fungus have been studied for their production. Some of these products can interact with the immune system. Cryptococcal polysaccharide is reported to be antiphagocytic and to induce B-cell tolerance [12,36,49], and yeast mannans can activate complement via the alternative pathway. Whether toxic metabolites, such as the aflatoxins or the toxins described by Iwata [29], will be detected routinely in the blood and other fluids remains for future research; however, it is clear that an understanding of the pathogenesis of fungal diseases must include studies on the biological activities of all the fungal metabolites that enter the body fluids, not only those that are useful diagnostically.

It is hard to predict what the future has in store, but we can expect that methods for the routine diagnosis of invasive and opportunistic fungal infections will be available in the not-too-distant future.

VI. RECENT DEVELOPMENTS

A. Nonantigenic Metabolites

1. Ethanol

A review of 24 persons from Japan, who suffered symptoms of drunkenness following their ingestion of nonalcoholic foods, has been made by Iwata [87]. Low acidity of the gastric juice, postoperative stenosis in the stomach or jejenum, and a naturally high susceptibility to alcoholic beverages were noted for several patients. In addition, the symptoms could be provoked by a glucose tolerance test. Several yeast species were implicated in the disorder, which could be treated with antifungal antibiotics with or without corrective surgery.

2. Oxalic Acid

A further report on oxalosis associated with *Aspergillus niger* aspergilloma has been presented [88]. Of four patients, only one was found with oxalosis. The oxalate crystals occurred in renal tissues and in the lung. More than 40 aspergillomas caused by *A. fumigatus* were reviewed but none was found to be associated with oxalosis. However, oxalic acid crystals were seen in postmortem tissues collected from parrots infected by *A. fumigatus* as well as by *A. oryzae* [105]. The conditions allowing oxalic acid production by *Aspergillus* in vivo need to be studied.

3. Arabitol

Several workers have confirmed the detection of arabitol in persons with invasive candidiasis [89-91]. As elevated levels of arabitol may be detected in persons having renal failure, the measurement of the arabitol/creatinine ratio in sera has been recommended for evaluating a patient's status [92]. Whether arabitol will be a useful marker for candidiasis caused by species of *Candida* other than *C. albicans* requires evaluation. Several species, including *C. tropicalis,* have been reported to release only small amounts of arabitol when grown in vitro [93].

4. Mannose

Using gas-liquid chromatography, Monson and Wilkinson [94] found that the level of free mannose in serum and CSF rose in patients with invasive candidiasis. Although poorly controlled diabetes is associated with elevated levels of serum mannose, this should not be a problem for the diagnosis of invasive candidiasis in many patients. Mannose had been observed previously [95] but had not been further evaluated.

5. Other Metabolites

Other researchers have reported the occurrence of abnormal peaks produced on chromatograms produced following GLC with samples of serum or cerebrospinal fluid taken from patients having cryptococcal meningitis [95,96] and aspergillosis [97].

B. Antigenic Metabolites

Recent research at several institutions has confirmed the importance of processing sera prior to their use in assays for the detection of antigen. Originally, alkalis or acids were added to bring about dissociation of fungal antigens from host serum components [64, 79], and such pretreatment procedures have been found to be effective for detecting, by RIA, *Aspergillus* antigen in bronchoalveolar lavage fluid and serum samples from rabbits which had been injected with conidia of *A. fumigatus* [98]. *Candida* mannan has been detected by ELISA following the pretreatment, with alkali, of sera from leukemic patients [99] or following the dissociation of the mannan complexes by heating sera in the presence of EDTA for 5 min at 121°C [100]. Even gentle heating (56°C, 30 min) has been found to increase the sensitivity of mannan detection in sera from rabbits in which candidal endocarditis had been induced [101]. Further standardization of the pretreatment methods has been made [103], resulting in a rapid, sensitive, sandwich ELISA procedure for the detection of mannan.

In spite of the findings of increased sensitivity following pretreatment of sera, others have been able to detect mannan in unprocessed sera from a number of leukemic patients using a passive hemagglutination inhibition technique [102]. This and the previous report [102] confirmed that assays for mannan, extracted from *C. albicans* of serotype B, were far less sensitive than the assays for mannan from *C. albicans* serotype A [41].

C. Biologically Active Metabolites

1. Enzymes

Yolken has suggested [104] that the presence of fungi in an infection could be diagnosed from their production of enzymes having activities not characteristic of human or bacterial enzymes. Such enzymes might include glucose oxidase and chitin synthetase, which are not produced by animals or bacteria. In his review, Yolken mentioned the detection of adenine deaminase in some human infections due to *A. fumigatus* and *C. tropicalis*, but a detailed description of the assay was not presented. This approach to diagnosis is exciting in that enzymes are often measurable in small concentrations. However, the antigenicity of the fungal enzymes probably will limit their detection in body fluids to conditions where they are present in the absence of antibody, or where they exist as soluble antigen-antibody complexes.

ACKNOWLEDGMENTS

Preparation of this chapter was supported in part by the Wellcome Trust (UK). I should like to thank Dr. J. B. Brooks for furnishing Figure 1, Drs. R. A. Cox and T. E. Kiehn for preprints of manuscripts, and Dr. E. M. Bernard for permission to quote unpublished findings.

REFERENCES

1. Avrameas, S., and T. Ternynck. 1969. The cross-linking of proteins with glutaraldehyde and its use for the preparation of immunoadsorbents. Immunochemistry 6: 53-66.
2. Axelsen, N. H., and C. H. Kirkpatrick. 1973. Simultaneous characterization of free *Candida* antigens and *Candida* precipitins in a patient's serum by means of cross immunoelectrophoresis with intermediate gel. J. Immunol. Methods 2: 245-249.
3. Barbara, J. A. J., P. J. Harrison, D. R. Howell, T. E. Cleghorn, D. S. Dane, M. Briggs, and C. H. Cameron. 1979. A single reverse passive haemagglutination test for detecting both HBsAg and anti-HBs. J. Clin. Pathol. 32: 1180-1183.
4. Bartnicki-Garcia, S. 1973. Fungal cell wall composition. In A. I. Laskin and H. A. Lechevalier (Eds.), *Handbook of Microbiology,* Vol. II: *Microbial Composition,* CRC Press, Cleveland, pp. 201-214.
5. Bennett, J. E., and J. W. Bailey. 1971. Control for rheumatoid factor in the latex test for cryptococcosis. Am. J. Clin. Pathol. 56: 360-365.
6. Bennett, J. E., and H. F. Hasenclever. 1965. *Cryptococcus neoformans* polysaccharide: studies of serologic properties and role in infection. J. Immunol. 94: 916-920.
7. Bloomfield, N., M. A. Gordon, and D. R. Elmendorf, Jr. 1963. Detection of *Cryptococcus neoformans* antigen in body fluids by latex particle agglutination. Proc. Soc. Exp. Biol. Med. 114: 64-67.
8. Blumenthal, H. J. 1976. Reserve carbohydrates in fungi. In J. E. Smith and D. R. Berry (Eds.), *The Filamentous Fungi,* Vol. II: *Biosynthesis and Metabolism,* Wiley, New York, pp. 292-307.
9. Brian, T. 1655. *The Pisse-Prophet, or Certain Pisse-Pot Lectures,* R. Thrale, London.
10. Brooks, J. B. 1977. Detection of bacterial metabolites in spent culture media and body fluids by electron capture gas-liquid chromatography. Adv. Chromatogr. 15: 1-31.
11. Bullock, W. E., R. P. Artz, D. Bhathena, and K. S. K. Tung. 1979. Histoplasmosis. Association with circulating immune complexes, eosinophilia, and mesangiopathic glomerulonephritis. Arch. Intern. Med. 139: 700-702.
12. Bulmer, G. S., and M. D. Sans. 1968. *Cryptococcus neoformans.* III. Inhibition of phagocytosis. J. Bacteriol. 95: 5-8.
12a. Chesney, R. W., S. O'Regan, H. J. Guyda, and K. N. Drummond. 1976. *Candida* endocrinopathy syndrome with membranoproliferative glomerulonephritis: demonstration of glomerular *Candida* antigen. Clin. Nephrol. 5: 232-238.
13. Cochrane, V. W. 1958. *Physiology of Fungi,* Wiley, New York.
14. Coombs, R. R. A., L. Edebo, A. Feinstein, and B. W. Gurner. 1978. The class of antibodies sensitizing bacteria measured by mixed reverse passive antiglobulin haemagglutination (M_RPAH). Immunology 34: 1037-1044.
15. Corry, J. E. L. 1978. A review. Possible sources of ethanol ante- and post-mortem: its relationship to the biochemistry and microbiology of decomposition. J. Appl. Bacteriol. 44: 1-56.

16. Craven, R. B., J. B. Brooks, D. C. Edman, J. D. Converse, J. Greenlee, D. Schlossberg, T. Furlow, J. M. Gwaltney, Jr., and W. F. Miner. 1977. Rapid diagnosis of lymphocytic meningitis by frequency-pulsed electron capture gas-liquid chromatography: differentiation of tuberculous, cryptococcal, and viral meningitis. J. Clin. Microbiol. *6*: 27-32.

17. Dawson, D. M., and A. Taghavy. 1963. A test for spinal-fluid alcohol in torula meningitis. N. Engl. J. Med. *269*: 1424-1425.

18. Dochez, A. R., and O. T. Avery. 1917. The elaboration of specific soluble substance by *Pneumococcus* during growth. J. Exp. Med. *26*: 477-493.

19. Fischer, A., J-J. Ballet, and C. Griscelli. 1978. Specific inhibition of in vitro *Candida*-induced lymphocyte proliferation by polysaccharidic antigens present in the serum of patients with chronic mucocutaneous candidiasis. J. Clin. Invest. *62*: 1005-1013.

20. Geha, R. S. 1977. Circulating immune complexes and activation of the complement sequence in acute allergic bronchopulmonary aspergillosis. J. Allergy Clin. Immunol. *60*: 357-359.

21. Gibson, I. A. S. 1953. Crown rot, a seedling disease of ground nuts caused by *Aspergillus niger*. Trans. Br. Mycol. Soc. *36*: 198-209.

22. Goodman, J. S., L. Kaufman, and M. G. Koenig. 1971. Diagnosis of cryptococcal meningitis. Value of immunologic detection of cryptococcal antigen. N. Engl. J. Med. *285*: 434-436.

23. Gordon, M. A., and E. W. Lapa. 1974. Elimination of rheumatoid factor in the latex test for cryptococcosis. Am. J. Clin. Pathol. *61*: 488-494.

24. Gordon, M. A., and D. K. Vedder. 1966. Serologic tests in diagnosis and prognosis of cryptococcosis. JAMA *197*: 961-967.

25. Goullier, A., Y. Ganansia, G. Decoux, A. Favier, J. Biguet, and P. Ambroise-Thomas. 1976. Apports de la chromatographie en phase gazeuse au diagnostic précoce des septicémies à *Candida*. Essais préliminaires. Bull. Soc. Fr. Mycol. Med. *5*: 103-107.

26. Hierholzer, J. C. 1976. Effects of sulfate concentration, electroendosmotic flow, and electrical resistance of agars and agaroses on counterimmunoelectrophoresis with adenovirus antigens and antisera. J. Immunol. Methods *11*: 63-76.

27. Hodgkinson, A. 1977. *Oxalic Acid in Biology and Medicine*, Academic, London.

28. Hunter, W. M. 1978. Radioimmunoassay. In D. M. Weir (Ed.), *Handbook of Experimental Immunology*, 3rd ed., Blackwell, Oxford, pp. 14.1-14.40.

29. Iwata, K. 1978. Fungal toxins as a parasitic factor responsible for the establishment of fungal infections. Mycopathologia *65*: 141-154.

30. Kaplan, W. M. 1975. Practical application of fluorescent antibody procedures in medical mycology. In *Mycoses,* Proceedings of the Third International Conference on the Mycoses, Pan Am. Health Organ. Sci. Publ. No. 304, Washington, D.C., pp. 178-185.

31. Kauffman, C. A., K. S. Israel, J. W. Smith, A. C. White, J. Schwartz, and G. F. Brooks. 1978. Histoplasmosis in immunosuppressed patients. Am. J. Med. *64*: 923-932.

32. Kaufman, L., and S. Blumer. 1968. Value and interpretation of serological tests for the diagnosis of cryptococcosis. Appl. Microbiol. *16*: 1907-1912.

33. Kerkering, T. M., A. Espinel-Ingroff, and S. Shadomy. 1979. Detection of *Candida* antigenemia by counterimmunoelectrophoresis in patients with invasive candidiasis. J. Infect. Dis. *140*: 659-664.

34. Kiehn, T. E., E. M. Bernard, J. W. M. Gold, and D. Armstrong. 1979. Candidiasis: detection by gas-liquid chromatography of D-arabinitol, a fungal metabolite, in human serum. Science *206*: 577-580.

35. Kozel, T. R., and J. Cazin, Jr. 1971. Nonencapsulated variant of *Cryptococcus neoformans*. I. Virulence studies and characterization of soluble polysaccharide. Infect. Immun. *3*: 287-294.

36. Kozel, T. R., and R. P. Mastroianni. 1976. Inhibition of phagocytosis by cryptococcal polysaccharide: dissociation of the attachment and ingestion phases of phagocytosis. Infect. Immun. *14*: 62-67.

37. Krick, J. A., and J. S. Remington. 1976. Opportunistic invasive fungal infections in patients with leukaemia and lymphoma. Clin. Haematol. *5*: 249-310.

38. Kurrein, F., G. H. Green, and S. L. Rowles. 1975. Localized deposition of calcium oxalate around a pulmonary *Aspergillus niger* fungus ball. Am. J. Clin. Pathol. *64*: 556-563.

39. Lehmann, P. F., and E. Reiss. 1978. Invasive aspergillosis: antiserum for circulating antigen produced after immunization with serum from infected rabbits. Infect. Immun. *20*: 570-572.

40. Lehmann, P. F., and E. Reiss. 1979. *Aspergillus fumigatus* antigenaemia: detection of antigen in mice and in human patients. Bull. Soc. Fr. Mycol. Med. *8*: 57-64.

41. Lehmann, P. F., and E. Reiss. 1980. Detection of *Candida albicans* mannan by immunodiffusion, counterimmunoelectrophoresis, and enzyme-linked immunoassay. Mycopathologia *70*: 83-88.

42. Lehmann, P. F., and L. O. White. 1975. Chitin assay used to demonstrate renal localization and cortisone-enhanced growth of *Aspergillus fumigatus* mycelium in mice. Infect. Immun. *12*: 987-992.

43. Lehmann, P. F., and L. O. White. 1976. Acquired immunity to *Aspergillus fumigatus*. Infect. Immun. *13*: 1296-1298.

44. Ligon, W. V., Jr. 1979. Molecular analysis by mass spectroscopy. Science *205*: 151-159.

45. Ling, N. R., S. Bishop, and R. Jefferis. 1977. Use of antibody-coated red cells for the sensitive detection of antigen and in rosette tests for cells bearing surface immunoglobulins. J. Immunol. Methods *15*: 279-289.

46. Maccani, J. E. 1977. Detection of cryptococcal polysaccharide using counterimmunoelectrophoresis. Am. J. Clin. Pathol. *68*: 39-44.

47. Melchers, F., M. Potter, and N. L. Warner (Eds.). 1978. Lymphocyte hybridomas. Curr. Top. Microbiol. Immunol. *81*: 1-246.

48. Miller, G. G., M. W. Witwer, A. I. Braude, and C. E. Davis. 1974. Rapid identification of *Candida albicans* septicemia in man by gas-liquid chromatography. J. Clin. Invest. *54*: 1235-1240.

49. Murphy, J. W., and G. C. Cozad. 1972. Immunological unresponsiveness induced by cryptococcal capsular polysaccharide assayed by the hemolytic plaque technique. Infect. Immun. *5*: 896-901.

50. Neill, J. M., and C. E. Kapros. 1950. Serological tests on soluble antigens from mice infected with *Cryptococcus neoformans* and *Sporotrichum schenckii*. Proc. Soc. Exp. Biol. Med. *73*: 557-559.

51. Neill, J. M., J. Y. Sugg, and D. W. McCauley. 1951. Serologically reactive material in spinal fluid, blood and urine from human case of cryptococcosis (torulosis). Proc. Soc. Exp. Biol. Med. *77*: 775-778.

52. Nime, F. A., and G. M. Hutchins. 1973. Oxalosis caused by *Aspergillus* infection. Johns Hopkins Med. J. *133*: 183-194.

53. Ouchterlony, O., and L.-A. Nilsson. 1978. Immunodiffusion and immunoelectrophoresis. In D. M. Weir (Ed.), *Handbook of Experimental Immunology*, 3rd ed., Blackwell, Oxford, pp. 19.1-19.44.

54. Palmer, D. F., L. Kaufman, W. Kaplan, and J. J. Cavallaro. 1977. *Serodiagnosis of Mycotic Diseases,* Thomas, Springfield, Ill.

55. Pappagianis, D., and R. Marovitz. 1966. Studies on ethanol production by *Cryptococcus neoformans.* Sabouraudia *4*: 250-255.

56. Poor, A. H., and J. E. Cutler. 1979. Partially purified antibodies used in a solid-phase radioimmunoassay for detecting candidal antigenemia. J. Clin. Microbiol. *9*: 362-368.

57. Rebers, P. A., S. A. Barker, M. Heidelberger, Z. Dische, and E. E. Evans. 1958. Precipitation of the specific polysaccharide of *Cryptococcus neoformans* A by types I and XIV antipneumococcal sera. J. Am. Chem. Soc. *80*: 1135-1137.

58. Reiss, E., and P. F. Lehmann. 1979. Galactomannan antigenemia in invasive aspergillosis. Infect. Immun. *25*: 357-365.

59. Richardson, M. D., L. O. White, and R. C. Warren. 1979. Detection of circulating antigen of *Aspergillus fumigatus* in sera of mice and rabbits by enzyme-linked immunosorbent assay. Mycopathologia *67*: 83-88.

60. Ride, J. P., and R. B. Drysdale. 1972. A rapid method for the chemical estimation of filamentous fungi in plant tissues. Physiol. Plant Pathol. *2*: 7-15.

61. Rippon, J. W., and D. N. Anderson. 1978. Experimental mycosis in immunosuppressed rabbits. I. Acute and chronic candidosis. Mycopathologia *64*: 91-96.

62. Rippon, J. W., and D. N. Anderson. 1978. Experimental mycosis in immunosuppressed rabbits. II. Acute and chronic aspergillosis. Mycopathologia *64*: 97-100.

63. Schlossberg, D., J. B. Brooks, and J. A. Shulman. 1976. Possibility of diagnosing meningitis by gas chromatography: cryptococcal meningitis. J. Clin. Microbiol. *3*: 239-245.

64. Segal, E., R. A. Berg, P. A. Pizzo, and J. E. Bennett. 1979. Detection of *Candida* antigen in sera of patients with candidiasis by an enzyme-linked immunosorbent assay-inhibition technique. J. Clin. Microbiol. *10*: 116-118.

65. Shaffer, P. J., G. S. Kobayashi, and G. Medoff. 1979. Demonstration of antigenemia in patients with invasive aspergillosis by solid phase (protein A-rich *Staphylococcus aureus*) radioimmunoassay. Am. J. Med. *67*: 627-630.

66. Shaffer, P. J., G. Medoff, and G. S. Kobayashi. 1979. Demonstration of antigenemia by radioimmunoassay in rabbits experimentally infected with *Aspergillus.* J. Infect. Dis. *139*: 313-319.

67. Smith, D., L. Muscatine, and D. Lewis. 1969. Carbohydrate movement from autotrophs to heterotrophs in parasitic and mutalistic symbiosis. Biol. Rev. Camb. Philos. Soc. *44*: 17-90.

68. Staib, F., J. Steffen, D. Krumhaar, G. Kapetanakis, C. Minck, and G. Grosse. 1979. Lokalisierte Aspergillose und Oxalose der Lunge durch *Aspergillus niger:* ein Beitrag zum Thema Topferde von Zimmerpflanzen als Aspergillen-Reservoir. Dtsch. Med. Wochenschr. *104*: 1176-1179.

69. Stoddart, R. W., and B. M. Herbertson. 1978. The use of fluorescein-labelled lectins in the detection and identification of fungi pathogenic for man: a preliminary study. J. Med. Microbiol. *11*: 315-324.

70. Theofilopoulos, A. N., and F. J. Dixon. 1979. The biology and detection of immune complexes. Adv. Immunol. *28*: 89-220.

71. Tyler, R. 1956. Spinal fluid alcohol in yeast meningitis. Am. J. Med. Sci. *232*: 560-561.

72. Voller, A., A. Bartlett, and D. E. Bidwell. 1978. Enzyme immunoassays with special reference to ELISA techniques. J. Clin. Pathol. *31*: 507-520.

73. Walters, B. A. J., J. E. D. Chick, and W. J. Halliday. 1974. Cell-mediated immunity

and serum blocking factors in patients with chronic dermatophytic infections. Int. Arch. Allergy Appl. Immunol. *46*: 849-857.

74.   Warren R. C., A. Bartlett, D. E. Bidwell, M. D. Richardson, A. Voller, and L. O. White. 1977. Diagnosis of invasive candidosis by enzyme immunoassay of serum antigen. Br. Med. J. *1*: 1183-1185.

75.   Warren, R. C., M. D. Richardson, and L. O. White. 1978. Enzyme-linked immuno-sorbent assay of antigens from *Candida albicans* circulating in infected mice and rabbits: the role of mannan. Mycopathologia *66*: 179-182.

76.   Warren, R. C., L. O. White, S. Mohan, and M. D. Richardson. 1979. The occurrence and treatment of false positive reactions in enzyme-linked immunosorbent assays (ELISA) for the detection of fungal antigens in clinical samples. J. Immunol. Methods *28*: 177-186.

77.   Weiner, M. H. 1980. Antigenemia detected by radioimmunoassay in systemic asper-gillosis. Ann. Intern. Med. *92*: 793-796.

78.   Weiner, M. H., and M. Coats-Stephen. 1979. Immunodiagnosis of systemic asper-gillosis. I. Antigenemia detected by radioimmunoassay in experimental infection. J. Lab. Clin. Med. *93*: 111-119.

79.   Weiner, M. H., and M. Coats-Stephen. 1979. Immunodiagnosis of systemic candi-diasis: mannan antigenemia detected by radioimmunoassay in experimental and human infections. J. Infect. Dis. *140*: 989-993.

80.   Weiner, M. H., and W. J. Yount. 1976. Mannan antigenemia in the diagnosis of in-vasive *Candida* infections. J. Clin. Invest. *58*: 1045-1053.

81.   White, L. O., E. Gibb, H. C. Newham, M. D. Richardson, and R. C. Warren. 1979. Comparison of the growth of virulent and attenuated strains of *Candida albicans* in the kidneys of normal and cortisone-treated mice by chitin assay. Mycopathologia *67*: 173-177.

82.   White, L. O., H. C. Newham, and J. P. Ride. 1978. Estimation of *Absidia ramosa* infection in the brain and kidneys of cortisone-treated mice by chitin assay. My-copathologia *63*: 177-179.

83.   White, L. O., M. D. Richardson, H. C. Newham, E. Gibb, and R. C. Warren. 1977. Circulating antigen of *Aspergillus fumigatus* in cortisone-treated mice challenged with conidia: detection by counterimmunoelectrophoresis. FEMS Microbiol. Lett. *2*: 153-156.

84.   White, R. W., D. B. Lindsay, and R. W. Ash. 1972. Ethanol production from glucose by *Torulopsis glabrata* occurring naturally in the stomachs of newborn animals. J. Appl. Bacteriol. *35*: 631-646.

85.   Wilson, D. E., T. W. Williams, and J. E. Bennett. 1966. Further experience with the alcohol test for cryptococcal meningitis. Am. J. Med. Sci. *252*: 532-536.

86.   Yoshinoya, S., R. A. Cox, and R. M. Pope. 1980. Circulating immune complexes in coccidioidomycosis. Detection and characterization. J. Clin. Invest. *66*: 655-663.

87.   Iwata, K. 1976. A review of the literature on drunken symptoms due to yeasts in the gastrointestinal tract. In K. Iwata (Ed.), *Yeasts and Yeast-like Microorganisms in Medical Science. Proceedings of the Second International Symposium on Yeasts, Tokyo, 1972.* University of Tokyo Press, Tokyo, pp. 260-268.

88.   Severo, L. C., A. T. Londero, G. R. Geyer, and P. D. Picon. 1981. Oxalosis associ-ated with an *Aspergillus niger* fungus ball. Report of a case. Mycopathologia *73*: 29-31.

89.   Roboz, J., R. Suzuki, and J. F. Holland. 1980. Quantification of arabinitol in serum by selected ion monitoring as a diagnostic technique in invasive candidiasis. J. Clin. Microbiol. *12*: 594-602.

90. Gold, J. W. M., E. M. Bernard, T. E. Kiehn, B. Wong, L. Borch, and D. Armstrong. 1980. Serum arabinitol levels in patients with invasive candidiasis. In *Current Chemotherapy and Infectious Disease. Proceedings of the 11th International Congress of Chemotherapy and the 19th Interscience Conference on Antimicrobial Agents and Chemotherapy.* American Society for Microbiology, Washington, D.C., pp. 978-980.

91. Eng, R. H. K., H. Chmel, and M. Buse. 1981. Serum levels of arabinitol in the detection of invasive candidiasis in animals and humans. J. Infect. Dis. *143*: 677-683.

92. Gold, J. W. M., B. Wong, E. M. Bernard, S. W. McKean, T. Kiehn, and D. Armstrong. 1981. 21st Interscience Conference on Antimicrobial Agents and Chemotherapy, Abstract 668. American Society for Microbiology, Washington, D.C.

93. Bernard, E. M., K. J. Christiansen, S.-F. Tsang, T. E. Kiehn, and D. Armstrong. 1981. Rate of arabinitol production by pathogenic yeast species. J. Clin. Microbiol. *14*: 189-194.

94. Monson, T. P., and K. P. Wilkinson. 1981. Mannose in body fluids as an indicator of invasive candidiasis. J. Clin. Microbiol. *14*: 557-562.

95. Davis, C. E., and R. A. McPherson. 1975. Rapid diagnosis of septicemia and meningitis by gas-liquid chromatography. In D. Schlessinger (Ed.), *Microbiology—1975.* American Society for Microbiology, Washington, D.C., pp. 55-63.

96. Amundson, S., A. I. Braude, and C. E. Davis. 1974. Rapid diagnosis of infection by gas-liquid chromatography: analysis of sugars in normal and infected cerebrospinal fluid. Appl. Microbiol. *28*: 298-302.

97. Ueda, S., K. Kimura, K. Miyamoto, T. Ito, K. Takahashi, H. Iida, and J. Wada. 1974. Diagnostic use of gas chromatography of serum carbohydrates for pulmonary aspergillosis. Jpn. J. Thoracic Dis. *12*: 301-309. (In Japanese with English abstract.)

98. Andrews, C. P., and M. H. Weiner. 1981. Immunodiagnosis of invasive pulmonary aspergillosis in rabbits. Fungal antigen detected by radioimmunoassay in bronchoalveolar lavage fluid. Am. Rev. Respir. Dis. *124*: 60-64.

99. Meckstroth, K. L., E. Reiss, J. W. Keller, and L. Kaufman. 1981. Detection of antibodies and antigenemia in leukemic patients with candidiasis by enzyme-linked immunosorbent assay. J. Infect. Dis. *144*: 24-32.

100. Lew, M. A., G. R. Siber, D. M. Donahue, and F. Maiorca. 1982. Enhanced detection with an enzyme-linked immunosorbent assay of *Candida* mannan in antibody-containing serum after heat extraction. J. Infect. Dis. *145*: 45-56.

101. Scheld, W. M., R. S. Brown, Jr., S. A. Harding, and M. A. Sande. 1980. Detection of circulating antigen in experimental *Candida albicans* endocarditis by an enzyme-linked immunosorbent assay. J. Clin. Microbiol. *12*: 679-683.

102. Reiss, E., L. Stockman, R. J. Kuykendall, and S. J. Smith. 1982. Dissociation of mannan-serum complexes and detection of *Candida albicans* mannan by enzyme immunoassay variations. Clin. Chem. *28*: 306-310.

103. Meunier-Carpentier, F., and D. Armstrong. 1981. *Candida* antigenemia, as detected by passive hemagglutination inhibition, in patients with disseminated candidiasis or *Candida* colonization. J. Clin. Microbiol. *13*: 10-14.

104. Yolken, R. H. 1981. Enzymic analysis for rapid detection of microbial infection in human body fluids: an overview. Clin. Chem. *27*: 1490-1498.

105. Kaplan, W., P. Arnstein, L. Ajello, F. Chandler, J. Watts, and M. Hicklin. 1975. Fatal aspergillosis in imported parrots. Mycopathologia *56*: 25-29.

7

Mode of Action of Antifungal Drugs

Gerald Medoff and George S. Kobayashi / Washington University School of Medicine,
St. Louis, Missouri

I. INTRODUCTION

The development of new agents for the treatment of mycotic infections has been slow, and there are relatively few available, compared to the number that have been developed for treatment of bacterial infections. There are several reasons for this:

1. Systemic fungal infections are much less common than bacterial infections, so that less effort has been placed in the development of new therapeutic agents.
2. The difficulties in establishing a reproducible animal model for the most frequent of the fungal infections, the dermatomycoses, have decreased the incentive for developing new agents (these infections are also non-life-threatening, cosmetic problems and in most cases trivial).
3. Since fungi are eucaryotes, it has been difficult to develop antifungal agents that specifically affect fungal structures and macromolecular synthesis without also affecting host cells.

In recent years, the increasing frequency of systemic fungal infections, particularly in the compromised host, has stimulated work on the development of several new antifungal agents. The most important of these affect cell membranes. The polyene macrolides and the more recently discovered imidazole derivatives are the two groups of antibiotics in this class. The other important antifungal agents are 5-fluorocytosine, which probably affects RNA and DNA synthesis, and griseofulvin, which interacts with receptors on tubulin to inactivate the free subunits and inhibits cell division or other microtubular functions in a variety of different cell types.

The recent increase in the number of antifungal agents and the discovery that some of the older ones have properties which allow them to be used in new ways have led to the development of several new therapeutic regimens. Therefore, it is important to learn the mode of action of these agents, their advantages, and their limitations because this information will be helpful in determining the indications for use of one regimen over the others. In this chapter we summarize what is known about the antifungal agents that are most important clinically.

II. POLYENE ANTIBIOTICS

Among the long list of polyenes that have been isolated and described are some of the most effective antifungal agents known. Problems associated with solubility, stability, absorption, and toxicity, however, have made only a few of these agents therapeutically useful (see Table 1).

A. Chemical and Physical Properties of the Polyene Antibiotics

All of the polyene antibiotics analyzed to date have certain common structural features in addition to a conjugated double-bond system (Fig. 1). They are all characterized by a macrolide ring of carbon atoms closed by the formation of an internal ester or lactone. The presence of the lactone confers a highly characteristic peak on the infrared spectrum of these compounds. The conjugated double-bond system is contained exclusively within the cyclic lactone and the degree of unsaturation ranges from trienes to heptaenes. Analysis of the characteristic ultraviolet (UV) spectrum of the polyene antibiotics and also x-ray

Table 1 Clinically Useful Polyene Antibiotics

Name	Producing organism	Chemical composition	Molecular weight
Amphotericin B	*Streptomyces nodosus*	$C_{47}H_{73}NO_{17}$	924
Nystatin	*S. albulus* or *S. noursei*	$C_{47}H_{75}NO_{17}$	926
Primaricin	*S. natalensis*	$C_{33}H_{47}NO_{14}$	666
Candicidin	*S. griseus*	$C_{63}H_{85}N_2O_{19}$	1200

crystalographic analysis of an N-iodoacetyl derivative of amphotericin B indicate that all the double bonds are in the trans conformation [34].

Another highly characteristic feature of the polyene antibiotics is the large number of hydroxyls present on the molecule, which are usually distributed along the macrolide ring on alternate carbon atoms (Fig. 1). The presence of such a large number of polar groups and the multiple hydrophobic double bonds on the opposite side of the macrolide ring confer on the polyene antibiotics the additional characteristic chemical property of being amphipathic. It is likely that this amphipathic feature plays an important role in the mode of action of these substances as they interact in various biological systems.

The most characteristic physical property of the polyene antibiotics is the UV absorption spectrum. The UV spectra of all the polyenes have a regular series of sharp peaks of absorption, which are separated by sharp troughs, all in the range 400-280 nm. Oroshnik and Mebane [89] have given an extensive tabulation of the exact absorption maximum for many of the polyene antibiotics and have shown that the characteristic UV absorbing pattern is due to the conjugated double bonds.

B. Mechanism of Action of the Polyene Antibiotics

In the early 1960s several laboratories independently presented evidence that polyene antibiotics could increase the cell membrane permeability of a number of organisms, thereby promoting a leakage of important cellular constituents and, ultimately, lysis and death of the cell [62,74]. Many other studies have substantiated these findings and have reinforced the belief that the other metabolic effects of the polyenes (inhibition of aerobic and anaerobic metabolism, etc.) were due to the leakage of vital cytoplasmic constituents resulting from the alteration of cell membrane permeability.

The antifungal effects of the polyene antibiotics are dependent on the binding of the drugs to cell membranes. Bacteria and protoplasts do not bind polyenes and are unaffected by them. These results have led to the notion that the polyene antibiotics act by binding to membrane sterols, which are constituents of eucaryotic cells but are not present in bacterial cell walls or membranes. Among the experiments suggesting a relationship between sterols and the action of polyenes, two series have been most convincing. First, all organisms susceptible to polyenes contain sterols (e.g., yeasts, algae, protozoans, flatworms, and mammalian cells) and all resistant organisms do not [33,44,129]. Second, sterols added to the medium protect polyene-susceptible cells from these antibiotics [36,71]. In this case, reversal of polyene effect is secondary to a physiological interaction between the polyene antibiotics and the added sterols which prevents the antibiotic from interacting with the sterols of the organisms under study.

In addition to this indirect evidence for the interaction between sterols and polyene

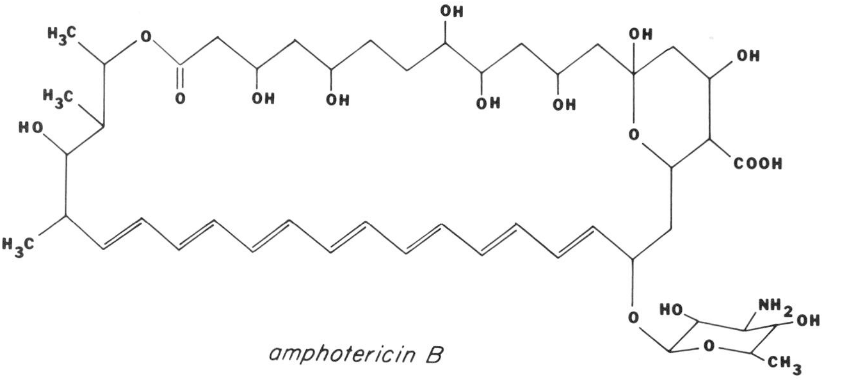

Figure 1 Chemical structures of amphotericin B, a heptaene; nystatin, a "degenerate heptaene"; and filipin, a pentaene.

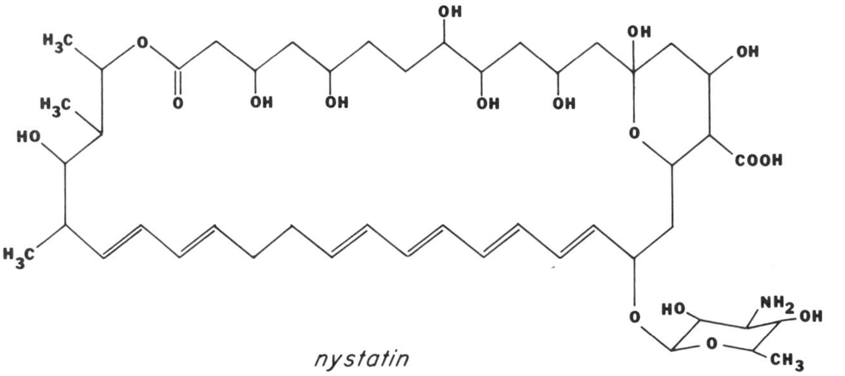

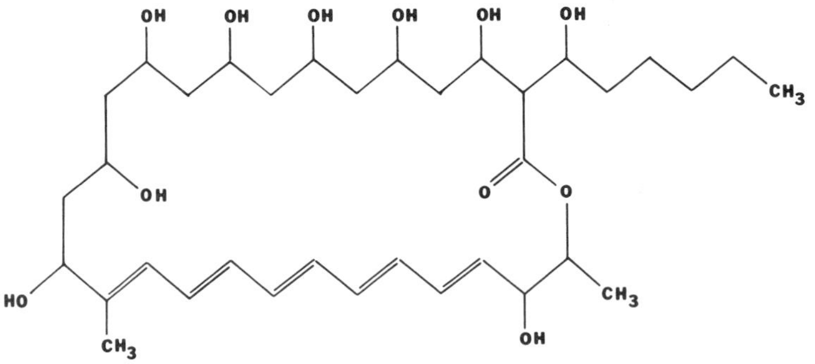

antibiotics, direct evidence is also available. We have already mentioned that polyene antibiotics have characteristic UV spectra. Lampen et al. [74] reported that when sterols are added to aqueous solutions of polyene antibiotics, UV absorbence decreases. This suggests that a direct interaction is occurring between the added sterol and the polyene. Several other groups have confirmed these findings [66,86,87,104] and more recently, Schroeder et al. [105] have presented additional evidence for binding between sterols and polyenes by use of a fluorometric technique.

The nature of the membrane lesion resulting from the polyene-sterol interaction has also been studied extensively using natural and artificial membranes. Freeze-etch electron microscopy of erythrocytes and *Acholeplasma laidlawii* treated with polyenes has shown structural alterations such as pits, doughnut-shaped craters, and protrusions in the membrane. It has been assumed, but not proven, that these structural changes are responsible for the enhanced permeability of the membrane. Studies have also been carried out using several different model membranes, including monolayer systems, and they have essentially confirmed the work in natural systems. Freeze-etch electron microscopy of *Epidermophyton floccosum* [88] has shown that amphotericin B induced in this fungus profound ultrastructural changes which consist of aggregation of membrane-associated particles, 85 Å in size, with the formation of depressions or craters on the inner face of the membranes. No pores or holes were observed, which is in agreement with the other studies using *A. laidlawii* and artificial membranes. The absence of pores or holes in the surface membranes after polyene treatment supports the notion that the interaction of the polyene antibiotics with the sterols in the membrane induces dramatic changes in the physical properties of the membranes.

C. In Vitro Effects of the Polyenes

The polyenes are inactive against bacteria. They exert various degrees of growth inhibition and killing against many species of fungi. Many polyenes alone are toxic to protozoans of medical importance, such as trichomonads, entamebae, naeglariae, and trypanosomes. Nystatin and amphotericin B are toxic to *Leishmania donovani.* This effect has been shown to be a result of the loss of intracellular constituents because of alteration of membrane permeability and eventual cell lysis. Nystatin is rapidly bound to these cells and is preferentially bound to those fractions containing the highest sterol content. As with fungi, the binding is inhibited by adding digitonin or cholesterol to the incubation medium. Seneca and Bergendahl [110] and Johnson et al. [58] have found that polyenes are toxic to snails and planaria.

As a general rule, the antifungal activity of the polyenes increases as the number of double bonds increases. Utahara et al. [121] were among the first to show that tetraenes and a pentaene (eurocidin) had about one-fourth of the activity, on a weight-for-weight basis, of the hexaene, mediocidin, and various heptaenes against *Candida albicans.* The heptaenes are more than one order of magnitude more active than pimaricin, a tetraene, and filipin, a pentaene, against *Candida,* whereas nystatin, a "degenerate" heptaene, is about twice as active as the latter two compounds.

The statement on the antifungal activities of the various polyene antibiotics is made despite the fact that accurate figures to support this widely held opinion are difficult to come by because experimentally determined values of minimum inhibitory concentration (MIC) vary so much among different laboratories. Among the factors that may affect the MIC are inoculum size, temperature and duration of incubation, and medium composition [65]. Therefore, susceptibility data from different centers can be compared only if strictly standardized conditions have been adhered to.

In the interpretation of susceptibility testing, the question arises whether an in vitro test is relevant to the in vivo effects of the drug. For example, are the concentrations of amphotericin B required for in vitro killing or growth inhibition the same as those needed for a clinical response, or are more subtle effects of the antibiotics combined with the host response adequate to manage the infection? The latter possibility is especially important for amphotericin B. Its significant toxicity makes the use of low levels attractive, if they are effective; and the correlation between serum or cerebrospinal levels and clinical response in patients is poor. Amphotericin B also has immunoadjuvant properties which may influence its in vivo effects and play a role in therapy (see Sec. II.D).

Several studies have tested which in vitro property of amphotericin B is most related to its clinical effect. Initial studies measured the binding of amphotericin B to cell membranes and the effects on membrane integrity. Kotler-Brajtburg et al. [69] have shown that there are two types of binding of amphotericin B to fungi. One is nonenergy dependent and reversible, and the other is irreversible and energy dependent. Only the latter irreversible binding is associated with lethality. The former type of binding is associated with low concentrations of amphotericin B and increases the permeability of eucaryotic cell membranes as measured by potassium leakage and/or sodium flux without killing the fungus [70]. It is possible to separate the permeabilizing and lethal effects of amphotericin B by several methods (J. Kotler-Brajtburg et al., unpublished report), and it is probable that the two actions of the drug result from independent mechanisms. HsuChen and Feingold [54] also inferred this by finding and studying fungal mutants that were resistant to the permeabilizing effects of amphotericin B but not to its lethal effects.

The permeabilizing, nonlethal effects of amphotericin B have been exploited to increase the therapeutic effectiveness of several second agents on fungi. This is discussed in more detail later in the chapter. An unsettled question is whether the permeabilizing effects of amphotericin B on fungi in vivo are sufficient to account for its therapeutic effectiveness. Based on reports which show that amphotericin B has immunoadjuvant properties (see Sec. II.D), studies have been initiated to look at the in vitro effects of amphotericin B on isolated cells of the immune system. Amphotericin B affects the phagocytic capacity of peritoneal macrophages and these effects are extremely dose dependent (J. Kotler-Brajtburg et al., unpublished report). At low concentrations, amphotericin B increases phagocytosis and higher concentrations result in a decreasing phagocytic capacity until a lethal concentration is reached. Therefore, low concentrations of amphotericin B "activate" or "prime" the cell membranes of macrophages and increase permeability to cations. These effects may complement the lethal effects of amphotericin B on certain subpopulations of lymphocytes (suppressor T-cells) and produce an immunoadjuvant effect in vivo [79].

Thus far Kotler-Brajtburg et al. [68] have studied 14 polyene antibiotics and six of their derivatives for permeabilizing, lethal, and immunoadjuvant effects. Although the antibiotics and derivatives vary greatly in potency, these researchers have categorized the polyene antibiotics into two groups according to the pattern of response of fungi to these agents (Fig. 2):

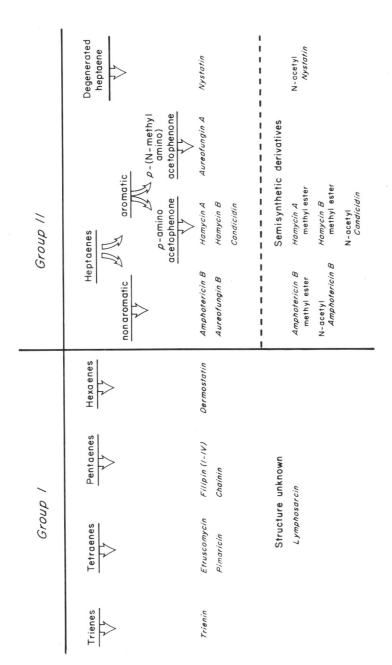

Figure 2 Classification of polyene macrolide antibiotics according to chemical and biological effects. (From Ref. 68.)

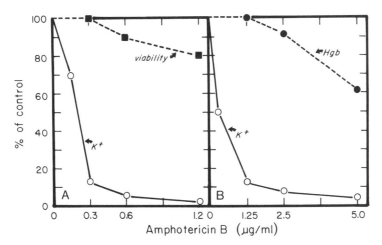

Figure 3 Effects of amphotericin B on (A) yeast and (B) red blood cells. (From Ref. 68.)

1. Those inducing considerable potassium leakage at lower concentrations and cell death at higher concentration (Fig. 3). These agents possess a fungistatic effect and a fungicidal effect, both of which are clearly separable. Depending on concentration, the former could occur in the absence of the latter. Included in this group are the heptaenes amphotericin B, candicidin, aureofungin A and B, hamycin A and B, nystatin (considered a "degenerate" heptaene), and six semisynthetic derivatives (amphotericin B methyl ester; N-acetyl amphotericin B; hamycin A and B methyl esters; N-acetyl candicidin; and N-acetyl nystatin).

2. Those causing little or no detectable potassium leakage prior to cell death or hemolysis (Fig. 4). There is no separation between fungicidal and fungistatic effects in this group. Included are a triene (trienin), tetraenes (pimaricin and etruscomycin), pentaenes (filipin and chainin), a hexaene (dermostatin), and one polyene with unknown chemical structure (lymphosarcin).

According to these measurements, there is a clear correlation between the type of biologic action and the size of the polyene macrolide ring. This is in accord with the notions of Lampen and Arnow [73]. These data would therefore predict that the search for polyenes with permeabilizing properties capable of acting synergistically with other drugs against fungi should be conducted among the amphotericin B-like group of drugs. In addition, Hammarström and Smith [47] have shown (and J. R. Little, in unpublished studies has confirmed) that the amphotericin B-like drugs are immunoadjuvants, but the filipin-like drugs are not. Therefore, adjuvanticity appears to be linked to molecular size and may also result from a preferential permeabilizing effect of the polyenes on certain cells of the immune system. The search for new immunoadjuvants should, therefore, also be conducted among heptaene polyene antibiotics. In contrast, when direct lethality to fungi is sought, antibiotics from the filipin class may be more effective.

D. In Vivo Effects of the Polyenes

At present, the only polyene antibiotic sufficiently nontoxic and tolerated parenterally in therapeutic doses in experimental animals is amphotericin B. In general, the in vivo ex-

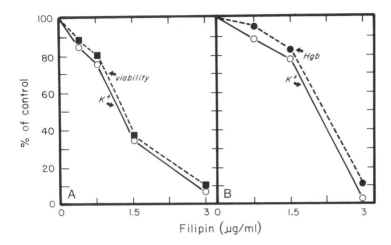

Figure 4   Effect of filipin on (A) yeast and (B) red blood cells. (From Ref. 68.)

periments in animals have confirmed the in vitro effectiveness of amphotericin B against a
variety of fungi at tolerable drug levels. In fact, inordinately low doses of amphotericin
B, resulting in nondetectable blood levels, have been effective therapeutically in several
animal models of infection. This observation has led to the hypothesis that amphotericin
B might have immunoadjuvant properties. This notion has been confirmed by Little et al.
[78] by finding that single intraperitoneal injections of amphotericin B significantly enhance the frequency of antibody-producing cells in the spleens and lymph nodes of many
mouse strains (Table 2). In general, the magnitude of immunopotentiation is much greater
for the secondary response than for the primary response, and amphotericin B showed a
potency similar to the adjuvants BCG, complete Freund's, or Pertussis-Maalox.

The effect of amphotericin B on the switch from IgM to IgG, its reduced effectiveness
in thymectomized mice, and the dramatic toxicity of amphotericin B in vitro for thymocytes and a subpopulation of spleen T-cells have led to the hypothesis that the immunoadjuvant effects of amphotericin B may require interaction with a particular T-cell population. The immunostimulant effects of amphotericin B methyl ester on murine responses
to three T-cell-independent antigens, TNP-*Brucella*, pneumococcal polysaccharide type
III, and the phosphocholine (PC) determinant group, have recently been studied by Little
et al. (unpublished results). Consistent with the hypothesis that amphotericin B methyl
ester causes immunostimulation via T-cells, no adjuvant effects were observed with any of
these T-cell-independent antigens.

In addition to its effect on humoral immunity, amphotericin B can augment delayed-
type hypersensitivity reactions and cell-mediated immunity. The earliest experiments
which suggested that amphotericin B could augment cell-mediated immune reactions
utilized the graft versus host reactions (GVHR). An extensive study has recently been
completed of the stimulation of contact sensitivity responses to 2,4-dinitrofluorobenzene
(DNFB) in AKR female mice by amphotericin B [115]. Amphotericin B was shown to
augment contact sensitivity responses to DNFB throughout the 3-week period following
skin sensitization. Since contact sensitivity in mice has been shown to be mediated by

Table 2 Effect of Amphotericin B on the Secondary Response in BALB/c Mice

Primary immunization	Secondary immunization	Days after primary immunization	Days after secondary immunization	PFC/1×10^6 spleen cells	
				IgM	IgG
AmB + TNP-HSA	TNP-HSA	20	7	25.1 ± 6.1	6.8 ± 12.0
TNP-HSA	TNP-HSA	20	7	18.0 ± 2.9	0
AmB	TNP-HSA	20	7	11.0 ± 4.3	0
AmB + TNP-HSA	TNP-HSA	27	7	30.2 ± 8.2	12.7 ± 8.8
TNP-HSA	TNP-HSA	27	7	21.6 ± 6.7	0
AmB	TNP-HSA	27	7	20.3 ± 6.0	0
AmB + TNP-HSA	TNP-HSA	34	7	37.4 ± 9.2	18.1 ± 11.6
TNP-HSA	TNP-HSA	34	7	20.6 ± 4.9	0
AmB	TNP-HSA	34	7	12.1 ± 4.3	0

Source: Ref. 8a.

T-cells, these data provide additional support for the argument that the adjuvant effects of amphotericin B are mediated via interaction with T-cells or certain T-cell subsets. Furthermore, it seems likely that suppressor T-cells or their precursors may be readily inhibited or intoxicated by amphotericin B. This conclusion is supported by further experiments with the contact sensitivity system showing that amphotericin B can largely or completely ablate the induction of immune tolerance produced in mice by intravenous administration of dinitrofluorobenzosulfonate [116]. This explanation for the mechanism of the adjuvanticity of amphotericin B is especially interesting because of the evidence that the anergy of disseminated fungal infection is due to activity of suppressor T-cells [118]. It may be that amphotericin B, the therapeutic agent for this kind of infection, has both a direct effect on the infecting organism and also an interaction with the suppressor T-cells generated by the infection.

E. Clinical Studies

Parenteral amphotericin B has been used to treat both systemic fungal infections and severe superficial mycoses. Successful results have been obtained in many fungal infections, including histoplasmosis, coccidioidomycosis, blastomycosis, cryptococcal meningitis, and disseminated *Candida* or *Aspergillus* infections.

The drug is marketed as a bile salt complex that forms a colloidal dispersion when hydrated. If the solution is too acid or contains sufficient electrolytes, the colloid aggregates and the solution becomes cloudy. Intravial dilution is accomplished with 10 ml of sterile water without preservatives. The final intravenous infusion mixture, obtained by adding the intravial mixture to 500-1000 ml of an intravenous solution of 5% dextrose in water, should be about 0.1 mg of amphotericin B per milliliter of 5% dextrose in water solution.

Most infections are successfully treated with amphotericin B provided that therapy is instituted soon after onset of the disease. A blood level of 1-2 μg/ml can be obtained on maximum dosages of the drug (0.5-1 mg/kg per day). On the other hand, some infections, such as disseminated aspergillosis and fungal endocarditis, may not respond to amphotericin B therapy, possibly because the causative organism is resistant as determined by in vitro susceptibility tests. Although fungal resistance to amphotericin B has been thought to be a rare phenomenon, a recent report indicates that it may be more frequent than realized [92].

Occasionally, there may be a lack of clinical response even when the organism is susceptible to the drug. It is also true that some infections respond to amphotericin B treatment even if susceptibility tests indicate that the MIC of amphotericin B is higher than what is thought to be a clinically achievable blood level of the drug. These poor correlations between the in vitro susceptibilities of the organisms to amphotericin B and the clinical response of the patients may be due to several factors:

1. The special nature of the disease within the host. For example, the large clots on prosthetic valves seen in *Candida* or *Aspergillus* endocarditis make the disease unresponsive to medical therapy [100], even though the organisms are susceptible in vitro.

2. The inability of amphotericin B to penetrate closed body cavities such as the central nervous system lowers the cure rate of cryptococcal meningitis to about 50%, even though *Cryptococcus neoformans* is almost always susceptible to amphotericin B [102].

3. The nature of the host contracting a systemic fungal infection. The patient with leukemia, when treated with drugs that suppress bone marrow and immune functions,

frequently cannot handle any kind of infection well, even though potent drugs are available for therapy.

4. As has already been noted, in vitro susceptibility tests do not measure the effect of amphotericin B on the host, and this may be an important determinant of some of the in vitro and in vivo discrepancies. Permeabilizing, nonlethal effects of amphotericin B, which occur at very low concentrations, as well as its adjuvant properties, may be important in the outcome of treatment of fungal infections.

5. Susceptibility tests for fungi are variable, largely because fungal infections are uncommon compared to bacterial infections, and the morphology, growth rate, optimal conditions of fungal growth are variable, and standardization is difficult. The poor solubility and stability of the most important antifungal agents have also complicated the assay procedures. For example, the levels of susceptibility of fungi to amphotericin B are dependent on the medium, inoculum size, temperature of incubation, and the length of incubation [64]. A low-dose inoculum, or an incubation temperature of 25-30°C, give a lower MIC than does a higher inoculum and an incubation temperature of 37°C. The half-life of amphotericin B activity in media is 24 h [19], so that a falsely high MIC is obtained with prolonged incubation. All these factors must be considered in choosing the susceptibility testing procedure most relevant to the clinical situation. Because of the variabilities in methodology, discrepancies will probably persist until a standard method of susceptibility testing is developed.

The dosage schedules for amphotericin B have been arrived at empirically, and are based on limits of toxicity, convenience, and what seems to have worked in practice. The critical factors in dosage are not well defined. There are also many disagreements regarding the details of administration of amphotericin B, and recommendations are often made purely on the basis of anecdotal clinical experience. The initial dose of amphotericin B should be 1-5 mg given over a period of 2-6 h, and on each subsequent day the dosage can be doubled until the desired maintenance dose is reached. It is seldom necessary to use more than 0.5 mg/kg per day.

Occasionally, patients will not even tolerate the 1-mg test dose, and chills, fever, and hypotension may occur. Initial doses may have to be lowered to 0.1 mg and then increased slowly as tolerance allows. These dosages may cause chills and fever and require the addition of 25-50 mg of hydrocortisone sodium succinate to the infusion bottle to control the reactions. It may also be necessary to premedicate the patient with aspirin, benadryl, and/or compazine to control nausea.

Recently a low-dose regimen has been recommended for certain types of infections caused by *Candida* [82]. According to this protocol, the patient receives a total dose of 100-200 mg of amphotericin B over a period of 7-10 days. Although experience with the regimen has been limited, it appears to be particularly successful in patients with mucocutaneous, esophageal, or urinary tract infections. Again, the basis for the effectiveness of this low-dose regimen may be a result of the nonlethal effects of amphotericin B on fungi and/or the immunoadjuvant properties of amphotericin B which act on the host.

The dosage of amphotericin B should not be reduced in patients with preexisting renal dysfunction, because comparatively little active amphotericin B is excreted in the urine and reduced renal function has no effect on the serum levels of the drug. On intravenous injection of the amphotericin B, roughly 10% of the bioactivity is retained in the plasma strongly bound to plasma proteins (probably lipoproteins). Serum levels obtained with therapeutic doses peak at about 0.5-2.0 μg/ml. The affinity with which amphotericin

B binds to cholesterol suggests that most of the administered dose is bound to sterol-containing membranes in different tissues. Details of the tissue distribution and excretion of the drug have not been clear, and the literature is not consistent on this point [56,80].

Intrathecal as well as intravenous amphotericin B is indicated in the treatment of coccidioidal meningitis and in some resistant cases of cryptococcal meningitis. Injection may be made into the lumbar theca, cisterna magna, or by use of a prosthesis into a lateral ventricle. Before initiating this mode of therapy, the physician should be thoroughly familiar with the technical details of administration and the formidable problems that may arise during this kind of therapy with amphotericin B [25]. Intra-articular injections of amphotericin B (5-15 mg) may be useful in the treatment of coccidioidal arthritis and in refractory cases of articular sporotrichosis [111]. Irritation commonly follows local injections of the drug, but systemic reactions are unusual.

Renal dysfunction is the most important toxic effect of amphotericin B, but it is usually reversible if it is recognized early and appropriate measures are taken [18]. Soon after the beginning of amphotericin B therapy, the glomerular filtration rate (GFR) falls to roughly 40% of normal in nearly every patient. After repeated doses, the GFR seems to stabilize at 20-60% of normal and remains at this level throughout the course of therapy. If renal function continues to deteriorate beyond this level, a safe and expeditious way to manage the nephrotoxicity of amphotericin B is to discontinue the drug for 3-5 days. Renal function almost always improves with the brief cessation of therapy. Then treatment with the previously administered dose can be reinstituted, and the renal function usually stabilizes. Despite the fact that nephrotoxicity has been reduced in laboratory animals by concommitant administration of intravenous mannitol or oral sodium bicarbonate, these have not worked in humans [37,50].

In most patients, creatinine and inulin clearances and the blood urea nitrogen (BUN) level return to nearly normal some months after cessation of amphotericin B therapy, but the degree of permanent GFR loss, if any, has not been firmly documented in a large series. A few patients, who cannot be singled out before therapy, sustain as much as a 50% loss of GFR for many months.

The monitoring of amphotericin B toxicity requires measurements of hematocrit, serum potassium, BUN, creatinine, and CO_2 values, and also a urinalysis. These tests should be obtained as a baseline before treatment, twice weekly for 4 weeks and then weekly until amphotericin B is discontinued.

Renal abnormalities other than azotemia are frequent but are not used as a guide to dosage. These include cylinduria, mild renal tubular acidosis, and other disturbances of renal tubular function. Hypokalemia occurs in about one-fourth of patients and may require potassium chloride supplementation. The hematocrit frequently falls to a stable value of 22-35% because of decreased erythrocyte production. Thrombocytopenia, hepatic dysfunction, and allergic reactions are rare toxic manifestations of amphotericin B therapy.

Recently, Schaffner and Mechlinski [103] and Lawrence and Hoeprich [75] have shown that esterification of the free carboxyl group of amphotericin B or its N-acetylated derivatives increased water solubility and caused only slight loss of biological activity. The methyl ester of amphotericin B appeared to be less toxic in animal models of infections and in a small number of patients. It is too early to know if these preparations will offer a therapeutic advantage, but the initial results in humans have raised some questions about central nervous system toxicity due to this preparation of amphotericin B [12].

Nystatin is the most frequently used topical or oral polyene antibiotic. Originally called fungicidin, this antibiotic is the product of a soil survey conducted by members of the staff of the New York State Department of Health, and its present name reflects the origin of its discovery. Nystatin is made by *Streptomyces noursei.* Its in vitro activity is highest against yeast-like fungi. Although it is stated to be too toxic parenterally, there are no data to support this contention. As with the other polyenes that have been characterized, there is little or no absorption when given orally. At the present time its most important use is to treat topical or superficial infections by direct application. The local action of nystatin and its use in controlling vaginal candidiasis or *Candida* infections involving any part of the alimentary tract from the mouth to the anus is now so generally familiar that it does not require further discussion.

Several other polyene antibiotics, notably pimaricin, hamycin, trichomycin, and candicidin, have been used to a limited extent, topically, orally, or by inhalation or injection for the treatment of localized and systemic mycotic infections. A colloidal preparation of hamycin when given orally [7] was found to be effective against some systemic fungal infections in animals and humans. However, the absorption of this polyene from the gastrointestinal tract has not been consistent enough to depend on it as a reliable systemic chemotherapeutic agent. Pimaricin, trichomycin, and candicidin have been used to treat local fungal infections of skin, eye, and vagina by topical application or injection into infected tissue. The fact that some of these agents combine activity against yeasts and trichomonas makes them particularly attractive for use in vaginitis, but it is difficult to judge if they offer any real advantages over amphotericin B or nystatin.

III. IMIDAZOLES

Imidazoles are broad-spectrum antimicrobial agents which act at the level of the cell membrane. With the exception of gram-negative bacteria, all microbes pathogenic to humans are covered by the antimicrobial spectrum of the imidazole derivatives [52]. However, their action against gram-positive bacteria is too limited to be of clinical relevance.

The structures of the three imidazoles that have been used clinically to treat fungal infections are shown in Fig. 5.

A. Clotrimazole

Clotrimazole has an extraordinarily wide range of activity against dermatophytes, yeasts, and filamentous and dimorphic fungi [95,122]. The drug has been shown to be fungistatic at 10 μg/ml and fungicidal at higher concentrations [84]. As with many other antibiotics, the MIC of clotrimazole depends on the organism being tested, size of inoculum, length of incubation, temperature, and type of medium employed in the assay procedure [53]. The drug is rapidly eliminated in the feces and urine of rats after oral or intravenous administration [119], and there is minimal absorption of the drug or its metabolites through the skin after topical application. Prolonged oral administration produces hepatic and adrenal changes, resulting in increased liver weight and cellular hyperplasia. In addition, limited clinical studies have revealed little efficacy and much toxicity [77]. The former appears to be due to an induction of enzymes in the host liver which degrade clotrimazole. The latter is manifested by gastrointestinal symptoms, especially nausea, vomiting, and diarrhea. Hallucinations and disorientation have also been reported with the use of this agent. Because of these negative results, the drug is used almost exclusively in the

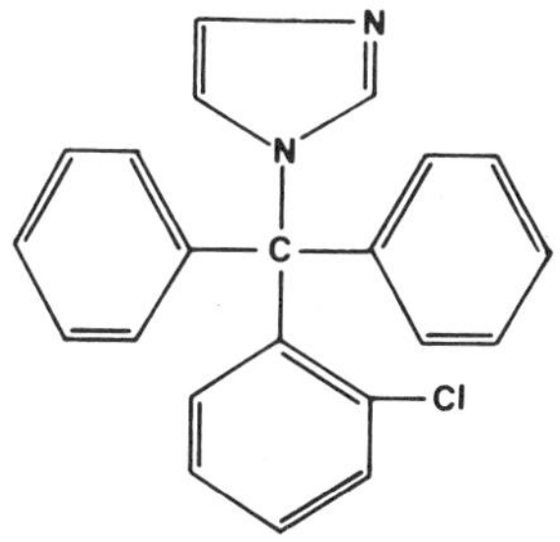

clotrimazole

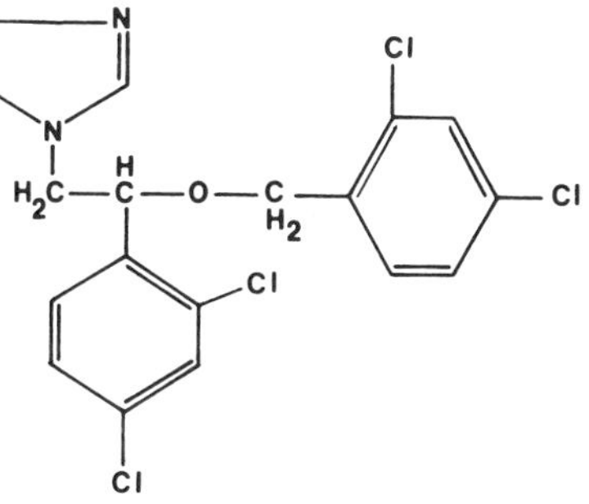

miconazole

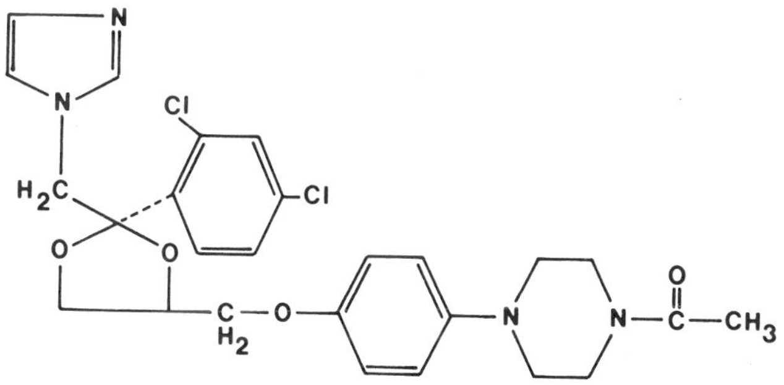

ketoconazole

Figure 5   Structures of imidazoles.

form of a 1% solution or cream as a topical antifungal agent in dermatophyte infections and at present there are no clinical indications for its use to treat systemic fungal infections.

B. Miconazole

Miconazole is a β-substituted 1-phenethylimidazole derivative with a broad spectrum of antifungal and antibacterial activity [124]. It is thought that miconazole works by interacting with the cell membranes of fungi and causing leakage of cytoplasmic contents [125].

Most of the experimental work with miconazole in animal models of infection have been done with murine models of coccidioidomycosis. It has been shown to be effective against this infection and against *Cryptococcus neoformans* infection in mice [76].

Clinical experience with miconazole has shown that it is effective in the treatment of coccidioidomycosis, candidiasis, paracoccidioidomycosis, and cryptococcal meningitis [24]. However, its short half-life requires frequent administration and its poor penetration in the CSF has necessitated intrathecal therapy for fungal meningitis.

Miconazole is metabolized in the liver to an inactive metabolite. Therefore, its dosage should not be altered in patients with compromised renal function. Its suggested dosage ranges from 400 to 1200 mg given every 8 h. The duration of therapy has not been established, but 3-12 weeks or longer have been suggested, depending on the type of infection being treated and the response of the patient. In meningitis, 20 mg is given every 1-3 days intraventricularly, intracisternally, or at lumbar injection sites.

The reported toxicity of miconazole has involved immediate problems during infusion such as anaphylaxis, tachycardia, arrhythmias, and fever and chills, nausea, and phlebitis. Pruritis, anemia, and hyponatremia are relatively common, and leukopenia, thrombocytopenia, and elevated liver enzymes have also been reported with its use. Increased serum cholesterol and triglyceride levels secondary to the Cremophor carrier have also been reported.

Studies with coccidioidomycosis and paracoccodioidomycosis suggest that it may be efficacious in these infections, but it requires long periods of treatment and intrathecal administration in patients with meningitis. The role of miconazole in the therapy of blastomycosis, histoplasmosis, sporotrichosis, and cryptococcal meningitis remains to be defined. For these reasons, at the present time it should not be considered as a first-line antifungal agent and should be reserved for patients who cannot tolerate amphotericin B.

C. Ketoconazole

Ketoconazole (R41,400) is the most recently developed imidazole derivative and is presently undergoing clinical testing. It is water soluble and well absorbed from the gastrointestinal tract. Its spectrum is similar to miconazole but appears to be about five times more active in vitro against *Coccidioides immitis* [28]. It is a potent inhibitor of ergosterol biosynthesis in *Candida albicans* and interferes with removal of the 14-α-methyl group from ergosterol [126]. This has made it an attractive agent for use in humans since cholesterol biosynthesis is unaffected by the mode of action of this imidazole.

Preliminary results in human infections have shown good results in the treatment of paracoccidioidomycosis, coccidioidomycosis, and histoplasmosis. The drug is supplied as 200-mg tablets and the suggested dose is 200-400 mg p.o. daily. Duration of therapy required for cure of the systemic fungal infections is unknown.

Ketoconazole is metabolized by the liver and inactive metabolites appear in bile, feces, and urine. The dosage should not be adjusted in patients with compromised renal function. Toxicity studies are incomplete but include pruritis and elevated liver function tests.

The added attractiveness of ketoconazole is that it may allow oral therapy for systemic fungal infection. However, additional studies are needed for its role to be defined. The lack of penetration into the CSF may present the same difficulties with the therapy of fungal meningitis as with miconazole.

IV. 5-FLUOROCYTOSINE

5-Fluorocytosine (5FC) is a synthetic oral antimycotic agent that is effective in the treatment of some systemic fungal infections, in particular those due to the yeasts.

5FC as an antimetabolite (Fig. 6) is probably of less importance, in regard to its antifungal properties, than the effect resulting from conversion to the antimetabolite 5-fluorouracil (5FU) by organisms possessing the necessary deaminases. Absence or low activity of cytosine deaminase activity in animal cells has been considered the key to the low toxicity of 5FC to the mammalian host. The evidence for this is that the antifungal effects of 5FC correlate directly with the amount of replacement of uracil by 5FU in RNA of the sensitive organisms. It is assumed that RNA heavily substituted with 5FU functions poorly and abnormal proteins are synthesized, which leads to cell death [96].

More recently, evidence suggests that the major site of action of 5FC is interruption of DNA synthesis [26]. This pathway would also depend on the deamination of 5FC to 5FU, which would then be covered to 5-fluoro-2'-deoxyuridine monophosphate (FdUMP). FdUMP would then inhibit thymidylate synthetase, which is the enzyme utilized to form thymidine needed for DNA synthesis (Fig. 7).

Either of the mechanisms described above would explain the observed antimicrobial spectrum of 5FC. 5FC resistance may be produced by deficiency of an enzyme at any step of the pathway or by a surplus of normal compounds present in the environment or synthesized de novo, which compete with the fluorinated antimetabolite. Several types of resistant organisms have been described which are deficient in permease, cytosine deaminase, or in uridine monophosphate (UMP) pyrophosphorylase [59]. According to the mechanism involved, the degree of fungal resistance to 5FC can differ considerably and vary from levels of partial resistance to complete resistance, the latter not affected by the highest concentration of 5FC. 5FC-resistant mutants are produced spontaneously and vary in frequency in different fungi. Selection of these mutants by 5FC during therapy has limited its efficacy as a primary antifungal agent.

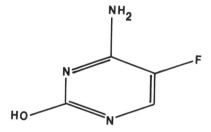

Figure 6 Structure of 5-fluorocytosine.

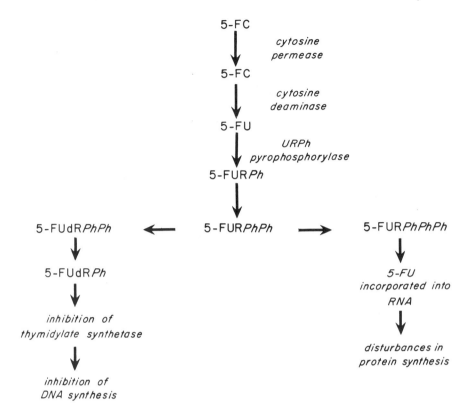

Figure 7 Pathway of metabolism of 5-fluorocytosine by fungi.

A. In Vitro Studies

Antifungal activity of 5FC is best against yeasts, including *Cryptococcus neoformans* and species of *Candida* and *Torulopsis;* somewhat less but still considerable activity is present against the dematiaceous fungi causing chromomycosis, and species of *Aspergillus,* when the susceptibility test is performed on appropriate culture media [5]. The MIC values vary from 0.1 to 25 µg/ml, the latter being the usually accepted cutoff between susceptible and resistant fungi. It is crucial in in vitro susceptibility tests that the media be free of competitive pyrimidines which the fungi can use selectively in preference to 5FC. 5FC is both fungistatic and fungicidal, with the effect dependent on the concentration of drug and time of incubation [94].

   Within 5FC-sensitive species, resistant strains occur which have had no previous contact with the drug. Such instances of primary resistance have occurred in 7-8% of pre-treatment isolates of *C. albicans,* unspeciated *Candida,* and *Torulopsis glabrata* [29]. In *Cryptococcus neoformans* the resistance is lower (1-2%) [107], but it is higher in *Candida* species other than *C. albicans* (22%) due to the prevalence of generally less sensitive species, such as *C. tropicalis* and *C. krusei.* Primary resistance is also observed with *Aspergillus* and dematiaceous fungi, but published information on this is still sparse. In *C. albicans* only about one-third of the isolates recorded as resistant were completely resistant [97].

This is an important consideration in the use of combination therapy with amphotericin B.

Development of resistant organisms during therapy is also a notable feature clinically. Almost two-thirds of isolates cultivated during or after therapy with 5FC from patients receiving the drug have been resistant [11].

B. In Vivo Studies

Chemotherapeutic effects of 5FC in experimental mouse infections have been demonstrated for *Candida albicans* [42,93] and *Cryptococcus neoformans* [9,46,112]. Murine models have shown only a marginal beneficial effect with *Sporothrix schenckii* [10] and *Aspergillus* infections [106]. No efficacy at all was found with mouse infections due to *Histoplasma capsulatum, Blastomyces dermatitidis,* and two agents of mucormycosis, *Rhizopus oryzae* and *Absidia corymbifera.*

C. Clinical Studies

The currently recommended dose of 5FC is 150 mg/kg per day divided into doses at intervals of 6 h. Absorption from the gastrointestinal tract is rapid and virtually complete. With normal kidney function, the half-life in serum is 3-5 h. Penetration into CSF and other body fluids and tissue is excellent. The exact distribution of 5FC in body fluids and organs is not known, but approximately 80-90% of the oral dose is excreted unchanged in the urine. Therefore, reduced doses have to be employed when renal function is impaired. One recommendation [23] is to administer the drug at intervals determined by the creatinine clearance: one dose every 6 h for clearance of at least 40 ml/min, every 12 h for clearance of at least 20-40 ml/min, and every 24 h for clearance of 10-20 ml/min. The drug is removed by dialysis, and in patients with end-stage renal disease a 20-mg/kg dose should be given after each dialysis [20]. Because levels can vary in these patients, it is always wise to monitor plasma concentrations by one of several available methods [60, 61].

5FC has been well tolerated by patients, even when given in large doses over extremely long periods of time. Gastrointestinal toxicity has been nausea, vomiting, diarrhea, and rarely bowel perforation. Disturbed hepatic function has been indicated primarily by elevated titers of transaminases and alkaline phosphatase levels, and bone marrow depression by neutropenia and thrombocytopenia. Usually, these toxic manifestations are mild and reversible when the dosage is reduced or the administration of the drug is stopped.

Many of the toxic manifestations, particularly the hematologic, appear to be related to serum concentration of 5FC in excess of 100 μg/ml [27]. This has been particularly true when 5FC has been given in combination with amphotericin B and may be related to a potentiation of toxicity by amphotericin B or to elevated blood levels of 5FC caused by the amphotericin B-induced renal impairment. In any case, it is useful to monitor 5FC blood levels, particularly in patients with impaired renal function and to keep peak blood levels between 50 and 75 μg/ml and certainly below 100 μg/ml.

There are few indications for therapy with 5FC alone. It can be used for *Candida* infections of the urinary, respiratory, and gastrointestinal tract. It can also be used for chromomycosis. In most instances, however, 5FC should be used in combination with amphotericin B to treat *Candida, Cryptococcus neoformans,* and perhaps *Aspergillus* infections.

V. IODIDE

Iodides are the drug of choice for lymphoreticular sporotrichosis although their mechanism of action is unknown [101]. A saturated solution of potassium iodide (1 g/ml) is taken orally three times a day. Adults begin with 3 ml/day and increase the amount slowly to 9 to 12 ml/day, as tolerance permits. Lower doses are recommended in children. Pregnant women should not be treated with iodide. Therapy is continued for about 6 weeks after healing. Several months of treatment are usually required. Most commonly encountered side effects include bitter taste, gastrointestinal discomfort, excessive lacrimation, swollen salivary glands, and rash.

VI. COMBINATION THERAPY

Until recently it has been customary in the treatment of fungal infections to use a dosage schedule for amphotericin B that results in blood levels maximally lethal to fungi. Unfortunately, at these blood levels, toxicity to the host almost always results because the specificity of amphotericin B for fungi compared to animal cells is not absolute.

The two types of binding of amphotericin B to fungal cell membranes and the permeabilizing and lethal effects on the cells have already been described [68,71]. The therapeutic implication of the nonlethal permeabilizing effects of amphotericin B alone is not known, but the success of the low-dose amphotericin B regimen in the treatment of *Candida* infections indicates that it may be important in some instances [82].

A. In Vitro Studies

One result from several in vitro studies is the concept that low, nonlethal, nontoxic permeabilizing concentrations of amphotericin B could be exploited to introduce other drugs or macromolecules into fungi and other eucaryotic cells. There are several antibiotics that can interfere with RNA or protein synthesis in fungal cell extracts but do not affect fungi, probably because the intact organism does not take them up into the cell. Nonlethal low-dose amphotericin B could be used to facilitate the entry of some of these agents into fungi, thus potentiating their effect. Thus far, it has been shown that in combination with concentrations of amphotericin B well below its MIC for each test organism, the antifungal effects of 5FC can be increased against *Cryptococcus neoformans,* various species of *Candida,* and *Aspergillus fumigatus* [72,81]. The same concentrations of amphotericin B could also enhance the effects of rifampin against *C. albicans, H. capsulatum,* and *B. dermatitidis* [4,67,83].

Similar in vitro antifungal effects of polymyxin B and synergism of polymyxin B in combination with 5FC or tetracycline against *Candida* and *Cryptococcus neoformans* have also been demonstrated [108].

It is important to note that not all combinations work against all fungi. For example, whereas amphotericin B and 5FC are synergistic against *C. neoformans,* they are not synergistic against *H. capsulatum.* In addition, some combinations may be antagonistic. Miconazole appears to decrease the effectiveness of amphotericin B because of its inhibitory effect on ergosterol synthesis, thereby decreasing binding of amphotericin B to cell membranes [21].

B. In Vivo Studies

Although it has been difficult to prove synergism in vivo, some investigators have shown that amphotericin B in combination with 5FC has at least an additive, and perhaps a synergistic, effect against *C. neoformans, Candida albicans,* and *A. fumigatus* infections in mice [9,120]. Others have reported that the combination of amphotericin B and tetracycline is synergistic against *Coccidioides immitis* infection in mice [55], and that the combination of amphotericin B and rifampin is synergistic against *H. capsulatum* and *B. dermatitidis* infections in mice [63].

The most important test of any new therapy is its effect on human infection. There have been several anecdotal reports of the efficacy of the combination of amphotericin B and 5FC against several different fungal infections in humans [123], but such uncontrolled studies are not adequate. Recently, a randomized controlled study has been published comparing amphotericin B alone and combined with 5FC in the treatment of cryptococcal meningitis [6]. Even though the combination regimen was given for only 6 weeks and amphotericin B for 10 weeks, the combination cured or improved more patients and produced fewer failures or relapses, more rapid sterilization of the CSF, and less nephrotoxicity than the prolonged course of amphotericin B alone. The participants in the study concluded that combined therapy was the regimen of choice for cryptococcal meningitis.

VII. GRISEOFULVIN

Griseofulvin, 7-chloro-2',4,6-trimethoxy-6'-methylspiro[benzofuran-2(3H),1'-[2]cyclohexene]-3,4'-dione, is a metabolic product of several species of *Penicillium* [13,16,57] and has been used for the past 20 years as the systemic drug of choice in the treatment of chronic dermatophytoses (tinea, ringworm). It was discovered and its chemistry was partially characterized in 1939 [90], and in 1947 the chemical structure, $C_{17}H_{17}ClO_6$ (Fig. 8), was completely elucidated. The compound has UV absorption maxima at 324, 291, 252, and 236 nm [39-41], and it is possible to assay this compound spectrophotometrically in biological specimens [3,32,131].

The biological properties of griseofulvin were first described by Brian and his colleagues [13-15]. It was termed "curling factor" because of its ability to induce stunted and aberrant growth of mycelium of *Botrytis alli.* Gentles [35] demonstrated in 1958 that experimental ringworm infection in the guinea pig could be effectively treated by oral administration of this drug. Therapeutic trials in humans followed, with dramatic responses reported [8,132].

The exact mechanism by which this antibiotic produces its effect against dermatophytes is unknown, but addition of griseofulvin to actively growing cultures of *Microsporum gypseum* or *Trichophyton mentagrophytes* temporarily halted protein and nucleic acid synthesis [31]. El-Nakeeb and Lampen [30] demonstrated that the uptake of griseofulvin by the dermatophyte *M. gypseum* involved a two-step process similar to the process

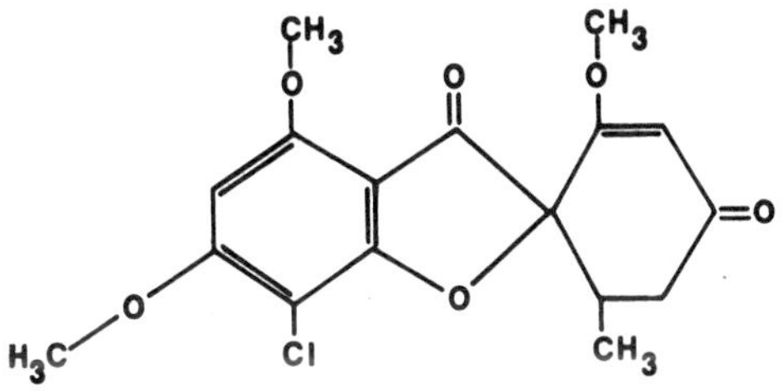

Figure 8   Structure of griseofulvin.

similar to the process seen in plants [22]. The first is an immediate uptake of small amounts of the antibiotic from the medium and does not require energy. This probably represents a simple absorption of the antibiotic to the lipids of the fungus. The second phase of uptake is prolonged and requires an energy source.

Although little is known about the antifungal mechanisms of griseofulvin on dermatophytes, there is a growing fund of information on the effects of griseofulvin on microtubules of animal cells. In 1958, Paget and Walpole [91] described the striking arrest of mitosis of bone marrow cells and intestinal epithelial cells in metaphase caused by intravenous injection of griseofulvin into rats. They also reported that griseofulvin caused spindle disorientation and chromosome scattering and that the drug inhibited chromosome movement at anaphase. Although these antimitotic properties are similar to antitubulins, such as colchicine, podophilotoxin, isopropyl-N-phenyl carbamate, and vinblastine sulfate [133], the molecular action of griseofulvin is thought to be different from the other compounds that bind to the receptors on tubulin and inactivate the free subunits [38]. The evidence for this is that griseofulvin does not interfere with tubulin polymerization in vitro and that HeLa cells arrested in metaphase by griseofulvin show morphologically normal microtubules [38]. It was proposed that the drug altered the function of intact microtubules in cells rather than causing them to depolymerize [134]. There are data that show, however, that griseofulvin interferes with tubulin polymerization in vitro and that the observed effects of griseofulvin on microtubular assembly are dependent on concentration of the antibiotic [130]. At 10^{-5} M concentration of griseofulvin, 3T3 cells are arrested in metaphase and exhibit normal microtubules, but when the concentration of drug is increased to 5×10^{-5} M, the cytoplasmic microtubules are destroyed.

In addition to its antimitotic properties and in vitro action on tubulin, griseofulvin shares other properties with those of colchicine and the vinca alkaloids [17]. It has been used in the treatment of gout [117,127] and has some anti-inflammatory activity [70]. In in vitro studies with Boyden chambers [1,2], clinically achievable concentrations of griseofulvin (0.1-1.0 μg/ml) inhibited chemotaxis of human polymorphonuclear leukocytes.

A. In Vitro Studies

Although much information has been gained about the effects of griseofulvin on microtubular structure and function in human cells, the mechanism of its action on dermatophytes remains to be elucidated. The recent isolation of tubulin from yeast [128] and filamentous fungi [114] may provide new insights into the understanding of the action of griseofulvin on fungi. Added impetus should be given to such studies in view of reports that griseofulvin specifically inhibits the mitotic spindle in various species of fungus [43-45].

The MIC of griseofulvin against various dermatophytes ranges between 0.14 and 0.6 μg/ml, but the antibiotic has little or no effect against yeasts and bacteria [99]. Microscopic studies on susceptible fungi have shown that, at concentrations of 0.1-0.2 μg/ml, grossly distorted growth of hyphae results [13].

B. In Vivo Studies

In recent years, a great deal of work has been done on the absorption, distribution, metabolism, and elimination of griseofulvin in humans and animals. Griseofulvin appears in the stratum corneum within 4-8 h after oral administration [113]. The degree of absorption from the gastrointestinal tract varies and depends on the particle size of the drug, fat and lipid intake, dissolution rate in intestinal fluids, and dosage regimen. In humans, the major metabolites of griseofulvin are 6-desmethylgriseofulvin and 6-desmethylgriseofulvin glucuronide. The absorption and metabolism of griseofulvin have been extensively reviewed by Lin and Symchowicz [77].

C. Clinical Studies

Griseofulvin has been effectively used in the treatment of dermatophyte infections of the scalp and hair, nails, glabrous skin, groin, and interdigital areas of the feet caused by *E. floccosum,* and various species in the genera *Microsporum* and *Trichophyton.* Hildick-Smith et al. [51] have extensively described its pharmacological, chemical, and antimicrobial properties. It is thermostable and unaltered by autoclaving at 15 lb/in.² for 15 min unless it is in solution, when it may lose potency on heating at autoclave temperatures. The usual oral dose in adults is 500 mg/day, but this will vary according to the severity of the infection and may require starting doses as high as 1.0 g/day. In children weighing 30-50 lb, a dosage of 10 mg/kg is usually adequate. Medication must be continued until the fungus is completely eradicated, as indicated by appropriate mycologic culture of specimens taken from the affected area. The duration of therapy generally depends on the time required for normal replacement of the infected tissues and varies from 3 weeks of therapy in uncomplicated tinea infections to 12 months in the case of toenail infections.

VIII. TOPICAL AGENTS

Several topical antifungal agents are currently available for the therapeutic management of superficial mycoses (i.e., pityriasis versicolor) and the dermatophytoses (i.e., ringworm). In general, there is a paucity of information available concerning the mechanism of action of these drugs.

A. Tolnaftate

Tolnaftate, m,N-dimethylthiocarbanilic acid O-2 naphthyl ester, synthesized and developed by Noguchi et al. [85], is a colorless, odorless compound with the chemical formula $C_{19}H_{17}NOS$ (Fig. 9). The antibiotic is applied topically twice a day, as a 1% solution or cream, to the affected area of skin. Tolnaftate is not very effective against dermatophyte infections of hair and nails. It is soluble in polyethylene glycol and chloroform, slightly soluble in ether and alcohol, and insoluble in water. Tolnaftate is effective against *M. gypseum, M. canis, M. audouinii, T. mentagrophytes, T. rubrum, E. floccosum,* and *P. orbiculare* but not against *C. albicans, A. fumigatus,* or *P. notatum* [98]. Although the drug has been used successfully in the treatment of cutaneous fungal infections, little has been published on the mechanism of its activity.

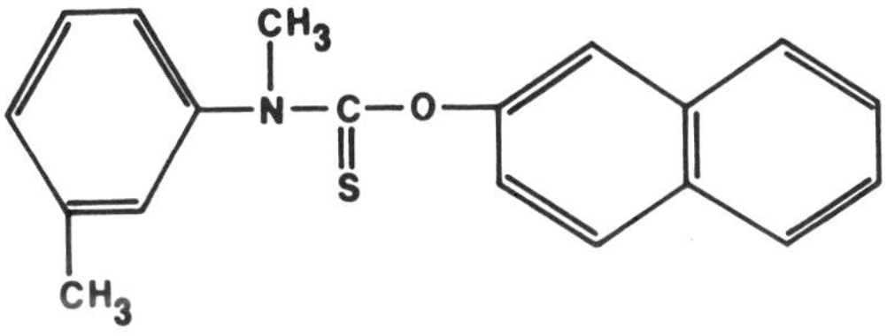

Figure 9   Structure of tolnaftate.

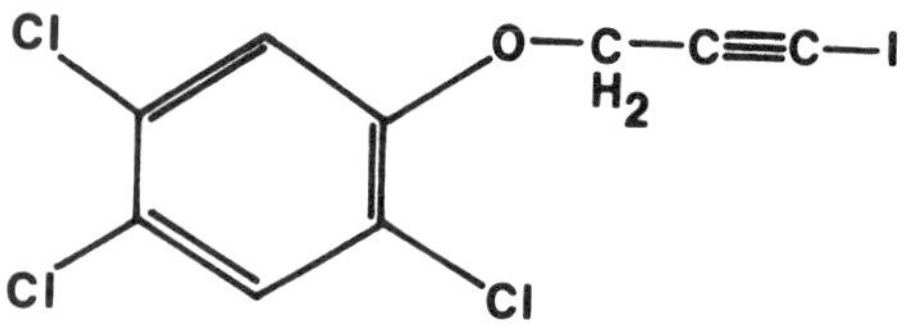

Figure 10 Structure of haloprigin.

B. Haloprigin

Another topically active antifungal agent is haloprigin, 3-iodo-2-propynl-2,4,5-trichloro-phenyl ether, which was synthesized by Seki et al. [109]. It is used as a 1% solution or cream in the treatment of dermatophyte infections of the skin. It is a γ-iodopropargyl aryl ether with the formula $C_9H_4Cl_3IO$ (Fig. 10). This synthetic compound is as active against dermatophytes as tolnaftate. Haloprigin, however, has a wider spectrum of action than tolnaftate because it is effective against species of *Allescheria, Alternaria, Aspergillus, Monosporium, Nigrospora,* and *Penicillium* [49]. Little is known about the mechanism of action of haloprigin, but yeast cells of *C. albicans* treated with it showed an inhibition of respiration and disruption of cell membrane integrity [48].

IX. CONCLUDING REMARKS

It is apparent that only a relatively small number of antibiotics are currently available for the treatment of fungus infections. Only amphotericin B is sufficiently nontoxic so that it can be given parenterally. It is presently most useful antifungal antibiotic for systemic fungal infections. 5FC, given orally, is most effective when given with amphotericin B, but its use is limited to selected fungal infections. Griseofulvin, the other oral drug, and several different topical agents are limited to the treatment of cutaneous fungus infections. The discovery of the antifungal activity of various imidazole derivatives has provided additional encouragement for the development of newer agents.

With the exception of the polyenes and 5FC, there is little known about the mechanism of action of the antifungal agents. The increasing occurrence of systemic fungal infections, particularly in the compromised host, should add impetus to the search for less toxic and more effective agents and perhaps to a greater understanding of the mechanisms of action of the ones presently available. This should result in more effective use of these agents and improvement of therapy.

ACKNOWLEDGMENTS

Portions of the work described in this chapter were supported by NIH Grants AI-10622, AI-00459, AI-107015, and CA-15665 and grants from the John A. Hartford Foundation, Inc. and the Brown Hazen Fund of the Research Corporation of America.

REFERENCES

1. Bandmann, U., B. Norberg, and L. Rydgren. 1974. Polymorphonuclear leucocyte chemotaxis in Boyden chambers. Scand. J. Haematol. *13*: 305-312.
2. Bandmann, U., B. Norberg, and G. Simmingskold. 1975. Griseofulvin inhibition of polymorphonuclear leucocyte chemotaxis in Boyden chambers. Scand. J. Haematol. *15*: 81-87.
3. Bedford, C., K. J. Chile, and E. G. Tomich. 1959. Spectrophotofluorometric assay of griseofulvin. Nature (Lond.) *184* (Suppl. 6): 364-365.
4. Beggs, W. H., G. A. Sarosi, and F. A. Andrews. 1974. Synergistic action of amphotericin B and rifampin on *Candida albicans*. Am. Rev. Respir. Dis. *110*: 671-673.
5. Bennett, J. E. 1977. Flucytosine. Ann. Intern. Med. *86*: 319-322.
6. Bennett, J. E., W. E. Dismukes, R. J. Duma, G. Medoff, M. A. Sande, H. Gallis, J. Leonard, B. T. Fields, M. Bradshaw, H. Haywood, Z. A. McGee, T. R. Cate, C. G. Cobbs, J. F. Warner, and D. W. Alling. 1979. A comparison of amphotericin B alone and combined with flucytosine in the treatment of cryptococcal meningitis. N. Engl. J. Med. *301*: 126-131.
7. Bennett, J. E., T. W. Williams, W. Piggot, and C. W. Emmons. 1964. The in vivo activity of colloidal hamycin, an antifungal antibiotic. Proc. Soc. Exp. Biol. Med. *117*: 166-170.
8. Blank, H., and F. J. Roth, Jr. 1959. The treatment of dermatomycoses with orally administered griseofulvin. Arch. Dermatol. *79*: 259-266.
8a. Blanke, T. J., J. R. Little, S. F. Shirley, and R. G. Lynch. 1977. Augmentation of murine immune responses by amphotericin B. Cell. Immunol. *33*: 180-190.
9. Block, E. R., and J. E. Bennett. 1973. The combined effect of 5-fluorocytosine and amphotericin B in therapy of murine cryptococcosis. Proc. Soc. Exp. Biol. Med. *142*: 476-480.
10. Block, E. R., A. E. Jennings, and J. E. Bennett. 1973. Experimental therapy of cladosporiosis and sporotrichosis with 5-fluorocytosine. Antimicrob. Agents Chemother. *3*: 95-98.
11. Block, E. R., A. E. Jennings, and J. E. Bennett. 1973. 5-fluorocytosine resistance in *Cryptococcus neoformans*. Antimicrob. Agents Chemother. *3*: 649-656.
12. Botter, A. A. 1971. Topical treatment of nail and skin infections with miconazole, a new broad-spectrum antimycotic. Mykosen *14*: 187-191.
13. Brian, P. W. 1949. Studies on the biological activity of griseofulvin. Ann. Bot. (Lond.) *133*: 59-78.
14. Brian, P. W. 1949. Curling-factor: a substance causing abnormal development of fungal hyphae, produced by *Penicillium janczewskii* Zal. In *Proceedings of the Fourth International Congress for Microbiology*, Rosenkilde and Bagger, Copenhagen, pp. 153-154.
15. Brian, P. W., P. J. Curtis, and H. G. Hemming. 1946. A substance causing abnormal development of fungal hyphae produced by *Penicillium janczewskii* Zal. I. Biological assay, production, and isolation of "curling factor." Trans. Br. Mycol. Soc. *29*: 173-187,
16. Brugmans, J. P., J. M. van Cutsem, and D. C. Thienpont. 1970. Treatment of long-term tinea pedis with miconazole. Double-blind clinical evaluation. Arch. Dermatol. *102*: 428-432.
17. Burnside, B. 1975. The form and arrangement of microtubules: an historical, primarily morphological, review. Ann. N.Y. Acad. Sci. *253*: 14-26.

18. Butler, W. T., J. E. Bennett, D. W. Alling, P. T. Wertlake, J. P. Utz, and G. J. Hill, II. 1964. Nephrotoxicity of amphotericin B; early and late effects in 81 patients. Ann. Intern. Med. *61*: 175-187.

19. Cheung, S. C., G. Medoff, D. Schlessinger, and G. S. Kobayashi. 1975. Stability of amphotericin B in fungal culture media. Antimicrob. Agents Chemother. *8*: 426-428.

20. Christopher, T. G., A. D. Blair, A. W. Forrey, and R. E. Cutler. 1976. Hemodialyzer clearances of gentamicin, kanamycin, tobramycin, amikacin, ethambutol, procainamide, and flucytosine, with a technique for planning theory. J. Pharmecokinet. Biopharm. *4*: 427-441.

21. Cosgrove, R. F., A. E. Beezer, and R. J. Miles. 1978. In vitro studies of amphotericin B in combination with the imidazole antifungal compounds clotrimazole and miconazole. J. Infect. Dis. *138*: 681-685.

22. Crowdy, S. H., J. F. Grove, H. G. Hemming, and K. C. Robinson. 1956. The translocation of antibiotics in higher plants. II. The movement of griseofulvin in broad bean and tomato. J. Exp. Bot. *7*: 42-64.

23. Cutler, R. E., A. D. Blair, and M. R. Kelly. 1978. Flucytosine kinetics in subjects with normal and impaired renal function. Clin. Pharmacol. Ther. *24*: 333-342.

24. Deresinski, S. C., R. B. Lilly, H. B. Levine, J. N. Galgiani, and D. A. Stevens. 1977. Treatment of fungal meningitis with miconazole. Arch. Intern. Med. *137*: 1180-1185.

25. Diamond, R. D., and J. E. Bennett. 1973. A subcutaneous reservoir for intrathecal therapy of fungal meningitis. N. Engl. J. Med. *288*: 186-188.

26. Diasio, R. B., J. E. Bennett, and C. E. Meyers. 1978. Mode of action of 5-fluorocytosine. Biochem. Pharmacol. *27*: 703-707.

27. Diasio, R. B., D. E. Lakings, and J. E. Bennett. 1978. Evidence for conversion of 5-fluorocytosine to 5-fluorouracil in humans: possible factor in 5-fluorocytosine clinical toxicity. Antimicrob. Agents Chemother. *14*: 903-908.

28. Dixon, D., S. Shadomy, H. J. Shadomy, A. Espinel-Ingroff, and T. M. Kerkering. 1978. Comparison of the in vitro antifungal activities of miconazole and a new midazole R41,400. J. Infect. Dis. *138*: 245-248.

29. Drouhet, E., L. Mercier-Soucy, and S. Montplaisir. 1975. Sensibilité et résistance des levures pathogènes aux 5-fluoropyrimidines I. Relation entre les phénotypes de résistance à la 5-fluorocytosine, le sérotype de *Candida albicans* et l'écologie de différentes espèces de *Candida* d'origine humaine. Ann. Microbiol. (Inst. Pasteur) *126*B: 25-39.

30. el-Nakeeb, M. A., and J. O. Lampen. 1965. Uptake of griseofulvin by the sensitive dermatophyte, *Microsporum gypseum*. J. Bacteriol. *89*: 564-569.

31. el-Nakeeb, M. A., W. L. McLellan, Jr., and J. O. Lampen. 1965. Antibiotic action of griseofulvin on dermatophytes. J. Bacteriol. *89*: 557-563.

32. Epstein, W. L., V. P. Shah, and S. Riegelman. 1972. Griseofulvin levels in stratum corneum. Arch. Dermatol. *106*: 344-348.

33. Feingold, D. S. 1965. The action of amphotericin B on *Mycoplasma laidlawii*. Biochem. Biophys. Res. Commun. *19*: 261-267.

34. Ganis, P., G. Avitabile, W. Mechlinski, and C. P. Schaffner. 1971. Polyene macrolide antibiotic amphotericin B. Crystal structure of the N-iodoacetyl derivative. J. Am. Chem. Soc. *93*: 4560-4564.

35. Gentles, J. C. 1958. Experimental ringworm in guinea pigs: oral treatment with griseofulvin. Nature (Lond.) *182*: 476-477.

36. Gottlieb, H. E., J. H. Carter, J. H. Sloneker, and A. Ammann. 1958. Protection of fungi against polyene antibiotics by sterols. Science *128*: 361.

37. Gouge, T. H., and V. T. Andriole. 1971. An experimental model of amphotericin B nephrotoxicity with renal tubular acidosis. J. Lab. Clin. Med. *78*: 713-724.

38. Grisham, L. M., L. Wilson, and K. G. Bensch. 1973. Antimitotic action of griseofulvin does not involve disruption of microtubules. Nature (Lond.) *244*: 294-296.

39. Grove, J. F., J. MacMillan, T. P. C. Mulholland, and M. A. T. Rogers. 1952. Griseofulvin Part I. J. Chem. Soc. *12*: 3949-3958.

40. Grove, J. F., and J. C. McGowan. 1947. Identity of griseofulvin and "curling factor." Nature (Lond.) *160*: 574.

41. Grove, J. F., D. Ismay, J. MacMillan, T. P. C. Mulholland, and M. A. T. Rogers. 1951. The structure of griseofulvin. Chem. Ind. No. 11, pp. 219-220.

42. Grunberg, E., E. Titsworth, and M. Bennett. 1963. Chemotherapeutic activity of 5-fluorocytosine. Antimicrob. Agents Chemother., pp. 566-568.

43. Gull, K., and A. P. Trinci. 1974. Effects of griseofulvin on the mitotic cycle of the fungus *Basidiobolus ranarum*. Arch. Microbiol. *95*: 57-65.

44. Gull, K., and A. P. Trinci. 1973. Griseofulvin inhibits fungal mitosis. Nature (Lond.) *244*: 292-294.

45. Gull, K., and A. P. J. Trinci. 1974. Ultrastructural effects of griseofulvin on the myxomycete *Physarum polycephalum*. Inhibition of mitosis and the production of microtubule crystals. Protoplasma *81*: 37-48.

46. Hamilton, J. D., and D. M. Elliott. 1975. Combined activity of amphotericin B and 5-fluorocytosine against *Cryptococcus neoformans* in vitro and in vivo in mice. J. Infect. Dis. *131*: 129-137.

47. Hammarström, L., and C. I. E. Smith. 1977. In vitro activating properties of polyene antibiotics for murine lymphocytes. Acta Pathol. Microbiol. Scand., Sect. C *85*: 277-283.

48. Harrison, E. F., and W. A. Zygmunt. 1974. Haloprogin: mode of action studies in *Candida albicans*. Can. J. Microbiol. *20*: 1241-1245.

49. Harrison, E. F., P. Zwadyk, Jr., R. J. Bequette, E. E. Hamlow, P. A. Tavormina, and W. A. Zygmunt. 1970. Haloprogin: a topical antifungal agent. Appl. Microbiol. *19*: 746-750.

50. Hellebusch, A. A., F. Salama, and E. Eadie. 1972. The use of mannitol to reduce the nephrotoxicity of amphotericin B. Surg. Gynecol. Obstet. *134*: 241-243.

51. Hildick-Smith, G., H. Blank, and I. Sarkany. 1964. *Fungus Diseases and Their Treatment*, Little, Brown, Boston.

52. Hoeprich, P. D., M. M. Kawachi, K. K. Lee, and C. P. Schaffner. 1979. Neuropsychiatric effects of amphotericin B methyl ester (AME) derivatives. In *Prog. Abstr. 11th International Congress of Chemotherapy and 19th Interscience Conference on Antimicrobial Agents and Chemotherapy*, No. 157, Washington, D.C.

53. Hoeprich, P. D., and A. C. Huston. 1975. Susceptibility of *Coccidioides immitis, Candida albicans,* and *Cryptococcus neoformans* to amphotericin B, flucytosine, and clotrimazole. J. Infect. Dis. *132*: 133-141.

54. HsuChen, C. C., and D. S. Feingold. 1974. Two types of resistance to polyene antibiotics in *Candida albicans*. Nature (Lond.) *251*: 656-659.

55. Huppert, M., S. H. Sun, and K. R. Vukovich. 1974. Combined amphotericin B-tetracycline therapy for experimental coccidioidomycosis. Antimicrob. Agents Chemother. *5*: 473-478.

56. Jagdis, F. A., N. Monji, R. M. Lawrence, P. D. Hoeprich, and C. P. Schaffner. 1976. Distribution of radiolabelled amphotericin B methyl ester and amphotericin B in non-human primates (abstr.). Clin. Res. *24*: 453.

57. Jeffreys, E. G., P. W. Brian, H. G. Hemming, and D. Lowe. 1953. Antibiotic production by the microfungi of acid heath soils. J. Gen Microbiol. *9*: 314-341.

58. Johnson, W. H., C. A. Miller, and J. E. Brumbaugh. 1962. Induced loss of pigment in planarians. Physiol. Zool. *35*: 18-26.

59. Jund, R., and F. Lacroute. 1970. Genetic and physiological aspects of resistance to 5-fluoropyrimidines in *Saccharomyces cerevisiae*. J. Bacteriol. *102*: 607-615.

60. Kauffman, C. A., J. A. Carleton, and P. T. Frame. 1976. Simple assay for 5-fluorocytosine in the presence of amphotericin B. Antimicrob. Agents Chemother. *9*: 381-383.

61. Kauffman, C. A., and P. T. Frame. 1977. Bone marrow toxicity associated with 5-fluorocytosine therapy. Antimicrob. Agents Chemother. *11*: 244-247.

62. Kinsky, S. C. 1961. Alterations in the permeability of *Neurospora crassa* due to polyene antibiotics. J. Bacteriol. *82*: 889-897.

63. Kitahara, M., G. S. Kobayashi, and G. Medoff. 1976. Enhanced efficacy of amphotericin B and rifampicin combined in treatment of murine histoplasmosis and blastomycosis. J. Infect. Dis. *133*: 663-668.

64. Kitahara, M., V. K. Seth, G. Medoff, and G. S. Kobayashi. 1976. Activity of amphotericin B, 5-fluorocytosine, and rifampin against six clinical isolates of *Aspergillus*. Antimicrob. Agents Chemother. *9*: 915-919.

65. Kitahara, M., V. K. Seth, G. Medoff, and G. S. Kobayashi. 1976. Antimicrobial susceptibility testing of six clinical isolates of *Aspergillis*. Antimicrob. Agents Chemother. *9*: 908-914.

66. Kleinschmidt, M. G., K. S. Chough, and J. B. Mudd. 1972. Effect of filipin on liposomes prepared with different types of steroids. Plant Physiol. *49*: 852-856.

67. Kobayashi, G. S., G. Medoff, D. Schlessinger, C. N. Kwan, and W. E. Musser. 1972. Amphotericin B potentiation of rifampicin as an antifungal agent against the yeast phase of *Histoplasma capsulatum*. Science *177*: 709-710.

68. Kotler-Brajtburg, J., G. Medoff, G. S. Kobayashi, S. Boggs, D. Schlessinger, R. C. Pandey, and K. L. Rinehart, Jr. 1979. Classification of polyene antibiotics according to chemical structure and biological effects. Antimicrob. Agents Chemother. *15*: 716-722.

69. Kotler-Brajtburg, J., G. Medoff, D. Schlessinger, and G. S. Kobayashi. 1974. Characterization of the binding of amphotericin B to *Saccharomyces cerevisiae* and relationship to the antifungal effect. Antimicrob. Agents Chemother. *6*: 770-776.

70. Kotler-Brajtburg, J., G. Medoff, D. Schlessinger, and G. S. Kobayashi. 1977. Amphotericin B and filipin effects on L and HeLa cells: dose response. Antimicrob. Agents Chemother. *11*: 803-808.

71. Kotler-Brajtburg, J., H. D. Price, G. Medoff, D. Schlessinger, and G. S. Kobayashi. 1974. Molecular basis for the selective toxicity of amphotericin B for yeast and filipin for animal cells. Antimicrob. Agents Chemother. *5*: 377-382.

72. Kwan, C. N., G. Medoff, G. S. Kobayashi, D. Schlessinger, and H. F. Raskas. 1972. Potentiation of the antifungal effects of antibiotics by amphotericin B. Antimicrob. Agents Chemother. *2*: 61-65.

73. Lampen, J. O., and P. M. Arnow. 1973. Differences in action of large and small polyene antifungal antibiotics. Bull. Res. Counc. Isr., Sect. A *11*: 286-291.

74. Lampen, J. O., P. M. Arnow, and R. S. Safferman. 1960. Mechanism of protection by sterols against polyene antibiotics. J. Bacteriol. *80*: 200-206.

75. Lawrence, R. M., and P. D. Hoeprich. 1976. Comparison of amphotericin B and amphotericin B methyl ester: efficacy in murine coccidioidomycosis and toxicity. J. Infect. Dis. *133*: 168-174.

76. Levine, H. B., D. A. Stevens, J. M. Cobb, and A. E. Gebhardt. 1975. Miconazole in coccidioidomycosis. I. Assays of activity in mice and in vitro. J. Infect. Dis. *132*: 407-414.

77. Lin, C., and S. Symchowicz. 1975. Absorption, distribution, metabolism, and excretion of griseofulvin in man and animals. Drug Metab. Rev. *4*: 75-95.

78. Little, J. R., T. J. Blanke, F. Valeriote, and G. Medoff. 1978. Immunoadjuvant and antitumor properties of amphotericin B. In M. A. Chirigos (Ed.), *Immune Modulation and Control of Neoplasia by Adjuvant Therapy,* Raven, New York, pp. 381-387.

79. Little, J. R., R. J. Plut, J. Kotler-Brajtburg, G. Medoff, and G. S. Kobayashi. 1978. Relationship between the antibiotic and immunoadjuvant effects of amphotericin B methyl ester. Immunochemistry *15*: 219-224.

80. Louria, D. B. 1958. Some aspects of the absorption, distribution, and excretion of amphotericin B in man. Antibiot. Med. Clin. Ther. *5*: 295-301.

81. Medoff, G., M. Comfort, and G. S. Kobayashi. 1971. Synergistic action of amphotericin B and 5-fluorocytosine against yeast-like organisms. Proc. Soc. Exp. Biol. Med. *138*: 571-574.

82. Medoff, G., W. E. Dismukes, R. H. Meade, and J. M. Moses. 1972. A new therapeutic approach to *Candida* infections. A preliminary report. Arch. Intern. Med. *130*: 241-245.

83. Medoff, G., G. S. Kobayashi, C. N. Kwan, D. Schlessinger, and P. Venkov. 1972. Potentiation of rifampicin and 5-fluorocytosine as antifungal antibiotics by amphotericin B. Proc. Natl. Acad. Sci. U.S.A. *69*: 196-199.

84. Meinhof, W., and D. Gunther. 1972. Treatment of chronic mucocutaneous candidiasis of children (*Candida* granuloma) with clotrimazole. Arch. Dermatol. Forsch. *242*: 293-308.

85. Noguchi, T., A. Kaji, Y. Igarashi, A. Shigematsu, and K. Taniguchi. 1962. Antitrichophyton activity of naphthiomates. Antimicrob. Agents Chemother., pp. 259-267.

86. Norman, A. W., R. A. Demel, B. De Kruyff, W. S. M. Geurts van Kessel, and L. L. M. Van Deenen. 1972. Studies on the biological properties of polyene antibiotics: comparison of other polyenes with filipin in their ability to interact specifically with sterol. Biochim. Biophys. Acta *290*: 1-14.

87. Norman, A. W., R. A. Demel, B. De Kruyff, and L. L. M. Van Deenen. 1972. Studies on the biological properties of polyene antibiotics. J. Biol. Chem. *247*: 1918-1929.

88. Nozawa, Y., Y. Kitajima, T. Sekiya, and Y. Ito. 1974. Ultrastructural alterations induced by amphotericin B in plasma membranes of *Epidermophyton floccosum* as revealed by freeze-etch electron microscopy. Biochim. Biophys. Acta *367*: 32-38.

89. Oroshnik, W., and A. D. Mebane. 1963. The polyene antifungal antibiotics. Fortschr. Chem. Org. Naturst. *21*: 17-79.

90. Oxford, A. E., H. Raistrick, and P. Simonart. 1939. Studies in biochemistry of micro-organisms. LX. Griseofulvin, $C_{17}H_{17}O_6C$, a metabolic product of *Penicillium griseo-fulvum* dierckx. Biochem. J. *33*: 240-248.

91. Paget, G. E., and A. L. Walpole. 1958. Some cytological effects of griseofulvin. Nature (Lond.) *182*: 1320-1321.

92. Pappagianis, D., M. S. Collins, R. Hector, and J. Remington. 1979. Development of resistance to amphotericin B in *Candida lusitaniae* infecting a human. Antimicrob. Agents Chemother. *16*: 123-126.

93. Pittillo, R. F., and B. J. Ray. 1969. Chemotherapeutic activity of 5-fluorocytosine against a lethal *Candida albicans* infection in mice. Appl. Microbiol. *17*: 773-774.

94. Plempel, M., K. Bartmann, K. H. Büchel, and E. Regel. 1969. Experimentelle Befunde über ein neues, oral wirksames Antimykotikum mit breitem Wirkungsspektrum. Dtsch. Med. Woshenschr. *94*: 1356-1364.

95. Plempel, M., K. H. Buchel, K. Bartmann, and E. Regel. 1974. Antimycotic properties of clotrimazole. Postgrad. Med. J. *50* (Suppl. 1): 11-12.

96. Polak, A., and H. J. Scholer. 1975. Mode of action of 5-fluorocytosine and mechanisms of resistance. Chemotherapy *21*: 113-130.

97. Polak, A., and H. J. Scholer. 1976. Combination of amphotericin B and 5-fluorocytosine. In J. D. Williams and A. M. Geddes (Eds.), *Chemotherapy,* Vol. 6, Plenum, New York, pp. 1371-1372.

98. Robinson, H. M., Jr., and J. Raskin. 1964. Tolnaftate therapy of mycotic infections. Preliminary report. J. Invest. Dermatol. *42*: 185-187.

99. Roth, F. J., Jr., B. Sallman, and H. Blank. 1959. In vitro studies of the antifungal antibiotic griseofulvin. J. Invest. Dermatol. *33*: 403-418.

100. Rubenstein, E., E. R. Noriega, M. S. Simberkoff, R. Holzman, and J. J. Rahal, Jr. 1975. Fungal endocarditis—analysis of 24 cases and review of literature. Medicine *54*: 331-334.

101. Sanders, E. 1971. Cutaneous sporotrichosis; beer, bricks, and bumps. Arch. Intern. Med. *127*: 482-483.

102. Sarosi, G. A., J. D. Parker, I. L. Doto, and F. E. Tosh. 1969. Amphotericin B in cryptococcal meningitis. Long-term results of treatment. Ann. Intern. Med. *71*: 1079-1087.

103. Schaffner, C. P., and W. Mechlinski. 1972. Polyene macrolide derivatives. II. Physical-chemical properties of polyene macrolide esters and their water soluble salts. J. Antibiot. *25*: 259-260.

104. Schroeder, F., J. F. Holland, and L. L. Bieber. 1973. Reversible interconversions of sterol-binding and sterol-nonbinding forms of filipin as determined by fluorimetric and light scattering properties. Biochemistry *12*: 4785-4789.

105. Schroeder, F., J. F. Holland, and L. L. Bieber. 1971. Fluorometric evidence for the binding of cholesterol to the filipin complex. J. Antibiot. *24*: 846-849.

106. Scholer, H. J. 1970. Antimykoticum 5-fluorocytosin (oral antimycotic agent, 5-fluorocytosine). Mykosen (Berlin) *13*: 179-188.

107. Scholer, H. J. 1976. Grundlagen und Ergebnisse der antimykotischen Chemotherapie mit 5-fluorocytosin. Chemotherapy *22* (Suppl. 1): 103-146.

108. Schwartz, S. N., G. Medoff, G. S. Kobayashi, C. N. Kwan, and D. Schlessinger. 1972. Antifungal properties of polymyxin B and its potentiation of tetracycline as an antifungal agent. Antimicrob. Agents Chemother. *2*: 36-40.

109. Seki, S., K. Nishihata, T. Nakayama, and H. Ogawa. 1963. Syntheses of halophenyl-alpha-iodopropargyl ethers having high antifungal activity. Agric. Biol. Chem. *27*: 150-151.

110. Seneca, H., and E. Bergendahl. 1955. Toxicity of antibiotics to snails. Antibiot. Chemother. *5*: 737-741.

111. Serstock, D. S., and H. H. Zinneman. 1975. Pulmonary and articular sporotrichosis. Report of two cases. JAMA *233*: 1291-1293.

112. Shadomy, S., H. J. Shadomy, and J. P. Utz. 1970. In vivo susceptibility of *Cryptococcus neoformans* to hamycin, amphotericin B, and 5-fluorocytosine. Infect. Immun. *1*: 128-134.

113. Shah, V. P., W. L. Epstein, and S. Riegelman. 1974. Role of sweat in accumulation of orally administered griseofulvin in skin. J. Clin. Invest. *53*: 1673-1678.

114. Sheir-Neiss, G., R. V. Nardi, M. A. Gealt, and N. R. Morris. 1976. Tubulin-like protein from *Aspergillus nidulans.* Biochem. Biophys. Res. Commun. *69*: 285-290.

115. Shirley, S. F., and J. R. Little. 1979. Immunopotentiating effects of amphotericin B. I. Enhanced contact sensitivity in mice. J. Immunol. *123*: 2878-2882.

116. Shirley, S. F., and J. R. Little. 1979. Immunopotentiating effects of amphotericin B. II. Enhanced in vitro proliferative responses of murine lymphocytes. J. Immunol. *123*: 2883-2889.

117. Slonim, R. R., D. S. Howell, and H. E. Brown, Jr. 1962. Influence of griseofulvin upon acute gouty arthritis. Arthritis Rheum. *5*: 397-404.

118. Stobo, J. D., S. Paul, R. E. Van Scoy, and P. E. Hermans. 1976. Suppressor thymus-derived lymphocytes in fungal infections. J. Clin. Invest. *57*: 319-328.

119. Tettenborn, D. 1974. Toxicity of clotrimazole. Postgrad. Med. J. *50* (Suppl. 1): 17-20.

120. Titsworth, E., and E. Grunberg. 1973. The combined effect of 5-fluorocytosine and amphotericin B in the therapy of murine cryptococcosis. Antimicrob. Agents Chemother. *4*: 306-308.

121. Utahara, R., Y. Okami, S. Nakamura, and H. Umezawa. 1954. On a new antifungal substance, mediocidin, and other antifungal substances of *Streptomyces* with three characteristic absorption maxima. J. Antibiot. *7A*: 120-124.

122. Utz, J. P. 1975. New drugs for the systemic mycoses: flucytosine and clotrimazole. Bull. N.Y. Acad. Med. *51*: 1103-1108.

123. Utz, J. P., I. L. Garriques, M. A. Sande, J. F. Warner, G. L. Mandell, R. F. McGehee, R. J. Duma, and J. Shadomy. 1975. Therapy of cryptococcosis with a combination of flucytosine and amphotericin B. J. Infect. Dis. *132*: 368-373.

124. van Cutsem, J. M., and D. Thienpont. 1972. Miconazole, a broad-spectrum antimycotic agent with antibacterial activity. Chemotherapy *17*: 392-404.

125. Van den Bossche, H. 1974. Biochemical effects of miconazole on fungi. I. Effects on the uptake and/or utilization of purines, pyrimidines, nucleosides, amino acids, and glucose by *Candida albicans*. Biochem. Pharmacol. *23*: 887-899.

126. Van den Bossche, H., G. Willemsens, and W. Cools. 1978. Janssen Pharmacentica, Preclinical Research Report R41400/20, August.

127. Wallace, S. L., and A. W. Nissen. 1962. Griseofulvin in acute gout. N. Engl. J. Med. *266*: 1099-1101.

128. Water, R. D., and L. J. Kleinsmith. 1976. Identification of α and β tubulin in yeast. Biochem. Biophys. Res. Commun. *70*: 704-708.

129. Weber, M. M., and S. C. Kinsky. 1965. Effect of cholesterol on the sensitivity of *Mycoplasma laidlawii* to the polyene antibiotic filipin. J. Bacteriol. *89*: 306-312.

130. Weber, K., J. Wehland, and W. Herzog. 1976. Griseofulvin interacts with microtubules both in vivo and in vitro. J. Mol. Biol. *102*: 817-829.

131. Weinstein, G. D., and H. Blank. 1960. Quantitative determination of griseofulvin by a spectrophotofluorometric assay. Arch. Dermatol. *81*: 746–749.

132. Williams, D. L., R. H. Martin, and I. Sarkany. 1958. Oral treatment of ringworm with griseofulvin. Lancet 2: 1212-1213.

133. Wilson, L. 1975. Action of drugs on microtubules. Life Sci. *17*: 303-310.

134. Wilson, L., J. R. Bamburg, S. B. Mizel, L. M. Grisham, and K. M. Creswell. 1974. Interaction of drugs with microtubule proteins. Fed. Proc. *33*: 158-166.

8

Measurement of Activity of Antifungal Drugs

George S. Kobayashi and Gerald Medoff / Washington University School of Medicine, St. Louis, Missouri

I. INTRODUCTION

Formerly, routine susceptibility testing of fungi was not considered necessary because there were only a limited number of useful antifungal agents and resistance was rarely encountered. However, in the past few years, more antifungals, such as the imidazoles and 5-fluorocytosine, have been developed [13,50], and resistance to 5-fluorocytosine and amphotericin B have been encountered [5,11,27,42,43,47,60,64,65,71,76,84,85]. These developments have changed our ideas on the need for susceptibility testing of fungi and we now believe that such testing may be necessary for optimal management of patients with systemic fungal infections.

The principles of susceptibility testing of fungi are essentially the same as those for bacteria [20,61,66,75]. However, testing with fungi must deal with the fact that interpretation of the results is complicated by inherent differences in fungal morphology, growth rate, and optimal culture conditions. For example, the vegetative phase of medically important fungi may be yeast-like or filamentous and the organism may have the phenotypic ability to exist in either morphologic state, depending on the conditions of culture. In the latter case the morphologic phase of the organism used in the test may have an important effect on the test results.

Since yeasts are unicellular organisms with uniform growth characteristics, the methods used for testing their susceptibility are readily adaptable from bacteria. These methods include dilution in broth [10,14,28,32,38,55,57,73,78] or agar [15,21,22,25,55,70,74,77, 82,83], disk diffusion [17,30,34,55,72,80], and cylinder and agar well [4,35,40,41,52, 53,86]. The minimum inhibitory concentration (MIC) values of a drug are those that result in predetermined changes in turbidity or optical density, or in zones of inhibition of growth. The minimum fungicidal concentration (MFC) is determined by subculturing from broth tubes showing no visible growth onto media without antibiotics and scoring for subsequent growth. The MFC is that concentration of drug which results in complete killing in the culture and therefore no growth on subcultures.

Because of their morphological complexity, the filamentous fungi do not lend themselves readily to laboratory procedures applied to the yeast. With filamentous organisms, reproducibility is a problem, and perhaps the single most difficult task confronting those who attempt to do studies with these fungi is the preparation of a standardized uniform suspension of hyphal cells. To circumvent this problem, suspensions of spores, when available, have been used as inocula. The criticism of the use of spores has been that the net effects of the antifungal agent measures the inhibition of germination rather than inhibition of an actively growing phase of the organism. For this reason, some investigators have employed single-cell spore suspensions that have been allowed to germinate prior to being dispensed in the test system. The susceptibility data generated thus far from comparative studies using both germinated and ungerminated spores indicate no differences in the MIC values, but more information is needed [48].

Several other factors could adversely affect the test results and must be considered in the design of susceptibility testing of fungi. These include the stability and solubility characteristics of the antifungal agent, characteristics of the medium, pH of the test system, the inoculum, and temperature and length of incubation. It is obvious when the present data on fungal susceptibility testing are reviewed that much more work on standardization of techniques and interpretation of results is necessary. For these reasons, we believe that the performance of these tests, at the present time, should be restricted to the few laboratories that have the facilities, experience, and interest to judge and to evaluate properly the variables that exist.

In this chapter we describe selected methods of susceptibility testing and assay of antifungal drug levels in body fluids that are being evaluated in several laboratories. We will deal with those done on the three classes of antifungal drugs: the polyenes, the imidazole derivatives, and the pyrimidine analog, 5-fluorocytosine. The methods to be discussed represent the state of the art at the present time and undoubtedly will be changed as more information becomes available and newer procedures are developed.

II. SUSCEPTIBILITY TESTING

Dilution of an antifungal agent in either broth or agar provides a simple in vitro assay procedure for the quantitative assessment of its MIC for a particular test organism. The dilution procedures are conventionally done in culture tubes or plates but can be miniaturized and performed in microtiter dishes [29,31,56]. The MIC values obtained from these tests will vary with the type of end point determination used to generate the results. The visual index of growth requires a subjective judgment of the lowest concentration of drug that inhibits growth. Several factors will influence the results of these assays, including size of inoculum, temperature of incubation, and duration of incubation [10,15,48]. Although turbidity is the most accepted method of determining the end point for drug susceptibility, dry weight measurements and radiometric measurements of RNA and protein synthesis have also been used [7,44,46,48,49,51,67].

Below we list the critical factors in susceptibility testing and discuss the problems and limitations of each. In addition to those listed, it is important to note that none of the in vitro susceptibility test procedures used at present includes any consideration of host effects. This is a critical point because we have learned over the past few years that sub-inhibitory concentrations of antimicrobial agents coupled with host cellular and humoral responses may have positive therapeutic effects [58]. In addition, the most potent anti-fungal agents, the polyenes, are active immunoadjuvants which boost both humoral and delayed-type hypersensitivity responses of the host to a variety of antigens [54]. This may have important effects on the outcome of infection.

A. Test Medium

The medium should be simple enough to support good growth of the test organism and, from a practical point of view, should mimic as closely as possible the situation as it exists in the host. This may be impossible, but at the very least the test medium should not grossly affect the action of the drugs and should yield reproducible results.

Comparative studies have been performed to determine the suitability of various media for antifungal susceptibility testing [40,41,48]. Sabouraud's glucose broth adjusted to pH 6.8-7.0 has generally been the medium of choice for studies dealing with the polyene macro-lide antibiotics. Chemically defined media such as Bacto yeast nitrogen base (YNB, Difco) supplemented with dextrose and Bacto yeast morphology agar (YMA, Difco) are not suit-able for polyene antifungal susceptibility studies because the pH of these media is 4.5, and acidic pH adversely affects the antifungal activity of amphotericin B and the other poly-enes [36]. Unbuffered media containing a high concentration of dextrose also should not be used when evaluating polyene antibiotics because of the acid produced by luxuriant growth of most fungi.

Because amphotericin B and the other polyene macrolides are polyunsaturated, they are highly susceptible to inactivation by autoxidation [69] or peroxidation [2]. Cheung et al. [18] found that amphotericin B is unstable in certain culture media with a half-life of 18-24 h. Others have confirmed these observations and shown that various antioxidants added to the culture medium stabilize amphotericin B and prolong its anti-fungal activity [3].

On the other hand, media such as Sabouraud's, brain heart infusion, trypticase soy, and others, containing complex undefined nitrogenous materials such as pyrimidines, purines, and some other nucleosides, are not recommended for susceptibility studies using 5-fluorocytosine and certain of the imidazole derivatives, because they compete with and decrease the activity of the test compound. Chemically defined media such as YNB or YMA supplemented with dextrose and asparagine are acceptable since they do not con-tain compounds that competitively inhibit the activity of 5-fluorocytosine [73].

It would be optimal to have one medium of simple composition which is capable of supporting the growth of all the pathogenic fungi and which does not interfere with the antifungal activities of the test drug. Unfortunately, this medium does not presently exist and most laboratories use a different medium to test each of the drugs.

B. Drug Preparation

Antifungal susceptibility testing must deal with the variable solubility and stability of the most important antifungal agents. With the exception of 5-fluorocytosine, most of the antifungal drugs are insoluble in water, and one must take into account when assessing the results what effects the organic solvent used to dissolve the drugs has on the test organism. For this reason, when solvents such as dimethylformamide, dimethyl sulfoxide, alcohol, acetone, and polyethylene glycol are used to solubilize the antibiotic, appropriate solvent controls without antibiotic must be included in the test system.

Three major categories of antifungal antibiotics (i.e., the polyenes, the imidazoles, and 5-fluorocytosine) are currently available for use in the treatment of systemic diseases. Concentrated working solutions of the antifungal agents can be prepared and stored without appreciable loss of activity for periods of time ranging from 1 week at $4°C$ for amphotericin B to up to 1 year at $-30°C$ for miconazole (Table 1). Once diluted, however, the test drugs should be used immediately. This practice may be difficult and expensive for laboratories that have a low volume of requests for susceptibility tests.

C. Inoculum

Unfortunately, there has been no standard inoculum developed for susceptibility testing of fungi. In general, the inoculum should be vigorously viable and at a concentration that results in a visibly turbid culture after one or two cell divisions of the organisms. Unlike yeasts, the filamentous fungi pose a special problem in the preparation of uniform suspensions of cells. The hydrophobic property of spores coupled with variability in morphology, growth rate, and optimal conditions of growth present other problems in developing a standard inoculum. In addition, the concentration, volume, and the age of the culture from which the inoculum is derived also influence the test results.

1. Yeasts

For yeast, a standard inoculum has been derived by the following methods. A 24-h-old culture of cells in log phase growth is suspended in the test medium to yield a final concentration of about 5×10^5 cells/ml. The total number of particles is estimated from hemocytometer counts and the number of viable particles is calculated from plate counts of suitable dilutions. The spectrophotometric method and the McFarland nephelometer barium standard method are alternative procedures used to estimate density of cells in suspension by turbidity measurements. They both require construction of standard curves before use. In the spectrophotometric method, the yeasts are diluted to known concentrations which are read in a spectrophotometer at $\lambda = 530$ nm. An arithmetic plot of optical density (O.D.) to number of organisms is then constructed and in subsequent tests, when a yeast suspension is to be prepared, the desired concentration of cells per unit volume is calculated from the O.D. reading at λ 530 nm. It is important to remember that only those values obeying Beer's law can be used with confidence. Viability counts should always be checked to ensure that the correct number of colony-forming units were used.

In the McFarland nephelometer barium standard method, the suspension of yeasts is

Table 1 Solubility Characteristics of Selected Antifungal Agents

Classification	Compound	Solubility properties	Stock solutions
Polyene macrolides	Amphotericin B	Insoluble in water; soluble in dimethyl sulfoxide (DMSO) and dimethylformamide (DMF)	5 mg/ml may be prepared in DMSO or DMF and stored in the dark at 4°C for 1 week
	Nystatin	Insoluble in water; soluble in DMSO and DMF	5 mg/ml may be prepared in DMSO or DMF and stored in the dark at 4°C for 1 week
	Amphotericin B—methyl ester	Soluble in water	Make up fresh
	Nystatin—methyl ester	Soluble in water	Make up fresh
Imidazole derivatives	Miconazole	Insoluble in water; soluble in polyethylene glycol, DMSO, and DMF	10 mg/ml may be prepared in DMSO and stored at –30°C for 1 year
	Clotrimazole	Insoluble in water; soluble in polyethylene glycol, DMSO, and DMF	10 mg/ml may be prepared in DMSO and stored at –30°C for 1 year
	Ketoconazole	Insoluble in water; soluble in polyethylene glycol, DMSO, and DMF	10 mg/ml may be prepared in DMSO and stored at –30°C for 1 year
Pyrimidine analog	5-Fluorocytosine	Soluble in water	10 mg/ml may be prepared in water and stored at –20°C for 6 months

diluted to read the same turbidity as an 0.5 McFarland barium sulfate standard (10^5 particles/ml). The standard is prepared by adding 0.1 ml of a 1% $BaCl_2$ solution to 9.9 ml of a 1% solution of H_2SO_4. Suspensions of yeasts that give a visual turbidity comparable to this standard will have approximately 1×10^7 cells/ml. This must be confirmed by hemocytometer and viability counts.

2. Filamentous Fungi

For filamentous fungi, inocula consist of either macerated hyphal fragments or spores uniformly suspended in saline or a suitable buffer. For organisms that produce spores in abundance (e.g., *Aspergillus fumigatus, Paecilomyces variotii, Mucor,* etc.) the procedure involves flooding the surface of well-sporulating cultures grown on agar with saline or buffer and gently agitating the culture to dislodge the spores. The dislodged spores and fluid are transferred to a sterile screw-capped tube or bottle containing glass Ballotini beads and the suspension is vigorously agitated to disperse the hydrophobic fungal spores. The spores can also be suspended in an aqueous menstrum without the aid of beads if the sterile saline contains 0.05% Tween 80. The concentration of spores is then estimated by hemocytometer count and adjusted to the desired concentration. It is always best to confirm the spore number by performing viability counts on the suspension.

To prepare suspensions of filamentous fungi that do not produce spores in abundance, colonies or mycelial mats from the surface of agar are harvested in a small volume of sterile physiological saline. The preparation is transferred to a sterile Tenbroeck tissue grinder and gently ground by hand to disperse the particles of fungi. The suspension is allowed to stand for 5-10 min so that the larger mycelial fragments settle. The supernatant fluid containing suspended fungal particles is carefully removed from the more rapidly sedimenting debris and examined microscopically to determine its characteristics. If the suspension appears homogeneous, it is diluted in saline and a particle count obtained by procedures similar to those described above for suspensions of spores. As in the case of spore and yeast suspensions, once a satisfactory suspension of mycelial fragments is obtained, the cell number can be confirmed by turbidity and viability studies.

D. Conditions of Incubation

The choice of temperature and length of incubation employed in susceptibility testing will be determined by the growth properties of the organism and the stability of the antifungal agent being evaluated. In studies with clinical isolates of *Aspergillus,* Kitahara et al. [48] found that the MIC of amphotericin B, 5-fluorocytosine, and rifampin were markedly lower at incubation temperatures of 25°C than at 37°C. Block et al. [10], on the other hand, reported an increased susceptibility of *Cryptococcus neoformans* for 5-fluorocytosine when the assay was conducted at 37°C compared to incubation at 25°C. In the dimorphic fungus, *Histoplasma capsulatum,* Cheung et al. [19] found that the two phases have selective drug susceptibilities. The yeast phase of *H. capsulatum,* grown at 37°C, was much more resistant to amphotericin B than the mycelial phase grown at 25°C. Therefore, fluctuations in the temperature of incubation dramatically influence susceptibility patterns of an organism to certain antifungal agents.

E. Types of Assay Procedures

Dilution of antifungal agents in broth or agar are two procedures commonly used in assessing the MIC of a given compound for a specific fungus. Only the broth dilution method can be used to determine the MFC of the agent. The agar dilution method is more efficient, particularly when several organisms are to be tested simultaneously against different concentrations of one drug since multiple inocula can be placed on each agar plate [6,15, 24,83]. Disk- and agar well-diffusion tests with antifungal compounds have limited usefulness at the present time because of problems with reproducibility and generally poor solubility and stability properties of most antifungal antibiotics in water.

The following procedures for broth dilution and agar dilution techniques describe performance of the tests in general terms. These tests may be scaled up or down according to the needs of the laboratory.

1. Broth Dilution Method

The type of medium employed in susceptibility testing will depend on the drug that is being evaluated (see Sec. II.A). All tests should be performed in duplicate and should include both a growth control and a formaldehyde-killed control. The former serves as an index of viability of the inoculum and the latter as an inoculum control. Since fungi are particulate, they rapidly sediment and when resuspended, may contribute a hazy appearance to the broth even when they are not viable. For this reason, the formaldehyde-killed control serves as an index of comparison for those concentrations of drug that only partially inhibit the growth of the organism.

The broth dilution procedure is easily performed, provides a uniform inoculum for each of the test drug levels, and can be adapted to a scale that fits the needs of the laboratory. In general, a duplicate set of serial twofold dilutions of the test compound are dispensed into suitable broth medium which is inoculated to contain 10^5 viable organisms/ml. The test system is incubated at $30°C$ for 24-48 h in the case of polyenes and imidazoles and for at least 48 h in the case of 5-fluorocytosine. For each of the test assays to be valid, there must be visible growth in the control tubes. After the recommended period of incubation, the contents of each tube are resuspended and examined. The MIC is recorded as that concentration of drug which clearly inhibits growth when compared to the control tubes. The imidazoles are not readily soluble in water and for this reason impart slight turbidity to the tubes containing the drug. To circumvent this problem, uninoculated media containing varying concentrations of the imidazole can be incorporated in the test system for purposes of comparison. To determine the MFC, tubes showing no growth and the growth control tubes are subcultured onto Sabouraud's agar medium without antibiotic and incubated at $30°C$. The cultures are examined until growth from the control tubes is apparent. The MFC is established as the lowest concentration of drug from which subcultures were negative. With slight modifications this method can be adapted to procedures of susceptibility testing which measure dry weight, radioisotopic incorporation, and RNA and protein synthesis [1,67,71].

Table 2 lists a range of sample MIC values reported for amphotericin B, miconazole, clotrimazole, and 5-fluorocytosine against various selected fungi. Since there is no standard method of fungal susceptibility testing against antifungal agents, each laboratory should employ control organisms of known susceptibility for comparative purposes, such as *Saccharomyces cerevisiae* (ATCC 9683) when 5-fluorocytosine is being evaluated, *Candida albicans* (ATCC 10231) when polyenes are tested, and *C. stellatoideae* (ATCC 26232) when imidazoles are evaluated. Susceptibility tests on these controls should yield consistently reproducible MIC and MFC values for each of the respective classes of compounds.

Table 2 Range of Values of Minimum Inhibitory Concentration Reported for Selected Fungi

Organism	Amphotericin B (μg/ml)	Clotrimazole (μg/ml)	Miconazole (μg/ml)	5-Fluorocytosine (μg/ml)
Cryptococcus neoformans	0.03–0.86	1.0–4.0	0.4–2	0.2–3.9
Candida spp.	0.05–3.7	0.02–16	0.0–2	0.13–16
Torulopsis glabrata	0.1–0.4	0.1–16	0.5–2	0.2–16
Aspergillus spp.	0.14–8.0	0.1–10		0.2–500
Histoplasma capsulatum	0.04–1.0	0.2–3.13	0.01–1	Resistant
Blastomyces dermatitidis	0.05–0.5		0.01–1	Resistant
Coccidioides immitis	0.2–0.8	0.1–3.13	0.3–2.0	Resistant

2. Agar Dilution Method

Dilution of antifungal agents in agar for susceptibility studies on fungi enables one to determine easily the MIC for multiple fungal isolates. This procedure is not suitable for all types of compounds or fungi, but under certain conditions it has been used successfully to evaluate the in vitro susceptibility of several yeast isolates to various antifungal agents [25]. Mechanical replicators, such as the Steers type, allow multiple inocula to be reproducibly placed onto the surface of antibiotic-containing agar media. The results of careful studies have shown that assessments based on the agar dilution procedure with ketoconazole and 5-fluorocytosine are highly dependent on inoculum size and length of incubation, and tend to be divergent when compared to those obtained by the broth dilution method [15]. This variability is not seen with the polyenes. At the present time, routine use of the method should be limited to those organisms and antifungal compounds that yield reproducible results. For this procedure, serial dilutions of the test compound are incorporated into suitable medium containing agar at a concentration of 1.5%. Since some of the media and antifungal compounds deteriorate at autoclave temperature [12,36], it is necessary to mix sterile liquified double strength agar and double-strength filter-sterilized media plus antifungal agent in order to preserve the stability of the nutrients and test compounds. The agar medium is dispensed into sterile petri dishes and allowed to gel. For the Steer's replicator method [83], test inocula are prepared at concentrations of about 10^6 viable particles per milliliter. The inoculating pegs of the Steer's replicator, with each surface area milled to deliver a volume of 0.001-0.003 ml, are dipped into wells containing the test organism and then transferred to agar plates containing varying concentrations of drug. Each test should include growth plates without added antibiotics. The inoculated plates are incubated at 30°C and the results recorded when the control plates show growth.

The agar dilution method susceptibility testing is not routinely performed in most laboratories mainly because there is not a great need, at the present time, to perform large-scale testing on a large number of clinical isolates. An additional problem is the variability of results obtained by this method compared to those obtained by the broth dilution method. Under ideal conditions the MIC results obtained for polyenes by the agar dilution method should agree with those obtained by the broth dilution method This relationship does not hold up for the imidazoles and 5-fluorocytosine [15].

III. MEASUREMENT OF DRUG LEVELS IN BODY FLUIDS

Monitoring the concentration of antifungal agents in serum and cerebrospinal fluid during treatment may, in theory, assist in the clinical management of a patient. At least this has been the assumption in many bacterial infections. However, in fungal diseases, the assays that are presently available for measurement of levels of polyenes, imidazoles, and 5-fluorocytosine in biological fluids are not standardized, are time consuming, and show considerable variability. In addition, there has been poor correlation with the results of these assays and the clinical response of the patients, e.g., levels of amphotericin B in cerebrospinal fluid and response of cryptococcal meningitis to therapy [8,59]. The method that has been employed the longest to measure drug levels has been a bioassay that utilizes an organism of known susceptibility to a specific agent. More recently, several different physicochemical assays of antifungal drug levels have been developed. These include fluorometry [68,81], spectrophotometry [39], gas-liquid chromatography [37], pyrolysis gas-liquid chromatography [16], high-pressure liquid chromatography [9,24,63], and thin layer chromatography [79]. In general, these newer methods are specific, sensitive, and can be performed more rapidly than the bioassay technique, but they may require an initial extraction procedure followed by further separation and analysis with specialized instrumentation not routinely available in many laboratories. In addition, the physical methods identify the compound or its derivatives but do not evaluate the biological activity of the agent.

A. Biological Methods

The microbiological assay of amphotericin B levels in body fluids, which involves growth inhibition of *P. variotii,* was first described by Green et al. [33] and was later modified by Bindschadler and Bennett [8]. The procedure is based on the radial growth inhibition of the test fungus, *P. variotii* (ATCC 36257), brought about by the diffusion of the antifungal compound from a fixed reservoir cut into an agar medium which supports the growth of the organism. The standard curve derived from zones of inhibition for amphotericin B plotted on semilogarithmic paper should be linear from 0.03 to 1.0 μg/ml. All test samples that exceed these values should be diluted accordingly and retested.

Bioassays can also be done on body fluids from patients who are being treated with 5-fluorocytosine or miconazole [66,75]. In the case of 5-fluorocytosine, *S. cerevisiae* (ATCC 9763) is seeded into the cooled agar at a final concentration of 10^5 viable particles/ml. The standard curve for 5-fluorocytosine should be linear from 2.5 through 80 μg/ml. When miconazole levels are being evaluated in body fluids, *C. stellatoideae* is employed at the same final concentration and the standard dose-response curve of inhibition should be linear from 0.5 through 16 μg/ml. Plates are usually held for 48 h at 30°C for the 5-fluorocytosine and miconazole assays. When sera from patients receiving both 5-fluorocytosine and amphotericin B in combination are being studied to determine levels from one of these agents in the presence of the other, additional procedures are necessary. *Chrysosporium pruinosum* (ATCC 36374) can be used as the bioassay indicator organism to determine the levels of amphotericin B in the presence of 5-fluorocytosine since this fungus is totally resistant to 5-fluorocytosine [63]. To ensure proper performance of the assay, the medium should be supplemented with cytosine.

Levels of 5-fluorocytosine can be assayed in the presence of amphotericin B by using a sensitive organism (*S. cerevisiae,* ATCC 9763) and a medium without purines or pyrimidines. The sera are heated to 90°C for 30 min to inactivate the biological activity of amphotericin B before the test is run [36].

B. Physical Methods

Although the physical methods for determining levels of drug in biological specimens are more rapid, sensitive, and specific than the bioassay procedures, the number of manipulations and care required to process the samples may discourage most laboratories from using them. Moreover, the sensitivity of such procedures often results in the detection of interfering substances and thus will yield unreliable results unless this fact is carefully monitored. Because of this problem, most physical procedures use an extraction step followed by chromatography performed on a suitable matrix with a solvent system capable of separating out the specific compounds being analyzed. Several methods, such as paper, thin layer, and gas chromatography, have been used for this purpose. Chromatography involves separation of mixtures into their individual components by placing the sample in a tube or column that is tightly packed with a solid matrix (the stationary phase) and allowing the sample to be carried by a flowing liquid (the mobile phase) through the solid matrix. The dissolved components move through the column at different rates and because of their characteristic physical and chemical interaction with the stationary phase, they exit the column separately. The efficiency of separation is achieved by virtue of a combination of the composition of the matrix, the physical properties of the compound, and the solvent system. In high-pressure liquid chromatography, the rate of flow of the liquid phase through the densely packed microparticulate column is controlled by pressures often in the range of several thousand pounds per square inch. A rapid separation of mixtures into their component parts is the result.

High-pressure liquid chromatography has been used to quantitate levels of amphotericin B [63] and 5-fluorocytosine [9,24] in biological specimens. In the case of amphotericin B, the recovery range has been 97.3-100% from seeded samples of cerebrospinal fluid and 98.6-100% from serum. Accurate determinations of amphotericin B concentrations as low as 0.02 μg/ml of serum have been achieved [63]. Until more information is available, data derived from physical methods have to be correlated with those obtained with the bioassays so that we may know what the relationship of the levels obtained by the former has to the active drug.

C. Other Methods

In addition to the bioassay and method of high-pressure liquid chromatography, several other procedures have been described for determination of levels of antifungal agents in biological fluids. In general, claims of rapidity, sensitivity, and specificity are made for each. Many, however, are not practical at the present time since they require sophisticated instrumentation not available in most laboratories. For example, many antifungal agents have characteristic fluorometric or spectral properties in the ultraviolet region and for this reason the compounds can be detected in the extracts of serum, cerebrospinal fluids, or urine by direct measurement in a spectrophotometer in the case of polyenes [39] or a fluorometer in the case of 5-fluorocytosine [68]. Determinations that are made by these methods must be interpreted with caution. The direct fluorometric and spectrophotometric methods will also detect interfering substances that have similar spectral properties and the values derived by these procedures may therefore be misleading.

Two radiometric methods have been described for determining antifungal concentrations in biological fluids. One of these measures the production of radioactive CO_2 by a susceptible yeast grown in medium containing the patient's serum and randomly labeled [^{14}C] glucose as substrate [45]. The radiometric bioassay procedure for determining the levels of drug in body fluids has little usefulness at the present time because of its relative insensitivity compared to other procedures.

Another radiometric procedure to estimate the levels of polyenes measures the efflux of rubidium (Rb$^+$) ions or radioactive rubidium (^{86}Rb$^+$) efflux from ^{86}Rb-loaded yeast cells [26]. This method is based on the fact that polyenes bind to the sterols of the fungal membranes and cause cell membrane damage, which leads to efflux of ions from the cells. The radioactive analog of K$^+$, ^{86}Rb$^+$, can be incorporated into target fungi such as susceptible *Candida tropicalis*. When such cells are exposed to polyenes, ^{86}Rb$^+$ leaks from the cells, and the release can be measured quantitatively. This method is sensitive but necessitates the use of atomic absorption spectroscopy [23,62].

IV. SUMMARY

In general, therapeutic regimens used in the treatment of mycotic infections have been empirically derived. However, clinical experience frequently offers no basis for judging whether the appropriate drug or its dosage was given to the patient. This is because the nature of the patient population most prone to life-threatening fungal infections is such that a large number of treatment failures occur despite the fact that the organism was sensitive to the drug and appropriate doses were being administered.

In theory, antifungal susceptibility testing and measurements of drug levels in biological fluids should define the minimal amounts of an antibiotic which will either inhibit the growth of the fungus or kill it and document whether this level is present in the patient. However, there are no universally accepted standard assay procedures and clinical correlations are incomplete, so that the clinical relevance of these tests is hard to interpret. It is apparent that much more work is needed in this area, particularly since these infections are becoming more prevalent and resistant organisms are being isolated with increasing frequency.

REFERENCES

1. Allen, P. M., and D. Gottlieb. 1970. Mechanism of action of fungicide thiabendazole, 2-(4'-thiazolyl)benzimidazole. Appl. Microbiol. *20*: 919-926.
2. Andrews, F. A., W. H. Beggs, and G. A. Sarosi. 1977. Influence of antioxidants on the bioactivity of amphotericin B. Antimicrob. Agents Chemother. *11*: 615-618.
3. Andrews, F. A., G. A. Sarosi, and W. H. Beggs. 1979. Enhancement of amphotericin B activity by a series of compounds related to phenolic antioxidants. J. Antimicrob. Chemother. *5*: 173-177.
4. Arai, M., K. Ishibashi, and H. Okazaki. 1970. Siccanin, a new antifungal antibiotic. Antimicrob. Agents Chemother. *1969*: 247-252.
5. Auger, P., C. Dumas, and J. Joly. 1979. A study of 666 strains of *Candida albicans*: correlation between serotype and susceptibility to 5-fluorocytosine. J. Infect. Dis. *139*: 590-594.
6. Bannatyne, R. M., and R. Cheung. 1978. Susceptibility of *Candida albicans* to miconazole. Antimicrob. Agents Chemother. *13*: 1040-1041.
7. Benitez, P., G. Medoff, and G. S. Kobayashi. 1974. Rapid radiometric method of testing susceptibility of mycobacteria and slow-growing fungi to antimicrobial agents. Antimicrob. Agents Chemother. *6*: 29-33.
8. Bindschadler, D. D., and J. E. Bennett. 1969. A pharmacologic guide to the clinical use of amphotericin B. J. Infect. Dis. *120*: 427-436.
9. Blair, A. D., A. W. Forrey, B. T. Meijsen, and R. E. Cutler. 1975. Assay of flucytosine and furosemide by high-pressure liquid chromatography. J. Pharm. Sci. *64*: 1334-1339.
10. Block, E. R., A. E. Jennings, and J. E. Bennett. 1973. Variables influencing susceptibility testing of *Cryptococcus neoformans* to 5-fluorocytosine. Antimicrob. Agents Chemother. *4*: 392-395.
11. Bodenhoff, J. 1968. Resistance studies of *Candida albicans* with special reference to two patients subjected to prolonged antibiotic treatment. Odontol. Tidskr. *76*: 279-294.
12. Bonner, D. P., W. Mechlinski, and C. P. Schaffner. 1975. Stability studies with amphotericin B and amphotericin methyl ester. J. Antibiot. *28*: 132-135.
13. Borgers, M. 1980. Mechanism of action of antifungal drugs with special reference to the imidazole derivatives. Rev. Infect. Dis. *2*: 520-534.

14. Brandsberg, J. W., and M. E. French. 1972. In vitro susceptibility of isolates of *Aspergillus fumigatus* and *Sporothrix schenkii* to amphotericin B. Antimicrob. Agents Chemother. *2*: 402-404.
15. Brass, C., J. Z. Shainhouse, and D. A. Stevens. 1979. Variability of agar dilution-replicator method for yeast susceptibility testing. Antimicrob. Agents Chemother. *15*: 763-768.
16. Calam, D. H. 1974. Chromatographic analysis of heptaene macrolide antibiotics. J. Chromatogr. Sci. *12*: 613-616.
17. Casals, J. B. 1979. Tablet sensitivity testing of pathogenic fungi. J. Clin. Pathol. *32*: 719-722.
18. Cheung, S. C., G. Medoff, D. Schlessinger, and G. S. Kobayashi. 1975. Stability of amphotericin B in fungal culture media. Antimicrob. Agents Chemother. *8*: 426-428.
19. Cheung, S. C., G. Medoff, D. Schlessinger, and G. S. Kobayashi. 1975. Response of yeast and mycelial phases of *Histoplasma capsulatum* to amphotericin B and actinomycin D. Antimicrob. Agents Chemother. *8*: 498-503.
20. Chmel, H., and D. B. Louria. 1980. Antifungal antibiotics: mechanism of action, resistance, susceptibility testing, and assays of activity in biological fluids. In V. Lorian (Ed.), *Antibiotics in Laboratory Medicine*, Williams & Wilkins, Baltimore, pp. 170-192.
21. Collins, M. S. 1975. Inhibition of growth of *Coccidioides immitis* on Sabouraud medium containing polymyxin B. J. Clin. Microbiol. *1*: 335-336.
22. Collins, M. S., and D. Pappagianis. 1976. Treatment of murine coccidioidomycosis with polymyxin B. Antimicrob. Agents Chemother. *10*: 318-321.
23. Cosgrove, R. F., and J. E. Fairbrother. 1977. Bioassay for polyene antibiotics based on the measurement of rubidium efflux from rubidium loaded yeast cells. Antimicrob. Agents Chemother. *11*: 31-33.
24. Diasio, R. B., M. E. Wilburn, S. Shadomy, and A. Espinel-Ingroff. 1978. Rapid determination of serum 5-fluorocytosine levels by high-performance liquid chromatography. Antimicrob. Agents Chemother. *13*: 500-504.
25. Dixon, D. M., G. E. Wagner, S. Shadomy, and H. J. Shadomy. 1978. In vitro comparison of the antifungal activities of R34000, miconazole and amphotericin B. Chemotherapy *24*: 364-367.
26. Drazin, R. E., and R. I. Lehrer. 1976. [86]Rubidium release: a rapid and sensitive assay for amphotericin B. J. Infect. Dis. *134*: 238-244.
27. Drutz, D. J., and R. I. Lehrer. 1978. Development of amphotericin B-resistant *Candida tropicalis* in a patient with defective leukocyte function. Am. J. Med. Sci. *276*: 77-91.
28. Drutz, D. J., A. Spickard, D. E. Rogers, and M. G. Koenig. 1968. Treatment of disseminated mycotic infections. Am. J. Med. *45*: 405-418.
29. Edwards, J. E., Jr., J. Morrison, D. K. Henderson, and J. Z. Montgomerie. 1980. Combined effect of amphotericin B and rifampin on *Candida* sp. Antimicrob. Agents Chemother. *17*: 484-487.
30. Esterhuizen, B., and K. J. V. D. Merwe. 1977. The antifungal activity of bacillomycin S. Mycologia *69*: 975-979.
31. Fisher, B. D., and D. Armstrong. 1977. Rapid microdilution colorimetric assay for yeast susceptibility to fluorocytosine. Antimicrob. Agents Chemother. *12*: 614-617.
32. Gordee, R. S., and T. R. Matthews. 1969. Systemic antifungal activity of pyrrolnitrin. Appl. Microbiol. *17*: 690-694.
33. Green, W. R., J. E. Bennett, and R. D. Goos. 1965. Ocular penetration of amphotericin B. Arch. Ophthalmol. *73*: 769-775.

34. Grendahl, J. G., and J. P. Sung. 1978. Quantitation of imidazoles by agar-disk diffusion. Antimicrob. Agents Chemother. *14*: 509-513.

35. Hamilton, J. D., and D. M. Elliott. 1975. Combined activity of amphotericin B and 5-fluorocytosine against *Cryptococcus neoformans* in vitro and in vivo. J. Infect. Dis. *131*: 129-137.

36. Hamilton-Miller, J. M. T. 1973. The effect of pH and of temperature on the stability and bioactivity of nystatin and amphotericin B. J. Pharm. Pharmacol. *25*: 401-407.

37. Harding, S. A., G. F. Johnson, and H. M. Solomon. 1976. Gas-chromatographic determination of 5-fluorocytosine in human serum. Clin. Chem. *22*: 772-776.

38. Harrison, E. F., P. Zwadyk, Jr., R. J. Bequette, E. E. Hamlow, P. A. Tavormina, and W. A. Zygmunt. 1970. Haloprigin: a topical antifungal agent. Appl. Microbiol. *19*: 746-750.

39. Herd, P. A., and H. F. Martin. 1974. A semi-quantitative spectrophotometric assay for amphotericin B in blood. Clin. Biochem. *7*: 359-365.

40. Hoeprich. P. D., and P. D. Finn. 1972. Obfuscation of the activity of antifungal antimicrobics by culture media. J. Infect. Dis. *126*: 353-361.

41. Hoeprich, P. D., and A. C. Huston. 1976. Effect of culture media on the antifungal activity of miconazole and amphotericin B methyl ester. J. Infect. Dis. *134*: 336-341.

42. Hoeprich, P. D., J. L. Ingraham, E. Klecker, and M. J. Winship. 1974. Development of resistance to 5-fluorocytosine in *Candida parapsilosis* during therapy. J. Infect. Dis. *130*: 112-118.

43. Holt, R. J., and R. L. Newman. 1973. The antimycotic activity of 5-fluorocytosine. J. Clin. Pathol. *26*: 167-174.

44. Hopfer, R. L., and D. Gröschel. 1977. Amphotericin B susceptibility testing of yeast with a Bactec radiometric system. Antimicrob. Agents Chemother. *11*: 277-280.

45. Hopfer, R. L., and D. Gröschel. 1977. Radiometric determination of the concentration of amphotericin B in body fluids. Antimicrob. Agents Chemother. *12*: 733-735.

46. Hopfer, R. L., K. Mills, and D. Gröschel. 1979. Radiometric method for determining the susceptibility of yeasts to 5-fluorocytosine. Antimicrob. Agents Chemother. *15*: 313-314.

47. Kim, S. J., J. Kwon-Chung, G. W. A. Milne, W. B. Hill, and G. Patterson. 1975. Relationship between polyene resistance and sterol compositions in *Cryptococcus neoformans.* Antimicrob. Agents Chemother. *7*: 99-106.

48. Kitahara, M., V. K. Seth, G. Medoff, and G. S. Kobayashi. 1976. Antimicrobial sensitivity testing of six clinical isolates of *Aspergillus.* Antimicrob. Agents Chemother. *9*: 908-914.

49. Kitahara, M., V. K. Seth, G. Medoff, and G. S. Kobayashi. 1976. Activity of amphotericin B, 5-fluorocytosine, and rifampin against six clinical isolates of *Aspergillus.* Antimicrob. Agents Chemother. *9*: 915-919.

50. Kobayashi, G. S., and G. Medoff. 1977. Antifungal agents: recent developments. Annu. Rev. Microbiol. *31*: 291-308.

51. Kobayashi, G. S., G. Medoff, D. Schlessinger, C. N. Kwan, and W. B. Musser. 1972. Amphotericin B potentiation of rifampicin as an antifungal agent against yeast phase of *Histoplasma capsulatum.* Science *177*: 709-710.

52. Levine, H. B., and J. M. Cobb. 1978. Oral therapy for experimental coccidioidomycosis with R 41400 (ketoconazole) a new imidazole. Am. Rev. Resp. Dis. *118*: 715-721.

53. Levine, H. B., D. A. Stevens, J. M. Cobb, and A. E. Gebhardt. 1975. Miconazole in coccidioidomycosis: I. Assays of activity in mice and in vitro. J. Infect. Dis. *132*: 407-414.

54. Little, J. R., T. J. Blanke, F. Valeriote, and G. Medoff. 1978. Immunoadjuvant and antitumor properties of amphotericin B. In M. A. Chirigos (Ed.), *Immune Modulation and Control of Neoplasia by Adjuvant Therapy*, Raven, New York, pp. 381-387.

55. Marks, M. I., and T. C. Eickhoff. 1971. Application of four methods of the susceptibility of yeast to 5-fluorocytosine. Antimicrob. Agents Chemother. *1970*: 491-493.

56. Mazens, M. F., G. P. Andrews, and K. C. Bartlett. 1979. Comparison of microdilution and broth dilution techniques for the susceptibility testing of yeasts to 5-fluorocytosine and amphotericin B. Antimicrob. Agents Chemother. *15*: 475-477.

57. Medoff, G., M. Comfort, and G. S. Kobayashi. 1971. Synergistic action of amphotericin B and 5-fluorocytosine against yeast-like organisms. Proc. Soc. Exp. Biol. Med. *138*: 571-574.

58. Medoff, G., W. E. Dismukes, R. H. Meade, III, and J. M. Moses. 1972. A new therapeutic approach to *Candida* infections: a preliminary report. Arch. Int. Med. *130*: 241-245.

59. Medoff, G., and G. S. Kobayashi. 1980. Strategies in the treatment of systemic fungal infections. N. Engl. J. Med. *302*: 145-155.

60. Merz, W. G., and G. R. Sanford. 1979. Isolation and characterization of a polyene-resistant variant of *Candida tropicalis*. J. Clin. Microbiol. *9*: 677-680.

61. Moore, G. S., and D. M. Jaciow. 1979. *Mycology for the Clinical Laboratory*, Reston, Reston, Va., pp. 257-278.

62. Ndzinge, I., S. D. Peters, and A. H. Thomas. 1977. Assay of nystatin base on the measurement of potassium released from *Saccharomyces cerevisiae*. Analyst *102*: 328-332.

63. Nilsson-Ehle, I., T. T. Yoshikawa, J. E. Edwards, M. C. Schotz, and L. B. Guze. 1977. Quantitation of amphotericin B with use of high-pressure liquid chromatography. J. Infect. Dis. *135*: 414-422.

64. Pappagianis, D., M. S. Collins, R. Hector, and J. Remington. 1979. Development of resistance to amphotericin B in *Candida lusitaniae* infecting a human. Antimicrob. Agents Chemother. *16*: 123-126.

65. Pierce, A. M., H. D. Pierce, Jr., A. M. Unrau, and A. C. Oehlschlager. 1976. Lipid composition and polyene antibiotic resistance of *Candida albicans* mutants. Can. J. Biochem. *56*: 135-142.

66. Proctor, A. G. J., and D. W. R. Mackenzie. 1980. Laboratory control. In D. C. E. Speller (Ed.), *Antifungal Chemotherapy*, Wiley, New York, pp. 407-436.

67. Qualman, S. J., H. E. Jones, and W. M. Artis. 1976. An automated radiometric microassay of fungal growth: quantitation of growth of *T. mentagrophytes*. Sabouraudia *14*: 287-297.

68. Richardson, R. A. 1975. Rapid fluorometric determination of 5-fluorocytosine in serum. Clin. Chim. Acta *63*: 109-114.

69. Rickards, R. W., R. M. Smith, and B. T. Golding. 1970. Macrolide antibiotic studies. XV. The autoxication of the polyenes of the filipin complex and lagosin. J. Antibiot. *23*: 603-612.

70. Robinson, H. J., H. F. Phares, and O. E. Graessle. 1964. Antimycotic properties of thiabendazole. J. Invest. Dermatol. *42*: 479-482.

71. Safe, L. M., S. H. Safe, R. E. Subden, and D. J. Morris. 1977. Sterol content and polyene antibiotic resistance in isolates of *Candida krusei, Candida parakrusei,* and *Candida tropicalis*. Can. J. Microbiol. *23*: 398-401.

72.   Saubolle, M. A., and P. D. Hoeprich. 1978. Disk agar diffusion susceptibility testing of yeast. Antimicrob. Agents Chemother. *14*: 517-530.

73.   Shadomy, S. 1969. In vitro studies with 5-fluorocytosine. Appl. Microbiol. *17*: 871-877.

74.   Shadomy, S., D. M. Dixon, A. Espinel-Ingroff, G. E. Wagner, H. P. Yu, and H. J. Shadomy. 1978. In vitro studies with armbrutiein, a new antifungal antibiotic. Antimicrob. Agents Chemother. *14*: 99-104.

75.   Shadomy, S., and A. Espinel-Ingroff. 1980. Susceptibility testing of antifungal agents. In E. Lennette, A. Balows, W. J. Hauser, Jr., and J. P. Truant (Eds.), *Manual of Clinical Microbiology,* American Society for Microbiology, Washington, D. C., pp. 647-653.

76.   Slim, A., E. Bergogne-Berezin, and D. Courtoux. 1978. Sensibilité in vitro des levures opportunistes (*Candida albicans* et *Torulopsis glabrata*) à la flucytosine et courbes de concordance. Ann. Biol. Clin. *36*: 69-74.

77.   Smith, R. F., S. L. Dayton, D. Blasi, and D. D. Chipps. 1975. Incidence of yeast and influence of nystatin on their control in a group of burned children. Mycopathologia *55*: 115-120.

78.   Sreedhara Swamy, K. H., M. Sirsi, and G. Ramanda Rao. 1974. Studies on the mechanism of action of miconazole: effect of miconazole on respiration and cell permeability of *Candida albicans.* Antimicrob. Agents Chemother. *5*: 420-425.

79.   Thomas, A. H. 1976. Analysis and assay of polyene antifungal antibiotics. Analyst *101*: 321-340.

80.   Utz, C. J., and S. Shadomy. 1977. Antifungal activity of 5-fluorocytosine as measured by disk diffusion susceptibility testing. J. Infect. Dis. *135*: 970-974.

81.   Wade, D. N., and G. Sudlow. 1973. Fluorometric measurement of 5-fluorocytosine in biological fluids. J. Pharm. Sci. *62*: 828-289.

82.   Wagner, G. E., and S. Shadomy. 1977. Effects of purines and pyrimidines on the fungistatic activity of 5-fluorocytosine in *Aspergillus* species. Antimicrob. Agents Chemother. *11*: 229-233.

83.   Wagner, G., S. Shadomy, L. P. Praxton, and A. Espinel-Ingroff. 1975. New method for susceptibility testing with antifungal agents. Antimicrob. Agents. Chemother. *8*: 107-109.

84.   Warner, J. F., R. J. Duma, R. F. McGehee, S. Shadomy, and J. P. Utz. 1971. 5-fluorocytosine in human candidiasis. Antimicrob. Agents Chemother. *1970*: 473-475.

85.   Woods, R. A., M. Bard, I. E. Jackson, and D. J. Drutz. 1974. Resistance to polyene antibiotics and correlated sterol changes in two isolates of *Candida tropicalis* from a patient with an amphotericin B-resistant funguria. J. Infect. Dis. *129*: 53-58.

86.   Yamaguchi, H. 1977. Antagonistic action of lipid components of membranes from *Candida albicans* and various other lipids on two imidazole antimycotics, clotrimazole and miconazole. Antimicrob. Agents Chemother. *12*: 16-25.

9

Poisonous Fungi: Mushrooms

Paul P. Vergeer / Richmond, California

I. INTRODUCTION

A growing interest in collection and consumption of wild mushrooms over the last decade
has led to an increase in the number of poisoning cases. Such intoxications present physicians with unique problems not encountered in other types of food poisoning, yet information on mushroom poisoning in the medical literature is scant and some of the advice
on treatment may be incorrect.

In poisoning cases, suspected fungi should be expertly identified whenever possible,
because mushrooms may contain one or a combination of toxins, each requiring a specific
treatment procedure. However, fungal material is not always available. In such cases,
vomitus, gastric contents, or stools can be examined for spores which, by their size and
shape, may indicate the genus involved. Methods for laboratory analysis have been published [68,97a]. When suspected mushrooms are available, they may be tentatively identified by comparing them to color photographs in guidebooks, such as *Mushrooms of
North America* [97]. Other recommended books are listed in the references [54,71,86,
104,124-126]. Brief descriptions of 24 toxic species are included in Secs. III to IX.

Other complications confronting physicians are panic reactions which hinder proper
diagnosis, idiosyncratic responses, and drug interactions, as shown in poisonings of the
group VIII type. Age and prior health of the patient, as well as the age, preparation, and
condition of the mushrooms themselves, are all factors to be taken into consideration.

A diagnosis based on time of onset of symptoms may be difficult when several different species have been consumed. The fast-acting toxins (groups IV, V, VII, and VIII)
usually cause short-term intoxications with a favorable prognosis, sometimes even in the
absence of treatment. The opposite is true for intoxications beginning after a latent period
of 6 h or more (groups III and VI). These are by far the most serious and require swift
medical attention and hospitalization.

II. MUSHROOM POISONING–GENERAL MEASURES

Specific treatments are discussed under the various groups. General measures that may be
taken in cases of mushroom poisoning include:

A. Reduced absorption
1. If within 4–6 h of ingestion and patient alert, consider induction of emesis with
15 cc of ipecac syrup followed by warm water.
2. If not alert, lavage stomach with large-bore gastric tube using 10-30 liters of iso-
tonic saline (unless contraindicated for cardiac reasons). Most safely performed
with an endotracheal tube in place.
3. Leave activated charcoal 30-60 g in 30-60 cc of water or saline in stomach.
4. Enemas to evacuate distal colon and possible cathartics if diarrhea not already
present.

B. Increased excretion
   1. Forced diuresis with furosemide, ethacrynic acid, and/or mannitol. Careful intake and output record.
   2. Hemodialysis, peritoneal dialysis, or charcoal hemoperfusion as indicated.
C. Symptomatic measures
   1. Maintenance of the airways and arterial blood gases as indicated.
   2. Cardiovascular monitoring depending on seriousness of poisoning. May range from vital signs only to cardiac monitor, central venous pressure, or Swan-Ganz catheter.
   3. Antishock measures
      a. Plasma expansion if hypovolemic.
      b. Vasoconstrictors or vasodilators as indicated by stage of shock and current theory.
      c. Adjuvant corticosteroids in massive doses.
   4. Renal function monitoring, including indices to distinguish prerenal azotemia from renal failure (urine:plasma ratios of urea or creatinine, renal failure index, fractional excretion of sodium).
   5. Hepatic function monitoring
      a. Measures to reduce ammonia (lactulose, antibiotics, protein restriction).
      b. Adequate glucose, electrolytes, blood volume, and coagulation factors.
      c. Consider extracorporeal circulation through charcoal filter for the very ill (if available).

III. TOXIC CYCLOPEPTIDES

A. Fungi

Mushrooms containing toxic cyclopeptides are found in the genera *Amanita, Lepiota, Conocybe,* and *Galerina.* Of these, the amanitas are the most dangerous. They are widely distributed, attractive-looking fungi which because of their large size, contain large doses of toxin per mushroom. The other genera have predominantly small species which are not readily gathered for the table.

Amanita spp. usually have white spores and gills. An important characteristic of this genus is the "universal veil," a membrane that totally envelops the young fruiting bodies or buttons. At this stage of development, amanitas can be easily confused with common puffballs. A sliced amanita button, however, shows the outline of a stem and cap, whereas young puffballs are solid throughout. When the mushroom develops, the universal veil does not expand, but separates from the cap edge. The top of the veil usually adheres to the surface of the cap in the form of warts or patches, and the lower part is left as a membranous, sac-like structure around the bottom of the stem. It may easily break loose from the base when the mushroom is pulled out of the ground. This bottom part of the universal veil is called the volva. A second, or partial veil runs from the stem to the edge of the cap and protects the young gills while maturing. When the fruit body expands, the partial veil breaks loose from the cap margin and remains hanging from the apex or midportion of the stem, forming a ring or annulus. Both annulus and warts or patches may be washed off by heavy rains, making identification difficult. The toxic species of *Amanita* are:

Amanita phalloides (Fig. 1): *Cap* 7-15 cm broad, olive-green to copper-brown, convex, with or without patches. *Stipe* 8-14 cm long, 10-20 mm thick, white, smooth, at times enlarged at the base; volva membranous, persistent, sac-like, and free from the stipe. *Spores* White, round, smooth, 8-10 μm. *Habitat* usually under oaks; mainly Pacific Coast states but rapidly spreading.

Figure 1 *Amanita phalloides*. (From Ref. 34.)

Amanita verna, A. virosa, and *A. bisporigera:* *Cap* 3-10 cm broad, pure white, con-vex, patches rare. *Stipe* 14-24 cm long, 10-23 mm thick, white, finely fibrous above annu-lus, hairy below; volva as in *A. phalloides. Spores* white, round, smooth, 7-10 μm. *Habi-tat* under hardwoods or mixed woods; eastern United States.

Amanita ocreata: *Cap* 5-12 cm broad; white at first, turning pinkish to buff; convex; patches rare. *Stipe* 8-10 cm long, 15-20 mm thick, white, sometimes staining pale brown when handled, surface dry, finely fibrillose, scaly above annulus; volva loose with floccose patches, white to creamy tan. *Spores* white, ovate, smooth, 9-14 × 7-10 μm. *Habitat* under live oak (*Quercus agrifolia*); solitary to gregarious; known from California only. KOH will stain the cap and stipe of fresh material bright yellow. Tests with $FeSO_4$, HNO_3, and HCl negative.

The genus *Lepiota* has white spores and somewhat resembles *Amanita*, but can be separated from the latter by the absence of a volva and surface structure of the cap. Lepiotas have a bulbous base; the caps often have a small raised mound or boss on the center, called the umbo. The cap surface is often tomentose (covered with minute tufts of hair) or squamulose (scaly). Unlike the warts of amanitas, the hairs and scales of *Lepiota* caps are an integral part of the covering tissue, or cuticle, of the cap.

European studies have confirmed the presence of ama- and phallotoxins in the following species [50]: *L. helveola, L. castanea, L. griseovirens, L. rufescens, L. subincarnata, L. brunneoincarnata, L. heimii, L. clypeolarioides, L. brunneolilacea,* and *L. ochraceofulva;* α and γ amanitin have been found in *L. helveola* from California [11a].

Conocybe and *Galerina* species have brown spores and are small brownish mushrooms. Conocybes are usually found on or near dung, in grassy areas, greenhouses, or around natural or artificial compost piles. Galerinas are found in moss beds or growing out of decayed wood. Species of both genera have an annulus and lack a volva. The known toxic members are *Conocybe filaris, Galerina autumnalis, G. marginata,* and *G. venenata* [15, 34,86,99].

Conocybe filaris: Cap 0.5-2 cm broad, ochraceous to pale yellow, campanulate. *Stipe* 1-4 cm long, 0.5-1.5 mm thick, pale yellowish white, faintly striate, thin veil when young. *Spores* brown, smooth, ellipsoid, 10.5-12.5 X 5.5-6.8 μm for specimens with two-spored basidia, 8-10 X 4.5-5.5 μm for specimens with four-spored basidia. *Habitat* on lawns or other grassy areas; widely distributed.

Galerina autumnalis: Cap 2.5-6 cm broad, dark brown when moist, light tan when dry, convex, viscid. *Stipe* 3-8 cm long, 3-9 mm thick, light brown to fawn-colored, streaked with whitish fibrils, thin hairy veil on upper portion of stipe. *Spores* rusty brown, elliptical, minutely wrinkled, 8.5-10.5 X 5-6.5 μm. *Habitat* on well-decayed hardwood or conifer logs; scattered to abundant in groups, widely distributed.

Galerina marginata: Cap 1.5-4 cm broad, ochraceous brown to yellow-brown, convex to flat. *Stipe* 2-6 cm long, 2-9 mm thick, ochraceous brown to honey-colored, darker brown below. *Spores* rusty brown, egg-shaped, wrinkled, 8-10 X 5-6 μm. *Habitat* on dead conifer wood.

Galerina venenata differs from *G. autumnalis* in its habitat (grassy soil instead of wood) and the absence of a viscid cap cuticle. Galerinas can be mistaken for the psychoactive psilocybes, with whom they often share habitat. Psilocybes, however, have smooth spores, never roughened or warted as in *Galerina.* The spore color of *Psilocybe* is purple-brown and brown in *Galerina.*

B. Toxins

The two groups of potentially toxic substances that have been isolated from these mushrooms, the phallotoxins and amatoxins, differ basically in their chemical composition and pharmacological reactions. Animal tests have shown that the phallotoxins are fast-acting agents; when injected intraperitoneally, phalloidin will kill within the first 2 h. In contrast, amatoxins, even in very high doses, never kill within the first 15 h despite the fact that α-amanitin is known to be 10-20 times more toxic than any of the phallotoxins [93,141, 145].

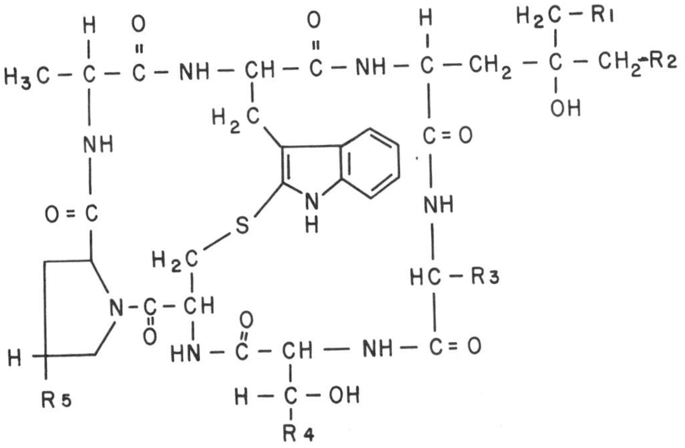

	R_1	R_2	R_3	R_4	R_5
Phalloidin	-OH	-H	-CH$_3$	-CH$_3$	-OH
Phalloin	-H	-H	-CH$_3$	-CH$_3$	-OH
Phallisin	-OH	-OH	-CH$_3$	-CH$_3$	-OH
Phallacidin	-OH	-H	-CH(CH$_3$)$_2$	-COOH	-OH
Phallacin	-H	-H	-CH(CH$_3$)$_2$	-COOH	-OH
Phallisacin	-OH	-OH	-CH(CH$_3$)$_2$	-COOH	-OH
Phallin B	-H	-H	-CH$_2$C$_6$H$_5$	-CH	-H

Figure 2 Phallotoxins.

Several methods of separation, identification, and quantitative determination of the toxins have been published, using various chromatographic methods [4,12,148]. For an evaluation of these methods, see Palyza and Kulhánek [106]. Palyza [107] more recently published a thin layer chromatographic procedure which permits detection of α-amanitin in quantities as little as 0.025 μg, the amount of toxin in approximately 0.5 mg of fresh mushroom tissue. But perhaps the most sensitive detection method developed so far has been described by Preston et al. [108]. Based on RNA polymerase II inhibition, amounts as low as 0.05 ng of α-amanitin can be detected during quantitative analysis.

Both phallotoxins and amatoxins are cyclic peptides. Phallotoxins (Fig. 2) are heptapeptides; their molecules consist of a ring of seven amino acids joined together by peptide bonds. A sulfur atom bridges the side chains of two amino acids on opposite sides of the ring (cysteine and tryptophan for phallotoxins, cysteine and hydroxylated tryptophan for amatoxins, respectively). At the present time, seven phallotoxins have been isolated: phalloidin, phalloin, phallisin, phallacidin, phallacin, phallisacin, and phallin B [78,146].

Amatoxins (Fig. 3A) are cyclic octapeptides; their molecules contain eight amino acids as well as the sulfur bridge. Members of this group include α-, β-, γ-, and ε-amanitin; amanin; amaninamide; and amanullin [42,121]. Amanullin is the only nontoxic member of the group.

Amatoxins are among the most lethal toxins found in nature; it has been determined

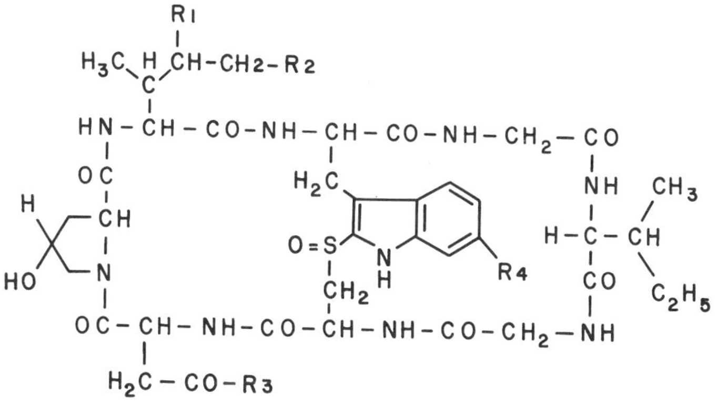

The chemical structure with substituent table:

	R_1	R_2	R_3	R_4
α-Amanitin	-OH	-OH	-NH$_2$	-OH
β-Amanitin	-OH	-OH	-OH	-OH
γ-Amanitin	-OH	-H	-NH$_2$	-OH
ϵ-Amanitin	-OH	-H	-OH	-OH
Amanin	-OH	-OH	-OH	-H
Amanullin	-H	-H	-NH$_2$	-OH
Amaninamide	-OH	-OH	-NH$_2$	-H

Figure 3A Amatoxins.

that the average 100 g of fresh European *Amanita phalloides* contains approximately 8 mg of α-amanitin, 5 mg of β-amanitin, and 0.5 of γ-amanitin [145]. When we consider the lethal dose to be about 0.1 mg/kg body weight, it then follows that a single mushroom weighing 50 g can be fatal for a 150-lb human adult [47a,140,142,143,145].

Other toxic or potentially toxic agents have been found in *Amanita* spp., for instance, the hemolytic cytotoxin, phallin, discovered in 1891 by Kobert [74]. Phallin has been further studied by Seeger and Scharrer [117], who renamed it phallosin, and by Faulstich et al. [41]. This compound is extremely toxic, causing death exclusively by erythrocyte hemolysis when injected into test animals [78]. Phallosin is, however, thermolabile and acid labile, in contrast to the amatoxins and phallotoxins, and is therefore rapidly removed when cooked or otherwise destroyed by the digestive juices. In 1980, Faulstich et al. reported the discovery of a unique toxin group found exclusively in *Amanita virosa* [42a]. Virotoxins are actin-binding cyclic monopeptides, whose structures and biological activity resemble in part the phallotoxins. They differ, however, in containing D-serine instead of L-cysteine, and lack the sulfur atom bridge found in the amatoxin and phallotoxin peptide rings. Two other amino acids, not found in nature before, were detected: 2,3-trans-3,4-dihydroxy-L-proline and 2'-(methylsulfonyl)-L-tryptophan. Virotoxin peptides have amino acids with both D and L configurations. It is speculated that these stereo configurations enable binding of the toxins to take place through "induced fit," the ability to adopt the active conformation only on contact with actin. So far, only the

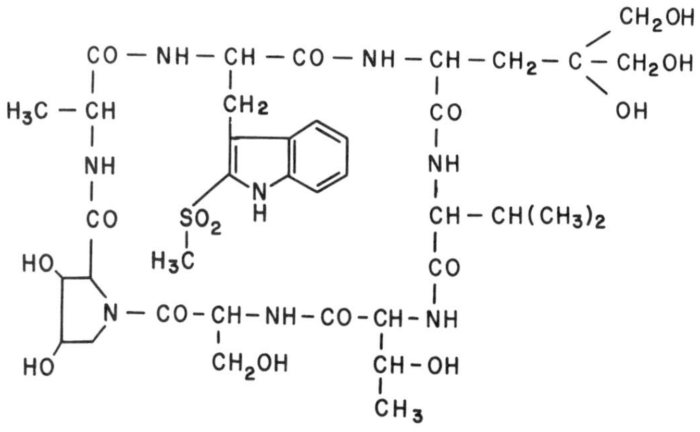

Figure 3B   Viroisin.

main component of the virotoxin group, viroisin, has been fully analyzed and its structure elucidated (Fig. 3B).

C. Mechanisms of Action

Several organs, including liver and kidneys, are damaged by amatoxins, whereas the action of phallotoxins is limited to the liver. There is a question, still not fully answered, as to whether phallotoxins are involved in human poisoning. Some authors believe that these toxins are eliminated from the body before harm can be done. Perhaps this hypothesis is based in part on laboratory observations. It has been found that when phalloidin was injected into rats and mice in doses of 5 mg/kg body weight, death occurred within 2-5 h. When the toxin was administered orally, no poisoning symptoms were observed within the first 8 h, nor has death been reported with oral doses [93]. In the absence of amatoxins, certain phallotoxins have also been found in a number of edible mushrooms without having caused human poisoning [1].

There is also a naturally occurring antagonist, antamanide, present in *A. phalloides* [144]. In a laboratory study, antamanide in doses of 0.5 mg/kg body weight administered to white mice proved sufficient to counteract lethal doses of phalloidin given simultaneously [146]. However, the amount of antamanide actually present in the mushroom is too low to be of any clinical significance.

Laboratory studies have shown that phallotoxins quickly attach themselves to the membrane of isolated rat liver cells (Figs. 4 and 5), where they cause swelling and rapid deformations of the surface within 10 min [49,52,53,138]. This is followed by a marked loss of potassium and lysosomal enzymes [48]. Loss of potassium is a result of the formation of potassium channels and not of inhibition of enzymes of a cation pump. Uptake of extracellular fluids into the phallotoxin-poisoned hepatocyte occurs by way of the endoplasmic reticulum [52,138].

Whereas action of the phallotoxins is limited to cells of the liver, action of the amatoxins shows up at multiple sites. Organs that may be affected include the liver, kidneys, pancreas, adrenal glands, muscles, and the brain [2,46,47,79,80,145]. Of the amatoxins,

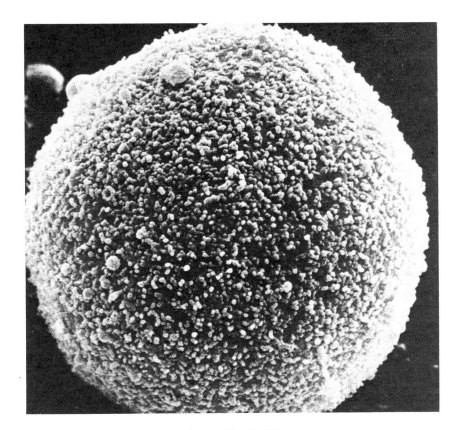

Figure 4   Isolated rat liver cell (control). 4950X.

α- and β-amanitin are the most toxic [140]. Both show similar mechanisms of action [47]. Cells exposed to α-amanitin will not show damage to the cell membranes, since the toxins do not bind there; instead, amanitins quickly penetrate the cell and immediately attack the nucleus [93]. The damage here has been summarized by Fiume [45] and includes: "chromatin condensation, breakup of nucleoli with separation of the fibrils from granules, followed by a progressive fall in nucleolar RNA content, temporary increase in the number of perichromatin granules, and accumulation of interchromatic granules at the center of the nucleus."

Alpha-amanitin inhibits, by as much as 80%, the action of the enzyme RNA polymerase II [88,101,102]. The toxin interferes with the action of the enzyme, but is not bound to it. The inhibition is furthermore limited to liver cells. Such specific and "limited" action of this toxin has made it a valuable tool in molecular and biological research. RNA polymerase II catalyzes RNA synthesis. When this enzyme is inhibited by amanitin, the inhibition prevents successful transcription of DNA, the fundamental process by which information of the cell is "read," and this inhibition then results in the termination of protein synthesis and cell death [47].

When the toxin enters the kidneys it is not completely filtered out of the body, but rather attacks the convoluted tubules and is then reintroduced into the bloodstream, to

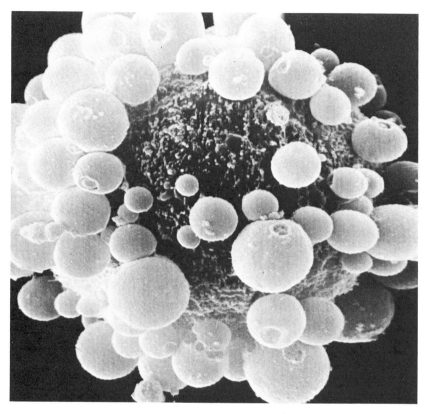

Figure 5 Isolated rat liver cell treated with 1.2×10^{-5} M phalloidin (10 min at 37°C). 4050X.

return eventually to the kidneys. Much of the damage to organs is caused by this recycling process [93].

D. Clinical Picture

Three phases can be distinguished during the course of intoxication by mushrooms of this toxin type: latent, gastrointestinal, and hepatorenal [34,93,146].

The latent phase covers the period between mushroom ingestion and onset of the first poisoning symptoms, and is characteristic for this type of poisoning. The time lapse may vary between 5 and 24 h, with the majority of patients reporting onset of symptoms during the twelfth hour [35]. There appears to be a correlation between time of onset and the number of fatalities, with most recorded deaths occurring in patients whose symptoms started during the first 12 h [73,82]. A long "time-effect" sequence may therefore indicate a favorable prognosis.

The gastrointestinal phase usually begins with acute abdominal pain, followed by diarrhea and vomiting. The diarrhea is usually profuse and watery, often painful and bloody, but in a few patients it is painless or minimal [35]. It was first reported that agents other than the cyclopeptides were responsible for the gastrointestinal symptoms

[34,146]. This hypothesis was based on results obtained from tests of the toxin on mice, rats, and guinea pigs. It does not take into account the fact that these animals never experience vomiting or diarrhea when they have been given the toxins, either orally or by injection [145]. Dogs, on the other hand, do experience such symptoms [31], and for this reason Fiume and co-workers used dogs in their amanitin experiments, with positive results [46]. The gastrointestinal phase started around the tenth hour for most of the dogs. Postmortem sections revealed shrinkage and rupture of cell nuclei in the mucosa of the stomach and intestinal tract. At the same time, mice used in the same experiment showed none of these lesions. It was therefore concluded that the "difference between the effects of α-amanitin on dogs and mice is probably due to its being able to penetrate into some of the cells in the gastro-intestinal canal in the former but not the latter. In the dog the probable order of events is: 1. α-amanitin enters into some of the cells in the gastric and intestinal mucosa, damaging them. 2. This damage leads to gastro-enteritis, with vomiting and diarrhea" [46]. Perhaps a similar mechanism of action takes place in human poisoning.

Other symptoms and signs noted during this second phase are nausea, dehydration, hypotension, oliguria, and hypokalemia. Occasionally, sweating, mental confusion, and muscle cramps (especially in the legs) are experienced as well [35]. Death can occur during the second phase; however, patients usually experience a short remission of symptoms instead, between the second and fourth day, only to relapse and enter the final hepatorenal phase.

During the hepatorenal phase, internal bleeding will increase. A decline in blood pressure can be expected. Other symptoms may include development of icterus of rapidly increasing severity, recrudescent diarrhea and abdominal pain, oliguria or anuria, and tonic-clonic spasms. When the victim does not respond to medication, blood pressure will continue to drop to shock level, at which time the patient will become comatose. Death, mostly from hepatic coma and/or renal failure, occurs between the fourth and seventh day.

E. Treatment

Conservative measures alone will reduce the mortality rate to well under 50%. In fact, the success of such measures has made it difficult to judge the effect, if any, of newly proposed "specific" treatments whose value has been compared only to historical controls.

Ordinarily, amanitin poisoning is discovered too late for gastric or colonic lavage to be useful, but these might be of some help up to 6-8 h after mushroom ingestion.

Hydration with replacement of water, Na^+, K^+, Ca^{2+}, Cl^-, and HCO_3^- are essential. In very ill persons consideration should be given to central venous pressure monitoring (or Swan-Ganz catheter), especially if dextran or transfusions are indicated for shock or depletion of coagulation factors. Serum glutamic-oxaloacetic transaminase (SGOT), serum glutamic-pyruvic transaminase (SGPT), blood ammonia, complete blood count, electrolytes, and prothrombin time should be checked frequently in order to follow the patient's course and to indicate any need for additional treatment. Most of the therapeutic maneuvers are fairly standard: $NaHCO_3$ or tromethamine for metabolic acidosis, intestinal antibiotics and protein restriction for hepatic failure, 5-25% mannitol for impending oliguria, intravenous glucose for prevention of hypoglycemia, and dopamine and similar agents for euvolemic shock.

Renal dialysis is unlikely to remove much of the toxin and is best reserved for patients who have progressed from prerenal azotemia to renal failure. Charcoal hemoperfusion has been tried, but the number of cases has been too small to evaluate its usefulness definitively [35]. There is at this time no good evidence that exchange transfusions are helpful. Although ventricular arrhythmias are rare, they have been documented, and seriously ill persons should be on a cardiac monitor. Intravenous lidocaine seems a reasonably safe drug in this circumstance.

Various antibiotic and antibacterial agents have been used for suppression of ammonia-producing bacteria in the colon. These agents include neomycin, kanamycin, tetracycline, and nonabsorbable sulfa drugs. All are potentially toxic, especially in the presence of damaged bowel, where toxic amounts can easily be absorbed.

A more debatable but also nonspecific treatment is that of corticosteroids. Hydrocortisone and its analogs have been shown to stabilize the lysosomal membranes in cells. Since the enzymes within lysosomes may be destructive to other cells, stabilization of the membrane of these intracellular organelles may provide protection against additional cytotoxicity. Large doses of corticosteroids (e.g., 20-40 mg of dexamethasone or even larger if shock is present) may be given intravenously to very ill patients.

A number of specific antidotes have been recommended, including cytochrome c and uridine diphosphogluconate. Only two types will be discussed here: protein conjugation blockers and thioctic acid.

In experimental animals protection against cyclic polypeptides is afforded by giving (before or together with the toxin) a variety of compounds that prevent conjugation of amanitins to tissue and red cell protein by competition for the binding sites. Many drugs in fact bind to such receptors, including penicillin, rifampin, chloromycetin, sulfonamides, and phenylbutazone. No evidence has been shown to date that these drugs help when given after the onset of symptoms. Nonetheless, on the basis of a tenuously postulated recirculation of protein-bound toxin, high doses of penicillin G (250 mg/kg per day) have been given.

Thioctic acid, which is highly controversial, at least has no known side effects except hypoglycemia (easily remedied by 5-10% glucose solutions). Thioctic or α-lipoic acid (Fig. 6) is a sulfated derivative of octanoic acid, and acts as a coenzyme in some oxidative decarboxylation reactions found in liver tissue. Reduction with formation of sulfhydryl groups could in theory bind toxic polypeptides. Results of its use in many liver disorders were marginal, until Kubička reported on its success in the treatment of *A. phalloides* poisoning [76]: of 40 patients treated, 39 survived; SGOT, SGPT, and serum bilirubin were measured twice daily, and used to determine dosage levels of thioctic acid to be employed. Initially, thioctic acid was started in a dose of 75 mg/day as a continuous infusion. As soon as an increase in the transaminases was noted, the amount of thioctic acid was increased (300 mg/day) and maintained at the higher dose level until the SGOT and SGPT started to decrease.

The difficulties of evaluating this type of treatment are: (1) conflicting reports on its

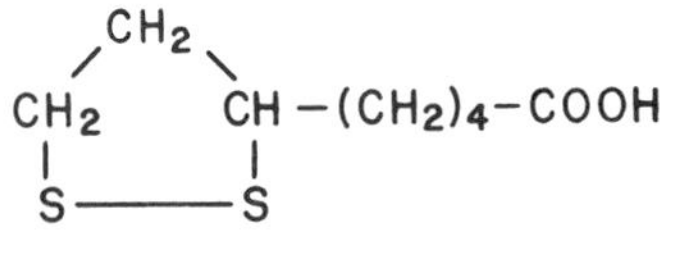

$$
\begin{array}{c}
\quad\;\; CH_2 \\
\diagup \qquad \diagdown \\
CH_2 \qquad\;\; CH-(CH_2)_4-COOH \\
\mid \qquad\qquad \mid \\
S \rule{1.5cm}{0.4pt} S
\end{array}
$$

Figure 6   Thioctic acid.

effectiveness [27,28,32,33,105]; (2) no control group has ever been employed; and (3) thioctic acid is always used in conjunction with a large number of other therapeutic drugs, making it practically impossible to judge its effectiveness.

Thioctic acid was first used in the United States in 1970 [44]. Becker et al. [4] have reported on its use in California. The drug has not been released for general use in this country but can be obtained from Dr. F. C. Bartter [Memorial Veterans Hospital, San Antonio, Texas 78284, (512) 696-9660 ext. 6463], who has the investigational new drug (IND) application of the Food and Drug Administration. The protocol must be followed carefully. It is important that the intravenous tubing and bottle be wrapped with aluminum foil, as thioctic acid decomposes readily in light.

IV. IBOTENIC ACID—MUSCIMOL

A. Fungi

Mushrooms belonging in this group are mostly amanitas: *Amanita cothurnata, A. crenulata, A. frostiana, A. muscaria, A. pantherina, A. smithiana,* and *A. strobiliformis* [71,86, 141]. *Tricholoma muscarium,* endemic to Japan, contains both ibotenic acid and tricholomic acid, a compound with insecticidal properties. Tricholomic acid is the erythro-2,3-dihydro derivative of ibotenic acid [141].

By far the most common and best known species are *A. muscaria* and *A. pantherina.* Ethnomycological studies on the use of *A. muscaria* as an inebriant in the rites of primitive peoples have been published [62,115,137]. *Amanita muscaria* and *A. pantherina* are also used as recreational drugs in the United States [34,78].

Figure 7   *Amanita muscaria.* (From Ref. 34.)

Amanita muscaria (Fig. 7): *Cap* 8-24 cm broad, red, orange to yellow-orange (depending on variety), white to yellow-white warts, convex to flat in age. *Stipe* 5-18 cm long, 2-3 cm thick, white, bulbous base with remains of volva appearing as concentric rings around upper part of bulb. *Spores* white, ellipsoid, thin-walled, smooth, 9-11 × 6-9 μm. *Habitat* under hardwoods and conifers, scattered to abundant; widely distributed throughout the Northern Hemisphere.

Amanita pantherina: *Cap* 5-12 cm broad, brown, pointed pyramid-shaped warts, convex to flat in age. *Stipe* 6-12 cm long, 1-2.5 cm thick, white; volva adhering to round bulb at the base of the stipe, with upper part forming a prominent collar. *Spores* white- elliptical, thin-walled, smooth, 10-12 × 7-8 μm. *Habitat* under conifers or mixed woods, scattered to abundant; widely distributed from the Rocky Mountains westward.

A special problem concerning toxicity in *Amanita* arises when toxic and nontoxic species hybridize. An example of this are the hybridized color forms of *A. pantherina* and *A. gemmata, A. pantherina-gemmata.* Benedict et al. [6] found no toxins in specimens identified as "pure" *A. gemmata.* There were, however, a number of specimens with cap colors ranging from brown (*A. pantherina*) to yellow-orange (*A. gemmata*) that did possess quantities of toxins intermediate to the amounts found in *A. pantherina* and *A. gemmata.* This quantitative variation of toxins in the hybrids may account for the widely varying reactions reported after ingestion of mushrooms identified as *A. pantherina,* which, in its pure dark form, is very poisonous.

B. Toxins and Mechanisms of Action

The chemical substances found here can be loosely divided into two artificial groups: the gastrointestinal irritants and the psychoactive agents. The irritants have not been identified. The parasympathetic stimulant muscarine, although named after *A. muscaria,* is present only in trace amounts (0.00025% in fresh mushrooms). Occasionally, specimens of *A. muscaria* have caused muscarinic symptoms such as salivation and sweating; these possible chemical variants, however, have never been assayed. There are no reports of muscarinic symptoms from *A. pantherina.* Muscarine does not enter the brain and is therefore not responsible for the psychoactive or central neurological actions of these mushrooms [8].

The psychoactive agents that have been determined are ibotenic acid (Fig. 8), muscimol (Fig. 9), and muscazone (Fig. 10) [136]. Other amino acids which are sometimes present are stizolobic acid, stizolobinic acid, aminohexadienoic acid, and chlorocrotylglycine [23,25]; the last two compounds have been found in *A. smithiana* exclusively, and only from the state of Washington.

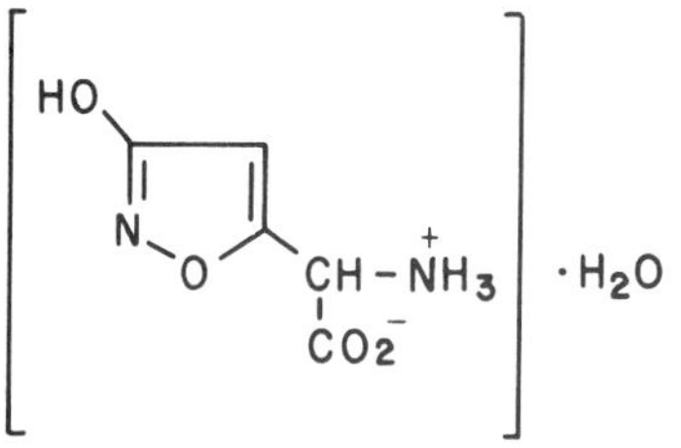

Figure 8   Ibotenic acid.

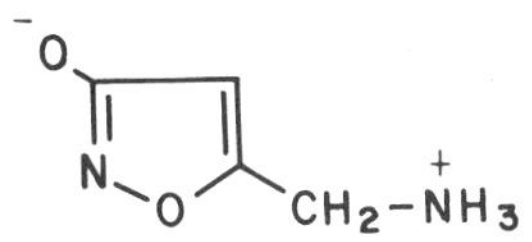

Figure 9 Muscimol.

Ibotenic acid derives its name from the Japanese common name ibotengu-take (*A. strobiliformis*); *ibo* = warted, *tengu* = long-nosed goblin, and *take* = mushroom. Ibotenic acid is an unstable and thermolabile amino acid which is readily decarboxylated to muscimol through loss of H_2O and CO_2. Muscazone has been found only in European specimens [8]. Muscazone is the ultraviolet irradiation rearrangement derivative of ibotenic acid [78]. The chemistry of these compounds was elucidated independently and simultaneously in Europe [13,14,100] and Japan [130,131].

There is a marked variation in the amount of toxins per specimen as well as distribution and location of the toxins inside the mushroom; however, it is believed that the highest concentration of ibotenic acid-muscimol are found directly below the cuticle, or covering layer of the cap [21]. In certain European countries, *A. muscaria* was consumed during times of famine; the mushrooms were cooked and the water discarded.

Quantitative estimates of total isoxazole extracts (ibotenic acid + muscimol) are facilitated by a distinctive color change in the ninhydrin-isoxazole complex. Extracts are prepared in the usual manner. The chromatograms, sprayed with ninhydrin and dried, will develop a bright yellow spot which slowly changes to purple [8]. Amounts thus estimated averaged about 0.18% for dried *A. muscaria* and 0.46% for dried *A. pantherina* [6].

Muscimol produces electroencephalographic alterations which resemble those evoked by anticholinergic drugs such as atropine [116]. However, muscimol does not possess peripheral atropine-like activity nor has physostigmine been found to have a significant effect on the EEG changes induced by muscimol, in complete contrast to its ability to counteract the change induced by atropine and atropine-like drugs [78]. Neither ibotenic acid nor muscimol appears to act on central nervous system acetylcholine, dopamine, or 5-hydroxytryptamine receptors. It has been suggested that ibotenic acid and muscimol act similarly by inhibiting the γ-aminobutyric GABA receptor [16,72,135].

C. Clinical Picture

The "pantherine syndrome" exemplified by the effects of *Amanita muscaria* and its allies usually appears within 20-90 min after ingestion, but may be delayed as long as 6 h. Poisonings from *A. pantherina* ingestion usually are more serious and complex than those resulting from *A. muscaria* consumption. Nevertheless, the reputation of these mushrooms as being "deadly" poisonous in either case is unwarranted.

Among the symptoms to be expected are drowsiness and nausea, at times followed by

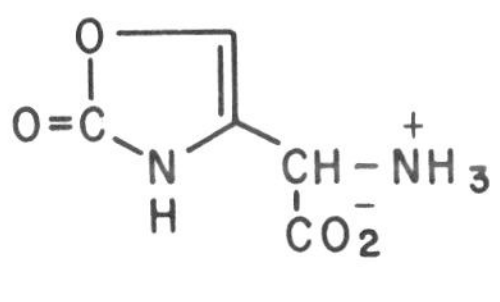

Figure 10 Muscazone.

gastroenteritis, elation, increased activities, and illusions. The illusions are often interpreted as hallucinations, but they are in fact misinterpretations of sensory stimuli, such as changes in color vision, echo images (seeing through walls), and identifying persons as divine figures. Subjective changes in bodily strength, distractibility, and retrograde amnesia are also noted at times. It is not unusual for patients to fall asleep for varying periods of time. According to a report of intentional ingestion of ibotenic acid, the most pronounced effects did not manifest themselves until the second or third hour, by which time most of the drug ingested had appeared in the urine [24]. Symptoms gradually subside over a period of 24-48 h.

D. Treatment

Within the first 4-6 h after ingestion (and without spontaneous vomiting), emesis may be induced by giving 1-2 tablespoons of ipecac syrup in adults, or an appropriate smaller amount in children. In the obtunded patient, copious lavage with a large bore gastric tube followed by 30-60 g of activated charcoal is useful. However, it is safer to precede such lavage with endotracheal intubation to prevent aspiration. Enemas or standard cathartics may be employed if diarrhea has not spontaneously occurred.

Supportive measures include the usual measures to maintain the airway, circulation, acid-base balance, and renal output. Very rarely is any other treatment required. Seizures, primarily seen in children, may be controlled by intravenous diazepam or diphenylhydantoin. A cautionary note relating to diazepam: two recent studies have found that small doses of diazepam, as well as phenobarbital, induced flaccid paralysis and an EEG pattern similar to deep anesthesia in muscimol-treated laboratory animals [116,132]. To date, this reaction has not been observed in humans. Chlorpromazine lowers the seizure threshold and is best avoided. Atropine is contraindicated.

V. MUSCARINE

A. Fungi

For almost a century, muscarine was believed to be the exclusive toxin in *Amanita muscaria;* however, this species contains only trace amounts of this compound (see Sec. IV. B). It is now known that muscarine is actually widespread in genera of the Agaricales and in certain boletes, but in many of these genera it is present in trace amounts only. On the other hand, most species of *Inocybe,* some *Clitocybe* spp., and two *Omphalotus* spp. contain toxin levels high enough to cause human poisoning. Stadelmann et al. [127] found muscarine and two of its isomers, epimuscarine and allomuscarine, in the following genera: *Amanita, Boletus, Clitocybe, Clitopilus, Collybia, Hygrocybe, Hypholoma, Lactarius, Mycena, Paxillus, Rhodophyllus, Russula, Tricholoma,* and *Tylopilus.*

Clitocybes have white to pinkish spores. The fruit bodies are firm, fleshy, and have decurrent gills (extending down the stipe). The mushrooms vary substantially in their sizes: caps 1.5-20 cm in width, stipes 1-10 cm in length and 0.1-4 cm in diameter. Clitocybes have a large color range, even within the same species; for instance, *C. aurantiaca* may range from nearly white to brown or orange-brown. The caps are convex to flat when young, becoming depressed in the center, with inrolled margins. Most of the species can be found growing directly on the ground.

Clitocybe dealbata: Cap 2-5 cm broad, pallid to dull white, convex to plane, center depressed in age. *Stipe* 4-7 cm long, 4-6 mm thick, white, smooth. *Spores* white, short elliptical, 4-6 × 3-4 μm. *Habitat* in grassy places or on leaves in open woods, solitary to abundant; smell and taste mealy.

Other suspected or muscarine containing species include *C. angustissima, C. aurantiaca, C. candicans, C. cerussata, C. ericetorum, C. festiva, C. gibba, C. hydrogramma, C. nebularis, C. phyllophila, C. rivulosa, C. suaveolens, C. truncicola,* and *C. vermicularis* [8,34, 86,99,127].

The genus *Omphalotus* closely resembles *Clitocybe*. The species of this genus can be found growing in large clusters at the base of hardwood trees. Because of their orange-like colors and decurrent gills, they have been mistaken for a look-alike edible species, *Cantharellus cibarius,* but this mushroom has decurrent ridges instead of gills.

Omphalotus olearius and the somewhat similar looking *O. olivascens* both contain muscarine. These species have been mistakenly called *Clitocybe illudens* in older books. *Omphalotus olivascens* is a Pacific Coast species which can be distinguished from the eastern and European *O. olearius* by its duller orange to yellow-orange colors, which have distinctive olive overtones or stains.

Omphalotus olearius: *Cap* 7-11 cm broad, bright orange to orange-yellow without olive overtones, convex to flat, depressed in center, often with a shallow knob, dry, smooth. *Stipe* 5-18 cm long, 5-22 mm thick, light orange, tapering to a narrow base, minutely downy or scaly in age. *Spores* white, round, smooth, 3-5 μm. *Habitat* at the base of hardwood stumps or from buried roots, growing in clusters. *Omphalotus* species are luminescent. This phenomenon can only be seen when the mushroom is taken into a totally dark closet.

The distribution of muscarine among species of *Inocybe* is widespread [96]. Detectable quantities of toxin vary from 0.01 to 0.8% in the fresh mushroom. Muscarine and its isomers can be found in at least 75% of all species tested [39]. Fatalities have not yet been reported from the United States, but *I. patouillardi* is listed in European literature as causing death [37].

Inocybes have umber-brown spores; spore dimensions range from small to large and their forms are quite variable, ranging from ovoid to ellipsoid to almost angular with nodulose swellings. Most inocybes are small to moderately sized mushrooms with conic, often umbonate caps (Fig. 11), 15-50 mm broad; most species have a cap surface which is fibrillose (covered with long hairs radiating from the center to the margin). The cap colors usually are some shade of brown, although a few are pure white to lilac to blackish brown. Inocybes are terrestrial and humicolous fungi and many species have an unpleasant chlorine-like odor.

B. Toxins and Mechanisms of Action

While searching for the toxic principle of *Amanita muscaria,* a compound was found in 1869 by Schmiedeberg and Koppe [114] which they named "muscarine," making this one of the oldest known mushroom toxins. Yet, further knowledge of the nature of this drug did not come till the 1950s, when Eugster and Waser [36], Kögl et al. [75], and Hardegger and Lohse [58] finally resolved the chemistry of this relatively simple amine. Muscarine chloride (Fig. 12) was obtained in crystalline form and its composition determined as $(C_9H_{20}NO_2)^+$. Three years later, using x-ray crystallographic techniques, the structure of the molecule was settled, showing it to be the oxoheterocyclic quarternary salt 2-methyl-3-hydroxy-5-trimethyl-ammoniummethyltetrahydrofuran chloride with S

Figure 11 *Inocybe sororia.*

configuration at position 2, R at position 3, and S at position 5 of the ring. Because of the three centers of asymmetry, four diastereoisomeric racemates and eight antipodes were anticipated and nearly all have now been obtained [8,141]. Eugster and Schleusener [40] have extracted allo-, epi-, and epiallomuscarine from fresh basidiocarps. Stadelmann et al. [128] found stereoisomeric muscarines (muscarine, epimuscarine, and allomuscarine) in the mycelium of cultured *Clitocybe rivulosa, C. dealbata, C. festiva,* and *Amanita muscaria.* Epiallomuscarine could not be detected in the mycelium of any of the fungi tested.

The parent compound, the L$^{(+)}$ isomer of muscarine, appears to be the most, and perhaps only, toxic alkaloid in human poisoning. It exhibits a highly specific activity at postganglionic parasympathetic effector sites which can be measured by the decrease of blood pressure in anesthetized cats with amounts of as little as 0.01 μg/kg.

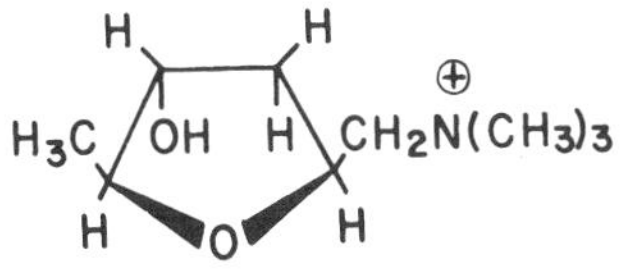

Figure 12   Muscarine.

In 1961, Malone et al. [95] developed a bioassay method, using albino rats, to determine the presence of stereoisomeric muscarines. The following year, Brown et al. [17] published on the detection of muscarine with the use of thin layer chromatography. The bioassay invariably shows higher muscarinic activity than does the chromatographic method. This apparent discrepancy is the result of measuring total muscarinic activity rather than individual muscarine-like compounds as is done with the TLC method. The bioassay method discloses physiological potentiators such as acetylcholine, whose structure is very similar to that of muscarine.

C. Clinical Picture

The onset of the symptoms usually occur within 15-30 min after ingestion, rarely beginning after 1 h. Muscarine is stable in boiling water and the method of preparation therefore has little effect on the severity of the intoxication. Muscarine acts like other drugs which stimulate the parasympathetic nervous system, but the quantity of ingested toxin will determine which symptoms are elicited. Symptoms include profuse sweating (most predominant and commonly reported feature), salivation, nausea, vomiting, and abdominal colic. In a few cases, probably due to higher concentrations of ingested muscarine, the following symptoms have also been reported: abnormally contracted pupils and blurred vision, diarrhea, a slowed heart beat, and dizziness.

D. Treatment

Since onset of symptoms occurs early, gastric lavage and activated charcoal may reduce the poison burden. Symptoms may then be titrated with atropine until all sweating and diarrhea have ceased. Pupillary constriction may disappear first. The initial dose in an adult is 2 mg intravenously, and in a child, 0.05 mg/kg body weight. Replacement of lost fluids and electrolytes is indicated, but great care must be taken in the presence of pulmonary edema to avoid fluid overload.

VI. MONOMETHYLHYDRAZINE

A. Fungi

Mushrooms in this group are found in the genus *Gyromitra* and in suspected species of the genera *Helvella, Disciotis,* and *Sarcosphaera* [86]. These genera are all members of the class Ascomycetes.

   Gyromitra and *Helvella,* or "false morels," have wavy or folded, often saddle-shaped caps; they lack gills. The colors of the caps are usually some shade of white, buff, or black. The species of these genera are all terrestrial fungi. A closely related genus, *Morchella* or "true morels," has geometrically pitted, rather than folded, caps. The caps, which are buff to brown, are conic to more or less rounded in shape. True morels are highly priced edibles, but even they should be well cooked. There have been some reports of poisonings involving raw morels; however, the toxic agent(s) have not been investigated or confirmed.

Gyromitra esculenta: Cap 2-8 cm broad, brown to red-brown, irregularly lobed, wrinkled, flesh brittle. *Stipe* 3-15 cm long, 10-20 mm thick, short and stout becoming hollow or chambered, surface slightly floccose, pale flesh-colored to grayish lavender. *Spores* white, smooth, elliptical, often containing two or more oil globules, 18-24 $\times$ 9-14 μm. *Habitat* under conifers, especially on sandy soil, solitary or in groups; widely distributed.

There is no doubt concerning the toxicity of this species (the toxic compound was originally extracted from *G. esculenta*); however, other species of the genus have been described as either toxic or edible in various mushroom guides. For instance, *Gyromitra gigas* is listed by Miller [97] as a good edible species, while Benedict [8] reports the existence of gyromitrin in this mushroom. The conflicting reports on toxicity may be the result of either misidentification, varying levels of toxin in identical species, or the occurrence of chemical variants in different geographical regions. The composition of the soil in which these mushrooms grow may also influence the chemicals found in the various specimens. Suspected *Gyromitra* species include *G. ambigua, G. brunnea, G. californica, G. caroliniana, G. fastigiata, G. infula,* and *G. sphaerospora* [86].

Helvella lacunosa, Disciotis venosa, and *Sarcosphaera crassa* have been reported as toxic when eaten raw; however, toxins have not yet been isolated from these species. It is speculated that these fungi may contain or can on occasion synthesize the same or similar toxins found in *Gyromitra.*

Helvella lacunosa: Cap 2-6 cm broad, dark brown to black, saddle-shaped with lobes, soon becoming undulate or convoluted, flesh brittle. *Stipe* 4-10 cm long, 15-20 mm thick, blackish to pallid, deeply ribbed, cylindrical or slightly swollen below, hollow or chambered. *Spores* white, elliptical, smooth, 17-20 $\times$ 11-13 μm. *Habitat* on soil, often on burnt areas, solitary or in groups.

Disciotis venosa: Cup large, up to 15 cm across when fully expanded, hymenial surface dark brown, in large specimens always more or less characterized by ribs and furrows radiating from the center, exterior whitish or cream-colored, minutely scurfy. *Stipe* short and stout, often sunk in the soil. *Spores* white, elliptical, 19-25 $\times$ 12-15 μm. *Habitat* on soil or on shady banks and lawns.

Sarcosphaera crassa: Cup of mature specimens can be large, up to 18 cm broad, develops under the soil surface as a hollow, smooth white ball, which splits open in a stellate manner to expose the violet-colored hymenium; flesh white, fragile, usually with fragments of soil attached. *Stipe* when present short and thick. *Spores* white, elliptical with abruptly truncated ends, 13-15 $\times$ 7-8 μm. *Habitat* under conifer or deciduous trees, solitary or clustered, widely distributed.

B. Toxins and Mechanisms of Action

The toxic component monomethylhydrazine (MMH) is not present in fresh mushroom tissue but is formed by hydrolysis of gyromitrin (Fig. 13), a volatile, acid- and base-labile, autooxidizable compound with the formula $C_4H_8N_2O$. Gyromitrin was first isolated from

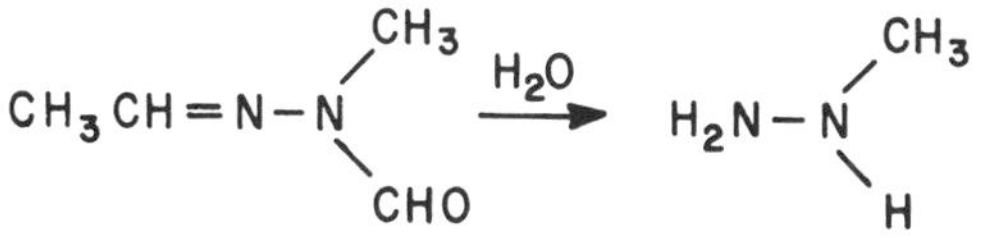

Figure 13 Conversion of gyromitrin to monomethylhydrazine by hydrolysis.

Gyromitra esculenta in 1967 by List and Luft [90,91]. One kilogram of fresh specimens was found to contain an average of 1.2-1.6 g of gyromitrin [92]. A chromatographic separation and spectral analysis of the eluent has been published [51].

Because of its volatile nature, gyromitrin is removed from the mushrooms by cooking or drying; however, sufficient amounts can be retained in the cooking liquid to cause intoxication when it is eaten with the mushrooms, as in stew or soups [34].

Consequences of MMH poisoning can be grouped as follows: gastrointestinal disturbance, liver and kidney damage, blood damage, central nervous system (CNS) disturbances and possible carcinogenesis. As a gastrointestinal irritant MMH will cause nausea, vomiting, diarrhea, and cramping. Liver damage is often quite variable in occurrence and difficult to assess. It may also be more severe in patients who are genetically disposed to metabolize hydrazines to acetyl derivatives [121]. Renal damage from MMH has been found in dogs and humans only, but it has not yet been determined whether the damage is caused directly by the toxin or as a secondary result of hemolysis [86]. Blood damage shows itself by methemoglobin formation and hemolysis. The early detection of methemoglobin and free hemoglobin in the blood occurs early and is therefore a good indicator of MMH exposure [86].

CNS disturbances have been reported with MMH intoxications. These include convulsions, coma, and respiratory failure. The exact mechanism of action is still not known. Simons [121] suggests that MMH acts on the CNS by interfering with the normal utilization and function of vitamin B_6 and hypothesizes that MMH would react with pyridoxal to form pyridoxal methylhydrazone. He goes on to say:

> The 5'-phosphate derivatives of the B_6-vitamers are cofactors in many important enzymatic reactions: transamination, decarboxylation, dehydration of serine, desulfurization of cysteine, etc. Pyridoxal methylhydrazone may preferentially bind to and inhibit the necessary enzymes. Furthermore, hydrazones inhibit phosphokinase, the enzyme catalyzing conversion of the B_6-vitamers to the corresponding phosphate esters. Thus an apparent B_6 deficiency can be induced because unphosphorylated B_6 reserves cannot be converted to the necessary 5'-phosphate esters.

Carcinogenicity of hydrazines in humans has not yet been reported, but has been shown to occur in animals. Toth and Shimizu [133] found that a low solution of methylhydrazine (0.01%) administered to 6-week-old hamsters resulted in the development of malignant histiocytomas in 32% of female and 54% of the male hamsters. No such lesions were found in the control group. Because of the latent period before tumors are discovered, it will probably be extremely difficult to ascertain a relationship between MMH and cancer in humans.

C. Clinical Picture

Initial symptoms usually occur after a latent period of between 6 and 12 h, but sometimes as early as 2 h or as late as 24 h. These symptoms include fatigue, dizziness, headache, and abdominal pain, often accompanied by watery diarrhea and vomiting. In the majority of cases this will be the extent of the intoxication with recovery in about 2-6 days. Occasionally, in severe cases, jaundice will become evident during the second day of poisoning. Initially, the liver becomes enlarged and tender. In fatal cases the acute liver degeneration is accompanied by convulsions, periods of delirium, periodic loss of consciousness, and sometimes kidney failure.

D. Treatment

The general treatment of liver and kidney failure due to MMH is similar to that of amanitin-type poisonings. However, renal failure is more likely to occur earlier, and use of fluids (furosemide and/or ethacrynic acid, mannitol) to promote a healthy renal output is mandatory. Large doses of pyridoxine HCl, (e.g., 25 mg/kg intravenously) have been suggested as a specific physiologic antagonist. Diphenylhydantoin or diazepam may be used to treat convulsions. Follow the free hemoglobin level in serum; if it rises, forced diuresis should be employed to prevent kidney damage.

VII. INDOLES

A. Fungi

Mushrooms belonging to this group are found in the genera *Conocybe, Copelandia, Gymnopilus, Naematoloma, Panaeolina, Panaeolus, Psilocybe,* and *Stropharia* [5,7,60,61,63,64, 122].

Psilocybe and *Stropharia* species have been used in shamanic rites of certain Mexican and Central American Indian tribes [62]. The use of hallucinogenic mushrooms as recreational drugs in Western society has been increasing, especially among the young. Books on identification and cultivation of these fungi, as well as the availability of spores and equipment through mail order suppliers, has contributed to the popularity of these mushrooms.

Identification of species in this group is difficult; the majority are small, brownish to grayish, look-alike fungi that often share habitat and features with species of many other genera. *Psilocybe* spp., for instance, have been found growing together with *Galerina autumnalis,* a similar-looking fungus containing Group III toxins. This sharing of habitat and appearance should certainly discourage indiscriminate collecting.

Naematoloma, Psilocybe, and *Stropharia,* genera of the family Strophariaceae, have purple-brown, smooth spores with a pore at the apex, a central stipe, and attached gills. The stipe may or may not have an annulus, but none has a volva. The mushrooms are small to medium in size, except for certain species in *Stropharia* which can be as tall as 20 cm and with a cap width of more than 15 cm. The overlapping features of these genera have resulted in many name changes whereby identical species have been placed in different genera by different authors, e.g., *Psilocybe cubensis* (Earle) Sing. and *Stropharia cubensis* Earle. Others, e.g., Shaffer [119], considered *Naematoloma* and *Stropharia* as subgenera of the genus *Psilocybe.* A revision of the taxonomic principles and classification of the Strophariaceae is needed.

Naematoloma popperianum is the only indole-containing species of that genus and has only been found in San Francisco, the type locality [123]. Guzmán and Vergeer [57] indexed the taxa of the genus *Psilocybe*, recognizing 180 species and varieties. They list 73 psilocybin- and/or psilocin-containing members in the genus. Perhaps the best known of these are *Psilocybe cubensis* and another common species found along the Pacific Coast, *P. cyanescens.*

Psilocybe cubensis (Fig. 14): *Cap* 2-20 cm broad, reddish cinnamon brown becoming yellowish brown to pale yellow in age, conic-campanulate at first, becoming convex to broadly convex to plane, surface viscid when moist, universal veil remnants adhering to the cap surface when young only; flesh firm, bruising blue when handled. *Stipe* 4-15 cm long, 5-15 mm thick, white, at times thickened at the base. *Spores* purple-brown, smooth elliptical, 11.5-17 × 8-11 μm. *Habitat* on horse or cow dung, scattered to gregarious; widespread in the southern United States, Mexico, Central America, Cuba, South America, and southeast Asia.

Psilocybe cyanescens: *Cap* 2-4 cm broad, chestnut-brown when young, becoming caramel-colored in age, convex to nearly plane with undulated or wavy margin, surface smooth and viscid when moist, hygrophanous (color fades upon drying). *Stipe* 6-8 cm long, 2.5-5 mm thick, white and readily bruising blue when handled, surface silky, sometimes covered with fine fibrils, base often with long white rhizomorphs (compacted mycelium strands). *Spores* purplish brown, smooth, elliptical, 9-12 × 5-8 μm. *Habitat* in humus enriched with lignin among leaves and twigs, on wood chips, or well-decayed conifer or eucalyptus substratum; scattered to gregarious.

Conocybe cyanopus and *C. smithii* are the two indole-containing species of this genus. Both are small mushrooms, rarely exceeding 2 cm in width; the color of the cap varies from reddish brown to cinnamon-brown, and the spores are rusty brown. Conocybes have fragile, whitish to brownish stipes, which display a faint bluing at the base of indole-containing species. They closely resemble galerinas and the fact that one species, *C. filaris,* contains toxic cyclopeptides, makes this a potentially dangerous genus.

Conocybe cyanopus: *Cap* 0.3-2.5 cm broad, reddish-brown to cinnamon-brown, convex to bell-shaped, translucent-striate. *Stipe* 1-3.5 cm long, 1-1.5 mm thick, whitish at first, becoming brown to grayish at the apex, greenish to bluish toward the base, fragile. *Spores* brown, smooth, ovoid-ellipsoid, thick-walled, 7-10 × 4-5 μm. *Habitat* in grassy areas, on lawns or on moss (*Polytrichum*).

Conocybe smithii: *Cap* 0.3-1.3 cm broad, ochraceous tawny to cinnamon-brown, obtusely conic, expanding to nearly plane with a distinct umbo, translucent-striate when moist. *Stipe* 1-6 cm long, 0.75-1.5 mm thick, whitish, equal to slightly enlarged at the base, surface covered with fine fibrils at first, becoming smooth in age, fragile. *Spores* brown, smooth, ovoid-ellipsoid, thick-walled, 7-9 × 4-4.5 μm. *Habitat* same as for *C. cyanopus.*

Panaeolus, Panaeolina, and *Copelandia* all have blackish spores. *Panaeolina* is a monotypic genus with *P. foenisecii* as the type species. *Panaeolina* is close to *Panaeolus* and might be combined with the latter as some authors propose. In *Panaeolina,* however, the color of the spores is off black (purplish fuscous to brownish black) and the spores are ornamented.

Panaeolina foenisecii: *Cap* 1-2.5 cm broad, smoky brown to reddish brown when moist, hygrophanous, narrowly conic, becoming bell-shaped in age. *Stipe* 4-10 cm long, 1.5-3 mm thick, whitish, equal, minutely hairy at the apex, brittle. *Spores* purplish fuscous, ellipsoid, 12-17 × 7-9 μm. *Habitat* on lawns or other grassy areas, scattered to abundant.

Figure 14 *Psilocybe cubensis.*

The genus *Panaeolus* contains approximately two dozen species and has been monographed by Ola'h [103]. Based on TLC and gas chromatography, Ola'h divided the species in three categories: psilocybian, latent psilocybian, and nonpsilocybian. At first it was believed that psilocybin-containing species grew exclusively in the eastern part of the United States, while western species contained serotonin instead; however, it is now known that *Panaeolus* spp. can vary greatly in their chemical content and are not geographically specific. The psilocybian species recognized by Ola'h are *P. ater, P. cambodginiensis, P. cyanescens, P. subbalteatus,* and *P. tropicalis.* The latent-psilocybian species: *P. africanus, P. castaneifolius, P. fimicola, P. foenisecii, P. microsporus,* and *P. sphinctrinus.*

Psathyrella is a large genus that closely resembles *Panaeolus* in appearance. Toxic species have not been found in *Psathyrella* to date. The differences between the two genera are: concentrated H_2SO_4 will discolor or bleach spores of *Psathyrella* species, no color change is noted in *Panaeolus* spores; all *Panaeolus* spp. grow on dung, *Psathyrella* spp. grow on humus, grass, or wood. A macroscopic feature of *Panaeolus* is the mottled appearance of the gills caused by irregularly maturing spores.

Gymnopilus spectabilis has been suspected since the mid-1960s to be hallucinogenic [19], but initial searches for psychoactive agents were unrewarding. Another accidental poisoning in Michigan in 1976 initiated further research of the genus *Gymnopilus* by Hatfield and Valdes [60]. Nineteen species were assayed and five were found to contain psilocybin: *G. spectabilis, G. luteus, G. aeruginosus, G. validipes,* and *G. viridans.* The earlier assays which showed negative results should not be accepted as unequivocal evidence that "hallucinogenic" collections of *G. spectabilis* were devoid of psilocybin, since the assay procedure utilized may have precluded the detection of this compound. Hatfield found that extracts used contained pigments that obscured the chromatogram and prevented unequivocal identification of detected compounds. This made it necessary to isolate the constituent via ion exchange and cellulose chromatography, a method not used in the previous research.

Gymnopilus spectabilis: *Cap* 5-20 cm broad, buff-yellow to yellow-orange, convex, becoming nearly flat in age. *Stipe* 3-20 cm long, 8-30 mm thick, concolorous with cap, enlarged in center or club-shaped. *Spores* bright rusty to ferruginous, verrucose roughened, elliptical, thick-walled, 7-10 × 4.5-6 µm. *Habitat* on ground growing from buried wood or directly from stumps, single or in cespitose clusters. Taste very bitter.

Gymnopilus luteus is almost identical with *G. spectabilis,* but differs in having radial hyphae (cells) in the cap trama (interwoven hyphae in *G. spectabilis*). In *G. validipes* the cap color is ochraceous, the stipe yellowish white, and it lacks the bitter taste. *Gymnopilus aeruginosus* is the only species of the genus with a blue-green to pinkish buff cap, rarely exceeding 5 cm in width. The stipe is thick and striate. The spores are cinnamon-colored and smaller than those ot other species in the genus (6-9 × 3.5-4.5 µm). *Gymnopilus viridans* has an ochraceous cap which becomes green-spotted where handled.

Alkaloids other than psilocybin/psilocin have been found in a variety of fungi, many of these belong to the family Polyporaceae [81]. Bis-noryangonin, a stimulant also found in the rootstock of the kava-kava plant (*Piper methysticum*), has been isolated from *Inonotus hispidus* and *Phaeolus schweinitzii.* Hordenine is found in *Boletus zelleri, Fomes pini, Bondarzewia berkeleyi, Grifola gigantea, Polyporus guttulatus,* and *Laetiporus sulphureus.* Tyramine is present in *Boletus zelleri, Bondarzewia berkeleyi, Albatrellus cristata,* and *Laetiporus sulphureus.* N-Methyltyramine is found in *Boletus zelleri, Bondar-*

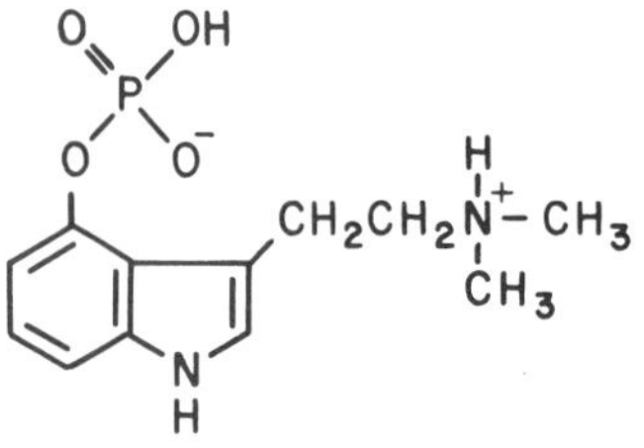

Figure 15   Psilocybin.

zewia berkeleyi, Grifola gigantea, Laetiporus sulphureus, and *Inonotus tomentosus.* Other alkaloids of unknown chemical composition have been reported from *Collybia confluens, Hygrophorus pusillus,* and *Amanita muscaria* [38,123]. Most of these compounds are present in trace amounts only, and it is doubtful that they play a role in human poisonings.

B.   Toxins and Mechanisms of Action

The original search for the chemical agents responsible for the psychoactive action of the mushrooms was conducted by Albert Hofmann of the Sandoz laboratories in Switzerland. The first active hallucinogen was isolated by him in 1958 from cultivated *Psilocybe mexicana* [67]. Hofmann was also responsible for the determination of its chemical structure and the synthetic preparation of this compound [68,69]. It was given the name psilocybin (Fig. 15). Degradation studies showed psilocybin to be 4-phosphoryloxy-N,N-dimethyltryptamine. A related substance sometimes present in fresh mushrooms was named psilocin (Fig. 16), 4-hydroxy-N,N-dimethyltryptamine. The chemical structures of psilocybin and psilocin were confirmed by synthesis [69].

The content of psilocybin in dried mushroom material varies from approximately 0.2% to 0.7%. Psilocin is only present in trace amounts. *Panaeolus* spp. usually contain smaller quantities of psilocybin per unit of dry weight than do *Psilocybe* spp.

Psilocybin is a fairly stable compound, readily soluble in water, and obtainable in colorless crystals. Psilocin, on the other hand, is highly sensitive to oxidation and not easily soluble in water. Psilocybin is the first natural indole derivative with a phosphoric acid radical. Psilocybin and psilocin are also unique in that they are substituted by a hydroxy group in position 4 of the indole structure [115]. Psilocybin, because of its stability, may be retained for long periods of time in dried mushroom material. Hatfield and Valdes [60] were able to detect psilocybin in a *Gymnopilus* sample more than 25 years old. On the other hand, it is known that certain psilocybes lose this indole quite rapidly. This apparent inconsistency needs further studying and may depend, in part, on the presence or absence of certain enzymes in the fungus.

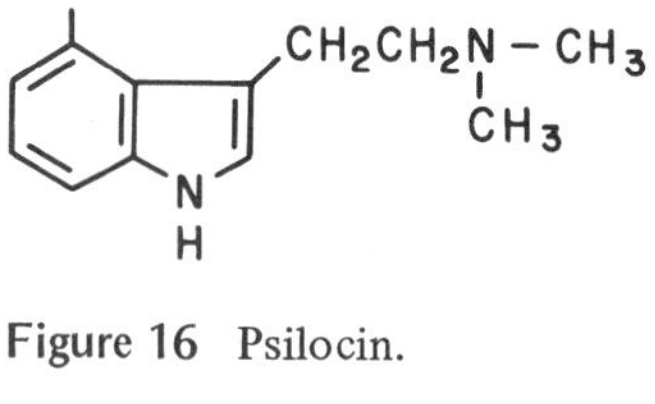

Figure 16   Psilocin.

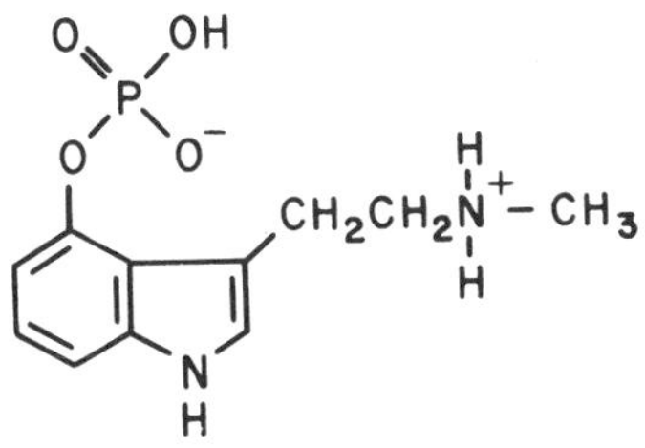

Figure 17 Baeocystin.

The bluing reaction of indole-containing mushrooms has been studied by Levine [85]. It was found that psilocybin could be dephosphorylated by homogenates of various mammalian tissues, as well as by purified alkaline phosphatase, to form psilocin. Psilocin is then readily oxidized by oxidants such as ferric chloride, cytochrome oxidase, or copper oxidase to a blue product, probably a quinone derivative. The oxidation is catalyzed by several enzymes, including some found in the mushrooms; however, the oxidative formation of the blue color can be elicited without enzymes in the presence of ferric ion, which acts as an electron acceptor. The bluing reaction has been suggested as a field indicator for indole-containing fungi. It should be pointed out, however, that many nonpsilocybian genera also display a bluing reaction when handled, for instance the genus *Boletus.*

In addition to psilocybin and psilocin, the monomethyl and nonalkylated analogs of psilocybin, baeocystin (Fig. 17) and norbaeocystin (Fig. 18), were isolated and so named by Leung and Paul [83,84] from *Psilocybe baeocystis,* believed to be the only species of *Psilocybe* to contain these compounds. Recently, however, Repke et al. [111] found that baeocystin is fairly common in hallucinogenic mushrooms; baeocystin was found in *Psilocybe cyanescens, P. cubensis, P. pelliculosa, P. semilanceata, P. silvatica, P. stuntzii, Conocybe cyanopus, C. smithii,* and *Panaeolus subbalteatus.*

C. Clinical Picture

Psilocybin, taken orally in doses of 4-8 mg, induces effects similar to the ingestion of about 20 g of fresh, or 2 g of dried mushrooms. Symptoms usually start within the first ½ h after ingestion. Reactions vary among individuals, ranging from a feeling of relaxation to one of tension, anxiety, or dizziness. Nausea and abdominal discomfort may be experienced, including vomiting and diarrhea. Two rare cases of paralysis of the limbs, lasting 24 h, have been reported [149]. Visual effects commence during the second ½ h, including sharpened

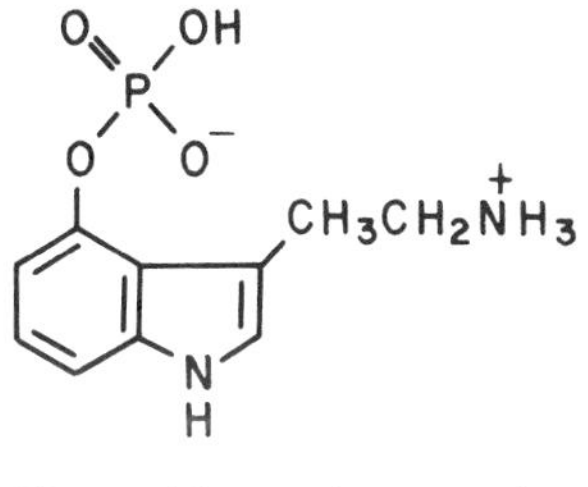

Figure 18 Norbaeocystin.

outlines of objects, heightened brilliance of colors, and visual imagery with closed eyes. True hallucinosis is rare, but has been reported with large doses (as much as 35 mg). Panic reactions have occurred, marked by fear of death or insanity and inability to distinguish reality from fantasy [34].

During the second hour the closed-eye visual effects increase and are marked by the perception of intense, brilliant colors, often in the form of geometric patterns, undulating, bursting forth like fireworks, and sometimes superimposed on dream-like shapes which rapidly appear and disappear. Such phenomena continue through the third hour, after which they gradually disappear. No "hangover" syndrome is experienced. Systemic effects of psilocybin may include pupillary dilation, rapid heart beat, rapid breathing, lowered body temperature, lowered blood sugar, and increased blood pressure, the effects being due mostly to stimulation of the central or sympathetic nervous system. In contrast to these symptoms of stimulation, motor activity tends to be slightly depressed [139].

Fatal poisonings resulting from eating *Psilocybe* or *Panaeolus* species are reported in the literature: for instance, the death of a 6-year-old girl in Oregon [94]. This, and an earlier report from Japan [70], should be questioned, since the mushroom was not positively identified; nevertheless, children are more susceptible to poisoning than are adults, because of the immaturity of their enzyme systems and their smaller size [77].

D. Treatment

Drugs should be avoided if possible and a quiet environment with the "talk-down" method used. If a sedative is needed, chlordiazepoxide or diazepam seem best at present. Chlorpromazine is variable in effectiveness, despite earlier recommendations for its use. Hyperpyrexia in children should be treated with fluids and cool sponging.

VIII. MUSHROOM-ALCOHOL SENSITIVITY

A. Fungi

The genus *Coprinus*, a common group of roadside and grassland fungi, has often been described as one of the safer edibles. *Coprinus* spp. are easily recognized by their characteristic oval or egg-shaped fruit bodies when young. The cap colors vary from grayish white to brownish gray; the gills are closely spaced and white at first, later turning to grayish blue and finally black, the color of the mature spores. At this stage, most *Coprinus* spp. start to deliquesce into an ink-like liquid, hence the common name "inky caps." Some inky caps, when consumed with alcohol, will cause a reaction similar to that of disulfiram (Antabuse), used in the treatment of alcoholism.

The following coprini are included in this toxin group: *C. atramentarius, C. insignis, C. quadrifidus,* and *C. variegatus* from North America and Europe; *C. erethistes* from Africa [62,97]. Also mentioned as toxic species in the literature, but doubtful, are *C. micaceus* [26,97], and *C. fuscescens* [112].

Coprinus atramentarius (Fig. 19): *Cap* 2-7 cm broad, gray to brownish gray, conic to bell-shaped in age, covered with small brownish scales when young, margin striate to wrinkled. *Stipe* 4-8 cm long, 6-10 mm thick, concolorous with cap, enlarged base with a thin volva-like fibrous ring. *Spores* black, elliptical, smooth, with pore at apex, 7.5-10 X 5-5.5 μm. *Habitat* usually in grass, in open fields or under trees, growing in clusters.

Another mushroom, *Clitocybe clavipes,* has also been reported capable of sensitizing

Figure 19 *Coprinus atramentarius.*

to alcohol in a manner similar to the coprini [29]. This species is widely distributed throughout the Northern Hemisphere, where it can often be found growing alone or in groups under conifers or mixed woods.

Clitocybe clavipes: Cap 2-9 cm broad, brown to grayish brown, flat or slightly depressed in the center, edge often wavy, gills extending down the stipe. *Stipe* 3.5-6 cm long, 4-10 mm thick, thickened toward the base (club-shaped). *Spores* white, egg-shaped, smooth, 5-7 × 3-4 μm. *Habitat* growing in conifer woods. Both cap and stipe have a fragrant odor.

Other reported mushroom-alcohol intoxications of "narrow-capped morels," possibly *Verpa bohemica* [55], and *Pholiota squarrosa* [118], belong to the gastrointestinal irritant type of poisoning. In neither of these cases did the intoxication mimic a disulfiram-like reaction. Nausea, vomiting, and diarrhea were present, and shock was observed in one of the poisonings [34,118].

B. Toxins and Mechanisms of Action

Simandl and Franc reported in 1956 [120] that they had isolated disulfiram from the mushrooms; however, subsequent research was unable to confirm this claim [59,89,147]. In 1975, a compound, coprine, was isolated simultaneously in the United States [59] and

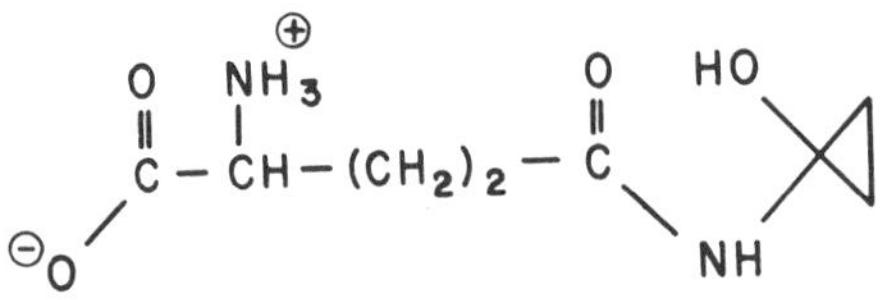

$$\underset{\ominus O}{\overset{O}{\underset{\|}{C}}} - \overset{\overset{\oplus}{NH_3}}{\underset{|}{CH}} - (CH_2)_2 - \overset{O}{\underset{\|}{C}} \diagdown \underset{NH}{\overset{HO}{\bowtie}}$$

Figure 20   Coprine.

Sweden [87]. Coprine and disulfiram have different chemical structures, but their mechanism of action is identical: the ability to inhibit the conversion of acetaldehyde, the initial metabolite of alcohol, into acetate. The result is a high level of acetaldehyde in the blood, which may then produce the characteristic vasomotor reaction [30,110].

Coprine (Fig. 20), is a γ-glutamyl conjugate of 1-aminocyclopropanol, formally the N,O-ketal of cyclopropanone. The structure of coprine has been confirmed by synthesis [87].

C. Clinical Picture

Duration as well as onset of symptoms can be quite varied from person to person. Prior alcohol consumption will instigate the alcohol-sensitizing effects only when sufficiently high blood alcohol levels are present. On the other hand, symptoms may recur if more alcohol is consumed during the next couple of days, although less vividly [34]. Sensitivity may also vary, and some people are able to consume these fungi with alcohol without effects.

Although gastrointestinal upsets have occurred with *Coprinus,* these symptoms have not yet been reported from *Clitocybe clavipes* in the United States. The classic syndrome is vasomotor, ensuing 10-30 min after ingestion. Typical symptoms include flushing, mydriasis, paresthesia, tachycardia, dyspnea, sweating, hypertension, and nausea. The effects will last about 3-4 h, after which they gradually disappear.

D. Treatment

Simple reassurance usually treats this type of poisoning adequately, since the unpleasant feelings disappear without aftereffects. Propranolol, +40 mg orally, may be given if symptoms are severe or supraventricular arrhythmias occur [110a]. Intravenous propranolol should be avoided except when cardiac arrhythmias are life threatening or when the poisoning is accompanied by coronary insufficiency. The usual intravenous dose is 1-2 mg; however, the rate of administration should not exceed 1 mg/min to diminish the possibility of lowered blood pressure and causing cardiac standstill. If sympathomimetic agents must be used for severe hypotension, minimal amounts of α-adrenergic drugs such as noradrenaline should be employed. Drugs with a β-adrenergic component may be expected to increase the tachycardia.

IX. MISCELLANEOUS TOXINS

This toxin group includes the gastrointestinal irritants and toxins of little known or unknown chemistry. The number of mushrooms capable of causing gastroenteritis is large and consists of many heterogeneous species. Certain fungi appear to vary in toxicity depending on location and habitat; *Laetiporus sulphureus* and the gray variant of *Leucoagaricus naucinus* may fall into this category. Others may be toxic when old, e.g., *Armillariella mellea,* or when raw, *Morchella* spp. and *Amanita vaginata.* Comprehensive references of gastrointestinal irritants, their pharmacology, chemistry, toxicology, and laboratory identification techniques have been published [78,86,134].

A biochemical cause of individual sensitivity has been suggested by Bergoz [10], Bergoz and Righetti [9], and Simons [121], involving the malabsorption of the disaccharide trehalose which has been found in many fungi, including the edible *Agaricus brunnescens,* which is sold in our markets [65]. Like other disaccharides, trehalose must be hydrolyzed before it can be absorbed from the intestinal tract. Individuals with a genetic deficiency in the intestine of the enzyme trehalase (an enzyme that accelerates the hydrolysis of trehalose) will suffer abdominal distress and diarrhea after mushroom ingestion.

The genus *Agaricus* may be recognized in the field by an annulus on the stem but no volva, free gills, and various color changes of the gills as the spores develop. Starting out white, they soon change to pink and finally become brown to blackish brown. Spores are brown under the microscope, smooth, and rarely larger than 10 μm [123]. The species usually are large and fleshy with white to brown to gray-brown caps which are often covered with appressed squamules (flattened scales). The cap surface is moist or dry but never viscid. Some of the species will instantly turn yellow when injured and some also have a phenol-like smell. *Agaricus hondensis* and *A. xanthodermus* belong to this group and have caused gastroenteritis in some individuals.

The genera *Russula* and *Lactarius* both contain a number of acrid-tasting species which cause gastroenteritis in some individuals but not in others. The toxic agents are unknown at the present time. The death of a 5-year-old girl after eating *Lactarius glaucescens* has been reported [22]; however, most intoxications are mild and of short duration.

Chlorophyllum molybdites is the mushroom most often misidentified with the lookalike and edible *Lepiota rachodes.* The gills are white at first, but later are colored slate-green by the developing spores. The unknown toxic component is soluble in ethanol and chloroform but not in ether and is destroyed when heated to 100°C. This species has caused poisonings in certain individuals, involving emesis, diarrhea, and occasionally mental confusion [8,11].

Entoloma lividum is a terrestrial species with salmon to pink spores and gray to pinkish gills. It is known to cause severe and prolonged gastrointestinal distress. The symptoms, starting during the first ½ h, include emesis, diarrhea, cramps, headaches, thirst, and general weakness. The toxic reactions may last for several days and may even be fatal in young children or individuals with a prior illness. Tyler [134] has reported liver damage in some patients. The toxic component is unknown at the present time. *Tricholoma pardinum* and *T. venenatum* have caused intoxications similar to *E. lividum.*

Gomphus floccosus somewhat resembles the edible *Cantharellus cibarius* with its orange color and ridges instead of gills. Ingestion of this species can cause nausea and diarrhea. The mushroom contains compounds related to agaricic and norcaperatic acids [98]. Carrano and Malone [20] found that both acids produced delayed effects of mydriasis, central nervous system depression, and skeletal muscle weakness in rats. They also inhibit the enzyme aconitase.

Hebeloma crustuliniforme is a cream-colored species with white to grayish gills and a bitter taste. It can often be found growing in lawns. The mushroom has caused severe attacks of abdominal cramps and profuse diarrhea, sometimes lasting for several days [109].

Naematoloma fasciculare is a species often mistaken for the nontoxic *N. sublateritium* or *N. capnoides. Naematoloma fasciculare*, however, has a very bitter taste, in contrast to the other two species. The cap and stipe color of *N. fasciculare* is yellow-orange, the gills are greenish yellow, and it is usually found growing in clusters from the stumps or bases of hardwoods. The onset of symptoms may be delayed for as long as 9 h and include vomiting, diarrhea, impaired vision, and sometimes paralysis of the limbs [86]. A death due to kidney failure has been documented [66]. The poisoning resembled the *Amanita phalloides* syndrome; however, no toxic cyclopeptides have been isolated from this fungus, and the toxic compound involved is unknown at the present time.

Lycoperdon and other true puffballs are edible when young and white throughout. The interior will darken as the mushroom matures and the species are then no longer edible. They are known to have caused mild gastrointestinal upset. In *Scleroderma* spp. the interior is never white but ranges from purplish to black. They should never be eaten. Stevenson and Benjamin [129] listed the following symptoms from a poisoning case involving *S. aurantium:* stomach pains, weakness, nausea without dizziness, and a tingling sensation over the entire body. The symptoms disappeared promptly after forced emesis.

Paxillus involutus is a potentially toxic species with a large convex to depressed cap varying in color from gray-brown to red-brown. Its decurrent gills are a dingy greenish yellow and turn brown when bruised. Consumption of this mushroom can result in massive hemolysis, but previous exposure to the fungus is probably required to cause production of the red blood cell-destroying antibodies [18,113]. Hemolytic crises following consumption of *Suillus luteus* and *Amanita vaginata* have also been reported [121]. Pharmacological and biochemical studies of the toxin(s) involved are needed.

Cortinarius orellanus has caused a number of fatalities in Europe [56]. The long latent period between consumption and first symptoms, average 8 days, is one of the reasons why this mushroom was never suspect until 1952, when a Polish physician, Stanislow Grzymala, began a search for the common factor in a strange "epidemic" affecting 102 patients, resulting in 11 fatalities. The common factor turned out to be consumption of *C. orellanus.* Grzymala obtained a toxin which he named orellanine [56], which is present in fresh specimens in amounts equal to approximately 1% of the weight of the mushroom. The molecule was recently described by Antkowiak and Gessner as being the bis-N-oxide of 3,3',-4,4'-tetrahydroxy-2,2'-bipyridyl, with the molecular formula $C_{10}H_8O_6N_2$ [2a]. Orellanine is partially soluble in water, acetone, and methanol, but is freely soluble in aqueous alkali, from which it reprecipitates on addition of acid. When heated at temperatures $>270°C$, it decomposes explosively, producing a cloud of greenish black smoke.

The hallmark of the clinical picture in orellanine-type poisoning is acute tubulointerstitial nephritis. Symptoms include diarrhea, vomiting, jaundice, fever, and sometimes rash [43]. Species known to be toxic are *C. orellanus, C. orellanoides, C. splendens,* and *C. speciosissimus.* Low concentrations of the toxin have also been found in *C. cinnamomeus, C. malicorius, C. phoeniceus,* and *C. sanguineus* [3]. It is possible that chemical variants will be found, especially in the latter group of species.

ACKNOWLEDGMENTS

The author wishes to thank the editor of this treatise for the invitation to contribute. He further thanks Dr. Thomas J. Duffy for his help with the medical sections, Mrs.

Dorothy B. Orr for editorial suggestions, and Dr. M. Frimmer for furnishing the micrograph illustrations (Figs. 4 and 5) of isolated rat liver cells.

REFERENCES

1. Abdel-Makal, S. H. 1974. Chemotaxonomic significance of alkaloids and cyclopeptides in *Amanita* species. Ph.D. dissertation, University of Maine, University Microfilms, No. 75-12, Ann Arbor, Mich., p. 410.

2. Abul-Haj, S. K., R. A. Ewald, and L. Kazyak. 1963. Fatal mushroom poisoning; report of a case confirmed by toxicologic analysis of tissue. N. Engl. J. Med. *269*: 223-227.

2a. Antkowiak, W. Z., and W. P. Gessner. 1978. Structure elucidation of orellanine. Symp. Pap., 11th IUPAC Int. Symp. Chem. Nat. Prod., Vol. 2, 45-47.

3. Arietti, N., and R. Tomasi. 1975. *I funghi velenosi*, Edizioni Agricole, Bologna.

4. Becker, C. E., T. G. Tong, F. Bartter, U. Boerner, R. L. Roe, R. T. A. Scott, and M. B. MacQuarrie. 1976. Diagnosis and treatment of *Amanita phalloides*-type poisoning. West. J. Med. *125*: 100-109.

5. Benedict, R. G., L. R. Brady, A. H. Smith, and V. E. Tyler, Jr. 1962. Occurrence of psilocybin and psilocin in certain *Conocybe* and *Psilocybe* species. Lloydia *25*: 156-159.

6. Benedict, R. G., V. E. Tyler, Jr., and L. R. Brady. 1966. Chemotaxonomic significance of isoxazole derivatives in *Amanita* species. Lloydia *30*: 333-342.

7. Benedict, R. G., V. E. Tyler, Jr., and R. Watling. 1967. Blueing in *Conocybe, Psilocybe,* and a *Stropharia* species and the detection of psilocybin. Lloydia *30*: 150-157.

8. Benedict, R. G. 1972. Mushroom toxins other than *Amanita*. In S. Kadis, A. Ciegler, and S. J. Ajl (Eds.), *Microbial Toxins*, Vol. 8, Academic, New York, pp. 281-320.

9. Bergoz, R., and A. Righetti. 1970. Intolérance aux champignons par malabsorption sélective du tréhalose: un syndrome rare et inédit. Sweiz. Med. Wochenschr. *100*: 1244-1245.

10. Bergoz, R. 1971. Trehalose malabsorption causing intolerance to mushrooms—report of a probable case. Gastroenterology *60*: 909-912.

11. Bessey, E. A. 1939. A case of poisoning by *Lepiota morgani*. Mycologia *31*: 109-110.

11a. Beutler, J. A., and P. P. Vergeer. 1980. Amatoxins in American mushrooms: evaluation of the Meixner test. Mycologia *72*: 1142-1149.

12. Block, S. S., R. L. Stephens, A. Barreto, and W. A. Murrill. 1955. Chemical identification of the *Amanita* toxin in mushrooms. Science *121*: 505-506.

13. Bowden, K., A. C. Drysdale, and G. A. Mogey. 1965. Constituents of *Amanita muscaria*. Nature (Lond.) *206*: 1359-1360.

14. Bowden, K., and A. C. Drysdale. 1965. A novel constituent of *Amanita muscaria*. Tetrahedron Lett. *12*: 727-728.

15. Brady, L. R., R. G. Benedict, V. E. Tyler, Jr., D. E. Stuntz, and M. H. Malone. 1975. Identification of *Conocybe filaris* as a toxic basidiomycete. Lloydia *38*: 172-173.

16. Brehm, L., H. Hjeds, and P. Krogsgaard-Larsen. 1972. The structure of muscimol, a GABA analogue of restricted configuration. Acta Chem. Scand. *26*: 1298-1299.

17. Brown, J. K., M. H. Malone, D. E. Stuntz, and V. E. Tyler, Jr. 1962. Paper chromatographic determination of muscarine in *Inocybe* species. J. Pharm. Sci. *51*: 853-855.

18. Bschor, F., and H. J. Mallach. 1963. Vergiftungen durch den kahlen Krempling (*Paxillus involutus*), eine geniessbare Pilzart. Arch. Toxikol. *20*: 82-95.

19. Buck, R. W. 1967. Psychedelic effects of *Pholiota spectabilis*. N. Engl. J. Med. *276*: 391-392.

20. Carrano, R. A., and M. H. Malone. 1967. Pharmacologic study of norcaperatic and agaric acids. J. Pharm. Sci. *56*: 1611-1614.

21. Catalfomo, P., and C. H. Eugster. 1970. *Amanita muscaria:* present understanding of its chemistry. Bull. Narc. (U.N.) *22*(4): 33-41.

22. Charles, V. K. 1942. Mushroom poisoning caused by *Lactarius glaucescens*. Mycologia *34*: 112-113.

23. Chilton, W. S., C. P. Hsu, and W. T. Zdybak. 1974. Stizolobic and stizolobinic acids in *Amanita pantherina*. Phytochemistry *13*: 1179-1181.

24. Chilton, W. S. 1975. The course of an intentional poisoning. McIlvainea *2*(1): 17-18.

25. Chilton, W. S., and J. Ott. 1976. Toxic metabolites of *Amanita pantherina, A. cothurnata, A. muscaria* and other *Amanita* species. Lloydia *39*: 150-157.

26. Christensen, C. M. 1975. *Molds, Mushrooms and Mycotoxins,* University of Minnesota Press, Minneapolis.

27. Ciocatto, E., U. Delfino, and M. A. Trompeo. 1970. Trattamento nell'intossicazione da *Amanita falloide* e contributo clinico. Parte seconda. Minerva Anestesiol. *36*: 636-653.

28. Ciucci, N., and A. Chiri. 1974. Considerazioni cliniche ed anatomopatologiche sui risultati a distanza ottenuti con l'acido tioctico nell'avvelemento da *Amanita phalloides*. Minerva Anestesiol. *40*: 61-70.

29. Cochran, K. W., and M. W. Cochran. 1978. *Clitocybe clavipes:* antabuse-like reaction to alcohol. Mycologia *70*: 1124-1126.

30. Coldwell, B. B., K. Genest, and D. W. Hughes. 1969. Effect of *Coprinus atramentarius* on the metabolism of ethanol in mice. J. Pharm. Pharmacol. *21*: 176-179.

31. Crisafulli, A., and C. M. Spagnolio. 1940. L'intossicazione da *Amanita phalloides;* azione tossica. Arch. "De Vecchi" *2*: 400-413.

32. Dabski, H., T. Drozd, and R. Bryc. 1970. Leczenie zatruć muchomorem sromotnikowym kwasem tioktanowym. Pol. Tyg. Lek. *25*(9): 338-340.

33. Delfino, U. 1970. L'intossicazione de *Amanita falloide*. Parte prima. Minerva Anestesiol. *36*: 629-635.

34. Duffy, T. J., and P. P. Vergeer. 1977. *California Toxic Fungi,* Toxicology Monograph No. 1, Toxicology Committee, Mycological Society of San Francisco, San Francisco.

35. Ebneter, K. 1976. Vergiftungen durch Knollenblätterpilze. Ph.D. dissertation, University of Zurich, Swiss Toxicologic Information Center, University of Zurich.

36. Eugster, C. H., and P. G. Waser. 1954. Zur Kenntnis des Muscarins. Experientia *10*: 298-300.

37. Eugster, C. H. 1957. Isolierung von Muscarine aus *Inocybe patouillardi*. Helv. Chim. Acta *40*: 886.

38. Eugster, C. H. 1967. Isolation, structure and synthesis of central-active compounds from *Amanita muscaria* (L. ex Fr.) Hooker. In D. H. Efron (Ed.), *Ethnopharmacologic Search for Psychoactive Drugs,* U.S. Public Health Service Publ. No. 1645, pp. 416-418.

39. Eugster, C. H. 1969. Chemie der Wirkstoffe aus dem Fliegenpilz (*Amanita muscaria*). In L. Zechmeister (Ed.), *Fortschritte der Chemie organischer Naturstoffe,* Vol. 27, Springer-Verlag, Vienna, pp. 261-321.

40. Eugster, C. H., and E. Schleusener. 1969. Stereomere Muscarine kommen in der

Natur vor. Gas-chromatographische Trennung der Norbasen. Helv. Chim. Acta *52*: 708-715.

41. Faulstich, H., S. Zobeley, and M. Weekauf-Bloching. 1974. Cytolytic properties of phallolysin. Hoppe-Seyler's Z. Physiol. Chem. *355*: 1495-1498.

42. Faulstich, H. 1975. Knollenblätterpilzvergiftung. Dtsch. Med. Wochenschr. *100*: 1714.

42a. Faulstich, H., A. Buku, H. Bodenmüller, and Th. Wieland. 1980. Virotoxins: actin-binding cyclic peptides of *Amanita virosa* mushrooms. Biochemistry *19*: 3334-3343.

43. Favre, H., M. Leski, P. Christeler, E. Vollenweider, and F. Chatelanat. 1976. Le *Cortinarius orellanus:* un champignon toxique provoquant une insuffisance rénale aiguë retardée. Schweiz. Med. Wochenschr. *106*: 1097-1102.

44. Finestone, A. J., R. Berman, B. Widmer, J. Markowitz, and U. J. Laquer. 1972. Thioctic acid treatment of acute mushroom poisoning. Pa. Med. *75*(7): 49-51.

45. Fiume, L. 1972. Pathogenesis of the cellular lesions produced by alpha-amanitin. In E. Farber (Ed.), *The Biochemistry of Disease,* Vol. 2, Dekker, New York, pp. 105-122.

46. Fiume, L., M. Derenzini, V. Marinozzi, F. Petazzi, and A. Testoni. 1973. Patho-genesis of gastro-intestinal symptomatology during poisoning by *Amanita phalloides*. Experientia *29*: 1520-1521.

47. Fiume, L. 1974. Meccanismo d'azione dei principali veleni della *Amanita phalloides* (Amanitine). Recenti Prog. Med. *56*: 547-553.

47a. Flammer, R. 1980. *Differentialdiagnose der Pilzvergiftungen,* G. Fischer, Stuttgart, New York.

48. Frimmer, M., J. Gries, D. Hegner, and B. Schnorr. 1967. Untersuchungen zum Wirkungsmechanismus des Phalloidens. Naunyn-Schmiedeberg's Arch. Pharmakol. Exp. Pathol. *258*: 197-214.

49. Frimmer, M., R. Kroker, and J. Porstendörfer. 1974. The mode of action of phal-loidin: demonstration of rapid deformation of isolated hepatocytes by scanning electron microscopy. Naunyn-Schmiedeberg's Arch. Pharmakol. *284*: 395-398.

50. Gérault, A., and L. Girre. 1975. Recherches toxicologiques sur le genre *Lepiota* Fries. C.R. Acad. Sci., Ser. D *280*: 2841-2843.

51. Giusti, G. V., and A. Carnevale. 1974. A case of fatal poisoning by *Gyromitra escu-lenta*. Arch. Toxicol. *33*: 49-54.

52. Govindan, V. M., H. Faulstich, Th. Wieland, B. Agostini, and W. Hasselbach. 1972. In vitro effect of phalloidin on a plasma membrane preparation from rat liver. Naturwissenschaften *59*: 521-522.

53. Govindan, V. M., G. Rohr, Th. Wieland, and B. Agostini. 1973. Binding of a phallo-toxin to protein filaments of plasma membrane of liver cell. Hoppe-Seyler's Z. Physiol. Chem. *354*: 1159-1161.

54. Groves, J. W. 1962. *Edible and Poisonous Mushrooms of Canada,* Can. Dept. Agric. Publ. No. 1112, Ottawa.

55. Groves, J. W. 1964. Poisoning by morels when taken with alcohol. Mycologia *56*: 779-780.

56. Grzymala, S. 1965. Etude clinique des intoxications par les champignons du genre *Cortinarius orellanus* Fr. Bull. Med. Leg. Toxicol. Med. *8*: 60-70.

57. Guzmán, G., and P. P. Vergeer. 1978. Index of taxa in the genus *Psilocybe*. Myco-taxon *6*: 464-476.

58. Hardegger, E., and F. Lohse. 1957. Synthese und absolute Konfiguration des Mus-carins. Helv. Chim. Acta *40*: 2383.

59. Hatfield, G. M., and J. P. Schaumberg. 1975. Isolation and structural studies of

coprine, the disulfiram-like constituent of *Coprinus atramentarius*. Lloydia *38*: 489-496.

60. Hatfield, G. M., and J. L. Valdes. 1978. The occurrence of psilocybin in *Gymno-pilus* species. Lloydia *41*: 140-144.

61. Heim, R., R. G. Wasson, and collaborators. 1958. *Les champignons hallucinogènes du Mexique*, Arch. Mus. Nat. Hist. Nat., 7e Ser. VI, Paris.

62. Heim, R. 1963. *Les champignons toxiques et hallucinogènes*, N. Boubée, Paris.

63. Heim, R., A. Hofmann, and H. Tscherter. 1966. Sur une intoxication collective à syndrome psilocybien causée en France par un *Copelandia*. C.R. Acad. Sci., Ser. D *262*: 519-523.

64. Heim, R., and collaborators. 1966. *Nouvelles investigations sur les champignons hallucinogènes*, Arch. Mus. Nat. Hist. Nat., 7e Ser. IX, Paris.

65. Hellerström, S. 1941. Sensitization to edible mushrooms. Acta Dermato-Venereol. *22*: 331-336.

66. Herbich, J., K. Lohwag, and R. Rotter. 1966. Tötliche Vergiftung mit dem grün-blättrigen Schwelkopf. Arch. Toxikol. *21*: 310-320.

67. Hofmann, A., R. Heim, A. Brack, and H. Kobel. 1958. Psilocybin, ein psychotroper Wirkstoff aus dem mexikanischen Rauschpilz *Psilocybe mexicana* Heim. Experientia *14*: 107-109.

68. Hofmann, A., A. Frey, H. Ott, Th. Petrzilka, and F. Troxler. 1958. Konstitution-aufklärung und Synthese von Psilocybin. Experientia *14*: 397-399.

69. Hofmann, A., R. Heim, A. Brack, H. Kobel, A. Frey, H. Ott, Th. Petrzilka, and F. Troxler. 1959. Psilocybin und Psilocin, zwei psychotrope Wirkstoffe aus mexikan-ischen Rauschpilzen. Helv. Chim. Acta *42*: 1557-1572.

70. Imai, S. 1932. On *Stropharia caerulescens*, a new species of poisonous toadstool. Trans. Sapp. Nat. Hist. Soc. *12*: 148-151.

71. Jenkins, D. T. 1977. A taxonomic and nomenclatural study of the genus *Amanita*, section *Amanita* for North America. Bibl. Mycol. *57*.

72. Johnston, G. A. R. 1971. Muscimol and the uptake of gamma-aminobutyric acid by rat brain slices. Psychopharmacologia *22*: 230-233.

73. Jubin, E., and S. Moeschlin. 1964. Die *Amanita*-Pilzvergiftung und Problematik ihrer Behandlung. Praxis *53*: 1502-1511.

74. Kobert, R. 1891. Ueber Pilzvergiftung. St. Petersburger Med. Wochenschr. *16*: 463-466, 471-474.

75. Kögl, F., C. A. Salemink, H. Schouten, and F. Jellinek. 1957. Ueber Muscarin. Recl. Trav. Chim. Pays-Bas Belge *76*: 109.

76. Kubička, J. 1963. Neue Möglichkeiten in der Behandlung von Vergiftungen mit dem Grünen Knollenblätterpilz—*Amanita phalloides*. Mykol. Mitteil. 7: 92-94.

77. Lamke, K. F. 1973. Mushroom poisoning in the young child. Paediatrician *2*: 83-89.

78. Lamke, K. F. 1978. Pharmacology and treatment of mushroom intoxication. In B. H. Rumack and E. Salzman (Eds.), *Mushroom Poisoning: Diagnosis and Treat-ment*, CRC Press, West Palm Beach, Fla., pp. 125-170.

79. Larcan, A. 1975. Actualités sur l'intoxication phalloidienne. Agressologie *16*: 257-261.

80. Larcan, A., M. Lamarche, H. Lambert, M. C. Laprevote-heully, and J. L. Patrer. 1975. Recherches expérimentales concernant un traitement de l'intoxication phal-loidienne. Agressologie *16*: 307-318.

81. Lee, T. M., L. G. West, J. L. McLaughlin, L. R. Brady, J. L. Lowe, and A. H. Smith. 1975. Screening for N-methylated tyramines in some higher fungi. Lloydia *38*: 450-452.

82. Lehman, L. 1963. Zur Klinik und Spätprognose der Vergiftung mit *Amanita phalloides*. Helv. Med. Acta *30*: 30-46.

83. Leung, A. Y., and A. G. Paul. 1967. Baeocystin, a mono-methyl analog of psilocybin from *Psilocybe baeocystis* saprophytic culture. J. Pharm. Sci. *56*: 146.

84. Leung, A. Y., and A. G. Paul. 1968. Baeocystin and norbaeocystin: new analogs of psilocybin from *Psilocybe baeocystis*. J. Pharm. Sci. *57*: 1667-1671.

85. Levine, W. G. 1967. Formation of blue oxidation product from psilocybin. Nature (Lond.) *215*: 1292-1293.

86. Lincoff, G., and D. H. Mitchel. 1977. *Toxic and Hallucinogenic Mushroom Poisoning*, Van Nostrand Reinhold, New York.

87. Lindberg, P., R. Bergman, and B. Wickberg. 1975. Isolation and structure of coprine, a novel physiologically active cyclopropane derivative from *Coprinus atramentarius* and its synthesis *via* 1-aminocyclopropanol. J. Chem. Soc. Chem. Commun. *1975*: 946.

88. Lindell, T. J., F. Weinberg, P. W. Morris, R. G. Roeder, and W. J. Rutter. 1970. Specific inhibition of nuclear RNA polymerase II by alpha-amanitin. Science *170*: 447-449.

89. List, P. H., and H. Reith. 1960. Der Faltentintling, *Coprinus atramentarius* Bull., und seine dem Tetraäthylthiuramdisulfid ähnliche Wirkung. Arzneim-Forsch. *10*: 34-40.

90. List, P. H., and P. Luft. 1967. Gyromitrin, das Gift der Frühjahrslorchel, *Gyromitra (Helvella) esculenta* Fr. Tetrahedron Lett. *20*: 1893-1894.

91. List, P. H., and P. Luft. 1968. Gyromitrin, das Gift der Frühjahrslorchel. Arch. Pharmazie *301*: 294-305.

92. List, P. H., and P. Luft. 1969. Nachweis und Gehaltsbestimmung von Gyromitrin in frischen Lorcheln. Arch. Pharm. *302*: 143-146.

93. Litten, W. 1975. The most poisonous mushrooms. Sci. Am. *232*(3): 90-101.

94. McCawley, E. L., R. E. Brummett, and G. W. Dana. 1962. Convulsions from *Psilocybe* mushroom poisoning. Proc. West. Pharmacol. Soc. *5*: 27-33.

95. Malone, M. H., R. C. Robichaud, V. E. Tyler, Jr., and L. R. Brady. 1961. A bioassay for muscarine and its detection in certain *Inocybe*. Lloydia *24*: 204-210.

96. Malone, M. H., R. C. Robichaud, V. E. Tyler, Jr., and L. R. Brady. 1962. Relative muscarinic potency of thirty *Inocybe* species. Lloydia *25*: 231-237.

97. Miller, O. K., Jr. 1972. *Mushrooms of North America*, E. P. Dutton, New York.

97a. Mitchel, D. H., and B. H. Rumack. 1978. Symptomatic diagnosis and treatment of mushroom poisoning. In B. H. Rumack and E. Salzman (Eds.), *Mushroom Poisoning: Diagnosis and Treatment*, CRC Press, West Palm Beach, Fla., pp. 171-186.

98. Miyata, J. T., V. E. Tyler, Jr., L. R. Brady, and M. H. Malone. 1966. The occurrence of norcaperatic acid in *Cantharellus floccosus*. Lloydia *29*: 43-49.

99. Moser, M. 1978. *Die Röhrlinge und Blätterpilze (Polyporales, Boletales, Agaricales, Russulales)*. *Kleine Kryptogamenflora*, Vol. IIb/2, 4th ed., G. Fischer, Stuttgart.

100. Müller, G. F. R., and C. H. Eugster. 1965. Muscimol, ein pharmakodynamisch wirksamer Stoff aus *Amanita muscaria*. Helv. Chim. Acta *48*: 910-926.

101. Novello, F., and F. Stirpe. 1969. Experimental conditions affecting ribonucleic acid polymerase in isolated rat liver nuclei. Biochem. J. *112*: 721-727.

102. Novello, F., L. Fiume, and F. Stirpe. 1970. Inhibition by alpha-amanitin of ribonucleic acid polymerase solubilized from rat liver nuclei. Biochem. J. *116*: 177-180.

103. Ola'h, G. M. 1970. *Le genre Panaeolus, essai taxinomique et physiologique*, Revue de Mycologie Mémoire Hors-Série No. 10, Laboratoire de Cryptogamie du Muséum National d'Histoire Naturelle, Paris.

104. Orr, R. T., and D. B. Orr. 1979. *Mushrooms of Western North America*, California Natural History Guides 42, University of California Press, Berkeley.

105. Paaso, B., and D. C. Harrison. 1975. A new look at an old problem: mushroom poisoning. Clinical presentations and new therapeutic approaches. Am. J. Med. *58*: 505-509.

106. Palyza, V., and V. Kulhánek. 1970. Über die chromatographische Analyse von Toxinen aus *Amanita phalloides*. J. Chromatogr. *53*: 545-558.

107. Palyza, V. 1974. Schnelle Identifizierung von Amanitinen in Pilzgeweben. Arch. Toxicol. *32*: 109-114.

108. Preston, J. F., H. J. Stark, and J. W. Kimbrough. 1975. Quantitation of amanitins in *Amanita verna* with calf thymus RNA polymerase B. Lloydia *38*: 153-161.

109. Prince, H. W. 1927. Mushroom poisoning due to *Hebeloma crustuliniforme*. Am. J. Dis. Child. *34*: 441-442.

110. Rappolt, R. T., and N. R. Rappolt. 1975. *C. atramentarius*, alcohol, beta blockers, and thee. Mycena News *26*(2): 3 (Mycological Society of San Francisco, San Francisco).

110a. Rappolt, T. R., Sr., G. Gay, D. S. Inaba, N. Rappolt, and R. T. Rappolt, Jr. 1978. Use of Inderal (propranolol-Ayerst) in I-a (early stimulative) and I-b (advanced stimulative) classification of cocaine and other sympathomimetic reactions. Clin. Toxicol. *13*(2): 325-332.

111. Repke, D. B., D. T. Leslie, and G. Guzmán. 1977. Baeocystin in *Psilocybe, Conocybe* and *Panaeolus*. Lloydia *40*: 566-578.

112. Rinaldi, A., and V. Tyndalo. 1974. *The Complete Book of Mushrooms*, Crown, New York.

113. Schmidt, J., W. Hartmann, A. Würstlin, and H. Deicher. 1971. Akutes Nierenversagen durch immunhämolytische Anämie nach Genuss des Kahlen Kremplings (*Paxillus involutus*). Dtsch. Med. Wochenschr. *96*: 1188-1191.

114. Schmiedeberg, O., and R. Koppe. 1869. *Das Muscarin, das giftige Alkaloid des Fliegenpilzes*, F. C. W. Vogel, Leipzig.

115. Schultes, R. E., and A. Hofmann. 1973. *The Botany and Chemistry of Hallucinogens*, Thomas, Springfield, Ill.

116. Scotti de Carolis, A., F. Lipparini, and V. G. Longo. 1969. Neuropharmacologic investigations on muscimol, a psychotropic drug extracted from *Amanita muscaria*. Psychopharmacologia *15*: 186-195.

117. Seeger, R., and H. Scharrer. 1972. Purification and characterization of phalloides hemolysin, a high molecular weight toxin from *Amanita phalloides*. Naunyn-Schmiedeberg's Arch. Pharmakol. *274*(Suppl.): R106.

118. Shaffer, R. L. 1965. Poisoning by *Pholiota squarrosa*. Mycologia *57*: 318-319.

119. Shaffer, R. L. 1968. *Keys to Genera of the Higher Fungi*, 2nd ed, The University of Michigan Biological Station, Ann Arbor.

120. Simandl, J., and J. Franc. 1956. Isolace tetraethylthiuramdisulfidu z hníku inkoustového (*Coprinus atramentarius*). Chem Listy *50*: 1862-1863.

121. Simons, D. M. 1978. The mushrooms toxins from a chemical point of view. Syllabus (Toxic Mushrooms, workshop), The New York Botanical Garden, The Bronx.

122. Singer, R., and A. H. Smith. 1958. Mycological investigations on Teonanácatl, the Mexican hallucinogenic mushroom. Part II. A taxonomic monograph of *Psilocybe*, section *Caerulescentes*. Mycologia *50*: 262-303.

123. Singer, R. 1975. *The Agaricales in Modern Taxonomy*, 3rd ed, J. Cramer, Vaduz.

124. Smith, A. H. 1949. *Mushrooms in Their Natural Habitats*, Sawyer's Inc., Portland.

125. Smith, A. H. 1966. *The Mushroom Hunter's Field Guide*, University of Michigan Press, Ann Arbor.

126. Smith, A. H. 1975. *A Field Guide to Western Mushrooms*, University of Michigan Press, Ann Arbor.

127. Stadelmann, R. J., E. Müller, and C. H. Eugster. 1976. Über die Verbreitung der stereomeren Muscarine innerhalb der Ordnung der Agaricales. Helv. Chim. Acta *59*: 2432-2436.

128. Stadelmann, R. J., E. Müller, and C. H. Eugster. 1977. Muscarine (Muscarin, epi- und allo-Muscarin) aus dem Mycel von *Amanita muscaria* und *Clitocybe* Arten. Sydowia *29*: 15-27.

129. Stevenson, J. A., and R. C. Benjamin. 1961. *Scleroderma* poisoning. Mycologia *53*: 438-439.

130. Takemoto, T., T. Nakajima, and T. Yokobe. 1964. Structure of ibotenic acid. Yakugaku Zasshi *84*: 1232-1233.

131. Takemoto, T., T. Nakajima, and R. Sakuma. 1964. Isolation of a flycidal constituent "ibotenic acid" from *Amanita muscaria* and *A. pantherina.* Yakugaku Zasshi *84*: 1233-1234.

132. Theobald, W., O. Büch, H. A. Kunz, P. Krupp, E. G. Stenger, and H. Heimann. 1968. Pharmakologische und experimental-psychologische Untersuchungen mit zwei Inhaltsstoffen des Fliegenpilzes (*Amanita muscaria*). Arzneim-Forsch. *18*: 311-315.

133. Toth, B., and H. Shimizu. 1973. Methylhydrazine tumorigenesis in Syrian golden hamsters and the morphology of malignant histiocytomas. Cancer Res. *33*: 2744-2753.

134. Tyler, V. E., Jr. 1963. Poisonous mushrooms. Prog. Chem. Toxicol. *1*: 339-384.

135. Walker, R. J., G. N. Woodruff, and G. A. Kerkut. 1971. The effect of ibotenic acid and muscimol on single neurons of the snail, *Helix aspersa.* Comp. Gen. Pharmacol. *2*: 168-174.

136. Waser, P. G. 1967. The pharmacology of *Amanita muscaria.* In D. H. Efron (Ed.), *Ethnopharmacologic Search for Psychoactive Drugs,* U.S. Public Health Service Publ. No. 1645, pp. 419-439.

137. Wasson, R. G. 1968. *Soma, Divine Mushroom of Immortality,* Harcourt Brace Jovanovich, New York.

138. Weiss, E., I. Sterz, M. Frimmer, and R. Kroker. 1973. Electron microscopy of isolated rat hepatocytes before and after treatment with phalloidin. Beitr. Pathol. *150*: 345-356.

139. Wiedmann, H., M. Taescher, and H. Konzett. 1958. Zur Pharmakologie von Psilocybin, einem Wirkstoff aus *Psilocybe mexicana* Heim. Experientia *14*: 378-379.

140. Wieland, Th. 1967. The toxic peptides of *Amanita phalloides.* Fortschr. Chem. Org. Naturst. *25*: 214-250.

141. Wieland, Th. 1968. Poisonous principles of mushrooms of the genus *Amanita.* Science *159*: 946-952.

142. Wieland, Th. 1972. Isolierung und chemische Bearbeitung der Inhaltsstoffe des grünen Knollenblätterpilzes (*Amanita phalloides*). Arzneim-Forsch. *22*: 2142-2146.

143. Wieland, Th. 1972. Struktur und Wirkung der Amatoxine. Naturwissenschaften *59*: 225-231.

144. Wieland, Th., G. Lüben, H. Ottenheym, J. Faesel, J. X. De Vries, A. Prox, and J. Schmid. 1968. The discovery, isolation, elucidation and synthesis of antamanide. Angew. Chem. Int. Ed. Engl. *7*: 204-208.

145. Wieland, Th., and O. Wieland. 1959. Chemistry and toxicology of the toxins of *Amanita phalloides.* Pharmacol. Rev. *11*: 87-109.

146. Wieland, Th., and O. Wieland. 1972. The toxic peptides of *Amanita* species. In S. Kadis, A. Ciegler, and S. J. Ajl (Eds.), *Microbial Toxins,* Vol. 8, Academic, New York, pp. 249-280.

147. Wier, J. K., and V. E. Tyler, Jr. 1960. An investigation of *Coprinus atramentarius* for the presence of disulfiram. J. Am. Pharm. Assoc. Sci. Ed. *49*: 426-429.
148. Yocum, R. R., and D. M. Simons. 1977. Amatoxins and phallotoxins in *Amanita* species of the northeastern United States. Lloydia *40*: 178-190.
149. Yokoyama, K. 1973. Poisoning by a hallucinogenic mushroom, *Psilocybe subcaerulipes* Hongo. Trans. Mycol. Soc. Jpn. *14*: 317-320.

10

Poisonous Fungi: Mycotoxins and Mycotoxicoses

A. Ciegler / Southern Regional Research Center, U.S. Department of Agriculture, New Orleans, Louisiana

H. R. Burmeister and R. F. Vesonder / Northern Regional Research Center, U.S. Department of Agriculture, Peoria, Illinois

I. INTRODUCTION

Mycotoxins are mold-produced, toxic, secondary metabolites that can cause harmful effects in animals or birds. The diseases caused by these metabolites, either by contact or by inadvertent ingestion of the toxin, are called mycotoxicoses. Mycotoxicoses differ from mycoses in that mold growth in the host is not directly involved. Mycoses usually involve a generalized invasion of living tissues by an actively growing fungus. This definition of a mycotoxin could also encompass some fungal metabolites that function as antibiotics; the difference appears to be one of degree rather than kind. Hence, under the proper circumstances, for example the strain of fungus, a particular physiochemical environment, a specific host, or the nutritional and health status of a host, many fungi can produce compounds that could effect a mycotoxicosis.

The problem is further exacerbated by the multiplicity of fungi that may coexist on an agricultural commodity or that may develop in sequence. Diverse fungal products can accumulate in a substrate either by simultaneous elaboration of several compounds by a single strain or by sequential development of several species and associated toxin production. Toxic interactions in which the cooperative effect of two or more substances elicits a total effect greater than the sum of the activities of individual agents (synergism) probably represents the norm in nature. This facet has been noted by many mycotoxin investigators, who usually cannot ascribe the symptoms observed in any given outbreak to any one specific toxin [20,47,196,330,389,439,440]. The topic of mycotoxin synergism has been reviewed in detail [208].

Certain diagnostic features characterize outbreaks of mycotoxicoses. Although outlined some time ago by Feuell [110], they still remain valid and pertinent:

1. Mycotoxicoses are not transmissible.
2. Drug and antibiotic treatment have little or no effect on the disease.
3. In field outbreaks, the trouble is often seasonal.
4. The outbreak is usually associated with a specific food or feedstuff.
5. The degree of toxicity is often influenced by the age, sex, and nutritional state of the host.
6. Examination of the suspected food or feed reveals signs of fungal activity.

A potential causal relationship between human consumption of moldy foods and various illnesses has been long suspected and was actually established with respect to the ergot alkaloids in the eighteenth century. Intoxications implicating moldy food in the nineteenth and twentieth centuries have been described recently [368].

Mycotoxicoses constitute a worldwide problem affecting both humans and animals. These diseases occasionally are manifested in an acute manner, but insidious effects probably constitute the current major portion of the problem, for example carcinogenesis in humans and loss of weight gains and feed efficiency in farm animals. Compounds probably involved or suspected of being involved in mycotoxicoses are shown in Tables 1 and 2.

The prevalence of mycotoxins in the environment is a difficult question to assess accurately. The simple approach used in most research laboratories, normally involving only a single fungus and a single toxin, can at best only approximate what occurs in the field. What toxins, then, are we to be concerned about and what are the conditions that lead to their occurrence? Some subjectivity is involved in any consideration of these points, for data are often lacking and all the variables are not well understood.

Table 1 Compounds Probably Involved in Mycotoxicoses

Toxin	Producing fungi	Susceptible host	Biological effects
Aflatoxins	*Aspergillus flavus,* *A. parasiticus*	Mammals, fish	Hepatotoxin, cancer
Penitrem A	*Penicillium palitans,* *P. crustosum*	Cattle, horses, sheep	Tremorgenic, convulsant
T-2	*Fusarium tricinctum*	Cattle, humans?	Dermal necrosis, hemorrhage
F-2	*Gibberella zeae*	Swine	Vulvovaginitis, abortion
Slaframine	*Rhizoctonia leguminicola*	Cattle	Excess salivation
Sporidesmins	*Pithomyces chartarum*	Swine, sheep	Hepatotoxin, facial eczema
Ochratoxin A	*Penicillium viridicatum,* *A. ochraceous*	Swine, humans?	Nephrotoxin
Psoralens	*Sclerotinia sclerotiorum*	Humans	Dermotoxin
Citrinin	*P. viridicatum,* *P. citrinum*	Swine	Nephrotoxin
Vomitoxin	*F. graminearum*	Swine, humans?	Vomiting
Maltoryzine	*A. oryzae*	Cattle	Death
Unidentified	*Phomopsis leptostromiformis*	Sheep	Hepatotoxin
Diplodiatoxin	*Diplodia maydis*	Cattle, sheep	Nephritis, mucoenteritis
Unidentified	*Phoma sorghina*	Humans	Hemorrhage, bullae mouth, rapid death
Secalonic acids D, F	*Penicillium oxalicum*	Humans	Death
Satratoxins	*Stachybotrys atra*	Horses, humans	Hemorrhage

Obviously, factors that influence mold growth on a commodity probably also influence toxin synthesis. Among such factors are moisture content, relative humidity, temperature, substrate composition, presence of competing microorganisms, and fungal strain. These points have been examined in detail in two reviews [146,227]. A current overview on the biochemistry and pathology of mycotoxins has been published by Uraguchi and Yamazaki [415]. Additional reviews that may be consulted are in Refs. 65, 67, 127, 314, 315, and 448.

An awareness of the relationship between moldy food and feeds and toxicoses was first established by the Japanese in the late nineteenth century with respect to "yellow rice disease" in Japan. Many deaths resulted from the consumption of moldy rice imported from southeast Asia [320]. In the USSR, reports of vomiting in humans who consumed moldy cereal grains had been recorded periodically since 1916, with a particularly violent outbreak in which thousands died occurring during World War II. Unfortunately, the literature describing the toxicoses in both Japan and the USSR was published in lan-

Table 2 Compounds Suspected of Being Mycotoxins

Toxin	Producing fungi	Possible host	Biological effects
Sterigmatocystin	*Aspergillus flavus*	Mammals	Carcinogen
Yellow rice toxins			
Luteoskyrin	*Penicillium islandicum*	Humans	Hepatotoxin
Cyclochlorotine	*P. islandicum*	Humans	Hepatotoxin
Citreoviridin	*P. citreoviride*	Humans	Neurotoxin
Rugulosin	*P. rugulosum*	Humans	Carcinogen
Rubratoxin	*P. rubrum*	Cattle	Hepatotoxin
Fusaranon-X	*Fusarium nivale*	Humans, swine	Vomiting
Nivalenol	*F. nivale*	Humans, swine	Vomiting
Cytochalasin E	*A. clavatus*	Humans	Death
PR toxin	*P. roqueforti*	Cattle	Abortion
Patulin	*P. urticae*	Cattle	Death
Penicillic acid	*Penicillium* spp.	Farm animals	No data

guages normally not read by Western scientists, so mycotoxicoses remained "neglected diseases" in the West [112]. In 1913, Alsberg and Black of the U.S. Department of Agriculture (USDA) concluded from their work on penicillic acid that products of mold growth could be involved in disease. The mycotoxin hypothesis was clearly stated in this classic publication [4].

The current era of mycotoxin research developed as a direct result of concurrent outbreaks of diseases in poultry and fish during 1960 in diverse locations. Most notoriety was given to the report of severe losses of turkey poults in Britain [31] as a result of an unknown etiological agent. The disorder was therefore called "turkey X disease." Acute manifestations of the disorder were loss of appetite, lethargy, wing weakness, and a distinctive attitude of the head and neck at time of death [360]. Histopathology revealed acute hepatic necrosis associated with bile duct proliferation [31]. Examination of the feed source showed that a common factor in the disease outbreaks was the utilization of a Brazilian groundnut meal in the rations [31,52].

At the same time, symptoms analogous to those of the turkey X syndrome were reported in outbreaks of other farm animals that had consumed feed whose groundnut component had originated from non-Brazilian sources [17]. Turkey X disease rapidly escalated from a relatively minor problem of concern only to veterinarians and poultry porducers to a problem of worldwide concern when it was determined subsequently that the toxic meal was also capable of producing hepatomas in laboratory animals [84,201].

Aspergillus flavus was soon implicated as the fungus involved in the turkey outbreak [324], and the aflatoxins were determined to be the secondary metabolites responsible both for toxicity and carcinogenicity in test animals [262,324,357,418]. The modern era of mycotoxin research was launched, for it was soon realized that the aflatoxins represented only one facet of an immense problem.

II. *ASPERGILLUS* TOXINS

A. Aflatoxin

1. Toxin-Producing Fungi

Since the initial implication of *A. flavus* in turkey X disease, aflatoxin production by this organism and by *A. parasiticus* has been well documented. *Aspergillus parasiticus* seems to be restricted primarily to tropical or semitropical regions, with all strains isolated to date being toxin producers [150,151]. However, not all strains of *A. flavus,* including the type culture (*A. flavus* NRRL 782), produce aflatoxin. The role played by aflatoxin as a selective or defensive factor for the fungus in nature has not been determined.

Other fungi reportedly produce aflatoxins, but none of these findings has been confirmed [138,139,155,198,248,329,341,392,421]. Evidence that only *A. flavus* or *A. parasiticus* can produce aflatoxins has been advanced by several laboratories [44,159,247, 274,439,440]. Within the past 10 years no further claims have been made for aflatoxin production by fungi other than *A. flavus* or *A. parasiticus.* Some of the apparently contradictory data may have arisen because various fungi produce fluorescent compounds that chromatograph with an R_f approximating that of the aflatoxins. Simple confirmatory tests now available can readily resolve this difficulty [10,14,302,362].

Since the initial discovery of the first four aflatoxins (aflatoxins B_1, B_2, G_1, G_2), additional members of this family of compounds have been isolated, many of them being the products of metabolism of aflatoxin B_1 by various birds or mammals.

2. Natural Occurrence

Aspergillus flavus seems to be ubiquitous and accordingly contaminates a wide variety of foods and feeds. In particular, corn and peanuts represent major potentially contaminated commodities. Soybeans are the major crop resistant to toxin contamination even though the fungus can grow readily upon the bean. Apparently, the chelation of zinc by phytic acid in soybeans prevents aflatoxin formation; zinc is necessary for aflatoxin synthesis [229].

Reports of laboratory formation of aflatoxin on various foods and feeds should be distinguished from the natural occurrence of the toxin. Nevertheless, the worldwide reports indicate that natural contamination is widespread (Table 3). Toxin levels found are not given in the table to avoid misinterpretation of the data as being absolute.

3. Production and Detection

Production may be divided into two aspects: production under natural conditions and production in the laboratory. Aflatoxin was initially believed to be a storage problem, but investigators at the Quaker Oats Co. [9] and the Northern Regional Research Center [209,212-214] determined that *A. flavus* can invade corn and produce aflatoxin before harvest. Similar findings were reported for cottonseed by Marsh et al. [225,226] and for peanuts by Pettit and Taber [289]. Factors affecting occurrence and production in the field are complex and not well understood. Insects seem to be vectors for the fungus, although the data are not definitive [15,108,137,209,237]. Other factors probably involved include cultivar and region in which planted [213,238,454]; moisture, drought, and irrigation stresses [87-89,137,290,319]; stage of fruit development and weather [211]; harvesting practice, crop sequence, and weather, including temperature cycles [289,373];

Table 3 Natural Occurrence of Aflatoxin in Agricultural Commodities

Aflatoxin B_1	Country	Reference
Corn (maize)	United States	354
Corn	Uganda	3
Corn	India	187
Corn	Thailand, Hong Kong	344
Corn	Philippines	51
Corn	Dominican Republic	203
Corn	Australia	74
Corn	France	200
Peanuts (groundnuts)	Thailand, Vietnam Mainland China	55
Peanut oil	Malaya	56
Peanuts	Nigeria	233
Peanuts	Uganda	218
Peanuts, expeller cake, peanut meal	Brazil, Nigeria, Sudan, Senegal, Indonesia, Kenya, Uganda, Argentina, Ghana, Congo	195
Peanuts	India	54
Peanuts	Thailand	344
Peanut butter	Philippines	51
Peanuts	Switzerland	184
Peanuts	Kenya	282
Peanut meal	Indonesia, United States, Brazil, India	224
Peanut plant parts	France	35
Farmer's stock peanuts	United States	85
Virginia peanuts	United States	90
Spanish peanuts	United States	289
Peanuts	Angola, Gambia, Ghana, Madagascar, Malawi, Mali, Mozambique, Nigeria, Bangladesh, Burma, India, Indonesia, Philippines, Taiwan, France, Poland, Spain	178
Peanut oil	India	95
Peanut oil, peanut products	Taiwan	216
Cottonseed meal	United States	169
Cottonseed and products	United States	319,415,435
Cottonseed meal, crush, cake	Brazil, Colombia, Guatemala, Nicaragua, El Salvador, Syria, Turkey, USSR	134

(continued)

Table 3   (continued)

Aflatoxin B_1	Country	Reference
Cottonseed	India	422
Pecans	United States	434
Pistachio	Turkey, Iran	81
Figs	Near East	11
Aleppo pine nuts	Tunisia	37
Hazelnut, walnut	Germany	126
Almonds	United States	328
Coconut and products	Ceylon	13
Walnut, almond	United States	118
Hazelnut, almonds, mixed nuts	Not stated	284
Harikot bean and other pulses	Egypt	133
Cheese	France	170
Dried milk products	Germany	265
Milk	Germany	185
Buffalo milk	India	449
Milk, milk products (barfi, khoa, paneer)	India	280
Milk	South Africa	222, 304
Milk	Iran	375
Milk, milk powder, yogurt, fresh cheese, hard cheese, camembert, processed cheese	Germany	297

strain of fungus; and composition of a mixed microbial milieu [16,209,259]. Additional undefined factors are undoubtedly involved, including the interplay of the variables.

On the other hand, laboratory production of aflatoxin falls under the procedures of fermentation technology. Two general approaches have been used: (1) those involving liquid media and (2) the "newer" approach in which solid substrates such as grains are fermented.

4. Analytical Procedures

Aflatoxin-detection procedures may be divided into three categories: (1) screening, (2) presumptive tests, and (3) confirmatory and analytical procedures.

Screening in cereals may employ a "black-light test." Under high-intensity ultraviolet light, *A. flavus*-infected grains exhibit a bright greenish yellow fluorescence (BGY) in the germ margin [352]. This fluorescence is not aflatoxin but is probably a kojic acid derivative whose structure has not yet been determined [225]. No quantitative relationship exists between BGY and aflatoxin, and, in general, only about half the samples of corn found BGY-positive also contain aflatoxin [210]. Usually, samples found positive are further analyzed by a minicolumn procedure.

A number of minicolumn procedures exist, but the principle for all of them is essentially the same. A solvent extract of the suspect sample is prepared. A portion is passed through a small glass column containing one or more layers of absorbent; an aflatoxin standard is absorbed onto a control column and the two columns observed under ultraviolet light; aflatoxin appears as a narrow blue-fluorescing band [156,157,299,348,423].

For quantitative analyses, the Official Methods of Analysis of the Association of Official Analytical Chemists (AOAC) are generally followed [18]. The AOAC procedures encompass a representative sampling technique (the amount varies from commodity to commodity), solvent extraction (one phase and two phase systems), precipitating agents (removal of protein and other interfering substances such as pigments), cleanup column chromatography (silica gel, cellulose), and thin layer chromatography (TLC) separation of the aflatoxins with measurement by fluorodensitometry or visual methods.

The AOAC procedures for aflatoxin assay have been applied with variations in precipitating reagents, extraction solvents, and chromatographic procedures to many commodities, including peanuts, peanut meal, cottonseeds, cottonseed meal, copra [18,19,99,235, 300,311,424], corn [350], roasted corn [298], mixed feeds, nuts [19,317], spices and herbs [332], cocoa beans [167], groundnuts and groundnut products [75], coffee beans [205], and cereal grains [149].

The accurate measurement of aflatoxins on thin layer chromatographic plates by either visual or fluorodensitometric methods depends on adequate sampling of the agricultural product. The nonuniform distribution of aflatoxin, as well as the relatively small amount present (usually 1-1000 ppb), necessitates that all of the commodity be evaluated to obtain the level of contamination in a unit lot. Because this is impractical, large representative samples from commodity lots are obtained by physical selection (probes, stream splitters). Product samples are ground and blended to give homogeneous toxin distribution with particle size correct to pass through a U.S. standard No. 20 sieve. Usually, 50 g

Table 4 LD_{50} of Aflatoxin for Laboratory Animals

Species	LD_{50} (single dose) (mg/kg body weight)
Duckling	0.3-0.6
Pig	0.6
Trout	0.8
Dog	1.0
Guinea pig	1.4-2.0
Sheep	2.0
Monkey	2.2
Rat	5.5-17.9
Chicken	6.3
Mouse	9.0

Table 5 Possible Cases of Fatal Acute Aflatoxicosis in Humans

Case	Country	Food source involved	Reference
Boy (15 yr)	Uganda	Cassava	343a
Child	Germany	Peanuts	32a
Girl (22 months)	Czechoslovakia	Rice product (?)	94
Girl (12 months)		Not known	
Girl (8 months)		Not known	
397 patients (106 deaths; 2:1 male:female)	India	Corn	187, 188
67 children (1-6 yrs)	Northeast Thailand	Not stated	36
23 children	North Thailand	Not stated	270
3 children	United States	Peanuts	A. W. Hays (personal communication)
Boy (22 months) Girl (8 months)	New Zealand	Rice (?)	26

of these homogeneous, ground samples are taken for toxin analyses. Samples after homogenization are analyzed for aflatoxins according to the CB or BF methods established for peanuts [18].

The use of high-pressure liquid chromatography (HPLC) as a possible alternative to the TLC fluorodensitometry and visual analytical procedures is attractive because of high-resolution potential, rapidity of separations, and potentially improved quantitative accuracy and precision [121,163,343]. However, HPLC does not seem to be as amenable as TLC for analysis of large sample numbers and, additionally, interfering substances contained in extracts obtained from naturally contaminated foods and feeds impede the measurement of aflatoxins by HPLC.

5. Biological Activity

The toxicity of aflatoxins varies according to the target host, dosage, and duration of exposure. Response also varies between individuals of a given species. The LD_{50} for a number of laboratory animals is given in Table 4.

The aflatoxins do not seem to represent a primary threat as an acute toxin to humans, although some suspect cases have been noted; representative findings are summarized in Table 5.

6. Possible Cases of Fatal Acute Aflatoxicoses in Humans

The outbreak in India and the cases of encephalopathy and fatty degeneration of the viscera (EFDV) in northeast Thailand noted in Table 5 [36,187,188,270,345] are worthy of further comment. In late 1974 and early 1975, over 200 villages in western India experienced an outbreak of a disease affecting humans and dogs [187,188]. The disease was characterized by jaundice, rapidly developing ascites, portal hypertension, and a high mortality rate (106 deaths among 397 cases), with death usually resulting from massive gastrointestinal tract bleeding. The disease was confined to the very poor, who were forced to consume badly molded corn containing aflatoxin between 6.5 and 15.6 ppm, an average daily intake of 2-6 mg of aflatoxin per person. Analyses of liver, sera, and urine were inconclusive, probably because of a lag between corn consumption and sample collection.

In northeast Thailand, where there is a high aflatoxin incidence in the food, EFDV is a common cause of death among children, in whom it resembles Reye's syndrome. Almost all cases occur in children from rural areas, with incidence increasing during the latter part of the rainy season. There is usually fever, vomiting, hypoglycemia, a rapid unexpected onset of convulsions, coma, and death; death usually occurs within 24 h of the onset of symptoms [36,270]. Several investigators have found aflatoxin in livers by biopsy [26,94,345]. In Missouri, aflatoxin has also been found in the liver of an adult who had carcinoma of the rectum and liver [291].

The cause of Reye's syndrome is unknown and may represent a response to a variety of insults, including a virus [153]. However, Becroft and Webster [26] and others have suggested an aflatoxin etiology; the facts are reviewed by Harwig et al. [142].

Although aflatoxin can function as an acute toxin, its potential carcinogenicity in humans is generally of greater concern. Aflatoxins are carcinogenic for a variety of laboratory animals, including primates [1,309,385], in whom the symptoms and pathology of aflatoxicosis closely resemble some types of human hepatic disease thought to be caused by aflatoxin. Obviously, experimentation on humans is impossible; hence, evidence for chronic toxicity in humans is based on epidemiological studies.

The dosage required to elicit toxicosis in primates [128] has been extrapolated to humans by Campbell and Stoloff [51]. They calculated that children eating food contaminated with 1.7 mg of aflatoxin per kilogram of body weight could develop serious liver damage within a short time. A single dose of 75 mg/kg could result in death, whereas 51.3 mg/kg would have no apparent effect. Daily doses of 0.34 mg/kg would also have no apparent effect. These figures should be accepted cautiously since they are extrapolations derived from primate data.

7. Cirrhosis

Aflatoxin has been implicated in cirrhosis in rhesus monkeys [223] and perhaps in Indian childhood cirrhosis. The disease seems to correlate with religious dietary habits of a middle-income group examined in the Mysore area [7]. The children involved consumed parboiled rice and crude peanut oil, which could have been high in aflatoxin because alkali refining is not used. In these studies, Amla et al. [6,7] reported that 20 children being treated for kwashiorkor were given 30-60 g/day of peanut meal for 5 days to 1 month; the meal was later found to contain 0.3 mg of aflatoxin B_1 per kilogram of meal. Three children on this supplement for 17 days showed liver cirrhosis. After withdrawal of the meal, there was a gradual transition from fatty to cirrhotic liver over a 1-year period, but characteristic clinical and histopathological signs were not noted before 6 months; histopathology seemed to correlate with duration of toxic meal consumption.

This study was subsequently expanded to include 255 cases of childhood cirrhosis in which urine [8] was examined for aflatoxins B_1 and M; B_1 but not M was found in 7% of the urine samples. Curiously, breast-milk samples from 25% of the mothers of affected children showed the presence of an aflatoxin with an R_f of B_1. This finding is in contrast to all other reports involving milk in which aflatoxin M but rarely B_1 is found. Yadagiri and Tulpule [449], in a similar study involving 16 cirrhotic children, did not find afla-toxin B_1 in the urine. Unfortunately, HCl had been added to the samples so that any B_1 would have been converted to B_{2a}, which was not reported. In an earlier study on relating cirrhosis and aflatoxin consumption in India, Robinson [312] also found a variety of violet-fluorescing spots in urine and mother's milk, but he lacked the technical facilities for positive identification.

In contrast to the data cited above, a study in the Philippines, where childhood cirrho-sis is not reported as endemic, although aflatoxin was consumed in peanut butter, revealed only aflatoxin M_1 in urine and breast milk [50]. Amla et al., in 1974 [8], noted neither cirrhosis in liver biopsy material nor clinical symptoms when peanut meals with an afla-toxin content averaging 17 μg/kg were used for kwashiorkor therapy. Additionally, van Rensburg [419] believes that the aflatoxin intakes reported by the Indian workers were too low to be responsible for the observed effects, on the basis that aflatoxin is not known to be a potent cirrhogen. However, this finding does not rule out a possible contributory role of aflatoxin in the disease.

It is difficult to resolve the differences shown between the studies in India and the Philippines, so the problem of aflatoxin acting as a cirrhogen must await further study.

8. Aflatoxin Carcinogenesis

The first evidence that aflatoxin is carcinogenic was published in 1961. Lancaster et al. [201] reported that 9 of 11 weanling rats fed for 7 months on a purified diet containing 20% toxic peanut meal developed multiple liver tumors, with lung metastases in two ani-mals. Since that report, aflatoxin has been shown to be carcinogenic to a variety of labor-atory animals, including nonhuman primates; the subject has been reviewed extensively [101,147,217,241,266,271].

In humans, evidence linking aflatoxin with carcinogenesis depends of necessity on epidemiological investigations. Epidemiological research requires accumulation of data that usually lack the rigorous controls demanded by most scientific investigations, so that "evidence" must be evaluated cautiously. But as van Rensburg [419] stated: "However laudable caution may be, there comes a point in the accumulation of evidence when it is more prudent to assume that the association is causal until proven otherwise. The *un-answerable* [italic added] question of whether the hypothesis is 'proven' is irrelevant." We believe this to be the current status implicating aflatoxin in human carcinogenesis.

An excellent summary of the epidemiological data up to 1974 has been published by Campbell and Stoloff [51]. Epidemiological investigations relating high incidence levels of hepatomas to aflatoxin ingestion have been reported for Mozambique [420] and the Philippines [43]. However, a study on varying incidence of esophageal cancer in the Caspian littoral of Iran showed only a low level of aflatoxin in the area [161].

Several tentative conclusions can be drawn:

1. Chronic ingestion levels more closely parallel cancer incidence than occasional high exposure [282,344,346,347,420].
2. Males appear to be more susceptible than females [184,282,283,442].
3. Climates that favor growth of *A. flavus* concomitant with poor food storage practices tend to favor increased hepatoma rates, indicating a cause-and-effect relationship [282, 344,346,347].

4. Over a wide range there is a linear relationship between cancer incidence and the
    logarithm of the level of aflatoxin intake, particularly in males [282,419].

A cause-and-effect relationship cannot be "proven" experimentally, but the weight
of epidemiological evidence implicates aflatoxin as a carcinogen in humans. In the absence
of contrary evidence, most progress can probably be made in solving the disease aspect by
at least a tentative acceptance of the aflatoxin hypothesis.

9. Aflatoxin Metabolism

Aflatoxin B_1 can be metabolized by higher animals and birds via several pathways, resulting in several transformation products. Various workers have indicated that aflatoxin B_1
requires activation to become metabolically active [120,122,123,380]. Garner [122]
postulated that of the various derivatives formed, aflatoxins B_{2a} and the epoxide represent the active forms responsible for acute toxicity and carcinogenicity, whereas the other
derivatives are probably detoxification products. However, some of these latter products,
e.g., aflatoxin M_1 and aflatoxin R_0, are toxic and carcinogenic.

Animal species vary in their susceptibility to aflatoxin B_1, presumably as a result of
differences in their metabolism of the toxin. However, Roebuck and Wogan [316], in in
vitro studies utilizing liver fractions from various species, including humans, were unable
to find a consistent pattern of metabolism that could be correlated with species differences
in response to aflatoxin B_1 toxicity or carcinogenicity. Some difficulty in correlation
may arise from the ability of various species, such as chickens, ducks, turkeys, rabbits, and
primates, to produce aflatoxicol (aflatoxin R_0) [278,323], although this derivative has
not been detected in the mouse, rat, or guinea pig [279]. Loveland et al. [219] presented
important evidence to show that R_0 could be converted back to B_1 by trout postmitochondrial enzymes; hence, R_0 could serve as a reservoir for the release of B_1 to target cells,
thereby enhancing the toxic effect of B_1. Hsieh et al. [164] have shown that a species'
ability to reduce B_1 to R_0 is related to the sensitivity of the species to acute aflatoxicosis
and that the ratio of reductive and oxidative activities (R_0/B_1) or (R_0/B_1 soluble metabolites) is an index of species susceptibility to the carcinogenic effect of aflatoxin B_1.

On the molecular level, considerable evidence indicates that aflatoxin B_1 requires
metabolism to aflatoxin B_1-2,3-oxide to exert its carcinogenic and mutagenic effects [120,
380]. Indirect evidence for the epoxidation is provided by the formation from aflatoxin
B_1 of toxic and mutagenic metabolites via cytochrome P_{450}-dependent microsomal oxidation of aflatoxins that contain 2,3 double bonds [120,232] and by the lower carcinogenicity of aflatoxin B_2 compared to B_1 [443]. Swenson et al. [377-379] presented
direct evidence on the isolation of 2,3-dihydro-2,3-dihydroxyaflatoxin B_1 from acid
hydrolysates of nucleic acid-aflatoxin from the livers of rats that had been given injections
of [^{3}H] aflatoxin B_1. Lin et al. [215] advanced further proof when they identified 2,3-
dihydro-2-(guan-7-yl)-3-hydroxyaflatoxin B_1 as the major acid hydrolysis product of aflatoxin B_1-DNA or -ribosomal RNA adducts formed in hepatic microsomal-mediated reactions with salmon sperm DNA and rat liver ribosomal RNA in vitro or from rat liver in
vivo. Two minor products were also isolated, one of which was identified as 2,3-dihydro-
2-(N^5-formyl-2,5,6,-triamino-4-oxopyrimidin-N^5-yl)-3-hydroxyaflatoxin B_1. However,
the role of these reactions, if any, in the mediation of mutagenesis of carcinogenesis has
yet to be elucidated.

10. Control Measures

Aflatoxin contamination of agricultural commodities was previously thought to result primarily from improper storage. Subsequent comprehensive studies have revealed that aflatoxin contamination, particularly of corn, the major commodity affected, can start in the field before harvest [9]. With respect to control, this finding has changed the entire concept of the problem. Current data indicate that aflatoxin contamination can occur at any part of the production, distribution, or usage cycle. This aspect, plus the wide variety of commodities affected, dictates a multiplicity of control approaches and measures.

In the field the obvious solution is the breeding of crops either resistant to the growth of the fungus or able to inhibit toxin biosynthesis. Some corn lines have shown promise, but an insufficient number of trials have been conducted to determine data reproducibility or the effect of climate or geography [453]. In peanuts, cultivars showing genetic resistance to invasion by A. flavus have been found, but, unfortunately, resistant genotypes give low peanut yields and cannot be used by the industry [249]. However, over the long range, the genetic approach to control will probably be the most fruitful.

For more immediate control other aspects of the problem have been investigated. For example, insects play a potential role as vectors in infection, carrying spores of toxin-producing fungi, although a cause-and-effect relationship has not been established [211, 436]. Unfortunately, use of insecticides has not precluded A. flavus infection of corn and subsequent toxin production.

Damage to grain and peanuts by mechanical harvesters render these commodities more susceptible to invasion by A. flavus. Obviously, more efficient units must be designed if control is to be achieved.

Moisture levels above 15% in commodities are conducive to contamination by A. flavus. Hence grain such as corn, which is often harvested at moisture levels around 30%, must be dried as rapidly as possible. A delay between harvest and drying can result in a considerable increase in toxin level. Adequate storage, although simple in concept, is often difficult in practice.

Various chemicals for prevention of mold growth on grains have been proposed, and a limited number of substances are in use commercially. These compounds, e.g., acetic and propionic acids and ammonia, have received attention as a means of preserving high-moisture corn [33,267]. At proper concentrations they immediately eliminate molds and yeasts, but in about 30 days, a secondary fungal growth ensues. Scopulariopsis predominates on ammonia-treated corn, and A. flavus is dominant on acetic-propionic-treated corn [33]. A possible solution to the problem has been achieved by Nofsinger et al. [267]. They intermittently applied small amounts of gaseous ammonia to high-moisture corn, permitting ambient air drying and preservation over 6 months without corn spoilage. The low percentage of ammonia (0.009-0.09% w/w) would not detoxify any aflatoxin present.

Physical sorting or culling is practical primarily in the peanut industry, where it is accomplished by screening at shelling plants, by hand removal of discolored kernels, and by electronic sorting devices that pass or reject on the basis of color. Detailed descriptions of aflatoxin control as practiced in the peanut industry have been published [86,384]. Culling is impractical for various reasons, primarily financial, when applied to corn, and is not practiced.

In the corn industry, the "black-light test" has been used as a preliminary screen to detect potentially contaminated lots. Under ultraviolet light, *A. flavus*-infected kernels exhibit a bright greenish yellow (BGY) fluorescence in the germ margin [352]. This fluorescence is not aflatoxin but is possibly an oxidized derivative of kojic acid [225, 226]. However, the presence of BGY is only an indicator of potential toxin contamination, and chemical assays should be conducted. Similar fluorescence has been observed in infected cottonseed, wheat, oats, barley, and sorghum, in which it also may be used as an aflatoxin presumptive test [34]. In cottonseed, removal of partially bald seed also helps control aflatoxin contamination [76].

A simple and obvious procedure would be the dilution of contaminated commodity with noncontaminated product to a safe toxin level or one that meets U.S. Food and Drug Administration (FDA) guidelines. Under the Delaney Amendment to the Federal Food, Drug and Cosmetics Act as revised in October 1962, Section 409c. 3A, a zero tolerance was established for the addition of any carcinogen to food or feed; aflatoxin falls under this act. However, from a practical standpoint, the FDA has permitted a guideline of 20 ppb in foods and feeds; this subsequently was lowered to 0.5 ppb for whole, skim, and low-fat milk. Legislation has been proposed by the FDA (*Federal Register,* December 6, 1974) to lower the guidelines in manufactured peanut products from 20 to 15 ppb.

Under these circumstances dilution of an aflatoxin-contaminated commodity is considered illegal by the FDA, and this interpretation has been upheld in the U.S. courts [314]. However, a one-time variance was granted for badly contaminated 1977-crop-year feed corn in Alabama, Florida, Georgia, Mississippi, North and South Carolina, and Virginia [107a]. In Europe, dilution is permitted for feedstuffs, but the maximum admissible toxin level varies based on the consuming animal [113]. In Germany, an ordinance that was effective March 1, 1977, limited the aflatoxin level in foodstuffs to 10 ppb aflatoxin $B_1 + B_2 + G_1 + G_2$, with the qualification that the aflatoxin B_1 level must not be higher than 5 ppb.

Solvent extraction of aflatoxins from contaminated agricultural products has been advocated. Most of the research has been conducted at the Southern Regional Research Center of the USDA in New Orleans, Louisiana. A review of their studies has recently been published, but commercialization has not taken place [308].

Chemical detoxification appears to be the approach currently receiving most research attention. This subject has been extensively reviewed and will be touched on only briefly here [25,63]. In India, hydrogen peroxide has been used commercially for detoxifying peanut protein for human food [361]. However, ammoniation for detoxification of oilseed meals and of corn seems to be the most promising approach. Gardner et al. [119] treated ton lots of heavily contaminated cottonseed meal with ammonia under 3 atm at 118°C for 30 min and reduced aflatoxin levels to well below FDA guidelines. A similar procedure has been successfully employed on corn, but without heat or pressure [39]. Both procedures are undergoing animal trials, and the data obtained to date are encouraging.

B. Ochratoxins

The ochratoxins constitute a family of seven closely related metabolites composed of a phenylalanine moiety amide-linked to a 3,4-dihydroisocoumarin system. Of the seven, only ochratoxin A is a serious natural contaminant in foods and feeds [58]. Van der Merwe et al. [417] initially isolated ochratoxin A from *A. ochraceus,* but these compounds were subsequently found in six additional species of *Aspergillus (A. alliaceus, A. ostianus, A. melleus, A. petrakii, A. sclerotiorum, A. sulphureus)* and in six species of *Penicillium (P. commune, P. cyclopium, P. palitans, P. purpurescens, P. variabile, P. viridicatum)* [61]. With respect to natural contamination of agricultural commodities, *P. viridicatum* seems to be primarily involved; hence, further discussion of the ochratoxins will be found in the *Penicillium* toxin section of this chapter.

C. *Aspergillus fumigatus* Toxins

This fungus has been isolated from moldy silage implicated in the deaths of beef cattle [73,356]. *Aspergillus fumigatus,* a known pathogen of animals and humans, seems to occur primarily as a storage fungus rather than as a field microorganism. Fumigillim [96] and fumigatoxin [168], compounds with antibiotic properties, have been isolated from *A. fumigatus,* but they have not been implicated in mycotoxicoses, although fumigatoxin does kill mice. Cole et al. [73] isolated two clavine alkaloids, fumigaclavine A and C, as well as several tremorgens belonging to the fumitremorgen group, from strains of *A. fumigatus* found on molded silage. The LD_{50} of fumigaclavine C was about 150 mg/kg oral dose in 1-day-old cockerels. Calves dosed with crude extracts of *A. fumigatus* cultures experienced severe diarrhea, irritability, and loss of appetite; postmortem examination showed serious enteritis and evidence of interstitial changes in the lungs but no abnormalities in any other tissues. Undoubtedly, *A. fumigatus* plays a role in mycotoxicoses, but the extent of this role has yet to be determined.

D. Miscellaneous *Aspergillus* Toxins

Numerous compounds were isolated from members of the genus *Aspergillus,* particularly during the 1940s, when a massive research effort was directed toward the discovery of new antibiotics. Many of the substances isolated, although possessing antibiotic properties, were too toxic for therapy; hence, they were forgotten. Undoubtedly, many of these substances may be involved in mycotoxicoses, but data on their toxicology or occurrence in nature are lacking. The current emphasis on mycotoxin research may revive interest in some of these compounds and perhaps uncover new applications.

E. Pyrazine-Ring Compounds

In addition to the aflatoxins, strains of *A. flavus* produce a number of substances possessing in common the pyrazine ring structure, including aspergillic acid ($C_{12}H_{20}O_3N_2$); hydroxyaspergillic acid ($C_{12}H_{20}O_3N_2$); flavacol ($C_{12}H_2ON_2$), produced by *A. sclerotiorum;* neohydroxyaspergillic acid ($C_{12}H_{20}O_3N_2$), also produced by *A. sclerotiorum;* mutaaspergillic acid ($C_{11}H_{18}O_3N_2$) from *A. oryzae;* and granegillin ($C_{12}H_{20}O_2N_2$) produced by *A. flavus* and *A. effusus* [381,437]. Few toxicological data are available on these substances, although in a limited study McDonald [234] demonstrated that some strains of *A. flavus* can synthesize both aflatoxin and aspergillic acid. Concomitant production of these two compounds raises the question of synergistic action, but this question has yet to be answered. McDonald found that the LD_{50} for neoaspergillic acid in mice is 125 mg/kg intraperitoneally (i.p.). Mice showed labored respiration, rigidity of limbs and trunk, followed by death from suffocation with no apparent lesions; aspergillic acid gave similar results.

F. Kojic Acid

Numerous species of the genus *Aspergillus,* for example *A. aliaceus, A. candidus, A. flavus, A. fumigatus, A. parasiticus,* and *A. wentii,* are capable of producing kojic acid, a γ-pyrone with the formula $C_6H_6O_4$. Kojic acid is not particularly toxic and its importance is probably as an indicator compound, possibly in an oxidized form, for the potential presence of aflatoxin as previously discussed.

G. β-Nitropropionic Acid

This relatively simple compound has been isolated from *A. flavus.* However, its role as a mycotoxin has not been elucidated, possibly because it has not been analyzed for. This same statement can be made for a host of fungal metabolites. Dairy cattle that had consumed the legume *Indigofera endecaphylla* exhibited severe toxic symptoms attributed to β-nitropropionic acid [253], but proof is lacking.

III. *PENICILLIUM* TOXINS

A. Introduction

The direct association of *Penicillium*-produced secondary metabolites with human illnesses has long been suspected but not established. The most noted intoxication, yellow rice disease, occurred in the nineteenth and twentieth centuries in Japan. Many Japanese became ill and died after eating moldy rice obtained from southeast Asia [186,320]. Several toxin-producing penicillia were isolated: *P. islandicum, P. citreoviride, P. rugulosum,* and *P. citrinum.* Although these fungi produce many different toxic substances, their role in the yellow rice syndrome, which involved symptoms of cardiac beriberi characterized by vomiting, convulsions, ascending paralysis, and respiratory arrest, is uncertain.

However, *P. citreoviride* produces citreoviridin, which produces yellow rice symptoms when administered to laboratory animals [394]. In 1921, the Japanese passed the "Rice Act," which called for strict rice inspection. These control measures eliminated the fatal beriberi that had characterized the yellow rice disease. *Penicillium islandicum* isolated from rice thought to cause edema of the legs in Japanese soldiers during World War II [391] produced luteoskyrin and islanditoxin (cyclochlorotine). These toxins caused hepatoxicoses and hepatocarcinogenesis in animals [102].

The nephrotoxin citrinin is a toxin produced by *P. citrinum.* This mold was isolated from yellow rice imported from Thailand to Japan in 1951, but its involvement in human ailments is unknown [320].

Evidence for ochratoxin A as an etiologic agent for the endemic disease known as Balkan nephropathy is only circumstantial, as is its association with other human mycotoxicoses. The Balkan disease, a slowly progressing renal failure, affected 20,000 humans in the areas adjacent to the Danube River and its tributaries [251]. Persons of middle age and over were involved. Because of the similarities between Balkan endemic nephropathy and ochratoxin A-induced porcine nephropathy [194], this toxin has been suggested as the possible disease determinant [189]. Patulin and penicillic acid have not been clearly established as potential causes of human intoxications. Their causal relationship to human diseases is suggested by their toxicity to various laboratory animals [62,438]. Because these two toxins are reported to be carcinogens, their importance is potentially enhanced. Mycotoxins produced by *Penicillium* species on various commodities are listed in Table 6.

Table 6 Natural Occurrence of Various Mycotoxins

Mycotoxin	Source	Country	Reference
Patulin	Apples	England	41
	Apple cider	France	92
	Apple juice, pears	Germany	107, 115
	Apples	Canada	141
	Grape juice	Canada	333
	Apple juice	Canada	338
	Bananas, pineapples, grapes, peaches, apricots	Germany	114
	Apple juice	Sweden	179
Penicillic acid	Corn, dried beans	United States	383
Ochratoxin	Corn, barley, wheat	Canada, Denmark	97, 191-193
	Oats, rye, green coffee beans	Finland, France	197
	Beans, peanuts, hay	Norway, Poland	191-193
	Animal products (pork and poultry)	Sweden, United Kingdom, United States, Yugoslavia	24
	Corn	Bulgaria	190
	Poultry feed, barley, oats	Denmark	354
	Corn	United States	136
	Barley, oats	Sweden	191
	Wheat	Canada	301
	Rice	Japan	393
	White beans, wheat, oats, peanuts, mixed feed	Canada	342
	Wheat	United States	353
Citrinin	Peanuts	India	374
	Wheat, oats, barley, rye	Canada	342

B. Patulin

Patulin is a carcinogenic lactone produced by many species of the genera *Penicillum, Aspergillus,* and *Byssochlamys* and given various names, e.g., clavicin, clavaitin, claviformin, expansin, leucopin, mycosin C, penicidin, and tercinin [441]. There are numerous reviews concerning patulin's biological and chemical properties [62,65,331,369,371]. Patulin poses a threat to human health because of its natural presence in foods, especially fruits (Table 7). The role of patulin in human mycotoxicoses is yet to be determined, even though much is known about its lethal toxicity to laboratory animals (Table 8).

The potential of patulin as a broad-spectrum antibiotic led to considerable clinical

Table 7 Physiochemical Properties of Penicillium Mycotoxins

Toxin	Formula	Molecular weight	Ultraviolet spectrum NM	E value	Infrared spectrum (cm^{-1})	NMR spectrum	Other	References
Patulin	$C_7H_6O_4$	154.12	276	14,450	3580, 3340, 1782, 1753	$CDCl_3$: 5.97 (complex), 4.73 dd, 3.46 d.	m.p. 110°C	181, 331
Penicillic acid	$C_8H_{10}O_4$	170.16	225	10,500	3335, 1757, 1645	$CDCl_3$: 1.75	m.p. 83–84°C	111, 307
Ochratoxin A	$C_{20}H_{18}O_6NCl$	403.83	213 332	36,800 6,400	3380, 1723 1678, 1655		m.p. 169°C; $(\alpha)_D +118$	417
Citrinin	$C_{13}H_{14}O_5$	250	222 253 319	22,280 8,279 4,710	3484, 2985 1675, 1639	1.25d, 1.38d, 2.03s, 3.04Q, 4.84q, 8.3s, 12.7, 15.2	m.p. 178–179°C; $(\alpha)_D -27.7$	196 260
Citreoviridin	$C_{23}H_{30}O_6$	402	204 234 286 sh. 294 388	17,000 19,200 24,600 27,100 48,000	3500, 1720 1689m, 654 1626	NMR spectra recorded in $CDCl_3$, C_6D_6	m.p. 210–214°C; $(\alpha)_D -107.8$	322
Rubratoxin A	$C_{26}H_{32}O_{11}$	520	204 252 sh.	31,900 4,430	3400, 1850 1815, 1770	Same as rub. B with 5.7 s for hemiacetal CHOH	m.p. 210–214°C; $(\alpha)_D 84$	256
Rubratoxin B	$C_{26}H_{30}O_{11}$	518	204 215	20,500 9,700	3520, 1858 1815, 1785 1755, 1705 1690	d-acetone; 0.84t, 1.3m, 2.0m, 2.45m, 2.7d, 2.8m, 3.18dd, 3.43d, 3.77dd, 4.40m, 4.65d, 5.62d, 5.88dt, 7.00dt	168–170°C; $(\alpha)_D +67$	256

Compound	Formula		UV	ε	IR	NMR	Physical properties	Ref.
Islanditoxin	$C_{25}H_{35}O_8N_5Cl_2$	602	257 292	292	3378, 1623		m.p. 250–251°Cd; $(\alpha)_D$ −49.7	228
Rugulson	$C_{30}H_{20}O_{10}$	542			3450, 1690, 1620	d-DMSO: 2.78d, 3.38br.s, 4.38br.d, 2.42s, 7.16d, 7.43d, 11.73d, 14.54s, 5.38	m.p. 290°Cdd; $(\alpha)_D$ +492	382
Luteoskyrin	$C_{30}H_{22}O_{12}$	574			3378, 1623	d-MSCO: 2.96d, 3.36 br.s, 4.53br.d, 2.28s, 7.27s, 11.28s, 12.38s, 14.53s, 5.48	m.p. 281°Cd; $(\alpha)_D$ −880	382
Erythroskyrine	$C_{25}H_{31}NO_5$	425	260 409	8,912 2,818	1693, 1635 1569, 1550	$CDCl_3$; 1.27d, 1.63d, 4.04, 4.25, 4.50, 4.79	m.p. 130–133°C; $(\alpha)_D$ +46.9	349
PR toxin	$C_{17}H_{20}O_6$	320	247	8,800	2945, 1735, 1720, 1680, 1620, 1460, 1435, 1380, 1245, 1035	$CDCl_3$; 1.03d, 1.45d, 1.49s, 2.16s, 3.65d, 3.96d, 5.16d, 6.43s, 9.75s, 1.6–2.3m	m.p. 155–157°C; 300 nm exc., 360 nm emx.	433
Roquefortine	$C_{22}H_{23}N_5O_2$	389	337	24,986			m.p. 192–202°C; $(\alpha)_D$ −733	333

Table 8 LD_{50} Values of *Penicillium* Mycotoxins (mg/kg)

Toxin	Host[a]
Patulin	Mouse: 8-10 s.c. [181]; 15 s.c., 25 i.v., 15 i.p. [42]; 5.7 i.p. [64]
	Rat: 15 s.c., 25.50 i.v. [42]; 25 s.c. [181]
	Chick: 170 p.c. [220]
	Chick embryo (4-day-old): 2.35 µg/embryo [64]
Penicillic acid	Mouse: 110 s.c., 250 i.v., 600 p.c. [257]
Citrinin	Mouse: 52 i.p., 110 p.o., 38 i.v. [30]
	Rat: 67 s.c., 67 i.p. [5]
Citreoviridin	Mouse: 72 i.p., 11 s.c., 29 p.o. [404]
	Rat: 3.6 s.c. [404]
Rubratoxin A	Mouse: 6.6 i.p. [318]
Rubratoxin B	Mouse: 2.6 i.p., 400 p.o. [444]
	Rat: 0.35 i.p. [444]
Islanditoxin	Mouse: 3 s.c. [228]
Ochratoxin A	Rat: 20 p.o. [303]
	Chick: 126 µg/chick [281]
Luteoskyrin	Mouse: 147 s.c., 6.6 i.v., 221 p.o. [382a]
	40.8 i.p., 147 s.c., 221 p.o., 665 i.v. [320]
	Brine
	Shrimp: 10 µg/disk 5% mortality [143]
Rugulson	Mouse: 55 i.p. [412]
	Rat: 44 i.p. [412]
Erythroskyrine	Mouse: 60 i.p. [320]
PR toxin	Mouse: 5 i.p., 20 s.c., 60 p.o. [433]
Roquefortine	Rat: 15-20 i.p. [337]

[a]s.c., subcutaneous; i.v., intravenous; i.p., intraperitoneal; p.c., per cutaneous; p.o., per

testing [12]. Oral administration to humans caused vomiting and stomach irritation [82, 116,428]. However, no intoxications were produced when 0.1 g of patulin was administered intravenously [82]. Patulin applied as a 1% ointment to human skin produced an irritation [77]. Carcinogenic properties were demonstrated by subcutaneous injection of 0.2 g in arachic oil twice a week to male rats. The rats developed tumors in 64-69 weeks at the site of injection [83]. However, no carcinogenicity was observed in mice and rats fed patulin.

Evidence of suspected patulin mycotoxicoses is often only circumstantial in livestock field outbreaks [62]. Moldy silage samples believed responsible for illness observed in sheep and cattle contained patulin [105,106].

1. Production and Detection of Patulin

Copious amounts of patulin (2.7 g/liter) are produced by *P. urticae* Banier on a potato dextrose broth at 25°C [268]. Other media for patulin production and the amount (g/liter) of toxin produced on them are: Czapek-Dox (1.3 g/liter by *P. patulum;* [370]); Raulin-Thom (1.26 g/liter [306]); and yeast-extract sucrose (0.48 g/liter [239]). At lower temperatures (1.7°C) patulin produced by species of *Penicillium* and *Aspergillus* is limited or inhibited but not at 12.8°C [220]. The ability of these two genera to produce patulin on various commodities has been tested; these commodities include fruits, stone fruits, meats, straw, bread [62], and fresh grapes [140].

The Association of Official Analytical Chemists method for analyses of patulin in apple juice consists of ethyl acetate extraction, elution of the extractate from a silica gel column with benzene-ethyl acetate (75:25), followed by thin layer chromatography (TLC) (toluene, ethyl acetate, 90% formic acid, 5:4:1). The limit of detection is 20 μg patulin per liter of apple juice [285,286,295]. Other analytical methods include TLC [294], liquid chromatography [431], and high-pressure liquid chromatography [372]. In a biological assay *Bacillus megaterium* is sensitive to 1.7 μg of patulin [370,371].

2. Mode of Action

Patulin is unstable in alkaline solution (pH 9.5), but stable at pH 2 [148]. It slowly decomposes in distilled water at room temperature [306]. The toxin is heat stable [294, 340], reacts with sulfur dioxide [294], and persists in apple juice and on dry corn for up to 14 days [294].

Perhaps the most perplexing problem patulin presents is its ability to bind with glutathione, cysteine, and in general with sulfhydryl groups and amino acids [29]. Identification of the products of such reactions with patulin has eluded chemists. Many of these adducts were noninhibitory to *Bacillus subtilis* and nontoxic to 4-day-old chick embryos [207], although some of them are teratogenic to chick embryos [64,180]. Patulin inhibits K^+ transport in erythrocytes [401] and Na^+-dependent transport of glycine in reticulocytes, indicating its potential for inactivating SH groups that are the active center of the nuclear transport system.

C. Penicillic Acid

The phenomenon of "blue eye" corn, wherein corn kernels have a blue-green discoloration, is associated with *Penicillium* species, many of which are capable of producing penicillic acid [199]. A survey of the 1972 corn crop by the FDA indicated penicillic acid in 7 of 20 samples of corn at levels of 5-213 ppb and in 5 of 20 samples of commercial dried beans at levels of 11-179 ppb [383]. The uncertain role of penicillic acid in human and animal mycotoxicoses may be due in part to the fact that it is no longer detectable after it interacts with amino acids and SH groups or perhaps with other components of grain [340] and meat [66,69,70]. Penicillic acid isolation from *P. puberlum* and its toxicity to mice and guinea pigs were reported in 1913 by Alsberg and Black [4]. Their work was documented subsequently by Birkenshaw et al. in 1936 [28]. This lactone is also a secondary metabolite of species of *Aspergillus* and *Paecilomyces ehrlichii.*

1. Detection and Production of Penicillic Acid

The physiochemical properties and lethal dosages of penicillic acid to laboratory animals
are shown in Tables 7 and 8.

Excellent production of the toxin (1.7 g/liter) is achieved on Raulin-Thom medium but
not on Czapek-Dox with *Penicillium cyclopium* NRRL 1888. On corn, 8.2 g of penicillic acid
per kilogram was produced at 20°C [199]. On yeast extract-sucrose broth, 44 cultures of
Penicillium of 422 isolated from sausages produced penicillic acid in yields of 0.1-5.0 mg
per 100 ml [69,70].

Penicillic acid can be extracted from corn with chloroform-methanol (7:3) and from
culture broths with ether, chloroform, or ethylacetate followed by acidification. The compound can be assayed by a variety of methods, including TLC [199], gas chromatography
(GC) [285,376], derivative formation [68,383], and GC-mass spectroscopy for confirmation.

Penicillic acid is stable at 100°C and at both acid and base pH [148]; however, it reacts with sulfhydryl-containing compounds [124], amines, and amino acids [273] at 37°C.
It is also reactive with arginine, histidine, and lysine [66,69,70].

D. Ochratoxins

Ochratoxin A occurs in many commodities throughout the world (Table 6). Discovered
in South Africa, ochratoxin A is produced by species of *P. frequentans, P. viridicatum, P.
cyclopium, P. commune, P. purpurescens,* and seven species of *Aspergillus* [190]. Other
ochratoxins have been described: ochratoxin B, a dechloro derivative of ochratoxin A, has
no toxicity [58,281]; ochratoxin C and D are, respectively, the methyl and ethyl esters of
ochratoxin A [80]. Chemical and physical properties are listed in Table 7 and toxicity
properties in Table 8.

Ochratoxin A synthesis has been described [310,365]. Elaboration of ochratoxin A
in submerged culture has been reported [109,335], but excellent yields can be attained
by solid substrate fermentation on wheat [58,364]. Toxin is recovered by liquid-liquid
extraction followed by high-pressure liquid chromatography [287]. Detection of ochratoxin A by TLC is based on its fluorescence in ultraviolet light [263,264,334,366,417].
Methods of detection involving multiple assays for ochratoxin A and other mycotoxins
have been developed [363]. A sensitive densitometric method with a lower limit of 0.5-
1 mg for ochratoxin on cereals has been reported [59].

1. Mode of Action and Control

Ochratoxin A is slowly hydrolyzed by acid [417] but more readily by carboxypeptidase
or chrymotrypsin [293] and bovine carboxypeptidase [91]. It binds to proteins and interacts with serum albumin [57] but does not bind with nucleic acid [292]. Ochratoxins
are relatively stable to steam sterilization (70% destroyed in 3 h) [390] and the canning
process [143] but are degraded under simulated coffee roasting conditions (200°C for
5 h) [206]. Breakage of the bond between the phenylalanine and the carboxyl moiety
of dehydroisocoumarin leads to biological inactivation.

E. Citrinin

Citrinin, a secondary metabolite of many species of *Penicillium* and *Aspergillus* [296,305],
was first isolated from *P. citrinum* [152]. Citrinin occurs naturally on many commodities
(Table 6) and is an antibiotic [182,386,387] active against some fungi and gram-positive

bacteria but not against gram-negative bacteria. Citrinin induced porcine nephropathy when fed at 20-40 mg/kg body weight over 1-9 weeks [117]. Data on the toxicity to various animals are summarized in Table 8.

Penicillium citrinum produces 1.75 g of citrinin per liter when grown on a 4% sucrose-2% yeast extract medium at 30°C in 21 days as still cultures [79]; its physiochemical properties are listed in Table 7. Available methods for citrinin analyses have been reviewed [260], with the most sensitive procedure being a fluorometric determination on TLC plates [342]. Ammonia changes citrinin to a green fluorescent compound, and its acetate derivative fluoresces blue under short-wavelength ultraviolet light. Its isolation from barley has been described [135].

Citrinin is sensitive to fluorescent light and is slightly heat labile under refluxing conditions [260]. It is thermally degraded under acid and base conditions [183]. It reportedly binds to albumin fractions of human serum [78], partially to plasma [429], and is inactivated by cysteine [53]. Its role as a carcinogen is unknown.

F. Citreoviridin

Citreoviridin is a neurotoxin produced by *P. citreoviride*. It acts at the level of the nervous system and causes paralysis in the extremities of laboratory animals, followed by convulsion and respiratory arrest [404]. The fungus had been isolated from yellow rice collected in Taiwan and Japan [320,321]. Penicillia reported to produce this toxin are *P. citreoviride, P. ochrosalmoneum, P. fellutanum,* and *P. pulvilarum* [322,394]. Its natural occurrence on cereals has not been reported. Physiochemical and toxicity data on citreoviridin are listed in Tables 7 and 8.

On maize meal, the toxin is produced in quantities of 7 g/kg of substrate by *P. pulvillorium* [258]. Purification procedures have been reported by Ueno and Ueno [404] and Lucci et al. [221]. Citreoviridin is usually detected on TLC plates by its yellow fluorescence in ultraviolet light and quantitated by its absorbence at a wavelength of 383 nm [258,404].

Citreoviridin is photochemically degraded in the presence of iodine to isocitreoviridin in a ratio of 7:3 but is heat stable in chloroform-methanol solution for 24 h [258]. This photochemical degradation product did not affect rats at levels of 100 mg/kg by subcutaneous injection. It has also been reported that citreoviridin affects energy-linked processes in mitochondria and submitochondrial proteins [26a].

G. Rubratoxin

The presence of rubratoxin A and rubratoxin B in nature has not been reported. These toxins were first isolated from *P. rubrum* cultured on a corn-sucrose medium [438] and have similar chemical structures [255,256,389]. Chemical and toxicity properties are listed in Tables 7 and 8.

Methods for bioproduction, extraction, and purification of rubratoxin have been reported [145]. Rubratoxin B is produced in corn in amounts of 1.8 g/kg of substrate and also on various natural substrates and semisynthetic media [98]. Rubratoxin is detected on silica gel plates by (1) formation of a 2,4-dinitrophenylhydrazone derivative [145]; (2) a fluorescent derivative formed by heating the preparation at 200°C for 10 min [144]; (3) using spray reagents such as alkaline $KMnO_4$, p-anisaldehyde, and 50% sulfuric acid [70,254]; or (4) fluorescence quenching of silica gel HF-254 under ultraviolet light [145].

Rubratoxin B can be quantitatively determined by ultraviolet absorption spectrometry [261]. Rubratoxin B is a potent hepatotoxin, is lethal to rats [444], and is also an embryocide and a teratogen [160].

H. Yellow Rice Toxins

Metabolites from the *Penicillium* species associated with toxic yellow rice in Japan include cyclopeptides, bianthraquinones, a fluorescent polyene, pyrans, and a tetramic acid. Some of these have been discussed, including citrinin, citreoviridin, and the rubratoxins. The chlorine-containing cyclopeptide, islanditoxin, causes liver damage in mice and is lethal (Table 8). Its toxicity is destroyed on dechlorination. Physiochemical properties are listed in Table 7.

Luteoskyrin is lipophilic and hepatocarcinogenic in mice by oral administration at a level of 30-100 ppm [100]. Biological effects include mitochondrial swelling [413] and binding to DNA in vitro [395,396]. DNA binding affects the activity of DNA-dependent RNA-polymerase [405,409,410]. Luteoskyrin is lethal to mice (Table 8), the LD_{50} value depending on the route of administration [32]. Its physiochemical properties are listed in Table 7.

Rugulosin is a hepatotoxic anthraquinoid produced by the mold *P. rugulosum* Thom [382]. Rugulosin causes acute liver damage in mice and rats [326] and its lethal dosage is about two or four times less than that of luteoskyrin for laboratory animals (Table 8). It is cytotoxic to protozoans [412], mutagenic to yeast [411], and attacks DNA [408]. Its physiochemical properties are listed in Table 7.

Erythroskyrine, a nitrogen-containing pigment, was isolated from *P. islandicum* [162], and was later obtained from the same mold isolated from rice grown in Spain [349]. More recently, *P. islandicum* was shown to produce 3.7 g of crude erythroskyrine on Czapek-Dox medium after 2 weeks at 25-27°C in still cultures [406]. Male mice given 60 mg/kg i.p. (Table 8) died between days 2 and 4 and showed marked hepatic damage, including centrolobular degeneration and necrosis of liver cells, nephrotic changes of the kidneys, and extensive cellular injuries with karyorrhexis in the lymph nodes, spleen, and thymus. The surviving mice showed less liver damage but an increase in mitotic cells and mobilization of the Kupffer cells of the liver and light hyperplasia of the lymph apparatus. The involvement of this mycotoxin pigment in liver damage of humans is still unknown, but implications are that it may be one of the yellow rice toxins. Its physiochemical properties are tabulated in Table 7.

I. *Penicillium roqueforti* Toxins

This mold is commonly used for the production of blue and Roquefort cheese. It is also a common contaminant on moldy silage [202]. The sequiterpenoid "PR" toxin was isolated from *P. roqueforti* cultured on various media [272,433]. It is a hepatotoxin in rats and mice. *Penicillium roqueforti* also elaborates alkaloids [269], one of which, roquefortine, has received attention because it is a neurotoxin that causes convulsive seizures in mice when administered intraperitoneally (Table 8) [337]. Its physiochemical properties are listed in Table 7.

IV. TREMORGENIC COMPOUNDS

This class of compounds, produced by *Penicillium* and *Aspergillus* species, is unique because they act on the central nervous system of vertebrate animals, causing a tremorgenic

Table 9 Tremogenic Mycotoxins: Physiochemical Data and Acute Toxicity

Toxin	Formula	Molecular weight	Ultraviolet spectrum		Other	Acute toxicity[a] (mg/kg)
			nm	E value		
Penitrem A	$C_{37}H_{33}NO_6Cl$	633	233 295	37,000 11,600	m.p. 237–239°C	Mouse: 1.1 i.p., 10 p.o.
Penitrem B	$C_{37}H_{35}NO_5$	583	227	38,450	m.p. 184–195°C	Mouse: 1.1 i.p., 10 p.o.
Penitrem C	Unknown					
Verruculogen	$C_{27}H_{33}N_3O_7$	511	224	47,500	m.p. 233–235°C	Mouse: 2.4 i.p., 126 p.o.
Fumitremorgen A	$C_{32}H_{41}N_3O_7$	579	225 275 295		m.p. 206–209°C; $(\alpha)_D$ + 61	Mouse: 5 i.p.
Fumitremorgen B	$C_{27}H_{33}N_3O_5$	479	Same as Fumitremorgen A		m.p. 211–212°C	
Paxilline	$C_{27}H_{33}NO_4$	435	230 281		m.p. 252°C	
Paspaline	$C_{28}H_{39}NO_2$	421	228–233 282 291 inf.	25,119 7,943 6,310	m.p. 264°C	
Tryptoquivaline	$C_{29}H_{30}N_4O_7$	546	228 275 305 317	37,000 8,550 3,800 3,040	m.p. 153–155°C; $(\alpha)_D$ + 142	
Tryptoquivalone	$C_{26}H_{24}N_4O_6$	488	234 292 320	34,950 9,550 6,300	m.p. 202–204°C; $(\alpha)_D$ + 254	

[a]i.p., intraperitoneal; p.o., per os.

response. Sixteen tremorgenic compounds have been reported to date. The physiochemical and toxic properties of some of them are listed in Table 9. These compounds have been reviewed extensively [71,72,73].

V. *FUSARIUM* TOXINS

A. Trichothecenes

1. Alimentary Toxic Aleukia and Trichothecene Toxicoses

The earliest and most complete review of the Soviet literature and the most detailed description of alimentary toxic aleukia (ATA) in the English language was presented by Mayer [230,231]; the brief description of the disease given here draws heavily on his coverage of this topic. ATA, a disease of the human blood and hematopoietic system, occurred as an epidemic in Siberia during the war years 1943-1945 but was never reported from another country. The disease was noticed in 1913 in eastern Siberia, but these observations were ignored until 1932, when it reappeared suddenly in western Siberia with fever, hemorrhagic rash on the skin, bleeding from the nose, throat and gums, and necrotic angina. In addition to the visible signs, deleterious effects of the disease were indicated by blood analyses.

A slight leucocytosis occurred shortly after the ingestion of the toxic cereals, followed by a disappearance of eosinophils, lymphopenia, and neutropenia. As more toxin was ingested, the bone marrow was further impaired, and just before the septic angina stage a critical low in the blood count was reached. Leucocyte counts dropped below 2000 per millimeters, the granulocytes almost completely disappeared, and the thrombocytes were in the range 3000 to 5000 per millimeter. Mortality rates varied from about 2% to as high as 80% as recorded in several areas of the USSR during the years between 1932 and 1945, but statistical data noting the number of persons involved have not been published, at least not in the Western literature.

This disease was given many names over the years of occurrence, including septic angina, alimentary hemorrhagic aleukia, alitoxicosis, agranulocytosis, and panmyelotoxicosis [230,231]. In 1943, the Public Health Service of the USSR officially named the disease "alimentary toxic aleukia." Investigations by Soviet scientists into the cause of this disease soon ruled out such factors as infectious agents, vitamin deficiencies, food spoilage, and inherent toxins in the cereals. After extensive testing of molds isolated from the toxic cereals, it was concluded that ATA was a result of toxic substances produced by molds of the genus *Fusarium* growing on overwintered cereals, especially millet, wheat, and barley [173-175,230]. Although no identifiable toxin was separated from the toxic cereals and even though the disease had never been reported outside the USSR, scientific interest was aroused in the West. Efforts to elucidate the case of ATA continued long after the last case of ATA was reported in 1947 [172,174-176,230,231].

The association of *Fusarium* with ATA was accepted by a large majority of scientists who studied the disease, although they were not able to identify the causative toxin. However, recent study of the mold cultures originating from the disease-causing cereals has revealed them to be producers of several potent mycotoxins. Comparison of the described symptoms of the disease with those induced by the *Fusarium* toxins in laboratory animals provide abundant circumstantial evidence linking ATA and the *Fusarium* toxins.

Early reports from the USSR, which could not be confirmed later, indicated that *F.*

sporotrichioides and *F. poae* recovered from inoculated grains produced the steroids sporo-
fusarin and poaefusarin and that each of the steroids produced symptoms of ATA in laboratory animals [27,172,204]. After the discovery and characterization of several of the
more than 40 trichothecenes [397], a family of closely related sesquiterpene compounds
and the development of analytical methods for the detection of these mycotoxins, Mirocha
and Pathre [243] obtained a sample of poaefusarin from Soviet scientists. They reported
that it contained 2.5% T-2 toxin and smaller amounts of T-2 tetraol, neosolaniol, and
zearalenone, all known metabolites of *Fusarium* species. The T-2 toxin content was sufficient to explain the toxicity of the sample, as determined by rat toxicity tests and rabbit
skin tests.

Studies of the fungi isolated from cereals involved in ATA support other findings that
the cause of this nutritional toxicosis was a toxin or toxins of *Fusarium* origin and most
likely the trichothecene T-2 toxin. An evaluation of the toxin-producing potential of 131
Fusarium strains taken from overwintered cereals that caused ATA in the USSR was conducted by Yagan and Joffe [450]. About three-fourths of the tested strains in the species
F. sporotrichioides and *F. poae* produced a strong skin reaction in rabbits. T-2 toxin from
cultures giving a positive skin test was identified by thin layer chromatography. Strains
of *Fusarium* obtained from overwintered cereals during the ATA endemics produced more
T-2 toxins than did strains from other sources [177]. Isolates of *F. sporotrichioides* and
F. poae brought to the West by Joffe have been intensively studied [177,400], but no
toxic steroids have been noted. At least one of these strains produces large amounts of
T-2 toxin and smaller amounts of HT-2 toxin, neosolaniol, and a butenolide [400,403].

Although mycotoxicologists generally accept that the products involved in ATA are
either T-2 toxin or a combination of T-2 toxin and other products of fungal origin [22,23],
the cause of this disease can be examined only retrospectively. No recent case of ATA has
occurred, and the implicated cereals are no longer available.

2. The Fungus

The genus *Fusarium* is a member of the Deuteromycotina (Fungi Imperfecti) in the family
Moniliaceae. Species of the genus have fusoid, curved septate macroconidia in slimy
masses (sporodochia) in branched conidiophores; septate microconidia with chlamydospores are common. The mycelia and spores are generally bright in color. A number of
species have perfect stages in the Hypocreales [2].

There is considerable confusion in *Fusarium* taxonomy because of the use of different
taxonomic systems in various countries. Thus, many of the strains studied in the United
States have been identified as *F. tricinctum* according to the system of Snyder and Hansen,
while in some other countries taxonomists consider this species to comprise at least four
distinct taxa. Similar confusion exists in the case of other fusaria, referred to in the literature as *F. roseum, F. nivale,* and *F. solani* [355]. As many as 1250 names have been proposed for the various isolates of *Fusarium* [2]. Among the more common classification
systems are those of Booth [32], Gorcon [129-131], Snyder and Hansen [358,359],
Toussan and Nelson [388], and Wollenweber and Reinking [445].

Fusarium species are widely distributed in soil and on organic substrates. They abound
in cultivated soils and are one of the most common fungi found on corn, barley, oat, and
wheat kernels. Fusaria are also involved in diseases of animals and humans and, as major
storage rots, often produce toxins that contaminate human and animal food. In the genus,
eight of the nine species in the Synder and Hansen system reportedly produce trichothe-

cenes. In addition to the genus *Fusarium*, trichothecenes are produced by species in the genera *Cephalosporium, Myrothecium, Trichoderma, Trichothecium, Stachybotrys*, and *Verticimonosporium* [395,396].

3. Natural Occurrence of Trichothecenes

Food poisoning cases caused by *Fusarium*-contaminated cereals were recognized many years ago [93] but only recently were the toxic metabolites from the cereals purified and studied in test animals. Symptoms described for humans and animals that had ingested cereals contaminated with *Fusarium* were in many respects like those observed in test animals given specific trichothecenes now known to be *Fusarium* metabolites. Despite the accumulation of evidence linking ATA to a *Fusarium* toxicosis, there has been no proof of human intoxication by these molds.

Several cases of trichothecene poisoning of animals, however, have recently been described. In a mycological study of moldy corn involved in a case of hemorrhagic syndrome in cattle, Gilgan et al. [125] isolated a highly toxic strain of *F. tricinctum* from which they crystallized a toxic chemical, diacetoxyscirpinol (DAS). This trichothecene was first isolated, characterized, and proven to be highly toxic to rats by Brian et al. [40]. Bamburg and his colleagues [21,22] continued the study of toxic fusaria and discovered another potent trichothecene which they characterized and named T-2 toxin. In 1972, Hsu et at. [165] were the first to identify a trichothecene from a naturally contaminated field sample. They associated the presence of T-2 toxin in corn silage with the death of 7 of 35 lactating cows.

Mirocha et al. [244], using highly sophisticated analytical procedures, examined feedstuffs obtained from sources where animals either refused the feeds or became ill after eating them. The researchers were able to detect either DAS, T-2 toxin, or vomitoxin and zearalenone in the nine toxic samples that caused problems for swine, cattle, and dogs. In Britain, Petrie [288] found T-2 toxin in stored brewers' grains fed to cows that subsequently showed a decreased appetite and weight yield, hair loss, bloody diarrhea, and multiple petechial hemorrhages of the mucous membrane.

Poultry also are affected by these toxins. Severe oral lesions induced by feeding rations to which a few parts per million T-2 toxin was added [446,447] have been observed also in birds that ate a moldy feed. The number of reported cases of proven trichothecene poisoning of animals is not large, but with improved analytical procedures, surveys of *Fusarium*-contaminated feedstuffs will provide needed information to assess the hazard of these toxins to animals and humans.

The less toxic trichothecene, vomitoxin, has not reportedly caused the death of farm animals. It is of interest, however, because it reduces the palatability of feeds, particularly for swine and because it has been more commonly encountered in the field than the other trichothecenes. A portion of the 1972 corn crop refused by swine was reexamined by Vesonder et al. [426]. Chemical analysis of this corn indicated the presence of no known mycotoxin, but a previously unknown trichothecene was extracted, purified, characterized, and named vomitoxin. *Fusarium graminearum* isolated from this corn produced vomitoxin on autoclaved grains that were also rejected by swine. When 7 mg of pure vomitoxin was given to a pig by intubation, vomiting occurred within 25 min [427].

At about the same time as Vesonder et al. [426] were identifying the refusal factors in corn, Morooka et al. [252] were characterizing trichothecenes from barley infected with *Fusarium*. They purified two new trichothecenes, nivalenol and deoxynivalenol. A *F. roseum* strain isolated from toxic barley produced deoxynivalenol, and it induced vomit-

ing in ducklings and dogs. Deoxynivalenol and vomitoxin have the same chemical structure. Like vomitoxin, the other *Fusarium*-produced trichothecenes, when tested in laboratory animals, caused vomiting. Although DAS and T-2 toxin are less frequently found
in field samples than vomitoxin, they are also refusal factors for swine [425]. It is highly
probably that all *Fusarium*-produced trichothecenes will reduce feed consumption of
swine.

4. Toxin Production

Although many toxin strains of *Fusarium* have been isolated from cereals or cereal products
and are capable of producing a variety of trichothecenes, only DAS, nivalenol, T-2 toxin,
and vomitoxin have been detected in naturally contaminated feedstuffs [165,244,246,288,
426,452]. Suitable conditions for growth of the fungus is a prerequisite for toxin formation. For production of maximum amounts of specific toxins, the species, strain, and the
environment are the determining factors. For the most part, fusaria do not grow on properly stored grains, but they may proliferate on grains before harvest and on moist cereals
not quickly dried. Because no studies to determine the conditions leading to the production of the various trichothecenes in natural environments have been reported, one can
only draw on information obtained from laboratory studies.

Yagan et al. [451] surveyed 131 isolates of *F. poae* and *F. sporotrichioides* taken
from overwintered cereals associated with ATA in the USSR. About three-fourths of the
isolates produced T-2. Toxin elaboration by *F. tricinctum* occurs at temperatures below
30°C, whereas production is poorer at higher temperatures [45,174]. Maximum yields
of other trichothecenes occur at more moderate temperatures. In fact, the production of
T-2 toxin by *F. tricinctum* strains seems to be the exception with respect to a low-tem-
perature requirement. Maximum yields of DAS [40], fusarenon-X [399,407], and vomitoxin [427] are produced at temperatures of 25°C or above. Laboratory conditions for
the production of some selected trichothecenes are given in Table 10.

5. Detection of Trichothecenes

Because of lack of a simple definitive test for the trichothecenes, only a few cases of their
natural occurrence have been reported. An abundance of presumptive biological tests for
trichothecenes in feedstuffs, such as animal skin tests [60,432], microbial inhibition and
phytotoxicity tests [46], brine shrimp lethality [143], and the inhibition of [14C]leucine
uptake by rabbit reticulocytes [398,402] have been used for detecting these toxins in
laboratory studies. Their value is limited, however, unless a positive biological response
is coupled to a definitive chemical test [103]. Of the biological tests, the skin test is the
most likely to detect small amounts of most trichothecenes, and it is a relatively inexpensive, simple procedure.

a. The Animal Skin Test: The cutaneous response of various animals to most of the
trichothecenes provides a simple, sensitive test that can be made semiquantitative. It is
the oldest and most used indicator of trichothecenes and other vesicants in extracts of
feedstuffs contaminated by fusaria or of active agents produced by *Fusarium* cultures in
the laboratory. According to Mayer [230,231], the rabbit skin test has been used since
1937 for the recognition of toxic grain in outbreaks of ATA. Toxic substances in the
grain caused hyperemia, leukocytosis, and necrosis of the skin. Joffe [171] employed
the rabbit skin test to advantage to determine the effects of growth temperatures on toxin
formation by *Fusarium* strains and other fungi. He was not able to show the nature of the

Table 10 Laboratory Production of Some Typical Trichothecenes

Trichothecene	Producing *Fusarium*	Producing medium	Temperature	Yield (mg/kg)	Reference
T-2 toxin	*F. tricinctum*	Corn	8°C	120	22
		White corn grits	15°C	2000	45
HT-2	*F. tricinctum*	Corn	24°C	20	22
Diacetoxyscirpinol	*F. scirpi*	GAN[a]	25°C	125	40
Monacetoxyscirpinol	*F. roseum* Gibbosum	Rice	25°C, 7 days	1000	277
Neosolaniol	*F. solani*	PSC[b]	25°C	10	403
Fusarenon-X	*F. nivale*	PSC[b]	25–27°C	40	407
Vomitoxin	*F. graminearum*	Rice	28°C	60	427

[a]Glucose ammonium nitrate broth.
[b]Peptone-supplemented Czapek's.

toxin before a number of trichothecene standards became available and physical-chemical confirmatory tests were developed. Some estimation of the total toxin in an unknown sample can be achieved by measuring the effects on an animal's skin and comparing it to the reaction caused by a known amount of toxin. Shaved rat skin developed reddish weals when only 0.05-0.1 μg of T-2 toxin was applied. Larger concentrations increased the severity of the skin reaction.

In a field study of the 1972 corn crop, Eppley et al. [104] applied the animal skin test to survey 173 samples for mycotoxins. They found that 93 of the samples caused a skin reaction, but the presence of the presumed T-2 toxin could not be confirmed by chemical tests. In summary, the animal skin test may eliminate negative samples for further analysis, but a positive test may result from any of a number of vesicants. Moreover, as of this writing, the only trichothecene found commonly in U.S. grain is vomitoxin, and it is not a potent skin irritant. Application of 100 μg of vomitoxin to the shaved backs of mice produces no visible vesication, and this toxin would probably go undetected in this test.

b. Physical-Chemical Analytical Methods: No sample, rapid, inexpensive method is available for the identification and quantitation of trichothecenes from natural substances. Thin layer chromatography, gas-liquid chromatography, and gas chromatography coupled to a mass spectrometer with computer analysis of ion peaks have been applied to toxin identification.

(1) Thin Layer Chromatography (TLC): The trichothecenes have no unique properties that provide easy identification. The pure compounds are colorless, have no fluorescent properties, and do not quench fluorescent indicators. Despite these unfavorable properties, TLC on silica gel is the most convenient and most used procedure for the identification of trichothecenes during purification work where the toxins are in high concentration and interfering substances are not abundant. Scott et al. [336] employed TLC on silica gel to separate trichothecenes from culture media and identified the toxins by comparing R_f values, color characteristics, and fluorescent properties after spraying the plates with sulfuric acid containing p-anisaldehyde and heating to 110-120°C. Many solvent systems for the TLC separation of trichothecenes have been published together with R_f values of some of the toxins [276]. TLC has not been adapted for the rapid, routine screening of foods or feedstuffs where relatively small amounts of toxins are present along with interfering substances.

(2) Gas-Liquid Chromatography (GLC): Because of lack of specificity, the inherent variability in biological assay methods, and the lack of spectral and chemical properties that permit detection of trichothecenes in small amount, Ikediobi et al. [166] developed a GLC method for this purpose. Compounds containing free hydroxyl groups were chromatographed as trimethylsilyl ethers. The GLC procedure was applied to known mixtures of the pure toxins and to the determination of T-2 toxin and HT-2 toxin in grain samples infected with *F. tricinctum*. Hsu et al. [165] successfully applied GLC in determining the presence of T-2 toxin in corn silage from a farm where the silage was associated with a lethal toxicosis in dairy cattle. The sensitivity of GLC methods for trichothecenes depends largely on the separation of trichothecenes from interfering components of the extract. Cleanup column chromatography and preparative TLC are usually required before GLC determinations can be made.

(3) Gas Chromatography-Mass Spectrometry (GC/MS): If the level of the trichothecene in a sample is less than 50 ppb, detection is extremely difficult, even by GLC.

The low levels of toxin encountered in mixed feed extracts plus nonspecific lipids that cochromatogram with trichothecenes exacerbate the problem of GLC separation and quantitation, e.g., interfering monoglycerides may be mistaken for trichothecenes when none are present [244]. Pathre and Mirocha [276] made analysis unambiguous by combining GC/MS and computer analysis of the mass spectral data. Fragmentation patterns of unknown trichothecenes in a sample were analyzed by the computer and compared to the fragmentation patterns of known trichothecenes. For each mycotoxin being analyzed intensities of a set of nine, characteristic-fragment ions were monitored as each component of the injected sample eluted from the GC column. The intensities of selected ions were used to identify the trichothecene. From feedstuffs associated with a disease syndrome in farm animals, these investigators were able to detect the presence of three trichothecenes at levels below 100 μg/kg of mixed feed.

6. Mode of Action

The LD_{50} dose of the more potent toxins, for example, DAS, T-2 toxin, monoacetoxyscirpinol, nivalenol, or fusarenon-X, for laboratory animals is under 10 mg/kg body weight [326,327]. Some of the other 12,13-epoxytrichothecenes, i.e., vomitoxin, with an LD_{50} of about 50 mg/kg body weight, may be less toxic but will cause feed refusal when present in animals feeds.

The response of animals, animal tissues, and cells to the action of *Fusarium* toxins has been studied intensively. For ATA, where the evidence weighs heavily in favor of T-2 toxin as the causative agent, the first symptoms of the disease following ingestion of toxic cereals were a burning sensation in the mouth, esophagus, and stomach. A couple of days after ingestion of the toxic cereals, the victim developed gastroenteritis with diarrhea, nausea, and vomiting. Local symptoms in the digestive tract disappeared, followed by a seemingly symptomless gap of 2 weeks to 2 months during which the ingested toxin acted upon the bone marrow. Bone marrow depletion is shown by a sharp reduction in leucocytes and a relative decrease in the percentage of granulocytes. At this point, the patient may develop necrotic angina, hemorrhagic diathesis, and hematological changes where the leucocyte count may drop to 1-2000 per millimeter, the granulocytes may disappear, thrombocytes decrease, and the platelets decrease to fewer than 30,000.

Because of availability, T-2 toxin and fusarenon-X have been the most studied of the trichothecenes. Symptoms produced in laboratory animals by these toxins are markedly similar to those described for ATA in humans. Sato et al. [325] reported that the major symptoms of cats administered T-2 toxin are vomiting, diarrhea, ataxia of the hind legs, discharge from the eyes, and ejection of hemorrhagic fluid. Administration of sublethal dosages of T-2 toxin over 17 days caused a reduction in circulating white blood cells and hemoglobin. On autopsy, extensive cellular damage to the bone marrow, intestines, spleen, and lymph nodes was noted. Ueno [395] reported that administration of T-2 toxin to mice caused a decline in circulating white blood cells and platelets, with a concomitant decrease in the weight of the thymus and the spleen. He felt that trichothecenes impair the hematopoietic system as well as the organs for immunological capacity. In experimental animals administered fusarenon-X, T-2 toxin, and 10 of 12 related trichothecenes, severe destruction and karyorrhexis were induced in the actively dividing cells of the bone marrow, mucosal epithelium of the small intesting, and other actively dividing tissue.

Comparative toxicology with various trichothecenes revealed that these toxins inhibit protein and DNA synthesis [321] and that this family of toxins contains the most potent small-molecule inhibitors of protein synthesis in eucaryotic cells [236]. An early effect of the trichothecenes on animal cells is to disaggregate the polysomal structure. Experiments with labeled T-2 toxin and fusarenon-X verify that the toxin binds to polysomes and ribosomes. The most likely action of trichothecenes on eucaryotic cells is their interference of active sites of peptidyl transferase of protein initiation and termination reactions. The action of these mycotoxins is probably related to their biochemical properties as protein synthesis inhibitors. The most toxic of the trichothecenes are probably those that inhibit the initiation and not the elongation or termination of protein synthesis on polyribosomes.

B. Zearalenone

1. Introduction

Zearalenone or F-2, a resorcylic acid lactone, is a natural metabolite produced by strains of *Fusarium.* No human cases of zearalenone intoxication have been reported. However, this fungal metabolite is of interest because it is often present in food and feed grains and because it has estrogenic activity in nonhuman primates [154]. Zearalenone is generally not lethal to the consuming animal, but it induces a hyperestrogenism when consumed by swine [240,242,245,275]. The visible symptoms of this mycotoxicosis include swelling and reddening of the vulva, increase in the size of the uterus, and growth and lactation in mammary glands. Estrogenic stimulation occurs even in males and gilts, as indicated by enlargement of the prepuce and mammary glands.

The association of moldy corn with estrogenism in swine was first noted in the late 1920s, and cases of the syndrome were reported periodically thereafter. *Gibberella zeae* [*F.* (sect. *Graminearum*) *roseum*] isolates taken from corn that caused uterotrophic activity in swine were cultured on sterile corn. When this corn was fed to immature pigs, vulvar and mammary enlargement was elicited within 4 days. From *G. zeae*-fermented corn, Stob et al. [367] were able to isolate, crystallize, and characterize the greenish blue fluorescent compound now known as zearalenone [416].

2. The Fungus and Natural Occurrence

The following species of *Fusarium,* according to the system of Snyder and Hansen, reportedly produce zearalenone: *F. roseum, F.* (sect. *Culmorum*) *roseum, F.* (sect. *Equiseti*) *roseum, F.* (sect. *Gibbosum*) *roseum, F.* (sect. *Graminearum*) *roseum, F. moniliforme, F. oxysporium, F. sporotrichioides,* and *F. tricinctum.* Of these species *F. roseum* produces the greatest amounts of zearalenone.

Reports of the estrogenic syndrome are sporadic but not infrequent and usually follow a season when wet grain is stored. Strains of *Fusarium* produce trichothecenes and zearalenone during storage if conditions are favorable, but little zearalenone is produced before harvest. Caldwell and Tuite [48] inoculated ears of field corn with zearalenone-producing fusaria to determine if toxin was accumulated before harvest. They found no production or an insignificant accumulation of zearalenone 100 days after inoculation at 90% of full silk with 8 strains of *F. tricinctum* and 12 strains of *F. roseum.* Production in the field never exceeded 5 ppm and was almost always less. Examination of naturally infected corn ears supported the findings that little zearalenone accumulates in freshly harvested corn [49]. In a comparable study, barley sprayed 6 weeks before harvest with spores of a zearalenone-producing strain of *F.* (sect. *Culmorum*) *roseum* had no detectable zearalenone at harvest. However, the moistened grain stored for 4 months at a low temperature developed significant amounts of the estrogen [132]. The greatest potential for zearalenone production is probably in grain stored in relatively open bins at high moisture concentrations; it has been found in barley, corn, mixed feeds, oats, wheat, and sorghum. Thus, the control of this mycotoxin seems possible by practicing proper grain drying and storage.

3. Production

In the laboratory, zearalenone is produced by inoculating a high-producing strain of *F. roseum* onto autoclaved, moistened, polished rice or corn kernels. The cultures are incubated for 7 days at 25°C followed by an incubation of 4-6 weeks at 12-14°C. Yields of 1-15 g/kg of substrate are commonly produced.

4. Detection

Eppley [103] developed a procedure for the detection of zearalenone which consists of column chromatography followed by TLC and identification under short-wavelength ultraviolet light. Interfering oils make the determinations of zearalenone difficult when concentrations present in the sample are below about 300 µg/kg [351]. Considerable variation in recovery values of collaborators occurred even when larger quantities of zearalenone were added to the samples. High-pressure liquid chromatography (HPLC) seems to be the most sensitive analytical method for the detection of zearalenone. Before samples are quantitated by HPLC, however, they require precleaning by column chromatography of thin layer chromatography or both. Thus, any of the described HPLC procedures are time consuming, but the method permits the detection of less than 20 µg/kg of cereal substrate of mixed animal feeds by ultraviolet or fluorescence detectors [158,250,339,430]. The coupling of chromatographic methods with GC/MS provides an excellent confirmation of an analysis.

Food samples can also be tested for estrogenic activity by feeding directly to 20-day-old female rats or by extracting the sample and applying it directly to the shaven skin of the rat. After 5 days the rats are sacrificed and the uterus is excised, weighed, and compared with control animals. An increase in uterine weight indicates that the test sample contained zearalenone.

5. Mode of Action

The estrogenic syndrome in swine involves development of swollen, edematous vulva in females, shrunken testes in young males, a dramatic increase in the weight of the uterine horn of gilt, enlarged mammary glands in the young of both sexes, and possible abortion in pregnant sows and gilts. In ovariectomized mice, zearalenone, like steroidal estrogens, accelerates the synthesis of uterine RNA, protein, and the permeability of uterine cells concomitant with proliferation of uterine muscle tissue and an increase in the uterine weight [414]. In vitro studies with cystosol preparations of mammary glands from lactating rats indicate that zearalenone inhibits the binding of estradiol by competing for binding sites of the estrogen receptors of target cells [38]. In rhesus monkeys zearalenone is nearly as effective as estradiol in the depression of gonadotrophin, luteinizing hormone, and follicle-stimulating hormone when administered by subcutaneous injection. It is not as effective by the oral route. Evidence to explain how zearalenone affects the anabolic and uterotrophic responses of animals is still relatively meager.

REFERENCES

1. Adamson, R. H., P. Correa, S. M. Sieber, K. R. McIntere, and D. W. Daigard. 1976. Carcinogenicity of aflatoxin B_1 in rhesus monkeys: two additional cases of primary liver cancer. J. Natl. Cancer Inst. *57*: 67-71.

2. Ainsworth, G. C. 1971. *Dictionary of the Fungi,* 6th ed, Commonwealth Mycological Institute, Kew, Surrey, England.

3. Alpert, M. E., M. S. R. Hutt, G. N. Wogan, and C. S. Davidson. 1971. Association between aflatoxin content of food and hepatoma frequency in Uganda. Cancer *28*: 253-260.

4. Alsberg, C. L., and O. F. Black. 1913. Contribution to the study of maize deterioration. Bio-chemical and toxicological investigations of *Penicillium puberulum* and *Penicillium stoloniferum.* Bur. Plant. Ind., U.S. Dept. Agric. Bull. 270, pp. 1-48.

5. Ambrose, A. M., and F. DeEds. 1946. Pharmacological properties of citrinin. J. Pharmacol. Exp. Ther. *88*: 173-186.

6. Amla, I., C. S. Kamala, G. S. Dopalakrishna, P. Jayaraj, V. Sreenivasa Murthy, and H. A. B. Parpia. 1971. Cirrhosis in children from peanut meal contaminated by aflatoxin. Am. J. Clin. Nutr. *24*: 609-614.

7. Amla, I. J., S. Kumari, V. Sreenivasa Murthy, P. Jayaraj, and H. A. B. Parpia. 1970. Role of aflatoxin in Indian childhood cirrhosis. Indian Pediatr. *7*: 262-268.

8. Amla, I. J., V. Sreenivasa Murthy, P. Jayaraj, and H. A. B. Parpia. 1974. Aflatoxin and Indian childhood cirrhosis—a review. J. Trop. Pediatr. Environ. Child Health *19*: 28-33.

9. Anderson, H. W., E. W. Nehring, and W. R. Wichser. 1975. Aflatoxin contamination of corn in the field. J. Agric. Food Chem. *23*: 775-782.

10. Andrellos, P. J., and G. R. Reid. 1964. Confirmatory tests for aflatoxin B_1. J. Assoc. Off. Anal. Chem. *47*: 801-803.

11. Anon. 1974. Imported figs recalled because of aflatoxin. Food Chem. News *16*: 17.

12. Anslow, W. K., H. Raistrick, and G. Smith. 1943. Antifungal substances from moulds. Part I. Patulin, a metabolic product of *Penicillium patulum* Banier and *Penicillium expansum* (Like) Thom. Trans. Soc. Chem. Ind. *62*: 236-238.

13. Arseculeratne, S. N., and L. M. DeSilva. 1971. Aflatoxin contamination of coconut products. Ceylon J. Med. Sci. *20*: 60-75.

14. Ashoor, S. H., and F. S. Chu. 1975. New confirmatory test for aflatoxins B_1 and B_2. J. Assoc. Off. Anal. Chem. *58*: 617-618.

15. Ashworth, L. J., Jr., R. E. Rice, J. L. McMeans, and C. M. Brown. 1971. The relationship of insects to infection of cotton bolls by *Aspergillus flavus.* Phytopathology *61*: 488-493.

16. Ashworth, L. J., Jr., H. W. Schroeder, and B. C. Langley. 1965. Aflatoxins: environmental factors governing occurrence in peanuts. Science *148*: 1228-1229.

17. Asplin, F. D., and R. B. A. Carnaghan. 1961. The toxicity of certain groundnut meals for poultry with special reference to their effect on ducklings and chickens. Vet. Rec. *73*: 1215-1219.

18. Association of Official Analytical Chemists. 1975. Natural poisons. In W. Horowitz (Ed.), *Official Methods of Analysis of AOAC,* 12th ed, AOAC, Arlington, Va., pp. 462-482.

19. Ayres, G. C., H. S. Lillard, and D. A. Lillard. 1970. Mycotoxins: detection in foods. Food Technol. *24*: 55-60.

20. Ayres, J. L., D. J. Lee, J. H. Wales, and R. O. Sinnhuber. 1971. Aflatoxin structure and hepatocarcinogenicity in rainbow trout. J. Natl. Cancer Inst. *46*: 561-564.

21. Bamburg, J. R., N. V. Riggs, and F. M. Strong. 1968. The structures of two toxins from two strains of *Fusarium trincinctum.* Tetrahedron *24*: 3329-3336.

22. Bamburg, J. R., and F. M. Strong. 1971. 12,13-Epoxytrichothecenes. In S. Kadis, A. Ciegler, and S. Ajl (Eds.), *Microbial Toxins*, Vol. 7, Academic, New York, pp. 207-292.

23. Bamburg, J. R., F. M. Strong, and E. B. Smalley. 1969. Toxins from moldy cereals. Agric. Food Chem. *17*: 443-450.

24. Barnes, J. M., R. L. Carter, P. K. C. Austwick, R. F. Flynn, and W. N. Aldridge. 1977. Balkan (endemic) nephropathy and a toxin-producing strain of *Penicillium verrucosum* var. *cyclopium*. An experimental model in rats. Lancet *1*: 671-675.

25. Beckwith, A. C., R. F. Vesonder, and A. Ciegler. 1976. Chemical methods investigated for detoxifying aflatoxins in foods and feeds. In J. V. Rodricks (Ed.), *Mycotoxins and Other Fungal Related Food Problems*. *Advances in Chemistry Series 149*, American Chemical Society, Washington, D.C., pp. 58-67.

26. Becroft, D. M. O., and D. R. Webster. 1972. Aflatoxins and Reye's syndrome. Br. Med. J. *4*: 117-118.

26a. Beechey, R. B., D. O. Osselton, H. Baum, P. E. Linnett, and A. D. Mitchell. 1974. Citreoviridin diacetate: a new inhibitor of the mitochondrial ATP-synthetase. In G. F. Azzone, M. E. Klingenberg, E. Quagliariello, and N. Siliprandi (Eds.), *Membrane Proteins in Transport and Phosphorylation*, Proceedings of the International Symposium on Membrane Proteins in Transport and Phosphorylation, North-Holland, Amsterdam, pp. 201-204.

27. Bilai, V. I. 1960. *Mycotoxicosis of Man and Agricultural Animals* (U.S.S.R.). Distributed by: Office of technical services, U.S. Dept. of Commerce, Washington, D.C. U.S. Joint Research Service, 1636 Connecticut Avenue, N.W., Washington, D.C.

28. Birkenshaw, J. H., A. E. Oxford, and H. Raistrick. 1936. Studies in the biochemistry of micro-organisms. XLVIII. Penicillic acid, a metabolic product of *Penicillium puberulum* Banier and *P. cyclopium* Westling. Biochem. J. *30*: 394-411.

29. Black, D. K. 1966. The addition of L-cysteine to unsaturated lactones and related compounds. J. Chem. Soc. (*C*): 1123-1127.

30. Blanpin, O. 1959. La citrinine. Nouvelles données sur l'action pharmacodynamique de cet antibiotique. Therapie *14*: 677-685.

31. Blount, W. P. 1961. Turkey "x" disease. Turkey *9*: 52, 55-58, 61-71.

32. Booth, C. 1971. *The Genus Fusarium*. Commonwealth Mycological Institute, Kew, Surrey, England.

32a. Bösenberg, H. 1972. Diagnostische Möglichkeiten zum Nachweis von Aflatoxin-Vergiftungen. Zentralbl. Bakteriol., Parasitenkd. Infektionskr. Hyg. 1, Abt: Orig. A *220*: 252-257.

33. Bothast, R. J., G. H. Adams, E. E. Hatfield, and E. B. Lancaster. 1975. Preservation of high moisture corn: a microbiological evaluation. J. Dairy Sci. *58*: 386-391.

34. Bothast, R. J., and C. W. Hesseltine. 1975. Bright greenish-yellow fluorescence and aflatoxin in agricultural commodities. Appl. Microbiol. *30*: 337-338.

35. Boudergues, R., H. Calvet, E. Discacciati, and M. Cliche. 1966. Note sur la présence d'aflatoxine dan les fanes d'arachides. Rev. Elev. Med. Vet. Pays Trop. *19*: 567-571.

36. Bourgeois, C., L. Olson, D. Comer, H. Evans, N. Keschamras, R. Cotton, R. Grossman, and T. Smith. 1971. Encephalopathy and fatty degeneration of the liver: a clinicopathologic analysis of 40 cases. Am. J. Clin. Pathol. *56*: 558-570.

37. Boutrif, E., M. Jemmali, A. E. Pohland, and A. D. Campbell. 1977. Aflatoxin in Tunisian aleppo pine nuts. J. Assoc. Off. Agric. Chem. *60*: 747-748.

38. Boyd, P. A., and J. L. Wittleff. 1978. Mechanisms of *Fusarium* mycotoxin action in mammary gland. J. Toxicol. Environ. Health *4*: 1-8.

39. Brekke, O. L., A. J. Peplinski, and E. B. Lancaster. 1977. Aflatoxin inactivation in corn by aqua ammonia. Trans. ASAE *20*: 1160-1168.

40. Brian, P. W., A. W. Dawkins, J. F. Grove, H. G. Hemming, D. Lowe, and G. L. F. Norris. 1961. Phytotoxic compounds produced by *Fusarium equiseti*. J. Exp. Bot. *12*: 1-12.

41. Brian, P. W., G. W. Elson, and D. Lowe. 1956. Production of patulin in apple fruits by *Penicillium expansum*. Nature (Lond.) *178*: 263-264.

42. Broom, W. A., E. Bulbring, C. J. Chapman, J. W. F. Hampton, and A. M. Thompson. 1944. The pharmacology of patulin. Br. J. Exp. Pathol. *25*: 195-207.

43. Bulatao-Jayme, J., E. M. Almero, and L. Salamat. 1976. Epidemiology of primary liver cancer in the Philippines with special consideration of a possible aflatoxin factor. J. Philipp. Med. Assoc. *52*: 129-150.

44. Bullerman, L. B., and J. C. Ayres. 1968. Aflatoxin producing potential of fungi isolated from cured and aged meats. Appl. Microbiol. *16*: 1945-1946.

45. Burmeister, H. R. 1971. T-2 toxin production by *Fusarium tricinctum* on solid substrate. Appl. Microbiol. *21*: 739-742.

46. Burmeister, H. R., and C. W. Hesseltine. 1970. Biological assays for two mycotoxins produced by *Fusarium tricinctum*. Appl. Microbiol. *20*: 437-440.

47. Burnside, J. E., W. L. Sippel, J. Forjacs, W. T. Carll, M. B. Atwood, and E. R. Coll. 1957. A disease of swine and cattle caused by eating moldy corn. Am. J. Vet. Res. *18*: 817-824.

48. Caldwell, R. W., and J. Tuite. 1970. Zearalenone production in field corn in Indiana. Phytopathology *60*: 1696-1701.

49. Caldwell, R. W., and J. Tuite. 1974. Zearalenone in freshly harvested corn. Phytopathology *64*: 752-753.

50. Campbell, T. C., J. P. Caedo, Jr., J. Butatao-Jayme, L. Salamat, and R. W. Engel. 1970. Aflatoxin M, in human urine. Nature (Lond.) *227*: 403-404.

51. Campbell, T. C., and L. Stoloff. 1974. Implications of mycotoxins for human health. J. Agric. Food Chem. *22*: 1006-1015.

52. Carnaghan, R. B. A., and K. Sargeant. 1961. The toxicity of certain groundnut meals to poultry. Vet. Rec. *73*: 726-727.

53. Cavallito, C. J., and J. H. Bailey. 1944. Preliminary note on the inactivation of antibiotics. Science *100*: 390.

54. Chandrasekhara, M. R., G. Ramas, and N. Leela. 1970. Processing of groundnut in India as a source of protein foods. Indian Food Packer *24*: 1-10.

55. Chong, Y. H. 1966. Aflatoxins in groundnuts and groundnut products. Far East Med. J. *2*: 228-230.

56. Chong, Y. H., and C. G. Beng. 1965. Aflatoxins in unrefined groundnut oil. Med. J. Malaya *20*: 49-50.

57. Chu, F. S. 1971. Interaction of ochratoxin A with bovine serum albumin. Arch. Biochem. Biophys. *147*: 359-366.

58. Chu, F. S. 1974. Studies on ochratoxins. Crit. Rev. Toxicol. *2*: 499-524.

59. Chu, F. S., and M. E. Butz. 1970. Mycotoxins: spectrophotofluorodensitometric measurements of ochratoxin A in cereals products. J. Assoc. Off. Anal. Chem. *53*: 1253-1257.

60. Chung, C. W., M. W. Trucksees, A. L. Giles, Jr., and L. Friedman. 1974. Rabbit skin test for estimating T-2 toxin and other skin-irritating toxins in contaminated corn. J. Assoc. Off. Anal. Chem. *57*: 1121-1127.

61. Ciegler, A. 1972. Bioproduction of ochratoxin A and penicillic acid by members of *Aspergillus ochraceous* Group. Can. J. Microbiol. *18*: 631-634.

62. Ciegler, A. 1977. Patulin. In J. V. Rodricks, C. W. Hesseltine, and M. A. Mehlman

(Eds.), *Mycotoxins in Human and Animal Health*, Pathotox, Park Forest South, Ill., pp. 609-624.

63. Ciegler, A. 1978. Detoxification of aflatoxin-contaminated agricultural commodities. In P. Rosenberg (Ed.), *Toxins: Animal, Plant and Microbial*, Pergamon, New York, pp. 729-738.

64. Ciegler, A., A. C. Beckwith, and L. K. Jackson. 1976. Teratogenicity of patulin and patulin adducts formed with cysteine. Appl. Enciron. Microbiol. *31*: 664-667.

65. Ciegler, A., R. W. Detroy, and E. B. Lillehoj. 1971. Patulin, penicillic acid and other carcinogenic lactones. In A. Ciegler, S. Kadis, and S. Ajl (Eds.), *Microbial Toxins*, Vol. 6, Academic, New York, pp. 409-434.

66. Ciegler, A., D. J. Fennell, H. J. Mintzlaff, and L. Leistner. 1972. Ochratoxin synthesis by *Penicillium* species. Naturwissenchaften *59*: 365-366.

67. Ciegler, A., S. Kadis, and S. Ajl. 1971. *Microbial Toxins*, Vols. 6-8, Academic, New York.

68. Ciegler, A., and C. P. Kurtzman. 1970. Fluorodensitometric assay of penicillic acid. J. Chromatogr. *51*: 511-516.

69. Ciegler, A., H. J. Mintzlaff, D. Weisleder, and L. Leistner. 1972. Potential production and detoxification of penicillic acid in mold fermented sausage (salami). Appl. Microbiol. *24*: 114-119.

70. Ciegler, A., H. J. Mintzlaff, W. Machnik, and L. Leistner. 1972. Unterschungen über das Toxinbildungsvermogen von Rohwursten Isolieter Schimmelpilze der Gattung *Penicillium*. Fleischwirtschaft *52*: 1311-1314, 1317-1318.

71. Ciegler, A., R. F. Vesonder, and R. J. Cole. 1976. Tremorgenic mycotoxins. In J. V. Rodricks (Ed.), *Mycotoxins and Other Fungal Related Food Problems. Advances in Chemistry Series 149*, American Chemical Society, Washington, D.C., pp. 163-177.

72. Cole, R. J. 1977. Tremorgenic mycotoxins. In J. V. Rodricks, C. W. Hesseltine, and M. A. Mehlman (Eds.), *Mycotoxins in Human and Animal Health*, Pathotox, Park Forest South, Ill., pp. 583-595.

73. Cole, R. J., J. W. Kirskey, J. W. Dorner, D. M. Wilson, J. C. Johnson, Jr., A. N. Johnson, D. M. Bedell, J. P. Springer, J. C. Clardy, and R. H. Cox. 1977. Mycotoxins produced by *Aspergillus fumigatus* species isolated from molded silage. J. Agric. Food Chem. *25*: 826-830.

74. Connole, M. D., and M. W. M. Hill. 1970. *Aspergillus flavus* contaminated sorghum grain as a possible cause of aflatoxicosis in pigs. Aust. Vet. J. *46*: 503-505.

75. Coomes, T. J., and J. C. Sanders. 1963. The detection and estimation of aflatoxin in groundnuts and groundnut material. Analyst (Lond.) *88*: 209-213.

76. Cucullu, A. F., L. S. Lee, and W. A. Pons, Jr. 1977. Relationship of physical appearance of individual mold-damaged cottonseed to aflatoxin content. J. Am. Oil Chem. Soc. *54*: 235A-237A.

77. Dalton, J. E. 1952. Keloid resulting from a positive patch test. Arch. Dermatol. Syphilol. *65*: 53-55.

78. Damodaran, C., C. S. Ramadoss, and E. R. B. Shanmugasundaram. 1973. A rapid procedure for the isolation, identification and estimation of citrinin. Anal. Biochem. *52*: 482-488.

79. Davis, N. D. 1975. Medium scale production of citrinin by *Penicillium citrinin* in a semisynthetic medium. Appl. Microbiol. *29*: 118-120.

80. Davis, N. D., G. A. Sansing, T. V. Ellenburg, and V. L. Diener. 1972. Medium-scale production and purification of ochratoxin A, a metabolite of *Aspergillus ochraceus*. Appl. Microbiol. *23*: 433-435.

81. Denizel, T., B. Jarvis, and E. Rolfe. 1976. A field survey of pistachio nut produc-

tion and storage in Turkey with particular reference to aflatoxin contamination. J. Sci. Food Agric. *27*: 1021-1026.

82. deRosnay, C. D., C. Martin-DuPont, and R. Jensen. 1952. Etude d'une substance antibiotique, la "Mycoine C." J. Med. Bordeaux *129*: 189-199.

83. Dickens, F., and H. E. H. Jones. Carcinogenic activity of a series of reactive lactones and related substances. Br. J. Cancer *15*: 85-100.

84. Dickens, F., and H. E. H. Jones. 1963. The carcinogenic action of aflatoxin after its subcutaneous injection in the rat. Br. J. Cancer *17*: 691-698.

85. Dickens, J. W. 1967. Survey of aflatoxin in farmers stock peanuts marketed in North Carolina during 1964-1966. In *Proceedings of the Mycotoxin Research Seminar*, U.S. Dept. of Agriculture, Washington, D.C., pp. 5-7.

86. Dickens, J. W. 1977. Aflatoxin control program for peanuts. J. Am. Oil Chem. Soc. *54*: 225A-228A.

87. Dickens, J. W., and H. E. Pattee. 1966. The effects of time, temperature and moisture on aflatoxin production in peanuts inoculated with a toxic strain of *Aspergillus flavus*. Trop. Sci. *8*: 11-22.

88. Dickens, J. W., and J. B. Satterwhite. 1971. Diversion program for farmers stock peanuts with high concentrations of aflatoxin. Oleagineux *26*: 321-328.

89. Diener, V. L., C. R. Jackson, W. E. Cooper, R. D. Stipes, and N. D. Davis. 1965. Invasion of peanut pods in the soil by *Aspergillus flavus*. Plant Dis. Rep. *49*: 931-935.

90. DiProssimo, V. P. 1976. Distribution of aflatoxins in some samples of peanuts. J. Assoc. Off. Anal. Chem. *59*: 941-944.

91. Doster, R. C., and R. O. Sinnhuber. 1972. Comparative ration of hydrolysis of ochratoxin A and B in vitro. Food Cosmet. Toxicol. *10*: 389-394.

92. Drilleau, J. F., and G. Bohuon. 1973. La patuline dans les produits cédricoles. C.R. Acad. Agric. Fr. *59*: 1031-1036.

93. Dounin, M. 1926. The fusariosis of cereal crops in European Russia in 1923. Phytopathology *16*: 305-308.

94. Dvorackova, I., F. Brodsky, and J. Cerman. 1974. Aflatoxin and encephalitic syndrome with fatty degeneration of viscera. Nutr. Rep. Int. *10*: 89-102.

95. Dwarakanath, C. T., V. Sreenivasa Murthy, and H. A. B. Parpia. 1969. Aflatoxin in Indian peanut oil. J. Food Sci. Technol. (Mysore) *6*: 107-109.

96. Eble, T. E., and F. R. Hanson. 1951. Fumagillin, an antibiotic from *Aspergillus fumigatus*. Antibiot. Chemother. *1*: 54-58.

97. Elling, F., B. Hald, C. Jacobsen, and P. Krogh. 1975. Spontaneous toxic nephropathy in poultry associated with ochratoxin A. Acta Pathol. Microbiol. Scand., Sect. A *83*: 739-741.

98. Emeh, C. O., and E. H. Marth. 1977. Methods to purify and determine rubratoxins. Z. Lebensm. Unters. Forsch. *163*: 115-120.

99. Englebrecht, R. H., J. L. Ayres, and R. O. Sinnhuber. 1965. Isolation and determination of aflatoxins B_1 in cottonseed meals. J. Assoc. Off. Anal. Chem. *48*: 815-818.

100. Enomoto, M. 1978. Carcinogenicity of mycotoxins. In K. Uraguchi and M. Yamayaki (Eds.), *Toxicology, Biochemistry and Pathology of Mycotoxins*, Halsted, New York, pp. 240-241.

101. Enomoto, M., and M. Saito. 1972. Carcinogens produced by fungi. Ann. Rev. Microbiol. *26*: 279-312.

102. Enomoto, M., and I. Ueno. 1974. *Penicillium islandicum* (toxic yellow rice) luteoskyrin-islanditoxin-cyclochlorotine. In I. F. H. Purchase (Ed.), *Mycotoxins*, Elsevier, Amsterdam, pp. 303-326.

103. Eppley, R. M. 1975. Methods for the detection of trichothecenes. J. Assoc. Off. Anal. Chem. *58*: 906-908.

104. Eppley, R. M., L. Stoloff, M. W. Trucksees, and C. W. Chung. 1974. Survey of corn for *Fusarium* toxins. J. Assoc. Off. Anal. Chem. *57*: 632-635.

105. Escoula, L. 1974. Moisissures toxinogènes des fourrages ensilés. I. Présence de patuline dans les fronts de coupe d'ensilages. Ann. Rech. Vet. *5*: 423-432.

106. Escoula, L. 1975. Contribution à l'étude écologique et toxicologique des moisissures des fourrages ensilés: présence de patuline dans les fronts de coupe. Thèse de 3e cycle, Université de Toulouse, 86 pp.

107. Eyrich, W. 1975. Zum Nachweis von Patulin in Apfelsaft. Chem. Mikrobiol. Technol. Lebensm. *4*: 17-19.

107a. Fed. Reg. *43*(65): 14122-14123.

108. Fennell, D. I., E. B. Lillehoj, and W. F. Kwokek. 1975. *Aspergillus flavus* and other fungi associated with insect damaged field corn. Cereal Chem. *52*: 314-321.

109. Ferreira, N. P. 1968. The effect of amino acids on the production of ochratoxin A in chemically defined media. Antonie Leeuwenhoek J. Microbiol. Serol. *34*: 433-440.

110. Feuell, A. J. 1966. Toxic factors of mold origin. Trop. Sci. *8*: 61-70.

111. Ford, J. H., A. R. Johnson, and J. W. Hinman. 1950. The structure of penicillic acid. J. Am. Chem. Soc. *72*: 4529-4531.

112. Forgacs, J. 1962. Mycotoxicoses—the neglected diseases. Feedstuffs *34*: 124-134.

113. Frank, H. K. 1977. Aflatoxin regulation in West Germany. In J. V. Rodricks, C. W. Hesseltine, and M. A. Mehlman (Eds.), *Mycotoxins in Human and Animal Health,* Pathotox, Park Forest South, Ill., pp. 759-764.

114. Frank, H. K., R. Orth, and A. Figge. 1977. Patulin in Lebensmitteln pflanzlicher Herkunft, 2. Verschiedene Obstarten, Gemuse und daraus hergestellte Produkte. Z. Lebensm. Unters. Forsch. *163*: 111-114.

115. Frank, H. K., R. Orth, and R. Herman. 1976. Patulin in Lebensmitteln pflanzlicher Herkunft. Kernobst und daraus hergestellte Produkte. Z. Lebensm. Unters. Forsch. *162*: 149-157.

116. Freerksen, E., and R. Bonicke. 1951. Die Inaktiverung des Patulins in vivo (Modell versuche zu Bestimmung des wertes antibaketerieller Substanzen für therapeutische Zwecke). Z. Hyg. Infektionskr. *132*: 274-291.

117. Fries, P., E. Hasselager, and P. Krogh. 1969. Isolation of citrinin and oxalic acid from *Penicillium viridicatum* West and their nephrotoxicity in rats and pigs. Acta Pathol. Microbiol. Scand. *77*: 559-560.

118. Fuller, F., W. W. Spooncer, A. D. King, Jr., J. Schade, and B. Mackey. 1977. Survey of aflatoxins in California tree nuts. J. Am. Oil Chem. Soc. *54*: 231A-234A.

119. Gardner, H. K., Jr., S. P. Koltun, F. G. Dollear, and E. T. Rayner. 1971. Inactivation of aflatoxins in peanut and cottonseed meals by ammoniation. J. Am. Oil Chem. Soc. *48*: 70-73.

120. Garner, R. C. 1973. Chemical evidence for the formation of a reactive aflatoxin B_1 metabolite by hamster liver microsomes. FEBS Lett. *36*: 261-264.

121. Garner, R. C. 1975. Aflatoxin separation by high pressure liquid chromatography. J. Chromatogr. *103*: 186-188.

122. Garner, R. C. 1976. The role of epoxides in bioactivation and carcinogenesis. In J. W. Bridges and L. F. Chassead (Eds.), *Progress in Drug Metabolism,* Vol. 1, Wiley, New York, pp. 77-128.

123. Garner, R. C., E. C. Miller, and J. A. Miller. 1972. Liver microsomal metabolism of aflatoxin B_1 to a reactive derivative toxic to *Salmonella typhimurium* TA 1530. Cancer Res. *32*: 2058-2066.

124. Geiger, W. B., and G. E. Conn. 1945. The mechanism of the antibiotic action of clavacin and penicillic acid. J. Am. Chem. Soc. *67*: 112-116.

125. Gilgan, M. W., E. B. Smalley, and F. M. Strong. 1966. Isolation and characterization of a toxin from *Fusarium tricinctum* on moldy corn. Arch. Biochem. Biophys. *114*: 1-3.

126. Giridhar, N., and G. V. Krishnamurthy. 1977. Studies on aflatoxin content of groundnut oil in Andhra Pradesh with reference to climatic conditions and seasonal variations. J. Food Sci. Technol. *14*: 84-85.

127. Goldblatt, L. A. 1969. *Aflatoxin*, Academic, New York.

128. Gopalan, C., P. G. Tulpule, and D. Krishnamurthi. 1972. Induction of hepatic carcinoma with aflatoxin in the rhesus monkey. Food Cosmet. Toxicol. *10*: 519-521.

129. Gordon, W. L. 1954. The occurrence of *Fusarium* species in Canada. III. Taxonomy of *Fusarium* species in the seed of vegetables, forage, and miscellaneous crops. Can. J. Bot. *32*: 576-590.

130. Gordon, W. L. 1956. The occurrence of *Fusarium* species in Canada. V. Taxonomic and geographic distribution of *Fusarium* species in soil. Can. J. Bot. *34*: 833-864.

131. Gordon, W. L. 1960. The taxonomy and habitats of *Fusarium* species from tropical and temperate regions. Can. J. Bot. *38*: 643-658.

132. Gross, V. J., and J. Robb. 1975. Zearalenone production in barley. Ann. Appl. Biol. *80*: 211-216.

133. Habish, H. A. 1972. Aflatoxin in harikot bean and other pulses. Exp. Agric. *8*: 135-137.

134. Hald, P., and P. Krogh. 1970. Occurrence of aflatoxin in imported cottonseed products. Nord. Veterinaermed. *22*: 39-47.

135. Hald, B., and P. Krogh. 1973. Analysis and chemical confirmation of citrinin in barley. J. Assoc. Off. Anal. Chem. *56*: 1440-1443.

136. Hamilton, P. B., W. E. Huff, J. R. Harris, and R. D. Wyatt. 1977. Outbreaks of ochratoxicosis in poultry. Abstr. Annu. Meet. Am. Soc. Microbiol., 1977, 021, p. 248.

137. Hamsa, T. A. P., and J. C. Ayres. 1977. Factors affecting aflatoxin contamination of cottonseed. I. Contamination of cottonseed with *Aspergillus flavus* at harvest and during storage. J. Am. Oil Chem. Soc. *54*: 219-224.

138. Hanssen, E. 1968. Ergebnisse der Untersuchung einiger Lebensmittel auf Aflatoxin B_1. Mitteilungsbl. GBDCh-Fachgruppe *22*: 83-88.

139. Hanssen, E. 1969. Schadigung von Lebensmitteln durch Aflatoxin B_1. Naturwissenschaften *56*: 90-92.

140. Harwig, J., B. J. Blanchfield, and P. M. Scott. 1978. Patulin production by *Penicillium roqueforti* from grape. Can. Inst. Food Sci. Technol. *11*: 149-151.

141. Harwig, J., Y. K. Chen, B. P. Kennedy, and P. M. Scott. 1973. Occurrence of patulin-producing strains of *Penicillium expansum* in natural rots of apple in Canada. Can. Inst. Food Sci. Technol. *6*: 22-25.

142. Harwig, J., W. Przybylski, and C. A. Moodie. 1975. A link between Reye's syndrome and aflatoxin? Can. Med. Assoc. J. *113*: 281.

143. Harwig, J., and P. M. Scott. 1971. Brine shrimp (*Artemia salina* L.) larvae as a screening system for fungal toxins. Appl. Microbiol. *21*: 1011-1016.

144. Hayes, A. W., and H. W. McCain. 1975. A procedure for the extraction and estimation of rubratoxin B from corn. Food Cosmet. Toxicol. *13*: 221-229.

145. Hayes, A. W., and B. Wilson. 1968. Bioproduction and purification of rubratoxin B. Appl. Microbiol. *16*: 1163-1167.

146. Heathcote, J. G., and J. R. Hibbert. 1978. *Aflatoxins: Chemical and Biological Aspects,* Elsevier, Amsterdam.

147. Heathcote, J. G., and J. R. Hibbert. 1974. Biological activity and electronic structure of the aflatoxins. Br. J. Cancer *29*: 470-476.

148. Heatley, N. G., and F. G. Philpot. 1947. The routine examination for antibiotics by moulds. J. Gen. Microbiol. *1*: 232-237.

149. Hesseltine, C. W. 1974. Natural occurrence of mycotoxins in cereals. Mycopathol. Mycol. Appl. *53*: 141-153.

150. Hesseltine, C. W., O. L. Shotwell, M. Smith, J. J. Ellis, E. Vandergraft, and G. Shannon. 1970. Production of various aflatoxins by strains of the *Aspergillus flavus* series. In *Toxic Micro-organisms,* Proceedings of the First U.S.-Japan Conference on Toxic Micro-organisms. UJNR Joint Panels on Toxic Micro-organisms and the U.S. Dept. of the Interior, Washington, D.C., pp. 202-210.

151. Hesseltine, C. W., W. G. Sorenson, and M. Smith. 1970. Taxonomic studies of the aflatoxin-producing strains in the *Aspergillus flavus* group. Mycologia *62*: 123-132.

152. Hetherington, A. C., and H. Raistrick. 1931. Studies on the biochemistry of microorganisms. Part SIV. On the production and chemical constitution of a new yellow colouring matter, citrinin, produced from glucose by *Penicillium citrinin* Thom. Philos. Trans. R. Soc. Lond., Ser. B *220*: 269-295.

153. Hilty, M. D. 1975. Etiology of Reye's syndrome. In J. D. Pollack (Ed.), *Reye's Syndrome,* Grune & Stratton, New York, pp. 383-385.

154. Hobson, W., J. Bailey, and G. B. Fuller. 1977. Hormone effects of zearalenone in nonhuman primates. J. Toxicol. Environ. Health *3*: 43-57.

155. Hodges, F. A., J. R. Zust, H. R. Smith, A. A. Nelson, B. H. Armbrecht, and A. D. Campbell. 1964. Mycotoxins: aflatoxin isolated from *Penicillium puberulum.* Science *145*: 1439.

156. Holaday, E. E. 1968. Rapid method for detecting aflatoxins in peanuts. J. Am. Oil Chem. Soc. *45*: 680-682.

157. Holaday, E. C., and J. Lansden. 1975. Rapid screening method for aflatoxin in a number of products. J. Agric. Food Chem. *23*: 1134-1136.

158. Holder, C. L., C. R. Nony, and M. C. Bowman. 1977. Trace analysis of zearalenone and/or zearalanol in animal chow by high pressure liquid chromatography and gas-liquid chromatography. J. Assoc. Off. Anal. Chem. *60*: 272-278.

159. Holker, J. S. E., and J. G. Underwood. 1964. A synthesis of a cyclopentenocoumarin structurally related to aflatoxin B. Chem. Ind. (Lond.) *45*: 1865-1866.

160. Hood, R. D., J. E. Innes, and A. W. Hayes. 1973. Effects of rubratoxin B on prenatal development in mice. Bull. Environ. Contam. Toxicol. *10*: 200-207.

161. Hormozdiari, H., N. E. Day, B. Aramesh, and E. Mahboubi. 1975. Dietary factors and esophageal cancer in the Caspian littoral of Iran. Cancer Res. *35*: 3493-3498.

162. Howard, B. H., and H. Raistrick. 1954. Studies in the biochemistry of microorganisms. 92. The colouring matters of *Penicillium islandicum* Sopp. Part 4. Iridoskyrin, rubroskyrin, and erythroskyrine. Biochem. J. *57*: 212-222.

163. Hsieh, D. P. H., D. L. Fitzell, J. L. Miller, and J. N. Seiber. 1976. High-pressure liquid chromatography of oxidative aflatoxin metabolites. J. Chromatogr. *117*: 474-479.

164. Hsieh, D. P. H., Z. A. Wong, J. J. Wong, C. Michas, and B. H. Ruebner. 1977. Comparative metabolism of aflatoxin. In J. V. Rodricks, C. W. Hesseltine, and M. A. Mehlman (Eds.), *Mycotoxins in Human and Animal Health,* Pathotox, Park Forest South, Ill., pp. 37-50.

165. Hsu, I. C., E. B. Smalley, F. M. Strong, and W. E. Ribelin. 1972. Identification

of T-2 toxin in moldy corn associated with a lethal toxicosis in dairy cattle. Appl. Microbiol. *24*: 684–690.

166. Ikediobi, C. O., I. C. Hus, J. R. Bamburg, and F. M. Strong. 1971. Gas-liquid chromatography of mycotoxins of the trichothecene group. Anal. Biochem. *43*: 327–340.

167. International Union of Pure and Applied Chemistry. 1973. Recommended method for aflatoxin analysis in cocoa beans. IUPAC Int. Bull. Tech. Rep. *8*: 1–12.

168. Iwata, K., T. Nagai, and M. Okudaira. 1969. Fumigatoxin, a new toxin from a strain of *Aspergillus fumigatus*. J. S. Afr. Chem. Inst. *22*: S131–S141.

169. Jackson, W. E., H. Wolf, and R. O. Sinnhuber. 1968. The relationship of hepatoma in rainbow trout to aflatoxin contamination and cottonseed meal. Cancer Res. *28*: 987–991.

170. Jacquet, J., P. Boutibonnes, and A. Teherani. 1970. Sur la présence des flavatoxines dans les aliments des animaux et dans les aliments d'origine animale destinés à l'homme. Bull. Acad. Vet. *43*: 36–41.

171. Joffe, A. Z. 1962. Biological properties of some toxic fungi isolated from over-wintered cereals. Mycopathol. Mycol. Appl. *16*: 201–221.

172. Joffe, A. Z. 1964. Toxin production by cereal fungi causing toxic alimentary aleukia in man. In G. N. Wogan (Ed.), *Mycotoxins in Foodstuffs*, MIT Press, Cambridge, Mass., pp. 77–85.

173. Joffe, A. Z. 1971. Alimentary toxic aleukia. In S. Kadis, A. Ciegler, and S. Ajl (Eds.), *Microbial Toxins*, Vol. 7, Academic, New York, pp. 139–189.

174. Joffe, A. Z. 1974. Growth and toxigenicity of fusaria of the Sporotrichiella section as related to environmental factors and culture substrates. Mycopathol. Mycol. Appl. *54*: 35–46.

175. Joffe, A. Z. 1974. Toxicity of *Fusarium poae* and *F. sporotrichioides* and its relation to alimentary toxic aleukia. In I. F. H. Purchase (Ed.), *Mycotoxins*, Elsevier, Amsterdam, pp. 229–262.

176. Joffe, A. Z. 1977. The genus *Fusarium*. In T. D. Wyllie and L. G. Morehouse (Eds.), *Mycotoxic Fungi, Mycotoxins, Mycotoxicoses*, Dekker, New York, pp. 59–82.

177. Joffe, A. Z., and B. Yagen. 1977. Comparative study of the yield of T-2 toxin produced by *F. poae* and *F. sporotrichioides* var. *tricinctum* strains from different sources. Mycopathologia *60*: 93–97.

178. Jones, E. D. 1974. Aflatoxin in feeding stuffs, its incidence, significance and control, Tropical Products Institute Conference, 1974, on Animal Feeds of Tropical and Subtropical Origin.

179. Josefsson, E., and A. Anderson. 1977. Patulin i. appledrycker. Sartryck ur Var fodu. *28*: 189–196.

180. Kahn, J. B., Jr. 1957. Effects of various lactones and related compounds on cation transfer in incubated cold storage erythrocytes. J. Pharmacol. Exp. Ther. *121*: 234–251.

181. Katzman, P. A., E. E. Hayes, C. K. Cain, J. J. van Wyk, F. J. Reithel, S. A. Thayer, E. A. Doisy, W. L. Gaby, C. J. Carroll, R. D. Muir, L. R. Jones, and J. J. Wase. 1944. Clavacin, an antibiotic substance from *Aspergillus clavatus*. J. Biol. Chem. *154*: 475–486.

182. Kavanagh, T. 1947. Activities of twenty-two antibacterial substances aganist nine species of bacteria. J. Bacteriol. *54*: 761–766.

183. Kawashiro, L., H. Tanabe, H. Tekauchi, and C. Nishimiera. 1965. Determination of citrinin contained in rice by fluorometry. Bull. Natl. Hyg. Lab. Tokyo *73*: 191–196.

184. Keen, P., and P. Martin. 1971. Is aflatoxin carcinogenic in man? The evidence in Swaziland. Trop. Geogr. Med. *23*: 44-53.

185. Kiermeier, F., and W. Mucke. 1972. Uber das Nachweis von Aflatoxin M in Milch. Z. Lebensm. Unters. Forsch. *150*: 137-143.

186. Kinosita, R., and T. Shibata. 1965. On toxic moldy rice. In G. N. Wogan (Ed.), *Mycotoxins in Foodstuffs*, MIT Press, Cambridge, Mass., pp. 111-132.

187. Krishnamachari, K. A., R. V. Bhat, V. Nagarajan, and T. B. G. Tilak. 1974. Investigations into an outbreak of hepatitis in parts of western India. Indian J. Med. Res. *63*: 1036-1040.

188. Krishnamachari, K. A., R. V. Bhat, V. Nagarajan, and T. B. G. Tilak. 1975. Hepatitis due to aflatoxicosis. An outbreak in western India. Lancet *1*: 1061-1062.

189. Krogh, P. 1972. Mycotoxic porcine nephropathy: a possible model for Balkan endemic nephropathy. In A. Puchlev, V. Dinev, B. Miley, and D. Doichinov (Eds.), *Endemic Nephropathy*, Bulgarian Academy of Sciences, Sofia, pp. 266-270.

190. Krogh, P. 1976. Mycotoxic nephropathy. In C. A. Brandly, C. E. Cornelius, and W. I. B. Beveridge (Eds.), *Advances in Veterinary Science and Comparative Medicine*, Vol. 20, Academic, New York, pp. 147-170.

191. Krogh, P. 1977. Natural occurrence of ochratoxin A: a kidney toxic mold metabolite in Scandinavian cereals. Zesz. Probl. Postepow Nauk Roln. *189*: 21-24.

192. Krogh, P. 1977. Ochratoxin A residues in tissue of slaughter pigs and nephropathy. Nord. Veterinaermed. *29*: 402-408.

193. Krogh, P. 1977. Ochratoxin in human and animal health. In J. V. Rodricks, C. W. Hesseltine, and M. A. Mehlman (Eds.), *Mycotoxins in Human and Animal Health*, Pathotox, Park Forest South, Ill., pp. 489-498.

194. Krogh, P., N. H. Axelsen, F. Elling, N. Gyrd-Hansen, B. Hald, J. Hyldgaard-Jensen, A. E. Larsen, A. Madsen, H. P. Mortensen, T. Moller, O. K. Petersen, U. Ravnskov, M. Rostgaard, and O. Aalund. 1974. Experimental porcine nephropathy: changes of renal function and structure induced by ochratoxin A-contaminated feed. Acta Pathol. Microbiol. Scand., Sect. A (Suppl.) 246.

195. Krogh, P., and B. Hald. 1969. Forekomst at aflatoksin i importerede jordnodprodukter. Nord. Veterinaermed. *21*: 398-407.

196. Krogh, P., B. Hald, and E. J. Pederson. 1973. Occurrence of ochratoxin A and citrinin in cereals associated with mycotoxic porcine nephropathy. Acta Pathol. Microbiol. Scand., Sect. B *81*: 686-695.

197. Krogh, P., B. Hald, R. Plestina, and S. Ceovic. 1977. Balkan (endemic) nephropathy and foodborn ochratoxin A: preliminary results of a survey of foodstuffs. Acta Pathol. Microbiol. Scand., Sect. B *85*: 238-240.

198. Kulik, M. M., and E. E. Holaday. 1966. Aflatoxin: a metabolic product of several fungi. Mycopathol. Mycol. Appl. *30*: 137-140.

199. Kurtzman, C. P., and A. Ciegler. 1970. Mycotoxin from a blue-eye mold of corn. Appl. Microbiol. *20*: 204-207.

200. Lafont, P., and J. Lafont. 1970. Contamination de produits céréaliers et d'aliments du bétail par l'aflatoxine. Food Cosmet. Toxicol. *8*: 403-407.

201. Lancaster, M. C., F. P. Jenkins, and J. McL. Philip. 1961. Toxicity associated with certain samples of groundnuts. Nature (Lond.) *192*: 1095-1096.

202. LeBars, J., and G. Escoula. 1974. Fungal contamination of animal feed. Toxicological aspects. Aliment. Vie *62*: 125-142.

203. Lebron, V. E., and A. M. Mejia. 1976. Determinación de aflatoxina B, por chromatographía de capa fina en productos y subproductos agricolas locales. Personal communication from INDOTEC, Dominical Republic.

204. Leonov, A. N. 1977. Current view of the chemical nature of factors responsible

for alimentary toxic aleukia.In J. V. Rodricks, C. W. Hesseltine, and M. A. Mehlman (Eds.), *Mycotoxins in Human and Animal Health,* Pathotox, Park Forest South, Ill., pp. 323-328.

205. Levi, C. P., and E. Borker. 1968. Survey of green coffee for potential aflatoxin contamination. J. Assoc. Off. Anal. Chem. *51*: 600-602.

206. Levi, C. P., H. L. Trenk, and H. K. Mohr. 1974. Study of the occurrence of ochratoxin A in green coffee beans. J. Assoc. Off. Anal. Chem. *57*: 866-870.

207. Lieu, F. Y., and L. B. Bullerman. 1978. Binding of patulin and penicillic acid to glutathione and cysteine and toxicity of the resulting adducts. Milchwissenschaft *33*: 16-20.

208. Lillehoj, E. B., and A. Ciegler. 1975. Mycotoxin synergism. In D. Schlessinger (Ed.), *Microbiology—1975,* American Society for Microbiology, Washington, D.C., pp. 344-358.

209. Lillehoj, E. B., D. I. Fennell, and W. F. Kwolek. 1976. *Aspergillus flavus* and aflatoxin in Iowa corn before harvest. Science *193*: 495-496.

210. Lillehoj, E. B., D. I. Fennell, and W. F. Kwolek. 1977. Aflatoxin and *Aspergillus flavus* occurrence in 1975 corn at harvest from a limited region of Iowa. Cereal Chem. *54*: 366-372.

211. Lillehoj, E. B., and C. W. Hesseltine. 1977. Aflatoxin control during plant growth and harvest of corn. In J. V. Rodricks, C. W. Hesseltine, and M. A. Mehlman (Eds.), *Mycotoxins in Human and Animal Health,* Pathotox, Park Forest South, Ill., pp. 107-120.

212. Lillehoj, E. B., W. F. Kwolek, R. E. Peterson, O. L. Shotwell, and C. W. Hesseltine. 1976. Aflatoxin contamination, fluorescence and insect damage in corn infected with *Aspergillus flavus* before harvest. Cereal Chem. *53*: 505-512.

213. Lillehoj, E. B., W. F. Kwolek, G. M. Shannon, O. L. Shotwell, and C. W. Hesseltine. 1975. Aflatoxin occurrence in 1973 corn at harvest. 1. A limited survey in the southeastern U.S. Cereal Chem. *52*: 603-611.

214. Lillehoj, E. B., W. F. Kwolek, E. E. Vandegraft, M. S. Zuber, O. H. Calvert, N. Widstrom, M. C. Futrell, and A. J. Bockholt. 1975. Aflatoxin production in *Aspergillus flavus* inoculated ears of corn grown at diverse locations. Crop Sci. *15*: 267-270.

215. Lin, J. K., J. A. Miller, and E. C. Miller. 1977. 2,3-Dihydro-2-(guan-7-yl)-3-hydroxyaflatoxin B, as a major hydrolysis product of aflatoxin B-DNA or -ribsomal RNA adducts formed in hepatic microsome-mediated reactions and in rat liver in vivo. Cancer Res. *37*: 4430-4438.

216. Ling, K.-H., C. M. Tung, P. Sheh, J.-J. Wong, and T.-C. Tung. 1969. Aflatoxin B, in unrefined peanut oil and peanut products in Taiwan. J. Formosan Med. Assoc. *67*: 309-314.

217. Linsell, C. A., and F. G. Peers. 1972. The aflatoxins and human liver cancer. In E. Grundmann and H. Tulinius (Eds.), *Current Problems in the Epidemiology of Cancer and Lymphomas,* Springer-Verlag, New York, pp. 193-207.

218. Lopez, A., and M. A. Crawford. 1967. Aflatoxin content of groundnuts sold for human consumption in Uganda. Lancet *2*: 1351-1354.

219. Lovelend, P. M., R. O. Sinnhuber, K. E. Berggren, L. H. Libbey, J. E. Nixon, and N. E. Pawlowski. 1977. Formation of aflatoxin B, from aflatoxicol by rainbow trout (*Salmo gairdneri*) in vitro. Res. Commun. Chem. Pathol. Pharmacol. *16*: 167-170.

220. Lovett, J., and R. G. Thompson, Jr. 1978. Patulin production by species of *Aspergillus* and *Penicillium* at 1.7, 7.2 and 12.8 C. J. Food Protect. *41*: 195-197.

221. Lucci, R., L. Merline, G. Nasine, and J. R. Locci. 1965. On a strain of *Penicillium fellutanum* Biourge producing a yellow fluorescent substance. J. Microbiol. *13*: 271-277.

222. Luck, H., M. Steyn, and F. C. Wehner. 1976. A survey of milk powder for aflatoxin content. S. Afr. J. Dairy Technol. *8*: 85-89.

223. Madhavan, T. V., P. G. Tulpule, and C. Gopalan. 1965. Aflatoxin-induced hepatic fibrosis in rhesus monkeys. Arch. Pathol. *79*: 466-469.

224. Manabe, M., O. Tsuruta, T. Sugiomoto, M. Minamisawa, and S. Matsuura. 1971. Aflatoxins in imported peanut meals. J. Food Sanit. *12*: 364-369.

225. Marsh, P. B., M. E. Simpson, R. J. Ferretti, T. C. Campbell, and J. Donoso. 1969. Relation of aflatoxins in cottonseeds at harvest to fluorescence in the fiber. J. Agric. Food Chem. *17*: 462-467.

226. Marsh, P. B., M. E. Simpson, R. J. Ferretti, G. V. Merola, J. Donoso, G. O. Craig, M. W. Trucksees, and P. S. Work. 1969. Mechanism of formation of a fluorescence in cotton fiber associated with aflatoxins in the seeds at harvest. J. Agric. Food Chem. *17*: 468-472.

227. Martin, P. M. D., and G. A. Gilman. 1976. A consideration of the mycotoxin hypothesis with special reference to the mycoflora of maize, sorghum, wheat and groundnuts. Tropical Products Institute Publication G 105, Grays Inn Road, London.

228. Marumo, S. 1955. Islanditoxin, a toxic metabolite produced by *Penicillium islandicum*. Bull. Agric. Chem. Soc. Jpn. *19*: 258-261.

229. Mateles, R. I., and J. C. Adye. 1965. Production of aflatoxins in submerged culture. Appl. Microbiol. *13*: 208-211.

230. Mayer, C. F. 1953. Endemic panmyelotoxicosis in the Russian grain belt. Part One: the clinical aspects of alimentary toxic aleukia (ATA). A comprehensive review. Mil. Surg. *113*: 173-189.

231. Mayer, C. F. 1953. Endemic panmyelotoxicosis in the Russian grain belt. Part Two: the botany, phytopathology, and toxicology of Russian cereal food. Mil. Surg. *113*: 295-315.

232. McCann, J., E. Choi, E. Yamasaki, and B. N. Ames. 1975. Detection of carcinogens as mutagens in the *Salmonella*/microsome test: assay of 300 chemicals. Proc. Natl. Acad. Sci. U.S.A. *72*: 5135-5139.

233. McDonald, D., and C. Harkness. 1967. Aflatoxin in the groundnut crop at harvest in northern Nigeria. Trop. Sci. *9*: 148-161.

234. McDonald, J. C. 1973. Toxicity, analysis and production of aspergillic acid and its analogues. Can. J. Biochem. *51*: 1311-1315.

235. McKinney, J. D. 1975. Use of zinc acetate in extract purification for aflatoxin assay of cottonseed products. J. Am. Oil Chem. Soc. *52*: 213.

236. McLaughlin, C. S., M. H. Vaughn, I. M. Campbell, C. M. Wei, M. E. Stafford, and B. S. Hansen. 1977. Inhibition of protein synthesis by trichothecenes. In J. V. Rodricks, C. W. Hesseltine, and M. A. Mehlman (Eds.), *Mycotoxins in Human and Animal Health*, Pathotox, Park Forest South, Ill., pp. 263-274.

237. McMeans, J. L., and C. M. Brown. 1975. Aflatoxin in cottonseed as affected by the pink bollworm. Crop Sci. *15*: 865-866.

238. McMeans, J. L., C. M. Brown, R. L. McDonald, and L. L. Parker. 1977. Aflatoxins in cottonseed: a comparison of two cultivans. Crop Sci. *17*: 707-709.

239. Mintzlaff, H. J., A. Ciegler, and L. Leistner. 1972. Potential mycotoxins in mould-fermented sausage. Z. Lebensm. Unter. Forsch. *50*: 135-137.

240. Mirocha, C. J., and C. M. Christensen. 1974. Oestrogenic mycotoxins synthesized by *Fusarium*. In I. F. H. Purchase (Ed.), *Mycotoxins*, Elsevier, Amsterdam, pp. 127-148.

241. Mirocha, C. J., C. M. Christensen, and G. H. Nelson. 1968. Toxic metabolites pro-
      duced by fungi implicated in mycotoxicoses. Biotechnol. Bioeng. *10*: 468-482.
242. Mirocha, C. J., C. M. Christensen, and G. H. Nelson. 1971. F-2 (zearalenone).
      Estrogenic mycotoxin from *Fusarium.* In A. Ciegler, S. Kadis, and S. Ajl (Eds.),
      Microbial Toxins, Vol. 7, Academic, New York, pp. 107-138.
243. Mirocha, C. J., and S. Pathre. 1973. Identification of the toxic principle in a sam-
      ple of poaefusarin. Appl. Microbiol. *26*: 719-724.
244. Mirocha, C. J., S. V. Pathre, and J. Behrens. 1976. Substances interfering with
      the gas-liquid chromatographic determination of T-2 mycotoxin. J. Assoc. Off.
      Anal. Chem. *59*: 221-223.
245. Mirocha, C. J., S. V. Pathre, and C. M. Christensen. 1977. Zearalenone. In J. V.
      Rodricks, C. W. Hesseltine, and M. A. Mehlman (Eds.), *Mycotoxins in Human and
      Animal Health,* Pathotox, Park Forest South, Ill., pp. 345-364.
246. Mirocha, C. J., S. V. Pathre, B. Schauerhammer, and C. M. Christensen. 1976.
      Natural occurrence of *Fusarium* toxins in feedstuffs. Appl. Microbiol. *32*: 553-
      556.
247. Mislivec, P. B., J. H. Hunter, and J. Tuite. 1968. Assay for aflatoxin production
      by the genera *Aspergillus* and *Penicillium.* Appl. Microbiol. *16*: 1053-1055.
248. Mishra, S. K., and H. S. R. Murthy. 1968. An extra fungal source of aflatoxins.
      Curr. Sci. *37*: 406.
249. Mixon, A. C. 1977. Influence of plant genetics on colonization by *Aspergillus
      flavus* and toxin production (peanuts). In J. V. Rodricks, C. W. Hesseltine, and
      M. A. Mehlman (Eds.), *Mycotoxins in Human and Animal Health,* Pathotox, Park
      Forest South, Ill., pp. 163-172.
250. Moller, T. E., and E. Josefsson. 1978. High pressure liquid chromatography of
      zearelenone in cereals. J. Assoc. Off. Anal. Chem. *61*: 789-792.
251. Moroeanu, S. B. 1967. Epidemiological observations on the endemic nephropathy
      in Rumania. In G. E. W. Wolstenholme and J. Knight (Eds.), *Balkan Nephropathy,*
      Little, Brown, Boston, pp. 4-13.
252. Morooka, N., N. Uratsuki, T. Yoshizawa, and H. Yamamoto. 1972. Studies on
      toxic substances in barley infected with *Fusarium* spp. J. Food Sanit. *13*: 368-
      375.
253. Morris, M. P., C. Pagan, and H. E. Warmke. 1954. Hiptagenic acid, a toxic com-
      ponent of *Indigofera endecaphylla.* Science *119*: 322-323.
254. Moss, M. O., and I. W. Hill. 1970. Strain variation in the production of rubratoxins
      by *Penicillium rubrum* Stoll. Mycopathol. Mycol. Appl. *40*: 81-88.
255. Moss, M. O., F. F. Robinson, A. B. Wood, H. M. Paisley, and J. Feeney. 1968.
      Rubratoxin B, a proposed structure for a bis-anhydride from *Penicillium rubrum*
      Stoll. Nature (Lond.) *220*: 767-770.
256. Moss, M. O., F. V. Robinson, A. B. Wood, and A. Morrison. 1967. Observations
      on the structure of the toxins from *Penicillium rubrum.* Chem. Ind. (Lond.) *2*:
      755-757.
257. Murnagham, M. F. 1946. The pharmacology of penicillic acid. J. Pharmacol. Exp.
      Ther. *88*: 119-132.
258. Nagel, D. W., P. S. Steyn, and D. B. Scott. 1972. Production of citreoviridin by
      Penicillium pulvillorum. Phytochemistry *11*: 627-630.
259. Naguib, M. M., and M. El-Khadem. 1976. Toxicity of *Aspergillus flavus* cultures
      isolated from Egypt. Zentralbl. Bakteriol. Parasitenkd. Infektionskr. Hyg., Abt.
      131: 506-509.
260. Neeley, W. C., S. P. Ellis, N. D. Davis, and V. L. Diener. 1972. Spectroanalytical
      parameters of fungal metabolites. I. Citrinin. J. Assoc. Off. Anal. Chem. *55*: 1122-
      1127.

261. Neeley, W. C., M. Y. Siraj, L. B. Smith, and A. W. Hayes. 1978. Spectroanalytical parameters of fungal metabolites. V. Rubratoxin B. J. Assoc. Off. Anal. Chem. *61*: 601-604.

262. Nesbitt, B. F., J. O'Kelly, K. Sargeant, and A. Sheridan. 1962. Toxic metabolites of *Aspergillus flavus.* Nature (Lond.) *195*: 1062-1063.

263. Nesheim, S. G. 1973. Analysis of ochratoxin A and B and their esters in barley, using partition and thin layer chromatography. II. Collaboration study. J. Assoc. Off. Anal. Chem. *56*: 822-826.

264. Nesheim, S. G., N. F. Hardin, O. G. Frances, and W. S. Langham. 1973. Analysis of ochratoxin A and B and their esters in barley, using partition thin layer chromatography. I. Development of method. J. Assoc. Off. Anal. Chem. *56*: 817-821.

265. Neumann-Kleinpaul, A., and G. Terplan. 1972. Zum Vorkommen von Aflatoxin M, in Trockenmilch-produkten. Arch. Lebensmittelhyg. *28*: 128-134.

266. Newberne, P. M., and A. E. Rogers. 1973. Animal model of human disease. Primary hepatocellular carcinoma. Am. J. Pathol. *72*: 137-141.

267. Nofsinger, G. W., R. J. Bothast, E. B. Lancaster, and E. B. Bagley. 1977. Ammonia-supplemented ambient temperature drying of high moisture corn. Trans. ASAE *20*: 1151-1154, 1159.

268. Norstadt, F. A., and T. M. McCalla. 1969. Patulin production by *Penicillium utricae* (Bainer) in batch culture. Appl. Microbiol. *17*: 193-196.

269. Ohmomo, S., T. Sato, T. Utagawa, and M. Abe. 1975. Isolation of festuclavine and three new indole alkaloids, roquefortine A, B and C from the cultures of *Penicillium roqueforti.* XII. Production of alkaloids and related substances by fungi. J. Agric. Chem. Soc. Jpn. *49*: 615-623.

270. Olson, L. C., C. H. Bourgeouis, R. B. Cotton, S. Harikul, R. A. Grossman, and T. J. Smith. 1971. Encephalopathy and fatty degeneration of the viscera in northeastern Thailand. Clinical syndrome and epidemiology. Pediatrics *47*: 707-716.

271. Ong, T.-M. 1975. Aflatoxin mutagenesis. Mutat. Res. *32*: 35-53.

272. Orth, R. 1976. PR Toxinbildung bei *Penicillium roqueforti*-Stammen. Z. Lebensm. Unters. Forsch. *160*: 131-136.

273. Oxford, A. E. 1942. On the chemical reactions occurring between certain substances which inhibit bacterial growth and the constituents of bacteriological media. Biochem. J. *36*: 438-444.

274. Parrish, F. W., B. J. Wiley, E. G. Simmons, and L. Long. 1966. Production of aflatoxin and kojic acid by species of *Aspergillus* and *Penicillium.* Appl. Microbiol. *14*: 139.

274. Pathre, S. V., and C. J. Mirocha. 1976. Zearalenone and related compounds. In J. V. Rodricks (Ed.), *Mycotoxins and Other Fungal Related Food Problems. Advances in Chemistry Series 149,* American Chemical Society, Washington, D.C., pp. 178-227.

276. Pathre, S. V., and C. J. Mirocha. 1977. Assay methods for trichothecenes and review of their occurrence. In J. V. Rodricks, C. W. Hesseltine, and M. A. Mehlman (Eds.), *Mycotoxins in Human and Animal Health,* Pathotox, Park Forest South, Ill., pp. 229-254.

277. Pathre, S. V., C. J. Mirocha, C. M. Christensen, and J. Behrens. 1976. Monoacetoxyscirpinol: a new mycotoxin produced by *Fusarium roseum* Gibbosum. J. Agric. Food Chem. *24*: 97-103.

278. Patterson, D. S. P., and B. A. Roberts. 1971. The in vitro reduction of aflatoxins B_1 and B_2 by soluble avian liver enzymes. Food Cosmet. Toxicol. *9*: 829-837.

279. Patterson, D. S. P., and B. A. Roberts. 1972. Aflatoxin metabolism in duck-liver

homogenates: the relative importance of reversible cyclopentenone reduction and hemiacetal formation. Food Cosmet. Toxicol. *10*: 501-512.

280. Paul, R., M. S. Kalra, and A. Singh. 1976. Incidence of aflatoxins in milk and milk products. Indian J. Dairy Sci. *29*: 318-321.

281. Peckham, J. C., B. Doupnik, Jr., and O. H. Jones, Jr. 1971. Acute toxicity of ochratoxin A and B in chicks. Appl. Microbiol. *21*: 492-494.

282. Peers, F. G., G. A. Gilman, and C. A. Linsell. 1976. Dietary aflatoxins and human liver cancer. A study in Swaziland. Int. J. Cancer *17*: 167-176.

283. Peers, F. G., and C. A. Linsell. 1973. Dietary aflatoxins and liver cancer. A population based study in Kenya. Br. J. Cancer *27*: 473-484.

284. Pensala, O., A. Niskanen, and S. Lahtinen. 1977. The occurrence of aflatoxin in nuts and nut products imported to Finland. Nord. Veterinaermed. *29*: 347-353.

285. Pero, R. W., D. Harvan, R. G. Owens, and J. P. Snow. 1972. A gas chromatographic method for the mycotoxin penicillic acid. J. Chromatogr. *65*: 501-506.

286. Pero, R. W., and D. Harvan. 1973. Simultaneous detection of metabolites from several toxigenic fungi. J. Chromatogr. *80*: 255-258.

287. Peterson, R. E., and A. Ciegler. 1978. Ochratoxin A. Isolation and subsequent purification by high pressure liquid chromatography. Appl. Environ. Microbiol. *36*: 613-614.

288. Petrie, L. 1977. The identification of T-2 toxin and its association with a haemorrhagic syndrome in cattle. Vet. Rec. *101*: 323-331.

289. Pettit, R. E., and R. A. Taber. 1968. Factors influencing aflatoxin accumulation in peanut kernels and the associated mycoflora. Appl. Microbiol. *16*: 1230-1234.

290. Pettit, R. E., R. A. Taber, H. W. Schroeder, and A. L. Harrison. 1971. Influence of fungicides and irrigation practice on aflatoxin in peanuts before digging. Appl. Microbiol. *22*: 629-934.

291. Phillips, D. L., D. M. Yourtree, and S. Searles. 1976. Presence of aflatoxin B in human liver in the United States. Toxicol. Appl. Pharmacol. *36*: 403-406.

292. Pitout, M. J. 1968. The effect of ochratoxin A on glycogen storage in the rat liver. Toxicol. Appl. Pharmacol. *13*: 299-306.

293. Pitout, M. J. 1969. A rapid spectrophotometric method for the assay of carboxypeptidase A. Biochem. Pharmacol. *18*: 1829-1836.

294. Pohland, A. E., and R. Allen. 1970. Analysis and chemical confirmation of patulin in grains. J. Assoc. Off. Anal. Chem. *53*: 686-687.

295. Pohland, A. E., K. Sanders, and C. W. Thorpe. 1970. Determination of patulin in apple juice. J. Assoc. Off. Anal. Chem. *53*: 692-695.

296. Pollock, A. V. 1947. Production of citrinin by five species of *Penicillium*. Nature (Lond.) *160*: 331–332.

297. Polzhofer, K. 1977. Aflatoxinbestimmung in Milch and Milchproduckten. Z. Lebensm. Unters. Forsch. *163*: 175-177.

298. Pons, W. A., Jr. 1976. Resolutions of aflatoxins B_1, B_2, G_1, and G_2 by high pressure liquid chromatography. J. Assoc. Off. Anal. Chem. *59*: 101-105.

299. Pons, W. A., Jr., A. F. Cucullu, A. O. Franz, Jr., L. S. Lee, and L. A. Goldblatt. 1973. Rapid detection of aflatoxin contamination in agricultural products. J. Assoc. Off. Anal. Chem. *56*: 803-807.

300. Pons, W. A., Jr., and L. A. Goldblatt. 1975. The determination of aflatoxins in cotton seed products. J. Am. Oil Chem. Soc. *42*: 471-475.

301. Prior, M. G. 1976. Mycotoxin determinations on animal feedstuffs and tissues in western Canada. Can. J. Comp. Med. *40*: 75-79.

302. Przybylski, W. 1975. Formation of aflatoxin derivatives on thin layer chromatographic plates. J. Assoc. Off. Anal. Chem. *58*: 163-164.

303. Purchase, I. F. H., and J. J. Theron. 1968. The acute toxicity of ochratoxin A to rats. Food Cosmet. Toxicol. *6*: 479-483.

304. Purchase, I. F. H., and L. J. Vorster. 1968. Aflatoxin in commercial milk samples. S. Afr. Med. J. *42*: 219.

305. Raistrick, H., and G. Smith. 1941. Antibacterial substances from mould. I. Citrinin, a metabolic product of *Penicillium citrinum* Thom. Chem. Ind. *60*: 828-831.

306. Raistrick, H., J. H. Birkenshaw, A. Bracken, S. E. Michael, W. A. Hopkins, and W. E. Cye. 1943. Patulin in common cold. Collaborative research on a derivative of *Penicillium patulum* Banier. Lancet *245*: 625-634.

307. Raphael, R. A. 1948. Compounds related to penicillic acid. Part III. Synthesis of penicillic acid. J. Chem. Soc. (Lond.), pp. 1508-1512.

308. Rayner, E. T., S. P. Koltun, and F. G. Dollear. 1977. Solvent extraction of aflatoxins from contaminated agricultural products. J. Am. Oil Chem. Soc. *54*: 242A-244A.

309. Reddy, J. K., D. J. Svoboda, and M. S. Rao. 1976. Induction of liver tumors by aflatoxin B_1 in the tree shrew (*Tupaia glis*), a nonhuman primate. Cancer Res. *36*: 151-160.

310. Roberts, J. C., and P. Woollven. 1970. Studies in mycological chemistry. Part XXIV. Synthesis of ochratoxin A, a metabolite of *Aspergillus ochraceous* Wilh. J. Chem. Soc. (*C*): 278-281.

311. Robertson, G. A., L. Lee, A. F. Cucullu, and L. A. Goldblatt. 1965. Assay of aflatoxin in peanuts and peanut products using acetone-hexane-water for extraction. J. Am. Oil Chem. Soc. *42*: 467-471.

312. Robinson, P. 1967. Infantile cirrhosis of the liver in India with special reference to probable aflatoxin etiology. Clin. Pediatr. (Philadelphia) *6*: 57-62.

313. Rodricks, J. V. (Ed.). 1976. *Mycotoxins and Other Fungal Related Food Problems. Advances in Chemistry Series 149.* American Chemical Society, Washington, D.C.

314. Rodricks, J. V., and H. R. Roberts. 1977. Mycotoxin regulation in the United States. In J. V. Rodricks, C. W. Hesseltine, and M. A. Mehlman (Eds.), *Mycotoxins in Human and Animal Health,* Pathotox, Park Forest South, Ill., pp. 753-757.

315. Rodricks, J. V., C. W. Hesseltine, and M. A. Mehlman. 1977. *Mycotoxins in Human and Animal Health,* Pathotox, Park Forest South, Ill.

316. Roebuck, B. D., and G. N. Wogan. 1977. Species comparison of aflatoxin B. Cancer Res. *37*: 1649-1656.

317. Romer, F. R. 1975. Screening method for the detection of aflatoxin in mixed feeds and other agricultural commodities with subsequent confirmation and quantitative measurement of aflatoxins in positive samples. J. Assoc. Off. Anal. Chem. *58*: 500-506.

318. Rose, H. M., and M. O. Moss. 1970. The effect of modifying the structure of rubratoxin B on the acute toxicity to mice. Biochem. Pharmacol. *19*: 612-615.

319. Russell, T. E., T. F. Watson, and G. F. Ryan. 1976. Field accumulation of aflatoxin in cottonseed as influenced by irrigation termination dates and pink bollworm infestation. Appl. Environ. Microbiol. *31*: 711-713.

320. Saito, M., M. Enomoto, and T. Tatsuno. 1971. Yellowed rice toxins. Luteoskyrin and related compounds. Chlorine-containing compounds and citrinin. In A. Ciegler, S. Kadis, and S. Ajl (Eds.), *Microbial Toxins,* Vol. 6, Academic, New York, pp. 299-380.

321. Saito, M., and T. Tatsuno. 1971. Toxins of *Fusarium nivale*. In S. Kadis, A. Ciegler, and S. Ajl (Eds.), *Microbial Toxins,* Vol. 7, Academic, New York, pp. 293-318.

322. Sakabe, N., T. Goto, and Y. Hirato. 1964. The structure of citreoviridin, a toxic compound produced by *Penicillium citreoviride* molded on rice. Tetrahedron Lett. 1825-1830.

323. Salhab, A. S., and D. P. H. Hsieh. 1975. Aflatoxin H_1: a major metabolite of aflatoxin B_1 produced by human and rhesus monkey livers in vitro. Res. Commun. Chem. Pathol. Pharmacol. *10*: 419-431.

324. Sargeant, K., A. Sheridan, J. O'Kelly, and R. B. A. Carnaghan. 1961. Toxicity associated with certain samples of groundnuts. Nature (Lond.) *192*: 1096-1097.

325. Sato, N., Y. Ueno, and M. Enomoto. 1975. Toxicological approaches to the toxic metabolites of fusaria. VIII. Acute and subacute toxicities of T-2 toxin in cats. Jpn. J. Pharmacol. *25*: 263-270.

326. Sato, N., and Y. Ueno. 1977. Hepatic injury and hepato accumulation of (+) rugulosin in mice. J. Toxicol. Sci. *2*: 261-271.

327. Sato, N., and Y. Ueno. 1977. Comparative toxicities of trichothecenes. In J. V. Rodricks, C. W. Hesseltine, and M. A. Mehlman (Eds.), *Mycotoxins in Human and Animal Health,* Pathotox, Park Forest South, Ill., pp. 295-308.

328. Schade, J. E., K. McGreevy, A. D. King, B. MacKey, and G. Fuller. 1975. Incidence of aflatoxin in California almonds. Appl. Microbiol. *29*: 48-53.

329. Schroeder, H. W., and M. J. Verrett. 1969. Production of aflatoxin by *Aspergillus wentii* Wehner. Can. J. Microbiol. *15*: 895-898.

330. Scott, De B. 1965. Toxigenic fungi isolated from cereal and legume products. Mycopathol. Mycol. Appl. *25*: 213-222.

331. Scott, P. M. 1974. Patulin. In I. F. H. Purchase (Ed.), *Mycotoxins*, Elsevier, Amsterdam, pp. 383-403.

332. Scott, P. M., P. C. Barry, and B. Kennedy. 1975. Analysis of spices and herbs for aflatoxins. Can. Inst. Food Sci. Technol. J. *8*: 124-126.

333. Scott, P. M., T. Fuleki, and J. Harwig. 1977. Patulin content of juice and wine produced from moldy grapes. J. Agric. Food Chem. *25*: 434-437.

334. Scott, P. M., and T. B. Hand. 1967. Method for detection and examination of ochratoxin A in some cereal products. J. Assoc. Off. Anal. Chem. *50*: 366-370.

335. Scott, P. M., B. Kennedy, and W. van Welbeck. 1971. Simplified procedure for the purification of ochratoxin A from extracts of *Penicillium viridicatum.* J. Assoc. Off. Anal. Chem. *54*: 1445-1447.

336. Scott, P. M., J. W. Lawrence, W. van Walbeck. 1970. Detection of mycotoxins by thin-layer chromatography: application to screening of fungal extracts. Appl. Microbiol. *20*: 839-842.

337. Scott, P. M., M. A. Merrein, and J. Polonsky. 1976. Roquefortine and isofumigaclavine A, metabolites from *Penicillium roqueforti.* Experientia *32*: 140-142.

338. Scott, G. M., W. F. Miles, P. Taft, and J. G. Dube. 1972. Occurrence of patulin in apple juice. J. Agric. Food Chem. *20*: 450-451.

339. Scott, P. M., T. Panalaks, S. Kanhere, and W. F. Miles. 1978. Determination of zearalenone in cornflakes and other corn-based foods by thin layer chromatography, high pressure liquid chromatography, and gas-liquid chromatography/high resolution mass spectrometry. J. Assoc. Off. Anal. Chem. *61*: 593-600.

340. Scott, P. M., and E. Sommers. 1968. Stability of patulin in fruit juices and flour. J. Agric. Food Chem. *16*: 483-485.

341. Scott, P. M., W. van Walbeck, and J. Forjacs. 1967. Formation of aflatoxins by *Aspergillus ostianus* Wehmer. Appl. Microbiol. *15*: 945.

342. Scott, P. M., W. van Walbeck, B. Kennedy, and D. Anyeti. 1972. Mycotoxins (ochratoxin A, citrinin and sterigmatocystin) and toxigenic fungi in grains and other agricultural products. J. Agric. Food Chem. *20*: 1103-1109.

343. Seitz, L. M. 1975. Comparison of methods for aflatoxin analysis by high pressure liquid chromatography. J. Chromatogr. *104*: 81-89.

343a. Serck-Hanssen, A. 1970. Aflatoxin-induced fatal hepatitis: a case report from Uganda. Arch. Environ. Health *20*: 729-731.

344. Shank, R. C., N. Bhamarapravati, J. E. Gordon, and G. N. Wogan. 1972. Dietary aflatoxins and human liver cancer. IV. Incidence of primary liver cancer in two municipal populations of Thailand. Food Cosmet. Toxicol. *10*: 171-179.

345. Shank, R. C., C. H. Bourgeous, N. Keschamras, and P. Chandavinol. 1971. Aflatoxins in autopsy specimens from Thai children with an acute disease of unknown aetiology. Food Cosmet. Toxicol. *9*: 501-507.

346. Shank, R. C., J. E. Gordon, G. N. Wogan, A. Nondasuta, and B. Subhamani. 1972. Dietary aflatoxins and human liver cancer. III. Field survey of rural Thai families for ingested aflatoxins. Food Cosmet. Toxicol. *10*: 71-84.

347. Shank, R. C., G. N. Wogan, J. B. Gibson, and A. Nondasuta. 1972. Dietary aflatoxins and human liver cancer. II. Aflatoxins in market foods and foodstuffs of Thailand and Hong Kong. Food Cosmet. Toxicol. *10*: 61-76.

348. Shannon, G. M., R. D. Stubblefield, and O. L. Shotwell. 1973. Modified rapid screening method for aflatoxin in corn. J. Assoc. Off. Anal. Chem. *56*: 1024-1025.

349. Shoji, J., and S. Shibata. 1964. The structure of erythroskyrine, a nitrogen-containing colouring matter of *Penicillium islandicum* Sopp. Chem. Ind. 419-421.

350. Shotwell, O. L. 1977. Aflatoxin in corn. J. Am. Oil Chem. Soc. *54*: 216A-224A.

351. Shotwell, O. L., M. L. Goulden, and G. A. Bennett. 1976. Determination of zearalenone in corn: collaborative study. J. Assoc. Off. Anal. Chem. *59*: 666-670.

352. Shotwell, O. L., M. L. Goulden, and C. W. Hesseltine. 1972. Aflatoxin contamination. Association with foreign material and characteristic fluorescence in damaged corn kernels. Cereal Chem. *49*: 458-465.

353. Shotwell, O. L., M. L. Gouldin, and C. W. Hesseltine. 1976. Survey of U.S. wheat for ochratoxin and aflatoxin. J. Assoc. Off. Anal. Chem. *59*: 122-124.

354. Shotwell, O. L., C. W. Hesseltine, H. R. Burmeister, W. F. Kwolek, G. H. Shannon, and H. H. Hall. 1969. Survey of cereal grains and soybeans for the presence of aflatoxin. II. Corn and soybeans. Cereal Chem. *46*: 454-463.

355. Smalley, E., A. Joffee, M. Palyusik, H. Kurata, and W. Marasas. 1977. Panel on trichothecene toxins. In J. V. Rodricks, C. W. Hesseltine, and M. A. Mehlman (Eds.), *Mycotoxins in Human and Animal Health*, Pathotox, Park Forest South, Ill., pp. 337-340.

356. Smith, D. F., and G. P. Lynch. 1973. *Aspergillus fumigatus* in samples of moldy silage. J. Dairy Sci. *56*: 828-829.

357. Smith, R. H., and W. McKernan. 1962. Hepatotoxic action of chromatographically separated fractions of *Aspergillus flavus* extracts. Nature (Lond.) *195*: 1301-1303.

358. Snyder, W. C., and H. N. Hansen. 1940. The species concept in *Fusarium*. Am. J. Bot. *27*: 64-67.

359. Snyder, W. C., and H. N. Hansen. 1945. The species concept in *Fusarium* with reference to discolor and other sections. Am. J. Bot. *32*: 657-666.

360. Spensley, P. C. 1963. Aflatoxin, the active principle in turkey "x" disease. Endeavour *22*: 75-79.

361. Sreenivasa Murthy, V., S. SriKanta, and H. A. B. Parpia. 1971. Removing aflatoxin from peanut seed meals. Indian Patent No. 120,257 (CA *77*: 4045f).

362. Stack, M. E., and A. E. Pohland. 1975. Collaborative study of a method for

chemical confirmation of the identity of aflatoxin. J. Assoc. Off. Anal. Chem. *58*: 110-113.

363. Steyn, P. S. 1969. The separation and detection of several mycotoxins by thin-layer chromatography. J. Chromatogr. *45*: 473-475.

364. Steyn, P. S. 1971. Ochratoxin and other dihydroisocoumarins. In A. Ciegler, S. Kadis, and S. J. Ajl (Eds.), *Microbial Toxins,* Vol. 6, Academic, New York, pp. 179-205.

365. Steyn, P. S., and C. W. Holzapfel. 1967. The synthesis of ochratoxins A and B, metabolites of *Aspergillus ochraceus* Wilh. Tetrahedron *23*: 4449-4461.

366. Steyn, P. S., and K. J. Van der Merwe. 1966. Detection and estimation of ochratoxin A. Nature (Lond.) *211*: 418.

367. Stob, M., R. S. Baldwin, J. Tuite, F. N. Andrews, and K. G. Gillette. 1962. Isolation of an anabolic uterotrophic compound from corn infected with *Gibberella zeae*. Nature (Lond.) *196*: 1318.

368. Stoloff, L. 1976. Occurrence of mycotoxins in foods and feeds. In J. V. Rodricks (Ed.), *Mycotoxins and Other Fungal Related Food Problems. Advances in Chemistry Series 149,* American Chemical Society, Washington, D.C., pp. 23-50.

369. Stott, W. T., and L. B. Bullerman. 1975. Patulin: a mycotoxin of potential concern in foods. J. Milk Food Technol. *38*: 495-705.

370. Stott, W. T., and L. B. Bullerman. 1975. Microbiological assay of patulin, using *Bacillus megaterium*. J. Assoc. Off. Anal. Chem. *61*: 497-499.

371. Stott, W. T., and L. B. Bullerman. 1975. Influence of carbohydrate and nitrogen source on patulin production by *Penicillium patulum*. Appl. Microbiol. *30*: 850-854.

372. Stray, H. 1978. High pressure liquid chromatographic determination of patulin in apple juice. J. Assoc. Off. Anal. Chem. *61*: 1359-1362.

373. Stutz, K. K., and P. H. Krumperman. 1976. Effect of temperature cycling on the production of aflatoxin by *Aspergillus parasiticus*. Appl. Environ. Microbiol. *32*: 327-332.

374. Subrahmanyam, P., and A. S. Rao. 1974. Occurrence of aflatoxins and citrinin in groundnuts (*Arachis hypogaea* L.) at harvest in relation to pod condition and kernel moisture content. Curr. Sci. *43*: 707-710.

375. Suzangar, M., A. Emami, and R. Barnett. 1976. Aflatoxin contamination of village milk in Isfahan, Iran. Trop. Sci. *18*: 155-159.

376. Suzuki, T., M. Takedo, and H. Tanabe. 1971. Studies on chemical analysis of mycotoxins. 11. Gas chromatography of penicillic acid and its derivatives and gas chromatographic analysis of penicillic acid in rice. Shokuhin Eiseigaku Zasshi *12*: 495-500.

377. Swenson, D. H., J. K. Lin, E. C. Miller, and J. A. Miller. 1977. Aflatoxin B_1-2,3-oxide as a probable intermediate in the covalent binding of aflatoxin B_1 and B_2 to rat liver DNA and ribosomal DNA in vivo. Cancer Res. *37*: 172-181.

378. Swenson, D. H., J. A. Miller, and E. C. Miller. 1973. 2,3-Dehydro-2,3-dihydro-aflatoxin B_1: an acid hydrolysis product of an RNA-aflatoxin B adduct formed by hamster and rat liver microsomes in vitro. Biochem. Biophys. Res. Commun. *53*: 1260-1267.

379. Swenson, D. H., E. C. Miller, and J. A. Miller. 1974. Aflatoxin B_1-2,3-oxide. Evidence for its formation in rat liver in vivo and by human liver microsomes in vitro. Biochem. Biophys. Res. Commun. *60*: 1036-1043.

380. Swenson, D. H., J. A. Miller, and E. C. Miller. 1975. The reactivity and carcinogenicity of aflatoxin B_1-2,3-dichloride, a model for the putative 2,3-oxide metabolite of aflatoxin B_1. Cancer Res. *35*: 3811-3823.

381. Taber, W. A., and L. C. Vining. 1959. Tryptophan as a precursor of the ergot alkaloids. Chem. Ind. (Lond.), Part 2, 1218-1219.

382. Takeda, N., S. Seo, Y. Ogihara, U. Sankawa, I. Iitaka, I. Kitagawa, and S. Shibata. 1973. Studies of fungal metabolites XXXI. Anthroquinoid colouring matters of *Penicillium islandicum* Sopp and some other fungi: (–) luteoskyrin, (–) rubroskyrin, (+) rugulosin and their related compounds. Tetrahedron *29*: 3703-3719.

382a. Tatsuno, T., M. Tsukioka, Y. Sakai, Y. Suzuki, and Y. Asami. 1955. Communications to the editor. Pharm. Bull. (Tokyo) *3*: 476-477.

383. Thorpe, C. W., and R. L. Johnson. 1974. Analysis of penicillic acid by gas-liquid chromatography. J. Assoc. Off. Anal. Chem. *57*: 861-865.

384. Tiemstra, P. J. 1977. Aflatoxin control during feed processing of peanuts. In J. V. Rodricks, C. W. Hesseltine, and M. A. Mehlman (Eds.), *Mycotoxins in Human and Animal Health,* Pathotox, Park Forest South, Ill., pp. 121-138.

385. Tilak, T. B. G. 1975. Induction of cholangiocarcinoma following treatment of a rhesus monkey with aflatoxin. Food Cosmet. Toxicol. *13*: 247-249.

386. Timonin, M. I. 1942. Another mold with anti-bacterial ability. Science *96*: 494.

387. Timonin, M. I. 1946. Activity of patulin against *Ustilago tritici* (Pers.) Jen. Sci. Agric. *26*: 358-368.

388. Toussan, T. A., and P. E. Nelson. 1968. *A Pictorial Guide to the Identification of Fusarium Species According to the Taxonomic System of Snyder and Hansen,* Pennsylvania State University Press, University Park, Pa.

389. Townsend, R. J., M. O. Moss, and H. M. Peck. 1966. Isolation and characterization of hepatotoxins from *Penicillium rubrum.* J. Pharm. Pharmacol. *18*: 471-473.

390. Trenk, H. L., M. E. Butz, and F. S. Chu. 1971. Production of ochratoxins in different cereals by *Aspergillus ochraceus.* Appl. Microbiol. *21*: 1032-1035.

391. Tsunoda, H. 1970. Microorganisms which deteriorate stored cereals and grains. In *Toxic Micro-organisms,* Proceedings of the First U.S.-Japan Conference on Toxic Micro-Organisms. UJNR Joint Panels on Toxic Micro-organisms and the U.S. Dept. of the Interior, Washington, D.C., p. 143.

392. Tung, T. C., and K. H. Ling. 1968. Study on aflatoxin of foodstuffs in Taiwan. J. Vitaminol. (Kyoto) *14*: 48-52.

393. Uchiyama, M., E. Isohata, and Y. Takeda. 1978. A case report on the detection of ochratoxin A from rice. J. Food Hyg. Soc. *17*: 103-104.

394. Ueno, Y. 1974. Citreoviridin from *Penicillium citreoviride* Biourge. In I. F. H. Purchase (Ed.), *Mycotoxins,* Elsevier, Amsterdam, pp. 283-302.

395. Ueno, Y. 1977. Mode of action of trichothecenes. Pure Appl. Chem. *49*: 1737-1745.

396. Ueno, Y. 1977. Trichothecenes: overview address. In J. V. Rodricks, C. W. Hesseltine, and M. A. Mehlman (Eds.), *Mycotoxins in Human and Animal Health,* Pathotox, Park Forest South, Ill., pp. 189-208.

397. Ueno, Y. 1978. Chemistry and biological activity of trichothecenes. Proc. Assoc. Mycotoxicol. *7*: 6-14.

398. Ueno, Y., M. Hosaya, and Y. Ishikawa. 1969. Inhibitory effects of mycotoxins on the protein synthesis in rabbit reticulocytes. J. Biochem. *66*: 419-421.

399. Ueno, Y., Y. Ishikawa, K. Saito-Amakai, and H. Tsundo. 1970. Environmental factors influencing the production of fusarenon-X, a cytotoxic mycoltoxin of *Fusarium nivale* Fn 2B. Chem. Pharm. Bull. *18*: 304-312.

400. Ueno, Y., K. Ishii, K. Sakai, S. Kanaeda, H. Tsunoda, T. Tanaka, and M. Enomoto. 1972. Toxicological approaches to the metabolites of fusaria. IV. Microbial survey on "bean hulls" poisoning of horses with the isolation of toxic trichothecenes, neosolaniol and T-2 toxin of *Fusarium nivale* M-1-1. Jpn. J. Exp. Med. *42*: 187-203.

401. Ueno, Y., H. Matsumoto, K. Ishii, and K. Kukita. 1976. Inhibitory effects of my-cotoxins on sodium-dependent transport of glycine in rabbit reticulocytes. Biochem. Pharmacol. *25*: 2091-2095.

402. Ueno, Y., M. Nakajima, K. Sakai, K. Ishii, N. Sato, and N. Shimada. 1973. Comparative toxicology of trichothec mycotoxins: inhibition of protein synthesis in animal cells. J. Biochem. *74*: 285-296.

403. Ueno, Y., N. Sato, K. Ishii, K. Sakai, and M. Enomoto. 1972. Toxicological approaches to the metabolites of fusaria. V. Neosolaniol, T-2 toxin and butenolide, toxic metabolites of *Fusarium sporotrichioides* NRRL 3510 and *Fusarium poae* NRRL3287. Jpn. J. Exp. Med. *42*: 461-472.

404. Ueno, Y., and I. Ueno. 1972. Isolation and acute toxicity of citreoviridin, a neurotoxic mycotoxin of *Penicillium citreoviride* Biourge. Jpn. J. Exp. Med. *42*: 91-105.

405. Ueno, Y., I. Ueno, K. Ito, and T. Tatsuno. 1967. Impairments of RNA synthesis in Ehrlich ascites tumors by luteoskyrin, a hepatotoxic pigment of *Penicillium islandicum* Sopp. Experientia *23*: 1001-1002.

406. Ueno, Y., Y. Kato, and M. Enomoto. 1975. Erythroskyrine, the third mycotoxin from *Penicillium islandicum* Sopp. Jpn. J. Exp. Med. *45*: 524-526.

407. Ueno, Y., I. Ueno, A. Kazue, Y. Ishikawa, H. Tsundo, K. Okubo, M. Saito, and M. Enomoto. 1971. Toxicological approaches to the metabolites of fusaria. II. Isolation of fusarenon-X from culture filtrates of *Fusarium nivale* Fn2B. Jpn. J. Exp. Med. *41*: 507-519.

408. Ueno, Y., and K. Kubata. 1976. DNA-attacking ability of carcinogenic mycotoxins in recombination deficient mutant cells of *Bacillus subtilis*. Cancer Res. *36*: 445-452.

409. Ueno, Y., I. Ueno, and K. Mizumoto. 1968. The mode of binding of luteoskyrin a hepatotoxic pigment of *Penicillium islandicum* Sopp to deoxyribonuclohistine. Jpn. J. Exp. Med. *38*: 47-64.

410. Ueno, Y., I. Ueno, K. Meyumoto, and T. Tatsuno. 1966. The bonding of luteo-skyrin, a hepatotoxic pigment of *P. islandicum* Sopp to deoxyribonucleic acid. Seikagaku. Tokyo. J. Jpn. Biochem. Soc. *38*: 687-699.

411. Ueno, Y., and M. Nakajima. 1974. Production of respiratory-deficient mutants of *Saccharomyces cerevisiae* by (–) luteoskyrin and (+) rugulosin. Chem. Pharm. Bull. *22*: 2258-2262.

412. Ueno, Y., I. Ueno, N. Sato, Y. Iitoi, M. Saito, M. Enomoto, and H. Tsunoda. 1971. Toxicological approach to (+) rugulosin, an anthraquinoid mycotoxin of *Penicillium rugulosum* Thom. Jpn. J. Exp. Med. *41*: 177-187.

413. Ueno, Y., I. Ueno, T. Tatsuno, and K. Uraguchi. 1964. Effects of luteoskyrin, a toxic substance of *Penicillium islandicum,* on mitochondrial swelling. Jpn. J. Exp. Med. *34*: 197-209.

414. Ueno, Y., and S. Yagasaki. 1975. Toxicological approaches to the metabolites of fusaria. X. Accelerating effect of zearalenone on RNA and protein synthesis in the uterus of ovariectomized mice. Jpn. J. Exp. Med. *45*: 199-205.

415. Uraguchi, X., and M. Yamazaki. 1978. *Toxicology: Biochemistry and Pathology of Mycotoxins,* Wiley, New York.

416. Urry, W. H., H. L. Wehrmeister, E. B. Hodge, and P. H. Hidy. 1966. The structure of zearalenone. Tetrahedron Lett. *27*: 3109-3114.

417. Van der Merwe, K. J., P. S. Steyn, and L. Fourie. 1965. Mycotoxins. Part II. The constitution of ochratoxins A, B and C, metabolites of *Aspergillus ochraceus* Wilh. J. Chem. Soc. *382*: 7083-7088.

418. van der Zidjen, A. S. M., W. A. A. B. Koekensmid, J. Boedingh, C. B. Barrett,

W. Q. Ord, and J. Philip. 1962. *Aspergillus flavus* and turkey x disease. Nature (Lond.) *195*: 1060-1062.

419. van Rensburg, S. J. 1977. Role of epidemiology in the elucidation of mycotoxin health risks. In J. V. Rodricks, C. W. Hesseltine, and M. A. Mehlman (Eds.), *Mycotoxins in Human and Animal Health*, Pathotox, Park Forest South, Ill., pp. 699-712.

420. van Rensburg, S. J., A. Kirsipuu, L. P. Continho, and J. J. van der Watt. 1975. Circumstances associated with the contamination of food by aflatoxin in a high primary liver cancer area. S. Afr. Med. J. *49*: 877-883.

421. van Walbeck, W., P. M. Scott, and F. S. Thatcher. 1968. Mycotoxins from foodborne fungi. Can. J. Microbiol. *14*: 131-137.

422. Vedanayagam, H. S., A. S. Indulkar, and S. R. Rao. 1971. Aflatoxins and *Aspergillus flavus* Link in Indian cottonseed. Indian J. Exp. Biol. *9*: 410-411.

423. Velasco, J. 1972. Detection of aflatoxin using small columns of florisil. J. Am. Oil Chem. Soc. *49*: 141-142.

424. Velasco, J., and T. B. Whitaker. 1975. Sampling cottonseed lots for aflatoxin contamination. J. Am. Oil Chem. Soc. *52*: 191-195.

425. Vesonder, R. F., A. Ciegler, H. R. Burmeister, and A. H. Jensen. Acceptance of corn amended with trichothecenes by swine and rats. Appl. Environ. Microbiol., in press.

426. Vesonder, R. F., A. Ciegler, and A. H. Jensen. 1973. Isolation of the emetic principle from *Fusarium*-infected corn. Appl. Microbiol. *26*: 1008-1010.

427. Vesonder, R. F., A. Ciegler, A. H. Jensen, W. K. Rohwedder, and D. Weisleder. 1976. Co-identity of the refusal and emetic principles from *Fusarium*-infected corn. Appl. Environ. Microbiol. *31*: 280-285.

428. Walker, K., and B. P. Wiesner. 1944. Patulin and clavicin. Lancet *246*: 294.

429. Wang, Y., H.-S. Ting, and W. Mann. 1950. Metabolism of citrinin. I. Determination of free citrinin in blood. Chem. J. Physiol. *17*: 259-270.

430. Ware, G. M., and C. W. Thorpe. 1978. Determination of zearalenone in corn by high pressure liquid chromatography and fluorescence detection. J. Assoc. Off. Anal. Chem. *61*: 1058-1062.

431. Ware, G. M., C. W. Thorpe, and A. E. Pohland. 1974. A liquid chromatographic method for the determination of patulin in apple juice. J. Assoc. Off. Anal. Chem. *57*: 1111-1113.

432. Wei, R., E. B. Smalley, and F. M. Strong. 1972. Improved skin test for detection of T-2 toxin. Appl. Microbiol. *23*: 1029-1030.

433. Wei, R. D., P. E. Still, E. B. Smalley, K. K. Schones, and F. M. Strong. 1973. Isolation and partial characterization of mycotoxin from *Penicillium roqueforti*. Appl. Microbiol. *25*: 111-114.

434. Wells, J. M., and J. A. Payne. 1976. Incidence of aflatoxin contamination on a sampling of southeastern pecans. Proc. Fla. State Hortic. Soc. *89*: 256-257.

435. Whitten, M. E. 1969. Screening cottonseed for aflatoxins. J. Am. Oil Chem. Soc. *46*: 39-41.

436. Widstrom, N. W., E. B. Lillehoj, A. N. Sparks, and W. F. Kwolek. 1976. Corn earworm damage and aflatoxin B_1 on corn ears protected with insecticide. J. Econ. Entomol. *69*: 677-679.

437. Wilson, B. J. 1971. Miscellaneous *Aspergillus* toxins. In A. Ciegler, S. Kadis, and S. Ajl (Eds.), *Microbial Toxins*, Vol. 6, Academic, New York, pp. 207-294.

438. Wilson, B. J., and C. H. Wilson. 1962. Extraction and preliminary characterization of a hepatotoxic substance from cultures of *Penicillium rubrum*. J. Bacteriol. *84*: 285-290.

439. Wilson, B. J., T. C. Campbell, A. W. Hayes, and R. T. Hanlin. 1968. Investigation of reported aflatoxin production by fungi outside the *Aspergillus flavus* group. Appl. Microbiol. *16*: 819-821.

440. Wilson, B. J., C. H. Wilson, and A. W. Hayes. 1968. Tremorgenic toxin from *Penicillium cyclopium* grown on food materials. Nature (Lond.) *220*: 77-78.

441. Wilson, D. M. 1976. Patulin and penicillic acid. In J. V. Rodricks (Ed.), *Advances in Chemistry Series 149*, American Chemical Society, Washington, D.C., pp. 90-109.

442. Wogan, G. N. 1965. Experimental toxicity and carcinogenicity of aflatoxins. In G. N. Wogan (Ed.), *Mycotoxins in Foodstuffs*, MIT Press, Cambridge, Mass., pp. 163-173.

443. Wogan, G. N. 1973. Aflatoxin carcinogenesis. Methods Cancer Res. *7*: 309-344.

444. Wogan, G. N., G. S. Edwards, and P. M. Newberne. 1971. Acute and chronic toxicology of rubratoxin B. Toxicol. Appl. Pharmacol. *19*: 712-720.

445. Wollenweber, H. W., and O. A. Reinking. 1935. *Die Fusarien ihre Beschreibung Schadwirking und Bekampfun*. Paul Parey, Berlin.

446. Wyatt, R. D., J. R. Harris, P. B. Hamilton, and H. R. Burmeister. 1972. Possible outbreaks of fusariotoxicosis in avians. Avian Dis. *16*: 1123-1130.

447. Wyatt, R. D., B. A. Weeks, P. B. Hamilton, and H. R. Burmeister. 1972b. Severe oral lesions in chickens caused by ingestion of dietary fusaric toxin T-2. Appl. Microbiol. *24*: 251-257.

448. Wyllie, T. D., and L. G. Morehouse. 1977. *Mycotoxic Fungi, Mycotoxins, Mycotoxicoses*, Vols. 1-3, Dekker, New York.

449. Yadagiri, B., and P. G. Tulpule. 1974. Aflatoxin in buffalo milk. Indian J. Dairy Sci. *27*: 293-297.

450. Yagan, B., and A. Z. Joffe. 1976. Screening of toxic isolates of *Fusarium poae* and *F. sporotrichioides* involved in causing alimentary toxic aleukia. Appl. Environ. Microbiol. *32*: 423-427.

451. Yagan, B., A. Z. Joffe, P. Horn, N. Mor, and I. I. Lutsky. 1977. Toxins from a strain involved in ATA. In J. V. Rodricks, C. W. Hesseltine, and M. A. Mehlman (Eds.), *Mycotoxins in Human and Animal Health*, Pathotox, Park Forest South, Ill., pp. 329-336.

452. Yoshizawa, T., and N. Morooka. 1977. Trichothecenes from mold infected cereal in Japan. In J. V. Rodricks, C. W. Hesseltine, and M. A. Mehlman (Eds.), *Mycotoxins in Human and Animal Health*, Pathotox, Park Forest South, Ill., pp. 309-322.

453. Zuber, M. S. 1977. Influence of plant genetics on toxin production in corn. In J. V. Rodricks, C. W. Hesseltine, and M. A. Mehlman (Eds.), *Mycotoxins in Human and Animal Health*. Pathotox, Park Forest South, Ill., pp. 173-179.

454. Zuber, M. S., O. H. Calvert, E. B. Lillehoj, and W. F. Kwolek. 1976. Pre-harvest development of aflatoxin B_1 in corn in the United States. Phytopathology *66*: 1120-1121.

Author Index

Numbers in brackets are reference numbers and indicate that an author's work is referred to although his name may not be cited in the text. Italicized numbers give the page on which the complete reference is listed.

A

Aalund, O., *456*
Aaronson, M., *113*
Abbott, P., 5[1], *39*
Abdel Kader, M. M., 10[280], *52*
Abdel-Makal, S. H., 380[1], *405*
Abdor, N. I., 82[1], *89*
Abdou, N. L., 21[68], *42*
Abe, M., 166[218], *205*, 436[269], *460*
Abel, T., 26[372], *56*, 245[449], 250 [450], *296*
Abernathy, R. S., 142[1], *194*, 226[1], *272*
Abrahams, I., 255[333], *290*
Abrami, P., 187[468], *217*
Abul-Haj, S. K., 380[2], *405*
Acquaviva, F. A., 77[44], *91*
Adam, A., 6[2], *39*
Adam, D., 146[214], 155[214], *205*
Adams, B. L., 68[134], 69[134], 71 [134], *95*
Adams, G. H., 425[33], *448*
Adamson, D. M., 71[2], 77[2], *89*
Adamson, R. H., 422[1], *447*
Adelstein, E. H., 101[76], *118*
Adler, H. E., 37[84], *43*
Adler, J. P., 161[176], *203*, 220[194], 229[194], 233[194], 234[194], *283*
Adler, W. H., 88[154], *97*
Adye, J. C., 417[229], *458*
Agarwal, S. C., 263[464,465,466], *297*
Agostini, B., 380[52,53], *407*
Aguilar-Torres, F. G., 161[2,3], *194*
Ainsworth, G. C., 439[2], *447*
Ajello, L., 7[3,4], 18[51], 21[429], *39, 41, 59*, 158[4,5], 170[6], 171 [386], *194, 213*, 230[2], 231[2], 262[188], *272, 283*

Ajl, S., 415[67], *450*
Aksoycan, N., 250[3], *272*
Al-Askari, S., 65[36], *90*
Albert, E. D., 76[3], *89*
Albornoz, M. B., 15[5], *39*, 177[472], *217*
Aldersley, T., 31[24], *40*
Al-Doory, Y., 181[139], *201*, 246[4], 258[4], 259[4,156], *272, 273, 281*
Aldridge, W. N., 429[24], *448*
Alford, R. H., 28[150], *46*, 71[65], 88 [64], *92*, 104[2], 108[2], 111 [108], *113, 120*, 256[166], *281*
Allen, P. M., 363[1], *368*
Allen, R., 433[294], *461*
Alling, D. W., 337[18], 345[6], *349, 350*
Allison, A. C., 28[87], *43*, 108[15], *114*
Almeida, F. P., 176[7], *194*
Almero, E. M., 423[43], *449*
Almy, R. E., 147[140], 156[140], 161 [140], 173[140], *201*, 246[157], *281*
Alpert, M. E., 418[3], *447*
Alsberg, C. L., 416[4], 433[4], *447*
Altemeis, W. A., 88[154], *97*
Alunurm, K., 191[332], *211*
Alvord, R. H., 77[4], *89*
Ambroise-Thomas, P., 100[3], *113*, 149 [72], 168[85], *197, 198*, 306[25], *320*
Ambrose, A. M., 432[5], *447*
Ames, B. N., 424[232], *458*
Amla, I. J., 422[6,7,8], *447*
Ammann, A., 327[36], *350*
Amos, D. B., 70[18], 86[18], *89*
Amos, W. M., 245[5], *273*
Amundson, S., 318[96], *324*
Anderes, E. A., 20[6,7], 25[7], *39*

471

N

Subject Index

Illustrations are indicated by italicized page numbers.

A